Chemical Sensors and Biosensors

Chemical Sensors and Biosensors

Fundamentals and Applications

Florinel-Gabriel Bănică

Department of Chemistry,
Norwegian University of Science and Technology (NTNU), Trondheim, Norway

Editorial Advisor
Professor Arnold George Fogg,
Visiting Professor, University of Bedfordshire

A John Wiley & Sons, Ltd., Publication

Library of Congress Cataloging-in-Publication Data

Banica, Florinel-Gabriel.
    Chemical sensors and biosensors : fundamentals and applications /
Florinel-Gabriel Banica, Department of Chemistry, Norwegian University of
Technology, Norway.
                pages cm
    Includes bibliographical references and index.
    ISBN 978-0-470-71066-1 (cloth) – ISBN 978-0-470-71067-8 (pbk.)   1.
Chemical detectors. 2.   Biosensors. I.   Title.
    TP159.C46B36 2012
    543–dc23
                                                                2012028141

A catalogue record for this book is available from the British Library.

HB ISBN: 9780470710661
PB ISBN: 9780470710678

Set in 10/12pt Times by Thomson Digital, Noida, India

Reprinted with corrections March 2014.

Dedicated to Ana and Irina

Summary Contents

Contents

PowerPoint slides for teaching purposes may be found online at http://booksupport.wiley.com by entering the author, title or ISBN and selecting the correct title. This will then allow you to access the slides for download.

Preface

As suggested by Marshal McLuhan, media (in the more general meaning of the term) act as extensions of the functions of the human body [1]. In the same way that the microphone acts as an extension of the ear, chemical sensors can be considered to be extensions of the organs of chemical perception that are the nose and the tongue.

The development of chemical sensors responds to the increasing demand of chemical data that characterize various systems of interest. Such a system can be the human body itself, whose physiological state can be assessed unequivocally by physical, chemical and biochemical parameters. The quality of the ambient and natural environment is characterized by measuring the content of noxious chemical species. No less important is the automatic control of certain industrial processes that depend on specific chemical parameters.

In general, standard analytical methods (e.g., chromatography, spectrometry and electrophoresis) can provide the same kind of information as that produced by chemical sensors. The advantage of the chemical sensor approach results from the fact that they are specialized, small size, portable and inexpensive devices that are suitable for *in situ* analysis and real-time monitoring of chemical parameters. Worthy of mention is the capability of dedicated chemical sensors to identify pathogen micro-organisms and viruses via characteristic compounds that are parts of the structure of the target species.

"There's plenty of room at the bottom" said Richard Feynman in a seminal lecture in 1959, that anticipated the advent of nanotechnology. This sentence can be paraphrased as follows: "There's plenty of new opportunities at the bottom". This applies well to the development of chemical sensors. Indeed, the most important trend in this area is the application of nanomaterials, either as substitutes for classical materials and reagents or in the implementation of completely new sensing and transduction methods. Of outstanding importance is the size compatibility of nanomaterials with biopolymer molecules, which allows fabrication of bionanocomposites with promising potential for application in the design of chemical sensors. New fabrication technologies, mostly inspired by microelectronic technology and nanotechnology, are expected to lead to an increase in the degree of integration in chemical-sensor arrays, thus prompting advances in production and application of artificial nose/tongue devices. Integration of chemical sensors with microfluidic systems is another promising trend since microfluidic systems allow extremely small sample volumes to be processed and analyzed automatically.

New books on chemical sensors are published regularly, but most of them are collective volumes profiling particular kinds of chemical sensor and particular applications of chemical sensors. A comprehensive overview of chemical sensors in one single book is needed for two reasons. First, such a book would serve as a useful teaching aid for use in courses covering the subject of chemical sensors. Secondly, an indepth introduction to the field of chemical sensors for scientists and engineers new to this subject would be advantageous. There are currently on the market a series of volumes that are intended to respond to the above aims. However, as the field progresses, a new book that covers recent advances is always welcome.

The development of a chemical sensor is very often a matter of material synthesis and processing. Synthetic materials (both inorganic and organic), materials of biological origin (proteins, nucleic acids, micro-organism and living cells), as well as biomimetic synthetic materials are widely used in the development of chemical sensors. Of equal importance is the fabrication technology, because the final goal in chemical-sensor research is the production of a marketable product. That is why the first eight chapters in this text introduce the main kinds of material used in the development of the chemical sensors, as well as typical processes and technologies involved in fabrication of chemical sensors. The next fourteen chapters present various classes of chemical sensors organized according to the transduction method. The final chapter is devoted to chemical sensors based on highly organized biological material such as micro-organisms and living cells.

This book has been designed mostly as an instruction manual in chemical sensors, with a particular attention on balancing classical topics with contemporary trends. Clearly, owing to its extent, the contents of this book cannot be covered in a normal course of lectures. However, the course instructor can select topics that fit the class level and the particular interest of the attending students. Moreover, the curriculum can be personalized by encouraging each student to explore more deeply into certain advanced topics. In addition, a study of chemical sensors is an enlightening excursion through various scientific and technological areas, thereby contributing substantially to the development of the student's scientific knowledge.

Additionally, this book will be useful to any scientist who needs an introduction into the field of chemical-sensor science and technology. As this is an interdisciplinary field, this book will be of interest to engineers, chemists, biochemists, microbiologists and physicists endeavoring to start up research work in the field of chemical sensors.

Nothing done by humans can be perfect, but, at least, it could be perfectible. Hence, any critical comment or suggestion is welcome.

1. McLuhan, M. (2003) *Understanding Media: The Extensions of Man*, Gingko Press, Corte Madera, Calif.

Acknowledgments

First, I would like to thank Professor Arnold Fogg, who kindly agreed to edit linguistically the initial draft text. Responsibility for the final text, however, lies with the author and the publishing editors. Also, I would like to acknowledge the assistance generously given by several colleagues at the Norwegian University of Science and Technology of Trondheim, Norway, who took the time to read certain chapters of the book and who made valuable comments and suggestions. These colleagues are: Professor Torbjørn Ljones, Professor David G. Nicholson, and Professor Kalbe Razi Naqvi. I also thank Dr. Alexandru Oprea (University of Tübingen, Germany) and Dr. Marian Florescu (University of Surrey, UK) for similar assistance.

Finally, I am grateful, in writing this book, to all those scientists who have contributed to the advance of chemical sensor science and technology. Many of these scientists are cited in the book, but, owing to space limitations, much valuable work in this area could not be included or cited.

List of Symbols

Roman Symbols

Symbol	Meaning	Section Reference
A	(a) surface area	4.2.2; 13.3.1
	(b) absorbance	18.3.1
	(c) amplitude of an electromagnetic wave	18.6.2
AC	subscript pertaining to alternating current	
a	(a) thermodynamic activity	10.2.1
	(b) sensor sensitivity	1.5
	(c) molar absorptivity	18.3.1
	(d) the exponent in the expression of CPE impedance	17.2.1
b	thickness of a light-absorbing layer	18.3.1
bipy	$2,2'$-bipyridine	
C	capacitance	11.1.3
C_{dl}	capacitance of the electric double layer	17.2.2
C_f	proportionality constant in the Sauerbrey equation	21.2.3
C_0	static capacitance in the equivalent circuit of a TSM oscillator	21.2.2
C_1	capacitance at the motional branch of the equivalent circuit of a TSM piezoelectric oscillator	21.2.2
c	(a) analyte concentration	1.5
	(b) concentration of the enzyme–substrate complex	3.6.1
	(c) concentration of the antibody–antigen complex	6.4
	(d) light velocity in vacuum	18.1
c^*	concentration of an excited-state species	18.3.6
c_A	concentration of the species A	10.4.2
c_{AR}	concentration of the analyte–receptor combination	18.4.1
c_B	concentration of the species B	10.4.2
$c_{O,b}$	concentration of the oxidized form of a redox couple in the bulk solution	11.3.1
$c_{O,i}$	concentration of the oxidized form of a redox couple at the electrode/electrolyte interface	11.3.1
c_Q	concentration of a fluorescence quencher	18.3.7
c_R	receptor concentration	18.4.1
$c_{R,t}$	total concentration of the receptor	18.4.1
$c_{R,b}$	concentration of the reduced form of a redox couple in the bulk solution	11.3.1
$c_{R,i}$	concentration of the reduced form of a redox couple at the electrode/electrolyte interface	11.3.1
D	diffusion coefficient	4.2.2; 13.3.1
Da	Damköhler number for internal diffusion in an immobilized enzyme layer	4.3.1
Da_M	mediator Damköhler number	15.2.2
Da_S	substrate Damköhler number	15.2.2
DC	subscript pertaining to direct current	
D_M	diffusion coefficient of a redox mediator	15.2.1
$D_{P,e}$	diffusion coefficient of the product within an immobilized enzyme layer	4.3.1
$D_{P,m}$	diffusion coefficient of the product in the external membrane of an enzymatic sensor	4.2.2
$D_{S,e}$	diffusion coefficient of the substrate within an immobilized enzyme layer	4.3.1
$D_{S,m}$	diffusion coefficient of the substrate in the external membrane of an enzymatic sensor	4.2.2
E	enzyme	3.6.1
E	(a) energy	11.1.2
	(b) electrode potential	10.2.2; 13.3.1
	(b) Young modulus	22.1.2

ΔE	difference between the actual electrode potential and the formal electrode potential	13.6.1
E_{cell}	cell voltage	10.2.2
E_{AC}	sine wave alternating potential	13.7.7
E_F	Fermi energy	11.1.2
EMF	electromotive force	10.2.2
E_{ph}	photon energy	18.1
E_{pzc}	potential of zero-charge	13.5.2
E_O	oxidase enzyme in the oxidized form	14.2.1
E_r	reference electrode potential	10.4.1
E_R	oxidase enzyme in the reduced form	14.2.1
ES	enzyme–substrate complex	3.6.1
$E_{1/2}$	half-wave potential	13.3.2
E^0	standard electrode potential	10.2.2
E_f^0	formal electrode potential	10.2.2; 13.3.1
E_0	prelogarithm constant in the response equation of a potentiometric ion sensor	10.4.1
e	(a) enzyme concentration	3.6.1
	(b) elementary charge	11.1.3
	(c) the base of natural logarithm	17.1
e_O	concentration of the oxidized form of an oxidase enzyme in an immobilized enzyme layer	15.1.1
e_R	concentration of the reduced form of an oxidase enzyme in an immobilized enzyme layer	15.1.1
e_t	total enzyme concentration	3.6.1
e_y	measurement error of the sensor response	1.5
e^-	electron	
F	(a) Faraday constant	
	(b) power of a fluorescence light beam	18.3.5
F_0	fluorescence power in the absence of a quencher	18.3.7
f	(a) F/RT	13.6.1
	(b) frequency	13.7.7; 21.2.2
f_e	enzyme loading factor	4.3.1
Δf_L	change in the resonance frequency due to liquid loading on a TSM piezoelectric resonator	21.2.5
Δf_m	change in the resonance frequency due to mass loading	21.2.3
f_O	(a) m_O/m_t	15.1.1
	(b) $m_{O,0}/m_t$	15.2.3
$\tilde{f}$	complex frequency of a TSM resonator	21.2.7
f_0	resonant frequency of an oscillator	21.2.1; 22.1.3
G	Gibbs free energy	10.2.1
ΔG	Gibbs free energy change in a chemical process	10.2.2
ΔG^*	activation energy of a chemical reaction	13.6.1
ΔG^0	standard Gibbs free energy change in a chemical process	10.2.2
GOD_{ox}	glucose oxidase, oxidized form	3.5.1
GOD_{red}	glucose oxidase, reduced form	3.5.1
ΔH	heat of reaction	Chapter 9
ΔH_r^0	standard enthalpy of reaction	Chapter 9
h	(a) Plank's constant	18.1
	(b) microcantilever thickness	22.1.1
I	(a) ionic strength	10.2.1
	(b) electric current	17.1
I_{AC}	sine-wave alternating current	17.1
I_D	drain current of a metal-insulator-semiconductor field effect transistor	11.1.4
I_{DC}	DC current	17.1
I_m	AC current amplitude	17.1
i	electrolytic current	13.3.1
i_a	anodic current	13.3.1
$i_{a,d}$	limiting, diffusion-controlled anodic current	13.3.2
i_C	capacitive current	13.5.3
i_c	cathodic current	13.3.1
$i_{c,d}$	limiting, diffusion-controlled cathodic current	13.3.1

i_f	Faradaic current	13.7.7
i_l	limiting current at an mediator-based amperometric enzyme sensor	15.1.3
i^*	the particular value of the limiting current recorded at $\alpha \gg 1$ and for $S = 1$	15.1.3
i_0	exchange current	13.6.4
J	diffusion flux	
J_d	limiting flux under first order kinetics and external diffusion control	15.1.2
J_l	flux limiting value	15.1.2
$J_{1,0}$	flux limiting value under zero-order kinetics	15.1.2
$J_{1,1}$	flux limiting value under first-order kinetics	15.1.2
J_M	mediator flux	15.2.1
J_P	product flux in an enzymatic sensor	4.2.1
J_S	substrate flux in an enzymatic sensor	4.2.1
$J_{P,m}$	product flux in the membrane of an enzymatic sensor	4.4.1
$J_{S,m}$	substrate flux in the membrane of an enzymatic sensor	4.4.1
J^*	the particular value of J_d for $s = K_M$	15.1.3
j	(a) current density	13.6.2
	(b) imaginary unit ($\sqrt{-1}$)	17.1; 21.2.2
j_a	anodic current density	13.6.2
$j_{a,d}$	limiting (diffusional) anodic current density	13.6.3
j_c	cathodic current density	13.6.2
$j_{c,d}$	limiting (diffusional) cathodic current density	13.6.3
j_0	exchange-current density	13.6.4
K_a	affinity constant	6.2.2
K_d	dissociation constant	18.4.1
K_e	equilibrium constant for the analyte–receptor interaction	18.4.1
K_{ex}	ion-exchange constant for a glass membrane	10.6.2
K_{exch}	ion-exchange constant for an ion-exchanger liquid membrane	10.8.3
K_M	Michaelis–Menten constant	3.6.1
K_p	partition coefficient	10.3.2
$K_{p,M}$	partition coefficient of the ion M	10.9.4
K_s	solubility constant of a sparingly soluble salt	10.5.2
k	spring constant of the microcantilever material	22.1.3
k_a	surface normalized pseudo-first-order rate constant for an enzymatic sensor	4.2.4
$k_{A,B}^{pot}$	potentiometric selectivity coefficient relative to ions A and B	10.4.2
k_B	Boltzmann constant	11.1.2
k_{cat}	turnover number of an enzyme	3.6.3
k_d	decay rate constant of an excited state species	18.3.6
k_e	(a) pseudo-first-order reaction rate for an enzyme-catalyzed reaction	6.2.2
	(b) excitation rate constant	18.3.7
k_H	proportionality coefficient in the Henry isotherm	11.3.5
k_m	mass-transfer coefficient	13.3.1
k_M'	reaction rate for enzyme regeneration by reaction with a redox mediator	15.1.1
$k_{M,N}^{pot}$	potentiometric selectivity coefficient relative to ions M and N	10.6.2
$k_{m,P}$	mass-transfer coefficient of the product	4.2.2
$k_{m,S}$	mass-transfer coefficient of the substrate	4.2.2
$k_{P,m}$	mass-transfer coefficients of the product in the membrane of an enzymatic sensor	4.4.1
k_s	standard rate constant of an electrochemical reaction	13.6.1
$k_{S,m}$	mass-transfer coefficients of the substrate in the membrane of an enzymatic sensor	4.4.1; 15.1.1
k_{SV}	Stern–Volmer constant	18.3.7
k_1	forward rate constant of the first step in the Michaelis–Menten mechanism	3.6.1
k_{-1}	backward rate constant of the first step in the Michaelis–Menten mechanism	3.6.1
k_2	rate constant for the second step in the Michaelis–Menten mechanism	3.6.1
L	(a) luminophores species	18.3.7
	(b) analyte-analog	18.4.2
L	electrical conductance	17.8.1
L^*	excited luminophores species	18.3.7
L_1	inductance at the motional branch of the equivalent circuit of a TSM piezoelectric oscillator	21.2.2
l	(a) distance between the plates of a capacitor	13.5.3
	(b) distance between the electrodes of an idealized conductometric cell	17.8.1
	(c) microcantilever length	22.1.1

Symbol	Description	Reference
l_e	thickness of an immobilized enzyme layer	4.2.1; 15.1.2
l_m	thickness of the external membrane in an enzymatic sensor	4.2.1
M_O	oxidized form of a redox mediator	14.2.1
M_R	reduced form of a redox mediator	14.2.1
m	(a) activity of an unspecified M^+ ion	10.5.2
	(b) mass	22.1.3
Δm	mass variation	21.2.3; 22.1.3
m^*	effective mass of a vibrating microcantilever	22.1.3
m_{aq}	activity of an unspecified M^+ ion in solution	10.3.2
m_m	activity of an unspecified M^+ ion in an ion-selective membrane	10.3.2
m_O	concentration of the oxidized form of a redox mediator	15.1.1
$m_{O,0}$	concentration of the oxidized mediator at the electrode surface	15.2.3
m_Q	mass of the vibrating zone of a TSM piezoelectric oscillator	21.2.3
m_R	concentration of the reduced form of a redox mediator	15.1.1
$m_{R,0}$	concentration of the reduced mediator form at the electrode surface	15.2.3
m_t	$m_O + m_R$	15.1.1
m_1	activity of an unspecified M^+ ion within the left-hand solution of an ion-selective membrane cell	10.5.2
m_2	activity of an unspecified M^+ ion within right-hand solution of an ion-selective membrane cell	10.5.2
N_O	number of moles of oxidized form of a redox couple	13.3.1
N_R	number of moles of reduced form of a redox couple	13.3.1
n	(a) number of moles	10.2.1
	(b) number of electrons in an electrochemical reaction	10.2.2; 13.3.1
	(c) activity of an unspecified N^+ ion	10.5.2
	(d) refractive index	18.2.1
	(e) overtone order	21.2.1
n_0	refractive index of the medium from which a light beam comes to an optical fiber	18.2.1
n_1	refractive index of the waveguide core	18.2.1
n_2	refractive index of the waveguide cladding	18.2.1
n_{eff}	effective refractive index	18.6.2
Ox	oxidized form of a redox couple	10.2.2; 13.3.1
O	subscript denoting quantities pertaining to the oxidized form of a redox couple	10.2.2; 13.3.1
P	reaction product	3.6.1
P	power of the transmitted light beam	18.3.1
P_e	dimensionless concentration of the reaction product in an immobilized enzyme layer (p_e/K_M)	4.2.5
P_0	power of the reference light beam	18.3.1
p	(a) concentration of a reaction product	4.2.1
	(b) partial pressure	10.2.2
p_e	concentration of the reaction product within an immobilized enzyme layer	4.2.1
$p_{e,0}$	concentration of the product at the transducer/immobilized enzyme layer interface	4.3.1
p_{CO_2}	partial pressure of carbon dioxide	10.17.4
pH	the negative logarithm (base 10) of hydrogen ion activity	
p_{H_2}	partial pressure of hydrogen	11.3.1
p_{O_2}	partial pressure of oxygen	10.17.2
$p_{m,i}$	product concentration at the membrane-enzyme layer interface	4.4.1
Q	(a) electrical charge	
	(b) quality factor of a resonator	21.2.7; 22.1.3
QY	fluorescence quantum yield	18.3.4
R	recognition receptor	
R	(a) ideal gas constant	
	(b) electrical resistance	9.1.1; 17.2.2
	(c) reflected light power	18.3.2
R	subscript pertaining to the reduced form of a redox couple	10.2.2; 13.3.1
R_{Air}	electrical resistance of a resistive gas sensor in contact with pure air	12.2.2
Red	reduced form of a redox couple	10.2.2; 13.3.1
R_{et}	electron-transfer resistance	13.6.5; 17.2.3
R_{Gas}	electrical resistance of a resistive gas sensor in contact with analyte-containing gas	12.2.2
R_{ref}	resistance of a resistive gas sensor in contact with a reference gas	12.2.2

RH	relative humidity	17.9.1
R_S	electrical resistance of a resistive gas sensor	12.2.2
R_s	resistance of an electrolyte solution	13.2; 17.2.1; 17.2.3
R_1	resistance at the motional branch of the equivalent circuit of a TSM piezoelectric oscillator	21.2.2
r	(a) radius	
	(b) receptor concentration	6.4
r_h	relative humidity	17.9.1
r_T	the turnover number of the substrate conversion relative to that of the enzyme reoxidation	15.1.2
r_0	total receptor concentration	6.4
S	enzyme substrate	3.6.1
S	dimensionless concentration of the substrate (s/K_M)	15.1.3
S_e	dimensionless concentration of the substrate within an enzymatic layer under external diffusion control (s_e/K_M)	4.2.5
s	substrate concentration	3.6.1
s_e	substrate concentration in an enzymatic layer	4.2.1; 15.1.1
$s_{e,0}$	substrate concentration at the transducer/immobilized enzyme layer interface	4.3.1
$s_{m,i}$	substrate concentration at the membrane/enzyme layer interface	4.4.1
T	(a) absolute temperature	
	(b) transmittance	18.3.1
T_d	dew point	17.9.1
T_{xy}	shear stress	21.2.4
t	time	
Δt	change in the thickness of a TSM piezoelectric oscillator	21.2.3
t_Q	thickness of a TSM piezoelectric oscillator	21.2.1
t_r	response time	4.3.2
u	ion mobility	10.3.1
V	(a) voltage	
	(b) volume of an enzyme layer	4.2.2
V_{AC}	sine wave alternating voltage	7.1
V_D	drain voltage for a metal-insulator-semiconductor field effect transistor	11.1.4
V_{DC}	DC voltage	17.1
V_G	gate voltage for a metal-insulator-semiconductor field effect transistor	11.1.4
V_{FB}	flat-band voltage	11.1.3
V_m	amplitude of sine-wave alternating voltage	17.1
V_T	threshold voltage of a metal-insulator-semiconductor device	11.1.3
V_T^*	threshold voltage of an electrolyte-insulator-semiconductor device	11.2.1
v	(a) velocity	21.2.4
	(b) potential scan rate	13.7.4
	(c) reaction rate	3.6.1
v'	reaction rate within an immobilized enzyme layer	4.2.2
v_a	surface-normalized reaction rate in an enzymatic sensor	4.2.2
v_e	reaction rate of an electrochemical reaction	13.3.1
$v_{e,a}$	velocity of an anodic reaction	13.6.1
$v_{e,c}$	velocity of a cathodic reaction	13.6.1
v_M	reaction rate of enzyme regeneration	15.1.1
v_m	maximum reaction rate of an enzyme-catalyzed reaction	3.6.1
v_S	reaction rate for the formation of an enzyme–substrate complex	15.1.1
v_{tr}	propagation velocity of a transverse wave	21.2.3
v_V	volume reaction rate within an immobilized enzyme layer	4.2.2
v_x	velocity along the x-axis	21.2.7
v_C	reaction rate of the substrate conversion in an enzyme–substrate complex	15.1.1
w	microcantilever width	22.1.1
X_C	capacitive reactance ($1/\omega C$)	21.2.2
X_L	inductive reactance (ωL)	21.2.2
x	distance	
Δx	microcantilever deflection	22.1.2
Y	admittance	17.1

Y_t	total admittance	17.1
y	response signal of a sensor	1.5
z	ion charge	10.2.1
Z	(a) electrical impedance	17.1
	(b) acoustic impedance	21.2.7
$|Z|$	impedance modulus	17.1
Z'	real part of the acoustic impedance of a TSM piezoelectric oscillator	21.2.2
Z''	imaginary part of the acoustic impedance of a TSM piezoelectric oscillator	21.2.2
Z_C	capacitive impedance	17.2.1
Z_F	Faradaic impedance	17.2.3
Z_{im}	imaginary part of electrical impedance	17.1
Z_m	motional impedance	21.2.7
Z_{m1}	motional impedance of an unloaded TSM piezoelectric resonator	21.2.2
Z_{m2}	motional impedance produced by loading a TSM piezoelectric resonator	21.2.7
$Z_{m,t}$	total motional impedance	21.2.7
Z_{re}	real part of electrical impedance	17.1
Z_s	mechanical impedance of a TSM resonator	21.2.7
Z_t	total impedance	17.1
Z_W	Warburg impedance	17.2.1

Greek Symbols

Symbol	Meaning	Section References
α	(a) substrate modulus for an enzymatic sensor under external diffusion control	4.2.4
	(b) transfer coefficient of a cathodic electrochemical reaction	13.3.3
β	(a) Biot number	4.4.1
	(b) transfer coefficient of an anodic electrochemical reaction	13.6.1
Γ	surface concentration	5.2
Γ_{max}	maximum surface concentration	5.2
γ	(a) activity coefficient	10.2.1
	(b) enzyme reoxidation capacity relative to the substrate conversion capacity in the absence of any diffusion limitation	15.2.2
δ	(a) thickness of the Nernst diffusion layer	13.3.1
	(b) charge fraction transferred in the interaction of a polar molecule with a semiconductor	11.3.5
	(c) Debye length	12.1.7
δ_{dl}	thickness of the electrical double layer	17.2.2
δ_S	partition coefficient of the substrate	4.4.1
δ_P	partition coefficient of the product	4.4.1
ε_d	dielectric constant	13.5.2
ε_{dl}	dielectric constant within the electrical double layer	17.2.2
η	(a) Da_M/Da_S	15.2.2
	(b) overvoltage (difference between the actual electrode potential and the equilibrium potential)	13.6.5
	(c) dynamic viscosity	21.2.4
η_L	dynamic viscosity of a liquid	21.2.5
θ	surface coverage degree	5.2
θ_e	internal lag factor	4.4.1
θ_c	critical incidence angle	18.2.1
θ_m	external lag factor	4.4.1
θ_1	incidence angle	18.2.1
θ_2	refraction angle	18.2.1
Λ	molar conductivity	17.8.1
Λ_i	molar conductivity of an ion i	17.8.1

λ	(a) electrical conductivity	17.8.1
	(b) light wavelength	18.1
	(c) acoustic wave wavelength	21.2.1; 21.2.3
	(d) the degree of completion of a combustion reaction	10.17.3; 13.10.5
μ	chemical potential	10.2.1
μ_i	chemical potential of a species i	10.2.1
$\bar{\mu}$	electrochemical potential	10.2.1
μ^0	standard chemical potential	10.2.1
μ_Q	shear elastic modulus of quartz	21.2.3
ν	(a) kinematic viscosity	13.7
	(b) frequency of electromagnetic radiation	18.1
	(c) Poisson ratio	22.1.2
$\tilde{\nu}$	wave number	18.1
π	the π number (ratio of a circle's circumference to its diameter)	
ρ	electrical resistivity	17.8.1
ρ_Q	quartz density	21.2.3
ρ_L	density of a liquid	21.2.5
σ	(a) Warburg constant	17.2.1
	(b) cell constant in conductometry	17.7.1
$\Delta\sigma$	difference in surface stress between the responsive and the passive surfaces of a cantilever	22.1.2
τ	fluorescence lifetime in the presence of a quencher	18.3.7
τ_d	diffusional time constant of an enzymatic sensor under internal diffusion conditions	4.3.2
τ_0	fluorescence lifetime	18.3.6
Φ_M	work function of a metal	11.1.3
Φ_S	work function of a semiconductor	11.1.3
ϕ	(a) Thiele modulus for internal diffusion in an immobilized enzyme layer	4.3.1
	(b) absolute electric potential	10.2.1
	(c) inner (Galvani) potential	10.2.1
	(d) phase of a sine wave signal	13.7.7; 17.1
$\Delta\phi_B$	boundary potential difference	10.3.2
ϕ_{CL}	chemiluminescence quantum yield	18.3.9
$\Delta\phi_D$	diffusion potential	10.3.1
ϕ_M	mediator Thiele modulus	15.2.2
$\Delta\phi_m$	membrane potential	10.4.1
ϕ_S	substrate Thiele modulus	15.2.2
χ	surface potential	10.2.1
ψ	outer potential	10.2.1
ω	angular velocity; angular frequency	17.1; 21.2.2

List of Acronyms

Acronym	Meaning	Section Reference
A	adenine	7.1
Ab	antibody	6.2.2
AC	alternating current	
AChE	acetylcholinesterase	3.4.3
ADH	adipic acid dihydrazide	5.3.2
Ag	antigen	6.2.2
ALP	alkaline phosphatase	3.4.4
APTES	(3-aminopropyl)-triethoxysilane	5.7
APTMS	(3-aminopropyl)-trimethoxysilane	5.7
ATP	adenosine triphosphate	
BOD	biological oxygen demand	23.2.1
BSA	bovine serum albumin	
C	cytosine	7.1
ChemFET	chemically sensitive field effect transistor	11.3.1
CDI	N,N.-carbonyldiimidazole	5.3.1
cDNA	complementary DNA	7.6
CNF	carbon nanofiber	8.3.5
CNT	carbon nanotube	8.3
Con A	Concanavalin A	19.5.2
CPE	constant phase element	17.2.1
CV	cyclic voltammetry	13.7.4
CVD	chemical vapor deposition	8.3.2
DC	direct current	
DDC	dicyclohexylcarbodiimide	5.3.1
DNA	deoxyribonucleic acid	7.1
DPV	differential pulse voltammetry	13.7.5
ds	double strand	7.1
ECL	electrochemically generated chemiluminescence	18.3.10
EDC	1-ethyl-3-(3-dimethylaminopropyl)carbodiimide	5.3.1
EDTA	ethylenediaminetetraacetic acid	
EIS	electrolyte-insulator-semiconductor structure	11.2.1
ELISA	enzyme-linked immunosorbent assay	6.2.6
EMF	electromotive force	10.2.2
EnFET	enzymatic field effect transistor biosensor	11.2.7
Fab	fragment, antigen binding region of an antibody molecule	6.2.1
FAD	flavin adenine dinucleotide	3.3
$FADH_2$	reduced FAD	3.3
FED	field effect device	11.1.3
FET	field effect transistor	Chapter 11
FIA	flow injection analysis	1.9
FPW	flexural plate wave	21.6.3
FRET	fluorescence resonance energy transfer	18.3.8
G	guanine	7.1
GFP	green fluorescent protein	23.5.3
GOPS	3-(glycidoxypropyl)-dimethyl-ethoxysilane	5.7
HOMO	highest occupied molecular orbital	
HOPG	highly ordered pyrolytic graphite	13.8.1
HPTS	8-hydroxypyrene-1,3,6-trisulfonic acid	19.9.1
HRP	horseradish peroxidase	3.4.1

HSA	human serum albumin	
HSAB	hard and soft acids and bases theory	10.9.2
IDT	interdigitated transducer	21.6.1
Ig	immunoglobulin	6.2.1
IgG	immunoglobulin G	6.2.1
ISE	ion-selective electrode	10.1
ISFET	ion selective field effect transistor	11.2.1
LAPS	light-addressable potentiometric sensor	11.2.5
LDH	lactate dehydrogenase	3.5.2
LOD	(a) limit of detection	1.6.3
	(b) lactate oxidase	3.5.2
LSV	linear scan voltammetry	13.7.4
LUMO	lowest unoccupied molecular orbital	
MEMS	microelectromechanical systems	Chapter 22
MIP	molecularly imprinted polymer	6.7.3
MIS	metal-insulator-semiconductor structure	11.1.3
MISFET	metal-insulator-semiconductor field effect transistor	11.1.4
MOS	metal-oxide-semiconductor structure	11.1.4
MOSFET	metal oxide-semiconductor field effect transistor	11.1.4
MPTS	(3-mercaptopropyl)-trimethoxysilane	5.7
mRNA	messenger RNA	7.6
MWCNT	multiwalled carbon nanotube	8.3
NAD^+	nicotinamide adenine dinucleotide	3.3
NADH	reduced nicotinamide adenine dinucleotide	3.3
$NADP^+$/NADPH	nicotinamide adenine dinucleotide phosphate	3.3
NASICON	$Na_3Zr_2Si_2PO_{12}$	10.17.4
NHS	N-hydroxysuccinimide	5.3.1
NPV	normal pulse voltammetry	13.7.5
OFET	organic field effect transistor	11.3.4
PB	Prussian Blue	13.9.4
PCR	polymerase chain reaction	7.5.3
PDMS	poly(dimethylsiloxane)	5.13.4
PEG	polyethylene glycol	
PHEMA	polyhydroxyethyl methacrylate	5.8.2
pH ISFET	pH-sensitive field effect transistor	11.2.2
PNA	peptide nucleic acid	7.2
PQQ	pyrroloquinoline quinone cofactor	3.3
PSA	prostate specific antigen	16.1.4; 21.4.2
PVA	poly(vinyl alcohol)	5.8.2
PVC	poly(vinyl chloride)	5.12.2
QCM	quartz crystal microbalance	Chap. 21
QCM-D	quartz crystal microbalance with dissipation monitoring	21.2.8
QD	quantum dot	20.1.1
RBM	radial breathing mode of a CNT	20.2.2
REFET	reference field effect transistor	11.2.6
RET	resonance energy transfer	18.3.8
RDCP	reciprocal derivative chronopotentiometry	13.7.8
RNA	ribonucleic acid	7.1
RTD	resistive thermal device	9.1
SAM	self-assembled monolayer	5.4.8
SAW	surface acoustic wave	21.6.2
SCE	saturated calomel electrode	10.2.2
SH-APM	shear horizontal acoustic plate mode	21.6.3
SHE	standard hydrogen electrode	10.2.2
SLAW	surface-launched acoustic wave	21.6.1
SMCC	(succinimidyl-4-(N-maleimidomethyl)cyclohexane-1-carboxylate)	5.3.2
SPR	surface plasmon resonance	18.6.1
STW	surface transverse wave	21.6.3
Sulfo-SMCC	sulfosuccinimidyl-4-(*N*-maleimidomethyl)cyclohexane-1-carboxylate	5.3.2
sulfo-NHS	N-hydroxysulfosuccinimide	5.3.1

ss	single strand	7.1
SWCNT	single-walled carbon nanotube	8.3
SWV	square-wave voltammetry	13.7.6
T	thymine	7.1
TCNQ	tetracyanoquinodimethane	14.2.3
TEOS	tetraethyl orthosilicate	5.7
THEOS	tetrakis(2-hydroxymethyl) orthosilicate	5.7
TISAB	total ionic strength adjustment buffer	10.4.4
TMOS	tetrakis(2-hydroxymethyl) orthosilicate	5.7
TMS	Teorell, Meyer and Sievers model	10.6.2
TOMEHE	thiazole orange methoxyhydroxyethyl	19.6.1
TSM	thickness-shear mode	21.2.1
TTF	tetrathiafulvalene	14.2.3
U	uracil	7.1
WGM	whispering-gallery modes	18.7.1
YSZ	yttrium-stabilized zirconia	10.17.1

1

What are Chemical Sensors?

1.1 Chemical Sensors: Definition and Components

The definition of an electrochemical biosensors given in ref. [1] can be adapted slightly to provide a general definition of the chemical sensor as follows. A chemical sensor is a self-contained device that is capable of providing real-time analytical information about a test sample. By chemical information we understand here the concentration of one or more chemical species in the sample. A target species is commonly termed the *analyte* or *determinand*. Besides chemical species, micro-organisms and viruses can be traced by means of specific biocompounds such their nucleic acid or membrane components. *Physical sensors* are devices used to measure physical quantities such as force, pressure, temperature, speed, and many others.

The first (and also best known) chemical sensor is the glass electrode for pH determination, which indicates the activity of the hydrogen ion in a solution.

When operated, a chemical sensor performs two functions, *recognition* and *transduction*, which are exemplified by the allegory in Figure 1.1. First, the analyte interacts in a more or less selective way with the *recognition (or sensing) element*, which shows affinity for the analyte. The sensing element may be composed of distinct molecular units called *recognition receptors*. Alternatively, the recognition element can be a material that includes in its composition certain *recognition sites*. Beyond this, the recognition element can be formed of a material with no distinct recognition sites, but capable of interacting with the analyte. In a chemical sensor, the recognition and transduction function are *integrated* in the same device. An analytical device with no recognition function included is not a chemical sensor but a concentration transducer.

Biosensors are chemical sensors in which the recognition system is based on biochemical or biological mechanisms. It is good to be aware of the fact that synthetic, biomimetic materials, that perform in the same way as materials of biological origin, have been developed and utilized in recognition elements.

As a result of the analyte interaction with the sensing element, certain physical or chemical properties of the sensing element vary as a function of the analyte concentration. In order to allow the user to assess this variation, a chemical sensor converts the above change into a measurable physical quantity. This process is called signal transduction (or, simply, *transduction*) or *signaling*. The word transduction is derived from the Latin "transducere," which means "to transfer or translate." A device that translates information from one kind of system (e.g., chemical) to another (e.g., physical) is called a *transducer* [2].

The sensing element and the transducer can be distinct components packaged together, in direct spatial contact, in the same unit. In certain types of chemical sensor, no physical distinction between the sensing element and the transducer can be made. Notwithstanding this, a distinction between the recognition and transduction functions as particular physical or chemical processes does exist.

The concept of the *molecular sensor* appears often in the literature. A molecular sensor is a molecule containing two distinct units. One of them is able to bind the analyte (e.g., an ion) in a selective way, while the second unit changes some physicochemical property (e.g., light absorption or emission) in response to binding of the analyte [3]. Therefore, recognition and signaling are performed by the same molecule.

By analogy with molecular sensors, the concept of *nanosensor* has been developed. A nanosensor is a sub-microscopic hybrid assembly including nanoparticles and molecular compounds featuring both recognition and signaling functions.

Molecular sensors and nanosensors are not chemical sensors in the sense of the above definition because the transducer is missing. Rather, such species can be considered as advanced analytical reagents. Notwithstanding this, molecular sensors and nanosensors can in principle be used to produce a true sensor upon integration with a transducer device.

The next two sections summarize the main recognition and transduction methods used in chemical sensors.

Chemical Sensors and Biosensors: Fundamentals and Applications, First Edition. Florinel-Gabriel Bănică.
© 2012 John Wiley & Sons, Ltd. Published 2012 by John Wiley & Sons, Ltd.

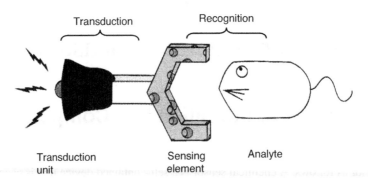

Figure 1.1 An allegory of a chemical sensor. A sensor is an assembly of a receptor and a signaling (transduction) unit. Adapted with permission from [3]. Copyright 1995 The Royal Society of Chemistry.

1.2 Recognition Methods

1.2.1 General Aspects

As a broad variety of recognition methods are utilized in chemical sensors, a general description of the recognition process is hardly achievable. Details on various recognition methods are given in the relevant chapters. Here, only a summary approach to this topic is presented.

A series of recognition processes occurs according to the following reaction scheme in which A is the analyte, R is the recognition receptor and P is a product of the analyte–receptor interaction:

$$A \quad + \quad R \quad \rightleftarrows \quad P$$
$$\text{Sample} \qquad \text{Sensing element} \tag{1.1}$$

The double arrow indicates that the recognition process is a reversible process at equilibrium. Reversibility of the recognition process arises from the fact that the product P involves noncovalent chemical bonds, such as ionic bonds, hydrogen bonds and van der Waals interactions. The recognition process can be characterized by its equilibrium constant K_A which is defined as:

$$K_A = \frac{c_P}{c_A c_R} \tag{1.2}$$

where symbols c represent concentrations of the species indicated by subscripts. This equilibrium constant indicates the affinity of the recognition receptor for the analyte. Great affinity results in a high value of the equilibrium constant. If the sensor response depends on the product concentration, the response will be determined by the concentration of the analyte in the sample.

An essential characteristic of the recognition process is its selectivity, which is the capacity of the sensor to respond preferentially to the analyte and not to another species B also present in the sample and acting as an interferent. The receptor–interferent interaction can be represented as follows:

$$B + R \rightleftarrows Q \tag{1.3}$$

The affinity of the receptor for the species B is indicated by the following equilibrium constant:

$$K_B = \frac{c_Q}{c_B c_R} \tag{1.4}$$

Sensor selectivity for the analyte is determined, in general, by the ratio of the above equilibrium constants and good selectivity is obtained if $K_A/K_B \gg 1$. More specific definitions of selectivity are given in the chapters addressing particular classes of chemical sensor.

It is important to note that certain recognition methods do not produce a well defined product, as shown in Scheme (1.1). In such cases, the interaction of the analyte with the sensing element is of a physical nature, such as gas sorption on a solid with no chemical reaction. In such cases, the monitoring of the recognition process can be performed by the measurement of a physical property of the sensing element, which depends on the analyte concentration in the sample.

The next sections present in summary form some of the common recognition methods used in chemical sensors.

1.2.2 Ion Recognition

Ion sensors were the first type of chemical sensors to be developed and produced on a large scale. The pH glass electrode was the first widely used ion sensor. It is based on the pioneering work of F. Haber and Z. Klemensiewicz (1908) and became commercially available by 1936 along with the Beckman pH-meter. Sensors for other ions (cations or anions) have been developed further.

Electric charge, which is the distinctive property of ions, is suitable for ion recognition. Therefore, ion recognition can be performed by various materials and reagents that have an electric charge opposite to that of the analyte ion.

Selectivity in electrostatic ion recognition arises from additional properties of the sensing material, such as the size of the ionic receptor site or some peculiarity of the analyte–receptor interaction, such as partial covalent character of the analyte–receptor bond.

Initially, ion sensors were based on solid materials, such as glass, or crystalline materials including selective recognition sites. By 1967–1968, molecular ion receptors were introduced. A molecular ion receptor can be a hydrophobic organic ion incorporated into a hydrophobic polymer membrane. As expected, this approach results in moderate selectivity. Superior selectivity has been obtained by using neutral ion receptors that interact with the analyte ion through a number of polarized atoms included in its structure.

Transduction in ion sensors is performed mostly by means of the effect of the ion charge upon the properties of the ion-recognition element. Typically, transduction in ion sensors is performed by potentiometric or optical methods.

1.2.3 Recognition by Affinity Interactions

Affinity interactions involve reversible multiple binding of two chemical species through noncovalent bonds, such as ionic bonds, hydrogen bonds, and van der Waals interactions. The product of an affinity interaction is a molecular association complex. In order for such a complex to form, the involved species should be complementary with respect to shape and chemical reactivity. For example, if one species displays a positively polarized hydrogen atom, the second one should display an electron donor atom placed such that it is able to form a hydrogen bond in the complex. The strength of the complex is indicated by its stability constant, which is similar to the equilibrium constant in Equation (1.2). Owing to the multiplicity of chemical bonding, association complexes of this type can be very stable.

Affinity interactions are very common in biological systems. An example of this type is represented by lectin proteins that recognize carbohydrates and form association complexes with such compounds.

A common type of affinity interaction is represented by the antibody–antigen interaction. *Antibodies* are glycoproteins produced by the immune system to identify and neutralize pathogen micro-organisms such as bacteria and viruses. The part of the pathogen that interacts with the specific antibody is called the antigen. The antibody–antigen interaction is an *immunochemical reaction*. Antibodies can be extracted from the blood of animals inoculated with an antigen, but can also be obtained from cell cultures.

In the clinical laboratory, immunochemical reactions are used for diagnostic purposes. Using specific antibodies as recognition receptors, pathogens can be identified. Conversely, using an antigen receptor, a specific antibody can be identified, which allows the detection of possible infection by a particular pathogen.

Besides pathogen or antibody detection, antibodies are used to recognize various protein molecules. Small organic molecules as such do not produce an immune response. However, an antibody specific for a small molecule is produced by an organism inoculated with a compound formed of this molecule attached to a protein. In this way, antibodies specific to certain small molecules can be obtained and used for analytical purposes.

Certain synthetic materials mimic the behavior of affinity reagents of biological origin. Molecularly imprinted polymers should be first mentioned in this respect. Molecularly imprinted polymers are polymeric material containing cavities with the size and shape matching the analyte molecule. In addition, the cavity includes functional groups that can bind reversibly to the analyte. Another class of synthetic affinity receptors are the nucleic acid aptamers that are synthetic nucleic acid molecules designed so as to form strong associations with certain small molecules or proteins.

All the above affinity recognition methods have found applications in the development of chemical sensors for a broad range of target species, including pathogenic micro-organisms, proteins and organic molecules.

1.2.4 Recognition by Nucleic Acids

In living organisms, nucleic acids function as supports for the storage and transfer of genetic information. In general, storage of genetic information is performed by deoxyribonucleic acids (DNAs) while transfer of information within cells is performed by ribonucleic acids (RNAs).

Nucleic acids are composed of a polymeric backbone onto which nucleobases are grafted. There are four nucleobases in DNA compositions, namely adenine (A), cytosine (C), guanine (G), and thymine (T). In RNAs, thymine is replaced by uracil (U). A sequence of three nucleobases codifies an amino acid and a sequence of nucleobase triplets codifies the primary structure of a protein.

Significantly, hydrogen bonds can only form between two distinct pairs of nucleobases, which are G-C and A-T in DNAs (or A-U in RNAs). This permits two complementary nucleic acids to form a double-strand association complex in a process called hybridization. Nucleic acid hybridization is a particular kind of affinity interaction that involves only hydrogen bonding between well-defined pairs of nucleobases.

Nucleic acid hybridization is the basis of the recognition process in nucleic acid sensors. A short nucleic acid forms the receptor (usually termed the capture probe) which is able to recognize by hybridization a particular nucleic acid sequence in the analyte nucleic acid.

Nucleic acid assays are of interest to clinical diagnosis, in the detection of genetic anomalies and also in the identification of pathogen micro-organisms. In forensic science, DNA testing assists in the identification of individuals by their particular DNA profiles.

1.2.5 Recognition by Enzymes

Enzymes are protein compounds that function as catalysts in chemical reactions occurring in living systems. The compound converted by the catalytic action of the enzyme is called the enzyme substrate. The catalytic property is selective to a particular substrate or to a particular functional group in a class of compounds.

Most chemical sensors rely on recognition processes at equilibrium as indicated by Scheme (1.1). In contrast, recognition by enzymes is a dynamic process which involves three main steps: First, the target compound (substrate) is bound to the active site of the enzyme to form a substrate–enzyme complex in a process similar to that in Scheme (1.1). The bound substrate undergoes a further chemical conversion, possibly with the participation of other coreagents. Finally, products are released and the active site of the enzyme returns to its initial state. This sequence is repeated with another substrate molecule as long as the substrate and coreagents are still present.

Many enzymes preserve their catalytic activity after isolation from a biological material and can be incorporated as recognition agents in the sensing element of a sensor. Transduction in enzymatic sensors can be achieved by measuring the steady-state concentration of a product or a coreagent involved in the enzymatic process.

There are various applications of enzymes in chemical sensors. First, enzymatic sensors can be designed for the purpose of substrate determination. Secondly, enzymatic sensors can be utilized in the determination of inhibitors, which are chemical species that affect the enzyme activity. Thirdly, enzymes can be employed as transduction labels in sensors based on affinity recognition. Hence, enzymes occupy a central role in the framework of chemical-sensor science.

1.2.6 Recognition by Cells and Tissues of Biological Origin

As shown before, enzymes form an important class of recognition receptors utilized in chemical sensors. Although isolated enzymes were initially used, it was soon realized that enzymes incorporated in biological materials (such as cells or tissues) can perform better due to the fact that they are in their natural environment. This leads to the development of a new class of chemical sensor in which the recognition is performed by cells or tissues of biological origin.

Application of biological cells and tissues is, however, much broader, as such entities can react to chemical stimuli by modifications in their metabolic processes. A metabolic modification leads to changes in the consumption of oxygen or to excretion of particular chemical species. Such modifications are exploited for transduction purposes.

1.2.7 Gas and Vapor Sorption

Determination of gases and vapors is a topic of great interest in various areas, including the monitoring of air quality, control of hazardous gases and vapors in industrial environmental and physiological investigations.

General recognition methods for gases and vapors are based on sorption either at the surface of (adsorption) or within (absorption) a solid material used for recognition. Depending on the target compound, various materials are used for gas and vapor recognition, including certain metals, polymeric materials or inorganic materials. Sorption can be a purely physical phenomenon or can be accompanied by chemical reactions that modify the chemical state of the analyte or that of the recognition material.

1.3 Transduction Methods

1.3.1 General Aspects

It is possible to distinguish two main transduction strategies, namely chemical transduction and physical transduction.

Chemical transduction is performed by monitoring the change in the chemical composition of the sensing element in response to the recognition process. In other words, the change in the concentration (or amount) of the product P is measured. If the primary product P is not detectable, one can resort to the monitoring of a coreagent or secondary product of the recognition process.

If none of the compounds involved in the recognition process is detectable, one can resort to product labeling by a detectable species called a *signaling label* (or a *transduction label*). The label can be a simple molecular species or a nanoparticle that can be detected by available physicochemical methods. Widely used labels are certain enzymes that allow indirect transduction. More specifically, an enzyme label catalyzes a chemical reaction that produces a readily detectable species.

Physical transduction focuses not on the chemical composition but on a specific physical property of the sensing element that is affected by its interaction with the analyte. Common physical transduction methods are based on the measurement of mass, refractive index, dielectric properties or electrical resistivity. Such methods are, as a rule, *label-free* transduction methods.

A brief overview of transduction methods applied in chemical sensors is given below.

1.3.2 Thermometric Transduction

A straightforward transduction method is based on the thermal effect of the recognition process, which leads to a change in the temperature. However, thermometric transduction is feasible only if the recognition is accompanied by a catalytic process, as in the case of enzyme-catalyzed reactions. Only catalytic processes generate sufficient heat to produce a measurable variation of the temperature. Thermometric transduction is also applicable in chemical sensors for combustible gases that react with oxygen at the surface of a suitable catalyst.

1.3.3 Transduction Based on Mechanical Effects

Recognition leads to a change in the overall mass of the sensing element. Mass change can be monitored by means of a mass transducer based on a vibrating piezoelectric crystal, known as the *quartz crystal microbalance*. The response signal of this transducer is the vibration frequency, which depends on the overall mass of the device.

More generally, propagation of mechanical vibration (acoustic waves) is affected by the change in the properties of the sensing element in response to the recognition process. For example, the speed of an acoustic wave can be modified as a result of analyte interaction with the sensing element.

Recently, a new class of mechanical transducer has been developed, namely microcantilevers. When integrated with a sensing element, a microcantilever undergoes bending as a function of the extent of the recognition process. Alternatively, vibrating microcantilevers provide information about mass change in a similar way to the quartz crystal microbalance.

1.3.4 Resistive and Capacitive Transduction

Analyte interaction with a properly selected recognition material can lead to changes in the electrical property of this material. Thus, interaction of combustible gases with semiconductor metal oxides causes the electrical resistivity to change as a function of the analyte concentration. This is the basis of *resistive transduction*.

Another electrical property that can be affected by the recognition process is the dielectric constant. The dielectric constant can be assessed by including the recognition material as a dielectric in the structure of a capacitor and measuring the capacitance of this capacitor. In this way, *capacitive transduction* is achieved.

1.3.5 Electrochemical Transduction

Sensors for aqueous solution samples can be based on electrochemical transduction methods. Electrochemistry deals with ion transport, ion distribution and electron-transfer reactions at the solution interface with a solid conductor (electrode). Besides electrolyte solutions, electrochemistry also addresses charge-transfer processes in systems involving ionic solids, which are also of relevance to certain types of chemical sensor.

Determination of ions can be achieved by means of sensors based on *potentiometric transduction*. The sensing element in potentiometric ion sensors is a membrane including ion-selective molecular receptors or receptor sites in a solid material. This membrane is placed between two solutions, one of them being the sample and the other one a solution containing the analyte ion at a constant concentration. Ion exchange at each side of the membrane leads to the development of a potential difference between the two sides of the membrane. This potential difference can be measured and related to the concentration of the analyte ion in the sample. Potentiometric ion sensors (commonly, but improperly designated ion-selective electrodes) form one of the main classes of chemical sensors.

An advance in potentiometric ion sensors was achieved by integrating ion-selective membranes with a semiconductor device of the field effect transistor type. In such sensors, the electric potential developed at the membrane–sample solution acts directly on the characteristics of the field effect semiconductor device.

An analogous principle is used in gas sensors based on field effect devices, with the notable exception that the gas-sensing element is formed of a metal with catalytic properties.

Potentiometric ion sensors have another important application in chemical sensors, namely they can act as transducers in sensors base on ion-generating recognition processes. Such applications refer to sensors for gases, such as carbon dioxide, ammonia, hydrogen cyanide and hydrogen fluoride that give rise to ions upon dissolution in aqueous solutions. Ion sensors are also widely used as transducers in enzymatic sensors as many enzyme reactions produce or consume certain ions (e.g., hydrogen or ammonium ions) or produce a detectable gas, such as carbon dioxide or ammonia.

Measurement of electric current forms another class of transduction method in electrochemical sensors, commonly known as *amperometric sensors*. The beginning of amperometric sensors is represented by the oxygen probe introduced by Leland C. Clark Jr. in 1956 [4]. This device indicates the concentration of dissolved oxygen using the electrochemical reduction of oxygen and the associated electrolytic current as the response signal. The discovery of the amperometric oxygen sensor opened the way to the development of amperometric enzymatic sensors, also pioneered by Leland C. Clark Jr. Amperometric enzymatic sensors are based on enzymes that catalyze oxidation–reduction reactions and involve a small, inorganic molecule as coreagent. Oxygen is the natural coreagent in such reactions, but it can be substituted by artificial coreagents in more advanced designs. Direct electron exchange between the working electrode of an electrochemical cell and the active site of an oxidase-type enzyme is an alternative transduction method in amperometric enzymatic sensors.

Amperometric transduction is also suited to affinity sensors provided that an electrochemically active compound is attached to the recognition product (P in reaction (1.1) and acts as an electrochemical label.

Some of the nucleobases included in the nucleic acid structure are electrochemically active and their electrochemical reactions are used to monitor the recognition by hybridization.

A series of electrochemical transduction methods are based on the concept of *electrochemical impedance*. The electrochemical impedance indicates the opposition to the flow of an alternating current through an electrochemical cell. Electrochemical impedance measurements provide a wealth of information about the physicochemical processes occurring in an electrochemical cell, such as ion migration, charge distribution at the electrode/electrolyte interface and the velocity of the electrochemical reaction. Each of the above processes can be related to the properties of a sensing element integrated with the electrochemical cell and used for transduction purposes.

1.3.6 Optical Transduction

Interaction of electromagnetic radiation with matter forms the basis of a broad range of analytical methods commonly known as spectrochemical methods of analysis. Commonly, electromagnetic radiation in the ultraviolet-visible-infrared domains is used for analytical purposes. Not surprisingly, a broad range of chemical sensors have been developed on the ground of interaction of the sensing element with electromagnetic radiation. Sensors based on this kind of transduction are termed optical sensors.

Optical transduction can be based on light emission or light absorption by the sensing element. Such processes are associated with transitions between energy levels of certain species (molecules or nanoparticles) included in the sensing element. The light-responsive species can be a transduction label, a coreagent or a product of the recognition process.

Optical transduction can also be achieved by monitoring a physical quantity connected to light propagation through the sensing layer, such as the refractive index. Light scattering provides additional methods for optical transduction.

1.4 Sensor Configuration and Fabrication

The final goal of sensor development is to obtain a marketable product. In order to achieve this goal, a sensor should be simple, robust and easy to use. Field applications require portable sensors, while biomedical applications often demand implantable sensors for *in vivo* monitoring of chemical species of physiological relevance. Miniaturization is, in this case, an essential condition. Miniaturization is also important for reducing the amount of sample required and for integration of multiple sensors in arrays in order to increase the throughput and to alleviate interferences (see Section 1.7).

Sensor miniaturization brings about an additional advantage, namely the possibility of constructing *smart sensors*. In a smart sensor, the sensor itself is integrated with microelectronic circuits that control the functioning parameters and perform data processing and interfacing with external readout equipment.

Good durability of a sensor is obtained, as a rule, at the price of using a more intricate fabrication technology and the consumption of expensive materials, which brings about a higher cost of the product. On the other hand, operation of a long-life sensor involves a preliminary calibration and some kind of conditioning after each run, which are not easily achievable in field or point-of-care applications. That is why it is preferable in certain cases to design cheap, disposable sensors for single-use application. As calibration of a disposable sensor is not feasible, it is essential that the fabrication technology secures very good batch reproducibility of the response characteristics.

Low cost and batch reproducibility can be obtained by excluding the utilization of hand work in the fabrication process. Advanced technologies for sensor fabrication are based on micromachining methods, initially developed in the area of microelectronic circuit technology. Micromachining allows for miniaturization and straightforward integration of multiple sensors in sensor arrays.

Chemical sensors can be shaped as a dip-in probe, similar to the well-known glass electrode for pH determination.

Very low volumes of samples can be tested by drop application onto a sensor with a flat surface, for example, a sensor formed as a thin layer on a plastic strip. This configuration is suitable for use in disposable sensors.

Facile sampling is provided by *capillary fill sensors*. Such sensors are formed of two sheets of glass held apart by a gap of capillary dimension. The sensor is formed as a thin layer onto the inner surface of a sheet. The sample enters the device with a reproducible volume by capillary ascension. An example of such a sensor is given in [5].

The principles of thin-layer chromatography have been applied to develop *lateral flow sensors*, which consists of a thin, porous layer deposited on a solid strip. Several distinct zones are formed in sequence on the strip; first a sample application pad, next, one or more zones containing reagents for chemical conditioning of the analyte, and, finally, the sensing-detection zone. When applied on the sampling pad, the sample drifts through capillary diffusion across the chemical conditioning zone and then, further to the detection zone where it is accumulated and generates the response signal.

Sequential analysis of multiple samples is best carried out by integration of the sensor in a flow-analysis system (see Section 1.8).

A *generic sensor* or *sensor platform* is a device that allows for straightforward integration of recognition receptors from a specific class in order to obtain sensors for various analytes belonging to the same class. As a rule, a generic sensor includes the transducer and additional elements (e.g., molecular linkers) that assist the integration of the receptor in an easy and rapid way. The generic sensor approach is convenient when the recognition element is not sufficiently stable. In this case, the receptor is integrated with the prefabricated generic sensor just prior to the test.

1.5 Sensor Calibration

In analytical chemistry, calibration aims at establishing an unequivocal mathematical relationship between the measured quantity and the analyte concentration [6,7].

The output of a chemical sensor is a measurable physical quantity called the *response signal*. The intensity of the signal (y) is correlated with the analyte concentration in the sample (c) by means of the following general relationship:

$$y = F(c) + e_y \tag{1.5}$$

Here, $F(c)$ represents the *calibration function* and e_y is the measurement error of the response. Hence, the concentration can be found from the inverse of the calibration function, which is called the *analytical function* or the *evaluation function*:

$$c = F^{-1}(y) \tag{1.6}$$

The form of the calibration function can be derived by mathematical modeling of the sensor or can be set as an empirical interpolation function. A common and very convenient calibration function is the direct proportionality relationship:

$$y = ac \tag{1.7}$$

where a indicates the sensor *sensitivity*. A direct proportionality function is characterized by constant sensitivity. More generally the sensitivity is defined by the following equation:

$$a = \frac{dy}{dc} \tag{1.8}$$

In the case of a nonlinear calibration function, the sensitivity is not a constant, but depends on the analyte concentration.

The calibration function may include constant parameters that are characteristic of the sensor but are independent of the sample properties. If these parameters can be derived from fundamental physicochemical laws and general constants (e.g., gas constant, Faraday constant), one has to deal with an *absolute analytical method*. This situation arises very infrequently.

Often, the response depends also on specific parameters of the analyte, such as the molar absorption coefficient in measurements based on light absorption. When a known specific parameter of the analyte comes into play, possibly along with other known empirical parameters, the analytical method is a *definitive measurement* method.

The most common situation is that in which one or more parameters in the calibration function cannot be derived a priori. If the mathematical form of the calibration function is known, the parameters in the calibration function are determined by measurements on samples with known concentration, commonly termed *reference samples* or *standard samples*. For example, in the case of a direct proportionality function, the sensitivity can be found using the measured responses and the concentration of a reference sample. In this case, the sensitivity is the quotient of the measured signal and the known concentration. More accurately, the sensitivity is obtained as the slope of the response–concentration relationship obtained by means of a series of reference samples. This approach is a *direct reference measurement*.

A common case is that in which the mathematical form of the calibration function is not known. In this case, the calibration function is obtained as an empirical interpolation function. This function is obtained by fitting y–c data produced by reference samples to a selected function, such as polynomials or another suitable function. A possible interpolation function is the linear function:

$$y = a_0 + a_1 c + e_y \tag{1.9}$$

where a_0 and a_1 are empirical parameters that are usually estimated by least square fitting; a_1 represents here the sensor sensitivity. If the linear function applies to concentrations near to zero, a_0 is the blank response. An analytical measurement based on these principles is an *indirect reference measurement*.

The quality of the calibration function should be validated by performing measurements on reference samples or by comparing the analysis results produced by the sensor with results obtained by an alternative analytical method. Reliable calibration is obtained by means of using reference samples with the chemical composition as close as possible to that of the unknown sample. In this way, the effect of the sample matrix on the response is corrected for.

The error of the measured concentration depends on measurement errors in both the calibration step and the sample analysis. When using a linear calibration function, the calibration error is minimal at the midpoint of the considered concentration range and increases with the distance from the midpoint. It is strongly recommended not to perform measurements outside the calibration range.

If suitable reference samples are not available, one can resort to the *standard addition method*, which is based on measurements on the plain sample and samples with the concentration modified in a controlled way. This method is applicable when the response is directly proportional to the analyte concentration. The response for a sample with modified concentration is:

$$y = a(c + \Delta c) \tag{1.10}$$

where c is the unknown concentration and Δc is the known variation in the concentration. A sequential increase in Δc gives a set of y–Δc data and the unknown concentration can be derived as the quotient of the intercept and the slope of the y–Δc straight line. Graphically, the unknown concentration can be obtained as $c = -(\Delta c)_{y=0}$, where $(\Delta c)_{y=0}$ is the Δc value at which the extended y–Δc line intersects the horizontal axis. As this method is essentially an extrapolation method, its accuracy is poorer than that of a measurement based on reference measurements.

It was assumed in the above approach that the response of the sensor depends on the concentration of one single species in the solution. This kind of sensor is known as a *zero-order sensor* and the calculation of the concentration is carried out by univariate calibration ((that is, single-component calibration). Higher-order sensors are introduced in Section 1.7.

1.6 Sensor Figures of Merit

The figures of merit of a sensor indicate how much a sensor fits the expected performances as far as the quality of the results, response stability, and ruggedness under storage and operation are concerned.

Being an analytical device, the performance characteristics of a chemical sensor can be defined by parameters used in the characterization of an analytical method. A systematic presentation of these parameters is given in ref. [8], which is recommended to the interested reader for more details. A number of performance characteristics are indicated by statistical parameters that are introduced in specialized texts (e.g., [6]).

An important statistical parameter used in the assessment of the quality of analytical results is the *confidence interval*, which indicates the scattering of measured values. The confidence interval for a series of replicate measurements with the average $\bar{c}$ and standard deviation of the mean $s_{\bar{c}}$ is:

$$\mathrm{cnf}(\bar{c}) = \bar{c} \pm \Delta\bar{c} \tag{1.11}$$

where $\Delta\bar{c}$ is the confidence limit given by:

$$\Delta\bar{c} = s_{\bar{c}}t_{1-\alpha,\nu} \tag{1.12}$$

Here, $t_{1-\alpha,\nu}$ is the quantile of the *t*-distribution at the level of significance α ($0 < \alpha < 1$) and for ν degrees of freedom. $P = 1 - \alpha$ indicates the probability that the true mean is likely to lie within the confidence interval. t values are tabulated as a function of $1 - \alpha$ and ν (see, e.g., ref, [6]). For example, if α is 0.05, there is a probability of 0.95 (that is, 95%) to find the mean value within the confidence interval. As indicated in Equation (1.12), a low standard deviation brings about a narrow confidence interval, which implies a low dispersion of data.

1.6.1 Reliability of the Measurement

The terms accuracy, precision and trueness define the reliability of the analytical measurement.

The *accuracy* indicates the degree of concordance between the concentration determined in a single test, and the true concentration (that is, the concentration in a certified reference material). The difference between the certified and measured concentration represents the *bias*:

$$\text{bias}(c) = c_{\text{test}} - c_{\text{true}} \tag{1.13}$$

A bias may be due to *systematic errors* produced by wrong calibration or improper operation of the sensor. Human errors, instrumental or computation errors are known as *gross errors* and also give rise to bias. An *outlier* is a result that appears to deviate markedly from other members of the data sample in which it occurs. Outliers should be discarded before proceeding to data analysis.

The accuracy of one single analytical result depends on the bias and the confidence limit and is quantified as follows:

$$\text{acc}(c) = 1 - \frac{\text{bias}(c)}{\Delta\bar{c}} \tag{1.14}$$

Trueness refers to a large number of replicate measurements on the same sample, giving the average concentration $\bar{c}$. Trueness is similar to the accuracy of the average concentration:

$$\text{trn}(c) = 1 - \frac{\text{bias}(\bar{c})}{\Delta\bar{c}} = \text{acc}(\bar{c}) \tag{1.15}$$

Precision indicates the degree of concordance between independent measurement results obtained under similar conditions. *Precision of an analytical procedure* is:

$$\text{prec}(c) = 1 - \frac{s_c}{\bar{c}} \tag{1.16}$$

where s_c is the standard deviation and $\bar{c}$ is the average of a series of replicate measurements. The numerical value of this parameter increases with decreasing s_c, that is, the error. For error-free measurements (that is, $s_c \rightarrow 0$) the precision becomes 1. The *precision of an analytical result* is determined by the relative confidence interval:

$$\text{prec}(\bar{c}) = 1 - \frac{\Delta\bar{c}}{\bar{c}} \tag{1.17}$$

Parameters defined in Equations (1.14)–(1.17) are graded on a scale extending from zero to one. For a good-quality sensor, each of the above parameter is close to one.

As a sensor may be used in the analysis of a series of samples, it is expected to maintain its calibration parameters at constant levels. If this condition is fulfilled, the sensor has a good *repeatability*. Sensor repeatability is checked by a sequence of measurements on identical reference samples carried out under the same conditions of measurement. Poor repeatability appears when one or more calibration parameters suffer a drift, that is a slow modification with time. The drift may be due to the alteration of the properties of the sensing element under repeated or prolonged contact with samples. Drift elimination is achieved by proper design and construction of the sensing element.

Repeatability should not be confused with *reproducibility*, which indicates the capacity of the sensor to give a similar result under different conditions, that is, different operators, different apparatus, different laboratories or after large intervals of time.

1.6.2 Selectivity and Specificity

An important figure of merit is the *selectivity* of the sensor, which indicates the extent to which a sensor can determine a particular analyte without interferences from other components of the sample. In order to account for selectivity, the calibration function should include one or more additional terms that express the interference of compounds (concomitants) accompanying the analyte in the sample. If the effect of such terms is below the accepted error level, the sensor has a satisfactory selectivity. Otherwise, the sensor delivers an incorrect result.

Interference arising from the interaction of a concomitant with the receptor site is a specific interference. If the concomitant affects the sensor response by interacting with other components of the sensing element, it produces nonspecific interference. Selectivity should be not confused with *specificity*, which designates ultimately selectivity and cannot be graded as it is an absolute term.

Sensor selectivity is mostly determined by the selectivity of the analyte–receptor interaction. In the early stages of chemical-sensor science, selectivity was a crucial issue and the main objective of research in this field was to find receptor materials as selective as possible for a particular analyte. As will be shown in Section 1.7, an assembly of poorly selective sensors can provide concentrations of a series of analytes by appropriate data processing methods. In this case, poor selectivity is an advantage rather than a drawback.

1.6.3 Detection and Quantification Capabilities

As a rule, sensors are expected to determine concentrations above a specified level. The lowest concentration level at which the sensor still provides reliable results represents the *limit of detection* (LOD). LOD is the lowest concentration or quantity of a substance that can be distinguished from the absence of that substance (a blank value) within a stated confidence limit. The meaning of terms included in the definition of the LOD is demonstrated in Figure 1.2. If y_b and s_b are the mean value and the standard deviation of the blank measurements, respectively, the signal giving the limit of detection is:

$$y_{LOD} = y_b + k s_b \tag{1.18}$$

where k is a numerical factor chosen according to the confidence level desired. As a rule, $k = 3$. The LOD in terms of concentration is obtained from the y_{LOD} value using the calibration function of the sensor. The definition of the LOD proves that this parameter depends on two factors: the blank response and its standard deviation. A good (that is, very low) LOD is obtained if the blank response is very low and the accompanying noise (which determine the s_b value) is also very low. Sensor LOD can depend on both the particular features of the recognition process and the intrinsic LOD of the transduction method.

More reliable analytical results are obtained when the concentration is above a higher limit, called the *limit of quantification* (LOQ). This limit corresponds to a signal that differs from the blank average by $10 s_b$.

It is clear that the detection limit depends on the fluctuations in the blank signal. Blank fluctuations may arise from random processes in the sensor functioning (sensor noise) and also from electronic noise in the readout equipment.

The sensor *sensitivity*, which has been already introduced in Section 1.5, reflects the change in the response produced by a unit variation in the concentration. Clearly, the LOD is connected with the sensitivity, as the LOD is obtained by multiplying the y_{LOD} value with the sensitivity. Often sensitivity is associated with the detection capability. This misuse is not recommended by IUPAC.

Every type of sensor can detect concentrations within a more or less wide range of concentrations, which is known as the *response range* (or concentration domain). The lower limit of the response range is the LOD introduced earlier; a more accurate designation of this parameter is the lower limit of detection. The upper limit of the

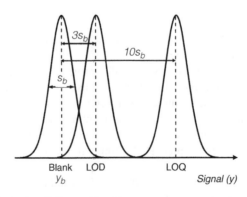

Figure 1.2 Definition of the limit of detection (LOD) and limit of quantification (LOQ). Curves represent the normal (Gaussian) distribution of errors.

concentration range is the concentration at which the response deviates significantly from the assumed calibration function. For example, a marked deviation from a linear calibration function determines the upper limit of detection. This deviation can be determined by various factors, such as the saturation of the receptor sites with analyte. The ratio between the upper and lower detection limits represents the *dynamic range* of the sensor. A response range of about one order of magnitude (that is, a dynamic range of about 10) is a minimal requirement.

Another figure of merit connected to the sensitivity is the *resolution*. The resolution of a sensor is the smallest detectable change in analyte concentration. It depends on two factors: the noise and the sensitivity. The noise determines the smallest detectable change in the response and the resolution is the ratio between this response variation and the sensitivity. A high sensitivity and a low noise level brings about a high resolution. Although this definition is reminiscent of the LOD definition, the resolution is not similar to the LOD because the noise in the presence of the analyte can differ from the blank noise.

1.6.4 Response Time

As a rule, a sensor is used to carry out sequential analysis of a series of samples. The throughput of the analytical process is determined by the time required by additional operations, such as sensor cleaning, sample conditioning, and sample introduction. The *response time* of the sensor may also determine to a certain extent the throughput. The response time is the time elapsed since the analyte is added to a well-stirred, analyte free solution, to the moment when the sensor response attains a practically constant value. During this time interval, the sensor functions in the transitory regime. The response time can be determined by the velocity of the analyte diffusion to the receptor sites or by the rate of analyte interaction with the receptor, or both. A response time below 1 min is excellent, and a response time of about 10 min is still satisfactorily. Much longer response times, in the range of tens of minutes, are still accepted if the nature of the physicochemical processes in the sensor do not permit improvements.

1.7 Sensor Arrays

Sensor arrays are assemblies of multiple sensors. If all sensors in the array are similar and respond to one single analyte, the array is useful for parallel analysis of multiple samples, each sample being applied to one of the sensors.

Sensor arrays can be designed for performing simultaneous determinations of multiple analytes in a sample (multiplexing). If each sensor is selective to a particular analyte, it acts independently of the other ones and provides a particular analyte concentration. However, this needs advanced selectivity which can be achieved only with a limited number of recognition methods.

A more general approach is based on arrays composed of poorly selective (cross-sensitive) sensors. In this case, each sensor responds to more than one sample component and the response of each sensor is a summation of the effects exerted by a series of components.

1.7.1 Quantitative Analysis by Cross-Sensitive Sensor Arrays

In order to determine the concentration of each analyte, the composite response should be processed by statistical methods known as *multivariate data analysis methods* Application of such methods to analytical chemistry is one of the objectives of Chemometrics [9]. An alternative approach to the analysis of data produced by a sensor array is based on *artificial neural networks*.

This section introduces some elements of multivariate data analysis. Suppose that the array includes n cross-sensitive sensors that respond linearly to m analytes in a certain sample k. Further, suppose that the response of each sensor to a particular analyte is directly proportional to the concentration of that analyte (c_{jk}), the proportionality constant (called sensitivity) being a_{ij}. Finally, assume that the overall signal produced by one sensor (y_{ki}) results by the superposition of the response to each analyte ($a_{ij}c_{jk}$). Then, sensor responses are given by the following set of equations:

$$
\begin{aligned}
y_{1k} &= a_{11}c_{1k} + a_{12}c_{2k} + \cdots + a_{1j}c_{jk} + \cdots + a_{1m}c_{mk} \\
y_{2k} &= a_{21}c_{1k} + a_{22}c_{2k} + \cdots + a_{2j}c_{jk} + \cdots + a_{2m}c_{mk} \\
&\cdots\cdots\cdots\cdots\cdots\cdots\cdots\cdots\cdots\cdots\cdots\cdots\cdots\cdots\cdots \\
y_{ik} &= a_{i1}c_{1k} + a_{i2}c_{2k} + \cdots + a_{ij}c_{jk} + \cdots + a_{im}c_{mk} \\
&\cdots\cdots\cdots\cdots\cdots\cdots\cdots\cdots\cdots\cdots\cdots\cdots\cdots\cdots\cdots \\
y_{nk} &= a_{n1}c_{1k} + a_{n2}c_{2k} + \cdots + a_{nj}c_{jk} + \cdots + a_{nm}c_{mk}
\end{aligned}
\tag{1.19}
$$

If N samples are analyzed with the same array, N similar systems of equations, with $k = 1, 2, \cdots N$ should be written. This assembly of systems of equations, which represents the mathematical model of the array response, can be written in a condensed form using matrix notation, as follows:

$$\underset{(n \times N)}{Y} = \underset{(n \times m)}{A} \times \underset{(m \times N)}{C} \tag{1.20}$$

Here, Y is the matrix including the response of n sensors to N samples. In this matrix, each column includes the responses associated with a particular analyte in a given sample, while each row represents the response of individual sensors to different analytes in the same sample. A is the *sensitivity matrix*, composed of the a_{ij} terms and C is the concentration matrix which contains the known concentrations of m analytes (rows) in N samples (columns). Equation (1.20) is used in the calibration stage to obtain the sensitivity matrix using samples of known concentrations. Concentrations in a set of N' unknown samples can be predicted from the matrix of the sensor responses for the unknown samples Y^s by the following matrix operation:

$$\underset{(m \times N')}{C^s} = \underset{(m \times n)}{B} \times \underset{(n \times N')}{Y^s} + \underset{(m \times N')}{e} \tag{1.21}$$

Here, e is the residual matrix, a residual being the difference between measured and model data. B is a *prediction matrix*. When the number of sensors in the array exceeds the number of analytes present in samples (that is, $n > m$), the B matrix is obtained from the sensitivity matrix A as:

$$B = A^T (AA^T)^{-1} \tag{1.22}$$

where A^T is the transpose of the sensitivity matrix. Although sensor arrays with $n = m$ can be used, arrays with $n > m$ allow for more reliable data processing.

Equation (1.21) can be handled by multilinear regression analysis (MLR), but more reliable methods are based on data compression. In such methods, the number of variables is reduced by forming new variables as linear combinations of original variables. The new variables are called *principal components* and the pth principal component can be expressed as:

$$X_p = \alpha_{1p} y_{1j} + \alpha_{2p} y_{2j} + \cdots + \alpha_{ip} y_{ij} + \cdots + \alpha_{np} y_{nj} \tag{1.23}$$

The coefficients α_{ip} are selected so that X_p complies to certain statistical conditions. Methods of this type are the *principal component regression* (PCR), and the *partial least square regression* (PLS).

As the array response to one single sample is a single-dimensional matrix (that is, a vector) a sensor array forms a *first-order sensor*. A *second-order sensor* can be obtained by recording each sensor signal as a function of the time and the response to one single sample is collected in the form of two-dimensional matrix. In a three-dimensional representation, the array response appears as a surface on which each point is determined by two variables: the numerical identifier of the sensor and the time of the measurement. Therefore, such a sensor array is operated in the transient regime. When used in the equilibrium state, the array functions as a first-order sensor. Compared with zero-order sensors (that is, one single sensor responding to one single analyte) higher-order sensors provide a considerably larger amount of information.

It is of interest to point out the advantages of multivariate data analysis over the univariate analysis method [10]. In the univariate method it is assumed that each sensor responds to one single analyte. However, the response can be affected by interferences and noise. In contrast, multivariate analysis provides interference-free results even if interferences in the recognition process do occur. This characteristic relieves the sensor from stringent selectivity, which is often difficult to achieve. In addition, multivariate analysis contributes to noise reduction and allows outliers to be detected. Last but not least, multivariate analysis permits simultaneous determination of a number of analytes in a mixture. A comprehensive overview of sensor array calibration is presented in ref. [11].

1.7.2 Qualitative Analysis by Cross-Sensitive Sensor Arrays

Arrays of cross-sensitive sensors can be used to identify particular mixtures of analytes in accordance with the array response to the mixture. Such a device can mimic the smell (*artificial nose* [12,13]) or the taste (*artificial tongue* [14,15]). In other words, an artificial nose or an artificial tongue performs the classification of a series of samples according to their similarity in chemical composition.

As an example, classification of mineral waters using an array of nonselective ion sensors is introduced next. Sensor responses are used to calculate 3 principal components (PC1, PC2, and PC3). In a three-dimensional plot, each sample is represented by a point defined by the values of the characteristic principal components, as shown in Figure 1.3A. The degree of similarity of two samples is expressed by the distance between the respective points. It is evident that various samples of the same trademark are grouped in well-defined clusters. If the sample is contaminated by a minute amount of organic matter, the cluster shifts, as indicated by the arrow in Figure 1.3A and contaminated samples form new clusters, labeled by the prefix c in front of the product name.

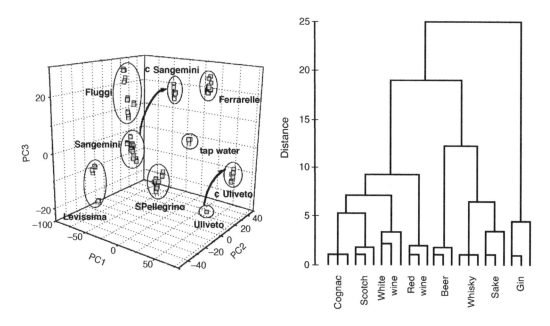

Figure 1.3 Applications of artificial nose/tongue devices. (A) Classification of different kinds of Italian mineral waters using the artificial tongue. Arrows indicate the shift of clusters as a result of contamination with organic compounds. (B) Classification of liquors (3 samples of each brand) by hierarchical cluster analysis. (A) Reproduced with permission from [16]. Copyright 2000 Elsevier. (B) Reproduced with permission from [17]. Copyright 1991 Elsevier.

Principal component analysis is advantageous in that a small number of variables are representative for a large set of original data. This allows for easy visualization and interpretation of data and reliable classification and identification of samples.

Another possible representation of data produced by the artificial nose or tongue is demonstrated in Figure 1.3B. This figure presents the results obtained by sensing vapors evolved by various liquors. An array of 6 nonspecific gas sensors has been used to this end. For each sample, the principal components are derived and used to calculate the distance from other samples. The array output is represented by means of *hierarchical cluster analysis* that can sort the samples in the form of a tree (dendrogram) based on the distances between them. Similar information can be obtained by pattern-recognition analysis of gas-chromatography data, but the artificial nose approach is simpler and faster.

Data-processing methods presented earlier belong to the class of *pattern-recognition* methods [18]. Pattern-recognition methods perform classification by assigning each sample to a particular class.

1.7.3 Artificial Neural Network Applications in the Artificial Nose/Tongue

An artificial neural network is a mathematical model or computational model that is inspired by the structure and functional features of the brain [18,19]. A neural network consists of an interconnected group of artificial neurons. The transmission line between two neurons is a synapse. A single neuron receives a series of input variables that are weighed along the synapses and then added. The resulting sum is transformed by application of a transfer function (e.g., multiplication with a constant or application of a nonlinear function). The result of the transformation forms the output signal of the neuron.

In order to obtain a neural network, artificial neurons are assembled in a sequence of layers, so that the output signals from a layer are fed to the neurons in the next layer. The last layer in the sequence provides the network output.

In artificial nose applications, the input signals are provided by the array's sensors. Before being used, a neural network is trained by means of a number of known compounds. In this process weighting factors are adjusted so as to minimize the difference between the actual output and the predetermined one. After this learning step, the network is able to determine these compounds.

Compared with the methods introduced earlier, artificial neural networks have a number of possible advantages. Artificial neural networks are adaptive systems that change their structure based on external or internal information that flows through the network during the learning phase. They are usually used to model complex relationships between inputs and outputs or to find patterns in data. A disadvantage of neural networks is the long training time needed by large networks. Notwithstanding this, artificial neural networks are very valuable tools for data processing in artificial olfaction and tasting.

1.7.4   Outlook

Sensor arrays allow for multiplexing the sample analysis. Particularly advantageous are the arrays of cross-sensitive sensors that provide more accurate and reliable results as compared with sensors dedicated to a particular analyte.

Arrays of cross-sensitive sensors can be operated in the artificial nose/tongue mode. Rather than detecting specific components, the artificial nose/tongue assigns the sample a characteristic fingerprint that depends on the overall chemical composition of the sample. In this way, it is possible to perform sample classification and identification.

There are a broad range of applications of the artificial nose/tongue in various areas [13]. Quality and organoleptic properties of foodstuff and beverages as well as aging or adulteration can be assessed by the assay of a characteristic assembly of chemical components. In medicine, the artificial nose can be useful for noninvasive diagnostic by the assay of volatile compound mixtures in breath, urine or sweat. Monitoring of pollution of air or water is another promising application of artificial olfaction. In the safety area, the artificial nose is a promising tool for the detection of chemical or biological warfare agents and detection of drugs or explosives.

Details of recognition and transduction methods in the artificial nose/tongue are presented in a further chapter dealing with various kinds of chemical sensors. The physicochemical principles of various types of gas sensor used in artificial noses are surveyed in refs. [12,20]. Overviews of the calculation methods for data processing in artificial nose application is given in refs. [12,21].

1.8   Sensors in Flow Analysis Systems

As very often a sensor or a sensor array is used in the analysis of a number of samples, automation of the operation sequence is essential for obtaining high sample throughput and reduction in personnel cost.

A common automation method in the analysis of liquid samples is flow injection analysis (FIA) [22,23]. In this method, the sensor is installed in a flow cell through which a fluid is continuously pumped. The sample is injected by means of a liquid chromatography valve (rotary valve) and is carried in the form of a plug inserted into the flowing carrier stream. Owing to diffusion, the sample is dispersed laterally, producing a variation in the concentration along the plug. When it reaches the sensor cell, the sample generates a signal that varies with time from the background level to a maximum value and then decreases as the sample plug leaves the cell. Either the peak height or the area of the signal–time curve can be correlated with the analyte concentration.

FIA produces reliable results if the fluid flow is laminar, which requires tubing diameter to be about 0.5 mm. A high degree of automation is achieved in FIA by integrating units for sample introduction and preliminary treatment of the sample. Moreover, FIA offers the possibility of running sequentially various steps such as sample assay, cell cleaning, and sensor regeneration.

A sensor designed for FIA application can be flat shaped or in the form of a flow-through channel with the sensing element on the inner wall.

Advanced methods for flow analysis are based on microfluidics systems [24]. Microfluidics deals with control and manipulation of fluid volumes in the submicroliter region that are constrained to very small size channels. Fluid flow can be prompted by applied pressure or electrokinetic phenomena. What distinguishes microfluidic systems from traditional flow analysis systems is the integration of a large network of channels and other micro-devices (such as actuators and valves) on a small chip. Therefore, microfluidics allows for a large-scale parallelization and multiplexing of flow analysis. An essential advantage of microfluidics is its compatibility with micromachining technology.

The advent of microfluidics prompted a revolution in the field of automatic analytical chemistry by the development of micro-total analytical systems (μTASs), also known as lab-on-a-chip systems. [25,26]. μTASs allow miniaturization and integration of complex functions, including physical and chemical conditioning of the sample, separations and analyte detection within the confines of a single chip. These systems can be accurate, reliable, rugged, and inexpensive, and, therefore, are well suited to point-of-care testing or field applications.

1.9   Applications of Chemical Sensors

In general, chemical sensors have been developed to provide alternatives to standard analytical methods based on techniques such as spectrometry, chromatography, biochemical or microbiological techniques. A chemical sensor can provide an inexpensive solution to a particular analytical problem without the need for expensive, multi-functional analytical equipment.

In addition, chemical sensors are suitable for field chemical analysis in environmental investigations. In medicinal applications, chemical sensors are useful in decentralized clinical investigations (point-of-care applications).

Chemical sensors offer the possibility of monitoring chemical parameters in industrial, environmental and medical applications. Of great interest is the application of chemical sensors to the *in vivo* determination and monitoring of chemical species of physiological relevance.

The use of sensors is faster than are conventional chemical, biochemical or microbiological assays. Therefore, it is not surprising that chemical sensors have found a broad range of applications in various areas such as monitoring of environment quality, clinical investigations, food technology and detection of warfare agents. The next sections outline certain typical analytical problems that can be tackled by means of chemical sensors. Surveys of chemical-sensor applications are available in refs. [27,28].

1.9.1 Environmental Applications of Chemical Sensors

Environmental applications of chemical sensors focus mainly on assessing water quality and air pollution [29–31].

Air pollution by automotive traffic and industrial activities is caused by toxic gases (sulfur, nitrogen and carbon oxides, hydrogen cyanide, etc.), toxic inorganic vapors (e.g., mercury) and other toxic vapors. Of particular relevance is the control of industrial environmental pollution caused by hazardous gases and vapors, such as those which are toxic, flammable or explosive. The quality of indoor air can be assessed by means of sensors for carbon dioxide and water vapor (humidity).

Water pollution directly affects aquatic organisms and, more generally, any organisms that need water for survival. The main water pollutants addressed by chemical sensors are toxic ions (e.g., mercury, lead, cadmium, and cyanide ions) and ions originating from agriculture activities. The use of fertilizers can lead to contamination of water sources by nitrate and phosphate ions that can disrupt aquatic ecological systems. Agriculture is also a source of water pollution by toxic pesticide residues. In addition to the general environmental impact, water quality is also a crucial issue in the supply of drinking water.

Various toxic compounds in water can be assessed by means of their inhibiting effects in enzyme-catalyzed reactions [32]. Organic pollutant sensors have also been developed using specific antibodies as recognition reagents.

Ion determinations can be achieved by standard potentiometric ion sensors, but, due to their high limits of detection, such sensors are suited only for the analysis of heavily polluted waters. However, recent progress in this field has led to the development of ion sensors with a very low limit of detection that can tackle determinations of metal ions below the concentration limit imposed by legal regulations for drinking water quality.

Sensors for toxins have been developed using micro-organisms as the sensing elements. Micro-organism metabolism is affected by toxins in the sample, which allow the monitoring of toxin concentrations by means of oxygen consumption in micro-organism respiration. The same principle has been used to develop sensors for the total content of dissolved organic material in water samples. In this case, dissolved organics stimulate the metabolism and hence oxygen consumption. Genotoxicity of environmental samples can be assessed by means of nucleic acid sensors [33].

The determination of possible pathogen micro-organisms in water is another important application that can be addressed by means of antibody-based sensors [34,35].

A problem of a great concern is acid rain caused by energy production by using fossil-fuel combustion. Common acidity sources are nitrogen and sulfur oxides that lead to increased acidity upon dissolution in atmospheric water. Increased acidity can be detected in an indirect way by monitoring the content of specific anions such as nitrite and sulfate.

A series of analytical problems in Marine Science are well suited to the application of chemical sensors [36,37]. Typical examples are the control of eutrophication due to increased concentration of nitrate and phosphate ions from fertilizers or sewage, monitoring of pollution by pesticides or by water diverted from oil extraction platforms, and determination of trace metals.

1.9.2 Healthcare Applications of Chemical Sensors

One of the main fields of application of chemical sensors is healthcare in which chemical sensors are utilized for *in vitro* or *in vivo* determination of chemical species of physiological relevance [38]. The functioning of *in vivo* sensors depends to a large extent to their biocompatibility [39]. As *in vivo* applications are very common, miniaturized sensors have been designed to this end. Often, multiple sensors are employed in order to determine simultaneously a series of various target analytes. Moreover, chemical sensors are used to detect pathogenic micro-organisms in clinical investigations.

Alkali and alkali-earth ions, as well as inorganic gases (dissolved oxygen and carbon dioxide, nitrogen monoxide), can be determined by means of dedicated sensors. Often, multiple sensors are employed in order to determine a series of target analytes.

Glucose determination in blood is very important in diabetic health care [40,41]. Currently, glucose sensors for self-monitoring of glucose in blood are widely available and intensive research efforts are devoted to the development and improvement of *in vivo* glucose sensors [42]. The next step in this area is the integration of glucose sensors with glucose delivery systems in order to maintain automatically the glucose level in blood at the normal level.

Sensors for a great number of biogenic compounds have also been developed. Among many target compounds, L-lactate, pyruvate, urea, uric and oxalic acids, histidine and histamine, phenolic compounds (L-dopa, dopamine and adrenalin), superoxide and sulfated bile acids can be mentioned. Dedicated sensors allow the monitoring of drugs in blood or urine.

Detection of pathogenic bacteria and viruses is another application of chemical sensors in clinical investigations. Pathogens can be detected by either immunological sensors or by nucleic acid-based sensors [35].

Normal biological processes, pathogenic processes, or pharmacologic responses to a therapeutic intervention can be assessed by means of *biomarkers* that are substances used as indicators of pathological states. Chemical sensors for biomarkers have been developed for the diagnostic of various forms of cancer, cardiovascular diseases, and hormone-related health problems [43].

1.9.3 Application of Chemical Sensors in the Food Industry, Agriculture and Biotechnology

Various chemical sensors have been developed in order to assess the quality of foodstuff and beverage quality and also for monitoring industrial processes in the food industry and biotechnology. Applications of enzymatic sensors in the food and drink industry are surveyed in refs. [44,45].

Food quality depends to a large extent on the content of nutrients and vitamins. Saccharides (such as glucose, fructose, sucrose, and lactose) can be determined by means of enzymatic sensors based on specific enzymes that produce chemical conversion of the target compound.

Other important components of foodstuff are lactic acid, malic acid, citric acid, and glutamic acid. Various enzymatic sensors for such compounds have been developed using relevant enzymes.

An important quality parameter of foodstuff is its freshness. Foodstuff freshness can be assessed by measuring the concentration of typical products of the spoilage process. Meat spoilage can be assessed by means of enzymatic sensors for putrescine ($NH_2(CH_2)_4NH_2$) and hypoxanthine (a purine derivative). Fish freshness can be assessed by determining a series of spoilage products such as inoxine-5-phosphate, inosine, and hypoxanthine [46].

Sensors for other compounds of relevance to food quality (such as cholesterol, fatty acids, lecithin, choline and polyphenols) have also been designed.

Of particular importance in the food industry is the control of pathogenic micro-organisms and microbial toxins in foodstuffs [34,35,47]. Chemical sensors for pathogens can be developed using either antibody–antigen recognition or by detection of the specific DNA.

In the beverage industry, dedicated chemical sensors can be used to determine ethanol and also glycerol, the latter being the main secondary product of alcoholic fermentation. Enzymatic sensors for both the above species have been produced. The content of sulfite ion in wine can also be determined by means of enzymatic sensors.

In agriculture, chemical sensors are employed in monitoring of macronutrients such as nitrate, phosphate and potassium ions [48].

Biotechnology uses biological systems, living organisms, or derivatives thereof (e.g., enzymes or living cells) to process raw materials. Various chemical sensors are used to monitor process parameters such as pH, dissolved oxygen, carbon dioxide, and various bio-organic compounds such as saccharides and amino acids [49,50]. Typical applications of chemical sensors in biotechnology are found in fermentation industry and production of certain antibiotics.

1.9.4 Chemical Sensors in Defense Applications

Defense in general, and defense against terror attacks in particular, is currently a matter of great concern that has prompted the development of chemical sensors for explosives and warfare agents such as pathogenic micro-organisms and toxic gases. Fast, *in situ*, detection of such matters is conveniently handled by means of chemical sensors.

Explosives can be traced using sensors specific to the explosive vapors. Such sensors have been developed using natural and synthetic affinity recognition reagents, enzymes and whole cells [51].

Biological warfare agents include living organisms, including viruses or infectious material derived from them, which could be used for hostile purposes. The targets of biological warfare agents can be humans, animals and crop plants. Such agents can multiply in the attacked organism and cause disease or death. Pathogenic bacteria, viruses and certain fungi are typical biological warfare agents.

Various types of chemical sensor for detection of biological warfare agents have been developed using various recognition mechanisms, such as affinity recognition by antibodies or synthetic materials, recognition by enzymes or whole cells, and the tracing of the pathogen DNA by means of a complementary DNA sequence [52,53]. Application of nanomaterials in sensors for biowarfare agents is currently arising much interest [54].

1.10 Literature on Chemical Sensors and Biosensors

As there is a vast amount of literature dealing with chemical sensors and biosensors, this section introduces only several general texts and certain relevant journals. References dealing with particular kinds of sensors are included in pertinent chapters.

The first monograph on chemical sensors was issued by J. Janata in 1989 [55]. A revised edition of this text become available recently [56].

By 1990, the field of biosensors attained the maturity and has been surveyed in a series of monographs published by E. A. H. Hall [57], F. W. Scheller and F. Schubert [58], and A. J Cunningham [59].

The status of chemical-sensor science and technology by about 1985–1995 is comprehensively surveyed in several collective volumes (refs. [60–64]). More recently, comprehensive collective volumes dedicated to biosensors [65,66] and electrochemical sensors [67] have been published.

Overviews of recent advances in chemical sensors are available in the continuing Springer Series on Chemical Sensors and Biosensors edited by G. Urban. Progresses in materials for chemical sensors as well as chemical sensors technology are amply surveyed in the series entitled Chemical Sensors edited by G. Korotcenkov and issued by Momentum Press (2010–2012).

Chemical sensors are amply represented in encyclopedic texts on physical and chemical sensors such as ref. [68].

Of great utility are collections of commented protocols for chemical sensor fabrication. Protocols for various kinds of biosensors are available in refs. [69–71]. Collections of protocols focusing on particular kinds of chemical sensors are referred to in pertinent chapters.

The development and expansion of chemical-sensor applications prompted the publication of several texts suitable for educational purposes (e.g., refs. [72–75]). A comprehensive overview of various kinds or recognition receptor used in chemical sensors is available in ref [76].

Research papers and reviews on chemical sensors are currently published in various journal. Two journals specialized on chemical sensors should be first mentioned, namely:

Biosensors and Bioelectronics (Elsevier)
Sensors and Actuators B: Chemical (Elsevier).

There are also some journals profiled on both chemical and physical sensors; for example:

IEEE Sensors Journal (Institute of Electrical and Electronics Engineers)
Sensors-Open Access Journal (MDPI – Open Access Publishing).

A great number of chemical sensor papers appear in journals profiled on general analytical chemistry such as those listed bellow:

Analytica Chimica Acta (Elsevier)
Analytical Chemistry (American Chemical Society)
Analytical and Bioanalytical Chemistry (Springer)
Analytical Letters (Taylor & Francis)
Analyst (The Royal Society of Chemistry)
Talanta (Elsevier)
TrAC Trends in Analytical Chemistry (Elsevier). This journal publishes only review papers.

The following journals profiled on electrochemistry and electroanalytical chemistry publish frequently papers on electrochemical sensors:

Bioelectrochemistry (Elsevier)
Electroanalysis (Wiley-WCH)
Electrochemistry Communications (Elsevier)
Journal of Electroanalytical Chemistry (Elsevier).

1.11 Organization of the Text

Chapter 1 introduces general terms and concepts in chemical-sensor science and technology. Further, Chapters 2–8 address a series of general recognition methods and materials commonly used in chemical sensors, as well as methods and technologies used in the manufacturing of such devices.

The next chapters focus on various kinds of chemical sensors classified according to the transduction method. As a rule, the physical or physicochemical principles of the addressed transduction method are first introduced. Further, various types of sensors obtained by combining the considered transduction method with diverse recognition schemes are presented.

Proteins are widely used in sensors as recognition receptors, either as enzymes, antibodies or biological receptors. For this reason the second chapter is dedicated to a short overview of protein structure and properties.

Chapter 3 introduces enzymes and their applications in chemical sensors for the determination of the enzyme substrate and enzyme inhibitors or as signaling labels in affinity-based sensors. The main transduction strategies in enzymatic sensors are also presented in this chapter. Understanding of the functioning of enzymatic sensors depends to a great extent on the mathematical modeling of such sensors. This topic is presented in Chapter 4.

An understanding of protein and enzyme chemistry is sufficient to grasp typical methods of fabrication of chemical sensors, which form the subject of Chapter 5. General methods for the immobilization of recognition receptors, materials used as the support for immobilization, other materials with relevance in sensor technology, as well as microfabrication technologies with applications in sensor fabrication are introduced in this chapter. Other methods and materials that are specific to particular classes of chemical sensors are addressed in subsequent chapters.

Chapters 6 and 7 address recognition methods based on affinity reactions. Chapter 6 introduces recognition methods based on antibodies and other natural and synthetic affinity reagents. A particular class of affinity interaction involves nucleic acids (Chapter 7). In this case, the analyte–receptor interaction is based on only two types of hydrogen bonding between complementary nucleobases. In both Chapters 6 and 7, general transduction strategies specific to each kind of recognition material, are presented.

As the application of nanomaterials in sensor technology is currently a topic of huge interest, Chapter 8 addresses the structure and properties of important classes of nanomaterials and the possibilities of application of such materials in sensor fabrication.

The following chapters introduce a series of typical transduction methods and present various chemical sensors based on the addressed transduction method.

Chapter 9 describes sensors in which transduction is based on a general property of chemical reactions, that is, the thermal effect. Two kinds of sensor are addressed in this chapter, namely enzyme-based sensors and sensors based on catalytic oxidation of combustible gases.

Chapter 10 presents ion-recognition methods and potentiometric ion sensors. Also, this chapter introduces a series of chemical sensors based on recognition processes leading to the formation of ions, such as sensors based on enzymatic reactions or gas dissolution in aqueous solutions. Such sensors make use of an ion sensors as transduction elements. Chapter 11 addresses an advanced class of potentiometric sensors that are based on the integration of electronic semiconductor devices with suitable sensing elements.

A particular but very important application of inorganic and organic semiconductor materials refers to resistive gas sensing, a topic that is dealt with in Chapter 12. The common feature of Chapters 11 and 12 is the application of semiconductor materials for recognition and transduction purposes.

The next five chapters address electrochemical sensors based on dynamic electrochemistry transduction. The fundamental principles of dynamic electrochemistry methods is the object of Chapter 13. Chapters 14 and 15 address amperometric enzymatic sensors; the former introduces the design principles of such sensors, while the latter gives an account of mathematical modeling of amperometric enzymatic sensors.

Applications of dynamic electrochemistry methods to affinity sensors and nucleic acid sensors are accounted for in Chapter 16. As in both cases the recognition is based on the formation of association complexes, there are many common features as far as the electrochemical transduction is involved.

A comprehensive examination of the dynamics of electrochemical processes is possible by application of electrochemical impedance spectrometry. This general method, as well as more specialized methods derived from it (such as conductometry and capacitance measurements), is presented in Chapter 17.

The extent of the space dedicated to electrochemical sensors is due to the central position occupied by this kind of sensor in the context of chemical-sensor science.

Optical sensors are addressed in Chapters 18–20. Chapter 18 presents the general principles of chemical sensing based on light waveguides. Chapter 19 addresses the design of optical sensors in conjunction with various recognition methods and emergent applications. Chapter 20 deals with an important contemporary trend in the development of optical sensors, namely the application of certain nanomaterials in optical sensing.

Chapter 21 deals with sensors based on a particular type of mechanical phenomenon, namely the acoustic wave, in other words, vibration and sound propagation in solid, liquid and viscoelastic materials.

Currently, much interest is aroused by the application of microelectromechanical systems (MEMS) in chemical sensing, a topic that is addressed in Chapter 22. As in the case of acoustic wave sensor, the MEMS sensor performs transduction by mechanical effects produced by the recognition event.

The last chapter, Chapter 23, introduces sensors based on living material (cells, tissue) as recognition receptors. This chapter has been placed at the end because living material sensors employ a series of transduction methods introduced in previous chapters.

Most of the chapters, and also certain sections, have been organized so as to introduce first the essential principles of the type of sensor addressed. Further, advanced topics are presented so that the interested reader can acquire a deeper knowledge of the field.

Each chapter gives some key references in order to help the interested reader to get a first contact with the relevant literature.

References

1. Thevenot, D.R., Toth, K., Durst, R.A. *et al.* (1999) Electrochemical biosensors: Recommended definitions and classification – (Technical Report). *Pure Appl. Chem.*, **71**, 2333–2348.
2. Sinclair, I.R. (2001) *Sensors and Transducers*, Newnes, Boston.
3. Fabbrizzi, L. and Poggi, A. (1995) Sensors and switches from supramolecular chemistry. *Chem. Soc. Rev.*, **24**, 197–202.
4. Heineman, W.R. and Jensen, W.B. (2006) Leland C. Clark Jr. (1918–2005), *Biosens. Bioelectron.*, **21**, 1403–1404.
5. Fogg, A.G., Scullion, S.P., Edmonds, T.E. *et al.* (1990) Adaptation of online reactions developed for use with flow-injection with amperometric detection for use in disposable sensor devices – reductive determination of phosphate as preformed 12-molybdophosphate in a capillary-fill device. *Analyst*, **115**, 1277–1281.
6. Miller, J.N. and Miller, J.C. (2010) *Statistics and Chemometrics for Analytical Chemistry*, Pearson Prentice Hall, Harlow.
7. Danzer, K. and Currie, L.A. (1998) Guidelines for calibration in analytical chemistry – Part 1. Fundamentals and single component calibration (IUPAC recommendations 1998). *Pure Appl. Chem.*, **70**, 993–1014.
8. Danzer, K. (2007) *Analytical Chemistry: Theoretical and Metrological Fundamentals*, Springer-Verlag, Berlin, Heidelberg.
9. Otto, M. (2007) *Chemometrics: Statistics and Computer Application in Analytical Chemistry*, Wiley-VCH, Weinheim.
10. Bro, R. (2003) Multivariate calibration – What is in chemometrics for the analytical chemist? *Anal. Chim. Acta*, **500**, 185–194.
11. Carey, W.P. and Kowalski, B.R. (1996) Sensor and sensor array calibration, in *Handbook of Chemical and Biological Sensors* (eds R.S. Taylor and J.S. Schultz), Institute of Physics Publ., Bristol, pp. 287–347.
12. Gardner, J.W. and Bartlett, P.N. (1999) *Electronic Noses: Principles and Applications*, Oxford University Press, Oxford.
13. Pearce, T.C., Schiffman, S.S., Troy Nagle, H. *et al.* (eds) (2003) *Handbook of Machine Olfaction: Electronic Nose Technology*, Wiley-VCH, Weinheim.
14. Citterio, D. and Suzuki, K. (2008) Smart taste sensors. *Anal. Chem.*, **80**, 3965–3972.
15. Zeravik, J., Hlavacek, A., Lacina, K. *et al.* (2009) State of the art in the field of electronic and bioelectronic tongues – towards the analysis of wines. *Electroanalysis*, **21**, 2509–2520.
16. Legin, A., Rudnitskaya, A., Vlasov, Y. *et al.* (2000) Application of electronic tongue for qualitative and quantitative analysis of complex liquid media. *Sens. Actuators B-Chem.*, **65**, 232–234.
17. Aishima, T. (1991) Discrimination of liquor aromas by pattern-recognition analysis of responses from a gas sensor array. *Anal. Chim. Acta*, **243**, 293–300.
18. Brereton, R. G. (2009) *Chemometrics for pattern recognition*, Wiley, Chichester.
19. Zupan, J. and Gasteiger, J. (1999) *Neural Networks in Chemistry and Drug Design*, Wiley-VCH, Weinheim.
20. James, D., Scott, S.M., Ali, Z. *et al.* (2005) Chemical sensors for electronic nose systems. *Microchim. Acta*, **149**, 1–17.
21. Hines, E.L., Llobet, E., and Gardner, J.W. (1999) Electronic noses: a review of signal processing techniques. *IEE Proc.-Circuit Device Syst.*, **146**, 297–310.
22. Růžička, J. and Hansen, E.H. (1988) *Flow Injection Analysis*, Wiley, New York.
23. Trojanowicz, M. (2000) *Flow Injection Analysis: Instrumentation and Applications*, World Scientific, Singapore.
24. Ohno, K., Tachikawa, K., and Manz, A. (2008) Microfluidics: Applications for analytical purposes in chemistry and biochemistry. *Electrophoresis*, **29**, 4443–4453.
25. Ghallab, Y.H. and Badawy, W. (2010) *Lab-on-a-Chip: Techniques, Circuits, and Biomedical Applications*, Artech House, Norwood, MA.
26. Li, P.C.H. (2010) *Fundamentals of Microfluidics and Lab on a Chip for Biological Analysis and Discovery*, CRC Press, Boca Raton, Fla.
27. Korotcenkov, G. (ed.) (2011) *Chemical Sensors Applications*, Momentum Press, Highland Park, N.J.
28. Ramsay, G. (1998) *Commercial Biosensors: Applications to Clinical, Bioprocess, and Environmental Samples*, J. Wiley, New York.
29. Lieberzeit, P.A. and Dickert, F.L. (2007) Sensor technology and its application in environmental analysis. *Anal. Bioanal. Chem.*, **387**, 237–247.
30. Badihi-Mossberg, M., Buchner, V., and Rishpon, J. (2007) Electrochemical biosensors for pollutants in the environment. *Electroanalysis*, **19**, 2015–2028.
31. Wanekaya, A.K., Chen, W., and Mulchandani, A. (2008) Recent biosensing developments in environmental security. *J. Environ. Monit.*, **10**, 703–712.
32. Trojanowicz, M. (2002) Determination of pesticides using electrochemical enzymatic biosensors. *Electroanalysis*, **14**, 1311–1328.
33. Palchetti, I. and Mascini, M. (2008) Nucleic acid biosensors for environmental pollution monitoring. *Analyst*, **133**, 846–854.
34. Leonard, P., Hearty, S., Brennan, J. *et al.* (2003) Advances in biosensors for detection of pathogens in food and water. *Enzyme Microb. Technol.*, **32**, 3–13.

35. Nayak, M., Kotian, A., Marathe, S. *et al.* (2009) Detection of microorganisms using biosensors-A smarter way towards detection techniques. *Biosens. Bioelectron.*, **25**, 661–667.

36. Varney, M.S. (2000) *Chemical Sensors in Oceanography*, Gordon and Breach Science Publishers, Amsterdam.

37. Kroger, S. and Law, R.J. (2005) Biosensors for marine applications – We all need the sea, but does the sea need biosensors? *Biosens. Bioelectron.*, **20**, 1903–1913.

38. D'Orazio, P. (2003) Biosensors in clinical chemistry. *Clin. Chim. Acta*, **334**, 41–69.

39. Vadgama, P. (2007) Sensor biocompatibility: Final frontier in bioanalytical measurement. *Analyst*, **132**, 495–499.

40. Wang, J. (2001) Glucose biosensors: 40 years of advances and challenges. *Electroanalysis*, **13**, 983–988.

41. Wang, J. (2008) Electrochemical glucose biosensors. *Chem. Rev.*, **108**, 814–825.

42. Cunningham, D.D. and Stenken, J.A. (eds) (2010) *In Vivo Glucose Sensing*, John Wiley and Sons, Hooken N.J.

43. Mascini, M. and Tombelli, S. (2008) Biosensors for biomarkers in medical diagnostics. *Biomarkers*, **13**, 637–657.

44. Prodromidis, M.I. (2010) Impedimetric immunosensors-A review. *Electrochim. Acta*, **55**, 4227–4233.

45. Mello, L.D. and Kubota, L.T. (2002) Review of the use of biosensors as analytical tools in the food and drink industries. *Food Chem.*, **77**, 237–256.

46. Venugopal, V. (2002) Biosensors in fish production and quality control. *Biosens. Bioelectron.*, **17**, 147–157.

47. Palchetti, I. and Mascini, M. (2008) Electroanalytical biosensors and their potential for food pathogen and toxin detection. *Anal. Bioanal. Chem.*, **391**, 455–471.

48. Sinfield, J.V., Fagerman, D., and Colic, O. (2010) Evaluation of sensing technologies for on-the-go detection of macronutrients in cultivated soils. *Comput. Electron. Agric.*, **70**, 1–18.

49. Freitag, R. (ed.) (1996) *Biosensors in Analytical Biotechnology*, Academic Press, San Diego, Calif.

50. Mulchandani, A. and Bassi, A.S. (1995) Principles and applications of biosensors for bioprocess monitoring and control. *Crit. Rev. Biotechnol.*, **15**, 105–124.

51. Smith, R.G., D'Souza, N., and Nicklin, S. (2008) A review of biosensors and biologically-inspired systems for explosives detection. *Analyst*, **133**, 571–584.

52. Gooding, J.J. (2006) Biosensor technology for detecting biological warfare agents: Recent progress and future trends. *Anal. Chim. Acta*, **559**, 137–151.

53. Shah, J. and Wilkins, E. (2003) Electrochemical biosensors for detection of biological warfare agents. *Electroanalysis*, **15**, 157–167.

54. Tok, J.B.H. (ed.) (2008) *Nano and Microsensors for Chemical and Biological Terrorism Surveillance*, RSC Publishing, Cambridge.

55. Janata, J. (1989) *Principles of Chemical Sensors*, Plenum Press, New York.

56. Janata, J. (2009) *Principles of Chemical Sensors*, Springer-Verlag US, Boston, MA.

57. Hall, E.A.H. (1990) *Biosensors*, Open University Press, Milton Keynes, England.

58. Scheller, F.W. and Schubert, F. (1992) *Biosensors*, Elsevier, Amsterdam.

59. Cunningham, A.J. (1998) *Introduction to Bioanalytical Sensors*, Wiley, New York.

60. Turner, A.P.F., Karube, I., and Wilson, G.S. (eds) (1987) *Biosensors: Fundamentals and Applications*, Oxford University Press, Oxford.

61. Edmonds, T.E. (ed.) (1988) *Chemical Sensors*, Blackie, Glasgow.

62. Göpel, W. (ed.) (1991) *Chemical and Biochemical Sensors*, vol. **1, 2**, VCH Verlagsgesellschaft, Weinheim.

63. Taylor, R.F. and Schultz, J.S. (eds) (1996) *Handbook of Chemical and Biological Sensors*, Institute of Physics Publ., Bristol.

64. Kress-Rogers, E. (ed.) (1997) *Handbook of Biosensors and Electronic Noses: Medicine, Food, and the Environment*, CRC Press, Boca Raton, Fla.

65. Gorton, L. (ed.) (2005) *Biosensors and Modern Biospecific Analytical Techniques*, Elsevier, Amsterdam.

66. Knopf, G.K. and Bassi, A.S. (eds) (2007) *Smart Biosensor Technology*, CRC Press, Boca Raton.

67. Alegret, S. and Merkoçi, A. (eds) (2007) *Electrochemical Sensor Analysis*, Elsevier, Amsterdam.

68. Grimes, C.A., Dickey, E.C., and Pishko, M.V. (eds) (2006) *Encyclopedia of Sensors*, American Scientific Publishers, Stevenson Ranch, Calif.

69. Cass, A.E.G. (ed.) (1990) *Biosensors: A Practical Approach*, IRL Press, Oxford.

70. Cooper, J.M. and Cass, A.E.G. (eds) (2004) *Biosensors: A Practical Approach*, Oxford University Press, New York.

71. Rasooly, A. and Herold, K.E. (eds) (2009) *Biosensors and Biodetection: Methods and Protocols*, Humana Press, New York.

72. Eggins, B.R. (1996) *Biosensors: An Introduction*, J. Wiley, Chichester.

73. Eggins, B.R. (2002) *Chemical Sensors and Biosensors*, J. Wiley, Chichester.

74. Diamond, D. (ed.) (1998) *Principles of Chemical and Biological Sensors*, Wiley, New York.

75. Gründler, P. (2007) *Chemical Sensors: An Introduction for Scientists and Engineers*, Springer-Verlag, Berlin.

76. Zourob, M. (ed.) (2010) *Recognition Receptors in Biosensors*, Springer, New York.

2

Protein Structure and Properties

Biosensors are chemical sensors that recognize target molecules by means of materials of biological origin, including proteins. Two categories of protein are of particular relevance to biosensors, namely enzymes and antibodies. Another type of biomaterial with broad applications in biosensors is polynucleotides that are also known as nucleic acids.

Enzymes, antibodies and nucleic acids have been designed by nature to perform particular tasks that rely on specific interactions with other chemical species. Such compounds can be advantageously used to impart selectivity to biomaterial-based sensors.

In this chapter a short survey of the structure and properties of proteins is given to aid understanding of their behavior in sensor applications. Enzymes, antibodies and nucleic acids are presented in subsequent chapters.

Protein molecules are polymers consisting of a sequence of α-amino acid residues. Such molecules display a high degree of organization, involving weak and strong bonding interactions between different regions along the length of the molecule as well as bonding interaction between distinct molecules [1,2].

2.1 Amino Acids

α-amino acids have the general structure shown in Figure 2.1. Each molecule has a hydrogen atom, a carboxyl group and an amino group attached to the central carbon atom. The amino acids differ from each other in the nature of the fourth group, R (side chain) attached to this carbon atom. The simplest amino acid is glycine in which R is a hydrogen atom. In other amino acids R can be an aliphatic group (e.g., $-CH_3$ in alanine), an acidic group (e.g., $-CH_2COOH$ in aspartic acid), a basic group (e.g. $-(CH_2)_4NH_2$ in lysine), a nonionic polar group (e.g., $-CH_2OH$ in serine; $-CH_2CONH_2$ in asparagine) or a hydrophobic aromatic group (in tyrosine and tryptophan). The particular side chain gives each amino acid a specific size and shape, and introduces properties such as hydrophobicity or hydrophilicity, acidic or basic character, positive or negative charge, or a polar nature of the side chain.

With the exception of glycine, α-amino acids contain an asymmetrically substituted carbon atom and display optical isomery. All natural α-amino acids have the L-configuration.

As each amino acid molecule contains at least two groups capable of undergoing deprotonation or protonation ($-COOH$ and $-NH_2$), its protonation state in solution depends on the pH of the solution (Figure 2.1). The ionization constant (pK_a) of the carboxyl group ($-COOH$) is about 2, whereas the ionization constant of the protonated amino group ($-NH_3^+$) is between 9 and 10. Due to inductive electronic effects, ionization constants depend to some extent on the structure of the side chain. As a consequence of these protonation/dissociation equilibria, at pH > 2 the carboxyl is ionized to $-COO^-$ whereas the amino group is protonated to $-NH_3^+$ at pH values < 10. In neutral solutions, both groups are in the ionized form and most of the molecules are in the form of a hybrid ion (*zwitterion*) with zero total charge. The pH at which all amino acid molecules are in the zwitterion form is known as the *isoelectric point* (pI). Additional protonation equilibria come into play if the side chain contains ionizable groups.

An amino acid that is particularly important in determining protein structure is cysteine for which $R = -CH_2-SH$. (Figure 2.2). Two cysteine molecules, can bind together to form a disulfide bond ($-S-S-$) by means of an oxidation reaction. Such a bridge between distant cysteine residues in a protein backbone contributes to the stabilization of the three-dimensional configuration of the protein molecule. In addition, a disulfide bridge can be the means by which two separate and individual protein macromolecular chains can be connected.

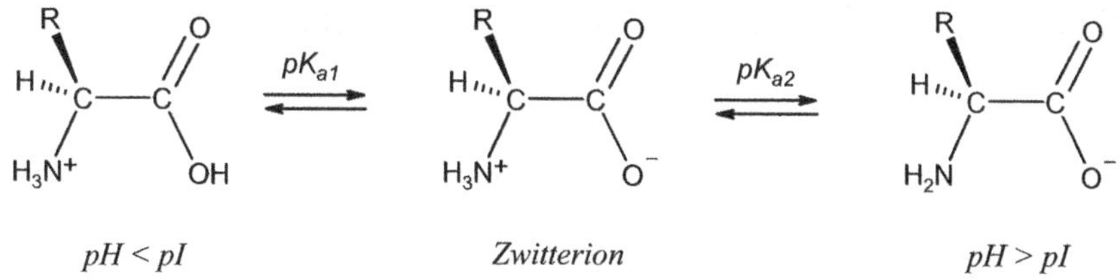

Figure 2.1 An α-amino acid and its protonation equilibria.

Chemical Sensors and Biosensors: Fundamentals and Applications, First Edition. Florinel-Gabriel Bănică.
© 2012 John Wiley & Sons, Ltd. Published 2012 by John Wiley & Sons, Ltd.

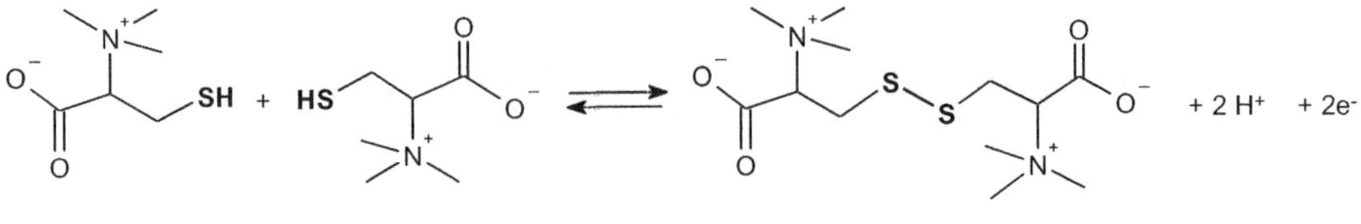

Figure 2.2   Cysteine oxidative dimerization yielding a cystine molecule (disulfide bridge).

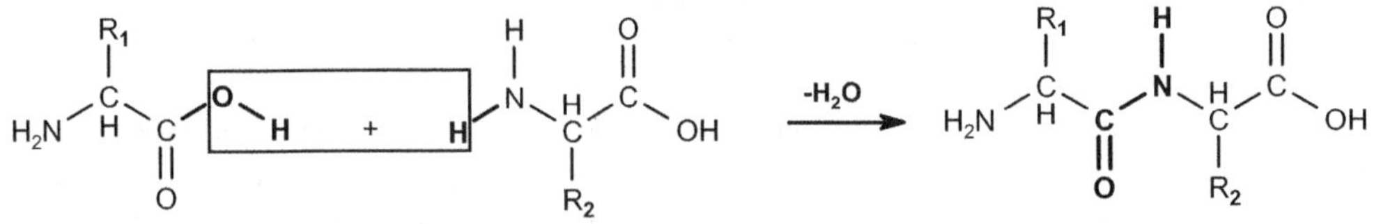

Figure 2.3   Condensation of two amino acids molecules to yield a dipeptide.

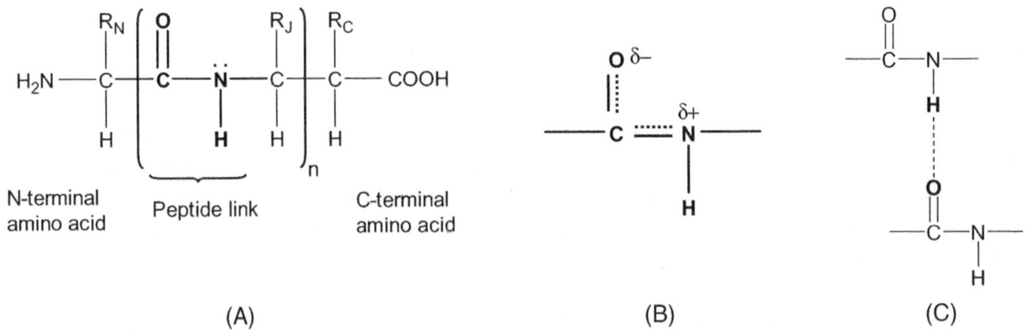

Figure 2.4   (A) Polymeric structure of a polypeptide; (B) the resonance structure of the peptide link; (C) a hydrogen bond between two peptide links.

2.2 Chemical Structure of Proteins

Proteins are assembled from amino acids using information encoded in genes. Each protein has its own unique amino acid sequence that is specified by the nucleotide sequence of the gene encoding this protein. Bonding of two amino acid molecules proceeds by a condensation reaction as shown in Figure 2.3 and a product called a *dipeptide* forms in this way. This dipeptide can then bind another amino acid to form a tripetide.

A sequence of reactions like that in Figure 2.3 leads to the formation of *polypeptides* (Figure 2.4A). In polypeptides and proteins the individual amino acids are joined together by *peptide links* (that is, the $-CO-NH-$ group). In the $-CO-NH-$ group, the lone electron pair at the nitrogen atom is actually delocalized giving the $C-N$ bond a partial double bond character (Figure 2.4B). This restricts rotation around the $C-N$ bond and imparts the peptide link a planar configuration.

2.3 Conformation of Protein Macromolecules

Several levels of organization can be distinguished in the configuration of the protein molecule. The primary level depends simply on the amino acids sequence. Higher levels of organization give the protein its characteristic three-dimensional shape.

The *primary structure* of proteins is given by the single-dimensional distribution of amino acids along the polymer chain.

Of crucial importance to protein conformation is the ability of the peptide links to form hydrogen bonds between them, as shown in Figure 2.4C. Hydrogen bonds between peptide links assist the protein strand in undergoing a self-organization process that gives the protein its *secondary structure*. The most important configurations of this type are the *α-helix* and the *β-sheet*.

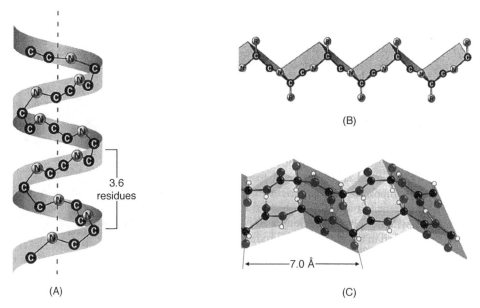

Figure 2.5   Elements of secondary protein structure. (A) Conformation of the α-helix; (B) a β-strand; (C) a β-sheet. Adapted with permission from [3]. Copyright 2010 Wiley, Hoboken.

In order to form an α-helix the polypeptide backbone coils as a right-handed screw, which permits side chains to aim to the outside of the chain (Figure 2.5A). The stability of this conformation is secured by hydrogen bonds between peptide links. An α-helix is usually represented by a folded ribbon (Figure 2.6), a flat strip or a tube. The α-helix has 3.6 amino acids per turn, which places the $-C=O$ group of an amino acid exactly in line with the $-N-H$ group of the next 4th amino acid down the molecule which allows hydrogen bonds to form.

A *β-strand* (Figure 2.5B) is a sequence of amino acids whose peptide backbones are almost fully extended. They are usually represented by an arrow pointing toward the C-terminus (Figure 2.6). An assembly of such strands that are hydrogen bonded to each other forms a β-sheet. (Figure 2.5C). Two or more adjacent β strands can aggregate in antiparallel, parallel, or mixed arrangements. The *β-strand* shows a pronounced trend to twisting and bending, thus deviating from the perfect parallel arrangement such as that shown in Figure 2.5C. Due to the elasticity of hydrogen bonds, twisting and bending do not hamper association of β- strands into β-sheets.

Additional interactions between distant regions in the polypeptide prompt the secondary structure elements to aggregate themselves in a well-defined arrangement that represents the *tertiary structure* of the protein (Figure 2.6). Such a structure can be stabilized by disulfide bridges and noncovalent bonds involving side-chain functionalities. Noncovalent chemical bonding in proteins is discussed in Section 2.4.

In the tertiary structure, connection between organized regions is provided by disorganized chain sections (loops) or by small turns stabilized by noncovalent bonds.

An example of protein tertiary structure with typical, secondary structural elements is shown in Figure 2.7A. This particular protein also includes small, nonprotein compounds (dihydrofolate and NADP$^+$) attached by noncovalent bonds.

Some protein molecules consist of two or more polypeptide chains assembled by noncovalent bonds and sometimes by disulfide bridges. Portions of distinct polypeptides can also gather into intermolecular β-sheets. The conformation adopted by such an assembly of polypeptide molecules form the *quaternary protein structure*. As an

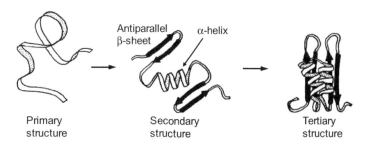

Figure 2.6   The folding of a polypeptide strand illustrating the stepwise self-organization of protein from primary through secondary to tertiary structure. Secondary structure is stabilized by hydrogen bonds between N and O atoms in peptide links. Tertiary structure is stabilized by chemical bonds involving side chains. Adapted with permission from [2]. Copyright 2000 John Wiley & Sons, Inc.

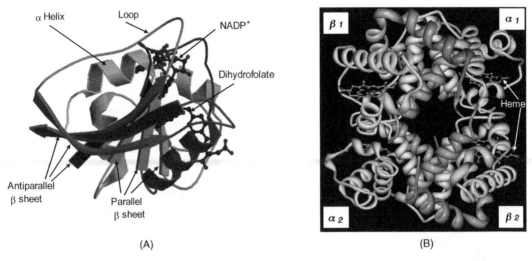

(A)   (B)

Figure 2.7   (A) Tertiary structures of Escherichia coli dihydrofolate reductase as a ternary complex with the substrate (dihydrofolate) and the NADP$^+$ coenzyme. Adapted from http://www.rcsb.org/pdb/explore/explore.do?structureId=1DRE. (B) Quaternary structure of hemoglobin. Iron-heme groups are also displayed. Adapted from http://www.rcsb.org/pdb/explore/explore.do?structureId=1A3N. Last accessed 15/05/2012.

example, Figure 2.7B shows the quaternary structure of hemoglobin, a tetramer consisting of four distinct chains: two similar α chains and two similar β chains.

Many proteins occur in mixed aggregates with polysaccharides (glycoproteins) or lipids (lipoproteins).

2.4 Noncovalent Chemical Bonds in Protein Molecules

Figure 2.8 summarizes the main kinds of noncovalent bonds that determine the folding of protein macromolecules [4]. That are hydrogen bonds (A), ionic bonds (B) and hydrophobic interactions based on van der Waals forces (C). van der Waals forces are attributed to electromagnetic interactions occurring as a consequence of fluctuations of charge distribution within atoms. In contrast to covalent bonds, the indicated noncovalent bonds have no fixed length and strength.

A central role in enzyme conformation is also played by water. Water can stabilize some conformations by forming hydrogen bonds that link two different sites (Figure 2.8D). Moreover, binding of water to polar and ionic groups causes proteins to be solvated and imparts the protein solubility in water when such groups are available at the external surface of the molecule.

Some metal ions can also produce a specific protein conformation (Figure 2.9) by interaction with side chains. Thus, alkali and alkali-earth ions are bound by electrostatic forces involving anions (such as carboxylate in aspartic acid) and negative ends of dipoles (such as $-C=O$ in asparagine and $=NH$ in arginine, Figure 2.9A). As electrostatic interactions are not spatially oriented, formation of electrostatic bonds experiences little restriction as far as the orientation and length are concerned. The length and strength of such bonds can vary within a wide range.

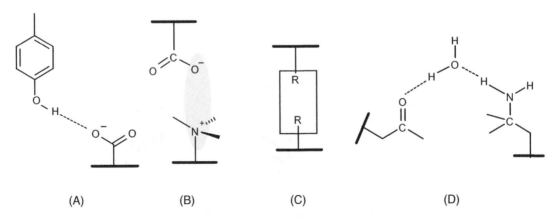

(A)   (B)   (C)   (D)

Figure 2.8   Examples of noncovalent chemical bonds involving side chains. (A) hydrogen bond; (B) ionic bond (salt bridge); (C) Hydrophobic bond (R are hydrophobic side chains); (D) hydrogen-bonded water bridge. Thick lines represent protein backbones.

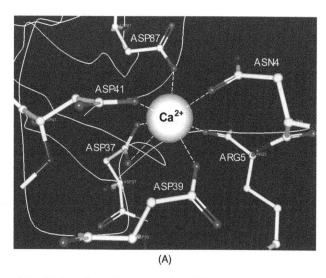
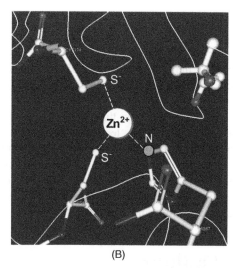

(A) (B)

Figure 2.9 Binding of metal ions in proteins. (A) A calcium ion held by electrostatic bonds in Cadherin 23. Adapted from http://www.rcsb.org/pdb/explore/explore.do?structureId=2WBX. (B) Zn^{2+} in alcohol dehydrogenase. Metal coordination is performed by two sulfur atoms from cysteine residues and a nitrogen from the imidazole ring in a histidine residue. Curved lines represent the protein backbone. Adapted from http://www.rcsb.org/pdb/explore/explore.do?structureId=1SBY. Last accessed 15/05/2012.

Transition-metal ions form coordination bonds that are spatially oriented strictly in accord with the disposition of molecular orbitals (Figure 2.9B). Also, the length and strength of coordination bonds is limited by quantum-mechanical rules. Therefore, transition-metal ions impart a rigid configuration.

2.5 Recognition Processes Involving Proteins

In living organisms many proteins function as *receptors* by linking to a specific *ligand* to form an aggregate called a *complex* (not to be confused with coordination compounds of metal ions). Binding in such a complex occurs via non-covalent bonds that impart stability by their great number. An example of a tertiary complex is shown in Figure 2.7A, in which a protein is bound to two nonprotein compounds to form a ternary complex. Figure 2.10 shows the binding details in the complex of a small sugar molecule to the protein avidin. The resulting complex is stabilized by hydrophobic and hydrogen bonds. The high specificity achieved in the formation of such complexes is due to the partner molecules being chemically and sterically complementary. However, perfect steric matching is not always essential as the receptor site can often undergo some conformational rearrangement in order to accommodate the ligand.

Of particular relevance to biosensor applications are enzymes and antibodies. Enzymes function as catalysts in biochemical reactions. The first step in an enzymatic reaction consists of specific binding of a ligand to the active site of the enzyme. Once the complex has been formed, the ligand undergoes a chemical reaction. The reaction

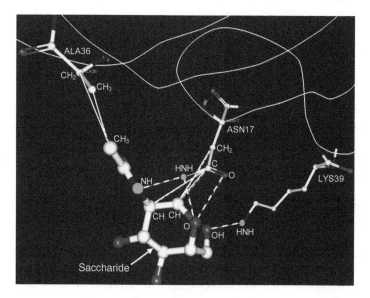

Figure 2.10 Binding of a saccharide (2-(acetylamino)-2-deoxy-a-D-glucopyranose) to avidin. Full straight lines: hydrophobic bonds; dashed lines: hydrogen bonds. Adapted from http://www.rcsb.org/pdb/explore/explore.do?structureId=2AVI Last accessed 15/05/2012.

products are released leaving the receptor site free to accept a further ligand molecule. The catalytic cycle is repeated as long as the ligand is available. A compound that undergoes catalytic conversion under the effect of an enzyme is called *enzyme substrate* (shortly, *substrate*). Clearly, enzyme–substrate interactions are not at equilibrium but are governed by the rules of chemical kinetics. When an enzyme is used as a recognition element in a biosensor, any physical or chemical consequence of the reaction can in principle be used in order to monitor the progress of the reaction and effect the transduction.

An antibody is a protein that is produced by the immune system of an organism to protect the organism against potentially harmful agents termed antigens. The antibody binds the antigen selectively by noncovalent bonds. This kind of interaction results in chemical equilibrium, and, therefore, the stability of the complex can be described in terms of thermodynamic functions. In biosensors, an antibody can function as a receptor for a ligand that is the analyte. Alternatively, the antigen can be used as ligand to determine the antibody. Physical effects (such as mass change) or signaling tags (e.g., fluorescent labels attached to one of the reactants) allow the complex to be detected in order to perform signal transduction.

2.6 Outlook

Amino acids can form by condensation reactions polymers called proteins. A protein molecule consist of a linear sequence of amino acids joined to each other by peptide links, with a free amino group at one end of the molecule and a carboxyl group at the other. The sequence of amino acid residues in protein is determined by the genetic code of the organism. This sequence forms the primary structure of the protein molecule. Hydrogen bonds between $-NH$ and $-C=O$ groups in the backbone stabilize particular conformations such as α-helices and β-sheets that represent the next organization degree known as the secondary structure. Further organization of the protein involves spatial disposition of these secondary structure elements. This constitutes the tertiary structure of the protein, and is the result of the formation of chemical bonds involving side chains. Some protein molecules result by aggregation of several distinct polypeptide chains, which is brought about by noncovalent bonding, and, sometimes by the formation of disulfide bridges. This final conformational arrangment is termed quaternary structure.

Comprehensive information on protein structure is available in specialized data bases, such as the Protein Data Bank at http://www.pdb.org/pdb/home/home.do. Each structure in this data base is identified by a four character index (for example, 2AVI for the protein in Figure 2.10).

The primary structure relies on stable, covalent bonds. In contrast, the higher organization levels are mostly based on noncovalent bonds that are susceptible to alteration under the effect of various factors such as pH, ionic strength, temperature and solvent composition. Alterations in the higher-level organization cause protein *denaturation* that results in the loss of the natural properties of the protein. That is why proteins are relatively sensitive materials when used in an artificial environment. Depending on circumstances, denaturation can be a reversible or irreversible process.

Application of proteins in biosensors relies on the property of some molecules of this type to bind to a specific ligands by multiple noncovalent bonds to yield a complex. Any physical or chemical consequence of complex formation can be exploited for the purpose of transduction.

Questions and Exercises

1 What is the general structure of an α-amino acid? Give an overview of general properties of amino acids. Comment on the particular characteristics of some amino acids and explain what is the reason for their particular behavior.

2 Use Internet resources and prepare a list of proteinogenic α-amino acid grouped according to their side-chain characteristics.

3 Write the chemical formula of a peptide that consists of a sequence of glycine, cysteine and lysine and that of another one that consists of asparagine, glycine, aspartic acid and cysteine. Is it possible to link the above peptides by a covalent bond?

4 Why does a protein undergo denaturation if the pH is shifted from that of its natural environment?

5 What may happen when an aqueous protein solution is mixed with a nonaqueous solvent?

6 A water-soluble protein can be precipitated by an inert salt (such as ammonium sulfate) due to the salting-out effect. What are the transformations at the molecular level leading to the modification of solubility?

7 Is there some receptor–ligand interaction included in Figure 2.7A? What is the receptor and what are the ligands.

8 Go to the Protein Data Bank web site, open the files for the structure of the complexes indicated in Figures 2.9 and 2.10 and check the interatomic distances. Make a correlation between the nature of the chemical bond and its length.

9 Search the structure PDB ID: 2WDO in the Protein Data Bank and examine the binding of glycerol and Mg^{2+} to this protein molecule.

10 Open the structure PDB ID: 1E9Z (the urease enzyme) in the Protein Data Bank and download the Ni(II) complex structure via the Ligand Chemical Component menu. Comment on the binding of nickel ions in this enzyme.

References

1. Whitford, D. (2005) *Proteins: Structure and Function*, John Wiley & Sons, Chichester.
2. Copeland, R.A. (2000) *Enzymes: A Practical Introduction to Structure, Mechanism, and Data Analysis*, Wiley-VCH, New York.
3. Karp, G. (2010) *Cell Biology*, John Wiley & Sons, Hoboken, N.J.
4. Chang, R. and Chang, R. (2000) Intermolecular forces, in *Physical Chemistry for the Chemical and Biological Sciences*, University Science Books, Sausalito, Calif, pp. 669–700.

3

Enzymes and Enzymatic Sensors

3.1 General

Enzymes are protein compounds that are specifically structured to bind to and act on a *substrate* (reactant molecule) to convert it by a catalytic mechanism, that is, by lowering the activation energy of the reaction with no effect on the chemical equilibrium [1–4].

Enzyme-catalyzed reactions rely on the formation of an intermediate involving both a shape and structure match of the substrate with the active site on the enzyme (Figure 3.1). In the resulting complex, substrate conversion is facilitated by various means. Thus, the enzyme–substrate interaction can cause a key chemical bond in the substrate to become weaker and prone to further alteration. Furthermore, the enzyme provides favorable conditions for stabilizing a reaction intermediate and preventing its reconversion to the initial form. When more than one reactant is involved, the enzyme, by specific chemical bindings, can gather all of them in a state that stimulates the reaction to proceed. In some cases, the enzyme active site can shuttle particles such as electrons or hydrogen ions that are needed in the reaction. Although Figure 3.1 shows a single-substrate reaction, many enzyme reactions involve two or more reactants (*cosubstrates*).

Enzymatic methods are widely used in bioanalytical chemistry in order to determine the enzyme itself or its substrate [5]. To this end, either the reaction rate or the concentration of a reactant or a product is assessed by a suitable analytical method.

As an extension of this, the application of enzymes in biosensor design relies on their specificity for a substrate-analyte that cannot be detected in a direct way. Therefore, in order to build up an enzyme sensor, the enzyme should be immobilized as part of a recognition layer at the surface of a suitable transducer so as to gage the concentration of a detectable species (Figure 3.2). The substrate and any additional reactants undergo first of all diffusion from the sample solution into the recognition layer where the enzyme reaction takes place. The transducer allows the monitoring of the course of the reaction by detecting either a product, or the excess of reactant that escaped the enzyme reaction. In addition, pure physical effects (such as heat evolution) are suitable for monitoring the rate of the enzyme reaction and, implicitly, the substrate concentration. Perm-selective membranes are often incorporated at interfaces in order to control the diffusion of certain reactants. Also, diffusion of some interfering compound from the sample to the recognition element can be hindered by a suitable external membrane.

As well as isolated enzymes, enzymes included in living entities (such as micro-organisms or living tissues) can be used directly in order to perform analyte recognition and conversion (see Chapter 23). A biosensor based on an enzyme, either isolated or incorporated in living materials, is often termed a *metabolism sensor*.

Moreover, any inhibitor (for example, an organic compound or a metal ion) of the enzyme catalytic activity can be determined by its slowing down effect on the substrate conversion rate. Further, an enzyme can act as a transduction tag if it is attached to a nondetectable species. By its action on a suitable substrate, a detectable product forms that enables indirect detection of the target compound. As a large amount of product results in the presence of a minute amount of enzyme-tagged compound, this allows for extremely sensitive detection.

For a long time, the availability of enzymes was limited to those produced by natural living organisms. Recently, genetic engineering has made it possible to extend the range of enzyme sources and to create new enzymes that are able to meet specific requirements.

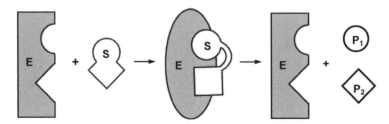

Figure 3.1 Mechanism of enzyme-catalyzed substrate conversion. E is the enzyme, S is the substrate, P_1 and P_2 are product. The intermediate state is an enzyme–substrate complex.

Chemical Sensors and Biosensors: Fundamentals and Applications, First Edition. Florinel-Gabriel Bănică.
© 2012 John Wiley & Sons, Ltd. Published 2012 by John Wiley & Sons, Ltd.

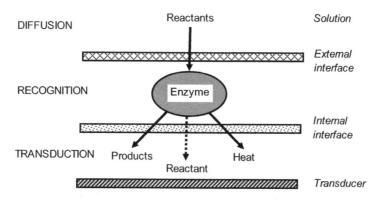

Figure 3.2 Typical configuration of an enzymatic biosensor.

3.2 Enzyme Nomenclature and Classification

Many enzymes have been named by adding the suffix "-ase" to the substrate name. Thus, urease catalyzes the hydrolysis of urea to ammonia and carbon dioxide, whereas phosphatase catalyzes the hydrolysis of phosphate esters. However, in order to avoid confusion, enzymes are classified into six major classes and a series of sub-classes, according to the kind of catalyzed reaction (Table 3.1). At the same time, rules for unambiguous identification of each enzyme were recommended by the Nomenclature Committee of the International Union of Biochemistry and Molecular Biology [6].

According to this systematic nomenclature, each enzyme is ascribed a *recommended name*, suitable for common use, a *systematic name,* which denotes the reaction it catalyzes, and a four digit *classification number* (preceded by the EC acronym that stands for Enzyme Commission). For example, glucose oxidase (that catalyzes the oxidation of glucose by oxygen) is denoted EC 1.1.3.4, and has the systematic name β-D-Glucose:oxygen 1-oxidoreductase, (see the scheme in Figure 3.3).

Table 3.1 Enzyme classes

Enzyme class	Reaction catalyzed	Systematic name	Recommended name
1. Oxidoreductases • Deydrogenases • Oxidases (O_2 is the electron acceptor)	Electron transfer (oxidation/ reduction) Two-substrate reactions	Donor:acceptor oxidoreductase	Donor dehyrogenase Donor oxidase (only if O_2 is the acceptor)
2. Transferases	Transfer of an atom or group (X) between two molecules: $AX + B \rightarrow A + BX$ Two-substrate reactions	Donor:acceptor group transferase	Acceptor (or donor) group transferase
3. Hydrolases	Hydrolysis reactions: $AB + H_2O \rightarrow A-OH + BH$ Two-substrate reactions; one of these is water	Substrate X-hydrolase (X is the group removed by hydrolysis)	Substrate + suffix -*ase*
4. Lyases	Nonhydrolytic bond cleavage One-substrate reactions (bond breaking) Two-substrate reactions (bond formation)	Substrate group-lyase	Reaction + suffix −*ase* (for example, dehydratase)
5. Isomerases	Isomerization reactions One-substrate reactions	Reaction type + suffix -*ase* (for example, racemase)	
6. Ligases	Bond -formation reactions Two-substrate reactions (ATP[a] as cosubstrate)	A:B ligase (A and B denote the substrates)	

[a]ATP stands for adenosine triphosphate.

Main class:	Oxidoreductase	– 1
Subclass:	Acting on -CHOH group	– – – – – – – – – – – – – – – – – – 1
Sub-subclass:	O₂ as electron acceptor	– – – – – – – – – – – – – – – – – 3
Systematic name:	β-D-glucose- O₂-1-oxidoreductase	– – – – – – – – – – 4

Figure 3.3   Assigning the classification number to the glucose oxidase enzyme (EC 1.1.3.4).

3.3 Enzyme Components and Cofactors

Although enzymes are protein-type compounds, many of them need a nonprotein *cofactor* in order to fulfill the biological function (Figure 3.4). The cofactor can be an organic molecule acting as a *coenzyme*. Some coenzymes are tightly bound to the protein structure as a *prosthetic group*. If the cofactor is removed, the remaining substance is called an *apoenzyme*, whilst the whole enzyme is called a *holoenzyme*. In other cases, the cofactor is an independent substance (*cosubstrate*) that is bound temporarily to the enzyme in order to take part in the catalytic process. By free diffusion, coenzymes contribute to the transport of electrons, atoms or groups of atoms between molecules.

The simplest cofactors are metal ions attached to the protein side chains by coordination or electrostatic bonds. Thus, an *activator* metal ion obliges the active site to adopt a favorable configuration or can be involved in binding the substrate to the active site. In some cases, a transition-metal ion acts as an electron conveyor. Tightly bound metal ions (such as nickel in urease) are structural constituents of the enzyme active site.

An example of a tightly bound cofactor is flavin adenine dinucleotide (FAD) that occurs as a prosthetic group in various oxidoreductases. It conveys electrons and hydrogen ions between an enzyme and its substrate by undergoing a redox reaction similar to that depicted in Figure 3.5. Electrochemical oxidation of the $FADH_2$ form may be the basis for electrochemical transduction in some biosensors. However, as the FAD center is surrounded by the protein backbone, direct electron transfer can be achieved only when electrodes composed of particular materials are used. Alternatively, the transduction can be performed by detecting a redox mediator that conveys electrons between the $FADH_2$ center and a metal or graphite electrode.

An example of a coenzyme is nicotinamide adenine dinucleotide (NAD^+) that consists of two nucleotides linked by two phosphate groups, with nicotinamide attached to one end position and adenine to the other end position. It can undergo the redox reaction shown in Figure 3.6 and act as an electron and proton conveyer in reactions catalyzed by some oxidoreductases. In an amperometric enzyme sensor, the electron acceptor/donor role in a reaction like that in

```
                        Cofactors
                       /         \
              Coenzymes           Metal ions
              /       \            /        \
  Prosthetic group  Cosubstrate  Activator ions  Active site ions
  (tightly bound)  (weakly bound) (weakly bound)  (tightly bound)
```

Figure 3.4   Different cofactors and their interaction with proteins.

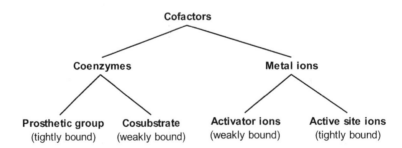

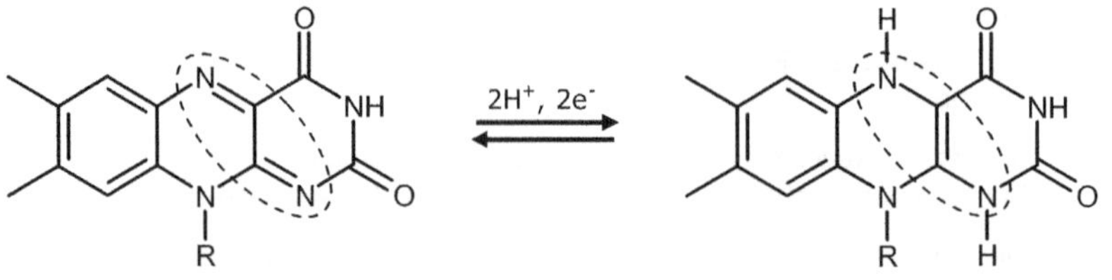

FAD   2H⁺, 2e⁻   **FADH₂**

Figure 3.5   The redox reaction of the active part of FAD. R denotes the remaining part of the FAD/FADH₂ unit.

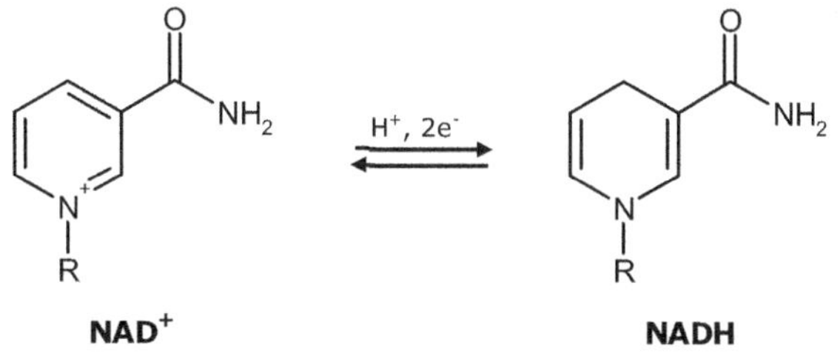

Figure 3.6 The redox reaction of nicotinamide adenine dinucleotide. The featured reacting moiety is nicotinamide; R denotes the remaining part of the coenzyme molecule.

Figure 3.6 can be assumed by an electrode in a suitable electrochemical cell. In such circumstances, the $NAD^+/NADH$ couple performs as an electron shuttle between a substrate and the electrode, giving rise to the response current.

Nicotinamide adenine dinucleotide phosphate ($NADP^+/NADPH$), which is a phosphorylated derivative of NAD/NADH, performs similar functions in various enzyme-catalyzed reactions.

Pyrroloquinoline quinone cofactor (PQQ, (Figure 3.7)) occurs in some bacterial enzymes and plays a role similar to that of NAD^+. This cofactor is attached to a protein carboxylate via a calcium ion, which, in addition, allows temporary binding of the substrate (for example, an alcohol) to the active site.

A common prosthetic group in some oxidoreductases is of the *heme* type (Figure 3.8). It consists of an iron atom coordinated in the center of a large heterocyclic porphyrin ring. As the iron atom can swing between several oxidation states, it acts as an electron donor/acceptor in redox reactions.

The above discussion has been limited to several cofactors that have wide applications in enzyme biosensor. Many other cofactors occur in living organisms.

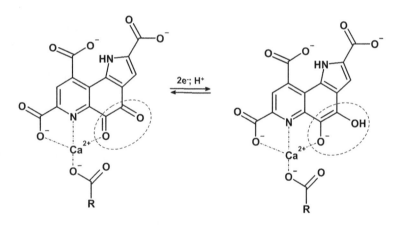

Figure 3.7 The pyrroloquinoline quinone cofactor (PQQ). R represents the protein backbone.

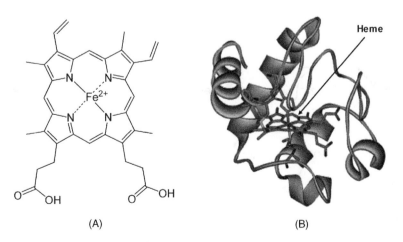

Figure 3.8 (A) Heme B, a typical heme group. (B) Cytochrome *c*, a heme protein. Adapted from http://www.rcsb.org/pdb/explore/explore.do?structureId=1HRC Last accessed 16/05/2012.

Table 3.2 Some flavoenzyme oxidases with applications in biosensors. Systematic names are given in parentheses

Enzyme	Reaction	Applications
Glucose oxidase (β-D-glucose:oxygen 1-oxidoreductase) EC 1.1.3.4	Figure 3.9	Clinical; Food industry
Galactose oxidase (D-galactose:oxygen 6-oxidoreductase) EC 1.1.3.9	D-galactose $+ O_2 \rightarrow$ D-galacto-hexodialdose $+ H_2O_2$	Food industry
Cholesterol oxidase (Cholesterol:oxygen oxidoreductase) EC 1.1.3.6	Cholesterol $+ O_2 \rightarrow$ Cholest-4-en-3-one $+ H_2O_2$	Clinical
Monoamine oxidase (amine:oxygen oxidoreductase (deaminating)) EC 1.4.3.4	$RCH_2NHR' + H_2O + O_2 \rightarrow RCHO + R'NH_2 + H_2O_2$	Clinical; Food industry
L-amino-acid oxidase (L-amino-acid:oxygen oxidoreductase (deaminating)) EC 1.4.3.3	L-amino acid $+ H_2O + O_2 \rightarrow$ a 2-oxo acid $+ NH_3 + H_2O_2$	Clinical
Lactate oxidase ((S)-lactate:oxygen 2-oxidoreductase (decarboxylating)) EC 1.13.12.4	Lactate $+ O_2 \rightarrow$ Acetate $+ CO_2 + H_2O$	Clinical Food industry

3.4 Some Enzymes with Relevance to Biosensors

Among the vast number of enzymes that occur in living organisms, several have proved useful for bioanalytical purposes and also for designing enzyme biosensors. The following section includes a short outline of enzymes of particular relevance from this standpoint. A comprehensive overview of enzymes application for biosensor design and construction is available in ref. [7].

3.4.1 Oxidases

FAD-oxidases use molecular oxygen as the electron and hydrogen ion acceptor in the catalytic cycle. A typical example involves glucose oxidation catalyzed by glucose oxidase using the FAD prosthetic group as an electron and hydrogen ion conveyer (Figure 3.9). Glucose oxidase from *Aspergillus niger* is broadly employed in bioanalytical chemistry, either as an analytical reagent or as recognition material in glucose biosensors [8–12]. This is because glucose is a compound of great analytical concern in diabetes monitoring as well as in the food industry. In addition, the relatively low price and good stability makes the glucose/glucose oxidase system a very suitable model for method development in biosensor science.

Other FAD-dependent oxidases with wide applications in sensor science are summarized in Table 3.2. The last enzyme in this table (lactate oxidase) relies on the flavin mononucleotide prosthetic group (FMD). Flavin-oxidases use oxygen as a cosubstrate and give rise to hydrogen peroxide. Monitoring of one of the above compounds is a general transduction method for FAD-oxidase-based sensors. Oxygen, however, can be substituted by an artificial electron acceptor allowing oxygen-independent operation. As shown in Table 3.2, some flavoenzyme oxidases give rise to inorganic gases such as ammonia or carbon dioxide. Monitoring of such gases or their hydrolysis products (for example, NH_4^+ or H^+) represents another convenient transduction strategy.

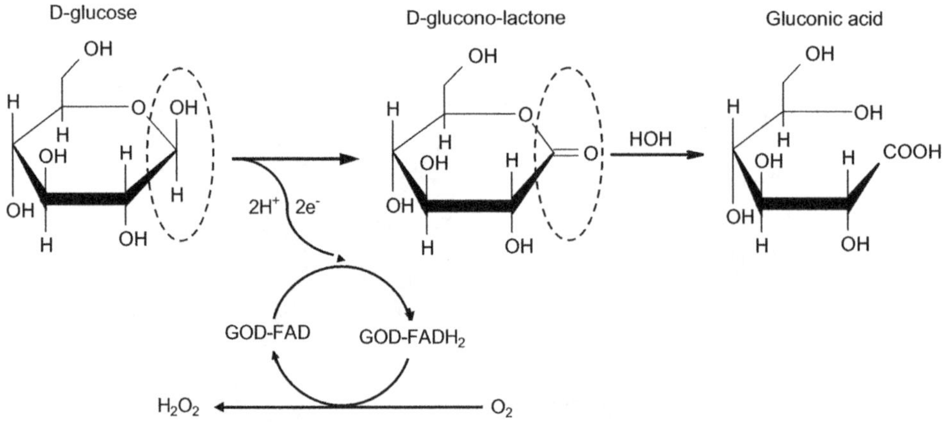

Figure 3.9 Glucose oxidation catalyzed by glucose oxidase (GOD). FAD and FADH$_2$ are the oxidized and reduced forms, respectively, of the prosthetic group. Glucose is oxidized upon electron and hydrogen ion transfer to the oxidized form of the prosthetic group (FAD) that is reduced to FADH$_2$. FAD is regenerated by the oxidation of FADH$_2$ with molecular oxygen. Hydrogen peroxide forms as a byproduct of this reaction.

Table 3.3 Copper-containing oxidases with applications in biosensors

Enzyme	Reaction	Applications
L-ascorbate oxidase (L-ascorbate:oxygen oxidoreductase) EC 1.10.3.3	L-ascorbate $+ \frac{1}{2}O_2 \rightarrow$ dehydroascorbate $+ H_2O$	Food industry
Tyrosinase (monophenol,L-dopa:oxygen oxidoreductase) EC 1.14.18.1	L-tyrosine $+$ L-dopa $+ O_2 \rightarrow$ L-dopa $+$ Dopaquinone $+ H_2O$	Clinical Environment
Catechol oxidase (1,2-benzenediol:oxygen oxidoreductase) EC 1.10.3.1	Catechol $+ \frac{1}{2}O_2 \rightarrow$ (1,2-Benzoquinone) $+ H_2O$	Clinical Environment
Laccase (benzenediol:oxygen oxidoreductase) EC 1.10.3.2	4 (Benzenediol) $+ \frac{1}{2}O_2 \rightarrow$ 4 (Benzosemiquinone) $+ H_2O$	Environment

A series of *copper-containing oxidases* (Table 3.3) are also used extensively in biosensors for biologically active compounds or phenolic pollutants [13,14]. The product of phenols oxidation is a reducible quinone, which can be determined by electrochemistry. Copper enzymes can perform direct electron transfer to an electrode, which can provide an electrochemical transduction method [15]. Laccase, that catalyses the oxidation of benzenediol, is used in sensors for monitoring the environment pollution [16].

Another class of oxidases is represented by *peroxidases* that can shuttle electrons to hydrogen peroxide (or small organic peroxides) from electron donors. The prosthetic group in many peroxidases is of a heme type (Figure 3.8). The most common member of this class is *horseradish peroxidase* (HRP) that can accept electrons from virtually any reducing agent, for example, ferrocyanide, phenol, hydroquinones, ortho- and para-phenylenediamine, ascorbate, iodide or ferrocene. The reaction mechanism for a peroxidase-catalyzed reaction can be described by the following steps [17]:

$$\text{Native enzyme}(Fe^{3+}) + H_2O_2 \rightarrow \text{Compound-I} \quad (3.1)$$

$$\text{Compound-I} + AH_2 \rightarrow \text{Compound-II} + AH^{\bullet} \quad (3.2)$$

$$\text{Compound-II} + AH_2 \rightarrow \text{Native enzyme}(Fe^{3+}) + AH^{\bullet} + H_2O \quad (3.3)$$

In the first step, the ferriheme prosthetic group in the enzyme undergoes a two-electron oxidation by hydrogen peroxide or organic hydroperoxides. This reaction results in the formation of the compound-I (oxidation state $+5$), consisting of oxyferryl ion ($Fe^{4+} = O$) and a porphyrin π cation radical. In the next reaction, compound-I accept one electron from the donor substrate molecule AH_2 and forms the compound-II (oxidation state $+4$). In the third step, the compound-II undergoes additional one-electron reduction by reaction with a second AH_2 molecule, whereby the enzyme is restored to its native state. Therefore, the overall reaction is:

$$\text{RO-OR}' + 2AH_2 \xrightarrow{\text{Peroxidase}} \text{ROH} + \text{R}'\text{OH} + 2AH^{\bullet} \quad (3.4)$$

where R and R′ represent organic residues or hydrogen atoms.

The final product depends on the nature of the substrate. Organic electron donors such as aromatic amines and phenolic compounds are oxidized to free radicals, $AH^{\bullet}$. Inorganic substrates like hexacyanoferrate(II) are simply oxidized by withdrawing one electron.

Reactions (3.1) and (3.2) can also proceed as electrochemical reactions (in which the cell cathode acts as electron donor) either by direct electron transfer or by means of redox mediators. In both cases, the electrolytic current is correlated to the concentration of peroxide in the solution.

Peroxidase sensors can be used for determining either the peroxide or the electron-donor substrate and also inhibitors such as CN^- and F^-. Peroxidases are also widely used as transduction tags in affinity sensors.

3.4.2 Dehydrogenases

Dehydrogenase performs the transfer of a hydride ion (H^-) between a substrate containing a –CHOH group and a suitable cofactor [18]. Such a reaction is equivalent to the transfer of one proton and two electrons. Thus, NAD^+- (or $NADP^+$-) dependent hydrogenases convert an alcohol to a carbonyl compound as follows:

$$\text{RR}'\text{CH} - \text{OH} + NAD^+ \underset{\xrightarrow{\text{Dehydrogenase}}}{\rightleftarrows} \text{RR}'\text{C} = O + NADH + H^+ \quad (3.5)$$

Table 3.4 Some NAD^+-dependent dehydrogenase enzymes with applications in biosensors

Enzyme	Reaction	Applications
Alcohol dehydrogenase (alcohol:NAD^+ oxidoreductase) EC 1.1.1.1	Reaction (3.5)	Fermentation Food industry
Glucose dehydrogenase (β-D-glucose: $NAD(P)^+$ 1-oxidoreductase) EC 1.1.1.118	β-D-glucose + NAD^+ → D-glucono-1,5-lactone + NADH + H^+	Clinical Food industry
Lactate dehydrogenase ((S)-lactate:NAD^+ oxidoreductase) EC 1.1.1.28	Lactate + NAD^+ → Pyruvate + NADH + H^+	Clinical Food industry
L-amino acid dehydrogenase (L-amino-acid:NAD^+ oxidoreductase) EC 1.4.1.5	L-amino acid + H_2O + NAD^+ → 2-oxo acid + NH_3 + NADH + H^+	Clinical Food industry
Glutamate dehydrogenase (L-glutamate:NAD^+ oxidoreductase) EC 1.4.1.3	L-glutamate + H_2O + NAD^+ → 2-oxoglutarate + NH_3 + NADH + H^+	Fermentation Food industry

Over 250 enzymes belong to this class, thus offering a broad range of analytical applications. Several dehydrogenases with relevance to biosensor applications are included in Table 3.4.

In dehydrogenase-based biosensors transduction can in principle be performed by monitoring the cofactor in either the oxidized or reduced form by electrochemical reactions or light absorption/emission. As the hydrogen ion is involved in such reactions, pH monitoring could also be a straightforward transduction method. In the case of amino acid dehydrogenases, determination of ammonia (or ammonium) by a suitable probe provides an additional transduction method.

Some bacterial dehydrogenases rely on the PQQ cofactor (Figure 3.7), which is relatively tightly bound to the enzyme. Such an enzyme can in some instances be a convenient alternative to NAD^+-dependent dehydrogenases because the NAD^+/NADH system is a soluble and freely diffusing species. In natural systems, PQQ-enzymes perform the same task as NAD^+-dependent dehydrogenases (that is, catalysis of hydride ion transfer) but use quinone as the electron acceptor. Quinoprotein alcohol dehydrogenase (alcohol:quinone oxidoreductase, EC 1.1.5.5) and quinoprotein glucose dehydrogenase (D-glucose:ubiquinone oxidoreductase, EC 1.1.5.2) are examples of PQQ-dehydrogenase already investigated for biosensor applications.

3.4.3 Hydrolases

A hydrolase is an enzyme that catalyzes the hydrolysis of a chemical bond such as the ester bond (esterases) or the peptide bond (proteases).

Several enzymes in the hydrolase class deserve a particular mention. Thus, *acetylcholinesterase* (AChE, EC 3.1.1.7) is essential to nerve cell function through its capacity to break down the neurotransmitter acetylcholine into its constituents (Figure 3.10). Various pesticide or war gases are AChE inhibitors that renders a inhibition-based AChE sensor useful for detecting such harmful compounds [19].

As hydrogen ions result from hydrolysis, transduction in AChE sensors can be performed simply by assessing the change in pH. For amperometric sensor applications, the natural substrate is substituted by a derivative that gives rise to an easily detectable product (such as acetylthiocoline). Thiocoline ($HS(CH_2)_2N^+(CH_3)_3$), which results from the hydrolysis reaction, can be monitored by its electrochemical oxidation.

Urease (urea amidohydrolase EC 3.5.1.5) catalyzes the hydrolysis of urea (Figure 3.11) and is widely used for both urea determination and also for determining enzyme inhibitors [20]. At a suitable pH, either ammonia or carbon dioxide form and probes for such gases are suitable for transduction.

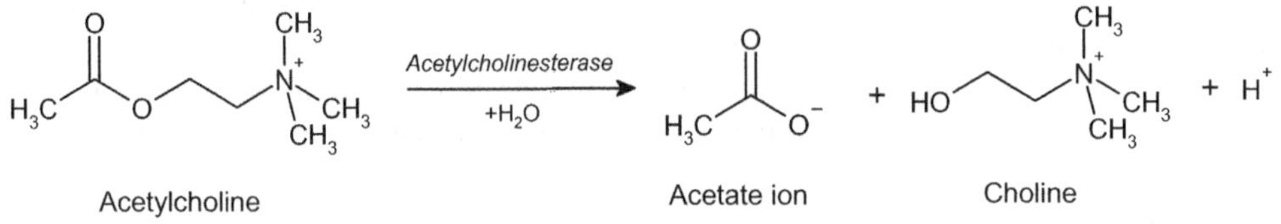

Figure 3.10 Enzymatic hydrolysis of acetylcholine.

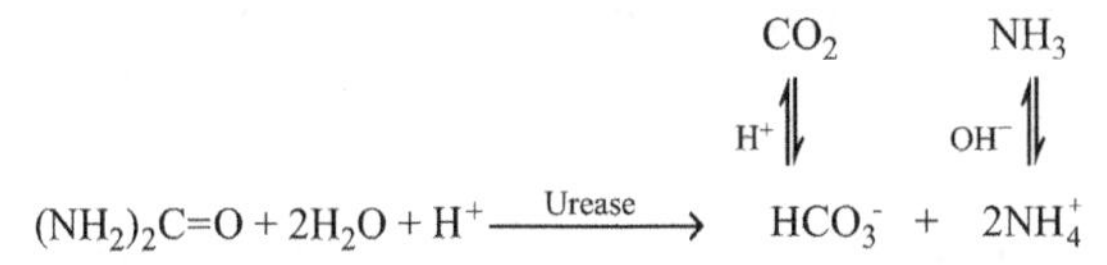

Figure 3.11 Enzymatic hydrolysis of urea.

Alkaline phosphatase (ALP) (phosphate-monoester phosphohydrolase (alkaline optimum), EC 3.1.3.1) is responsible for removing phosphate groups from many types of phosphate esters, including nucleotides, proteins, and alkaloids. ALP-catalyzed dephosphorylation proceed as shown in reaction (3.6).

$$\text{R-O-PO}_3\text{H}_2 + \text{H}_2\text{O} \xrightarrow{\text{ALP}} \text{R-OH} + \text{H}_3\text{PO}_4 \tag{3.6}$$

ALP as an isolated enzyme or as a component of living micro-organisms is useful for pollutant determination by inhibition of the enzymatic reaction. ALP is also employed as a transduction tag in affinity sensors.

Sensors used in the antibiotics industry have been developed by means of *penicillase*, which converts the substrate (for example, penicillin) into a substituted β-amino acid. This brings about a change in the pH that is utilized as a response signal.

Some amino derivatives undergo enzymatic hydrolysis with formation of ammonia, which can be determined by an ammonia probe included in a biosensor. In this way, clinically relevant compounds such as creatinine and adenosine can be determined by means of sensors based on relevant enzymes (creatinase and adenosine deaminase, respectively).

3.4.4 Lyases

Lyases are enzymes that catalyse the breaking of various chemical bonds by means other than hydrolysis and oxidation.

Clinically relevant oxalic acid can be determined by sensors based on *oxalate decarboxylase*. This enzyme converts oxalic acid to formic acid and carbon dioxide (reaction (3.7)). The transduction can be performed by means of a carbon dioxide probe.

$$\text{HOOC-COOH} \xrightarrow{\text{Oxalate decarboxylase}} \text{HCOOH} + \text{CO}_2 \tag{3.7}$$

Another lyase of interest is *L-aspartase* (aspartate ammonia-lyase) that converts aspartate to fumarate yielding ammonia, which can be monitored by an ammonia probe.

3.4.5 Outlook

Enzymes are protein compounds that perform as specific catalysts in living organisms. Enzyme specificity is imparted by the chemical structure and the configuration of the active site, which allows the enzyme to bind the substrate and give rise to an enzyme–substrate complex. In this state, the substrate undergoes chemical conversion to products that are released, leaving the active site. At the end of this process, the enzyme is able to bind and convert another substrate molecule.

Certain enzymes are plain proteins, whilst other enzymes rely on an active site formed of a nonprotein species (cofactors). A cofactor can be bound to the protein backbone or can be an independent molecule that acts as a conveyor of electrons, atoms or groups of atoms.

Enzymatic sensors are produced by integrating an enzyme layer with a transducer that monitors the concentration of a reactant or product involved in the enzymatic reaction. The response signal thus generated is in relation with the substrate concentration in the sample.

The selectivity of an enzymatic sensor depends on the enzyme selectivity. In this respect, it should be borne in mind that only a few enzymes exhibit absolute selectivity that is, they will catalyze only the conversion of one particular compound. Other enzymes will be specific for a particular type of chemical bond or functional group.

Besides the substrate determination, enzymatic sensor can be used to determine enzyme inhibitors. Moreover, enzymes are utilized as signaling labels in certain types of sensors such as immunosensors or nucleic acid sensors.

Questions and Exercises (Sections 3.1–3.4)

1 What is the role of enzymes in living organism and how do enzymes perform this function?
2 How can enzymes be used in biosensors? What kind of analytes can be targeted by enzyme biosensors?
3 What is the role of an enzyme cofactor?
4 Point out the similarities and differences between cofactors introduced in Section 3.3, taking into account their structure, function and mode of action.
5 What are the main differences between FAD-oxidases and peroxidases from the standpoint of structure and chemical reaction? Expand this comparison by including copper-containing oxidases.

6 Devise a general transduction procedure for the copper enzymes in Table 3.3.

7 Point out several enzymes with applications in sensors for environmental applications.

8 Suggest several different enzymes that are used in glucose biosensors. For each of them, write the chemical reaction and mention possible transduction methods.

9 Answer the same question for the case of amino acid biosensors.

10 List several enzymes with applications in sensors for biomedical sciences and use Internet resources to find out the biological relevance of the target substrate.

11 Use Internet resources to find out (a) the systematic name and the EC number for penicillase, creatinase, adenosine deaminase, oxalate decarboxylase and L-aspartase, and (b) the chemical reaction catalyzed by each of the above enzyme.

3.5 Transduction Methods in Enzymatic Biosensors

3.5.1 Transduction Methods

The transduction strategy is chosen with regard to the enzyme reaction by taking into account the methods available for gaging the physical and chemical consequences of the recognition process.

Purely physical effects such as heat evolution or change in ionic conductivity represent general transduction methods that can, in principle, be put into operation with any kind of enzyme sensor. However, such transduction methods are not selective and should be used cautiously.

Chemical transduction relies on monitoring the concentration of any possible compound that is involved in the enzyme reaction. The main restriction in this case is the availability of a suitable monitoring method. If such a method is missing, additional reactions can be integrated into the transduction scheme, in order to generate a detectable compound.

The general principles of transduction in enzymatic sensors will be illustrated here for the particular case of the glucose oxidase-based glucose sensor. Figure 3.12 illustrates the main course of glucose oxidation in the presence of glucose oxidase and the subsequent strategies for performing transduction in a glucose sensor. The substrate–enzyme electron transfer takes place within an intermediate complex formed by the substrate and the oxidixed form of the enzyme (GOD_{ox}). Then, the complex undergoes a chemical conversion that releases the product and the enzyme in its reduced form (GOD_{red}). In order for the enzyme to fulfill further its function this form should be changed back to GOD_{ox} by an electron-transfer reaction. In natural media, the electron donor in this step is dissolved oxygen and hydrogen peroxide results as the product (path I). Hence, transduction can be performed by monitoring either the oxygen or the hydrogen peroxide concentration within the sensing layer by means of an appropriate probe. However, in this method both oxygen and a pH buffer should be added to the sample as auxiliary reagents. A reagentless glucose biosensor can be obtained if GOD_{ox} is regenerated by direct electrochemical oxidation of GOD_{red} with the electrolytic current acting as the response signal (path II). Another method relies on a redox mediator (M) that is included in the sensor structure along with the enzyme itself (path III). GOD_{red} is reoxidized to GOD_{ox} by electron transfer to the oxidized form of the mediator (M_{ox}). The resulting reduced mediator (M_{red}) is converted back to M_{ox} by an electrochemical reaction that provides the response current. In addition,

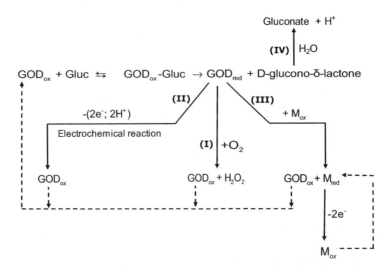

Figure 3.12 Possible transduction methods for glucose oxidase-based sensors. GOD and Gluc stands for glucose oxidase and glucose, respectively.

hydrolysis of the primary product (gluconolactone) to the gluconate anion (path IV) induces a pH change that can be monitored for achieving transduction.

Additional reactions can be integrated with the transduction scheme in order to generate products that can be detected by means of available probes. Thus, hydrogen peroxide can react with the iodide ion in the presence of a suitable catalyst, which allows transduction to be performed by means of a potentiometric iodide sensor.

It is clear from the above discussion that, for a given enzyme–substrate couple, different transduction approaches can be feasible. A selective transduction method, which responds specifically to a particular reaction product, is preferable, but perfect specificity is difficult to achieve. Endogenous components in the sample may in many cases interfere with the transduction step. In such cases, corrections should be applied to the response signal in order to avoid errors. Thus, a reference sensor, containing no enzyme, can be used in order to obtain the response background due to interferences and this signal has to be subtracted from the overall response provided by the main sensor. Alternatively, perm-selective membranes should be selected such as to prevent sample interferents from reaching the transduction zone. So, a key problem in enzyme sensor research is finding the most suitable transduction procedure for a particular recognition reaction in order to obtain a reliable sensor that fits specific requirements imposed by the sample composition and the operational conditions.

Certain enzymatic reactions releases gases such as ammonia or carbon dioxide. These gases can be monitored by specific probes in order to generate the response signal.

3.5.2 Multienzyme Sensors

In enzyme assays, a reaction leading to a detectable product is termed an *indicator reaction*. If such a reaction is not available for a specific substrate, the initial product can be further converted into a detectable compound (also called an *indicator species*) by means of a second enzyme-catalyzed reaction. In such cases the primary reaction is termed an *auxiliary reaction*, whereas the second reaction acts as the actual indicator reaction.

Application of multienzyme systems has also proved to be successful in developing enzymatic sensors. Thus, the glucose sensor can be adapted in order to build up a sucrose sensor by integrating glucose oxidase with invertase (EC 3.2.1.26) within the sensing layer. Invertase catalyzes the hydrolysis of sucrose to D-fructose and a mixture of α- and β-glucose and the latter product can be monitored by means of a glucose sensor. This scheme has been improved by including glucose mutarotase (EC 5.1.3.3), which converts α-glucose into β-glucose. Such an enzymatic sensor is termed a *sequence sensor*. It includes two enzyme layers: the first one contains the auxiliary enzyme and is in contact with the sample solution, whereas the next layer, which contains the indicator enzyme, communicates with the transducer.

An additional enzyme in the biosensor configuration can be included in order to remove interferents. Thus, the above-mentioned sucrose biosensor responds not only to sucrose itself but also to endogenous glucose from biological and foods samples. In order to remove glucose interference, an additional glucose oxidase layer is incorporated in front of the invertase layer. Within this layer, glucose is oxidized and is thus removed from the flux of reactants before reaching the indicator enzyme layer.

A large increase in sensitivity can be achieved by means of a second enzyme that performs substrate recycling. This principle is demonstrated in Figure 3.13 for the case of a lactate sensor. Here, the lactate ion is oxidized to pyruvate by means of oxygen under catalysis by lactate oxidase (LOD). Next, lactate dehydrogenase (LDH) restores the initial analyte molecule and thus assists a recycling process which enhances the response signal arising from either oxygen depletion or peroxide formation. What results is a *chemical amplification* of the response signal.

Clearly, a great expansion of the field of application of enzyme sensor can be achieved by integrating more enzymes into the sensing layer [7,21]. However, such an approach should be implemented with caution because a multienzyme sensor is more susceptible to interferences and other disturbances caused by various factors affecting each of the enzymes (such as enzyme degradation, inhibition or sensitivity to pH changes).

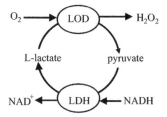

Figure 3.13 Chemical amplification by substrate recycling in a lactate sensor.

Questions and Exercises (Section 3.5)

1 Draw the sketch of a sucrose sensor including a layer for removing endogenous glucose and indicate by arrows the diffusion flux for each reactant and product.

2 Potato contains acid phosphatase (phosphate-monoester phosphohydrolase (acid optimum), EC 3.1.3.2). Draw up the configuration of a possible glucose-6-phosphate sensor that consists of a glucose sensor complemented with a thin potato slice. Indicate by arrows the flux of each reactant/product and indicate the change in flux intensity by means of the arrow thickness.

3 Phosphate and fluoride anions are inhibitors of acid phosphatase. Could this property be useful for determining such anions? Draw up the expected response–concentration graph for the cases in which transduction is performed by monitoring either oxygen or hydrogen peroxide.

4 Conversion of glucose to glucono-lactone by glucose dehydrogenase is a reversible reaction. Could this property be used for performing substrate recycling in a glucose sensor? Write the relevant chemical reactions and draw a scheme of the sensor configuration.

5 Endogenous glucose interference with a bi-enzyme sucrose sensor can be eliminated by means of an additional glucose oxidase layer. If glucose concentration is too high, oxygen would be consumed to a high degree in the external layer, which impairs the functioning of the adjacent glucose sensor. Could this be relieved by catalase (EC 1.11.1.6)? Write the relevant reactions. *Hint*: Catalase is an enzyme that catalyzes the decomposition of hydrogen peroxide to water and oxygen.

3.6 Kinetics of Enzyme Reactions

The application of enzymes in analytical chemistry is based on their ability to act as specific catalysts for a broad range of biochemical reactions [3,22,23]. When an enzyme and its substrate are present in a solution and certain physicochemical conditions are fulfilled, substrate conversion proceeds steadily. Clearly, such a system is not at equilibrium and its behavior is dominated by kinetic factors. Hence, a preliminary examination of the enzyme kinetics is essential for understanding the underlying principles of enzyme biosensors. The goal of the kinetic approach is to derive the rate equation, that is, an equation giving the reaction rate as a function of relevant concentrations. The reaction rate indicates how fast a reaction proceeds and it represents the number of moles consumed or produced per time unit and volume unit (in mol dm^{-3} s^{-1}). In order to derive the rate equation, a reaction mechanism based on experimental investigation of the reaction system should be assumed.

3.6.1 The Michaelis–Menten Mechanism

A large diversity of reaction mechanisms is possible for enzyme-catalyzed reactions, as emphasized in standard texts [24–26]. The present approach will refer to the *Michaelis–Menten mechanism* that is a representative one and that highlights some essential features of enzyme kinetics. Denoting by E, S, and P the enzyme, the substrate, and the product, respectively, the Michaelis–Menten mechanism can be formulated as a two-step process (reaction (3.8)), the k symbols being assigned to relevant reaction rate constants:

$$E + S \underset{k_{-1}}{\overset{k_1}{\rightleftharpoons}} ES \overset{k_2}{\rightarrow} E + P \tag{3.8}$$

Accordingly, the first step is an equilibrium reaction driven by the affinity of the substrate for the catalytic site, which yields an enzyme–substrate complex, ES. After the substrate has been converted irreversibly into the product P, the enzyme is released and is able to convert another substrate molecule. The symbols e, s, c, and p will be further used to denote the concentration of the free enzyme, substrate, enzyme–substrate complex, and product, respectively. It will be assumed for simplicity that the substrate concentration remains approximately constant. According to the rules of chemical kinetics, the reaction rate with reference to formation of the product depends on complex concentration as:

$$v = \frac{dp}{dt} = k_2 c \tag{3.9}$$

As c is an unknown, it will be expresses as a function of known parameters using the steady-state assumption that states that c is time independent (Briggs–Haldane assumption). Accordingly, the rate of ES formation in the first step above should be equal to the rate of ES decay:

$$k_1 es = k_2 c + k_{-1} c \tag{3.10}$$

This equation can be rearranged in order to derive the expression for the *Michaelis–Menten constant* (K_M) as follows:

$$K_M = \frac{es}{c} = \frac{k_2 + k_{-1}}{k_1} \tag{3.11}$$

Now, the concentration of the complex ES can be expressed as:

$$c = e\frac{s}{K_M} \tag{3.12}$$

Therefore, the reaction rate equation becomes:

$$v = \frac{k_2}{K_M} es \tag{3.13}$$

As some enzyme molecules are bound in ES, the concentration of the free enzyme is:

$$e = e_t - c \tag{3.14}$$

where e_t stands for the total enzyme concentration and c is given by Equation (3.12). By combining Equations 3.11, 3.12 and 3.13, one obtains:

$$v = \frac{k_2 e_t s}{K_M + s} \tag{3.15}$$

This equation proves that the reaction rate increases with substrate concentration and attains a constant, limiting value (v_m) if $s \gg K_M$:

$$v_m = k_2 e_t \tag{3.16}$$

Under these conditions, Equation (3.12) combined with Equation (3.14) proves that all enzyme molecules will be in the form of ES complex, which correspond to the *enzyme saturation* state. The reaction rate is in this case independent of substrate concentration (*zero-order kinetics*) but it is proportional to the enzyme concentration. Using Equation (3.16), Equation (3.15) becomes:

$$v = \frac{v_m s}{K_M + s} \tag{3.17}$$

Equation (3.17) is known as the *Michaelis–Menten equation*. It was assumed above that the substrate concentration is constant. In order to fulfill this condition, the experimentally determined reaction rate should be extrapolated to initial moment. The resulting value (that is, *initial rate of reaction*, v_0) should be plotted as a function of the initial substrate concentration (s_0) in order to determine K_M and v_m (Figure 3.14). The $v_0 - s_0$ curve is a hyperbola that tends towards a limit (v_m) at substrate concentrations very high with respect to K_M.

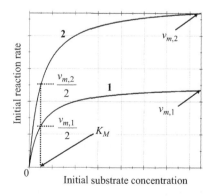

Figure 3.14 A plot of enzyme kinetics data according to the Michaelis–Menten equation. Enzyme concentration for curve 2 is twice as high as that for curve 1. Notice that K_M equals the substrate concentration yielding a reaction rate that is half of the maximum value. Substrate determination is best achieved within the quasilinear range ($s \ll K_M$), whereas enzyme determination should be performed within the quasihorizontal region that corresponds to the $s \gg K_M$ condition.

At the opposite extreme ($s \ll K_M$), the reaction rate is proportional to the substrate concentration (*first-order kinetics*) but is independent of the concentration of the enzyme:

$$v = \frac{k_2 e_t}{K_M} s = \frac{v_m}{K_M} s \tag{3.18}$$

The first-order kinetics limit represents the ideal condition for performing substrate determination. When designing an enzyme biosensor, it is therefore crucial to secure a sufficiently low substrate concentration within the enzyme layer. The pre-s factor in Equation (3.18) is the pseudofirst-order rate constant (k_e) that depends on the total concentration of the enzyme:

$$k_e = \frac{k_2 e_t}{K_M} = \frac{v_m}{K_M} \tag{3.19}$$

An accurate determination of K_M and v_m can be obtained by means of a linearized form of Equation (3.17) which is known as the Lineweaver–Burk equation:

$$\frac{1}{v} = \frac{1}{v_m} + \frac{K_M}{v_m} \frac{1}{s} \tag{3.20}$$

According to the Lineweaver–Burk equation, a plot of $1/v$ vs. $1/s$ yields a straight line whose intercept is $1/v_m$ and the slope is K_M/v_m; this enables determining both v_m and K_M. However, this method is statistically inconsistent and the best alternative is the nonlinear regression based on Equation (3.17).

It should be emphasized that K_M is not an equilibrium constant because it is associated with a dynamic system under steady-state conditions. However, if the backconversion of the complex is very fast relative to the second step ($k_{-1} \ll k_2$), K_M will reduce to:

$$K_M \approx k_{-1}/k_1 \tag{3.21}$$

Only under these special circumstances does the formation of EC occurs under near-equilibrium conditions and K_M then represents the dissociation constant of EC. K_M is, in this case, a measure of the enzyme *affinity* for substrate; the higher the value of K_M, the lower is the affinity.

In conclusion, the reaction rate of an enzymatic reaction can follow three different kinetic laws, depending on the concentration of the substrate relative to the Michaelis–Menten constant. If $s \ll K_M$, the reaction follows a first order kinetic law with the pseudorate constant defined in Equation (3.19). At the other extreme ($s \gg K_M$), the reaction rate is independent on substrate concentration (Equation (3.16)) and the reaction proceeds according to the zero-order kinetics. In the intermediate case ($0.1 K_M < s < 10 K_M$), a second-order kinetics operates, according to Equation (3.13).

3.6.2 Other Mechanisms

So far, enzyme-catalyzed reactions involving only one substrate have been considered. However, in a large number of cases two or more substrates are involved. For example, the conversion of the enzyme–substrate complex to product may involve a second substrate, W, which causes the second step in reaction (3.8) to be formulated as follows:

$$ES + W \xrightarrow{k_3} E + P \tag{3.22}$$

It is easy to prove that, in the steady state, the reaction rate is:

$$v = \frac{k_1 k_3 e_t s w}{k_1 s + k_{-1} + k_3 w} \tag{3.23}$$

Here, w is the concentration of W. Setting $k_{cat} = k_3 w$ (for a high excess of W, its concentration is almost constant), $v_m = k_{cat} e_t$, and $K_M = (k_{cat} + k_{-1})/k_1$ causes Equation (3.23) to assume the form of the Michaelis–Menten equation. Therefore, although the overall reaction involves three reactants (S, E, and W), at a high excess of W the reaction is described fairly by a pseudosecond-order kinetic equation.

Multisubstrate reactions are often represented by schemes proposed by Cleland [27] such as that presented in Figure 3.15 for the double displacement mechanism (also termed the *ping-pong mechanism*). This mechanism operates in reactions involving coenzymes. During such a process, the enzyme alternates between two states that are, for example, the oxidized and reduced forms, in the case of an oxidase.

In Figure 3.15, below the horizontal line, the enzyme itself, along with the enzyme complexes are shown. The arrows pointing downwards or upwards indicate substrate binding an product release, respectively. Although the

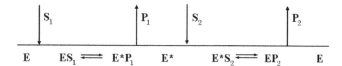

Figure 3.15 Cleland scheme for a two-substrate reaction that obeys the ping-pong mechanism.

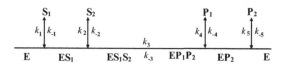

Figure 3.16 Schematics of the single displacement ordered mechanism for a two-substrate reaction.

$$NAD^+ + CH_3CH_2OH \leftrightarrows NADH + CH_3CHO$$

$$S_1 \qquad S_2 \qquad\quad P_1 \qquad P_2$$

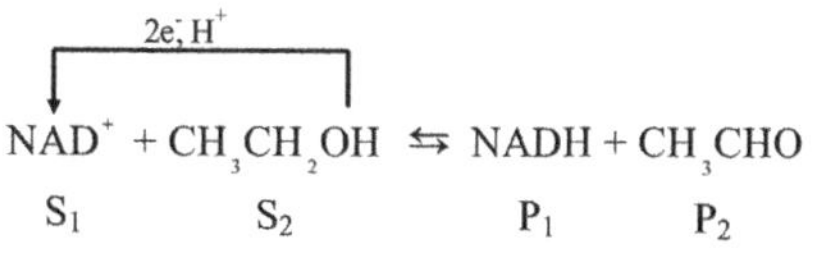

Figure 3.17 An example of single displacement ordered reaction: dehydrogenase-catalyzed conversion of ethanol to acetaldehyde.

arrows are single-headed, they symbolize as a rule reversible reactions. Often, the relevant rate constants are included on the left or right side of the arrow. Thus, in the scheme in Fig. 3.15, the substrate S_1 binds first to the enzyme to form the ES_1 complex. After ES_1 conversion to E^*P_1, the product P_1 and the altered enzyme E^* are released. Next, the second substrate S_2 gives the E^*S_2 complex that undergoes conversion to EP_2, in which the enzyme is restored to the initial state. Finally, the dissociation of EP_2 yields the product P_2 and the free enzyme. In the particular case of glucose oxidation, E and E^* are the enzyme with the FAD group in the oxidized or reduced form, respectively. S_1 and S_2 represent glucose and oxygen, respectively, whereas P_1 and P_2 represent glucono-lactone and hydrogen peroxide, respectively.

The ping-pong mechanism is characterized by the following kinetic equation:

$$v = v_m \left(1 + \frac{K_{M1}}{s_1} + \frac{K_{M2}}{s_2} \right)^{-1} \tag{3.24}$$

Here, K_{M1} and K_{M2} are constant specific to each of the substrates. If the concentration of S_2 is sufficiently high such that $K_{M2}/s_2 \ll K_{M1}/s_1$, the above kinetic equation reduces to the Michaelis–Menten equation in s_1, which allows the determination of the K_{M1} constant. If the above inequality is reversed, K_{M2} can be determined in a similar way. These constants, as well as v_m depend on the relevant rate constants in the reaction scheme in Figure 3.15.

Another kind of two-substrate reaction pathway is the single displacement ordered mechanism represented in Figure 3.16. In this case, a ternary complex $((ES_1S_2))$ forms by successive reactions of the enzyme with substrates, followed by the conversion of the substrates to products to yield a second ternary complex (EP_1P_2), followed by the stepwise release of products.

An example of this type is the conversion of ethanol to acetaldehyde by the action of alcohol dehydrogenase, with the coenzyme NAD^+ acting as electron and proton acceptor (Figure 3.17). This reaction is used in ethanol enzymatic sensors.

In summary, many enzymatic reactions involve more than one substrates. Under particular conditions, the kinetic equation of a multisubstrate reaction can be reduced to the Michaelis–Menten equation.

3.6.3 Expressing the Enzyme Activity

The kinetic approach in the previous sections allows the derivation of several physical parameters that characterize the enzyme under study. As in some cases the conversion of ES to product involves several distinct steps, it would be more correct to replace k_2 with an overall constant k_{cat}, which can be calculated by means of experimental data as follows:

$$k_{cat} = \frac{v_m}{e_t} \tag{3.25}$$

Hence, k_{cat} is the rate constant at enzyme saturation. This constant is often referred to as the *turnover number* of the enzyme and is expressed in reciprocal time units (s^{-1}). It represents the number of catalytic cycles that each active site undertakes per unit time under saturation conditions. So, if $k_{cat} = 10\,s^{-1}$, the catalytic cycle occurs 10 times a second.

In many real situations, the substrate concentration is very low ($s \ll K_M$); this implies that the free enzyme concentration is approximately equal to the total enzyme concentration ($e \approx e_t$). Taking into account this approximation, and also substituting k_{cat} for k_2 in Equation (3.13), this equation assumes the following form:

$$v = \frac{k_{cat}}{K_M} e_t s \qquad (3.26)$$

The second-order rate constant defined by the ratio k_{cat}/K_M (dimensions: $M^{-1} s^{-1}$) in the above equation is called the *specificity constant* and is a quantitative measure of the substrate specificity of an enzyme.

Very often, the experimenter has to deal not with pure enzyme but with enzyme samples containing other (inactive) protein species. Besides, some enzymes contain more catalytic sites per molecule and the turnover number, as defined above, cannot be unambiguously determined. A suitable alternative in such instances is to express the quantity of enzyme in an empirical way, by means of the *enzyme activity*. Enzyme activity represents that amount of the enzyme preparation that catalyzes the conversion of a given quantity of substrate per unit time, under specified conditions (such as temperature and pH):

$$\text{Activity} = \frac{\Delta m_S}{\Delta t} = \frac{V \Delta c_s}{\Delta t} \qquad (3.27)$$

where m_S is the amount of substrate, V is the sample volume, and c_s is the concentration of the substrate. This definition implies that the activity is determined close to the zero-order kinetics region, such as to render it practically independent of substrate concentration. The SI unit for enzyme activity is the *katal* that represents the amount of enzyme preparation that converts 1 mole of substrate per second (1 katal = mole/s). For example, one katal of trypsin is that amount of trypsin that breaks a mole of peptide bonds per second under specified conditions.

A commonly used value is the *enzyme unit* (U) that denotes the amount of enzyme that converts 1 μmol of substrate per minute (units: [μmol/min]. 1 U corresponds to 16.67 nanokatals. From the kinetic standpoint, enzyme activity represents an approximation of the maximum reaction rate.

Activity is an extensive quantity. Hence, in order to characterize the enzyme activity independently of its amount, the term *specific activity* was introduced; it is defined as follows:

$$\text{Specific activity} = \frac{\text{Activity}}{\text{Amount of enzyme material}} \qquad (3.28)$$

The SI unit for specific activity is katal kg^{-1}, but a more practical unit is $\mu mol \, mg^{-1} \, min^{-1}$. The specific activity is an indicator of enzyme purity:

$$\% \, \text{purity} = \frac{\text{Specific activity of enzyme sample}}{\text{Specific activity of pure enzyme}} \times 100 \qquad (3.29)$$

Clearly, an impure sample has lower specific activity because some of its content is not an actual enzyme.

3.6.4   pH Effect on Enzyme Reactions

Acid or base groups can be present at the enzyme catalytic site. Consequently, their ionization state can influence the k_{cat} and K_M values and, therefore, the reaction rate. The pH can determine the protonation state of a site involved in the binding of substrate and affect the stability of the enzyme–substrate complex with an immediate effect on the K_M value. On the other hand, the protonation state of a particular site can be essential for the progress of the reaction within this complex. If a change in the protonation state removes the catalytic activity completely, only a fraction of the total enzyme amount will remain active. As the reaction rate is measured with respect to the total enzyme concentration, this change results in a decrease in the k_{cat} value. Often, the combined pH effect on both K_M and k_{cat} causes the reaction rate–pH curve to display a maximum at a specific pH.

Charged substrates may also influence the pH dependence of the reaction rate either directly or indirectly, as they affect the local ionic strength and thereby the thermodynamic activity of the reactants.

Moreover, the transduction process itself can be affected by pH such as in the case of NH_3- or CO_2-linked transduction. At a high pH, protonation of the above compounds leads to the formation of NH_4^+ or HCO^{3-}, respectively. Consequently, the response generated by a gas-sensitive transducer is diminished with respect to the response obtained at the optimum pH.

In summary, an enzyme reaction and, implicitly, the response of an enzymatic sensor can be strongly dependent on pH. Hence, this parameter should be controlled by means of a pH buffer system whose components do not participate in the enzyme-catalyzed reaction.

3.6.5 Temperature Effect on Enzyme Reactions

Like any chemical reaction, the rate of an enzyme-catalyzed reaction increases as the temperature is raised. The Michaelis–Menten equation includes a reaction rate constant (k_{cat}) and a pseudoequilibrium constant (K_M) that are both temperature dependent. The effect of temperature on the rate constant is given by the Arrhenius Equation (3.30) that demonstrates an exponential increase in the rate constant with the temperature, the steepness of the variation being determined by the activation energy ($\Delta G^{\#}$). In this equation, A is a constant factor R is the ideal gas constant and T is the absolute temperature.

$$k_{cat} = A e^{-\frac{\Delta G^{\#}}{RT}} \tag{3.30}$$

The effect of temperature on K_M can be expressed by the van't Hoff equation, which includes the reaction standard Gibbs energy for the formation of the enzyme–substrate complex (ΔG_f).

$$K_M = B e^{\frac{\Delta G_f}{RT}} \tag{3.31}$$

where B is a constant. $\Delta G^{\#}$ is always positive, whereas ΔG_f can be either positive or negative. Consequently, k_{cat} always increases with temperature, whereas K_M can increase or decrease with this parameter. As $|\Delta G^{\#}| > |\Delta G_f|$, the reaction rate will generally increase with temperature up to a maximum value. Further increase is prevented by enzyme denaturation, which leads to a sharp decrease in enzyme activity. As a rule, enzyme processes should be carried out about 10–20 °C below the maximum activity temperature in order to avoid enzyme denaturation.

3.6.6 Outlook

Enzyme-catalyzed reactions can follow a large variety of pathways. A plot of the initial reaction rate vs. substrate concentration (as in Figure 3.14) is the first test for assessing the reaction mechanism. A double-reciprocal plot according to Equation (3.20) is actually much more informative at this stage. Any deviation from the Michaelis–Menten behavior suggests that additional factors come into play (such as inhibition by substrate or product) or the mechanism does not match the Michaelis–Menten assumptions. Under particular conditions, a more complicated mechanism can be described by the Michaelis–Menten equation provided that the characteristic constants are properly defined in accordance with mechanism details.

The Michaelis–Menten equation puts into evidence two limiting cases, according to the substrate concentration relative to the Michaelis–Menten constant. If the substrate concentration is very low with respect to this constant, the reaction rate is proportional to the substrate concentration. This case represents the first-order kinetics limit. Enzymatic sensors designed for substrate determination should operate under these conditions. The opposite case, of a high substrate concentration, corresponds to the zero-order kinetics regime, in which the reaction rate is proportional to enzyme concentration but independent of substrate concentration. Conditions close to this limit are convenient for performing enzyme quantifications when the enzyme is used as a label tag or when the sensor is designed for inhibitor determination.

Questions and Exercises (Section 3.6)

1 What are the key assumptions in the Michaelis–Menten mechanism?

2 What is the meaning of the Michaelis–Menten constant?

3 In what way can the substrate concentration affect the kinetics of an enzyme reaction? What conditions are suited for determining the substrate?

4 Comment on possible two-substrate mechanisms. In what conditions does the Michaelis–Menten equation represent a fair description of the reaction kinetics for such mechanisms?

5 Summarize the quantities used to assess the catalytic activity of an enzyme and point out the specific conditions for each of them.

6 What is the effect of pH and temperature on enzyme reactions?

7 The following kinetic data were obtained for an enzyme-catalyzed reaction.

Initial substrate concentration (mmol/L)	0.1	0.3	0.5	1.5	3.0	4.0
Initial reaction rate (μmol/L min)	0.018	0.031	0.045	0.059	0.063	0.065

a. Plot the above data according to the Michaelis–Menten equation and estimate the values of v_m and K_M from the curve.
b. Process the Michaelis–Menten equation in order to derive the following linear relationships (1) $1/v$ vs. $1/s$ (Lineweaver–Burk equation); (2) s/v vs. s (Hanes equation); (3) v vs. v/s (Eady–Hofstee equation).
c. Plot the kinetic data according to the previously derived equations and calculate the values of v_m and K_M using the slopes and intercepts of the plotted lines.

Hints and answers: (c) v_m: $= 0.069$ μmol/L min; $K_M = 0.29$ mM. Small differences between the values determined by each of the above plots may occur due to the effect of error propagation when calculating the new variables. These differences are insignificant if the experimental data are accurate enough. The most reliable results are obtained if experimental data are processed by nonlinear regression according to the Michaelis–Menten equation.

8 An enzymatic reaction occurs as follows: $S + NAD^+ = P + NADH$. In order to assess the enzyme, the progress of the reaction was determined by recording NADH absorbance at 340 nm (molar absorptivity: $a = 6200 \, L \, mole^{-1} \, cm^{-1}$) in a $l = 1$ cm cell containing $V = 3$ ml solution. The absorbance (A) varied linearly with time at the beginning and changed by $\Delta A = 0.26$ units within the time interval $\Delta t = 590$ s.

a. What is the role of NAD^+ in this reaction?
b. Calculate the enzyme activity in katal and enzyme units.
c. Calculate the specific enzyme activity assuming that the solution contains 0.02 mg enzyme.

Hints and answers: Convert all quantities into SI units. Δs is proportional to ΔA, in accordance to Beer's law. Activity $= 0.84$ nanokatal $= 0.05$ U. Specific activity $= 42$ μkatal/s kg $= 2.5$ U/mg.

9 In order to determine the enzyme activity, the substrate concentration should be selected such that the reaction rate is as close as possible to its maximum value. For an enzyme with $K_M = 3.1$ mM, calculate the lowest substrate concentration that leads to a deviation of the reaction rate not lower than 95% of the maximum reaction rate.

Hints and answers: Rearrange the Michaelis–Menten equation such as to obtain the v/v_m ratio as a function of s, equate this ratio to 0.95 and solve for s.

Answer: $s \geq 23$ mM.

3.7 Enzyme Inhibition

Some compounds termed *effectors* can bind to an enzyme molecule and thereby diminish or enhance its activity. In the first case, such a compound acts as an *inhibitor*; in the second one, it performs as an *activator* [24,28]. Enzyme inhibition allows the quantification of inhibitor concentration by standard nonsensor assay methods but also forms the basis for inhibitor determination by means of enzyme biosensors [29–31]. This section reviews first the principles of enzyme inhibition and then presents the basics of inhibition-based biosensors.

3.7.1 Reversible Inhibition

Reversible inhibition involves a reversible reaction between the inhibitor and the enzyme (E); the enzyme can be in the free state or can be part of an enzyme–substrate (ES) complex. Thus, the inhibitor (I) may form one of two kinds of inactive complex: a binary (EI) or a ternary (ESI) complex. Therefore, in order to derive kinetic equations for inhibited catalysis, the following reactions should be considered:

- Enzyme reaction in the absence of the inhibitor:

$$E + S \underset{k_{-1}}{\overset{k_1}{\rightleftharpoons}} ES \overset{k_2}{\longrightarrow} E + P \qquad (3.32)$$

- Reversible binding of the inhibitor to the free enzyme:

$$E + I \underset{}{\overset{K_i}{\rightleftharpoons}} EI \qquad (3.33)$$

- Reversible formation of an enzyme-substrate-inhibitor ternary complex

$$\text{ES} + \text{I} \xrightleftharpoons{K_i} \text{ESI} \tag{3.34}$$

$$\text{EI} + \text{S} \xrightleftharpoons{K_i} \text{ESI} \tag{3.35}$$

Reversible inhibition is characterized by the *inhibition constant*, K_i, which is the dissociation constant for each of reactions (3.33) to (3.35). The reaction rates in the presence (v_i) and in the absence (v) of the inhibitor are conveniently compared using the *degree of inhibition*, d_i:

$$d_i = \frac{v - v_i}{v} \tag{3.36}$$

Various types of reversible inhibition can be defined by combinations of the above reactions (see Table 3.5). The reaction rate for the inhibited reaction can be derived under the steady-state condition using a procedure similar to that employed in Section 3.6.1. It can be proved in this way that the kinetic equation in the presence of reversible inhibition has the same form as the Michaelis–Menten equation but either v_m or K_M (or both) are replaced by the apparent parameters v_m^i and K_M^i. v_m^i and K_M^i may be dependent on both inhibitor concentration (i) and the inhibition constant. The combined effect of these quantities is indicated by the over-unity parameter q:

$$q = 1 + \frac{i}{K_i} \tag{3.37}$$

Competitive inhibition occurs if the inhibitor binds to the free enzyme (reaction (3.33)), either at the active site or at another (distant) site on the enzyme, with destructive effects on catalytic activity. As a result, the concentration of the enzyme available to the substrate is reduced. Thus, hydroxyurea, a urea-analog, is a competitive inhibitor of urea hydrolysis in the presence of soya bean urease. As in the case of urea, it forms a coordination bond with a nickel atom at the active site but yields an inactive complex. It is worth noting that fluoride ions also cause inhibition by bonding to the nickel atoms in the enzyme [32].

Competitive inhibition has no effect on v_{max} but K_M is replaced by the greater parameter qK_M. Due to the competition of substrate and inhibitor for the same site, the degree of inhibition decreases with increasing the substrate/inhibitor concentration ratio.

In *uncompetitive inhibition* an inhibitor binds only to the enzyme–substrate complex (reaction (3.34)) but not to the free enzyme. The formation of the ternary complex prevents the substrate from being converted to the final product. Both v_{max} and K_M are lowered by uncompetitive inhibition. As there is no competition between substrate and inhibitor for the same binding site, an increase in the substrate concentration does not overcome the uncompetitive inhibition. Conversely, at a constant inhibitor concentration, the degree of inhibition may increase with the substrate concentration because the chance of forming the inert EIS species is thus enhanced.

In *noncompetitive inhibition* the inhibitor can combine with both the free enzyme and the enzyme–substrate complex (reactions (3.33) and (3.34)) in order to give a dead-end complex. This implies that the inhibitor binds at a site other than the active one and brings about a conformational change that removes enzyme affinity for substrate. By virtue of this two-site feature, the substrate can also combine with the EI complex (reaction (3.35)). For the same reason, noncompetitive inhibition is not alleviated by increased substrate concentration. A specific inhibition constant should be defined in this case for each of reactions (3.34) and (3.35).

The effect of inhibition on reaction parameters is illustrated in Table 3.5 for the particular case in which the above constants are equal. If this condition is not fulfilled, v_{max} decreases, while K_M may increase or decrease with increasing inhibitor concentration. Often, the term *mixed inhibition* is assigned to this general case.

The most frequent type of inhibition is the competitive one. Uncompetitive and noncompetitive inhibition are rare in nature [2].

Table 3.5 Selected mechanisms of reversible inhibition

Inhibition type	Reactions	v_m^i	K_M^i
Competitive	(3.32) and (3.33)	v_m	qK_M
Uncompetitive	(3.32) and (3.34)	v_m/q	K_M/q
Noncompetitive	(3.32), (3.33), (3.34) and (3.35)	v_m/q	K_M

3.7.2 Irreversible Inhibition

Irreversible inhibitors interact with enzymes so strongly that the concentration of active enzyme is reduced from an initial value e_t to $e_t - i$, according to a 1-to-1 stoichiometry. Consequently, the maximum reaction rate in the presence of the inhibitor is always less than the normal value:

$$v_{max}^i = v_{max}[1 - (i/e_t)] \tag{3.38}$$

A well-known example is represented by organophosphorus derivatives (widely employed as pesticides) that react with –OH groups in serine side chains of some enzymes such as acetylcholine esterase.

In contrast to reversible inhibition, which is a fast, diffusion-controlled process, irreversible inhibition is a slow reaction that needs an incubation time in order to be completed. During this stage, enzyme activity decreases exponentially.

3.7.3 Enzymatic Sensors for Inhibitors: Design and Operation

In order to detect an enzyme inhibitor, the sensor should be operated under zero-order kinetics, as pointed out in Section 3.6.1. Inhibitor determination is therefore prone to interference by any factor affecting the enzyme activity, such as pH, temperature and enzyme degradation.

Determination of the inhibitor can be performed under *steady-state* conditions by means of a calibration function that expresses the *percentage of inhibition* (%*I*, Equation (3.39)) vs. inhibitor concentration. Here y_o and y_i stand for sensor response in the absence and in the presence of inhibitor, respectively.

$$\%I = 100(y_o - y_i)/y_o \tag{3.39}$$

Clearly, this equation is formally analogous to Equation (3.36) defined for the inhibition of a dissolved enzyme.

The calibration line is, as a rule, nonlinear but, under properly optimized conditions, a quasilinear region can be identified. A log plot can sometimes display a better linear feature. In some instances, a plot of the residual activity vs. inhibitor concentration yields a convenient calibration graph. The main parameters to be optimized are the immobilization procedure, the enzyme loading, substrate concentration, pH and incubation time.

Alternatively, a *kinetic* approach can be implemented. In this case, after the reaction attains the steady state with the substrate alone, the inhibitor is added and the rate of inhibition is assessed as the slope of the signal–time curve (dy/dt), which is a function of inhibitor concentration.

In principle, any transduction method that is compatible with the enzyme reaction can be implemented. Much research work in this respect has been devoted to electrochemical methodologies [33–35].

A critical issue with inhibitor assay is the limited selectivity. As the inhibition phenomenon can be caused by very different type of compounds (for example, metal cations and various inorganic or organic species) the selectivity is poor [36]. That is why an inhibitor sensor should be appraised in natural samples in order to assess how useful it is for solving real problems. Selectivity of enzyme-inhibition-based biosensors can be enhanced by using an array of sensors, each of them based on a different enzyme, with subsequent chemometric treatment of data. Alternatively, the assay can include a separation step in order to get rid of interferences. Limited selectivity, however, can be an advantage if the sensor is used for screening the toxicity of environmental samples.

Another critical aspect is the reactivation of the enzyme after being exposed to an inhibitor-containing sample. Reactivation should be designed according to the chemistry of each enzyme–inhibitor couple. So, reversible inhibition by metal ions can be removed by means of a strong chelating agent, such as EDTA. In many instances, reversible inhibition by organic compounds can be eliminated by soaking the sensor in a buffer solution at the optimum pH of the enzyme. As the reactivation is not always total, the calibration graph should be checked regularly. Reactivation is much more difficult in the case of irreversible inhibition, but even in this instance, suitable procedures may be devised. If reactivation proves not to be feasible, it is possible to resort to the disposable-sensor approach.

The detection limit for an inhibitor sensor depends on both the substrate concentration and enzyme loading. In the case of competitive inhibition, the limit of detection decreases as the substrate and enzyme concentration decrease but is limited by the characteristic detection limit of the transduction process.

As the analysis by means of inhibition-based biosensors involves several steps (such as pH adjustment, incubation and regeneration) it is convenient to include the sensor in an automatic flow-analysis system.

A comprehensive overview of the performance of electrochemical biosensors developed for the determination of inhibiting species is given in [37].

3.7.4 Applications of Enzyme-Inhibition Sensors

Enzyme-inhibition methods have been applied mostly to determine hazardous substances such as pesticides and heavy-metal ions. The choice of a particular enzyme is made with reference to the properties of the targeted pollutant. Widely used are choline esterase enzymes [18,38,39]. Thus, acetylcholine esterase (AChE, EC 3.1.1.7) is an enzyme that catalyzes the hydrolysis of the neurotransmitter acetylcholine, producing choline and acetate (see also Figure 3.10):

$$R\text{-}COO(CH_2)_2N^+(CH_3)_3 + H_2O \xrightarrow{AChE} HO(CH_2)_2N^+(CH_3)_3 + R\text{-}COO^- + H^+ \qquad (3.40)$$

Similarly, butyrylcholine esterase (BuChE, EC 3.1.1.8) acts on the synthetic substrate butyrylcholine. In the above reactions, carbamate pesticides induce reversible, competitive inhibition, while organophosphorus derivatives are irreversible inhibitors via esterification to a hydroxyl group in serine.

As an acid product results in reaction (3.40), transduction can be performed simply by means of a potentiometric or optical pH probe. Properly selected synthetic substrates can enhance the sensitivity. Thus, a nonfluorescent substrate that yields a fluorescent product enables a very low limit of detection to be achieved owing to the intrinsic sensitivity of fluorimetry. Alternatively, the substrate can be selected so as to give an electrochemically active product (such as thiocholine), which allows amperometric transduction. On the other hand, choline oxidase can be added to the sensing layer in order to convert choline (resulted from reaction (3.40)) into betaine aldehyde with oxygen consumption and hydrogen peroxide formation. With such a bi-enzyme sensor, transduction can be performed by the electrochemical monitoring the concentration of either oxygen or hydrogen peroxide.

Also useful for pesticide determination are oxidoreductase enzymes (such as glucose oxidase and tyrosinase) and hyrolases (such as alkaline phosphatase and urease).

Heavy-metal ions (Hg^{2+}, Cu^{2+}, Pb^{2+}, Cd^{2+}, Ag^+) cause enzyme inhibition by strong interaction with thiol groups near the active site of the enzyme. Urease, choline esterases and oxidases are among the enzymes mostly employed in developing heavy-metal ion sensors based on this principle. Selectivity is, as a rule, very poor, and determination of an individual metal ion in the presence of similar interfering species is hardly achievable.

In conclusion, enzyme inhibition represents a convenient approach for performing the determination of organic and inorganic pollutants in environment samples as well as in foods. Application of nanomaterials is expected to bring about further progress in this field [40]. However, it should be kept in mind that the selectivity of inhibition sensors could be problematic.

Although much research effort has been exerted in this field, there are still many concerns regarding the reliability of the sensor in applications to real samples.

Exercises (Section 3.7)

1 Download the structure 1IE7 in the Protein Data Bank (http://www.pdb.org/pdb/home/home.do) and examine the binding of the phosphate ion inhibitor to the active center of urease.

2 In the absence of inhibition, K_M of an enzyme was of 8 μM. In the presence of 3 μM inhibitor, the apparent Michaelis–Menten constant (K_M^i) was 12 μM. Calculate the inhibition constant.

Answer: 6 μM

3 Derive equations relating d_i to substrate and inhibitor concentrations for each kind of inhibition in Table 3.5 and discuss the effect of inhibitor and substrate concentration on d_i. Prove that, for competitive inhibition, d_i increases if the i/s ratio decreases at a constant inhibitor concentration.

Hint: use dimensionless concentrations $\iota = i/K_i$; $\sigma = s/K_M$.

Answer: Competitive: $\iota/(1 + \iota + \sigma)$; uncompetitive: $\sigma\iota/(1 + \sigma + \sigma\iota)$; noncompetitive: $i/(1 + \iota)$.

4 The concentration of the mercury ion can be determined by means of a urea sensor consisting of a urease layer attached to a CO_2 probe (log-type transduction). Using calibration data in the table below, plot the percentage of inhibition vs. log $[Hg^{2+}]$, define the linear response range, calculate the sensitivity within the linear region and calculate the Hg^{2+} concentration in two samples yielding $\%I = 38$ and 67, respectively.

c/μM	0	0.01	0.03	0.05	0.1	0.3	0.5	1.0	3.0
Signal	55	53	45	40	30	18	11	6	1

Hint. A linear calibration function can be identified within the intermediate concentration range.

Answer: 0.07 and 0.27 μM.

5 An urea sensor, which was used to determine the fluoride ion by inhibition, was checked for the effect of the enzyme activity in the sensing layer. To this end, the effect of fluoride ion concentration on the percentage of inhibition was determined at two different enzyme activities, as shown in the Table below. Plot the percentage of inhibition vs. the fluoride concentration and comment on the effect of enzyme activity on the working range and limit of detection.

Hint. Assume that the limit of detection is the fluoride concentration at which the percentage of inhibition is 10%.

$[F^-]$/mM	1.0	1.2	1.3	1.5	1.7	2.0	2.3	2.5	2.7
%I (0.05 U urease)	18	32	40	55	68				
%I (0.50 U urease)					10	29	47	60	73

3.8 Concluding Remarks

Enzymes are proteins that catalyze chemical reactions in biological systems. Under suitable conditions, enzymes isolated from biological preparations display catalytic activity in artificial systems as well. This property is utilized in the development of enzymatic sensors. An enzymatic sensor is obtained by integration of an enzyme with a transduction device that indicates the concentration of a reactant or product of the enzymatic reaction. Integration of the enzyme and the transducer is achieved by enzyme immobilization at the surface of the transducer.

As long as the enzyme substrate is available, an enzyme reaction continues to proceed. Therefore, the process occurring in an enzymatic sensor is a dynamic process, characterized by its reaction rate. Substrate supply occurs by diffusion from the solution to the active part of the sensor. As diffusion has a limited rate, a steady state is reached; in the steady state, the concentration of reactants and products at the transducer surface is constant. As a result, the sensor response is also constant and is determined either by the diffusion rate or by the reaction rate or both. The next chapter demonstrates that the best conditions for the determination of the substrate are achieved when the enzymatic reaction is of the first order with respect to the substrate and the rate of the overall process is determined by diffusion.

Application of enzymes in chemical sensing is not limited to the determination of the enzyme substrate. Other applications are based on the measurement of the activity of the enzyme incorporated in an enzymatic sensor.

Enzyme activity can be affected by inhibitors such as toxic metal ions or certain organic compounds. Therefore, enzymatic sensors can be used for the determination of enzyme inhibitors. In such applications, the sensor should be operated such that the enzyme reaction occurs close to the zero-order kinetic regime. Under these conditions, the response is independent on the substrate concentration but is dependent on the enzyme activity.

Monitoring of enzyme activity is also of great interest in the application of enzymes as signaling labels in affinity sensors. In such applications, an enzyme-tagged species is allowed to interact with the sensing layer such that the amount of enzyme incorporated in this layer is a function of the analyte concentration. In the presence of added substrate, the sensor functions as an enzymatic sensor acting as an indicator of the amount of incorporated enzyme. The optimal kinetic regime in this case is the zero-order regime in which the response depends on the concentration of the incorporated enzyme. In order to obtain good sensitivity, enzymes with a high turnover number are used as labels.

In conclusion, owing to their broad range of applications, enzymes occupy a central role in chemical-sensor science and technology.

References

1. Palmer, T. (2001) *Enzymes: Biochemistry, Biotechnology, and Clinical Chemistry*, Horwood Publ., Chichester.
2. Bugg, T. (2004) *Introduction to Enzyme and Coenzyme Chemistry*, Blackwell, Oxford.
3. Copeland, R.A. (2000) *Enzymes: A Practical Introduction to Structure, Mechanism, and Data Analysis*, Wiley-VCH, New York.
4. Price, N.C. and Stevens, L. (1999) *Fundamentals of Enzymology: The Cell and Molecular Biology of Catalytic Proteins*, Oxford University Press, Oxford.
5. Guilbault, G.G. (1976) *Handbook of Enzymatic Methods of Analysis*, M. Dekker, New York.
6. Webb, E.C. (1992) *Enzyme Nomenclature 1992: Recommendations of the Nomenclature Committee of the International Union of Biochemistry and Molecular Biology on the Nomenclature and Classification of Enzymes*, Academic Press, San Diego.
7. Scheller, F.W. and Schubert, F. (1992) *Biosensors*, Elsevier, Amsterdam.

8. Bankar, S.B., Bule, M.V., Singhal, R.S. *et al.* (2009) Glucose oxidase - an overview. *Biotechnol. Adv.*, **27**, 489–501.

9. Wilson, R. and Turner, A.P.F. (1992) Glucose-oxidase - an ideal enzyme. *Biosens. Bioelectron.*, **7**, 165–185.

10. Raba, J. and Mottola, H.A. (1995) Glucose-oxidase as an analytical reagent. *Crit. Rev. Anal. Chem.*, **25**, 1–42.

11. Wang, J. (2008) Electrochemical glucose biosensors. *Chem. Rev.*, **108**, 814–825.

12. Wang, J. (2001) Glucose biosensors: 40 years of advances and challenges. *Electroanalysis*, **13**, 983–988.

13. Marko-Varga, G., Emneus, J., Gorton, L. *et al.* (1995) Development of enzyme-based amperometric sensors for the determination of phenolic compounds. *TrAC-Trends Anal. Chem.*, **14**, 319–328.

14. Peter, M.G. and Wollenberger, U. (1997) Phenol-oxidizing enzymes: Mechanism and applications in biosensors, in *Frontiers in Biosensorics* (eds F.W. Scheller, F. Schubert, and J. Fedrowitz), Birkhäuser, Basel, pp. 63–81.

15. Shleev, S., Tkac, J., Christenson, A. *et al.* (2005) Direct electron transfer between copper-containing proteins and electrodes. *Biosens. Bioelectron.*, **20**, 2517–2554.

16. Riva, S. (2006) Laccases: Blue enzymes for green chemistry. *Trends Biotechnol.*, **24**, 219–226.

17. Ruzgas, T., Csoregi, E., Emneus, J. *et al.* (1996) Peroxidase-modified electrodes: Fundamentals and application. *Anal. Chim. Acta*, **330**, 123–138.

18. Fitzpatrick, P.F. (2001) Substrate dehydrogenation by flavoproteins. *Accounts Chem. Res.*, **34**, 299–307.

19. Andreescu, S. and Marty, J.L. (2006) Twenty years research in cholinesterase biosensors: From basic research to practical applications. *Biomol. Eng.*, **23**, 1–15.

20. Singh, M., Verma, N., Garg, A.K. *et al.* (2008) Urea biosensors. *Sens. Actuators B-Chem.*, **134**, 345–351.

21. Scheller, F., Renneberg, R., and Schubert, F. (1988) Coupled enzyme reactions in enzyme electrodes, in *Methods in Enzymology*, vol. **137d** (ed. K. Mosbach), Academic Press New York, pp. 29–43.

22. Carr, P.W. and Bowers, L.D. (1980) *Immobilized Enzymes in Analytical and Clinical Chemistry: Fundamentals and Applications*, John Wiley & Sons, New York.

23. Holme, D.J. and Peck, H. (1993) *Analytical Biochemistry*, Longman Scientific & Technical, Harlow.

24. Leskovac, V. (2003) *Comprehensive Enzyme Kinetics*, Kluwer Academic/Plenum Pub., New York.

25. Bisswanger, H. (2002) *Enzyme Kinetics: Principles and Methods*, John Wiley & Sons, Weinheim, pp. 108–120.

26. Grunwald, P. (2009) *Biocatalysis: Biochemical Fundamentals and Applications*, Imperial College Press, London.

27. Cleland, W.W. (1963) Kinetics of enzyme-catalyzed reactions with two or more substrates or products. 1. Nomenclature and rate equations. *Biochim. Biophys. Acta*, **67**, 104–137.

28. Bisswanger, H. (2002) *Enzyme Kinetics: Principles and Methods*, Wiley-VCH, Weinheim.

29. Domínguez, E. and Narváez, A. (2005) Non-affinity sensing technology: The exploitation of biocatalytic events for environmental analysis, in *Biosensors and Modern Biospecific Analytical Techniques* (ed. L. Gorton), Elsevier, Amsterdam, pp. 429–537.

30. Amine, A., Mohammadi, H., Bourais, I. *et al.* (2006) Enzyme inhibition-based biosensors for food safety and environmental monitoring. *Biosens. Bioelectron.*, **21**, 1405–1423.

31. Marques, P. and Yamanaka, H. (2008) Biosensors based on the enzymatic inhibition process. *Quim. Nova*, **31**, 1791–1799.

32. Tarun, E. I., Rubinov, D. B. and Metelitza, D. I. (2004) Inhibition of urease by cyclic beta-triketones and fluoride ions. *Appl. Biochem. Microbiol.*, **40**, 337–344.

33. Sole, S., Merkoci, A. and Alegret, S. (2003) Determination of toxic substances based on enzyme inhibition. Part I. Electrochemical biosensors for the determination of pesticides using batch procedures. *Crit. Rev. Anal. Chem.*, **33**, 89–126.

34. Sole, S., Merkoci, A. and Alegret, S. (2003) Determination of toxic substances based on enzyme inhibition. Part II. Electrochemical biosensors for the determination of pesticides using flow systems. *Crit. Rev. Anal. Chem.*, **33**, 127–143.

35. Trojanowicz, M. (2002) Determination of pesticides using electrochemical enzymatic biosensors. *Electroanalysis* **14**, 1311–1328.

36. de Castro, M. D. L. and Herrera, M. C. (2003) Enzyme inhibition-based biosensors and biosensing systems: questionable analytical devices. *Biosens. Bioelectron.*, **18**, 279–294.

37. Evtugyn, G. A., Budnikov, H. C. and Nikolskaya, E. B. (1998) Sensitivity and selectivity of electrochemical enzyme sensors for inhibitor determination. *Talanta*, **46**, 465–484.

38. Skladal, P. (1996) Biosensors based on cholinesterase for detection of pesticides. *Food Technol. Biotechnol.*, **34**, 43–49.

39. Schulze, H., Vorlova, S., Villatte, F. *et al.* (2003) Design of acetylcholinesterases for biosensor applications. *Biosens. Bioelectron.*, **18**, 201–209.

40. Periasamy, A. P., Umasankar, Y. and Chen, S. M. (2009) Nanomaterials – Acetylcholinesterase Enzyme Matrices for Organophosphorus Pesticides Electrochemical Sensors: A Review. *Sensors*, **9**, 4034–4055.

4

Mathematical Modeling of Enzymatic Sensors

4.1 Introduction

A particular feature of enzymatic biosensors is the dynamic character of the sensor processes. Once the substrate-containing solution comes into contact with the sensor, conversion of substrate under the catalytic action of the enzyme begins and induces concentration gradients. As a result, diffusion of both reactant and product is sustained as long as the substrate is available. Consequently, modeling of enzymatic biosensors is a problem of mass transfer coupled with a chemical reaction that is catalyzed by an immobilized catalyst. From this standpoint, an enzymatic sensor can be viewed as a small enzyme reactor integrated with a transduction device.

The expected output of the modeling approach is the response function that gives the correlation between the sensor response and the substrate concentration in the test solution. Modeling reveals the essential working parameters and is of particular importance for the rational design of the sensor. Also, it allows predicting important features, such as the extent of the linear calibration range, the limit of detection, the response time and possible interferences.

The key assumption in the present approach is that the transduction response is a function of the concentration of a product at the transducer interface. Therefore, the main goal of the mathematical modeling is to find a relationship between the product concentration at the transducer surface and the substrate concentration in the test solution.

It will be assumed that the transduction process occurs without consumption of the product. This assumption is reasonably fulfilled if the transduction is performed by potentiometric, conductometric or optical methods. Amperometric enzyme sensor do not always fulfill this condition and modeling methods for such sensors are therefore left for a later chapter (Chapter 15).

Clearly, a stable, time-independent response, results only under steady-state conditions, in which the concentration of any reactant and product is time-independent at any point. The steady-state condition is therefore a main assumption in further derivations. Mathematically, it is expressed by equating the rates of the chemical reaction and mass-transfer process relative to both substrate and product. As is typical of a sequential process in the steady state, the overall reaction rate is determined by the slowest step, which can be either the enzymatic reaction or a diffusion step.

Modeling of immobilized enzyme reactors is of outstanding interest in biotechnology and is comprehensively dealt with in various texts (e.g., [1–3]). A good introduction to the modeling of enzymatic sensors is available in ref. [4].

Functioning of an enzymatic sensor is determined to a great extent by the diffusion processes. That is why this chapter is organized in accordance to the diffusion conditions. The first two sections deal with limiting cases in which diffusion is localised either out of the enzyme layer (external diffusion) or inside the enzyme layer (internal diffusion). The final section addresses the case in which diffusion in both the solution and the enzyme layer are relevant to the sensor functioning.

4.2 The Enzymatic Sensor under External Diffusion Conditions

4.2.1 The Physical Model

The physical model of an enzymatic sensor functioning under external diffusion conditions is shown in Figure 4.1. This sensor consists of a homogeneous enzyme layer intercalated between the transducer surface and a membrane permeable to the substrate and the product. It is assumed that the substrate and product concentrations are constant within both the enzyme layer and in the test solution. An even concentration distribution within the solution phase is secured by stirring. In order to obtain an even distribution of concentration within the enzyme layer, the enzyme loading should be high and the diffusion through the membrane should be sluggish. A more accurate definition of the external diffusion conditions is given in Section 4.4.

Under the above conditions, concentration gradients occur only within the membrane. It is assumed for simplicity that concentration profiles within the membrane are linear.

In the framework of this model, the transduction is performed by detecting the product of the enzymatic reaction and the transduction process does not involve product consumption. The response signal depends on the product

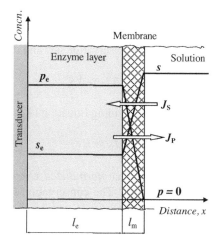

Figure 4.1 Concentration profiles and mass fluxes in an enzymatic sensor under external diffusion conditions. p and s denote the concentration of the substrate and product in the solution, respectively; the same symbols with the subscript e indicate substrate and product concentrations in the enzyme layer. J_S and J_P are diffusive fluxes of the substrate and product, respectively.

concentration within the enzyme layer and the goal of this approach is to derive an equation relating this parameter to the substrate concentration in the sample solution. Initially, the product is not present in the solution and the amount of product accumulated next into the solution is very low. Hence, the product concentration in the solution is negligible during the run. As the mass-transfer process is limited to the sensor region outside the enzyme layer, it is said that the sensor is operated under *external mass-transfer* conditions.

The physical model depicted in Figure 4.1 pertains to a membrane-covered enzyme layer. However, this model applies also to membraneless sensors. In this case, the diffusion gradient is localized in a solution film adjacent to the enzyme layer according to the Nernst diffusion model.

4.2.2 The Mathematical Model

The functioning of enzymatic sensors is based on a sequential two-step process involving diffusion and enzymatic conversion of the substrate. In the steady state, the velocity of the enzymatic reaction is equal to the diffusion rates of the substrate and the product. The reaction velocity will be expressed by the *volume reaction rate*, which is the product of the reaction rate within the enzyme layer (v') and the volume of the enzyme layer (V):

$$v_V = V v' \tag{4.1}$$

The volume reaction rate (in $mol\,s^{-1}$) indicates the number of moles of substrate converted within the whole enzyme layer per time unit. Assuming that the enzyme reaction follows Michaelis–Menten kinetics, the volume reaction rate is:

$$v_V = V \frac{k_2 e_t s_e}{K_M + s_e} \tag{4.2}$$

where e_t is the enzyme concentration in the enzyme layer, k_2 is the rate constant and K_M is the Michaelis–Menten constant. Each of the above constants refers to the immobilized enzyme and can be different from the relevant constant determined in the solution phase.

The diffusion rate is expressed by the *diffusive flux* (J) that indicates the number of moles crossing in unit time a unit area cross section perpendicular to the diffusion direction; the flux unit is $mol\,m^{-2}\,s^{-1}$. In the framework of the Nernst layer model, the diffusive flux is proportional to the concentration difference between the solution and the enzyme layer, the proportionality constant being termed the *mass-transfer coefficient*. Therefore, the substrate and product fluxes depend on concentration as follows:

$$J_S = -k_{m,S}(s - s_e); \quad J_P = -k_{m,P}(p - p_e) \tag{4.3}$$

where $k_{m,S}$ and $k_{m,P}$ are the mass-transfer coefficients of the substrate and the product, respectively and $p = 0$. The mass-transfer coefficient is the quotient of the diffusion coefficient and the thickness of the diffusion layer:

$$k_{m,S} = \frac{D_{S,m}}{l_m}; \quad k_{m,P} = \frac{D_{P,m}}{l_m} \tag{4.4}$$

where $D_{S,m}$ and $D_{P,m}$ are the diffusion coefficients of the substrate and product, respectively, in the membrane, and l_m is the thickness of the membrane.

The steady-state condition is derived from the mass-conservation law and reads:

$$A|J_s| = AJ_p = Vv' \tag{4.5}$$

where A is the surface area of the enzyme layer. Upon dividing Equation (4.5) by A one obtains:

$$|J_s| = J_p = l_e v' \tag{4.6}$$

The quantity $v_a = l_e v'$ (in mol m^{-2} s^{-1}) is the *surface-normalized reaction rate*. It represents the amount of substrate converted per unit time and unit surface area. The conservation Equation (4.6) can now be formulated as follows:

$$k_{m,S}(s - s_e) = k_{m,P}p_e = l_e \frac{k_2 e_t s_e}{K_M + s_e} \tag{4.7}$$

Two limiting kinetic cases are possible depending on the s_e/K_M ratio, that is, either $s_e \gg K_M$ or $s_e \ll K_M$.

4.2.3 The Zero-Order Kinetics Case

In the zero-order kinetic case, $s_e \gg K_M$ and, therefore, the Michaelis–Menten constant in the kinetic equation can be neglected. Thereby, an equation formed by the last two terms in Equation (4.7) gives:

$$p_e = \frac{l_e}{k_{m,P}} k_2 e_t = \frac{l_e}{k_{m,P}} v_m \tag{4.8}$$

where v_m is the maximum reaction rate for the immobilized enzyme. Equation (4.8) demonstrates that under zero-order kinetics conditions, p_e (and, hence, the response signal) is independent of the substrate concentration in the solution but depends on the total concentration of the enzyme. Therefore, the determination of the substrate concentration cannot be accomplished with an enzymatic sensor functioning under zero-order kinetics conditions. However, the sensor response depends on the total concentration of active enzyme within the biocatalytic layer. Such conditions are favorable when using the enzymatic sensor to determine an enzyme inhibitor. Under the effect of the inhibitor, a fraction of the total enzyme amount loses its activity and the sensor signal decreases as a function of the inhibitor concentration.

In addition, this case is suitable when an enzyme is utilized as a label tag for an otherwise nondetectable compound. The recognition process brings the labeled analyte near the transducer and, in the presence of the substrate the sensor generates a signal depending on the concentration of the enzyme label. This signal is directly related to the analyte concentration.

The $s_e \gg K_M$ condition required in the above two applications can be fulfilled by proper adjustment of the substrate concentration that has to be added to the sample solution.

4.2.4 The First-Order Kinetics Case

In the first-order kinetics case, $s_e \ll K_M$. In this instance, the limiting Michaelis–Menten equation for the first-order kinetics case will be used to define the surface-normalized reaction rate as follows:

$$v_a = k_a s_e \tag{4.9}$$

where

$$k_a = k_e l_e = \frac{k_2 e_t}{K_M} l_e \tag{4.10}$$

where $k_e = k_2 e_t$. At a constant enzyme concentration, k_a represents the *surface normalized pseudofirst-order rate constant*. Further, by equating the diffusional flux of the substrate with the reaction rate, it results:

$$k_a s_e = k_{m,S}(s - s_e) \tag{4.11}$$

This equation yields:

$$s_e = \frac{k_{m,S}}{k_{m,S} + k_a} s = \frac{1}{1 + (k_a/k_{m,S})} \qquad (4.12)$$

The above equation puts into evidence the following coefficient:

$$\alpha = \frac{k_a}{k_{m,S}} \qquad (4.13)$$

Using the α coefficient, Equation (4.12) becomes:

$$s_e = \frac{1}{1 + \alpha} s \qquad (4.14)$$

The dimensionless coefficient α (termed the *substrate modulus for external diffusion*) is a key parameter in the modeling of processes involving immobilized enzymes. By expanding the terms in Equation (4.13) one obtains an expression for α that shows the effect of the kinetic, diffusional and geometric parameters:

$$\alpha = \frac{v_m l_e}{K_M k_{m,S}} = \frac{k_2 e_t l_e}{K_M k_{m,S}} \qquad (4.15)$$

According to its definition in Equation (4.13), the α parameter depends on the rate constants of two consecutive processes: the transport of the substrate ($k_{m,S}$) and the enzymatic conversion of the substrate (k_a). Therefore, the physical meaning of the α parameter can be rationalized as follows:

$$\alpha = \frac{\text{Substrate conversion capacity}}{\text{Transport capacity (external diffusion)}} \qquad (4.16)$$

A value of α greater than 1 indicates a fast chemical reaction preceded by an intrinsically slow diffusion step. At the $\alpha \gg 1$ limit, diffusion is so slow that the supply of substrate is much below the conversion potential of the enzyme layer and the overall reaction rate is wholly determined by the rate of substrate diffusion. Under these circumstances, any substrate molecule reaching the enzyme layer is instantaneously converted into the product and the sensor operates under *diffusion control*. In the opposite case ($\alpha \ll 1$) the supply capacity greatly exceeds the conversion capability of the enzyme layer. Therefore, the concentration gradient adjusts itself to a very low value in order to allow diffusion to keep pace with the sluggish chemical reaction. In this situation, the sensor process occurs under *kinetic control*.

It is now possible to derive an expression for the p_e variable that determines the sensor response. To this end, the above expression for s_e (4.14) will be substituted in an equation formed by the first two terms in Equation (4.7). Upon solving it for p_e, one obtains:

$$p_e = \frac{k_{m,S}}{k_{m,P}} \frac{\alpha}{\alpha + 1} s \qquad (4.17)$$

Equation (4.17) proves that if $s_e \ll K_M$, the product concentration in the enzyme layer is proportional to the substrate concentration in the solution, the proportionality constant being dependent on the α coefficient. If $\alpha \gg 1$, the proportionality constant in Equation (4.17) reduces to $k_{m,S}/k_{m,P}$, which causes the kinetic parameters of the enzyme reaction to become irrelevant. Under such circumstances, the sensor functions under diffusion control. The p_e variable assumes in this situation its maximum possible value for a given substrate concentration and imparts to the sensor the maximum sensitivity.

It is important to notice that under diffusion control, the response is independent of any parameter of the enzyme layer, which renders the sensor stable under long-term storage and operation and less sensitive to fluctuations arising from the manufacturing process.

If the condition $\alpha \gg 1$ is not fulfilled, the sensitivity is lower. Also, the response depends not only on diffusion parameters but also on the enzyme concentration as well as the thickness of the enzyme layer. In this case, the sensor response is sensitive to changes in pH or temperature (which affect the reaction rate) as well as to possible enzyme inactivation by degradation or inhibition.

In conclusion, in the first-order kinetics case, the product concentration in the enzyme layer is proportional to the substrate concentration in solution. If, in addition, the sensor operates under diffusion control ($\alpha \gg 1$), the product concentration in the enzyme layer is independent on the parameters of this layer, which renders the sensor reliable and resilient.

4.2.5 The Dynamic Range and the Limit of Detection under External Diffusion Conditions

It has been proved in the previous sections that substrate determination can be performed with the sensor functioning under the first-order kinetics regime that implies that the condition $s_e \ll K_M$ is fulfilled. As the s_e value is out of direct control, it is important to define the limit of the linear range in terms of substrate concentration in the solution. To this end, the response equation should be derived with no restrictions in the value of the s_e/K_M ratio. The starting point in this derivation is the mass-conservation equation formulated as follows:

$$k_{m,S}(s - s_e) = \frac{v_{a,m} s_e}{K_M + s_e} \tag{4.18}$$

where $v_{a,m} = l_e v_m$ is the surface-normalized maximum reaction rate. Using dimensionless concentrations ($S_e = s_e/K_M$ and $S = s/K_M$) and solving for S_e, one obtains from Equation (4.18) the following expression:

$$S_e = \frac{1}{2}\left[-(1 + \alpha - S) + \sqrt{(1 + \alpha - S)^2 + 4S}\right] \tag{4.19}$$

where $\alpha = v_{a,m}/k_{m,S}K_M$ is similar to the coefficient defined in Equation (4.15). In order to derive the concentration of the product in the enzyme layer, the following conservation equation with respect to dimensionless concentrations will be used:

$$k_{m,P}P_e = k_{m,S}(S - S_e) \tag{4.20}$$

where $P_e = p_e/K_M$ is the dimensionless concentration of the product within the enzyme layer. By solving this equation, it follows that:

$$P_e = \frac{1}{2}\frac{k_{m,S}}{k_{m,P}}\left(1 + S + \alpha - \sqrt{4S + (1 - S + \alpha)^2}\right) \tag{4.21}$$

The above equation gives the product concentration in the enzyme layer as a function of the substrate concentration in the solution with no restrictions on the value of the s_e/K_M ratio. This equation has been used to generate the curves in Figure 4.2 that indicate the $P_e - S$ relationship in the first-order kinetics regime. The curves in this figure demonstrate clearly that the extent of the linear response region increases with increasing α coefficient. An estimation of the limit of the linear range can be obtained by setting $S_e = 1$ and solving Equation (4.19) for S. One obtains thus the approximate limit of the linear range at $S = 0.5\,\alpha + 1$. It is worth mentioning that the upper limit of the working range could be determined not only by the α modulus but also by the transduction process itself.

As far as the limit of detection is concerned, it is independent of α but is determined by the intrinsic limit of detection of the transduction method.

Sensor sensitivity is determined by the pre-s factor in Equation (4.17). Since the sensor is as a rule designed such that $\alpha \gg 1$, it follows that the sensitivity is practically independent of this factor. In agreement with this conclusion, the linear parts of the curves in Figure 4.2 do overlap.

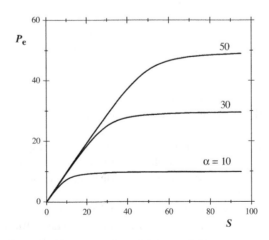

Figure 4.2 Correlation between the product concentration within the enzyme layer (P_e) and the substrate concentration in solution (S) in the case of external diffusion control. Curves generated using Equation (4.21) with $k_{m,S} = k_{m,P}$.

Recalling the definition of α, it is clear that this parameter can be tuned by adjusting the enzyme concentration in the biocatalytic layer, the thickness of the enzyme layer and the characteristics of the external membrane. The α parameter increases with decreasing mass-transfer coefficient through the membrane, that is, with increasing membrane thickness. However, a decrease in the mass-transfer coefficient is accompanied by an increase in the response time.

In conclusion, a high value of the α parameter is beneficial for the sensor response as well as for sensor resilience.

Questions and Exercises (Section 4.2)

1 What are the characteristics of an enzymatic sensor functioning under external diffusion control? What sensor parameters secure the external diffusion regime?

2 Try to find an alternative wording for the definition of α in Equation (4.16).

3 Substitute pertinent expressions for k_a and $k_{m,S}$ in Equation (4.13) in order to derive an equation for the α coefficient including all characteristic constants and geometrical parameters. Discuss the effect of each term in this equation on the α parameter and comment on possible methods for increasing the α value.

4 Using the Michaelis–Menten equation and Equation (4.12), derive a kinetic equation for the reaction rate in the enzymatic layer under the assumption of external diffusion control. Derive the limiting forms of this equation for (a) $\alpha \ll 1$, and (b) $\alpha \gg 1$ and comment on the effect of the substrate concentration in the solution. For case (b), define the condition needed in order for the reaction rate to be proportional to the substrate concentration in the solution.
Answer: $v = k_2 e_t s [s + K_M(1 + \alpha)]^{-1}$

5 Using the results in the previous exercise derive an equation for the *effectiveness factor* that is defined as follows:

$$\eta_E = \frac{\text{Reaction rate with external diffusion limitation}}{\text{Reaction rate with no diffusion limitation}} = \frac{v}{v|_{k_{m,S} \to \infty}} \qquad (4.22)$$

where v represents the reaction rate of the reaction proceeding within the enzyme layer. The effectiveness factor represents the degree of utilization of the enzyme. Using this equation, plot the η_E–S curve for $\alpha = 1$, 10, and 50 and comment on the degree of utilization of the enzyme under first-order kinetics and external diffusion control.

Answer: $\eta_E = (2\alpha S)^{-1}(1 + S)\left(1 + \alpha + S - \sqrt{(1 + \alpha - S)^2 + 4S}\right)$

6 Assuming that the transducer response y is proportional to the product concentration in the enzyme layer ($y = bp_e$), derive an equation relating the response to the substrate concentration in the solution. Comment on the effect of the α coefficient on the sensor sensitivity.

7 Derive the equation of the sensor response under the assumption that the transducer response is a logarithmic function of the product concentration in the enzyme layer ($y = a + b\log p_e$). Comment on the effect of the α coefficient on the sensor response and point out the advantage of a very high value of α.

4.3 The Enzymatic Sensor under Internal Diffusion Control

4.3.1 The Steady-State Response

In many instances, concentration gradients form inside the enzyme layer and the mass transfer within this region cannot be neglected. Such a situation arises if the enzyme layer is relatively thick or the diffusion coefficients within this layer are very small. The concentration profiles of the substrate and product in this case are shown in Figure 4.3. In order to simplify the mathematical treatment, it will be assumed that the concentration gradients out of the enzyme layer are negligible. This condition is fulfilled when the solution is well stirred and the membrane permeability is very high. Actually, this model fits best membraneless sensors.

The mathematical model will be obtained upon combining Fick's second law of diffusion with the expression of the reaction rate. Taking into account the conclusions of the previous section, it is convenient to consider the limiting

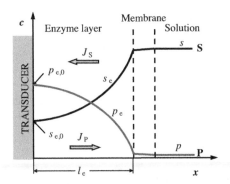

Figure 4.3 Concentration (c) profiles for substrate and product in the case of an internal diffusion enzymatic sensor. Concentration gradients out of the enzyme layer are negligible. Substrate and product partition at interfaces is neglected. s and p represent substrate and product concentrations in the solution, respectively; s_e and p_e represent similar concentration in the enzyme layer.

case of first-order kinetics ($s_e \ll K_M$). Under these conditions, the differential equations for diffusion combined with chemical reaction assume the following forms:

$$\frac{\partial s_e}{\partial t} = D_{S,e} \frac{\partial^2 s_e}{\partial x^2} - \frac{k_2 e_t s_e}{K_M} \tag{4.23}$$

$$\frac{\partial p_e}{\partial t} = D_{P,e} \frac{\partial^2 p_e}{\partial x^2} + \frac{k_2 e_t s_e}{K_M} \tag{4.24}$$

$D_{S,e}$ and $D_{P,e}$ denote the diffusion coefficients within the enzyme layer for the substrate and product, respectively, and x is the distance from the surface of the transducer. In the above equations, the left-hand term represents the local time variation of the concentration. This is caused by two processes: diffusion and the enzymatic reaction, whose contributions are expressed by the first and second terms on the right-hand side of the equation, respectively. The required mathematical solution of the above equations is the product concentration at the transducer surface ($x = 0$) that determines the transducer response.

The limiting conditions are, as follows: for $x = l_e$, $s_e = s$ and $p_e = 0$ (that is, product concentration in the test solution is negligible); for $t = 0$ and $0 \leq x \leq l_e$, $s_e = p_e = 0$ (that is, neither the substrate nor the product is initially present in the enzyme layer). Assuming that the transduction occurs without material consumption, concentration gradients at $x = 0$ are equal to zero.

In the steady state, concentrations are constant at any point within the enzyme layer. Therefore, the steady-state solution is obtained by nullifying the left-hand term in Equations (4.23) and (4.24). The solution of this problem demonstrates that the sensor response depends essentially on the dimensionless parameter ϕ defined in Equation (4.25) and identified as the *Thiele modulus* or the *substrate modulus for internal diffusion*.

$$\phi = l_e \sqrt{\frac{e_t}{K_M} \frac{k_2}{D_{S,e}}} \tag{4.25}$$

Therefore, the Thiele modulus groups all relevant parameters of the enzyme layer. Often, the square of ϕ is utilized instead of ϕ. ϕ^2 is called the *enzyme loading factor* (f_e) whereas chemical engineering identifies ϕ^2 as the *Damköhler number*, Da:

$$\phi^2 = f_e \equiv \text{Da} \tag{4.26}$$

The physical meaning of f_e can be inferred if the Equation (4.25) is rearranged such as to give the following relationship:

$$f_e = \frac{k_a}{k_{S,e}} \tag{4.27}$$

where $k_a = k_2 l_e$ is the surface-normalized rate constant and $k_{S,e} = D_{S,e}/l_e$ is the mass-transfer coefficient. According to Equation (4.27), f_e is the quotient of two rate constants, namely the rate constant of the enzymatic reaction and the mass-transfer coefficient. Therefore, Equation (4.27) can be expressed as follows:

$$f_e = \frac{\text{Substrate conversion capacity}}{\text{Transport capacity (internal diffusion)}} \tag{4.28}$$

It is clear that a high value of f_e implies slow mass transfer, that is, the mass-transfer step determines the overall rate of the process. In the case of a low f_e value, the overall reaction rate depends on kinetic parameters included in the k_a constant and is independent of the mass-transfer parameters. Therefore, the f_e parameter can be viewed as the degree of catalyst utilization.

Using the Thiele modulus, the differential equations for the steady-state conditions led to the following expression for the product concentration profile [4]:

$$p_e = \frac{D_{S,e}}{D_{P,e}} \left[1 - \left(\cosh \phi \frac{x}{l_e} \right) (\cosh \phi)^{-1} \right] s \tag{4.29}$$

This equation was used to plot the response factor (p_e/s) as a function of the dimensionless distance (x/l_e) for various values of the Thiele modulus, as shown in Figure 4.4A. This figure demonstrates that at low ϕ values the degree of conversion is rather low all over the enzyme layer, as a result of the sluggishness of the reaction. On the contrary, high ϕ values bring about a high conversion of substrate to product. At $\varphi = 10$, as a consequence of the rapid chemical reaction, most of the substrate is converted within the right-hand portion of the enzyme layer, that is, before it moves far into the membrane enzyme film.

It is possible now to account for the sensor response. As before, it will be assumed that the response is a function of the product concentration at the transducer surface, that is, at $x/l_e = 0$. For this particular condition, Equation (4.29) gives:

$$p_{e,0} = \frac{D_{S,e}}{D_{P,e}} \left(1 - \frac{1}{\cosh \phi} \right) s \tag{4.30}$$

This equation proves that the product concentration at the transducer surface ($p_{e,0}$) is proportional to the substrate concentration in solution. The proportionality constant depends on the ratio of diffusion coefficients. What is even more important is that the proportionality constant is also a function of the Thiele modulus. However, if $\phi > 5$, $\cosh \varphi$ becomes exceedingly greater than unity and the proportionality constant in Equation (4.30) reduces to the ratio of the diffusion coefficient. From the physical standpoint, this limit corresponds to a slow mass transfer, which makes the overall process occur under *internal diffusion control*.

Assuming that the response is proportional to p_{e0}, the profile of the sensor response will be the same as the straight lines in Figure 4.4B. This figure demonstrates an increase in sensitivity with increasing the ϕ value. Nevertheless, the sensitivity reaches a limit at $\phi \approx 5$ and any further increase of this parameter has no effect on the sensitivity. Besides, as $D_{S,e}$ and $D_{P,e}$ could be very different in some instances, the effect of the $D_{S,e}/D_{P,e}$ ratio on sensitivity should not be disregarded.

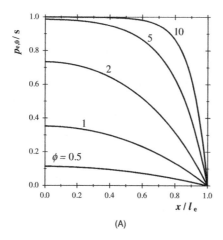
(A)

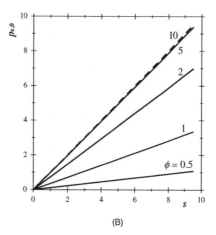
(B)

Figure 4.4 The enzymatic sensor under internal diffusion conditions and first-order kinetics. (**A**) Profile of the product concentration within the enzyme layer for various ϕ values. (**B**) Simulated response of the enzyme sensor under internal diffusion control. ($D_{S,e} = D_{P,e}$) for both graphs.

It should also be borne in mind that Equation (4.30) was derived under the assumption that $s_e \ll K_M$. If the substrate concentration in solution is high enough to render this condition no longer applicable, the sensor response will deviate from the linear trend displayed in Figure 4.4B.

As in the case of external diffusion control, no restriction on the limit of detection arises as far as the reaction kinetics are concerned. Consequently, the limit of detection depends mostly on the response characteristics of the transduction process. This latter can also impose the upper limit of the linear range, even if the above-mentioned kinetic features allow, in principle, the attainment of a higher limit.

In conclusion, the sensitivity of an enzymatic sensor functioning under internal diffusion conditions increases with increasing Thiele modulus ϕ. According to Equation (4.25), the value of the Thiele modulus depends on the enzyme concentration within the enzyme layer, the thickness of this layer and the diffusion coefficient of the substrate. In turn, the diffusion coefficient is determined by the method of enzyme immobilization. It should be borne in mind that an increase of the thickness and a decrease of the diffusion coefficient result also in an increase in the response time.

A high value of the Thiele modulus brings also advantages as far as the long-term sensor response, stability and reproducibility are concerned. Equation (4.25) demonstrates that for $\phi > 5$, this modulus drops out from the response function and the response becomes independent of the parameters of the enzyme layer. Therefore, to a certain degree, alteration of the enzyme activity (by degradation of inhibition) has no effect on the sensor response.

4.3.2 The Transient Regime and the Response Time under Internal Diffusion Conditions

The transient behavior of the sensor is noticeable immediately after initiating the sensing process. In order to investigate transient behavior, the substrate should be initially absent in both the solution and the enzyme layer. In order to trigger the sensing process, the substrate is added to the well-stirred solution at $t = 0$. It is clear that the response signal will vary with time before reaching a constant value, which is typical of the steady state.

The next treatment is adapted from ref. [5] in which the modeling of the transient behavior has been accomplished at the first-order kinetics limit ($s_e \ll K_M$) under the assumption that substrate and product have identical diffusion coefficients within the enzyme layer ($D_e = D_{S,e} = D_{P,e}$). By solving the time-dependent Equations (4.23) and (4.24), the product concentration in the enzyme layer has been obtained as a function of time and distance from the transducer surface (Equation (4.12) in [5]), which can be put in the following form:

$$\frac{p_e}{s} = \left(1 - \frac{\cosh\left(\frac{x}{l_e}\right)\sqrt{f_e}}{\cosh\sqrt{f_e}} \right) - f\left(\frac{t}{\tau_d}\right) \tag{4.31}$$

The first term in this equation represents the steady-state response (see Equation (4.29). The second term is a function of the dimensionless time variable t/τ_d, where t is the time and τ_d is the *diffusional time constant*, which is defined as:

$$\tau_d = \frac{l_e^2}{D_e} \tag{4.32}$$

The time-dependent term in Equation (4.31) also includes the Thiele modulus. This term decreases asymptotically with time and becomes close to zero when the steady state is approached.

Equation (4.31) has been used to simulate the profile of the product concentration within the enzyme layer at various stages after initiating the enzymatic reaction. The results of the simulation are shown in Figure 4.5. The curve for $t/\tau_d = 0.02$ represents the profile of the product concentration at an early stage after initiating the sensing process. Under these conditions, substrate molecules entering the enzyme layer are immediately converted and a product concentration gradient develops. Consequently, the product undergoes diffusion towards the interior of the enzyme layer. However, due to the inherent slowness of diffusion, the product concentration fades away at some distance from the interface. As the process proceeds further, the gradient of the product concentration increases. Thereby the diffusion rate is enhanced, prompting the product to spread further within the enzyme layer (see the intermediate curves in Figure 4.5A). When the time elapsed is such that $t/\tau_d \geq 2$, the second term in Equation (4.31) becomes negligible and the steady-state profile is set up. This situation is represented by the uppermost curve in Figure 4.5A, which overlaps the points plotted by means of the steady-state equation (4.29).

As the sensor response depends on the product concentration at the transducer surface, it is of interest to inspect the time variation of this variable. Equation (4.31) can be used to this end after setting $x = 0$. The simulated time variation of the p_{e0}/s ratio is displayed in Figure 4.5 B for $f_e = 25$ and several values of the diffusion time constant. At the selected value of f_e, the overall reaction rate is determined by the diffusion step. The curves in Figure 4.5 B demonstrate that the greater the value of the time constant, the longer is the time needed for the steady state to set up.

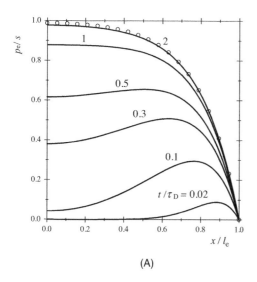

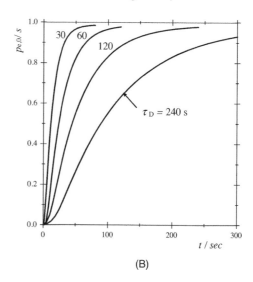

(A) (B)

Figure 4.5 The enzymatic sensor in the transient regime. (A) Product concentration profiles within the enzyme layer at various stages for $f_e = 25$. Circles indicate the steady-state profile according to Equation (4.29). (B) Transient response of an enzymatic sensor under internal diffusion control for $f_e = 25$ and various values of the diffusional time constant.

The response time of the sensor can be estimated by putting Equation (4.31) in an approximate form, which is valid only when the steady state is approached (that is, for $t/\tau_d > 1$). According to ref. [5], the time dependence under these conditions is represented by the following equation:

$$\left(\frac{p_{e,0}}{s}\right)_{t/\tau_d > 1} = \left(1 - \frac{1}{\cosh \sqrt{f_e}}\right) - \frac{4}{\pi} \exp\left(-\frac{\pi^2}{4}\frac{t}{\tau_d}\right) \tag{4.33}$$

This equation proves that shortly before the steady state is approached, the time variation of the response is independent of the enzyme loading. It depends only on the diffusional parameters included in the τ_d constant. An expression for the response time (t_r) can be derived by assuming that the response time is the time elapsed until the second term in Equation (4.33) drops to a negligible value such as 0.01, that is about 1% of the final p_e/s value. Based on this assumption, the response time is obtained as:

$$t_r \approx 2\tau_d = 2\frac{l_e^2}{D_e} \tag{4.34}$$

Accordingly, the steady state is practically set up when the time elapsed after triggering the sensor process reaches twice the value of the diffusional time constant. This conclusion is in good agreement with the plots in Figures 4.5A and B. Therefore, the response time can be modified by changing the diffusion-related parameters. Among them, the thickness of the enzyme layer is the most susceptible to adjustment. The diffusion coefficient in turn depends essentially on the enzyme immobilization method. It could be in the region of 10^{-5} cm^2 s^{-1} if an enzyme solution is entrapped between the transducer surface and a semipermeable membrane, but it is much lower if the enzyme is embedded in a gel or is incorporated in the pores of a solid material.

Summing up, under internal diffusion control, the transient response depends on both the enzyme loading factor and the diffusional time constant. Nevertheless, the effect of enzyme loading is manifest only at the beginning of the transient regime and the actual response time depends only on diffusion-related parameters, namely the thickness of the enzyme layer and the diffusion coefficient of the substrate within the enzyme layer.

It should also be kept in mind that the above results were derived under simplifying conditions. An actual sensor may deviate more or less from this model. Nevertheless despite its approximate character, this approach indicates the main factors affecting the response time and also the expected effects of these factors.

Questions and Exercises (Section 4.3)

1 Comment on the characteristics of an enzymatic sensor that operates under external diffusion control.
2 Demonstrate that the enzyme loading factor represents the quotient of the maximum possible reaction rate (which is achieved when $s_{e,0} = s$) and the maximum attainable mass transfer rate (which occurs when $s_{e,0} = 0$).

3 What advantages bring about a high value of the enzyme loading factor as far as the response characteristic and sensor reliability are concerned?

4 Suppose that after prolonged utilization of an enzymatic sensor, the enzyme activity dropped by 25% with respect to the initial value. Assuming that $f_e = 1$, 25, and 100, calculate the per cent variation of the sensor response due to the change in enzyme activity for each of the above values of f_e. Assume that $D_{S,e} = D_{P,e}$.

Answer: -35%; -3.4%; -0.1%, respectively.

5 What parameters determine the response time of an enzymatic sensor under internal diffusion control? What sensor parameters should be adjusted in order to find a compromise between the response time, on the one hand, and sensitivity and reliability, on the other hand?

4.4 The General Case

4.4.1 The Model

This section addresses the general problem of a catalytic reaction in an immobilized enzyme layer coupled with diffusion in both the enzyme layer and an adjacent layer, which could be a semipermeable membrane or a solution layer within which significant concentration gradients develop. The following approach is based on the theoretical treatment in ref. [6]. The physical model of the considered system is presented schematically in Figure 4.6.

This model takes into account the effect of partition at the interface between the catalytic and noncatalytic layers. Partition causes a sudden change in the concentration at this interface and can be quantified by the partition constants of the substrate (δ_S) and the product (δ_P) that are defined as follows:

$$\delta_S = \frac{s_{e,i}}{s_{m,i}}; \quad \delta_P = \frac{p_{e,i}}{p_{m,i}} \tag{4.35}$$

The meaning of symbols in these equations is given in Figure 4.6.

It is also assumed that the reaction obeys the first-order kinetics ($S_e \ll K_M$) and occurs according to a Michaelis–Menten-type mechanism in which each substrate molecule gives rise to n_P product molecules.

$$E + S \underset{k_{-1}}{\overset{k_1}{\rightleftarrows}} ES \overset{k_2}{\longrightarrow} E + n_P P \tag{4.36}$$

No provision is made for a possible pH change in the enzyme layer. Hence, it is assumed that the pH is kept constant by means of a pH buffer system.

Diffusion through the enzyme layer is described by modified Fick's Equations (4.23) and (4.24). At the enzyme layer/membrane interface, concentration profiles show discontinuities due to partition, which should be accounted

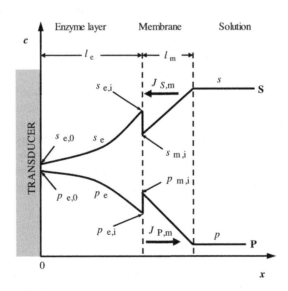

Figure 4.6 Concentration (c) profiles for a biocatalytic layer attached to the surface of a transducer. s and p denote concentrations of the substrate and product, respectively. Subscripts m and e to p and s indicate concentration in the membrane and the enzyme layer, respectively.

for in the relevant limiting conditions of Fick's equations. The transport in the membrane layer is modeled by means of the Nernst layer concept that gives the following expressions for the diffusional fluxes:

$$J_{S,m} = -k_{S,m}\left(s_{m,i} - s\right) \tag{4.37}$$

$$J_{P,m} = -k_{P,m}\left(p_{m,i} - p\right) \tag{4.38}$$

where $k_{S,m}$ and $k_{P,m}$ are the mass-transfer coefficients of the substrate and product, respectively, in the membrane.

Assuming that the product concentration in the solution is negligible, the above model leads to an equation relating the product concentration at the transducer surface ($p_{e,0}$) to the substrate concentration in the solution, as follows:

$$\frac{p_{e,0}}{s} = n_P \delta_S \frac{\delta_P \theta_m \dfrac{\phi}{\beta} \tanh \phi + \theta_e \left[1 - (\cosh \phi)^{-1}\right]}{\delta_S \dfrac{\phi}{\beta} \tanh \phi + 1} \tag{4.39}$$

The Thiele modulus is immediately recognizable in the above equation that includes several additional dimensionless parameters denoted by β, θ_m, and θ_e.

β is the *Biot number* that quantifies the relative preponderance of internal or external diffusion:

$$\beta = \frac{k_{S,m}}{k_{S,e}} = \frac{\text{Transport capacity in the membrane}}{\text{Transport capacity in the enzyme layer}} \tag{4.40}$$

where $k_{S,e} = D_{S,e}/l_e$.

Large β values imply $k_{S,m} \gg k_{S,e}$ and indicates that internal diffusion is very slow compared with the external diffusion. This is typical of an enzymatic sensor operating under internal diffusion control. The case of very low β values implies much slower diffusion within the membrane compared with the diffusion within the enzyme layer, which leads to the external diffusion control case. The Biot number allows therefore making of a clear-cut distinction between internal and external diffusion regimes as limiting cases. Using the definitions of α, ϕ and f_e coefficients, it is easy to prove that these coefficients are interconnected by means of the Biot number as follows:

$$\alpha = \frac{f_e}{\beta} = \frac{\phi^2}{\beta} \tag{4.41}$$

Each of the two θ coefficients in Equation (4.40) indicates the rapidity of the substrate diffusion relative to that of the product and will be termed *lag factors*. The *internal lag factor* is defined as:

$$\theta_e = \frac{D_{S,e}}{D_{P,e}} \tag{4.42}$$

A large value of θ_m results when the diffusion coefficient of the product in the enzyme layer is much lower that that of the substrate. Under these circumstances, product depletion by diffusion out of the enzyme layer is a slow process and the product can accumulate within the enzyme layer. This leads to an increase in sensitivity.

The *external lag factor* (θ_m) is identified as:

$$\theta_m = \frac{k_{S,m}}{k_{P,m}} \tag{4.43}$$

A large value of θ_m indicates a slow diffusion of the product across the membrane and leads consequently to an enhanced product concentration within the enzyme layer.

4.4.2 Effect of the Biot Number

Equation (4.39) will be used next to examine the effect of the Biot number and understand in which way the limiting situations of internal or external diffusion can be attained. To this end, it is convenient to made to following simplifying assumptions: $n_P = 1$, $\delta_S = \delta_P = 1$, $\theta_m = \theta_e = 1$. Thus, the effect of partition and unevenness of diffusion coefficients is removed and Equation (4.39) simplifies to:

$$\frac{p_{e,0}}{s} = 1 - \frac{1}{\left(\dfrac{\phi}{\beta} \tanh \phi + 1\right) \cosh \phi} \tag{4.44}$$

For the purpose of this discussion it is useful to mention that for $\phi \geq 0$ we have $0 \leq \tanh \phi \leq 1$ and for $\phi \geq 5$, $\tanh \phi = 1$.

As already mentioned, the internal diffusion regime is attained at very high β values, more precisely, if $\beta \gg \phi$, which implies that $(\phi/\beta)\tanh \phi \ll 1$. Therefore, under the internal diffusion regime, the response factor Equation (4.44) becomes:

$$\left(\frac{p_{e,0}}{s}\right)_{ID} = 1 - \frac{1}{\cosh \phi} \tag{4.45}$$

Taking into account the above simplifying assumptions, this equation is similar to Equation (4.30) derived under the hypothesis of a purely internal diffusion regime.

The external diffusion regime occurs if β is sufficiently low to satisfy the condition $(\phi/\beta)\tanh \phi \gg 1$. Taking into account this condition, a limiting form of Equation (4.44) can be derived. However, this limiting equation is somewhat ambiguous because it can yield negative values for the response factor. That is why Equation (4.17) which was derived under the assumption of purely external diffusion control will be further compared with the more general Equation (4.44) upon assuming that $\beta \ll 1$. To this end, a change of variable in Equation (4.17) will be operated in order to substitute the parameter α by its expression in Equation (4.40). As $\theta_m = 1$ and the concentration is evenly distributed across the enzyme layer (that is, $p_e = p_{e,0}$), one obtains the following expression for the response factor under external diffusion conditions:

$$\left(\frac{p_{e,0}}{s}\right)_{ED} = \frac{\phi^2}{\phi^2 + \beta} \tag{4.46}$$

In order to assess the interplay between internal and external diffusion, Equation (4.44) was used to plot the response factor vs. ϕ for selected values of the Biot number (solid lines in Figure 4.7). A typical curve for intermingled diffusion is that for $\beta = 1$ (curve 3). An increase in the Biot number results in a shift of the curve toward higher values of ϕ until the limit of the internal diffusion regime is reached for $\beta \geq 25$ (curve 4). For comparison, data obtained with the limiting Equation (4.45) are plotted as dots in Figure 4.7. It is clear that the general Equation (4.44) and the limiting Equation (4.45) lead to concordant results.

The external diffusion case occurs at subunity values of β and is represented by curve 1 in Figure 4.7, which is plotted for $\beta = 0.01$. This curve overlaps the circles plotted with the limiting Equation (4.46). For $\beta = 0.1$ (curve 2) small differences between the two series of data can be noticed.

The above discussion demonstrates clearly that it is the Biot number β that dictates the prevalence of internal or external diffusion. Internal diffusion control is dominating at $\beta \geq 25$, whereas external diffusion control operates at $\beta \leq 0.1$. The above figures can change if the effect of partition and lag coefficients are also taken into account, but the trends illustrated in Figure 4.7 are applicable even in such situations.

Keeping in mind that the maximum sensitivity is achieved when $p_{e,0}/s \approx 1$, Figure 4.7 demonstrates that for this limit to be achieved, a higher enzyme loading is needed with an increase in β. While the internal diffusion regime requires that $\phi \geq 5$ in order to attain the maximum sensitivity, this limit can be achieved at lower ϕ values in the case of external diffusion. This feature can be rationalized if we take into account the fact that the external membrane

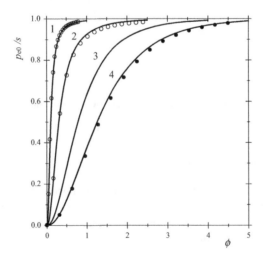

Figure 4.7 Effect of the Thiele modulus ϕ on the product concentration at the transducer surface for various values of the Biot number β. Solid lines were generated with Equation (4.44) for combined internal and external diffusion. Dots represent the limiting case of internal diffusion (Equation (4.45)). Circles were generated with the approximate Equation (4.46) for external diffusion. Biot number: (1) 0.01; (2) 0.1; (3) 1; (4) 25.

brings a limitation to the access of the substrate to the catalytic layer. Therefore, a lower catalytic potential is in this case required in order to reach full conversion of the substrate to product. Clearly, a high diffusion resistance in the external membrane does not affect the sensitivity if $\theta_m = 1$, but it is beyond any doubt that the lower the value of β the longer the response time. Figure 4.7 proves, in addition, that the limiting Equation (4.17) is in good agreement with the general Equation (4.44) if $\beta \geq 0.1$.

The above discussion has pointed out the importance of the Biot number in deciding the localization of concentration gradients either within the enzyme layer or within an external layer. In order to distinguish the actual parameters that determine the value of the Biot number, Equation (4.40) will be reformulated as follows:

$$\beta = \frac{D_{S,m}}{D_{S,e}} \frac{l_e}{l_m} \tag{4.47}$$

It is clear that the simplest way to alter the value of the Biot number relies on adjusting the layer thicknesses. Thus, a very thin enzyme layer favors the external diffusion regime, whereas internal diffusion is expected to dominate when the thickness of the external membrane is very low or substrate diffusivity within it is very high. If the enzyme layer is in direct contact with the solution, the thickness of the external diffusion layer is determined by the hydrodynamic conditions of the solution and vigorous stirring results in a reduced thickness.

4.4.3 Effect of Partition Constants and Diffusion Coefficients

The following discussion is devoted to an analysis of the effect of partition constants and diffusion parameters that were neglected in the previous section. This analysis can be conveniently performed if it is assumed that $\phi \geq 5$ because under such conditions the response factor approaches a value close to one (Figure 4.7). Since, in this instance, $(\cosh\phi)^{-1} \ll 1$ and $\tanh\phi = 1$, Equation (4.39) assumes the following form:

$$\left(\frac{p_{e,0}}{s}\right)_{\phi \geq 5} = \delta_P \theta_m \frac{\dfrac{\phi}{\beta} + \dfrac{1}{\delta_P}\dfrac{\theta_e}{\theta_m}}{\dfrac{\phi}{\beta} + \dfrac{1}{\delta_S}} \tag{4.48}$$

According to this equation, diffusional and partition parameters intermingle in such an intricate way that it is impossible to distinguish the effect of each of them. That is why only the extreme cases of internal and external diffusion will be considered next. The external diffusion regime is achieved if β is so small that $(\phi/\beta) \gg (\theta_e/\theta_m)/\delta_P$. If the condition $(\phi/\beta) \ll (1/\delta_S)$ is also fulfilled, Equation (4.48) becomes:

$$\left(\frac{p_{e,0}}{s}\right)^{ED}_{\phi \geq 5} = \delta_P \theta_m \tag{4.49}$$

For the internal diffusion regime to occur, β must be very high. In this case, $(\phi/\beta) \ll (\theta_e/\theta_m)/\delta_P$. Assuming that $(\phi/\beta) \gg (1/\delta_S)$, the following equation results from (4.48):

$$\left(\frac{p_{e,0}}{s}\right)^{ID}_{\phi \geq 5} = \delta_S \theta_e \tag{4.50}$$

The above equations prove that, in the case of the external diffusion regime, the response factor is proportional to the partition constant of the product, whereas the substrate partition affects the response only in the case of the internal diffusion regime. It is proved also that the response is proportional to the lag factor within the layer that determines the diffusion rate.

4.4.4 Experimental Tests for the Kinetic Regime of an Enzymatic Sensor

It was demonstrated above that the response of an enzymatic sensor is determined by the interplay of kinetic and diffusional factors. From the practical standpoint it is important to distinguish what kind of process represents the rate-determining step in the sensor process.

Kinetic control of the overall process is indicated by a strong increase of the sensor response with increasing temperature, provided that the transduction process is less affected by this parameter. This interpretation is based on the fact that the rate of a chemical reaction is strongly dependent on temperature (according to the Arrhenius equation) whereas the temperature has a much smaller effect on diffusion coefficients.

External diffusion control can be verified by checking the effect of the thickness of the external membrane. In the absence of the membrane, external diffusion is localized within a liquid film (Nernst layer) in which strong concentration gradients develop. If external diffusion control is present, the thickness of the Nernst layer can be altered by changing the hydrodynamic conditions such as the stirring rate in batch analysis or the fluid flow rate in flow analysis systems.

Questions and Exercises (Section 4.4)

1 Comment on the concentration profiles in an enzymatic sensor in the case in which external and internal mass-transfer capacities are comparable.
2 What dimensionless parameter indicates the interplay between internal and external diffusion? What values of this parameter indicate control by only one of the above types of diffusion?
3 How can the Biot number be modified in order to impose either internal or external diffusion control to a membrane-containing or a membraneless enzymatic sensor?
4 Comment on the effects of partition constants and diffusion coefficients on the sensitivity of an enzymatic sensor.

4.5 Outlook

The overall process occurring in an enzymatic sensor involves first the catalytic conversion of the substrate to products. As the enzyme is present in an immobilized form, diffusion of the substrate and product are coupled with the enzymatic reaction.

The sensor response depends to a large extent on the kinetics of the enzymatic reaction. Suitable conditions for substrate determination are provided by first-order kinetics. In this case, the sensor response is proportional to the substrate concentration in the sample.

The substrate diffuses from the solution to the enzyme layer, whilst the products travel from the enzyme layer to the solution. The enzyme layer may be coated with a semipermeable membrane or may be in direct contact with the test solution. In both cases, concentration gradients form in the film adjacent to the enzyme layer. Diffusion processes occur therefore in three distinct regions: the enzyme layer, the adjacent film and the bulk of the solution. The diffusion rate in each layer is characterized by the mass-transfer coefficient, which indicates the transport capacity. The overall diffusion rate is determined by the region of lowest transport capacity. Owing to the sluggishness of diffusion in such a region, high concentration gradients develop within it while in adjacent regions of high transport capacity, concentration gradients could be negligible. Therefore, two limiting cases can be distinguished: external diffusion and internal diffusion control. In the first case, the low transport capacity in the external (membrane) layer determines the overall mass-transfer rate, whilst in the second case it is the transport capacity in the enzyme layer that controls the overall mass-transfer process.

The interplay of internal and external diffusion is determined by the dimensionless Biot number. A high value of the Biot number implies sluggish internal diffusion that determines internal diffusion control. Conversely, low Biot number values correspond to sluggish external diffusion and lead to external diffusion control.

In general, the response of the enzymatic sensor depends on the enzyme concentration and the rapidity of the enzymatic reaction (through the k_2 rate constant). It depends also on the diffusion rate within the layer with the lowest transport capacity. Finally, the response depends on the thickness of the enzyme layer that determines the total amount of enzyme per surface area unit. The effect of all these factors is synthetically expressed by particular dimensionless parameters, namely the substrate modulus in the case of external diffusion control and the Thiele modulus in the case of internal diffusion control.

In the general case, which corresponds to near-unity values of the Biot number, diffusion in both the enzyme layer and in the membrane should be considered. As the substrate modulus and the Thiele modulus are interdependent, the response is determined in general by the Biot number and one of the above moduli.

The above treatment refers to the case in which the enzyme kinetics obeys the Michaelis–Menten mechanism and the sensor response is determined by the product concentration at the transducer surface. Despite its limitations, an examination of this particular model system allowed the functioning principles of an enzymatic sensor and the parameters that determine the sensor response to be determined. A comprehensive survey of enzyme sensor modeling is available in ref. [7].

References

1. Engasser, J.M. and Horvath, C. (1976) Diffusion and kinetics with immobilized enzymes, in *Immobilized Enzyme Principles* (eds E. Katchalski-Katzir, L.B. Wingard, and L. Goldstein), Academic Press, New York, pp. 127–220.
2. Goldstein, L. (1976) Kinetic behavior of immobilized enzyme systems, in *Immobilized Enzymes* (ed K. Mosbach), Academic Press, New York, pp. 397–443.

3. Illanes, A., Fernandez-Lafuente, R., Guisan, J.M. *et al.* (2008) Heterogeneous enzyme kinetics, in *Enzyme Biocatalysis: Principles and Applications*, Springer, Dordrecht, pp. 155–203.

4. Carr, P.W. and Bowers, L.D. (1980) *Immobilized Enzymes in Analytical and Clinical Chemistry: Fundamentals and Applications*, John Wiley & Sons, New York.

5. Carr, P.W. (1977) Fourier-analysis of transient-response of potentiometric enzyme electrodes. *Anal. Chem.*, **49**, 799–802.

6. Blaedel, W.J., Boguslaski, R.C., and Kissel, T.R. (1972) Kinetic behavior of enzymes immobilized in artificial membranes. *Anal. Chem.*, **44**, 2030–2037.

7. Baronas, R., Kulys, J., and Ivanauskas, F. (2009) *Mathematical Modeling of Biosensors: An Introduction for Chemists and Mathematicians*. Springer Netherlands, Dordrecht.

5

Materials and Methods in Chemical-Sensor Manufacturing

5.1 Introduction

Fabrication of a chemical sensor consists of integrating a transducer with the sensing element, which includes receptor sites. Often, this is largely a matter of immobilizing the receptor at the transducer surfaces. However, many sensors require additional active components. At the same time, the sensing element should be designed so as to be able to allow the access of the analyte to the receptor. In addition, implantable sensors should be compatible with living tissue. That is why, in general terms, building up the sensing part of the sensor is not simply a problem of receptor immobilization, although this is a key issue.

Commercialization of chemical sensors brings about additional requirements as far as the fabrication method is concerned. The ideal commercial sensor should be as simple as possible and suitable for mass production. The sensor characteristics should be stable for a sufficiently long period of time under operational and storage conditions. As far as stability under operation is concerned, the requirements are less stringent in the case of disposable sensors, but in this case an inexpensive fabrication technology is sought.

A broad variety of methods for assembling the sensing element has been developed [1]. For example, confinement of an enzyme solution between a perm-selective membrane and the transducer surface is a straightforward method that was widely used at the beginning of the development of enzymatic sensors. However, the stability of this system is rather poor and this method is clearly not suitable for mass production. More advanced approaches to sensor fabrication are summarized in Figure 5.1. In one of these methods the sensing part is assembled as a monolayer on the support surface by physical adsorption (A). More stable systems are obtained by forming strong linkages between the receptor compound and the support, either by covalent bonding or affinity interactions (B). Covalent bonding can also be applied for crosslinking protein molecules (C). Entrapment within a polymer network is another widely used approach (D). Finally, a solution of a biocompound can be encapsulated within vesicles (E). A broad spectrum of immobilization protocols has been developed in the framework of biotechnology, and biosensor science can greatly benefit from this readily available know-how [2].

The next sections refer mostly to the use of proteins as bioactive components. Although some of the methods described here are also convenient for nucleic acid immobilization, some particular features relating to the immobilization of nucleic acids are addressed in Chapter 7.

5.2 Noncovalent Immobilization at Solid Surfaces

Immobilization of biocompounds by adsorption involves noncovalent bonding (such as hydrophobic interactions, hydrogen bonding or electrostatic attraction). Such methods are gentle and uncomplicated. Generally, adsorption is achieved by prolonged contact of the support with a solution of the adsorbate. As monolayers are usually formed by such interactions, the analyte can access the sensing part with no restrictions. However, an adsorbed film is not particularly stable and can be damaged by desorption caused by changes in temperature, pH or ionic strength. Nevertheless, an adsorbed layer can be stabilized by crosslinking of adsorbed proteins.

Adsorption of protein molecules conform as a rule to the Langmuir isotherm that relates the surface concentration, Γ (in mole m^{-2}) with the concentration of the adsorbate in solution c:

$$\Gamma = \Gamma_{max}\frac{Kc}{1 + Kc}, \quad \text{or} \quad \theta = \frac{Kc}{1 + Kc}, \quad \text{where} \quad \theta = \frac{\Gamma}{\Gamma_{max}} \tag{5.1}$$

K is the equilibrium constant of the adsorption process, Γ_{max} is the maximum achievable surface concentration, and the θ ratio is termed the *surface coverage*. This equation was derived under the assumption that adsorption occurs by equilibrium interaction between the adsorbate and corresponding binding sites at the support surface. It is also assumed that the adsorbed molecules do not interact with each other and that only a single molecular layer forms.

Chemical Sensors and Biosensors: Fundamentals and Applications, First Edition. Florinel-Gabriel Bănică.
© 2012 John Wiley & Sons, Ltd. Published 2012 by John Wiley & Sons, Ltd.

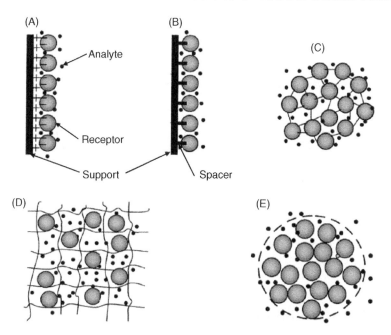

Figure 5.1 Several methods for building up the sensing layer. (A) Physical adsorption at a solid support; (B) Covalent bonding to the support surface; (C) Support-free crosslinking; (D) Entrapment in a polymer network; (E) encapsulation. The large circles represent biomacromolecules, the smaller circles represent free-diffusing small molecules or ions. Adapted with permission from [3]. Copyright 2011 Wiley-VCH Verlag GmBH & Co. KGaA.

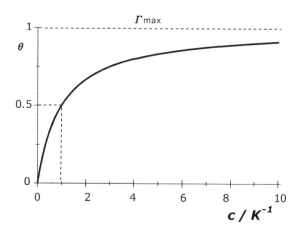

Figure 5.2 Graphical representation of the Langmuir isotherm. The concentration is indicated in K^{-1} units. Half-coverage ($\theta = 0.5$) is obtained when $c = K^{-1}$.

As shown in Figure 5.2, the coverage increases with increasing adsorbate concentration and tends asymptotically to a value close to 1 at high concentrations.

Despite its intrinsic limitations, adsorption is a method of choice in preliminary investigations, and also for fabricating disposable sensors that are exposed to sample solutions for only short periods of time. Various support materials have been used for adsorption, the most popular being silica, cellulose acetate, activated carbon, and synthetic polymers such as poly(vinyl chloride) (PVC) and polystyrene.

Ion exchangers represent another alternative for noncovalent immobilization. Such materials are porous polymers bearing charged groups that can capture protein molecules bearing the opposite charge (see Figure 5.1A). Electrostatic interactions are affected by the ionic strength of the solution, a quantity that depends on both ion concentrations and ion charges. At a high ionic strength, charged groups can be screened by counterions leading to the attenuation of the electrostatic forces.

Questions and Exercises (Section 5.2)

1 What kinds of interaction are involved in noncovalent immobilization at solid surfaces?
2 How can the pH and the ionic strength of the solution affect a surface-adsorbed molecular layer?
3 What are the limitations of noncovalent immobilization? What are the main applications of this method?

5.3 Covalent Conjugation

Outstanding stability of the sensing layer is obtained by chemical reactions resulting in covalent bonds [4,5]. However, this method can be time-consuming and may need expensive reagents.

In this section, several typical conjugation methods are described by assuming that two species, A and B have to be bound together by a covalent link. Both A and B can be biomacromolecules; alternatively, one of them is a macromolecule and the second is a small molecule or a reactive solid surface. Common reactive groups involved in bioconjugation are hydroxyl, primary amine, carboxyl and sulfhydryl groups that are all ubiquitous in proteins. Bioconjugation can be applied for various purpose such as immobilization on solid supports [5], attaching small molecular labels or linking enzyme or nanoparticle labels to biomacromolecules. The covalent bond should not involve groups located in the vicinity of the active site of the bioreceptor in order to preserve its biological activity.

If the conjugation partners cannot react directly (which is often the case), crosslinking is performed in two stages. In the first stage (*activation*), a conjugation reagent interacts with the species of interest to append a reactive group to it. This group then reacts with the second species.

A wealth of protocols for covalent crosslinking has been developed within the framework of various biological disciplines [3,6]. Chemical-sensor technology can benefit advantageously from this vast pool of expertise.

5.3.1 Zero-Length Crosslinkers

In this approach, one atom of a molecule is covalently attached to an atom of a second molecule in a condensation reaction that results in a two-atom linkage. For example, conjugation of carboxyl and amine derivatives can be performed by compounds of the carbodiimide class. Typical reagents of this type are EDC and DDC that are shown in Figure 5.3A. Such reagents react first with a carboxyl group to yield a reactive intermediate that reacts next with a

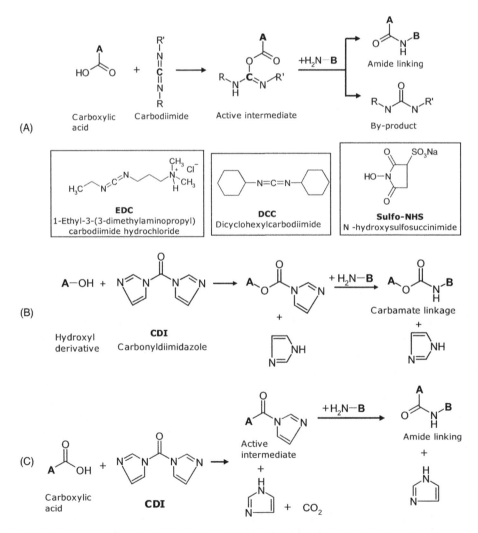

Figure 5.3 Crosslinking by means of zero-length crosslinkers. (A) Crosslinking of carboxyl and amino groups by means of carbodiimides; (B) crosslinking of alcohols and amines by means of N,N.-carbonyldiimidazole (CDI); (C) crosslinking of carboxyl and amino groups by means of CDI.

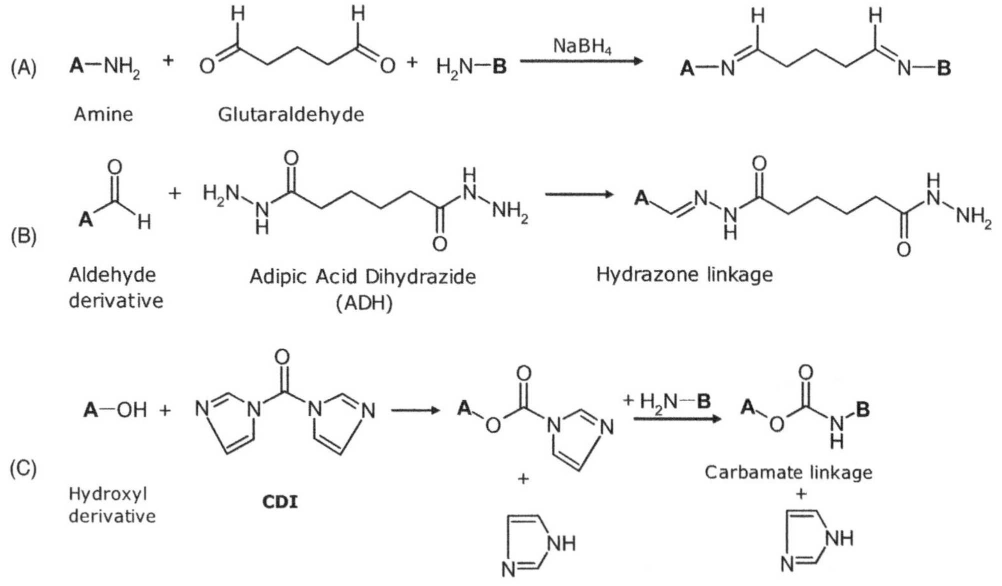

Figure 5.4 Crosslinking with homobifunctional crosslinkers. (A) Glutaraldehyde linking of two amine derivatives; (B) Crosslinking of aldehyde derivatives with adipic acid dihydrazide (ADH); (C) Crosslinking of hydroxyl and amine derivatives by means of CDI.

primary amine group giving an amide linkage. EDC coupling occurs best in the presence of sulfo-NHS (Figure 5.3A) that increases the solubility of the intermediate product and thus enhances the effectiveness of the amine attack. Amide linking can also be obtained by activation with CDI (Figure 5.3B and C).

5.3.2 Bifunctional Crosslinkers

Bifunctional crosslinkers have a reactive group at each end of a short molecule. By conjugation, such reagents link two species through a spacer arm. When immobilizing a biocompound on a solid support, the presence of a spacer arm has particular advantages. Thus, the immobilized molecule preserves sufficient mobility that results in enhanced reaction velocity of the recognition process due to less steric hindrance. In addition, the spacer secures some distance from the support that prevents disturbing effects of the interface microenvironment on the immobilized species.

Homobifunctional crosslinkers contain the same reactive group at each end and are used to conjugate similar functionalities. Thus, glutaraldehyde reacts with the amino group of lysine in proteins, and is widely used to perform crosslinking of protein molecules (Figure 5.4A). In this way, a crosslinked protein network forms. The reaction proceeds via a poorly stable Schiff base intermediate that, by reduction with $NaBH_4$ or $NaCNBH_3$, gives a hydrazone linkage. Enzyme immobilization by crosslinking with glutaraldehyde is a very popular method as it produces a very stable preparation. Crosslinking with glutaraldehyde is also a convenient method to append proteins to amine-functionalized solid supports.

Homobifunctional hydrazides (such as adipic acid dihydrazide, ADH) can be used to conjugate molecules that contain carbonyl or carboxyl groups. Such reagents also react spontaneously with aldehydes to form hydrazone linkages (Figure 5.4B). This reaction is used to attach amine groups to polysaccharides (for example, dextran) after oxidation of the saccharide with periodate, which creates aldehyde groups.

CDI acts as a bifunctional reagent for crosslinking of hydroxyl and amine derivatives via a carbamate spacer, as shown in Figure 5.4C.

Coupling of amine and thiol derivatives can be achieved by means of a *heterobifunctional* reagent such as SMCC (succinimidyl-4-(N-maleimidomethyl) cyclohexane-1-carboxylate), as shown in Figure 5.5. This molecule contains two different reactive groups: an N-hydroxysuccinimide (NHS) ester at one end and a maleimide residue at the opposite end. The first group reacts with a primary amine group to yield a reactive intermediate that reacts further via the maleimide fragment with a thiol forming a spacer-linked conjugated species. As antibodies contain a large number of thiol groups, this method is widely used to perform antibody conjugation with an enzymes label.

5.3.3 Immobilization by Protein Crosslinking

As is apparent from the above discussion, crosslinking reactions can be used to graft biomolecules to solid supports, and to link different molecules in the absence of a support. The second alternative is particularly suited for preparing enzyme gel layers. In order to conserve valuable enzyme, crosslinking is carried out in the presence of an inexpensive protein (such as bovine serum albumin). Glutaraldehyde is a commonly used reagent in this respect. In

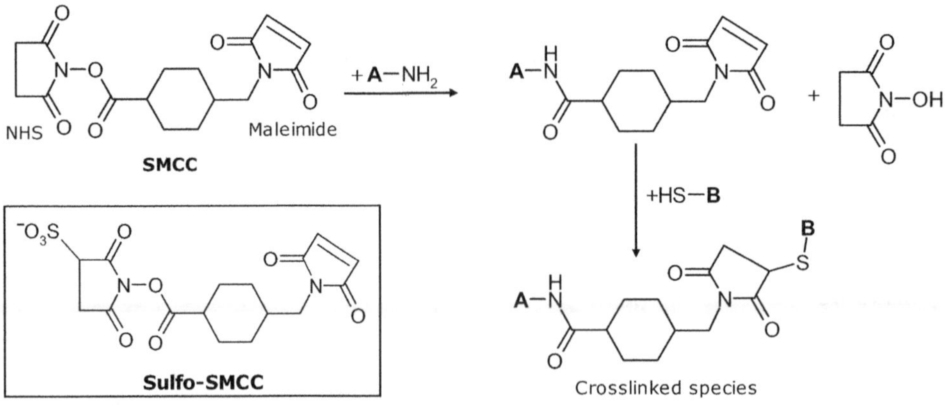

Figure 5.5 Conjugation by a heterobifunctional reagent. Coupling of an amine and a thiol by means of SMCC (succinimidyl-4-(N-maleimidomethyl)cyclohexane-1-carboxylate).

this case, a critical issue is preserving enzyme activity in the resulting preparation. An advanced version of support-free crosslinking is based on preliminary formation of enzyme aggregates by the addition of salts (salting out effect), organic solvents or polymers. This promotes the precipitation of the enzyme in the form of molecular aggregates under mild conditions that safeguard enzyme activity. Finally, enzyme aggregates are stabilized by crosslinking.

Questions and Exercises (Section 5.3)

1 What are the principles of covalent conjugation and what are its advantages and main applications?
2 Review the methods for crosslinking of (a) two amine derivatives; (b) amine and carboxyl derivatives; (c) alcohol and amine derivatives; (d) amine and thiol derivatives.
3 Suppose that you want to prepare an enzyme gel at a solid surface with no covalent bonding between the enzyme and the support. Design a suitable method and sketch the pertinent chemical reactions.

5.4 Supports and Support Modification

5.4.1 General Aspects

Protein immobilization on solid supports (carriers) is a topic of great interest in biotechnology [2,7], and methods developed in this field are also practical in biosensor research and technology. Many standard bioconjugation reactions can be adapted for grafting biocompounds to functional groups at the support surface either directly, when the support bears reactive groups, or after support activation [4,5]. This section reviews several suitable support materials for biocompound immobilization and presents typical reactions for support activation and biomolecule grafting to solid supports.

Supports can be made of nonporous or porous materials. Due to the high specific area, porous supports allow the achievement of a high density of immobilized compound, which is beneficial in various applications. Also, pores can act as a diffusion barrier that is favorable in an enzyme sensor as it contributes to expanding the linear response range. Among porous supports, hydrogels occupy a special position as they allow for immobilization by entrapment, and also by covalent grafting.

An important property of solid supports is their wettability, which represents the ability of a liquid to maintain contact with the surface. Wettability is assessed by the contact angle θ, which is the angle at which a liquid/gas interface meets a solid surface (Figure 5.6). A contact angle less than 90° indicates that wetting of the surface is

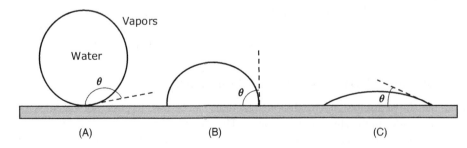

Figure 5.6 The contact angle of a water drop with a hydrophobic surface (A), an intermediate surface (B) and a hydrophilic surface (C).

favoured, and the fluid will spread over a large area of the surface. Contact angles greater than 90° indicate that wetting of the surface is not favoured and the liquid tends to minimize the contact area with the surface.

In the case of water, a wettable surface is also termed hydrophilic and a nonwettable surface is termed hydrophobic. Polar functional groups at a surface make it hydrophilic, whereas the absence of such groups renders the surface hydrophobic.

5.4.2 Natural Polymers

Two kinds of natural polymer are particularly suited to enzyme immobilization, namely, polysaccharides and proteins, that provide a hydrophilic, biocompatible environment to the immobilized species. Such materials are prepared in the form of gels and are useful for immobilization by entrapment, but can also be conveniently employed as supports for covalent grafting. Common polysaccharide materials are considered next.

Cellulose is a practical support for protein immobilization but it is susceptible to microbial degradation and is currently used to a lesser extent.

Dextran is a basically linear, water-soluble polysaccharide. It turns into an insoluble porous gel by crosslinking and in this form it exhibits size-exclusion perm-selectivity. Dextran materials are commercially available under the trade name Sephadex. Ionic derivatives of dextran are useful for protein coupling by electrostatic interactions.

Chitosan (Figure 5.7A) is a polysaccharide containing amino groups. It is obtained by processing chitin, which is a byproduct of fishing (crabs, shrimps) and fermentation industries. Chitosan in a soluble form can be mixed with a protein solution and a gel can be formed by glutaraldehyde crosslinking.

Algal polysaccharides include materials such as agarose and alginates (Figure 5.7B), both of them being able to form gels. Crosslinked agarose (Sepharose) is more satisfactory than crude agarose.

Alginate gels can be formed in the presence of calcium ions or other multivalent counterions. It is very stable between pH 5 and 10 and is mechanically stable in the presence of high concentrations of ions. Enzyme entrapment in alginate gels can be impaired by leakage, but it is a mild and versatile technique.

Carrageenans (Figure 5.7C and D) are extracted from red algae and consist of galactose units that are partially esterified with sulfuric acid. Carragenan gels are very stable at pH > 4.5 and tolerate heat-sterilization conditions. Gelation is induced by calcium and potassium ions but such gels are not stable in the presence of sodium ions.

A common feature of algal polysaccharides and chitosan is the presence of ionic groups such as $-COO^-$, $-SO_3^-$ and $-NH_3^+$. Gelation of such materials can be induced by multivalent ions that interact electrostatically with the charged groups in polysaccharides (ionotropic gelation [8]). A protein can be entrapped in the gel if it is present in the mother liquor.

Among protein materials, collagen and gelatin deserve particular mention.

Collagen is abundant in higher vertebrates as a constituent of flesh and connective tissues. It is hydrophilic, water insoluble and displays a high concentration of binding sites. In addition, its fibrous structure and high swellability in water are convenient for enzyme immobilization, which can be performed by adsorption, entrapment and covalent coupling.

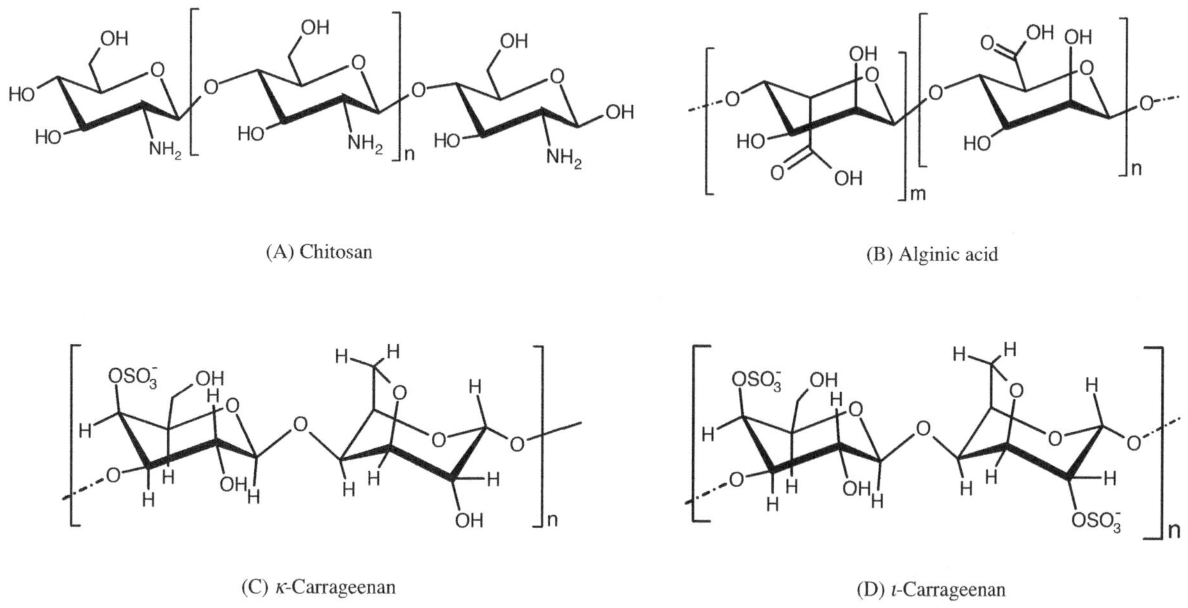

(A) Chitosan

(B) Alginic acid

(C) κ-Carrageenan

(D) ι-Carrageenan

Figure 5.7   Natural polysaccharides used for enzyme immobilization.

Gelatin is a mixture of peptides and proteins produced by partial hydrolysis of collagen on boiling with water. It forms a gel by cooling down the solution to about $40\,°C$, a temperature that is compatible with many enzymes. However, the gel strength is rather low and crosslinking of the entrapped enzyme is required.

5.4.3 Synthetic Polymers

Synthetic polymers present a large variety of immobilization opportunities due to the possibility of tailoring the chemical structure, morphology and physical properties for diverse applications. In addition, such materials are inert to microbial attack. Within the broad class of synthetic polymers it is possible to distinguish *active polymers* and *inactive polymers*. Compounds in the first class bear active groups that can react directly with functional groups in biocompounds in order to perform covalent coupling. Inactive polymers need activation with suitable reagents prior to coupling. Various polymers purposely designed for enzyme immobilization are commercially available.

Polystyrene includes benzene as a pending group that can be easily converted into an amino derivative by nitration and reduction. This can be further activated by diazotization. If the hydrophobicity of polystyrene raises compatibility problems, it is practical to use materials obtained by copolymerization of styrene with hydrophilic monomers such as acrylic acid. Polystyrene is inexpensive, readily available and, after functionalization, displays a high density of binding groups.

Acrylic polymers are available in different forms, each of them including specific reactive groups such as carboxyl, anhydride, amide, hydroxyl, nitrile or epoxide. Thus, polyacrylates and polymethacrylates contain carboxyls and form a negatively charged matrix. *Polyacrylamide* is widely used both as a matrix for gel entrapment and as a support for covalent coupling. As linear polyacrylamide is water soluble, gelation is induced by crosslinking.

Polyamides, known as *nylons* are a category of copolymers formed by condensation of α, ω-dicarboxylic acids and α, ω-diamines. Polyamides are available in a variety of physical forms, such as fibers, membranes, powders and tubes. These materials have a good mechanical strength, biological resistance and hydrophilicity. Terminal carboxyl and amino groups are possible reactive groups. In order to increase the density of binding groups, polyamides are subject to mild acid depolymerization that generates carboxyl and amino groups on the surface. The abundant amide groups can be converted into another group that is susceptible to subsequent conjugation with proteins.

5.4.4 Coupling to Active Polymers

Certain polymers contain active groups that can be conjugated with biocompounds without preliminary activation. This class includes polymers containing anhydride or epoxide groups or halogens.

The *epoxide* group (also known as oxirane or ethylene oxide group) is a three-member cycle formed of one oxygen and two carbon atoms (Figure 5.8A). Acid-catalyzed hydrolysis of an epoxide generates a glycol. The epoxide group reacts readily with nucleophile compounds (primary amine, hydroxyl and thiol) in a ring-opening process.

Polycarbonates contain $-O-(C=O)-O-$ groups and can be prepared in either a soluble or an insoluble form suitable for covalent grafting of proteins by reaction with the amino group (Figure 5.8B)

The *isocyanate* group $(-N=C=O)$ reacts easily with the amino group to produce a urethane linkage $(-NH-(C=O)-O-,$ also called an isourea linkage) that functions as a linker to biomolecules (Figure 5.8C). The isothyocyanate group $(-N=C=S)$ behaves similarly.

Certain *halogenated copolymers* are suitable for immobilization via reaction of the halogen with primary amines giving secondary amine bonds.

5.4.5 Coupling to Inactive Polymers

Polymers containing amino, hydroxyl and carboxyl groups are widely employed as immobilization supports, but coupling to such materials can only be carried out after prior activation.

Amine groups can be present in raw polymers or can be created by chemical conversion of other functional groups. It can be activated by diazotization, as shown in Figure 5.9A or by reaction with crosslinkers such as glutaraldehyde or trichlorotriazine (cyanuric acid).

Vicinal hydroxyl groups are found in polysaccharides as well as in polyvinyl alcohol. Oxidation of polysaccharides with periodate produces aldehyde groups that can be crosslinked according to Figure 5.4B. Activation with cyanogen bromide (CNBr) yields reactive imidocarbamates that react with amine groups in proteins to form a peptide bond (Figure 5.9B).

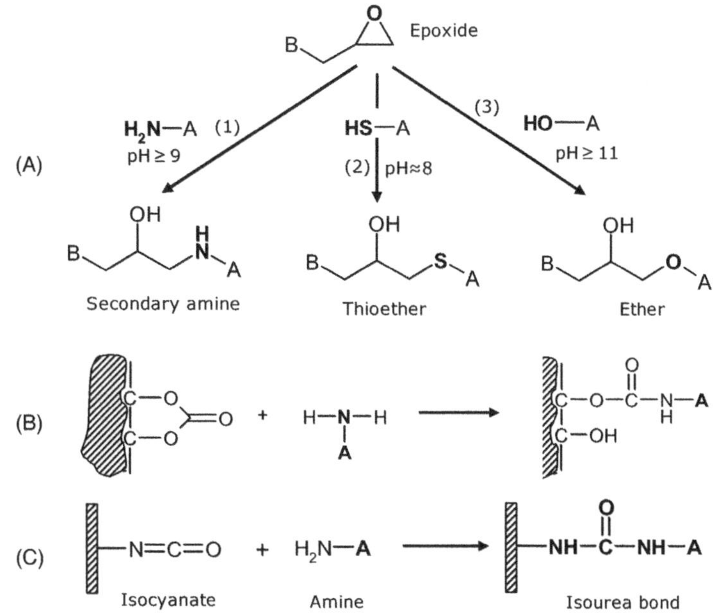

Figure 5.8   Coupling to synthetic polymers containing epoxide (A) carbonate (B) and isocyanate (C) groups.

Reactions with trialkoxysilanes provide a wealth of methods for attaching various functionalities (X) for subsequent coupling. (Figure 5.9C). Various groups can be attached in this way (including amino, carboxyl, thiol or epoxide) using suitable trialkoxysilanes. When affinity binding is sought as an immobilization method, the surface can be functionalized by a biotin-derivatized precursor. As thiols bind spontaneously to gold, gold surfaces can be covered with a silica layer by sol-gel chemistry using a thiol-derivatized trialkoxysilane.

Separate hydroxyl groups can be found in synthetic polyols, such as polyethylene glycol and related polymers. Such groups can be activated by reaction with CDI (Figure 5.4C) or with trichloro-s-triazine (Figure 5.9D). Functionalized epoxide derivatives react with hydroxyl as shown in Figure 5.8A.

Carboxyl groups can be activated via the reaction of carbodiimide in the presence of sulfo-NHS (Figure 5.3A). Also, reactions with CDI (Figure 5.3B), thionil chloride (Figure 5.9E) and the Woodward's reagent are commonly employed to this end.

The *amide group* is present in polyacrylamide and also in Nylon-type polyamides. This group reacts with hydrazine to form an acyl azide derivative that couples with amino groups in biomolecules (Figure 5.9F).

The *nitrile* group is present in poly(acrylonitrile) and related polymers. This group can be converted to carboxyl by acid or base catalysis. Nitrile activation can also be carried out by reduction to amine or by aminolysis leading to an amine side chain (Figure 5.9G). Further amine activation (e.g., by glutaraldehyde) allows various biocompounds to be grafted to the polymer. Remarkably, the conversion of nitrile groups into amino groups results in the formation of a polymeric gel with a high swelling factor.

5.4.6   Inorganic Supports

Inorganic materials such as controlled-pore glass, silica and metal oxides (Al_2O_3, TiO_2) are characterized by high chemical stability that imparts resilience to chemical and thermal treatments. Selection of a support from among such materials is mostly determined by their resilience at extreme pH values. Thus, raw controlled-pore glass is not stable at pH > 8 but its durability is increased by coating it with zirconium oxide. Al_2O_3 and TiO_2 resist alkaline media well, whereas silica is better suited to acidic solutions. As hydroxyl groups are present at the surface of these materials, their functionalization can be achieved by reaction with alkoxysilanes (Figure 5.9C).

5.4.7   Carbon Material Supports

Various forms of graphite and other carbon materials are commonly employed as electrode materials in electrochemical sensors [9]. Graphite consists of parallel sheets of sp^2-bonded carbon atoms held together by van der Waals interactions. The properties of the sheet edge differ essentialy from those of the basal plane (Figure 5.10). Thus, the basal plane is hydrophobic and resistant to chemical attack. Conversely, the unsaturated covalent bond at the edge

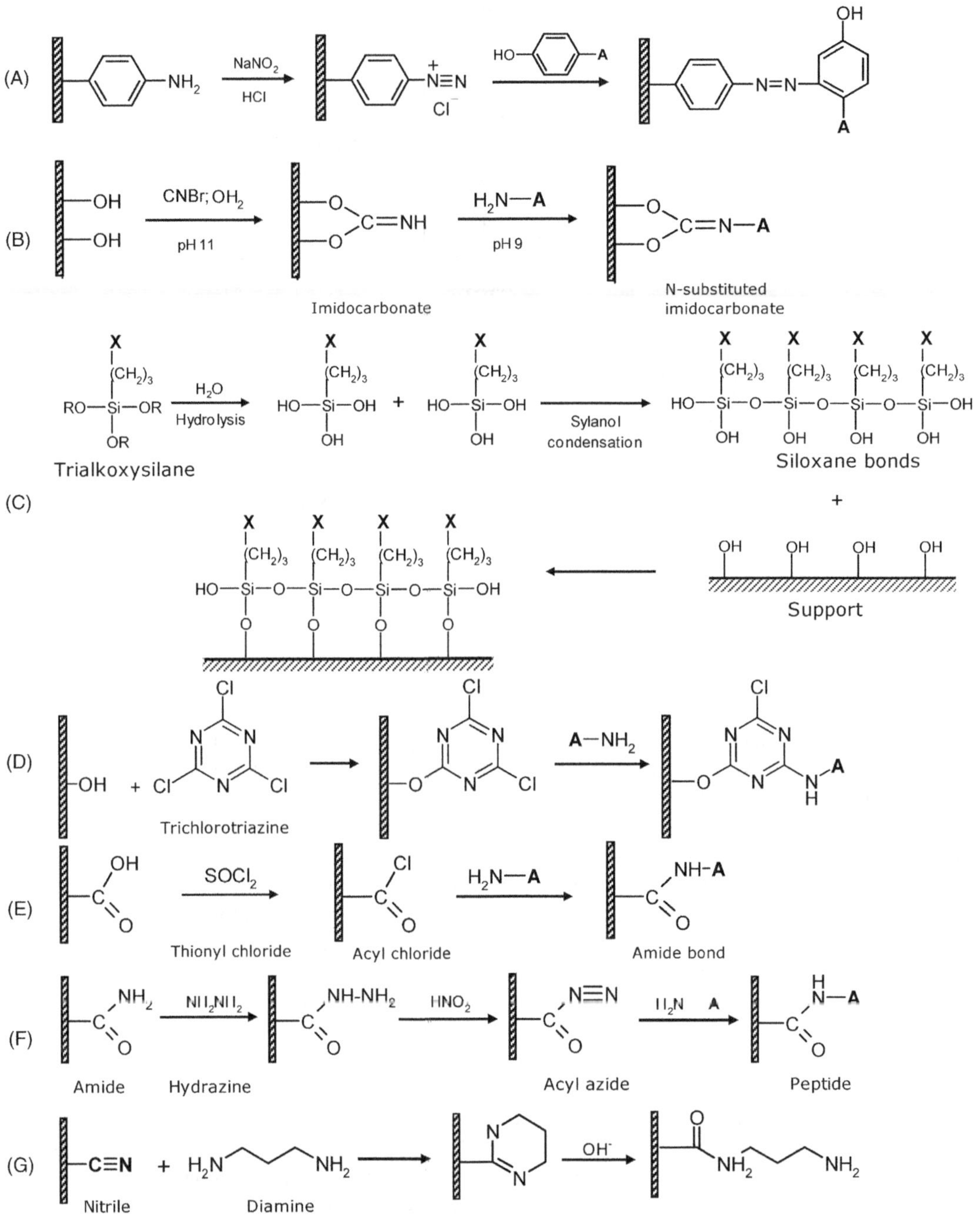

Figure 5.9 Common coupling reactions for inactive polymer supports. (A) Amine activation by diazotization; (B) activation of vicinal hydroxyls by cyanogen bromide; (C) trialkoxysilane functionalization; (D) activation of separate hydroxyl by trichlorotriazine (cyanuric chloride); (E) activation of carboxyl group by thionyl chloride; (F) coupling to amide group via acyl azide; (G) Conversion of nitrile group into amine.

imparts chemical reactivity and normally various oxygen functionalities are appended to marginal carbons. A high density of carboxyl groups at the edge can be obtained by chemical or electrochemical oxidation. Various forms of graphite materials are available and the immobilization method is selected in accordance with the surface constitution. Surfaces displaying preponderantly basal planes are suitable for adsorptive immobilization of hydrophobic compounds. Conversely, surfaces consisting mostly of edges are hydrophilic. Moreover, in neutral and alkaline solution, dissociation of carboxyl groups imparts to the surface a negative charge. Covalent bonding to such surfaces is achievable via suitable activation of the carboxyl group.

A very common method for integrating biocompounds with carbon materials is based on carbon paste materials [10]. The carbon paste consists of graphite powder mixed with an oil binder and a suitable modifier such as a synthetic or biological receptor. By proper selection of the binder, the modifier nature and the component percentage,

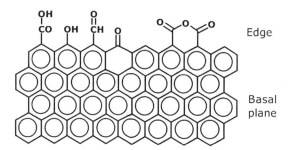

Figure 5.10 Graphite structure.

a huge variety of carbon paste compositions can be prepared for applications in sensors for inorganic, organic and biological species. Mass production of sensor including similar mixture as planar films can be achieved by screen-printing technology (Section 5.13.2).

Contemporary research in the field of carbon materials focuses on carbon nanomaterials such as graphene, fullerene and carbon nanotubes. This topic, which is of a high interest in chemical-sensor technology, is addressed in detail in Chapter 8.

5.4.8 Metal Supports

Due to their chemical inertness, noble metals such as gold and, to a lesser extent, silver, platinum and palladium are commonly used in various types of chemical sensor.

Molecular monolayers can be assembled at a metal surface by chemisorption of molecules containing a reactive head-group. A widely used method of this kind is based on self-assembly of thiols on gold or other metals.

Thiols (R—SH, also referred to as mercaptans) react spontaneously with a gold surface to form strong chemical bonds via the sulfur head group [11]. Under normal conditions, this process is irreversible, which results in high stability of the self-assembled monolayer. In general, the adsorbed alkylthiols molecules form close-packed monolayers with the alkane chain tilted at about 30° from the vertical (Figure 5.11A). van der Waals interactions between alkyl tails result in a regular pattern that is characteristic of a self-assembled monolayer (SAM). Thiol monolayers can also be formed on silver or platinum supports. Other sulfur derivatives, such as thiones ($R_2C=S$[12]) or thiocarboxylic acids (RC(=O)SH) behave similarly. Selenols (R—SeH) also form self-assembled monolayers on gold [13].

Mixed layers are made using mixtures of functionalized alkanethiols with unfunctionalized alkanethiols for diluting the functionalities at the modified surface.

Gold electrodes and gold surfaces are widely used in the design of chemical sensors. In such instances, modification by chemisorption of thiols is a method of choice for attaching functional molecules. This can be achieved simply by hydrophobic interaction of a target molecule with a nonfunctionalized thioalkane layer. A clean gold surface is hydrophobic but it turns hydrophilic by chemisorption of −OH terminated thiols. If the end group in the surface layer bears an electric charge (such as in –COO⁻ or –NH₃⁺), charged molecules can be attached by electrostatic interactions. A more firm immobilization is achieved by covalent grafting to terminal functionalities in the thiol molecules. Suitable linking groups can be attached to the surface by chemisorption of amino- or carboxyl derivatives such as cysteamine, and dihydrolipoic acid (Figure 5.1B). Remarkably,

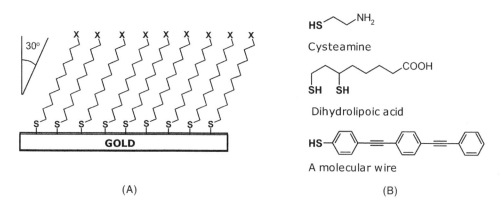

(A) (B)

Figure 5.11 (A) The conformation of an alkanethiol layer at a gold surface. X = H (in alkanethiols) or a functional group (e.g., −COOH, or −NH₂). (B) Common thiol reagents for functionalization of a gold surface by sulfur chemisorption.

molecular wires can also be appended to gold surfaces by thiol chemisorption. A molecular wire is a linear molecule containing a system of conjugated π-bonds [14]. Due to delocalization, π-electrons can flow along this molecule allowing for electric contact between the metal and redox species within the sensing layer.

Moreover, thiol-functionalized receptors (such as proteins or oligonucleotides) spontaneously form a chemisorbed surface layer on gold. Many protein molecules contain thiol groups in cysteine residues and react directly. If such groups are absent, one can resort to Traut's reagent (2-iminothiolanel), which reacts with amino groups in proteins and appends a short thiol tail. Other receptors can also be obtained as thiol-derivatives for immobilization by sulfur chemisorption on gold. In order to impart more stability to the surface layer, several thiol groups are appended to the receptor molecule so as to form multiple sulfur–gold linkages.

Chemisorbed close-packed layers only form with long linear thioalkanes. Large receptor molecules form a loose layer leaving a large fraction of the surface unoccupied. In order to prevent nonspecific interactions with this surface, a linear thioalkane is assembled in a second step to form a compact layer over the surface left free after receptor immobilization.

Thiol chemisorption is widely used in various kinds of chemical sensor [15–17] such as amperometric sensors, piezoelectric sensors and sensors based on the field-effect transistor. Gold nanoparticles can be crosslinked to various biomolecules by a similar approach. Thiol self-assembly is also an important method in microtechnology [18].

An alternative method for self-assembly on metal surfaces relies on aryl diazonium salts ($Ar-N^+\equiv N)A^-$, where Ar is an aromatic ring and A^- is an anion). By electrochemical reduction, such compounds yield a compact film that adheres well to the gold surface. Thus, 4-carboxyphenyl diazonium has been deposited on gold and used for cross-linking polypeptides as metal ion receptors [19]. Compared with thiol chemisorption, diazonium self-assembly produces a more stable and compact layer of short spacers that performs better as linkers to the metal surface.

5.4.9 Semiconductor Supports

Silicon is currently the standard material in microelectronics and is therefore widely available. In addition, well-established micromachining technologies allow for patterning the silicon surface in the form of sensor arrays. In addition, the control of the electronic properties of silicon by coating with bio-organic layers opens up new perspectives in both microelectronics and sensor technology [20,21].

Modification of the silicon surface begins with activation by a chemical treatment that produces reactive groups at the surface. Thus, hydroxyl groups can be formed by treatment with an oxygen plasma followed by hydrolysis in water. These groups react with trichlorosilanes in the gas phase as shown in Figure 5.12A. The resulting self-assembled monolayer is stabilized by siloxane bonds as well as van der Waals interactions between the alkane chains.

Another approach is based on prior formation of a hydride layer at the silicon surface, either by temperature treatment with H_2 in vacuum or by etching with HF or NH_4F solutions [22]. Thermally or ultraviolet-driven reactions with a 1-alkene molecule appends it by a Si–C bond. An alkyl layer covalently bound to the silicon surface forms in this way. The terminal methyl in these molecules can be activated by photochemical reactions. Direct linking of functionalized 1-alkenes is hampered by the reactivity of the functional group with the surface hydride. That is why the functional group should be protected prior to grafting and deprotected thereafter. Carboxyl, amino and thiol groups can be appended in this way to the silicon surface. Alternatively, crosslinking reagents can be first attached to silicon in order to perform subsequent grafting of functional groups of interest.

An alternative route relies on the reaction of surface hydride with diazonium salts, as shown in Figure 5.12B. Coupling to hydride is initiated by electrochemical oxidation of such salts yielding an aryl radical. Diazonium coupling proceeds spontaneously at room temperature certain compounds (including molecular wires) [23].

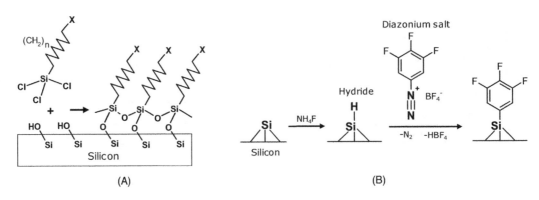

Figure 5.12 (a) Modification of silicon surface by trichlorosilanes. X can be an amino group that is suitable for subsequent crosslinking. (b) Silicon surface modification by self-assembly of diazonium salts.

A relatively new semiconductor material is doped diamond that is commercially available in the form of thin layers prepared by chemical vapor deposition on various supports. Its main advantages over silicon are the higher chemical stability and the large potential window for electrochemical reactions. Biomolecules can be covalently grafted to a hydrogen-terminated diamond surface [20].

Questions and Exercises (Section 5.4)

1 What kind of natural compounds are useful for enzyme immobilization? What kind of immobilization methods are suitable for such supports?
2 Review the active groups in polymers and their chemical reactions with biocompound functional groups.
3 Draft several methods for covalent coupling to polystyrene.
4 Outline several methods for covalent coupling of proteins to silica and glass surfaces using trialkoxysilanes as crosslinkers.
5 How can a protein be crosslinked with nylon polymers?
6 What procedures and reactions can be used for enzyme immobilization in poly(acrylamide)?
7 Give an account of chemical reactions involved in protein grafting to poly(acrylonitrile).
8 Review the main inorganic supports for biocompound immobilization and their pertinent properties.
9 What properties of graphite are relevant to sensor applications?
10 Draft several possible procedures for covalent grafting to a graphite surface.
11 Sketch several possible routes for direct and indirect grafting of biocompounds to gold surfaces.

5.5 Affinity Reactions

The high specificity of affinity interactions is advantageously exploited to link various species modified with affinity reagents. The most frequently used system of this kind is the avidin–biotin couple [24–26].

Avidin and *streptavidin* are proteins with different origin. Avidin is present in egg white, whereas streptavidin is found in the cell membrane of *Streptomyces* bacteria. These proteins share a common property, namely, both of them combine by affinity interactions with the small molecule biotin (also known as vitamin H or vitamin B_7, Figure 5.13A). This interaction involves the heterocyclic moiety of biotin and rests on multiple hydrophobic and hydrogen bondings that impart to the complex an outstanding stability (dissociation constant close to 10^{-15} M^{-1}). Avidin and streptavidin molecules are homotetramers, each of the four subunit being able to bind a biotin molecule.

Biotin can be conjugated with proteins by reaction of the carboxyl group with the amino group of lysine. Alternatively, it can be derivatized with reactive groups in order to perform crosslinking with the compound of interest [27,28]. The second alternative is more convenient because it produces a longer linker and avoids steric hindrance in the interaction with (strept)avidin. Biotin and its derivatives are widely used for conjugation of both proteins [29] and nucleic acids [30] with a (strept)avidin functionalized support (Figure 5.13B). Moreover, due to the multivalent character of (strept)avidin, up to four biotinylated species can be assembled with such a molecule (Figure 5.13C). This property allows the growing of intricate structures such as multilayers or dendrimers.

Avidin is a glycoprotein, whereas streptavidin is a genuine protein that makes it less prone to nonspecific interactions. By removing the saccharides from the avidin molecule, one obtains a product known as NeutrAvidin. As a result of carbohydrate elimination, NeutrAvidin is not prone to nonspecific interactions that can occur in the case of avidin. However, the biotin binding capacity is retained because the carbohydrate is not necessary for this interaction. NeutrAvidin has a near-neutral isoelectric point (pI = 6.3) and is therefore not electrically charged in neutral solutions. Consequently, nonspecific electrostatic interactions are also reduced to a minimal level.

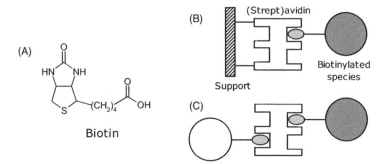

Figure 5.13 The biotin–avidin system. (A) the biotin molecule: (B) immobilization of a biotinylated species to a (strept)avidin modified support; (C) conjugation of two biotinylated species by a (strept)avidin bridge.

Coupling by affinity proceeds under gentle conditions with a high specificity and yields very stable products. However, the cost of materials can be prohibitive in some applications.

Questions and Exercises (Section 5.5)

1 What are the main compounds employed for affinity immobilization? Review their relevant properties.
2 Draft several possible methods for coupling streptavidin to polymer supports.

5.6 Thin Molecular Layers

Assembly of one or several molecular layers on a solid support is a practical method for building up the sensing part of the sensor and integrating it with a transducer [31]. This method allows the smart structuring of the sensing layer according to predefined conditions and also provides a biocompatible environment for biological receptors. Such layers can be obtained by *self-assembly*, a process in which a disordered system of pre-existing components forms an organized structure as a consequence of specific, local interactions among the components. A self-assembly process is spontaneous and is often reversible.

A self-assembled layer can also be obtained by strong interactions (such as chemisorption) with a solid surface. Hydrophobic interactions provide another alternative that is appropriate for assembling surfactants and other amphiphilic compounds. Electrostatic interaction between successive layers allows regular sequences of molecular layers to be assembled by the so-called layer-by-layer method. Organized sequences of molecular layers can also be obtained by affinity interactions.

Such methods belong to the field of Supramolecular Chemistry that addresses large and complex entities formed from distinct molecules assembled by noncovalent bonding [32]. Supramolecular chemistry is essential in living organisms as many biocompounds are built up according to supramolecular chemical principles and provide a rich source of inspiration for synthetic and semisynthetic chemistry. The quaternary structure of some proteins, as well as affinity interactions and formation of enzyme–substrate complexes, are typical examples of supramolecular interactions.

5.6.1 Self-Assembly of Amphiphilic Compounds

Typical amphiphilic compounds are surfactants, which are molecules with a hydrophilic head and a long hydrophobic tail. The hydrophilic head can be either an ionized group (such as $-SO_3^-$ or $(CH_3)_3N^+-$) or a polar, nonionic moiety, while the hydrophobic tail is a long hydrocarbon chain. Among various classes of surfactants, *phospholipids* (such as phosphatidylcholine, (Figure 5.14)) deserve a particular mention. Phospholipids are of interest because they are present in natural membranes and, therefore, are compatible with biocompounds.

When a small amount of surfactant is mixed with water it tends to accumulate at the surface with the hydrophobic tail sticking out from the water. However, if the surfactant concentration is over a certain level (termed the *critical micellar concentration*), its molecules assemble in small molecular aggregates called *micelles* in which the hydrophilic heads are oriented toward the aqueous phase (Figure 5.15A). Although this figure shows only a spherical micelle, micelles can also be assembled as rods. Rod-like micelles can further assemble into hexagonal arrangements looking like a submillimeter honeycomb.

A more advanced degree of organization is achieved in *liposomes* (Figure 5.15B). Liposomes are artificially prepared vesicles made of surfactant bilayers with a minute amount of aqueous phase occluded within.

Liposomes can be obtained by evaporation of the organic solvent of a phospholipid solution so that a thin film is formed at the bottom of round bottom flask. Then, a liposome suspension in an aqueous buffer solution is formed under stirring or sonication. Controlled-size liposomes form when using surfactant micelles or proteins as templates.

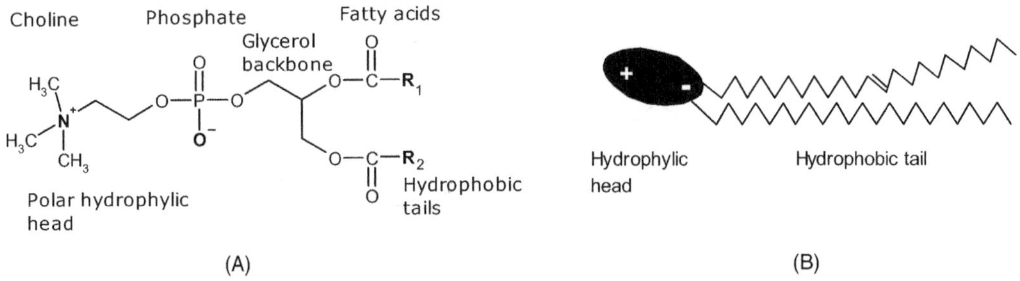

(A) (B)

Figure 5.14 Structure of phosphatidylcholine, a typical phospholipid (A) and its molecule conformation (B).

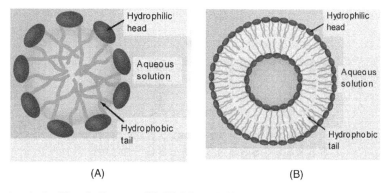

(A)

(B)

Figure 5.15 Structure of a micelle (A) and a liposome (B). (A) Adapted with permission from http://commons.wikimedia.org/wiki/File: Micelle_scheme-en.svg Last accessed 17/05/2012. (B) Adapted with permission from http://commons.wikimedia.org/wiki/File: Liposome_scheme-en.svg Last accessed 17/05/2012.

A liposome can consist of a single bilayer (Figure 5.15B) (unilamellar liposome) or of many bilayers stacked upon one another (multilmellar liposomes). Multilamellar liposomes are larger in size but of nonuniform size and undergo fast sedimentation, which may limit their application.

Small molecules, but also proteins, can be entrapped in liposomes when present in the reaction mixture. Enzyme entrapment in liposomes stabilizes the enzyme and provides a barrier to inhibitors such as small molecules and metal ions. Selective permeability can be imparted to liposomes by forming size-selective or charge-selective pores in the bilayer.

5.6.2 Bilayer Lipid Membranes

Bilayer lipid membrane are essential structure in living cells and can be easily prepared artificially using natural phospho-lipids [33–35]. A widely used method for preparing supported lipid bilayers is the Langmuir–Blodgett technique [36].

The Langmuir–Blodgett technique gives rise to molecular layers attached to an inert solid support by adsorption (Figure 5.16). In order to prepare a Langmuir–Blodgett film, a tiny concentration (far less than the critical micellar concentration) of amphiphilic molecules are spread at the air/water interface where they adjust themselves with the hydrophobic tail sticking out from water to form a disordered layer. Lateral compression of this layer produces a close-packed monolayer that is consolidated by mutual molecular interactions. This layer is then transferred to a solid surface by dipping followed by slow retraction. The hydrophilic end will be oriented towards the support surface when the surface is hydrophilic, whereas an opposite orientation will be adopted with hydrophobic supports.

Successive molecular layers can be assembled by the Langmuir–Blodgett technique by repeating the above procedure. In each new layer, molecules are oriented such as to fit the pertinent external moieties in the previous layer. Many applications rely on molecular bilayers. Bilayers formed in this way mimic natural membranes and are thus compatible with many proteins and other biocompounds.

A phospholipid bilayer does not always fulfill the stability requirements imposed by a chemical sensor. Better stability results when using a porous support such as an ultrafiltration membrane. In such a case, microlayers are supported in pores of 1–5 μm diameter. Bilayers formed on a smooth surface can be stabilized by additional treatments such as crosslinking of reactive terminal groups, or postassembly covalent grafting to the solid surface.

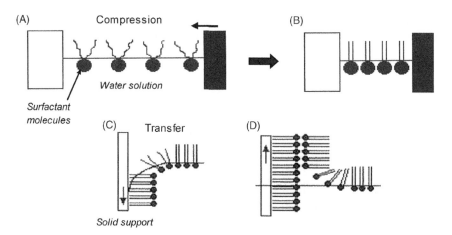

Figure 5.16 Deposition of a Langmuir–Blodgett film on a hydrophobic solid support. (A) random film at the surface of a diluted surfactant solution; (B) close-packed film formed by compression; (C) transfer of the film by dipping the support into the solution; (D) bilayer formation by retracting the support. Adapted with permission from [34]. Copyright 2008 John Wiley & Sons, Ltd.

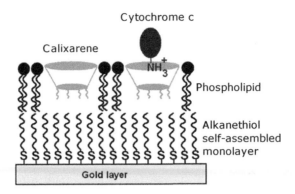

Figure 5.17 Sensing layer for cytochrome c detection using carboxylated calixarene as receptor. The receptor is embedded in a phospholipid layer attached by hydrophobic interaction on an alkanethiol layer assembled on gold. Adapted with permission from [38]. Copyright 2011 Wiley-VCH Verlag GmBH & Co. KGaA.

A sensing element can be assembled by incorporating a suitable receptor into a bilayer membrane [35,37]. The receptor can be introduced directly into the surfactant solution prior to the formation of the membrane. Alternatively, the receptor can be codeposited with the surfactant layer at the air/solution interface before the formation of the bilayer. Another method rests on the encapsulation of the receptor molecules in liposomes that are then left to fuse with a preformed bilayer.

Supported bilayers can also be prepared by assembling a phospholipid layer on a preformed alkane thiol self-assembled monolayer. An example is shown in Figure 5.17 that illustrates a method for immobilizing carboxyl-functionalized calixarene. This receptor interacts with cytochrome c by electrostatic interaction between cationic lysine groups in the protein and anionic carboxylate groups in the receptor.

Firm attachment of a bilayer to a support is achieved in tethered bilayer lipid membranes [39]. In this approach, the first layer is formed by chemisorption of a thiol-tethered lipid to a gold surface, whereas the second layer is obtained by self-assembly of the second layer over the first one.

5.6.3 Alternate Layer-by-Layer Assembly

Molecular multilayers can be assembled on solid supports by consecutive adsorption of alternate monolayers bearing opposite electric charges.

Layer-by-layer assembly does not provide oriented films as the Langmuir–Blodgett technique does, but it can be carried out using a very simple procedure. The solid support should bear ionic groups and the multilayer is assembled using solutions of polyanionic or polycationic compounds. Figure 5.18 illustrates this technique for the case of a positively charged support surface. The first layer is formed by immersing the support in a solution containing a polyanion. After rinsing, the support is immersed in a solution of the polycation to form the second layer. This sequence can be reiterated in order to add more layers. The layer-by-layer technique provides a high degree of control over the width of the layer, owing to the linear growth of the thickness with the number of bilayers.

Charged protein molecules can be assembled as multilayers simply by substituting the second surfactant by the pertinent protein. As the electric charge is the only restriction, different proteins can be assembled within the same film as consecutive layers (Figure 5.19). In this way, multi-enzyme sensors can be produced.

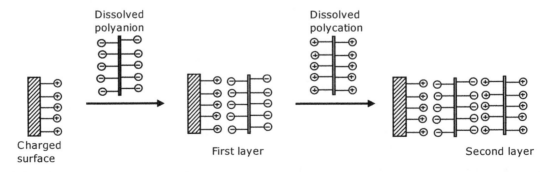

Figure 5.18 Build up of a multilayer assembly by sequential adsorption of anionic and cationic polyelectrolytes, for example, poly (vinyl sulfate) and poly(allylamine), a polyanion and a polycation, respectively.

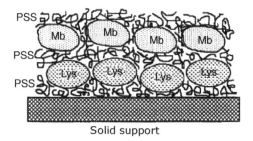

Solid support

Figure 5.19 Assembly of different proteins at a solid support using the layer-by-layer method. Mb is myoglobin; Lys is lysozyme; PSS is the poly(styrenesulfonate) polyanion. Adapted with permission from [40]. Copyright 1995 American Chemical Society.

Organized multilayers can also be prepared by exploiting the multivalent character of (strept)avidin in its interaction with biotinylated species. Molecules grafted with multiple biotin units can be assembled sequentially with avidin layers by strong affinity interactions.

The layer-by-layer technique is equally practical for integrating nanoparticles or living cells with the sensing layer. Nanostructured films can be fabricated in this way [41]. This method has proved effective for assembling the sensing layer in sensors based on various transduction methods, such as amperometric, optical and field effect semiconductor devices.

Questions and Exercises (Section 5.6)

1 What kinds of compound are used to prepare thin molecular layers?
2 What types of structure are produced by self-assembly of amphiphilic compounds dispersed in aqueous solutions? What are their applications in chemical sensors?
3 How can molecular multilayers be prepared at the surface of a solid support? Comment on the stability of such layers and possible methods for reinforcing them.
4 How can enzymes be integrated in a molecular multilayer? What are the advantages of multilayer assembly?

5.7 Sol-Gel Chemistry Methods

Sol-gel chemistry addresses the preparation of gels by chemical reactions of dissolved precursors [42,43]. The product is a loose network of polymer chains including an appreciable amount of solvent. Typical materials in this class are the silica gels that form by condensation of orthosilicic acid in acidic solutions of Na_2SiO_3. They consist of crosslinked polymer chains of $-O-Si-O-$ units. Sol-gel chemistry allows the preparation of similar materials that include organic substituents bound to either silicon or oxygen atoms. The synthesis relies on esters of orthosilicic acids (Figure 5.20) and proceeds by a sequence of hydrolysis (5.2) and condensation (5.3) reactions:

$$Si(OR)_4 + H_2O \rightarrow HO-Si(OR)_3 + R-OH \tag{5.2}$$

$$(OR)_3Si-OH + HO-Si(OR)_3 \rightarrow (OR)_3Si-O-Si(OR)_3 + H_2O \tag{5.3}$$

The above sequence produces linear polymer molecules but further hydrolysis and condensation result in a network of crosslinked polymer chains. It is important to note that siloxanes are not stable in an alkaline solutions.

The constitution of the final product depends on the balance of the reaction rates of hydrolysis and condensation, the solution pH being a key parameter. In neutral solutions, hydrolysis is very slow relative to condensation, but this situation is reversed in an alkaline solution. At the same time, if the available amount of water is limited, hydrolysis is much slower than condensation. Fast hydrolysis favors formation of colloidal particles (*sols*), whereas fast condensation promotes formation of large clusters that, by crosslinking, form a continuous three-dimensional network including water (a *gel*).

Entrapment of biomolecules in silica gel can be effected in two stages. In the first stage, a sol is produced by reaction of the precursor (for example, TEOS) in an acidic solution. Biopolymers are immobilized after shifting the pH into the neutral region where the structure of the biopolymer is not altered. Under these conditions, the condensation is accelerated yielding a gel that includes biopolymer molecules within its pores (Figure 5.21). However, the alcohol formed in reaction (5.2) can cause protein denaturation. The alcohol should therefore be removed before proceeding to biopolymer entrapment. Some additive, such as poly(ethylene glycol), polysaccharides, polyelectrolytes or ionic surfactants have beneficial effects on the stability of entrapped enzymes as the entrapped molecules are surrounded by additive molecules. In order to prevent precipitation, THEOS should be used as precursor. It is

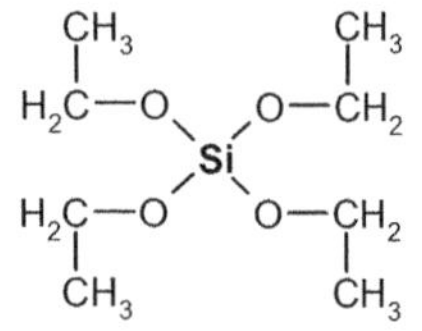

TEOS (Tetraethoxysilane or tetraethyl orthosilicate) **TMOS** is the acronym of the analogous methyl derivative.

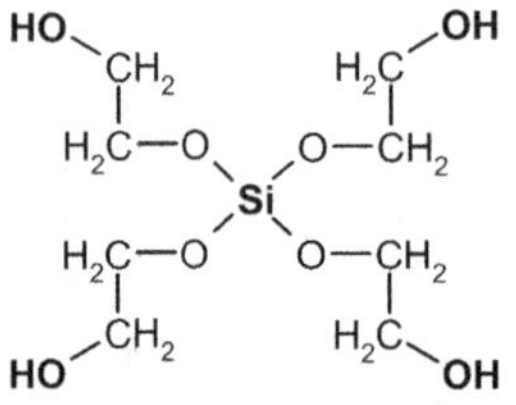

THEOS (Tetrakis (2-hydroxyethyl) orthosilicate))

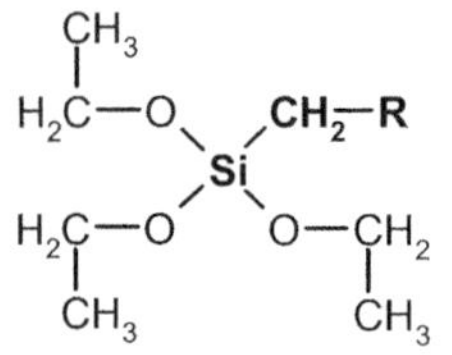

An organically modified precursor (methyltriethoxysilane if R is hydrogen).

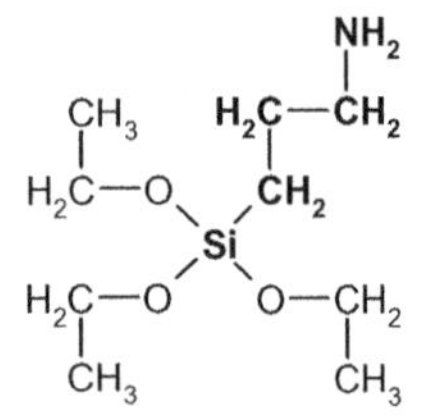

APTES ((3-aminopropyl)-triethoxysilane). **APTMS** is the acronym of the analogous methyl derivative

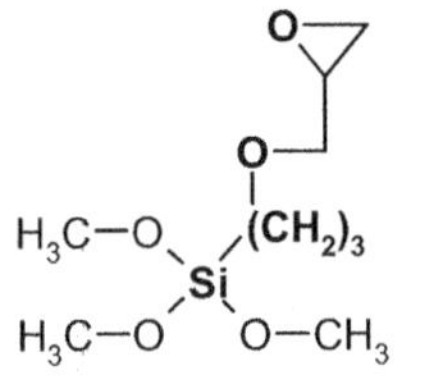

GOPS ((3-Glycidoxypropyl)-dimethyl-ethoxysilane)

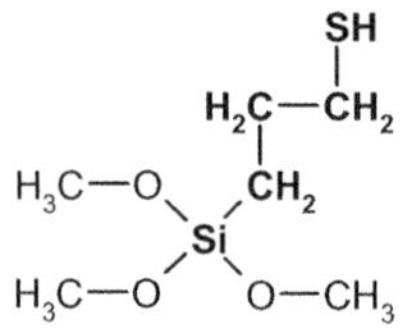

MPTS ((3-mercaptopropyl)-trimethoxysilane)

Figure 5.20 Typical precursors for the synthesis of silica materials by sol-gel chemistry.

completely miscible with water and, by hydrolysis, it produces ethylene glycol, which is a biocompatible organic solvent. By using THEOS and suitable additives, the immobilization can be effected in a single step consisting of gel formation in the presence of the guest molecules.

Due to the mild reaction conditions, the sol-gel method is a convenient method for the entrapment of enzymes and other biological materials including living cells and organisms [44,45]. For example, Figure 5.21 shows an enzyme encapsulated by the sol-gel method in the presence of a surfactant. Remarkably, the enzyme activity is preserved even if the pH in the aqueous phase is very far from the optimum pH of the enzyme. Note that silanol groups can be ionized at a suitable pH imparting negative charges to the gel pore surface. This property brings about some selectivity towards positively charged guest molecules.

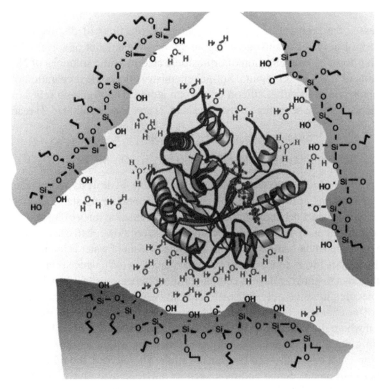

Figure 5.21 Entrapment of a phosphatase enzyme in a sol-gel silica network. Adapted with permission from [48]. Copyright 2005 American Chemical Society.

The properties of the gel pores can be engineered by using *organically modified precursors*. Such molecules include an organic group linked directly to a silicon via a carbon–silicon bond, as in methyltrietohysilane and APTES (Figure 5.20). A material obtained by the condensation of organically modified precursors is known as *ormosil* [46]. One or two C−Si bonded substituents can be introduced in the precursor molecule. Depending on its properties, the substituent imparts to the pore surfaces specific characteristics, such as hydrophobicity (through an alkyl substituent) or an electric charge (APTES with protonated amine groups). By noncovalent interactions, the substituents contribute to the stabilization of the immobilized molecule. The C−Si bonded arm can be designed such as to contain a reactive end group (Figure 5.20), such as amino (in APTES and APTMS), epoxide (in GOPS), or thiol (in MPTS). The reactive group serves as an anchor for covalent conjugation of proteins and another compounds.

Organically modified precursors can be used to activate certain support materials in view of further immobilization of receptor molecules. So, materials with exposed hydroxyls at the surface (glass, cellulose, pretreated carbon) can be modified with reactive groups by reaction with APTES, GOPS or MPTS. A similar treatment can be applied to silicon surfaces after preparing a silicon oxide layer by oxidation and hydrolysis to yield hydroxyl groups. Receptor molecules are then covalently conjugated to the silane derivative linker via the appended reactive group.

Sol-gel chemistry represents a very convenient method for enzyme entrapment due to the mild reaction conditions and the possibility of engineering the surface structure so as to impart specific properties such as hydrophilicity, hydrophobicity or electric charge [45]. Various immobilization protocols based on sol-gel chemistry are available [47].

In summary, sol-gel chemistry is a mild and straightforward method for immobilization of biopolymers and cells in biocompatible matrices. Also, small molecules can be integrated either by inclusion or covalent attachment to the gel network.

Sol-gel entrapment preserves the three-dimensional structure of proteins and hence their biological activity. However, protein denaturation is possible due to the alcohol byproducts. Nanomaterials can be included in the gel matrix in order to bring about additional functionality such as direct electron transfer from redox enzymes to the electrode in amperometric sensors.

This section has focused on silica sol-gel materials, but it is important to note that other materials (such as metal oxides) can be obtained in the gel form by analogous chemistry [49].

Questions and Exercises (Section 5.7)

1 What are the structural details of sol-gel chemistry precursors?
2 What is the mechanism of polymerization and how can the conformation of the product be controlled?
3 How can the surface properties of the product be engineered?
4 How can solid surfaces be functionalized by sol-gel chemistry?
5 What are the advantages of sol-gel immobilization?

5.8 Hydrogels

Hydrogels are solid–liquid systems in which a polymeric matrix forms a loose three-dimensional network including a high proportion of water, up to 90% of the total mass [50,51]. The polymer network results from crosslinking either by physical interactions between polymer molecules or by covalent bonding (Figure 5.22). A *xerogel* forms from a gel by drying with unhindered shrinkage. Xerogels retain high porosity and a very large specific surface area along with a very small pore size (1–10 nm). When the solvent is removed under supercritical conditions, the network does not shrink and a highly porous, low-density material known as an *aerogel* is produced.

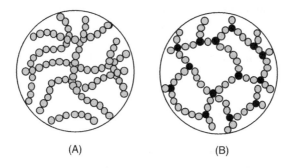

(A) (B)

Figure 5.22 Physically (A) and chemically (B) crosslinked hydrogels. The blank space is filled with water. Adapted from http://commons.wikimedia.org/wiki/File:Structures_of_macromolecules.png?uselang=fr. Last accessed 17/05/2012.

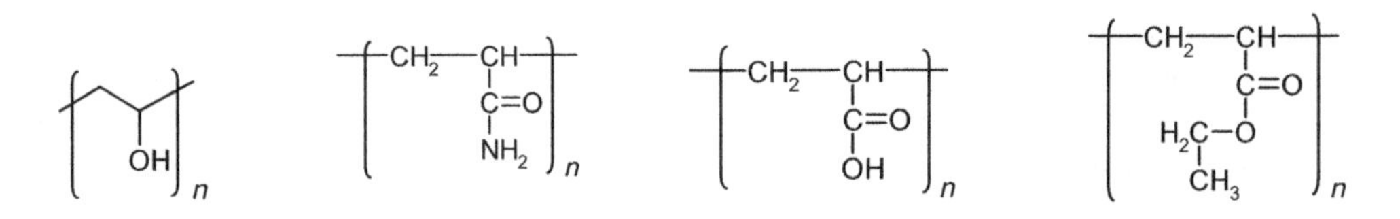

Polyvinyl alcohol (PVA)

Polyacrylamide

Polymethacrylic acid

Polyhydroxyethyl methacrylate (PHEMA)

Figure 5.23 Synthetic polymers for hydrogels.

5.8.1 Physically Crosslinked Hydrogels

As shown in Figure 5.22A, the stability of a physically crosslinked hydrogel is secured by noncovalent bonding such as hydrogen bonding, van der Waals forces and hydrophobic interactions. Such hydrogels can be obtained from natural polysaccharides (such as dextran or agarose). Certain natural polysaccharides include additional functionalities, such as the amine in chitosan, carboxyl in alginate and carboxyl and sulfonate in some carrageenans (Figure 5.7). Such groups impart particular properties such as an electric charge and can also be practical for conjugation with biocompounds. Protein materials, such as collagen and gelatin are alternative hydrogel materials. Physically crosslinked hydrogels can also be obtained from mixtures of components, such as PVA (Figure 5.23), carboxymethyl cellulose or sodium silicate.

A common procedure for obtaining physically crosslinked hydrogels consists of successive freezing and thawing of the precursor solution in water or a water–organic solvent mixture (freeze-thaw method). The product of this process is often termed a *cryogel*. During gel formation, growing ice crystals determine the shape and size of the pores that develop after defrosting of the sample. Biocompounds or cells are incorporated in the hydrogel if they are present in the system. Alternatively, the species of interest can be adsorbed onto a preformed gel surface or appended to reactive sites purposely included in the gel precursor molecule [52].

5.8.2 Chemically Crosslinked Hydrogels

Such hydrogels consist of networks of branched polymer molecules held together by covalent bonds (Figure 5.22B) and are obtained by polymerization in solution [53]. Typical polymers for such applications are summarized in Figure 5.23. Alternatively, hydrogels can be obtained from prepolymers crosslinked by addition of water (polyurethanes) or by ultraviolet irradiation. Incorporation of a biocomponent into the hydrogel can be effected by adding it to the reaction mixture.

Entrapment in hydrogels is a mild method for immobilization of biomacromolecules or cells. The gel is biocompatible and its matrix is sufficiently loose to allow for penetration of small analyte molecules such as enzyme substrates or oligonucleotides. This method is therefore widely used for the preparation of the sensing element in various types of chemical sensors.

The previous presentation addresses application of hydrogels as a passive matrix for the immobilization of a bioreceptor within the sensing part. However, certain functionalized hydrogels can perform as an active component in the sensing–transduction process. Such functions can be carried out by redox hydrogels, eletroconductive hydrogels or stimuli-responsive hydrogels.

5.8.3 Redox Hydrogels

Redox hydrogels include metal complexes of ruthenium or osmium appended to the main polymer chain [54,55]. As the metal ion is stable in two oxidation states, electron flow is possible by electron hopping from a reduced metal ion to the oxidized ion under the effect of an applied potential difference. Redox hydrogels are essential components of certain types of amperometric enzyme sensors as they allow direct electron transfer from a redox enzyme molecule to the electrode to be performed. Redox hydrogels have also led to excellent applications in amperometric biosensors based on redox enzyme labels, such as in immunosensors and in nucleic acid sensors. This topic is addressed in detail in Chapters 14 and 16.

5.8.4 Responsive Hydrogels

Certain hydrogels display the outstanding property of changing their volume dramatically under the effect of some physical or chemical stimulus, such as temperature or pH. Such materials are known as responsive hydrogels or *smart hydrogels* [56–59]. Thus, a response to pH change is secured by appending ionizable acidic groups to the polymer network. Ionization of this group results in a negative charge appearing within the gel. Within the pH range

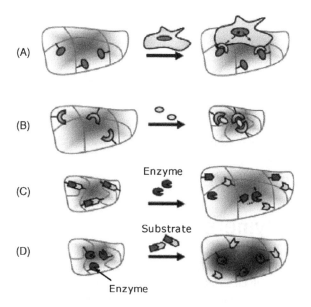

Figure 5.24 Swelling/collapse of a hydrogel under the effect of chemical stimuli. (A) Interaction of a multitopic target entity with gel-immobilized receptors; (B) bridging of receptors by interaction with the target entity; (C) enzyme detection by cleavage of chemical bonds in the immobilized substrate; (D) Substrate detection by enzymatic reaction with pending groups. Adapted with permission from [59]. Copyright 2007 Elsevier.

around the apparent pK_a of the acid the gel volume expands by electrostatic repulsion when the pH increases. The same effect is noted with basic group appended polymers, as the protonation of the base produces positive charges, for example, by conversion of a neutral $-NH_2$ group to $-NH_3^+$. Rigorously speaking, gel expansion is due to the rise in the osmotic pressure as a consequence of the accumulation of counterions within the ionized gel. As this process is reversible, the hydrogel responds to either an increase or a decrease in pH. This physicochemical effect has been exploited in developing pH sensors [60,61]. As many enzymatic reactions produce hydrogen ions and, consequently, bring about a local pH change, enzyme-containing responsive hydrogels are suitable for substrate determination [59,62].

Figure 5.24 summarizes several possible interactions between a responsive hydrogel and a target entity. Thus, living cells can adhere to specific receptors incorporated in the gel (A). Small molecules, ions and also antibodies or other proteins can trigger gel shrinkage by forming bridges between receptor sites within the gel (B). An enzyme that induces the cleavage of chemical bonds within a gel-included substrate causes gel swelling (C). This effect allows the detection of living cells using the effect of a secreted enzyme. A substrate can be sensed if a gel-immobilized enzyme induces reactions with reactive groups appended to the gel network (D).

The volume change can be reported by various methods [60,63,64]. If the gel is confined between a metal support and a metal bending plate, the volume variation causes a variation in the electrical capacity of the capacitor thus assembled. The deformation of a piezoresistive bending plate can be detected by the change in its electrical resistance. Other transduction methods are based on the variation in the electrical resistance of the gel itself in response to the volume change. The change in the gel volume is accompanied by a drastic modification of the refractive index, so, optical methods are particularly suited for transduction. The change in the gel thickness can also be reported by a piezoelectric oscillator. Clearly, an important advantage of smart gel-based sensors arises from the possibility of performing label-free transduction.

In conclusion, hydrogels are outstanding materials for immobilization of both biomacromolecules and living cells. In addition, certain hydrogels can perform as active components of the sensing element, either by providing electrical conductivity or by responding specifically to chemical stimuli. Integration of nanoparticles in hydrogels has led to new materials with very promising properties [65]. Biomolecule-responsive hydrogels appear to be particularly appealing for biosensor development [66].

Questions and Exercises (Section 5.8)

1 What is the structure of a hydrogel?
2 What kinds of material are suited for the preparation of hydrogels?
3 Explain why the calcium ion causes gelation of algal polysaccharides. What kind of interaction is involved in polymer crosslinking?
4 How can a chemically crosslinked hydrogel be obtained and what properties should the raw material feature?

5 How can receptors be incorporated into a hydrogel and what are the advantages of hydrogel immobilization?

6 What is the key difference between electroconductive hydrogels and redox hydrogels?

7 What structural details impart to a gel the ability to respond to a chemical stimulus? How does the gel react to such a stimulus?

8 Review several possible interactions of smart hydrogels with chemical and biological species.

9 How does the electrical resistance of a smart hydrogel vary in response to shrinking.

10 Design a sequence of reactions for grafting a carboxyl derivative to a chitosan gel network and for grafting an amine derivative to alginate.

11 Sketch possible routes for grafting proteins to polyacrylamide and polymethacrylic acid gels.

5.9 Conducting Polymers

Conducting organic polymers are compounds that consist of polyconjugated macromolecules. This particular chemical structure imparts these materials with electrical conductivity, in contrast to common polymers that are insulators. Conducting polymers have backbones of contiguous sp^2-hybridized carbon centers. One valence electron on each center resides in a p_z orbital, which is orthogonal to the other three sigma bonds. The electrons in these delocalized orbitals have high mobility when the material is doped by oxidation, which removes some of these delocalized electrons. Thus, the conjugated π-orbitals form a one-dimensional electronic band, and the electrons within this band become mobile when the band is partially emptied.

Electrochemical deposition is an advantageous method for the synthesis of conducting polymers. This is because it allows for thin films to be prepared at the surface of conducting materials, such as metals or carbon, for application as receptor entrapping matrix in chemical sensors. Alternatively, the synthesis can be conducted so as to obtain a material containing receptor molecules covalently bound to the polymer backbone. In addition, conducting polymers are used as sensing materials in gas sensors and as auxiliary materials in a series of electrochemical sensors. Comprehensive surveys of conducting polymer applications in chemical sensors can be found in refs. [55,67–69].

Typical conducting polymers are shown in Figure 5.25. Among the conducting polymers, polypyrrole (Figure 5.26A) is the most commonly used [70]; therefore, the subsequent discussion will focus on this compound.

The simplest method of inducing electrochemical polymerization consists of applying a sufficiently positive constant potential to an electrode immersed in an aqueous pyrrole solution. In its first stage, pyrrole oxidation results in radical cations that couple with each other and produce a polymer film. If the electrode potential exceeds a certain limit, the product is an overoxidized polymer.

Clearly, the support should be a conductor material (graphite, platinum) which should be stable under the electropolymerization conditions. Formation of oxides on metal electrodes is detrimental. As the polypyrrole chains bear a positive charge, anions are spontaneously integrated in the network in order to secure electrical neutrality.

According to the idealized structure (Figure 5.26A), the polypyrrole macromolecule features an extended π orbital system. In other words, the π electrons are delocalized over the entire macromolecule. Actually, due to the oxidative conditions in the synthesis, many π orbitals are not completely filled and such a partially filled orbital includes a positive vacancy. However, vacancies are delocalized as well and each unit in the chain assumes formally a fractional charge, $\delta+$ ($\delta < 1$). The electric charge on the polymer backbone is balanced by counterions A^- inserted within the polymer network. The fraction δ is called the *doping level* and the counterion is a *dopant*. A dopant can be a small inorganic ion or a large polyelectrolyte molecule, that is, a polymer with pendant ionic groups. Due to the charge balance, the doping level also reflects the amount of included counterion. It is worth noting that the type of dopant affects appreciably the properties of the polymer.

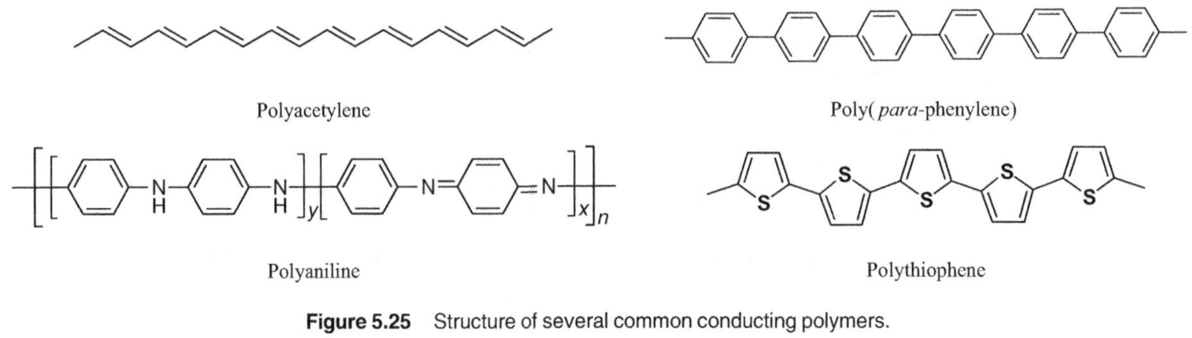

Polyacetylene

Poly(*para*-phenylene)

Polyaniline

Polythiophene

Figure 5.25 Structure of several common conducting polymers.

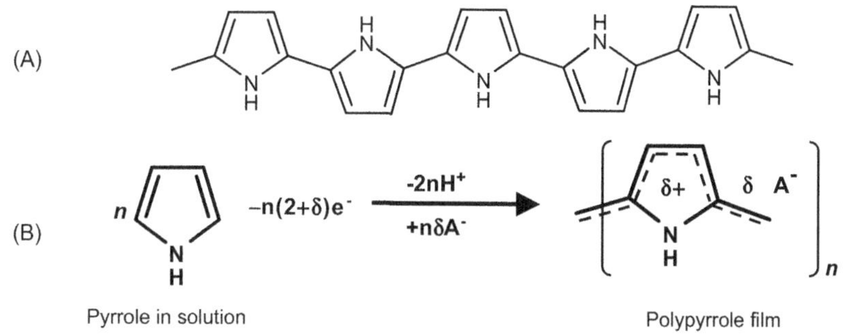

Figure 5.26 (A) Idealized structure of polypyrrole; (B) electrochemical polymerization of polypyrrole. $\delta+$ is the average fractional charge at the repeating unit of the polymer.

Once prepared, polypyrrole can be subjected to subsequent oxidation or reduction. Such transformations are accompanied by modifications in dopant concentration and consequent modifications of the physical properties. For example, overoxidized polypyrrole is less conductive and more porous.

The most striking property of polypyrrole is its electrical conductivity. From this standpoint, polypyrrole behaves as a semiconductor. Under the effect of an electric field, electrons jump from one molecule to the next one, this process being accompanied by dopant displacement in the opposite direction. The mobility of the dopant has a great effect on the conductivity of the polymer.

Pyrrole can be readily functionalized at the N atom and substituted pyrrole will also undergo electrochemical polymerization, but the arrangement of polymer chains is influenced by the steric effect induced by the substituent.

Taking into account the advantageous properties as well as the easy synthesis of polypyrrole, polypyrrole applications in chemical-sensor design encompasses a very large area [69,71]. Polypyrrole conductivity is sensitive to gas absorption, which is the basis of the application of polypyrrole in gas sensors [72].

Polypyrrole is widely employed as an entrapment matrix for biological receptors such as proteins and nucleic acids [73]. Physical entrapment is achieved by conducting the synthesis in a solution containing the pertinent receptor. Clearly, this is possible if the receptor molecule bears a negative charge that interacts electrostatically with the positive charge at the polymer backbone. The loose structure of the polymer film permits free diffusion of small analyte molecules and imparts some degree of size-exclusion selectivity.

More control over the structure of the sensing film can be achieved by covalent grafting onto the polymer backbone Thus, pyrrole grafted with a receptor can be copolymerized with crude pyrrole to form a receptor-functionalized film. If the polymerization conditions are not compatible with the receptor, one can resort to another method, namely post-polymerization grafting onto a preformed polymer films. In this approach, the film is prepared by copolymerization of pyrrole and a N-substituted pyrrole monomer. Receptors can then be covalently linked to the polymer through the reactive group at the second monomer unit. Co-polymerization of pyrrole with biotin-functionalized pyrrole allows for subsequent grafting of biotinylated receptors via avidin–biotin linkages. Besides imparting stability to the film, the avidin layer also prevents nonspecific adsorption of sample components.

The stability of a polypyrrole sensing film can be hampered by the hydrophobic character of the backbone that is not compatible with hydrophilic receptor molecules. This problem can be alleviated by synthesizing amphiphilic polymer matrices [74]. Such materials are obtained by the polymerization of monomers consisting of a hydrophilic fragment inserted between two terminal pyrrole residues.

It is also important to point out the use of polypyrrole in fabricating sensor microarrays by electropolymerization. The support in this case is an electrode array. If the electrode at each site is individually connected to the voltage source, particular sensing properties can be imparted to each component of the array by local electrochemical polymerization under specific conditions.

Composites of conducting polymers with various species such as nanoparticles, and organic and bio-organic materials provide various opportunities for chemical-sensor applications [75].

Polypyrrole is extensively employed as the immobilization matrix for various receptors such as enzymes [76–78], nucleic acids [79], and affinity receptors [80,81].

Other conducting polymers are shown in Figure 5.25. Polyacetylene and poly(*para*-phenylene) are hydrocarbon compounds, whereas poly(aniline) includes secondary amine bridges that can be protonated to secondary ammonium groups and this makes this material pH-responsive. Polythiophene, like polypyrrole, consists of chains of five-membered heterocycles.

It is worth noting that *insulator polymers* (such as polyphenol) can also be prepared by electrochemical synthesis [82]. Overoxidized polypyrrole is also a nonconducting polymer. Such polymers are useful as entrapment matrices for enzymes. Due to their insulator character, growing of insulator polymer films is limited to 10–100 nm thickness. Hence, substrate and product diffusion within the film is very fast and allows for a even concentration distribution within it. Electrochemically polymerized dielectric polymers are also employed as insulator layers over transducer surfaces.

Electropolymerization is achievable only at conducting solid surfaces. Plasma polymerization and photochemical polymerization are methods of choice for obtaining polymer coatings on insulator surfaces.

Semiconductor polymers addressed in this section are currently termed simply conducting polymers. This name is assigned also to redox hydrogels. However, the mechanisms of electrical conduction in these two classes of material are fundamentally different.

Questions and Exercises (Section 5.9)

1 What are the essential properties of semiconductor polymers?
2 How can biocompounds be integrated with polypyrrole?
3 Discuss the common features of polypyrrole and standard redox hydrogels.
4 Write a sequence of reactions for assembling proteins at a gold surface using a properly derivatized pyrrole as starting material.

5.10 Encapsulation

Encapsulation is an immobilization method that consists of enclosing the receptor solution inside a small vesicle (Figure 5.27). The surrounding membrane should be permeable to small molecules (substrate and product) but not to the receptor itself. The capsule size can vary from a few μm to a few hundred μm. This method is suitable for the immobilization of sensitive enzymes and whole cells as the native structure of the entrapped species is well preserved. If the membrane remains still permeable to the enzyme one can resort to crosslinking in order to prevent enzyme leakage (Figure 5.27). Crosslinked enzyme aggregates are much larger than individual molecules and are not susceptible to leakage. Crosslinking excludes the necessity to control rigorously the size of membrane pores. Multiple enzyme systems can also be obtained by encapsulation.

Polymer encapsulation can be achieved by interfacial polymerization. This process is conducted in emulsions formed by mixing an aqueous solution of enzyme with a monomer solution in a nonaqueous solvent. In this way, dispersed aqueous droplets form and the polymerization occurs only at the droplet surface producing vesicles that include enzyme solution. Alternatively, an emulsion can be formed by mixing an aqueous solution of monomer and enzyme with another monomer dissolved in a nonaqueous solvent. In this case, the membrane forms by copolymerization. Encapsulation can also be performed by coating liposome-containing enzymes with a polymer layer.

Other encapsulation methods are based on layer-by-layer technology, sol-gel reactions and hydrogel formation [83].

Questions and Exercises (Section 5.10)

1 Draw a flow chart for the process of enzyme encapsulation in polymer vesicles.
2 How can vesicles with uniform size be prepared?
3 Assume that an encapsulated enzyme can leak out through the vesicle pores. Draft several methods for preventing this leakage, and sketch the relevant chemical reactions.

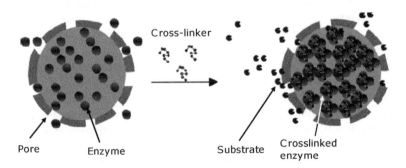

Figure 5.27 Schematics of enzyme encapsulation and postencapsulation crosslinking. Left: enzyme molecule escape through the pores. Right: bulky enzyme aggregates are tightly confined within the vesicle. Adapted with permission from [2]. Copyright 2005 Wiley-VCH Verlag GmBH & Co. KGaA.

5.11 Entrapment in Mesoporous Materials

Mesoporous materials are characterized by a regular spatial distribution of pores of similar diameter [84–86]. According to the IUPAC definition, mesoporous materials are porous materials with pore diameters in the range 2–50 nm, which is rather close to the size of proteins. For comparison, microporous materials (such as zeolites) have pore diameters of less than 2 nm, while macroporous materials have pore diameters greater than 50 nm.

Mesoporous silica particles are synthesized by reacting tetraethyl orthosilicate with a template made of micellar rods or another material (Figure 5.28). By reactions similar to that used in the sol-gel method a collection of nano-sized spheres or rods is produced that are filled with a regular arrangement of pores. The template can then be removed by a suitable chemical or physical treatment. Periodic mesoporous organosilicates are obtained when organic fragments are included as bridging groups. This modification provides numerous possibilities for tuning chemical and physical properties of the material by varying the structure of the precursors.

Functionalization of mesoporous materials can be achieving during the synthesis by using precursors substituted with selected functional groups. It is also possible to perform postsynthesis functionalization by grafting other groups onto the material network.

In addition to silica, many other materials can be obtained in the mesoporous form. Thus, nanoporous carbon can be obtained by conversion of sucrose to carbon inside the silica mesopores through mild carbonization. Mesoporous metal oxides and phosphates, as well as mesoporous ceramics, metals and semiconductors have also been obtained.

As suggestively stated in ref. [85], mesoporous materials are "holes filled with opportunities". Among these opportunities, application of such materials in biosensors is of interest, particularly for enzyme immobilization by entrapment [87,88]. Enzyme entrapment can be achieved simply by incubation of the mesoporous material with an enzyme solution. In order to prevent the enzyme from leaking, crosslinking with glutaraldehyde or capping of the pores are recommended. Compatibility of the enzyme with the inner pore surface is a key issue. Thus, depending on the protein characteristics, either a hydrophilic (silica) or hydrophobic (carbon) material should be selected as the support. The electrical conductivity of carbon makes it the material of choice for amperometric enzyme sensors.

As an example, Figure 5.29 shows the configuration of the sensing part of an amperometric glucose sensor based on the detection of hydrogen peroxide. The enzyme is entrapped in ordered mesoporous carbon that displays a hierarchical distribution of pore diameters. Larger mesopores (37 nm diameter) accommodate enzyme molecules, whereas the smaller pores (5.6 nm) secure the diffusion of substrate to the catalytic sites.

Mesoporous materials are also suitable for entrapping mixed assemblies including enzymes and nanoparticles. It is also interesting to note that certain mesoporous materials allow well-organized assemblies of chemical functionalities that elicit enzyme-like catalytic activity to be obtained. Such biomimetic materials function as artificial enzymes.

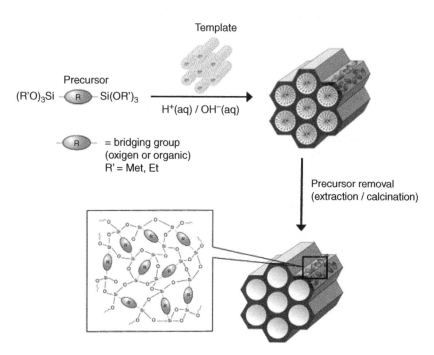

Figure 5.28 Schematics of mesoporous silica synthesis. Adapted with permission from [86]. Copyright 2006 Wiley-VCH Verlag GmBH & Co. KGaA.

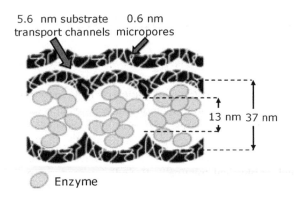

5.6 nm substrate   0.6 nm
transport channels micropores

13 nm 37 nm

Enzyme

Figure 5.29   Immobilization of glucose oxidase in ordered mesoporous carbon for application in an amperometric glucose sensor. Adapted with permission from [89]. Copyright 2009 Wiley-VCH Verlag GmBH & Co. KGaA.

Questions and Exercises (Section 5.11)

1 What are the specific properties of mesoporous materials?
2 What are the processes involved in fabrication of (a) silica and (b) carbon mesoporous structures?
3 How can enzymes be entrapped in mesoporous materials, and how can enzyme leakage be prevented?

5.12 Polymer Membranes

Polymer membranes can perform various functions in chemical sensors [90–92]. Thus, polymer membranes can be employed as matrices for physical entrapment of enzymes. In addition, some membranes feature selective permeability and are therefore useful to prevent interference from certain sample components. A perm-selective membranes can be used simply to contain an enzyme solution as a thin a layer at the transducer surface. Moreover, a polymer membrane can act as a diffusion barriers in enzymatic sensors. This membrane control the rate of substrate diffusion to the immobilized enzyme and allows obtaining a linear response function.

Applications of polymer supports and polymer hydrogels have been discussed before. This section addresses mainly compact polymers that are relevant to chemical sensors. It introduces first several methods for obtaining polymer films and then emphasizes typical applications of polymer membranes in sensor development.

5.12.1 Deposition of Polymers onto Solid Surfaces

Preformed polymers can be obtained in a well-characterized form before they are applied to the underlying surface. Deposition of preformed polymers is carried out using polymer solutions that are cast onto solid surfaces and leave a polymer film after the evaporation of the solvent. The surface should be selected so as to be wetted by the solution. If a hydrophobic receptor is added to the polymer solution, it will be entrapped in the polymer film. This immobilization method is widely used in ion sensors based on the use of macrocyclic compounds or liquid ion exchangers as receptors.

Dip coating is achieved by immersing the support in the polymer solution followed by solvent evaporation. The coverage depends on the immersion time and the polymer concentration. These parameters afford to control of the amount of deposited material.

In the *drop coating* technique the polymer solutions is applied as several microliter droplets onto the support surface. Spreading the solution out and evaporation results in a thin film. This method is limited to surface areas below 1 cm² and a uniform coverage cannot be guaranteed. However, it is a very simple technique and permits to control the amount of deposited polymer through the concentration and the volume of the polymer solution.

Spin coating relies on the deposition of polymer solution onto the support while the support is spun at a high speed. Spinning ensures an uniform spreading of the polymer solution that results in an even film thickness. The thickness of the film depends on the rotation speed and the viscosity of the polymer solution. Films as thick as 1 μm can be obtained in this way.

When preparing coated polymer films it is essential to obtain reproducible and compact, pinhole-free layers. This depends on the film thickness and the roughness of the underlying surface. After deposition, the film can be strengthened by crosslinking based on suitable chemical reactions.

A greater degree of control over the film characteristics at the molecular level is achieved by resorting to reactive deposition. Polymer layers covalently grafted onto the support can be obtained using precursors appended through reactive terminal groups that react with the support surface. Thus, triethoxysilanes serves for grafting polymer layers to oxide surfaces.

A very practical method is based on photochemical polymerization that allows the growing of polymer layers at predefined sites. In this approach, photosensitive precursors are first cast onto the surface and then crosslinked by irradiation through a shadow mask. Polymerization only occurs within the irradiated zones. By selective deposition of different polymers at preselected sites, this method allows to produce sensor arrays.

5.12.2 Perm-Selective Membranes

Certain polymer membranes provide selective permeability for some species and are used as selectivity-imparting coatings onto the sensing element in certain sensors. Selectivity can be imparted by the physical size of membrane pores, the solubility of particular compounds in the membrane or specific interactions between the polymer and the diffusing species.

Gas-permeable membranes are permeable to certain gases and are employed to contain an aqueous solution at the sensor surface or as selective coatings onto a solid-state sensing element. The membrane can be either compact or porous. In the case of compact membranes (for example, silicon rubber), gas diffusion through the membrane involves dissolved gas molecules. In the case of porous membranes (for example, Teflon), the gas diffuses through hydrophobic pores that are not permeable to water.

Size-exclusion perm-selective membranes feature micropores that allow the transport of only molecules smaller than the pore size. A typical material of this type is Nucleopore, which is available as about 10-μm thick polycarbonate membranes with uniform pore diameter that can range from 0.01 to 30 μm. Small molecules (for example, an enzyme substrate or a coreagent) can cross this membrane to reach the enzyme layer, whereas large proteins cannot. A typical application of such membranes is in the coating of the sensing element of certain glucose amperometric sensors. The membrane is permeable to glucose and oxygen but prevents the enzyme from leaking away.

A frequently used membrane is based on cellulose acetate (that is, the acetate ester of cellulose) which is available as dialysis membranes and hollow fibers. Cellulose acetate membranes are prepared by solvent casting from a solution of cellulose acetate in acetone onto a flat surface. The membrane is not permeable to anions and large molecules such as proteins. Such a membrane is used as a coating over the electrode surface in amperometric oxygen sensors and also as internal membrane in oxygen-linked enzyme sensors. However, hydrophobic polycarbonate membranes perform better as they have better permeability to oxygen.

Ion-exchanger membranes are made of polymers featuring ionic groups attached to the backbone and display selective permeability to ions bearing an opposite charge. Typical examples in this class are Nafion and poly(vinyl-pyridine) (Figure 5.30).

Nafion is a sulfonated perfluorinated polyether featuring anionic $-SO_3^-$ groups. Nafion can be dissolved and cast to form supported membranes. A particular characteristic of this material is the internal phase segregation. This polymer has a high density of hydrophilic sites ($-SO_3^-$ groups), long fragments of hydrophobic chains and is not cross-linked, which imparts to the backbone certain flexibility. In contact with aqueous solutions, the anionic sites separate themselves from the hydrophobic segments forming water-filled pockets and channels. Due to the anionic sites, Nafion behaves as a cation exchanger. Applications of Nafion membranes rest on their selective ionic conductivity. Indeed, under the effect of an electric potential difference, Nafion membrane allow cations to cross it through the water-filled pockets but it is not permeable to anions. This property is exploited in amperometric sensors in which Nafion coating over the sensing part prevents access of certain anionic species that might interfere with sensor functioning.

Anion perm-selective membranes can be prepared from the cation-exchanger poly(pyridine) (Figure 5.30B). Poly-pyrrole, with its positively charged backbone is also selectively permeable to anions.

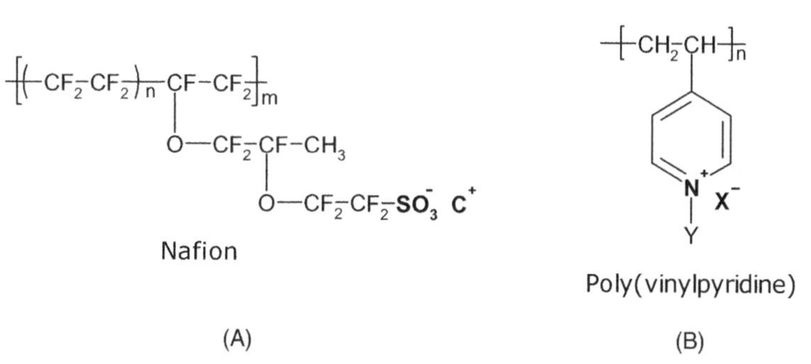

Figure 5.30 Chemical structure of two ion exchanger polymers: Nafion (A) and poly(vinylpyridine) (B). C^+ and X^- are countercations and counteranions, respectively.

Ion selective membranes can be obtained by incorporating ion exchangers in a poly(vinyl chloride) (PVC) matrix. Thin PVC membranes are prepared by solvent evaporation of PVC solutions on a flat surface. A lipophilic plasticizers, such as bis(2-ethylhexyl)phthalate (also called dioctyl phthalate) or isopropyl myristate, should be incorporated in the PVC solution. The membrane thus obtained occludes the plasticizer in anionic form and is permeable to cations.

Plasticized PVC membranes are widely used as matrices for immobilization of ion receptors in various types of ion sensors, the selectivity arising from the receptor characteristics. Such membranes are obtained by dissolving the receptors along with PVC and a plasticizer in tetrahydrofuran, casting this solution on a flat glass surface and letting the solvent evaporate slowly. Unplasticized PVC membranes that are also lipophilic in nature, elicit selective permeability to certain small organic ions (for example, urate, lactate, ascorbate and oxalate) probably due to residual void interstices in the polymer matrix. Details on ion selective polymer membranes can be found in Chapter 10.

Questions and Exercises (Section 5.12)

1 What is the role of a polymer membrane in a chemical sensor?
2 How can a polymer membrane be fabricated?
3 What kinds of membrane prevent enzyme leakage from the sensing layer while allowing the access of the substrate molecule? What sensor characteristics are improved by a perm-selective membranes?
4 Suppose that you want to prevent the access of anions to the sensing part of an enzyme sensor by means of a membrane. What material is suitable to prepare the membrane and what is the mechanism of selective permeability?
5 What materials are suitable for obtaining membranes that separate an enzyme layer from an electrode at which oxygen reduction is monitored?

5.13 Microfabrication Methods in Chemical-Sensor Technology

Previous sections pointed to the basic physicochemical methods for the integration of a sensing element with a solid support, which is, in many instances, a component of the transducer. On the laboratory scale, manual fabrication is a common practice. However, mass production of chemical sensors and biosensors needs a high throughput and more reliable fabrication methods. In addition, miniaturization and multiplexing of sensors require methods for accurate deposition of receptors as well-defined patterns at predetermined sites on solid surfaces. Basic methods of this kind are introduced in this section. These methods rely on technologies borrowed from the microelectronics industry, such as the structuring of the surface in the form of thick films ($>5\,\mu m$) or thin films ($<5\,\mu m$). Such methods are suitable for creating complex patterns involving one or more receptors and are very suitable for the fabrication of sensor microarray [93,94].

5.13.1 Spot Arraying

Biocompound spot arrays can be formed by the deposition of small solution volumes either manually, using micropipettes or by means of a microdispenser. Commercial printer ink-jet dispensers or commercial instruments used in the fabrication of DNA array can been used to this end. However, the high temperature and shearing stress involved with such devices can be detrimental to proteins. Electrospray devices have also been used to produce enzyme spots of $100-300\,\mu m$ diameter. However, as this method imparts to the molecule a large electric charge, it may induce enzyme denaturation.

A purposely designed microdispenser for enzyme solutions is shown in Figure 5.31. In this device, the solution is continuously fed in and droplets of about 100 pL are delivered under the impulse from a piezoelectric actuator that vibrates at a high frequency. Enzymes can be attached to the support surface either by sulfur chemisorption, cross-linking or entrapment in a polymer film. In the last two methods, the solution contains immobilization reagents in addition to the enzyme. This general method allows for deposition of mono- or multienzyme structures in the form of mono- or multimolecular layers as about 100-μm-sized patterns [94,95].

5.13.2 Thick-Film Technology

Thick-film technology is based on the principle of *screen printing* that is a general method for the deposition of patterned coatings onto flat surfaces [96]. Screen-printing is performed by coating a paste onto a flat support through openings in a screen (Figure 5.32). The paste is pressed onto the screen and the support by a squeegee and the screen

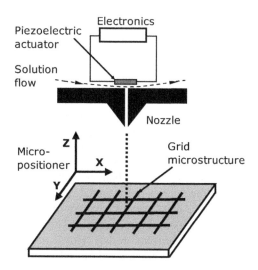

Figure 5.31 Building a biocompound grid microstructure by means of a flow-through microdispenser.

determines the pattern of the deposited layer on the support. The success of this operation depends on the viscosity of the paste and also on its thixotropy, a term that defines the capacity of a thick paste or gel to turn into a fluid form under mechanical stress. The paste is a mixture of the sensor materials to be applied, an organic additive (that regulates the thixotropy) and a solvent. After printing, the layer is left to dry and then is subjected to thermal treatment. During this stage, the organic carrier is burned off, and other chemical and physical processes impart adhesion. The paste composition is selected so as to undergo the expected transformation during the heat treatment. A standard mixture consists of silicon, lead, and bismuth oxides that form a dielectric glass at elevated temperatures. A conductive paste can be obtained by including copper oxide that yields copper by reduction. The firing-resistant support is made of ceramic flexible tape. When no high-temperature treatment is required, the support can by a plastic material (such as a polyester).

Typically, the screen is formed from a finely woven mesh of stainless steel or synthetic polymer mounted on a metal frame. The mesh is coated with an ultraviolet-sensitive emulsion, for example, polyvinyl acetate sensitized with a dichromate solution, which allows the required patterns to be produced by photolysis.

Screen printing allows the fabrication of 5- to 50-μm thick layers with a resolution of about 100 μm. The width of the sensors thus fabricated can vary from a few mm down to 0.5 mm, which allows the installation of such a sensor in a catheter for *in vivo* application. Different pastes are commercially available, including, conductive, resistive, dielectric and overglaze pastes. Custom-made pastes are designed for particular applications.

By sequential printing, various patterns with specific properties can be formed on the same support using different masks and different paste compositions. Screen printing can also be used to deposit a polymer film over selected areas of the sensor to function as an immobilization support, as a perm-selective membrane or as an insulator.

This technology is widely used in the fabrication of screen-printed electrodes for use in electrochemical sensors. The paste or ink in this case contains graphite powder and the resulting layer, which is electroconductive, functions as an electrode.

Enzymes can be attached to preformed screen-printed layers by common methods such as adsorption, crosslinking, or entrapment. Alternatively, a second layer can be formed by printing a paste containing an enzyme and, if necessary, a cofactor.

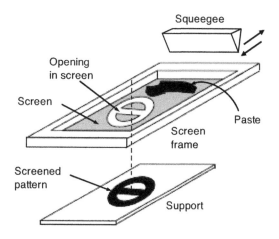

Figure 5.32 The screen-printing process. Adapted with permission from [97] Copyright 1973 John Wiley & Sons, Ltd.

5.13.3 Thin-Film Techniques

Microfabrication is the term used to denote methods for the fabrication of miniature patterns of micrometer size and smaller. This class of technologies is well established in the integrated circuit industry [98] and is also a valuable method in the fabrication of chemical sensors [99,100]. Microfabrication yields miniature devices and also assemblies of microsensors using readily available equipment. Moreover, electronic circuitry can be integrated with the sensor in order to provide an electronic interface for signal processing. The basic material in microfabrication is silicon; inexpensive silicon wafers (some 250 μm thick) are widely available. A comprehensive survey of microfabrication applications in the field of chemical sensors and biosensors is available in ref. [101].

Microfabrication encompasses a series of unit operations, the most important being patterning, deposition or growth of films, and etching. The thickness of the structure obtained by microfabrication is in the μm range.

Photolithography is a technology for creating micropatterns on flat surfaces (Figure 5.33A). In this technique, light irradiation of a photosensitive coating through a mask creates a pattern of exposed and covered domains that can be further processed so as to create functional units at the microscale. The process begins with spin coating of the support surface with a photosensitive polymer layer (photoresist). This is next irradiated through a photomask with transparent patterns identical to those that are to be transposed onto the support. With a positive photoresist, the irradiated zones are altered and can be dissolved away in the next stage called developing. The pattern on the surface is similar to that on the mask, but about 10 times smaller. In the case of a negative photoresist, photochemical reactions reinforce the exposed material while the shadowed areas remain susceptible to being dissolved away. Clearly, the pattern on the wafer surface is in this case the reverse of that on the mask.

Photomasks are typically transparent fused silica blanks coated with an opaque chromium layer. Chromium is etched by a laser or electron beam under computer control leaving a clear path for the light to travel though.

After irradiation and development, exposed areas on the surface are processed by etching or film deposition in order to obtain the expected configuration. Finally, the remaining photoresist is removed by attack with a liquid reagent or by ashing with oxygen-containing plasma.

In etching processes, the uppermost layer of the substrate in the areas not protected by photoresist is removed. Plasma or reagent solutions can be used to this end. Plasma etching can be made anisotropic, that is, asymmetric patterns can be etched in this way. Wet etching processes that rely on attack by chemical reagents are generally isotropic in nature, that is, give rise to symmetrical patterns.

Various layers (0.2 to 200 μm thick) can be grown on exposed areas by sputtering, physical vapor deposition or chemical vapor deposition. A metal layer can be formed by sputtering a metal target in vacuum with an argon ion beam. This beam conveys metal atoms to the surface.

In a typical chemical vapor deposition process, the substrate is exposed to one or more volatile precursors that react or decompose at high temperatures on the substrate surface to produce the expected deposit. An advanced version of chemical vapor deposition relies on chemical reactions taking place in a plasma, a highly ionized gas that is produced by electrical discharge or other methods (Figure 5.33B). In plasma, chemical reaction are very fast and this allows the deposition to be performed at lower temperatures, which is often beneficial in the manufacture of semiconductor devices.

After completing the patterning process, a fraction of silicon wafer surface still remains uncoated. The uncoated surface is next passivated by growing films of silicon dioxide or silicon nitride (Si_3N_4) on them. Silicon dioxide is formed by surface reaction in a damp oxygen atmosphere at a high temperature, or by various chemical vapor deposition processes. Better passivation is achieved with silicon nitride layers formed by the gas-phase reaction of dichlorosilane (SiH_2Cl_2) with ammonia.

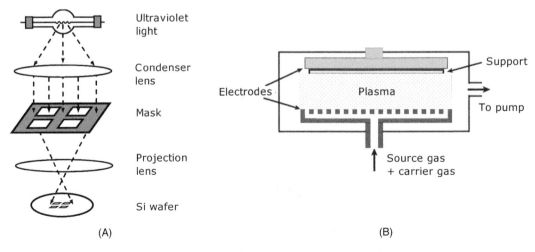

Figure 5.33 Methods of thin-film technology. (A) Photolithography; (B) plasma-enhanced chemical vapor deposition.

Thin-film methods are used in sensor technology to obtain patterned support surfaces to which the receptor layer is then attached. Gold spot arrays can be prepared by vapor deposition in order to function as support for receptor immobilization by thiol chemisorption. If no site-specific immobilization method is available, one can resort to photolithography to form an array of exposed support spots. Then, the immobilization is performed over the whole area. In this way, the receptor is thus applied on the photoresist-coated domains as well as on the exposed support spots. After removing the photoresist (along with the material deposited on it), only the uncoated spots preserve the sensing layer. This method should be carried out carefully in order to avoid degradation of the receptor by the solvent used to remove the photoresist. This drawback is absent in mask-less photolithography. In this approach, the support is coated with a photolabile layer containing a protein linker, such as biotin. This layer is patterned by irradiation through a mask and then removed from the irradiated zones. Subsequently, the receptor (for example, an avidin-tagged species) will be appended only to the spots that were shaded during the irradiation step.

The resolution of the photolithographic immobilization methods seems to be near 2 μm. Much better resolution can be achieved by soft lithography, which is presented in the next section.

5.13.4 Soft Lithography

In soft lithography a soft stamp (mold) is used to apply a patterned layer to the surface of a solid support [102,103]. As shown in Figure 5.34, a master is first fabricated on a photoresist-covered support by standard photolithography. The polymer stamp is then obtained by pouring a degassed polymer over the master. The stamp materials are elastic polymers, the most common one used being poly(dimethylsiloxane) (PDMS), a silicone-based elastomer. The polymer stamp is cured and then detached from the master. Once the stamp is fabricated, its surface is inked by contact with a thioalkane solution in ethanol. The thioalkane diffuses into the stamp and when the stamp is applied on the surface of a gold support, a thiol self-assembled monolayer (SAM) forms only in the regions of direct contact. The amount of thiol absorbed in the stamp is sufficient for multiple prints. This method is called *microcontact printing*.

Self-assembly of derivatized thiols impart to the surface various properties, depending on the end group of the thiol. Thus, a plain thioalkane renders the surface hydrophobic and chemically inert. A hydrophilic end group at the thiol end imparts hydrophilicity, whereas reactive end groups allow for crosslinking of compounds of interest. So, either adsorptive or covalent immobilization can be performed on SAMs formed by microcontact printing.

More rugged stamps can be obtained using the PDMS stamp as a master to mold a harder polymer stamp based on oligomers or prepolymers that do not penetrate into PDMS. This technology, which is referred to as replica molding, can produce numerous mold replicas using inexpensive materials that are suitable for mass production.

5.13.5 Microcontact Printing of Biocompounds

Based on the standard soft lithography, various methods for micro- and nanopatterning of surfaces with biocompounds have been developed [104–106].

The SAM microcontact printing technology (Figure 5.34) can be easily transposed to protein printing directly on the support surface by inking the stamp with a protein solution. During the printing process, proteins are transferred

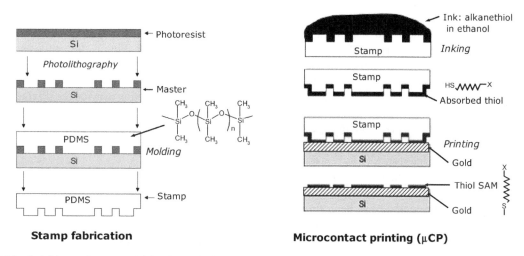

Stamp fabrication   **Microcontact printing (μCP)**

Figure 5.34 Soft lithography: stamp fabrication and microcontact printing. Adapted from http://commons.wikimedia.org/wiki/File: Creating_the_PDMS_Master.JPG (A) and http://commons.wikimedia.org/wiki/File:Inking_and_Contact_Process.JPG (B).

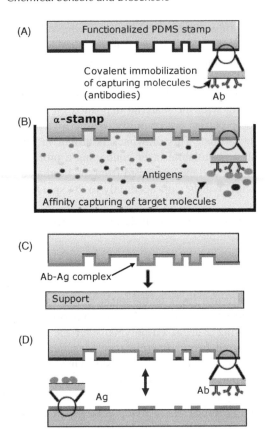

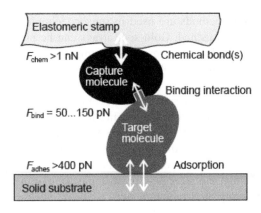

Figure 5.35   Affinity contact printing (α-CP). Left: main steps: (A) Covalent linking of capturing molecules on a functionalized stamp; (B) Inking of the α-stamp by affinity bonding of target molecules; (C) Printing on a solid support; (D) stamp removal. Right: Forces involved in the microcontact printing of proteins. Reprinted by permission from Macmillan Publishers Ltd: Nature Biotechnology [107]. Copyright 2001.

to the support surface by adsorption. Thiol-containing biomolecules undergo chemisorption at a gold surface via the sulfur group and protein monolayers can be formed in this way. The main limitation of this method is that the stamp should be inked after each application as the amount of adsorbed protein is limited.

The above problem can be circumvented by using hydrogel stamps swollen with a solution of protein. Such a stamp consists of two parts: a reservoir filled with protein dissolved in a biological buffer and a hydrogel that makes contact with the support and transfers proteins by diffusion.

An advanced method for biopatterning termed *affinity-contact printing* (α-CP) relies on selective inking with proteins appended to the stamp by affinity interactions (Figure 5.35 left). In this method, the stamp surface is activated by treatment with oxygen plasma. This forms a very thin silica-like layer that includes silanol groups ($-Si-OH$). Trialkoxysilane-bearing reactive groups are then crosslinked with the silanol groups and finally, capture molecules (e.g., an antibody) is covalently grafted to the surface. The result of this process is a so-called α-stamp. This stamp is then put in contact with a solution containing a mixture of proteins. Affinity interactions occur at the surface leading to the selective capture of target proteins, for example, an antigen. Finally, the antigens are transferred by contact printing to a solid support.

Figure 5.35 (right) shows the strengths of the binding interactions involved in the printing process as determined by atomic force microscopy. It is clear that multiple interactions involved in adsorption provide the driving force for protein transfer from the stamp to the support surface.

The strength of this method arises from the selectivity of the inking process. Only molecules in the ink that fit the capturing molecule are transferred to the stamp. This allows crude protein preparations to be used with no prior separation of the protein of interest.

Various affinity interactions (including avidin–biotin linking) are suitable for inking the stamp. This allows the printing of various proteins (including enzymes) and also nucleic acids and living cells [105,107].

The α-stamp fabrication method has been further refined in order to enhance the throughput and resolution (<1 μm) of the biomolecule microcontact printing process [108].

Microprinting of biocompounds by soft lithography has also been developed so as to allow printing of multiple receptor spots on the same support [109]. This is therefore a valuable technology for fabrication of protein and nucleic acid microarrays.

Questions and Exercises (Section 5.13)

1 Review briefly the methods for fabricating biocompound patterns on solid surfaces.
2 How can a biocompound solution be sputtered in a controlled way onto a predetermined surface area? Give a detailed account of methods for coupling the biocompound to a variety of support surfaces.
3 Draw a flow chart of the thick-film fabrication process. Comment on the characteristics of the materials and devices involved in this process.
4 Suppose that you have to fabricate a gold micropattern on a silicon surface. Draft a flow chart for this process and review the characteristics of the equipment used.
5 Point out the main steps in the process of micropatterning two different biocompounds, each of them at a specific site on a solid surface using an intermediate gold layer. Do the same for the case in which the patterning should be carried out at an oxidized silicon surface. Sketch the relevant reactions.
6 How can a biocompound be micropatterned directly onto a flat solid support? Review the main steps, devices and materials used to this end.
7 Draw a scheme of the device used for microcontact printing with a hydrogel stamp. Design a method for fabrication of such a stamp.
8 What are the key materials in affinity contact printing? Draft the flow chart of this process.

5.14 Concluding Remarks

A great variety of methods for the fabrication of the sensing element are available. Selection of a particular method depends on a series of factors. First, the fabrication method should be compatible with the receptor and preserve its activity for as long a time as possible. Testing of receptor stability is a key step in developing a chemical sensor, particularly when using sensitive biological materials such as enzymes. Methods for the characterization of immobilized enzymes are well established [110] and most of the standard approaches in this field also apply to enzymatic sensors.

Another important factor is the receptor loading in the sensing layer. Enzyme sensors require, as a rule, a high loading, whereas immunosensors are not particularly demanding as far as this feature is concerned. In general the immobilization should be performed such as to allow the access of the analyte to the binding sites.

The immobilization method should be compatible with the transduction method, which often imposes the restriction of using some particular material as the support for immobilization.

The long-term stability of the sensor depends not only on the stability of the receptor compound but also on the survival of the sensing layer overall. Greater long-time stability is achieved generally at the cost of using more intricate fabrication technology and more expensive reagents. Often, it is preferable to resort to simple and inexpensive methods to fabricate disposable sensors. In this case, the batch reproducibility of the sensor response is the essential issue.

Mass fabrication issue is essential when aiming at the commercialization of sensors. To this end, thick- or thin-film technologies are the methods of choice.

Various immobilization and fabrication protocols are available in the literature [111,112] and the interested reader can take the advantage of this large pool of expertise.

References

1. Cass, A.E.G., Cass, T., and Ligler, F.S. (1998) *Immobilized Biomolecules in Analysis*, Oxford University Press, Oxford.
2. Cao, L. (2005) *Carrier-Bound Immobilized Enzymes: Principles, Applications and Design*, Wiley-VCH, Weinheim.
3. Bisswanger, H. (2011) *Practical Enzymology*, Wiley-VCH, Weinheim.
4. Cabral, J.M.S. and Kennedy, J.F. (1991) Covalent and coordination immobilization of proteins, in *Protein Immobilization: Fundamentals and Applications* (ed. R.F. Taylor), M. Dekker, New York, pp. 73–138.
5. Dugas, V., Elaissari, A., and Chevalier, Y. (2010) Surface sensitization techniques and recognition receptors immobilization on biosensors and microarrays, in *Recognition Receptors in Biosensors* (ed. M. Zourob), Springer, New York, pp. 47–134.
6. Hermanson, G.T. (2008) *Bioconjugate Techniques*, Elsevier, Amsterdam.
7. Cao, L.Q. (2005) Immobilised enzymes: science or art? *Curr. Opin. Chem. Biol.*, **9**, 217–226.
8. Patil, J.S., Kamalapur, M.V., Marapur, S.C. *et al.* (2010), Ionotropic gelation and polyelectrolyte complexation: the novel techniques to design hydrogel particulate sustained, modulated drug delivery system: A review. *Dig. J. Nanomater. Biostruct.*, **5**, 241–248.

9. Lukaszewicz, J.P. (2006) Carbon materials for chemical sensors: A review. *Sens. Lett.*, **4**, 53–98.

10. Švancara, I., Walcarius, A., Kalcher, K. *et al.* (2009) Carbon paste electrodes in the new millennium. *Cent. Eur. J. Chem.*, **7**, 598–656.

11. Wink, T., van Zuilen, S.J., Bult, A. *et al.* (1997) Self-assembled monolayers for biosensors. *Analyst*, **122**, R43–R50.

12. Ion, A., Partali, V., Sliwka, H.R. *et al.* (2002) Electrochemistry of a carotenoid self-assembled monolayer. *Electrochem. Commun.*, **4**, 674–678.

13. Bănică, A., Culetu, A. *et al.* (2007) Electrochemical and EQCM investigation of L-selenomethionine in adsorbed state at gold electrodes. *J. Electroanal. Chem.*, **599**, 100–110.

14. Tour, J.M. (2003) *Molecular Electronics: Commercial Insights, Chemistry, Devices, Architecture And Programming*, World Scientific, River Edge.

15. Chaki, N.K. and Vijayamohanan, K. (2002) Self-assembled monolayers as a tunable platform for biosensor applications. *Biosens. Bioelectron.*, **17**, 1–12.

16. Gooding, J.J. and Hibbert, D.B. (1999) The application of alkanethiol self-assembled monolayers to enzyme electrodes. *TrAC-Trends Anal. Chem.*, **18**, 525–533.

17. Gooding, J.J., Mearns, F., Yang, W.R. *et al.* (2003) Self-assembled monolayers into the 21(st) century: Recent advances and applications. *Electroanalysis*, **15**, 81–96.

18. Love, J.C., Estroff, L.A., Kriebel, J.K. *et al.* (2005) Self-assembled monolayers of thiolates on metals as a form of nano-technology. *Chem. Rev.*, **105**, 1103–1169.

19. Liu, G.D. and Lin, Y.H. (2007) Nanomaterial labels in electrochemical immunosensors and immunoassays. *Talanta*, **74**, 308–317.

20. Stutzmann, M., Garrido, J.A., Eickhoff, M. *et al.* (2006) Direct biofunctionalization of semiconductors: A survey. *Phys. Status Solidi A-Appl. Mat.*, **203**, 3424–3437.

21. Ciampi, S., Harper, J.B., and Gooding, J.J. (2010) Wet chemical routes to the assembly of organic monolayers on silicon surfaces via the formation of Si-C bonds: surface preparation, passivation and functionalization. *Chem. Soc. Rev.*, **39**, 2158–2183.

22. Boukherroub, R. (2005) Chemical reactivity of hydrogen-terminated crystalline silicon surfaces. *Curr. Opin. Solid State Mater. Sci.*, **9**, 66–72.

23. Stewart, M.P., Maya, F., Kosynkin, D.V. *et al.* (2003) Direct covalent grafting of conjugated molecules onto Si, GaAs, and Pd surfaces from aryldiazonium salts. *J. Am. Chem. Soc.*, **126**, 370–378.

24. Wilchek, M. and Bayer, E.A. (1990) Introduction to avidin-biotin technology. *Method Enzymol.*, **184**, 5–13.

25. Diamandis, E.P. and Christopoulos, T.K. (1991) The biotin (strept)avidin system - principles and applications in bio-technology. *Clin. Chem.*, **37**, 625–636.

26. Anzai, J., Hoshi, T., and Osa, T. (2000) Avidin-biotin mediated biosensors, in *Biosensors and Their Application* (eds V.C. Yang and T.T. Ngo), Kluwer, New York, pp. 35–46.

27. Hermanson, G.T. (2008) *Biotinylation Reagents In Bioconjugate Techniques*, Elsevier, Amsterdam, pp. 506–545.

28. Wilchek, M. and Bayer, E.A. (1990) Biotin-containing reagents. *Method Enzymol.*, **184**, 123–138.

29. Bayer, E.A. and Wilchek, M. (1990) Protein biotinylation. *Method Enzymol.*, **184**, 138–160.

30. Smith, C.L., Milea, J.S., and Nguyen, G.H. (2005) Immobilization of nucleic acids using biotin-strept(avidin) systems, in *Immobilisation of DNA on Chips II* (ed. C. Wittmann), Springer, Berlin, pp. 63–90.

31. Davis, F. and Higson, S.P.J. (2005) Structured thin films as functional components within biosensors. *Biosens. Bioelectron.*, **21**, 1–20.

32. Ariga, K. and Kunitake, T. (2006) *Supramolecular Chemistry: Fundamentals and Applications*, Springer, Berlin.

33. Hianik, T. and Passechnik, V.I. (1995) *Bilayer Lipid Membranes: Structure and Mechanical Properties*, Kluwer, London.

34. Hianik, T. (2008) Biological membranes and membrane mimics, in *Bioelectrochemistry: Fundamentals, Experimental Techniques and Applications* (ed. P.N. Bartlett), John Wiley & Sons, Chichester, pp. 86–156.

35. Nikolelis, D.P., Krull, U.J., Ottova, A.L. *et al.* (1996) Bilayer lipid membranes and other lipid-based methods, in *Handbook of Chemical and Biological Sensors* (eds R.F. Taylor and J.S. Schultz), Institute of Physics Publishing, Bristol, pp. 221–256.

36. Petty, M.C. (1996) *Langmuir-Blodgett Films: An Introduction*, Cambridge University Press, Cambridge.

37. Trojanowicz, M. (2001) Miniaturized biochemical sensing devices based on planar bilayer lipid membranes. *Fresenius J. Anal. Chem.*, **371**, 246–260.

38. Mohsin, M.A., Bănică, F.G., Oshima, T. *et al.* (2011) Electrochemical impedance spectroscopy for assessing the recognition of cytochrome c by immobilized calixarenes. *Electroanalysis*, **23**, 1229–1235.

39. Knoll, W., Morigaki, K., Naumann, R. *et al.* (2004) Functional tethered bilayer lipid membranes, in *Ultrathin Electrochemical Chemo- and Biosensors: Technology and Performance* (ed. V.M. Mirsky), Springer, Berlin, pp. 239–252.

40. Lvov, Y., Ariga, K., Ichinose, I. *et al.* (1995) Assembly of multicomponent protein films by means of electrostatic layer-by-layer adsorption. *J. Am. Chem. Soc.*, **117**, 6117–6123.

41. Siqueira, J.R., Caseli, L., Crespilho, F.N. *et al.* (2010) Immobilization of biomolecules on nanostructured films for biosensing. *Biosens. Bioelectron.*, **25**, 1254–1263.

42. Wright, J.D. and Sommerdijk, N.A.J.M. (2003) *Sol-Gel Materials: Chemistry and Applications*, Taylor & Francis, London.

43. Lev, O. and Sampath, S. (2010) Sol-gel electrochemistry, in *Electroanalytical Chemistry: A Series of Advances*, vol. **23** (eds A.J. Bard and C.G. Zoski), CRC Press, Boca Raton, pp. 211–304.

44. Shchipunov, Y. (2008) Entrapment of biopolymers into sol-gel-derived silica nanocomposites, in *Bio-Inorganic Hybrid Nanomaterials: Strategies, Syntheses, Characterization and Applications* (eds E. Ruiz-Hitzky, K. Ariga, and Y. Lvov), Wiley-VCH, Weinheim, pp. 75–112.

45. Gupta, R. and Chaudhury, N.K. (2007) Entrapment of biomolecules in sol-gel matrix for applications in biosensors: Problems and future prospects. *Biosens. Bioelectron.*, **22**, 2387–2399.
46. Tripathi, V.S., Kandimalla, V.B., and Ju, H. (2006) Preparation of ormosil and its applications in the immobilizing biomolecules. *Sens. Actuators B*, **114**, 1071–1082.
47. Dave, B.C., Dunn, B., Valentine, J.S. *et al.* (1998) Sol-gel matrices for protein entrapment, in *Immobilized Biomolecules in Analysis* (eds A.E.G. Cass, T. Cass, and F.S. Ligler), Oxford University Press, Oxford, pp. 1–14.
48. Frenkel-Mullerad, H., and Avnir, D. (2005) Sol-gel materials as efficient enzyme protectors: Preserving the activity of phosphatases under extreme pH conditions. *J. Am. Chem. Soc.*, **127**, 8077–8081.
49. Niederberger, M. and Antonietti, M. (2007) Nonaqueous sol-gel routes to nanocrystalline metal oxides, in *Nanomaterials Chemistry: Recent Developments and New Directions* (eds C.N.R. Rao, A. Müller, and A.K. Cheetham), Wiley-VCH, Weinheim, pp. 119–138.
50. Jagur-Grodzinski, J. (2010) Polymeric gels and hydrogels for biomedical and pharmaceutical applications. *Polym. Adv. Technol.*, **21**, 27–47.
51. Omidian, H. and Park, K. (2010) Introduction to hydrogels, in *Biomedical Applications of Hydrogels Handbook* (eds R.M. Ottenbrite, K. Park, and T. Okano), Springer, New York, pp. 1–16.
52. Plieva, F.M., Galaev, I.Y., Noppe, W. *et al.* (2008) Cryogel applications in microbiology. *Trends Microbiol.*, **16**, 543–551.
53. Kuckling, D., Arndt, K.F., and Richter, A. (2009) Synthesis of hydrogels, in *Hydrogel Sensors and Actuators: Engineering and Technology* (eds G. Gerlach and K.-F. Arndt), Springer, Berlin, pp. 15–67.
54. Heller, A. (2006) Electron-conducting redox hydrogels: design, characteristics and synthesis. *Curr. Opin. Chem. Biol.*, **10**, 664–672.
55. Inzelt, G. (2008) *Conducting Polymers: A New Era in Electrochemistry*, Springer, Berlin.
56. van der Linden, H.J., Herber, S., Olthuis, W., and Bergveld, P. (2003) Stimulus-sensitive hydrogels and their applications in chemical (micro)analysis. *Analyst*, **128**, 325–331.
57. Kuckling, D. (2009) Responsive hydrogel layers – from synthesis to applications. *Colloid Polym. Sci.*, **287**, 881–891.
58. Tokarev, I. and Minko, S. (2009) Stimuli-responsive hydrogel thin films. *Soft Matter*, **5**, 511–524.
59. Ulijn, R.V., Bibi, N., Jayawarna, V. *et al.* (2007) Bioresponsive hydrogels. *Mater. Today*, **10**, 40–48.
60. Richter, A., Paschew, G., Klatt, S. *et al.* (2008) Review on hydrogel-based pH sensors and microsensors. *Sensors*, **8**, 561–581.
61. Guenther, M. and Gerlach, G. (2009) Hydrogels for chemical sensors, in *Hydrogel Sensors and Actuators: Engineering and Technology* (eds G. Gerlach and K.-F. Arndt), Springer, Berlin, pp. 165–195.
62. Urban, G.A. and Weiss, T. (2009) Hydrogels for biosensors, in *Hydrogel Sensors and Actuators: Engineering and Technology* (eds G. Gerlach and K.-F. Arndt), Springer, Berlin, pp. 197–219.
63. Deligkaris, K., Tadele, T.S., Olthuis, W. *et al.* (2010) Hydrogel-based devices for biomedical applications. *Sens. Actuators B-Chem.*, **147**, 765–774.
64. Gawel, K., Barriet, D., Sletmoen, M. *et al.* (2010) Responsive hydrogels for label-free signal transduction within biosensors. *Sensors*, **10**, 4381–4409.
65. Schexnailder, P. and Schmidt, G. (2009) Nanocomposite polymer hydrogels. *Colloid Polym. Sci.*, **287**, 1–11.
66. Miyata, T. (2010) Biomolecule-responsive hydrogels, in *Biomedical Applications of Hydrogels Handbook* (eds K. Park and T. Okano), Springer Science+Business Media, LLC, New York, NY, pp. 65–86.
67. Cosnier, S. and Holzinger, M. (2011), Electrosynthesized polymers for biosensing. *Chem. Soc. Rev.*, **40**, 2146–2156.
68. Chandrasekhar, P. (1999) *Conducting Polymers, Fundamentals and Applications: A Practical Approach*, Kluwer, Boston.
69. Lange, U., Roznyatouskaya, N.V., and Mirsky, V.M. (2008) Conducting polymers in chemical sensors and arrays. *Anal. Chim. Acta*, **614**, 1–26.
70. Ramanavicius, A., Ramanaviciene, A., and Malinauskas, A. (2006) Electrochemical sensors based on conducting polymer-polypyrrole. *Electrochim. Acta*, **51**, 6025–6037.
71. Giuseppi-Elie, A., Wallace, G.G., and Matsue, T. (1998) Chemical and biological sensors based on electrically conducting polymers, in *Handbook of Conducting Polymers* (eds T.A. Skotheim, R.L. Elsenbaumer, and J.R. Reynolds), Marcel Dekker, New York, pp. 963–991.
72. Potje-Kamloth, K. (2002) Chemical gas sensors based on organic semiconductor polypyrrole. *Crit. Rev. Anal. Chem.*, **32**, 121–140.
73. Cosnier, S. (2003) Biosensors based on electropolymerized films: New trends. *Anal. Bioanal. Chem.*, **377**, 507–520.
74. Cosnier, S. (1997) Electropolymerization of amphiphilic monomers for designing amperometric biosensors. *Electroanalysis*, **9**, 894–902.
75. Hatchett, D.W. and Josowicz, M. (2008) Composites of intrinsically conducting polymers as sensing nanomaterials. *Chem. Rev.*, **108**, 746–769.
76. Cosnier, S. (1999) Biomolecule immobilization on electrode surfaces by entrapment or attachment to electrochemically polymerized films. A Review. *Biosens. Bioelectron.*, **14**, 443–456.
77. Bartlett, P.N. and Cooper, J.M. (1993) A review of the immobilization of enzymes in electropolymerized films. *J. Electroanal. Chem.*, **362**, 1–12.
78. Schuhmann, W. (1995) Conducting polymer based amperometric enzyme electrodes. *Mikrochim. Acta*, **121**, 1–29.
79. Bidan, G., Billon, M., Calvo-Munoz, M.L. *et al.* (2004) Bio-assemblies onto conducting polymer support: Implementation of DNA-chips. *Mol. Cryst. Liquid Cryst.*, **418**, 983–998.
80. Cosnier, S. (2005) Affinity biosensors based on electropolymerized films. *Electroanalysis*, **17**, 1701–1715.
81. Sadik, O.A. (1999) Bioaffinity sensors based on conducting polymers: A short review. *Electroanalysis*, **11**, 839–844.

82. Miao, Y.Q., Chen, J.R., and Wu, X.H. (2004) Using electropolymerized non-conducting polymers to develop enzyme amperometric biosensors. *Trends Biotechnol.*, **22**, 227–231.

83. Park, B.W., Yoon, D.Y., and Kim, D.S. (2010) Recent progress in bio-sensing techniques with encapsulated enzymes. *Biosens. Bioelectron.*, **26**, 1–10.

84. Ozin, G.A., Arsenault, A.C., and Cademartiri, L. (2009) Microporous and mesoporous materials from soft building blocks, in *Nanochemistry: A Chemical Approach to Nanomaterials*, RSC Publ., Cambridge, pp. 521–592.

85. Bonifacio, L.D., Lotsch, B.V., Ozin, G.A. *et al.* (2010) Periodic mesoporous materials: Holes filled with opportunities, in *Comprehensive Nanoscience and Technology* (eds D. Andrews, G. Scholes, and G. Wiederrecht), Academic Press, Amsterdam, pp. 69–125.

86. Hoffmann F., Cornelius, M., Morell, J. *et al.* (2006) Silica-based mesoporous organic-inorganic hybrid materials. *Angew. Chem. Int. Edit.*, **45**, 3216–3251.

87. Vinu, A., Gokulakrishnan, N., Mori, T. *et al.* (2008) Immobilization of biomolecules on mesoporous structured materials, in *Bio-inorganic Hybrid Nanomaterials: Strategies, Syntheses, Characterization and Applications* (eds E. Ruiz-Hitzky, K. Ariga, and Y. Lvov), Wiley-VCH, Weinheim, pp. 113–157.

88. Lee, C.H., Lin, T.S., and Mou, C.Y. (2009) Mesoporous materials for encapsulating enzymes. *Nano Today*, **4**, 165–179.

89. Lee, D., Lee, J., Kim, J. *et al.* (2005) Simple fabrication of a highly sensitive and fast glucose biosensor using enzymes immobilized in mesocellular carbon foam. *Adv. Mater.*, **17**, 2828–2833.

90. Tierney, M.J. (1996) Practical examples of polymer-based chemical sensors, in *Handbook of Chemical and Biological Sensors* (eds R.F. Taylor and J.S. Schultz), Institute of Physics Publishing, Bristol, pp. 349–369.

91. Emr, S.A. and Yacynych, A.M. (1995) Use of polymer-films in amperometric biosensors. *Electroanalysis*, **7**, 913–923.

92. Reddy, S.M. and Vadgama, P.M. (1997) Membranes to improve amperometric sensor characteristics, in *Handbook of Biosensors and Electronic Noses: Medicine, Food, and the Environment* (ed. E. Kress-Rogers), CRC Press, Boca Raton, Fla, pp. 111–135.

93. Shoji S. (2000) Micromachining for biosensors and biosensing systems, in *Biosensors and Their Applications* (eds V.C. Yang and T.T. Ngo), Kluwer, New York, pp. 225–241.

94. Gaspar S., Schuhmann, W., Laurell, T. *et al.* (2002) Design, visualization, and utilization of enzyme microstructures built on solid surfaces. *Rev. Anal. Chem.*, **21**, 245–266.

95. Mosbach, M., Zimmermann, H., Laurell, T. *et al.* (2001) Picodroplet-deposition of enzymes on functionalized self-assembled monolayers as a basis for miniaturized multi-sensor structures. *Biosens. Bioelectron.*, **16**, 827–837.

96. Lambrechts, M. and Sansen, W. (1992) *Biosensors: Microelectrochemical Devices*, Institute of Physics Publishing, Bristol.

97. Rikoski, R.A. (1973) *Hybrid Microelectronic Circuits: The Thick Film*, Wiley New, York.

98. Licari, J.J. and Enlow, L.R. (1998) *Hybrid Microcircuit Technology Handbook: Materials, Processes, Design, Testing, and Production*, Noyes Publications, Westwood, N.J.

99. Hierlemann, A. (2005) *Integrated Chemical Microsensor Systems in CMOS Technology*, Springer-Verlag Berlin Heidelberg, Berlin, Heidelberg.

100. Madou, M. and Kim, H.L. (1996) Integrated circuit manufacturing techniques applied to microfabrication, in *Handbook of Chemical and Biological Sensors* (eds R.F. Taylor and J.S. Schultz,), Institute of Physics Publishing, Bristol, pp. 45–82.

101. Hierlemann A., Brand, O., Hagleitner, C. *et al.* (2003) Microfabrication techniques for chemical/biosensors. *Proc. IEEE*, **91**, 839–863.

102. Xia, Y. and Whitesides, G.M. (1998) Soft lithography. *Angew. Chem. Int. Ed.*, **37**, 550–575.

103. Gates, B.D., Xu, Q.B., Stewart, M. *et al.* (2005) New approaches to nanofabrication: Molding, printing, and other techniques. *Chem. Rev.*, **105**, 1171–1196.

104. Delamarche, E. (2004) Microcontact printing of proteins, in *Nanobiotechnology: Concepts, Applications and Perspectives* (eds C.M. Niemeyer and C.A. Mirkin), Wiley-VCH, Weinheim, pp. 31–52.

105. Truskett, V.N. and Watts, M.P.C. (2006) Trends in imprint lithography for biological applications. *Trends Biotechnol.*, **24**, 312–317.

106. Mendes, P.M., Yeung, C.L., and Preece, J.A. (2007) Bio-nanopatterning of surfaces. *Nanoscale Res. Lett.*, **2**, 373–384.

107. Bernard, A., Fitzli, D., Sonderegger, P. *et al.* (2001) Affinity capture of proteins from solution and their dissociation by contact printing. *Nature Biotechnol.*, **19**, 866–869.

108. Renault, J.P., Bernard, A., Juncker, D. *et al.* (2002) Fabricating microarrays of functional proteins using affinity contact printing. *Angew. Chem. Int. Ed.*, **41**, 2320–2323.

109. Ganesan, R., Kratz, K., and Lendlein, A. (2010) Multicomponent protein patterning of material surfaces. *J. Mater. Chem.*, **20**, 7322–7331.

110. Buchholtz, K. and Klein, J. (1987) Characterization of immobilized biocatalysts, in *Immobilized Enzymes and Cells* (ed. K. Mosbach), Academic Press, Orlando, pp. 3–30.

111. Bickerstaff, G.E. (1997) *Immobilization of Enzymes and Cells*, Humana, Riverview Drive.

112. Rasooly, A. and Herold, K.E. (2009) *Biosensors and Biodetection: Methods and Protocols*, Humana Press, New York.

6

Affinity-Based Recognition

6.1 General Principles

Affinity interactions are based on multiple noncovalent interactions between two species leading to the formation of a molecular aggregate usually called a complex. Such interactions are very common in living organism, being associated with important physiological functions. The strength of the complex arises from the multiplicity of the noncovalent bonds and is quantified by its stability constant, often termed the affinity constant. At the same time, the noncovalent character of the interaction imparts reversibility to the affinity association process. Finally, affinity interactions are very specific as a result of the steric and structural complementarity of the reactants.

The high specificity of biological affinity interactions prompted the development of a broad range of analytical applications in which one affinity reagent acts as a recognition receptor, whereas the second, target compound functions as the analyte. Inspired by natural affinity reactions, various synthetic receptors have been developed in the field of supramolecular chemistry [1,2]. Synthetic receptors are often more stable and cheaper than the natural ones but the latter are well established and will be further employed on a large scale.

From the standpoint of analytical applications, immunoassay is the most common analytical methods in this class. It is based on antibodies that are secreted by higher organisms in defense against pathogens. Biological receptors that are essential in the response of living cell to chemical stimuli form another group of natural compounds with relevance to analytical applications. In addition, a series of synthetic receptors or natural compounds with no receptor functions in living systems proved extremely useful from the analytical standpoint.

This chapter reviews first the principles of antibody-based chemical sensing, which is the most representative among the affinity-based analytical methods. Next, chemical sensing based on biological receptors is addressed, and finally several classes of synthetic receptors are dealt with.

6.2 Immunosensors

Immunoassay and immunosensors rely on species involved in the immune response of living organisms. In order to remove pathogens and other foreign species, the organism reacts by a series of physiological processes aiming at indentifying and annihilating the intruder. Immunoassay is currently a common diagnostic method [3,4]. In addition, immunochemical analysis methods are extremely useful in other areas of a great interest. The key compounds in immunoassay are the antibodies.

6.2.1 Antibodies: Structure and Function

An *antibody* is a large protein that is used by the immune system to identify and neutralize foreign objects such as bacteria and viruses. The antibody recognizes a unique part of the foreign target, called an *antigen*. Typical antibodies are immunoglobulins (Ig), a class of proteins abundant in blood plasma, the most prevalent one being immunoglobulin G (IgG). When a microorganism or a foreign protein enters the organism, the immune system produces antibodies that append to the antigen to form an *immunochemical complex*. This allows the immune system to recognize the antigen and trigger other processes leading to its removal from the organism. The immune response is elicited by *immunogens*, that are foreign compounds in the host organism. Immunogens can belong to certain classes of biopolymers such as proteins and polysaccharides.

Typically an antibody molecule is Y-shaped, as shown in Figure 6.1. An antibody molecule consists of four protein chains, two light chains with the molecular weight of about 25 kD and two heavy chains with the molecular weight of about 50 kD. These subunits are assembled by noncovalent bonds and sulfur bridges to form a quaternary structure. Within the antibody molecule there are *constant* (C) and *variable* (V) domains. In the constant domains, the amino acid sequence is similar in any antibody belonging to a specific class (e.g., IgG) while the sequence in the variable domains is adapted in order to allow for the recognition of a specific antigen. The Fc region ("*fragment, crystallizable*") ensures that each antibody generates an appropriate immune response for a given antigen, by binding

Chemical Sensors and Biosensors: Fundamentals and Applications, First Edition. Florinel-Gabriel Bănică.
© 2012 John Wiley & Sons, Ltd. Published 2012 by John Wiley & Sons, Ltd.

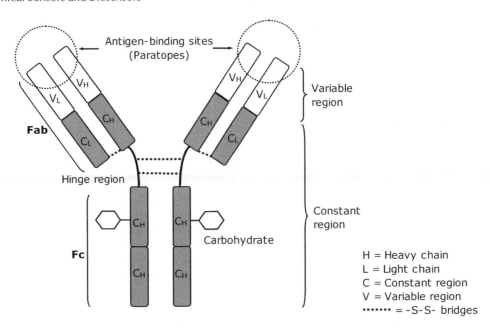

Figure 6.1 Schematic diagram of the ImmunoglobulinG (IgG) antibody.

to a specific class of Fc receptors, and other components of the immune system. Thus, an antibody acts as a label that triggers different physiological processes leading to the destruction of the antigen.

The arms of the Y contain the sites that can bind to an antigens (different or identical) and, therefore, recognize specific foreign species. This region of the antibody is called the Fab (Fab for "*fragment, antigen binding*"). The variable domains shape the *paratopes* of the antibody that is, the regions that interact directly with the antigen.

Antibodies are widely used in analytical immunochemistry for very sensitive detection and quantitation of various antigens. Figure 6.2 depicts the interaction of an antibody with a sample containing a mixture of antigens. Antibodies react only with small fragments of the antigen called *epitopes*. Only antigens featuring a fitting epitope interact with the antibody forming a strong antibody–antigen complex. Detection of this complex allows for identification and analytical determination of the target antigen.

Antibody preparations can be obtained by inoculating laboratory animals with an antigen. Blood serum from the inoculated animal contains a mixture of antibodies showing specificity to various epitopes on the antigen. Such a preparation contains *polyclonal antibodies*. In contrast, *monoclonal antibodies* that are produced from cell culture, bind to only one particular antigen epitope. As will be shown later, monoclonal antibodies present substantial advantages in immunoassay.

From the standpoint of the immunogenic behavior, two classes of antigens can be distinguished, namely *complete* and *partial* antigens. A complete antigen induces an immune response by itself and is usually a high molecular weight protein or polysaccharide. A complete antigen can feature several identical epitopes or contain different epitopes, each of them reacting with a specific antibody. Antigens in the second class are called *multideterminant* antigens. Organism inoculation with such antigens produces a polyclonal serum (Figure 6.3).

An partial antigen (also called a *hapten*) is a low molecular weight compound that cannot elicit an immune response alone but only if it is linked to a carrier-protein. Examples of haptens are hormones, drugs, allergens, and organic environment contaminants. When inoculated into an organism, the haptene-carrier adduct elicits immune response. Remarkably, the antihapten antibody thus produced binds to the hapten even when this one is free from

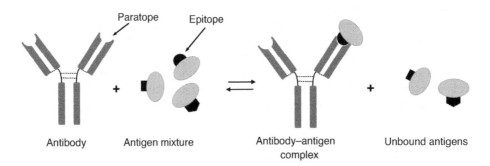

Figure 6.2 Specific interaction of an antibody with a mixture of antigens resulting in an antibody–antigen complex.

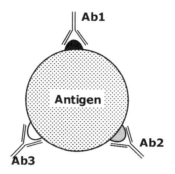

Figure 6.3 Interaction of a multideterminant antigen with polyclonal antibodies (Ab1, Ab2 and Ab3).

the carrier. This allows to develop very sensitive analytical methods for the determination of small-molecule compounds by interaction with a specific antibody. Such methods are commonly used for assessing pharmaceutical products and organic pollutants.

6.2.2 Antibody–Antigen Affinity and Avidity

The strength of the antibody–antigen interaction is expressed by the *association constant* that indicates the affinity of a paratope for a specific epitope. Adopting the symbol Ab for antibody, Ag for antigen and Ab:Ag for the complex, this reversible interaction can be formulated as:

$$Ab + Ag \rightleftarrows Ab\ Ag \tag{6.1}$$

The association constant K_a is the equilibrium constant of the above reaction:

$$K_a = \frac{[Ab : Ag]}{[Ab][Ag]} \tag{6.2}$$

where square parentheses indicate concentrations and K_a (also referred to as *affinity constant*) is quantified in reciprocal molar concentration units, M^{-1} or $L\,mol^{-1}$. K_a values can range from 10^5 to $10^{12}\,M^{-1}$, but only systems with $K_a > 10^8$ are useful for analytical purposes. A high affinity constant enables one to design assays with the limit of detection down to 10^{-9} to $10^{-12}\,M$.

The affinity constant has an unambiguous meaning only if it refers to monoclonal antibodies. When dealing with heterogeneous polyclonal antibodies, each component of the preparation features a particular affinity for its epitope. In such a case, K_a represents an average value that is termed *avidity*.

Association of an antibody and an antigen is an intricate process involving various kinds of interaction. First, the two partners are brought in close proximity and favorable positions by diffusion. When the epitope–paratope distance approaches some 10 nm, electrostatic attraction becomes effective. This reduces the distance between reactants and water molecules are excluded, allowing for hydrogen bonds to form between relevant groups in each partner. At a very short distance, van der Waals interactions come into play. In addition, nonpolar groups can aggregate with each other in the aqueous environment by hydrophobic interactions. The interplay of so many interactions, combined with steric complementarity explains the high selectivity of the antibody–antigen coupling.

6.2.3 Analytical Applications

Antibody–antigen reactions form the basis of a wide area of analytical methods with applications in various fields such as biomedicine, toxicology, food industry, and environment chemistry. The range of application encompasses detection of pathogenic micro-organisms, proteins, and small organic molecules that behave as haptens. Immunoassay kits are commercially available for application in clinical laboratories. However, such procedures require a series of operations such as washing out the sample or immunoreagents. The throughput is considerably enhanced by automation but even so there remain some problems such as field application of immunoassay or searching simultaneously for multiple analytes in a sample. Such problems are alleviated by combining immunoreaction with suitable transducers to form immunosensor that provide in real time the requested analytical information [5].

Immunoassay is equally concerned with the determination of either antigens or antibodies. Antibody detection is a valuable diagnostic tool as it allows an infection to which the organism elicited immune response to be revealed. So, according to circumstances, either an antibody or an antigen functions as a receptor, also called capture probe. The target analyte is also termed a ligand.

As the recognition is provided by the Fab domain, it is sometime convenient to resort to this fragment only instead of the whole antibody molecule. This fragment can be engineered such as to be more stable than the whole molecule.

A key issue is the regeneration of the sensor after each assay. This can be done either by shifting the pH to an extreme value, by an increase in ionic strength or by means of guanidinium chloride. Such a treatment breaks up noncovalent bonds involving hydrogen bonds or electrostatic interactions and releases the analyte. At the same time, the regeneration is potentially harmful to the binding site and may lead to a reduced lifetime of the sensor and a drift of the response signal. This problem can be alleviated by using a mutant antibody or an antibody fragment with enhanced stability. When possible, one can resort to an alternative regeneration methods. For example, when dealing with a small-molecule analyte, this one can be displaced from the complex by an excess of an analogous compound with a lower affinity. Regeneration conditions are less stringent in the case of antibody determination by means of immobilized antigens as receptors. The regeneration problem is irrelevant if the sensor is designed as a disposable device.

Nonspecific interactions of the analyte with the support surface could cause a disturbing background signal. In order to avoid such problems, it is recommended to resort to blocking agents such as proteins (gelatin, bovine serum albumin or lysosyme) or surfactants (Triton X 100, Tween 20) as coatings over the exposed support surface. Alternatively, one can rely on background subtraction using a nonspecific reference sensor that provide the correction signal.

In addition to the effect of the recognition process, the response can also be influenced by parameters unrelated to the target species. For example, uncontrolled variation of temperature, salt concentration and pH, as well as and nonspecific interactions can produce random variations in the response signal. To compensate for such effects, a reference sensor can be used. Ideally, no analyte binding occurs at the reference sensor but otherwise it elicits an identical response to the disturbing factors. The corrected signal is obtained as the difference between the total signal and the reference sensor response. If the working and reference sensors responds similarly to disturbing parameters, differential schemes can improve both selectivity and sensitivity of the assay.

Immunosensor operation involves several steps including addition of reagents and washing out. That is why it is used to operate an immunosensor in flow analysis systems that allow for automation of the run. Application of microfluidics in immunoassay brings about the noteworthy advantage of using very small amounts of sample and reagents [6].

6.2.4 Label-Free Transduction Methods in Immunosensors

It was already mentioned that immunoassay is based on the detection of the antibody–antigen complex. In certain immunoassays methods as well as in immunosensors, the receptor is immobilized at the surface of a solid support where the recognition event does occur. Therefore, any physical change caused by the formation of the complex can in principle be handy for transduction. The most striking alteration is the mass change that can be assessed by means of mass-sensitive devices such as the quartz crystal microbalance. On the other hand, the formation of the complex modifies drastically the behavior of the surface layer in its interaction with light beams. For example the refractive index of the sensing layer changes in response to complex formation and allows for the analyte concentration to be assessed. As antibodies and antigens are charged particles, their coupling dramatically changes the local charge distribution, an effect that can be monitored by electrochemical methods.

What limits the performance of a label-free affinity sensor is more often the limited selectivity of the recognition step due to nonspecific interaction of the analyte with the receptor or the support. Such interaction may drastically impair the sensor selectivity. Nonspecific binding of concomitants can prevent the analyte binding or cause a positive false signal.

Selectivity is particularly important when dealing with genuine biological samples where the analyte concentration can be much less that the concentration of similar concomitants. This problem can be resolved by preliminary separation of the analyte but separation may be a time-consuming step. To alleviate this problem, the sensor is often exposed to a blocking agent prior to the assay. Under favorable conditions, the blocking agent adsorbs on nonspecific binding sites, hopefully not occupying the receptor binding sites. Antifouling agents, such as polyethylene glycol, can also be placed on support areas surrounding the sensing layer to prevent analyte depletion by nonspecific binding. Similar effects can be obtained in the case of gold supports by chemisorption of alkanethiols or their derivatives

A differential approach can be used to subtract the nonspecific component of the sensor response. This scheme should be applied very cautiously as no general secure method for quantifying the nonspecific term can be worked out. Washing the sensor before reading out the signal can improve the selectivity by removing nonspecifically bound molecules that are, as a rule, more loosely attached than the analyte.

6.2.5 Label-Based Transduction Methods in Immunosensors

Very sensitive and reliable immunoassay methods are based on labeled immunoreagents. The antibody–antigen coupling brings the label into the complex and affords reporting the extent of the reaction. The most current labels are radioisotopes, luminescent compounds, and enzymes. Radioisotopes allow for extremely sensitive detection, but present, of course, serious hazards and are avoided whenever possible. Luminescent labels are in turn commonly used. Briefly, luminescence denotes light emission by excited molecules. Excited molecules can be generated by light absorption (photoluminescence, also called fluorescence) or as products of certain chemical (chemiluminescence) or biochemical (bioluminescence) reactions. See Section 19.5 for details.

Enzymes labels are characterized by exceptional sensitivity as the enzymatic reaction results in amplification of the signal.

It is obvious that a broad variety of transduction methods are prone to integration with immunorecognition reactions [7], as will be shown in detail in later chapters. The next section introduces the principles of enzyme labels that are suited for both standard immunoassay and immunosensors.

6.2.6 Enzyme Labels in Immunoassay

Enzymes are widely used labeling compound in standard immunoassay [8]. The principle of enzyme-label transduction is featured in Figure 6.4, which refers to the determination of an antigen. The antigen sample is incubated with the enzyme-labeled antibody in order to allow for complex formation. The complex should be next separated from the unreacted labeled antibody and the incubated with the substrate solution. The substrate is selected so as to form a detectable product by the enzyme-catalyzed reaction. The amount of complex, and implicitly the analyte concentration, can be estimated by various methods that are common to enzyme sensors.

In order to facilitate the separation of the complex, the receptor is immobilized at the surface of a solid support, such as a microtiter well. A method using immobilized antibodies and enzyme labels is usually termed *enzyme-linked immunosorbent assay* (shortly ELISA) [9,10].

It is obvious that the analyte concentration can be inferred from the amount of enzyme coupled to the immobilized complex. This one can be estimated by adding a substrate and allowing the enzymatic reaction to proceed for a predetermined time interval. The reaction is then stopped by shifting the pH from neutral to alkaline (by adding a concentrated base) and the product concentration is the determined by a suitable analytical method. The amount of detectable product must be proportional to the local enzyme concentration. In order to fulfill this requirement, the substrate concentration should be sufficiently high so as to allow the enzymatic reaction to proceed under zero-order kinetic conditions.

Under zero-order kinetics, the reaction rate of the enzymatic reaction is also an indicator of the amount of complex. The reaction-rate-based method is convenient for enzyme-linked immunosensors. Such a sensor is formally a combination of an immunorecognition system with an enzyme sensor that is designed so as to report on the total amount of enzyme linked to the Ab:Ag complexe.

A label-enzyme should feature high specific activity, stability and possibility of facile conjugation with immunoreagents. As the substrate should be selected so as to give a detectable product, it is convenient to use enzymes that are equally active with as many as possible substrates.

Next, several enzymes commonly used as immunoassay labels are briefly reviewed. More details on such enzymes are available in Chapter 3.

Peroxidases are heme proteins that catalyzes the oxidation of various substrates in the presence of a peroxide as electron acceptor. The most common enzyme of this type is horse radish peroxidase, which is stable and has a high specific activity. Care should be taken to avoid the presence of inhibitors such as cyanide and sulfide ions that inhibit the enzyme by covalent coupling to iron. Microperoxidases, that are catalytically active fragments of cytochrome *c*, function mostly in the same way except for the fact that they exhibit higher substrate selectivity. Catalase is also able to perform peroxidase-like reactions.

Alkaline phosphatase does catalyze the hydrolysis of orthophosphoric acid monoesters. A common substrate of this enzyme is *p*-nitrophenyl phosphate, which gives detectable *p*-nitrophenol.

β-Galactosidase catalyzes the hydrolysis of terminal *β*-D-galactose residues in a *β*-galactosides, in which the galactose residue is linked to a noncarbohydrate moiety. This moiety is selected so as to as yield a detectable product

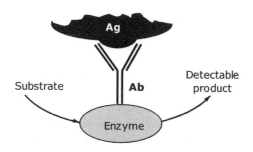

Figure 6.4 Detection of the antibody–antigen complex in enzyme-linked immunoassay.

by enzymatic hydrolysis. Thus, *p*-aminophenol forms when using 4-aminophenyl *β*-galactopyranoside as substrate. In contrast to alkaline phosphatase which requires a pH close to 10, *β*-galactosidase functions best at pH 6–8, which is also optimal for the immunorecognition processes.

Glucose oxidase catalyzes the oxidation of glucoses by dissolved oxygen yielding hydrogen peroxide as byproduct. Either oxygen or peroxide concentration can be assessed for transduction purposes. The specific activity is relatively low but mammalian biological fluids do not contain enzymes that perform similar reactions. Hence, interference by endogenous enzymes is not a problem. As glucose sensors based on this enzyme are well established, integration of glucose oxidase reaction with an immunorecognition process is rather straightforward.

Luciferases catalyze bioluminescent reactions (Section 18.3.9). Bacterial and firefly luciferases are commonly employed for optical monitoring in immunoassay.

Caution should be exercised when using enzyme labels in immunoassay. For example, the large size of enzyme molecules brings about some particular problems. The enzyme labels have the potential to hinder sterically the receptor-analyte interaction. It can also alter the tertiary structure of the antibody, thus disturbing the specific binding and increasing the probability of nonspecific interaction with concomitants. Conjugation of enzyme with immunoreagents should therefore be conducted so as to preserve the activity of the active site in each partner. On the other hand, enzyme–antibody conjugation can modify the transport behavior of the labeled entity causing slower diffusion and reduced permeability through certain support matrices. The presence of enzyme inhibitors or activators in the sample solution could be a serious impediment to proper functioning of the enzyme-based immunoassay system. Also, it is important to avoid the presence of sample enzymes that act similarly to the label enzyme.

6.3 Immobilization Methods in Immunosensors

Most of the general methods for protein immobilization (adsorption, entrapment, crosslinking and covalent binding) are also applicable in the case of immunoreagents if attention is paid to the particular aspects of the antibody structure and the constraint imposed by the recognition and transduction processes. Various immobilization methods such as absorption, entrapment, covalent grafting or affinity binding are achievable using electrochemically polymerized films [11].

A straightforward immobilization method relies on magnetic beads tagged to the compound of interest. This conjugate is then patterned on the sensor working area by a thick-film magnet deposited on the back side [12].

Certain transduction methods rely on changes in the properties of the sensor surface. This is the case in optical, mass-sensitive, and capacitance-sensitive methods. In such cases, the receptor should be immobilized as a monolayer at the support surface and care should be exerted to maintain the highest possible binding capacity by proper orientation of the receptor molecule at the interface. Best performances are obtained by antibody immobilization via the Fc region, that directs the Fab (antigen-binding) domains outward and faraway from the transducer surface. This could be achieved by linking the antigen via the carbohydrate residue in the antibody Fc region. Thus, periodate oxidation of the carbohydrate produces reactive aldehyde groups that can be crosslinked with the support by dihydrazide activation.

Antibody immobilization can be achieved by affinity interactions using *Protein A* as crosslinker [13]. Protein A is found in the cell wall of the *Staphylococcus aureus* bacteria. This protein binds to the Fc region of immunoglobulins through interaction with the heavy chain. It is therefore possible to append antibodies to Protein A functionalized supports.

The avidin–biotin system is also practical in antibody immobilization as a monolayer. In this approach, avidin is first attached to the support and then a biotinylated antibody is coupled by affinity interaction.

With certain transduction methods, antibody multi-layers function satisfactory and good sensitivity is achievable even with randomly oriented antibodies. In such cases, the receptor can be assembled in the form of 1- to 50-μm thick films obtained by entrapment or crosslinking with an inert protein.

6.4 Immunoassay Formats

Immunoassay can be performed either in homogeneous or heterogeneous format. The second alternative relies on receptors immobilized on a solid support and shares therefore certain common features with immunosensors. Several approaches in heterogeneous immunoassay are presented below.

Direct immunoassay is used to detect antigens and makes use of labeled antibodies that are added to the sample. The resulting labeled complex is detected after washing out the excess antibody.

Next, two indirect assay formats based on immobilized receptors and labeled reagents are introduced. It will be assumed that monoclonal antibodies are used in the assay, in accordance with the common practice.

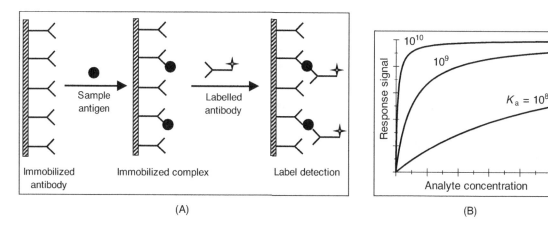

Figure 6.5 Excess reagent immunoassay (sandwich assay; two-site assay). (A) Assay sequence; (B) calibration graph for various values of K_a.

Noncompetitive assay (Figure 6.5A) (also known as *excess reagent* or *sandwich* assay or *two-site assay*) relies on two monoclonal antibodies, each of them being able to link to a particular epitope on the antigen. The first antibody is the immobilized receptor, while the second one is a labeled signaling antibody (secondary antibody). In the first step, the sample solution is incubated with the receptor and the Ab–Ag interaction is allowed to attain the equilibrium. The sample is then washed out and the signaling antibody solution is added. The signaling antibody links to the complex formed in the previous step, yielding a labeled ternary complex. After removing the above solution, the signal generated by the immobilized labels is read-out in order to infer the antigen concentration. The "excess reagent" designation of this technique arises from the fact that only a fraction of the total amount of receptor ("reagent") reacts with the analyte.

In order to derive the response function, the reversible binding of a receptor R and an analyte (ligand) A to yield the complex C is considered:

$$A + R \rightleftharpoons C \tag{6.3}$$

The binding process can be very fast, often close to the limit imposed by a diffusional encounter. Denoting by a, r and c the concentration of A, R and C, respectively, the reaction rates for the association (v_a) and the dissociation (v_d) processes are, respectively:

$$v_a = k_a a r \tag{6.4}$$

$$v_d = k_d c \tag{6.5}$$

where k_a and k_d are the rate constants for association and dissociation, respectively. At equilibrium, the rates of the above reactions are equal to each other. Therefore, by equating Equations (6.4) and (6.5) and rearranging gives an expression for the association constant K_a as:

$$K_a = \frac{c}{ar} = \frac{k_a}{k_d} = K_d^{-1} \tag{6.6}$$

K_a quantifies the strength of the complex while K_d, that is the reciprocal of the association constant, represents the dissociation constant of the complex. Concentrations involved in Equation (6.6) are equilibrium concentrations $(a$ and $r)$ that are generally different from initial concentrations $(a_0$ and r_0, respectively). However, under typical condition, only a small fraction of the available amount of analyte is bound to the receptor and the bulk concentration of the analyte can be assumed as invariable $(a \approx a_0)$. At the same time, the receptor concentration undergoes a noticeable variation and the equilibrium values becomes $r = r_0 - c$. Substituting this into Equation (6.6) gives, after rearrangement, an expression for the concentration of the complex as a function of the analyte concentration:

$$c = r_0 \frac{a}{K_d + a} \tag{6.7}$$

This equation (which is similar to the Langmuir isotherm) shows that the complex concentration increases nonlinearly with the analyte concentration and tends asymptotically to a limit at high concentrations (Figure 6.5B). An approximately linear range is available at very low analyte concentrations when $a \ll K_a$. As is evident from Figure 6.5B, the linear response range shifts to lower concentrations with increasing K_a, which demonstrates the advantage

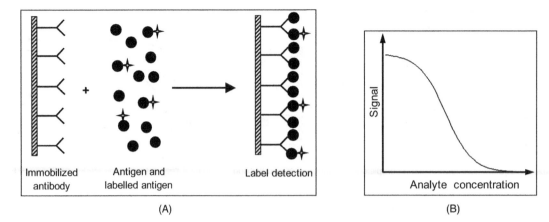

Figure 6.6   Competitive immunoassay. (A) assay sequence; (B) calibration graph.

of a very strong analyte–receptor interaction. When dealing with antibody–antigen systems, Equation (6.7) holds rigorously only when using a homogeneous, monoclonal antibody and the nonspecific binding is negligible. However at very low analyte concentrations, the $a \approx a_0$ approximation does not hold and the real curve deviates from the Equation (6.7). Nevertheless a calibration graph features as a rule an approximately linear region that is useful for analytical applications. As the analyte concentration could expand over a very broad range, the calibration graph (also called dose–response in immunoassay) is often plotted as a logarithmic-scale graph.

Readout of the immunosensors response should be performed when the system is close to equilibrium in order to secure best sensitivity. However, the equilibration of the recognition process can be rather slow and impose a long incubation time. Hence, fast response and high sensitivity are mutually exclusive. Satisfactory sensitivity is warranted if an antibody with a high affinity is used.

Competitive assay (Figure 6.6) makes use of a labeled analyte species. A fixed and known amount of labeled analyte is added to the sample to compete with the sample analyte for the limited number of receptor sites. The sample is then washed out and the signal produced by labels included in the immobilized complex is quantified. This method is also termed *limited reagent assay* since the amount of antigen (both labeled and nonlabeled) should be in excess with respect to the available amount of receptor in order to secure receptor saturation.

In competitive assay, both the plain analyte and the labeled one (A^*) compete with equal chances for the receptor binding site yielding two complexes, C and C^*, respectively:

$$A + R \rightleftarrows C \tag{6.8}$$

$$A^* + R \rightleftarrows C^* \tag{6.9}$$

As there are no differences in affinity, the equilibrium constant is the same for both of the above equilibria:

$$K_a = \frac{c}{ar} = \frac{c^*}{a^*r} \tag{6.10}$$

Assuming that receptor saturation is achieved, we have:

$$c + c^* = r_0 \tag{6.11}$$

The signal is determined by the C^* concentration that, in principle, can be derived from the above equations. However, the exact solution is intricate and it is more expedient to resort to some approximations. Thus, it will be assumed that the concentration of each ligand do not vary by binding to the receptor ($a \approx a_0$ and $a^* \approx a_0^*$) and, on the other hand, the receptor layer is saturated at equilibrium (Equation 6.11). Under these conditions, the concentration of the labeled complex is:

$$c^* = \frac{r_0}{1 + a_0(a^*)^{-1}} \tag{6.12}$$

Therefore, at a constant concentration of A^*, the signal decreases asymptotically with increasing analyte concentration and approaches zero at an excess of analyte. Under these conditions, the labeled ligand has almost no chance to bind to the receptor. On the other hand, at a very low analyte concentration, the $a \approx a_0$ approximation does not hold and a real response curve deviates from the shape predicted by Equation (6.12). Hence, a real response curve looks like the curve in Figure 6.6B, that displays a linear working range at intermediate concentrations.

It has been assumed in the above discussion that the antigen is to be determined. However, both above methods can be adapted to antibody determination using a suitable antigen as receptor.

6.5 Protein and Peptide Microarrays

A protein microarray is a collection of microscopic protein receptor spots attached to a solid surface. As each spot includes a particular receptor, a microarray allows the simultaneous detection of a large number of proteins in a complex sample ([14,15]. The most common protein microarray is the antibody microarray, where antibody spots of about 100 μm diameter are formed onto a chip of about 1 cm^2 a total area. Antibody microarrays are invaluable as diagnostic and prognostic tools for cancer and are also used in basic research applications in proteomics. If a transducer is integrated at each array site, the device functions as an immunosensor microarray.

Figure 6.7 shows an example of direct screening procedure based on an antibody microarray. In this particular case, each particular monoclonal antibody is assembled at two neighboring sites in order to improve the reliability of the result. Incubation with a sample allows each analyte to bind to a specific antibody on the array. Receptor-bound antigens are then labeled by a fluorescent marker and detected by means of an optical scanner that performs array imaging. An optical scanner is a device that excites fluorescence at each site by means of a laser beam and measures the intensity of the emitted light. The antigen is identified by its position on the microarray, whereas its concentration can be assessed from the intensity of the light arising from the specific spot.

In general, protein microarrays make use of various types of capture receptors for protein screening [16,17]. For example, antigen receptors are used for the detection of antibodies in serum. More recently, other types of capture molecules such as peptides or nucleic acid aptamers have been introduced.

Protein microarray formats falls into two main classes: forward phase array and reverse phase array. In a forward phase microarray, each spot consists of an assembly of identical receptors and is able to indicate a specific analyte in the sample (Figure 6.8). Hence, this format allows for multiplexed screening for various analytes in the same sample. According to the labeling alternative, a forward phase microarray can be designed for direct, competitive or sandwich assay.

In the reverse phase array, all proteins in a sample are attached at a specific spot on a microarray, which allows for screening simultaneously a great number of samples (Figure 6.9). The microarray is then incubated with a single specific antibody to detect the presence of the target protein across many samples. This method allows proteins levels in a large number of biological samples to be screened simultaneously. Cells or micro-organisms can also be screened in the same way.

In many cases, the biological activity of a protein (such as an antibody) can be mimicked by a shorter peptide, that is similar to a specific sequence in the protein molecule, although often with a partial loss of activity. The most impressive application of peptide receptors is in peptide microarrays that rely on synthetic peptides [18–20].

The fabrication of peptide microarrays can be performed according to two different alternatives: parallel, on-chip peptide synthesis or immobilization of previously synthesized peptides.

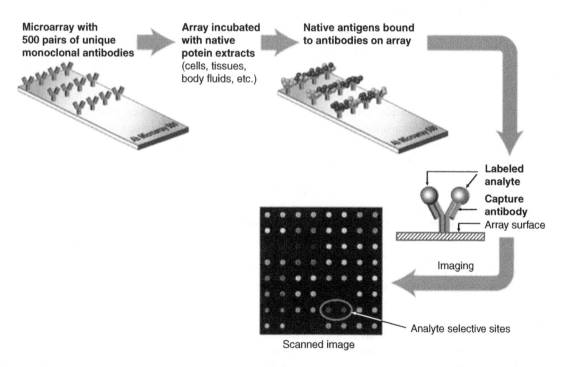

Figure 6.7 Schematics of an antibody microarray screening. A large number of identical antibody molecules are assembled at each spot. Adapted with permission from http://www.clontech.com/products/detail.asp?tabno=2&product_id=10576-. Copyright 2011 Clontech Laboratories, Inc.

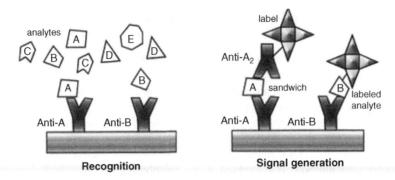

Figure 6.8   Forward phase protein microarray. At each spot on the microarray, a particular receptor (for example, an antibody) is immobilized in order to detect a particular analyte in the same sample. Adapted with permission from [15]. Copyright 2003 Elsevier Ltd.

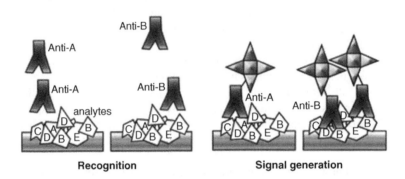

Figure 6.9   Reverse phase protein microarray. At each spot on the microarray, all proteins in a particular sample are immobilized. The target protein is then is detected by means of a specific signaling antibody. Adapted with permission from [15]. Copyright 2003 Elsevier Ltd.

The main advantage of peptide microarrays steams for their wide field of applications. Thus, antibody and nucleic acids arrays are limited to particular types of analyte. By contrast, peptide arrays allow for high throughout screening of a broad variety of interactions such as binding of proteins or ions, action of enzymes, cell adhesion and many others.

6.6 Biological Receptors

In biochemistry, a *receptor* is a protein molecule embedded in the plasma membrane or the cell cytoplasm, to which specific kinds of signaling molecules may attach. As the term "receptor" is used in chemical sensors to denote the recognition reagent or site, the term chemoreceptor will next denote a biochemical receptor that originates from living cells. A chemical species that binds to a chemoreceptor is called a ligand. Depending on the receptor, the ligand can be an ion or a small-molecule compound, such as neurotransmitters, hormones, pharmaceuticals, or toxins [21]. In living organisms, endogenous ligands (such as adrenalin) trigger a physiological process in response to the binding event. Ligands eliciting such an effect are called *agonists*. On the other hand, some ligands (called *antagonists*) merely block receptors without inducing a response. Many functions of the body are regulated by these chemoreceptors responding uniquely to specific agonists or antagonists that act as chemical messengers inside the body. Chemoreceptors are hence critical mediators in both intra- and extracellular communication.

Exogenous agonists and antagonists act like endogenous ligands causing stimulation or inhibition of a physiological response. Such an effect can be produced by pharmaceutical products, toxins or alkaloids. Binding of exogenous ligands is due to the fact that a chemoreceptor interacts with one or a few small atom groups in the ligand molecule but is relatively insensitive to the overall molecule structure. This behavior contrasts that of antibodies that are much more selective.

Ligand–chemoreceptor interactions are typical affinity reaction and follow in many cases the Equation (6.7). Deviations from this equation occur when the binding ratio differs from 1 : 1.

Chemoreceptors can be used in biosensors either as incorporated in a cell or after isolation from a biological preparation. The first alternative relies on the detection of the electric signal (action potential) developed by nerve cells under the effect of a chemical stimulus acting as an agonist. Alternatively chemical sensing of an agonist can be performed by monitoring the cellular response of micro-organisms caused by a chemical stimulus. Thus, a changes in the pH of the cell environment can be used for drug-screening purposes.

Isolated receptors can be attached to the transducer surface by common methods for protein immobilization (Chapter 5). Incorporation into phospholipid membranes or liposomes (that mimic cell membranes) is often

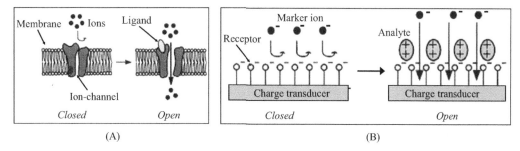

Figure 6.10 Principle of the ion-channel sensors. (A) Permeability modulation by channel blocking/opening (ligand-gated ion channel); (B) Configuration of a chemical sensor based on electrostatic modulation of membrane permeability. Adapted with permission from [23]. Copyright 2002 Chemical Society of Japan.

preferred. Direct or competitive determination of the ligand proceeds according to the immunoassay schemes described before.

Very promising chemoreceptors are the so-called *ion channels*, proteins that feature permeability to ions in response to the a triggering chemical stimulus (Figure 6.10) [22–24]. There are two different ion-channel response mechanisms. In the first one, the ligand blocks physically or opens the intramolecular channel (Figure 6.10A). The second mechanism is based on electrostatic attraction or repulsion that modulates the membrane permeability (Figure 6.10B). The recognition event can be detected by monitoring the ions crossing the membrane or the change in the local electrical state (voltage, current, or capacitance) in response to ion transfer. The great advantage of such receptors arises from the amplification caused by the channel opening that allow the transfer of a relatively large amount of ions.

The prototypic ligand-gated ion channel is the nicotinic acetylcholine receptor. It made up of five subunits, arranged symmetrically around a central pore. When acetylcholine binds to the receptor, it alters the receptor configuration and cause the pore to open. This pore allows Na^+ ions to flow down their electrochemical gradient.

Light-activated receptors provide unique transduction opportunities. Thus, the bacterial light receptor bacteriorhodopsin promotes hydrogen-ion pumping across the bacterial membrane. When incorporated in a lipid layer, light stimulation produces a change in the voltage across the membrane. If an ion carrier (e.g., valinomycin) is also included in the membrane, light can induce transport of ions, thus enabling for transduction. The presence of an ion in the sample affects the ion transport provided that this ion binds to the ion-carrier.

Inspired by natural chemoreceptors, certain synthetic systems with analyte-modulated ion permeability have been designed and applied to the development of chemical sensors [23].

Other compounds of biological origin are useful as recognition receptors in sensors although they do not act as biological receptor in living organism. Thus, lectins are proteins that bind to a specific sugar. The most representative in this class is *Concanavalin A* that has been employed as recognition receptor in sugar sensors.

6.7 Artificial Receptors

6.7.1 Cyclodextrins and Host–Guest Chemistry

Cyclodextrins are a class of compounds made up of sugar molecules (D-glucopyranoside) bound together in a ring (Figure 6.11A) [25,26]. Cyclodextrins are not found in nature but can be obtained by the enzymatic conversion of

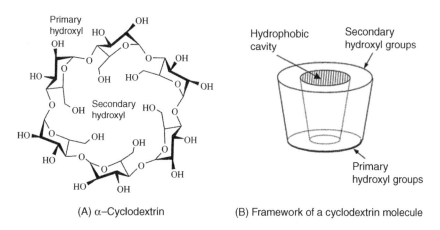

(A) α–Cyclodextrin　　　　(B) Framework of a cyclodextrin molecule

Figure 6.11 Cyclodextrins structure and shape. Adapted with permission from [1]. Copyright John Wiley & Sons, Ltd.

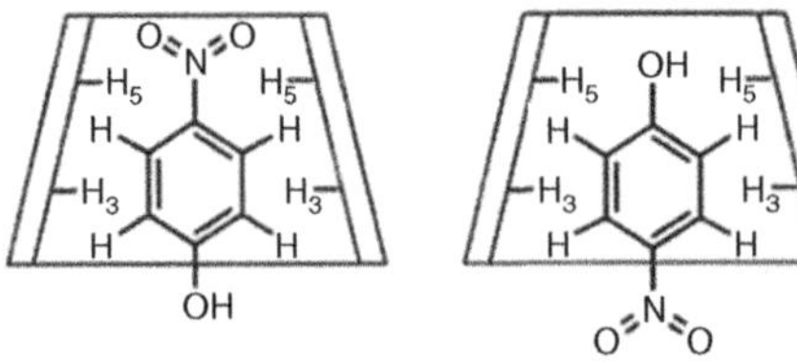

Figure 6.12   A nitrophenol guest molecule accommodated in the cavity of an α-cyclodextrin in the solution (left) and in the solid state (right). Reproduced with permission from [26]. Copyright 2006 Wiley-VCH Verlag GmBH & Co. KGaA.

starch. The main members of the group are α-, β- and γ-cyclodextrin, comprising six, seven or eight sugars, respectively, per molecule. The shape of the cyclodextrin molecule is often represented as a tapering cone displaying two different faces referred to as primary and secondary rims (Figure 6.11B). The narrower rim comprises the primary hydroxyl groups, whereas the wider rim contains the secondary hydroxyl groups ($-CH_2OH$). Cyclodextrins are generally water soluble but the solubility depends on the number of sugar molecules in the ring. Thus, β-cyclodextrin is less soluble than the α and γ analogs. The upper and lower rims display hydroxyl groups that can be derivatized.

Cyclodextrins are able to form host–guest complexes with hydrophobic molecules producing *inclusion complexes*. For example, Figure 6.12 shows the inclusion complex of 4-nitrophenol with α-cyclodextrin. In water solution, the hydrophobic moiety of the guest is accommodated within the cavity while the hydrophilic $-OH$ group is positioned in the hydrophilic environment provided by rime hydroxyls. In the solid state, the hydrophobic $-NO_2$ group is not subject to interactions with a polar solvent and hence points outwards from the cavity.

This class of interaction belongs to the host–guest chemistry which addresses complexes formed by inclusion of a host species in the cavity of a host molecule.

The stability of host–guest compounds is determined by noncovalent interactions. Due to the complex chemical structure of the cyclodextrin cavity, almost any kind of noncovalent interaction with the guest is possible, including hydrophobic effects, van de Waals interactions, hydrogen bonding, and dipole-dipole interactions. A guest molecules will be fully enclosed within the cyclodextrin molecule if it fits the cavity. With long molecules, cyclodextrins are capable to form channel structures in which several cyclodextrin cavities line up to form an extended hydrophobic cavity into which the guest can be threaded. Moreover, formation of a cyclodextrin host–guest complex can in some instance be followed by catalytic conversion of the guest, an effect similar to that produced by enzymes.

As usually with recognition receptors, cyclodextrins need to be interfaced with the transducer by immobilization. This can be done by covalent coupling via the hydroxyl groups or by incorporation in plasticized PVC membranes. Thiol-derivatized cyclodextrins can be assembled on a gold surface by sulfur chemisorption. It is also possible to synthesize polymers with cyclodextrins appended to the backbone.

Common transduction procedures in affinity sensors are also applicable in cyclodextrin-based sensors. Thus, charged analytes can be detected by assessing the charge density or distribution at interface. Transduction by mass-sensitive devices is also appropriate. The peculiar molecular shape of cyclodextrins allows certain specific transduction methods to be devised. For example, Figure 6.13 depicts an optical method based on a fluorescent label tagged to the cyclodextrin molecule. In the absence of the guest, the label will be entrapped in the cavity. If the guest analyte binds stronger than the label, the label is displaced from the hydrophobic cavity and gets in contact with the aqueous milieu that affects the intensity of fluorescence light.

Properly modified cyclodextrins imparts excellent recognition properties for neutral or charged organic and bio-organic compounds [27–29]. The capacity of cyclodextrins to elicit chiral recognition deserves particular mention. In other words, cyclodextrins interact preferentially with a specific enantiomer in an enantiomer mixture. This is not surprising if one takes into account the fact that the sugar building blocks in cyclodextrins are themselves chiral compounds.

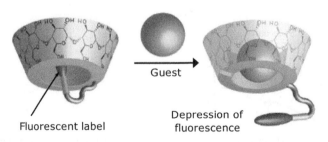

Guest

Fluorescent label

Depression of fluorescence

Figure 6.13   Turn-off fluorescent sensor using cyclodextrin as receptor and a fluorescent label inserted in the cavity. Label fluorescence is depressed after moving from the hydrophobic cavity into the aqueous milieu. Adapted from [27].

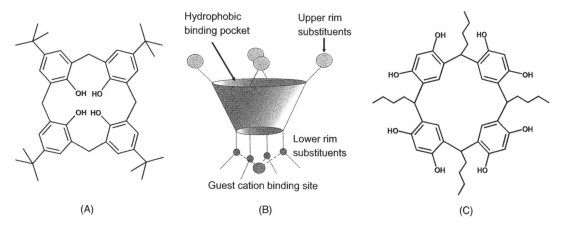

Figure 6.14 (A) Chemical structure of para-tert-butyl-calix[4]arene; (B) framework of a calixarene molecule in the cone conformation; (C) a resorcin[4]arene. Adapted with permission from [1]. Copyright John Wiley & Sons, Ltd.

6.7.2 Calixarenes

Calixarenes are macrocyclic compounds formed by condensation of *p*-substituted phenols with formaldehyde [30]. Figure 6.14A emphasizes a calixarene derived from *p*-tert-butylphenol; it consists of a four-phenol residues\. Calixarene molecules can be formed of more than four phenols, and the compound designation is done by including the phenol number. Thus, the compound in Figure 6.14a is a calix[4]arene, whereas calix[6]arene denotes a molecule including six phenol residues.

Calixarene molecules assume a three-dimensional basket shape (Figure 6.14B). Three distinct regions can be distinguished in a calixarene molecule: a wide upper rim, a narrow lower rim, and a central annulus. In calix[4]arenes, the internal volume is about $10\,nm^3$. In calixarenes derived from phenol, the four hydroxyl groups are intra-annular at the lower rim (Figure 6.14A). Conversely, in a resorcin[4]arene, eight hydroxyl groups are placed extra-annular at the upper ring (Figure 6.14C). Hydroxyl groups at the rims can act as binding sites for calixarene functionalization in order to impart selectivity for various guest species.

According to the properties of the guest species, its interaction with a calixarene can occur either at a rim or within the annular region. Recognition by the rim regions is secured by suitable substituents. Thus, hydroxyls groups at the rim are able to form multiple hydrogen bonds with guest molecules with a suitable shape and size. Conversion of calix[4]arene hydroxyls into ethers imparts the lower rim specificity to Na^+. Rim functionalization with charged groups allow for recognition of small charged proteins. Hydrophobic species of fitting size can be accommodated within the intracavity. At the same time, π-electrons in this region can interact with certain metal ions by coordinate bonding. Other recognition possibilities arise from building up receptor sites sandwiched between two calixarene fragments.

It is obvious that calixarene molecules can be tailored such as to recognize various types of analytes [31]. Most of the current applications refer to ion sensors [32,33] based on either potentiometric or optical transduction methods [34]. Sensing of biocompounds by means of calixarene receptors is also achievable. For example, a carboxyl-derivatized calixarene embedded in a lipid layer interacts with cytochrome c by electrostatic forces [35]. Potential applications of calixarenes in enantioselective sensing have also been envisaged.

6.7.3 Molecularly Imprinted Polymers (MIPs)

Recognition by natural receptors relies on the complementarity of the binding site and the analyte as far as both shape and interactive groups are concerned. This principle has been exploited to prepare synthetic receptors in the form of cavities in a polymer structure leading to the development of molecularly imprinted polymers (MIPs) [36–38].

Molecular imprinting is a method to produce specific recognition sites for a target molecule in a synthetic polymer matrix. Fabrication of MIPs involves several steps, as shown in Figure 6.15A. First, a template (the analyte or a similar derivative) is assembled with functionalized monomers linked to specific groups in the template. Next, a polymer matrix is formed by copolymerization in the presence of nonfunctionalized monomers. During this process, the template is incorporated into small cavities shaped after its form and also featuring overhanging functional groups. The polymer monolith is then mechanically ground to obtain micrometer-sized particles. Finally, the template is removed leaving free receptor sites imprinted in the rigid polymer. The imprint maintains a steric (size, shape) and chemical (special arrangement of complementary groups) memory of the template. The analyte binds reversibly to such sites (Figure 6.15B) with selectivity close to that of the antibody–antigen interaction.

Assembly of the template with the functional monomers can be carried out by either covalent or noncovalent binding. The covalent approach is more time consuming but the yield of binding sites is higher and their distribution

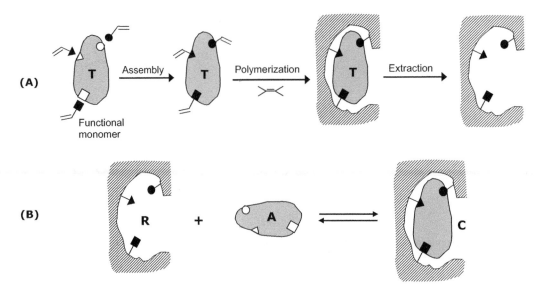

Figure 6.15 Synthesis (A) and recognition (B) by molecularly imprinted polymers. T is the template, A is the analyte and R is the receptor site.

is more homogeneous. Noncovalent assembly allows for more expedient fabrication but, due to the weak interactions between the template and functionalized monomers, both the yield and homogeneity may be somewhat inferior.

The most widely used MIPs materials are acrylic and vinyl polymers. There is a vast choice of suitable monomers to produce such materials and a selection is made in accordance with the expected properties of the binding site. By selecting a suitable monomer, the cavity surface can be tailored so as to be hydrophobic, electrically charged, or providing hydrogen bonding. In this way, one obtains an analyte-compatible microenvironment within the cavity.

The molecular imprinting technique can be applied to a wide range of target molecules. The imprinting of small molecules (for example, pharmaceuticals, pesticides, amino acids, peptides, nucleotide bases and sugars) is well established. Techniques for imprinting proteins and other macromolecular compounds have also been developed [39,40].

Molecularly imprinted polymer membranes are particularly advantageous in sensor applications. Such membranes can be prepared at a metal surface by electrochemical polymerization or at an insulator surface by chemical grafting or by using polymerization initiators linked to the surface. The membrane thickness should be below 100 nm in order to allow for free diffusion of analyte molecules to the receptor sites. Thicker membranes are also effective if the material has a porous structure.

Clearly, it is advantageous to use MIPs with a high density of imprinted sites. In order to quantify this parameter, the *imprinting efficiency* (*IE*) has been introduced [41]. This parameter refers to a polymer attached to the transducer surface and is defined by the following equation in which NMIP represents a nonimprinted polymer similar to the MIP:

$$IE = \frac{\text{Amount of bounded MIP}}{\text{Amount of bounded NMIP}} \tag{6.13}$$

This parameter cannot be easily measured but it can be estimated by the signal enhancement ratio (SE):

$$SE = \frac{\text{Signal with MIP}}{\text{Signal with NMIP}} \tag{6.14}$$

The term in the denominator of the above ratio is assessed by a control sensor whose surface is coated by the nonimprinted polymer.

Substantially improved characteristics have been noticed with MIP nanomaterials such as nanobeads or nanofibers [42]. Bare MIP nanoparticles provide much higher density of accessible receptor sites and faster analyte diffusion to the receptors. Alternatively, a very thin MIP layer can be formed on inert nanoparticles in the core–shell configuration. In this arrangement, the core particle can also act as support to reporting elements such as fluorescent labels. Prospects for MPSs application in the development of sensor microarrays are very promising.

As far as the transduction is concerned, common affinity sensor transduction methods are also suitable for the MIPs-based sensors. Widely used are the mass-sensitive transducers [43], but methods based on ionic conduction or luminescent labels are also currently applied [38,44].

MIPs are very promising as substitutes for natural receptors in affinity sensors [44]. The main benefits arise from facile receptor tailoring for a very broad range of target species, from ions and small molecules to macromolecular compounds. Also, the good stability and the possibility of obtaining MIPs in various physical forms such membranes or nanoparticle render these materials very attractive in various applications [38,44,45,46].

6.8 Outlook

A broad spectrum of recognition methods rely on affinity reactions, that is, strong noncovalent association resulting in the formation of a stable receptor–ligand complex. By far, the most advanced field in this class is represented by immunosensors that are used as diagnostic tools in clinical analysis [47,48] and also in environment monitoring [49,50] and food industry [51,52] for the detection of pathogens, allergens and organic pollutants.

Although chemoreceptors are less selective than antibodies, they have found a series of interesting applications. The reduced selectivity of chemoreceptors could be an advantage when screening for a particular class of analytes. Thus, the presence of illegal drugs in a sample can be expediently assessed by means of a chemoreceptor-based sensor, while the detection of a specific analyte within this class can be next performed by means of an antibody-based method. The advent of molecularly imprinted polymers and other receptors of nonbiological origin expands the field of application of affinity reactions and provides new possibilities of fabricating more rugged affinity sensors. At the same time, nanomaterials open new opportunities for improving the characteristics of affinity sensors [53]. A series of well-documented protocols for affinity sensors are available [54].

A particular mention deserve nucleic acid aptamers, that are synthetic nucleic acids that mimic the antibody behavior and hence interact selectively with proteins or small-molecules [55]. More details on nucleic acid aptamers are available in Chapter 7. Synthetic peptides (peptide aptamers) that emerged as a new class of recognition receptors show excellent potential for application in multiplexed assays.

Questions and Exercises

1 What are the main domains in an antibody molecule and what are their functions?
2 What kinds of antibodies are best suited for analytical applications and why?
3 Review the particular features of antibody immobilization.
4 Why is the two-site immunoassay name also ascribed to the excess reagent format?
5 Using Figures 6.5 and 6.6 as examples, sketch the reaction sequences in the antibody assay by means of antigen receptors.
6 What is the effect of K_a in competitive and noncompetitive assay?
7 Assume that in a competitive immunoassay the response is proportional to the c^*/r_0 ratio. Using Equation (6.12), calculate the a_0/a^* ratios that lead to c^*/r_0 ratios of 0.1 and 0.9, respectively. The above values indicate the limits of the working range. How can the linear response range in competitive assay be shifted into the sample concentration region?
8 What is the effect of the receptor total concentration on the response in competitive and noncompetitive assay?
9 Comment on the configuration and advantages of protein and peptide microarrays.
10 What are the particular features of cyclodextrin molecules?
11 How cyclodextrins interact with other molecules and what determines the specificity of such interactions?
12 Comment on possible immobilization and transduction methods in cyclodextrin-based sensors.
13 What are calixarenes? Review the main characteristics of calixarene molecules as far as the structure and the conformation are concerned.
14 What are the applications of calixarenes in chemical sensors? How can the calixarene selectivity be tuned in order to recognize a particular analyte?
15 Review the main steps and reagents in the synthesis of MIPs.
16 In what physical forms can MIPs be obtained? What are the most suitable forms for sensor applications and why?
17 Review the advantages and drawbacks of artificial receptors when compared with natural receptors.

References

1. Steed, J.W. and Atwood, J.L. (2009) *Supramolecular Chemistry*, John Wiley & Sons, Chichester.
2. Cragg, P.J. (2010) *Supramolecular Chemistry: From Biological Inspiration to Biomedical Applications*, Springer, Dordrecht.
3. Diamandis, E.P. and Christopoulos, T.K. (eds) (1996) *Immunoassay*, Academic Press, San Diego, Calif.
4. Liddel, E. (2005) Antibodies, in *The Immunoassay Handbook* (ed. D. Wild), Elsevier, Amsterdam, pp. 144–166.
5. Byrne, B., Stack, E., Gilmartin, N. *et al.* (2009) Antibody-based sensors: Principles, problems and potential for detection of pathogens and associated toxins. *Sensors*, **9**, 4407–4445.

6. Bange, A., Halsall, H.B., and Heineman, W.R. (2005) Microfluidic immunosensor systems. *Biosens. Bioelectron.*, **20**, 2488–2503.
7. Marquette, C.A. and Blum, L.J. (2006) State of the art and recent advances in immunoanalytical systems. *Biosens. Bioelectron.*, **21**, 1424–1433.
8. Walker, M.R., Stott, R.A., and Thorpf, G.H.G. (1992) Enzyme-labelled antibodies in bioassay, in *Bioanalytical Applications of Enzymes* (eds C.H. Suelter and L.J. Kricka), John Wiley & Sons, New York, pp. 179–207.
9. Crowther, J.A. (1995) *ELISA: Theory and Practice*, Humana Press, Totowa, N.J.
10. Deshpande, S.S. (1996) *Enzyme Immunoassays: From Concept to Product Development*, Chapman & Hall, New York.
11. Cosnier, S. (2005) Affinity biosensors based on electropolymerized films. *Electroanalysis.*, **17**, 1701–1715.
12. Varlan, A.R., Suls, J., Jacobs, P. *et al.* (1995) A new technique of enzyme entrapment for planar biosensors. *Biosens. Bioelectron.*, **10**, R15–R19.
13. Langone, J.J. (1982) Protein A of Staphylococcus aureus and related immunoglobulin receptors produced by Streptococci and Pneumococci. *Adv. Immunol.* **32**, 157–252
14. Donahue, A.C. and Albitar, M. (2010) Antibodies in biosensing, in *Recognition Receptors in Biosensors* (ed. M. Zourob), Springer Science, New York, pp. 221–248.
15. Liotta, L.A., Espina, V., Mehta, A.I. *et al.* (2003) Protein microarrays: Meeting analytical challenges for clinical applications. *Cancer Cell*, **3**, 317–325.
16. Herr, A.E. (2009) Protein microarrays for the detection of biothreats, in *Microarrays* (eds K. Dill, P. Grodzinski, and R.H. Liu), Springer, New York, pp. 169–190.
17. Kingsmore, S.F. (2006) Multiplexed protein measurement: Technologies and applications of protein and antibody arrays. *Nature Rev. Drug Discovery*, **5**, 310–320.
18. Min, D.H. and Mrksich, M. (2004) Peptide arrays: Towards routine implementation. *Curr. Opin. Chem. Biol.*, **8**, 554–558.
19. Panicker, R.C., Sun, H., Chen, G.Y.J. *et al.* (2009) Peptide-based microarray, in *Microarrays* (eds R.A. Potyailo, K., Dill, P. Grodzinski, and R.H. Liu), Springer, New York, pp. 139–167.
20. Tothill, I.E. (2010) Peptides as molecular receptors, in *Recognition Receptors in Biosensors* (ed. M. Zourob), Springer Science, New York, pp. 249–274.
21. Taylor, R.F. (1991) Immobilized antibody- and receptor-based biosensors, in *Protein Immobilization: Fundamentals and Applications* (ed. R.F. Taylor), M. Dekker, New York, pp. 263–303.
22. Cornell, B.A. (2008) Ion channel biosensors, in *Handbook of Biosensors and Biochips* (ed. R.S. Marks, D.C. Cullen, I. Karube, C.R. Lowe and H.H. Weetall), John Wiley & Sons, New York.
23. Sugawara, M., Hirano, A., Buhlmann, P. *et al.* (2002) Design and application of ion-channel sensors based on biological and artificial receptors. *Bull. Chem. Soc. Jpn.*, **75**, 187–201.
24. Luo, L.Q., Yang, X.R., and Wang, E.K. (1999) Ion channel sensor. *Anal. Lett.*, **32**, 1271–1286.
25. Szejtli, J. (1998) Introduction and overview of cyclodextrin chemistry. *Chem. Rev.*, **98**, 1743–1753.
26. Dodziuk, H. (ed.) (2006) *Cyclodextrins and Their Complexes: Chemistry, Analytical Methods, Applications*, Wiley-VCH, Weinheim.
27. Ogoshi, T. and Harada, A. (2008) Chemical sensors based on cyclodextrin derivatives. *Sensors*, **8**, 4961–4982.
28. Shahgaldian, P. and Pieles, U. (2006) Cyclodextrin derivatives as chiral supramolecular receptors for enantioselective sensing. *Sensors*, **6**, 593–615.
29. Hayashita, T., Yamauchi, A., Tong, A.J. *et al.* (2004) Design of supramolecular cyclodextrin complex sensors for ion and molecule recognition in water. *J. Incl. Phenom. Macrocycl. Chem.*, **50**, 87–94.
30. Gutsche, C.D. (1998) *Calixarenes Revisited*, Royal Society of Chemistry, Cambridge.
31. Diamond, D. and McKervey, M.A. (1996) Calixarene-based sensing agents. *Chem. Soc. Rev.*, **25**, 15–24.
32. Arora, V., Chawla, H.M., and Singh, S.P. (2007) Calixarenes as sensor materials for recognition and separation of metal ions. *Arkivoc.*, 172–200.
33. El Nashar, R.M., Wagdy, H.A.A., and Aboul-Enein, H.Y. (2009) Applications of calixarenes as potential ionophores for electrochemical sensors. *Curr. Anal. Chem.*, **5**, 249–270.
34. Leray, I. and Valeur, B. (2009) Calixarene-based fluorescent molecular sensors for toxic metals. *Eur. J. Inorg. Chem.*, 3525–3535.
35. Mohsin, M.A., Banica, F.G., Oshima, T. *et al.* (2010) Electrochemical impedance spectroscopy for assessing the recognition of cytochrome c by immobilized calixarenes. *Electroanalysis*, **23**, 1229–1235.
36. Komiyama, M. (2003) *Molecular Imprinting: From Fundamentals to Applications*, Wiley-VCH, Weinheim.
37. Haupt, K. (2003) Molecularly imprinted polymers: The next generation. *Anal. Chem.*, **75**, 376A–383A.
38. Haupt, K. (2001) Molecularly imprinted polymers in analytical chemistry. *Analyst*, **126**, 747–756.
39. Bossi, A., Bonini, F., Turner, A.P.F. *et al.* (2007) Molecularly imprinted polymers for the recognition of proteins: The state of the art. *Biosens. Bioelectron.*, **22**, 1131–1137.
40. Bergmann, N.M. and Peppas, N.A. (2008) Molecularly imprinted polymers with specific recognition for macromolecules and proteins. *Prog. Polym. Sci.*, **33**, 271–288.
41. Suryanarayanan, V., Wu, C.T., and Ho, K.C. (2010) Molecularly Imprinted Electrochemical Sensors. *Electroanalysis*, **22**, 1795–1811.
42. Tokonami, S., Shiigi, H., and Nagaoka, T. (2009) Review: Micro1- and nanosized molecularly imprinted polymers for high-throughput analytical applications. *Anal. Chim. Acta*, **641**, 7–13.
43. Uludag, Y., Piletsky, S.A., Turner, A.P.F. *et al.* (2007) Piezoelectric sensors based on molecular imprinted polymers for detection of low molecular mass analytes. *FEBS J.*, **274**, 5471–5480.
44. Danielsson, B. (2008) Artificial receptors, in *Biosensing for the 21st Century*, Springer, Berlin, pp. 97–122.

45. Bui, B.T.S. and Haupt, K. (2010) Molecularly imprinted polymers: synthetic receptors in bioanalysis. *Anal. Bioanal. Chem.*, **398**, 2481–2492.

46. Haupt, K. and Mosbach, K. (2000) Molecularly imprinted polymers and their use in biomimetic sensors. *Chem. Rev.*, **100**, 2495–2504.

47. Luppa, P.B., Sokoll, L.J., and Chan, D.W. (2001) Immunosensors-principles and applications to clinical chemistry. *Clin. Chim. Acta*, **314**, 1–26.

48. Aizawa, M. (1994) Immunosensors for clinical analysis. *Adv. Clin. Chem.*, **31**, 247–275.

49. Jiang, X.S., Li, D.Y., Xu, X. *et al.* (2008) Immunosensors for detection of pesticide residues. *Biosens. Bioelectron.*, **23**, 1577–1587.

50. Mallat, E., Barcelo, D., Barzen, C., *et al.* (2001) Immunosensors for pesticide determination in natural waters. *TrAC-Trends Anal. Chem.*, **20**, 124–132.

51. Yadav, R., Dwivedi, S., Kumar, S. *et al.* (2010) Trends and Perspectives of Biosensors for Food and Environmental Virology. *Food Environ. Virol.*, **2**, 53–63.

52. Rasooly, A. and Herold, K.E. (2006) Biosensors for the analysis of food- and waterborne pathogens and their toxins. *J. AOAC Int.*, **89**, 873–883.

53. Liu, G.D. and Lin, Y.H. (2007) Nanomaterial labels in electrochemical immunosensors and immunoassays. *Talanta*, **74**, 308–317.

54. Rogers, K.R. and Mulchandani, A. (eds) (1998) *Affinity Biosensors: Techniques and Protocols*, Humana Press, Totowa.

55. Crawford, M., Woodman, R., and Ferrigno, P.K. (2003) Peptide aptamers: Tools for biology and drug discovery. *Brief. Funct. Genom.*, **5**, 72–79.

7
Nucleic Acids in Chemical Sensors

Nucleic acids are essential compounds in all living organisms. They provide the means of storing and transferring genetic information. The determination of nucleic acid composition and content in biological samples is a key issue in biosciences and medicine. It is therefore not surprising that nucleic acid sensors are one of the main areas of the biosensor science and technology [1–4]. In such a sensor, nucleic acids can act as both receptors and analytes. Moreover, nucleic acids can also interact in a specific way with a number of small molecules and proteins. This property expands the field of application of nucleic acids as receptors in chemical sensors for such compounds. Applications of nucleic acid sensors encompass a very broad area including not only biomedical sciences but also environment monitoring, food industry and forensic science.

This chapter includes an overview of the main properties of nucleic acids as far as the chemical structure and reactivity are concerned. The applications of nucleic acids as recognition components in biosensors, as well as the transduction strategies in nucleic acid sensors are closely associated with the properties of these compounds. Transduction strategies and applications of nucleic acid sensors are considered later in the chapter.

7.1 Nucleic Acid Structure and Properties

Nucleic acids are biopolymers belonging to the polynucleotide class. The most common representative types of polynucleotide are the deoxyribonucleic acids (DNAs) and the ribonucleic acids (RNAs) that are involved in the storage and transfer of genetic information [5,6]. Both DNAs and RNAs are polymeric compounds that incorporate *nucleotides* as basic units. The nucleotide unit consists of three structural fragments: a heterocyclic *nucleobase* bound to a pentose sugar molecule that in turn is attached to a phosphate residue via an ester linkage at the 5′ position on the sugar (see Figure 7.1A). The canonical bases found in natural nucleic acids are adenine, guanine, cytosine thymine and uracil (Figure 7.1B). Adenine and guanine are *purines*, whereas cytosine thymine and uracil are *pyrimidines*. The first four bases occur in DNA; in RNA, uracil replaces thymine. In addition, a series of less-common nucleobases may occur in RNAs. An example of this kind is hypoxanthine (a purine derivative) that forms, in combination with a sugar, the inosine nucleoside.

A key structural difference between DNAs and RNAs is apparent from their names. This difference is in the sugar component of each classes of polynucleotide; *ribose* is the sugar in RNA and *2′-deoxyribose* is the sugar in DNA (Figure 7.1A). The sugar plus base combination forms a *nucleoside*. The sugar plus base plus phosphate combination is termed a *nucleotide*. Nucleotides are identified by the first letter (as a capital) in the name of the particular base (that is, A, G, C, T and U).

In the nucleic acid structure, nucleotides are associated in polymer chains through intermediate phosphate residues attached to 3′ and 5′ carbons in the sugar moiety (Figure 7.2). By convention, the direction of the chain is recorded from the 5′ to the 3′ end as shown by the arrow in Figure 7.2. This sequence of nucleobases in a polynucleotide gives the *primary structure* of a nucleic acid. Shorthand notation of the primary structure consists of the sequence of nucleotide acronyms (starting from the 5′ end) proceeded by the symbol d or r for DNA and RNA, respectively. Sometimes, the symbol p is inserted between base symbols to denote the phosphate linkage. Thus, a *single-strand* structures (*ss-nucleic acid*) can be symbolized as:

5′CpGpCpGpApApTpTpCpGpCpG

or more simply as:

d(CGCGAATTCGCG) or, r(TCGCGCGCGAAT)

Two complementary polynucleotide chains can form a duplex by association involving hydrogen bonds between A–T and C–G pairs (Watson–Crick model) as shown in Figure 7.3A. The set of interactions between complementary

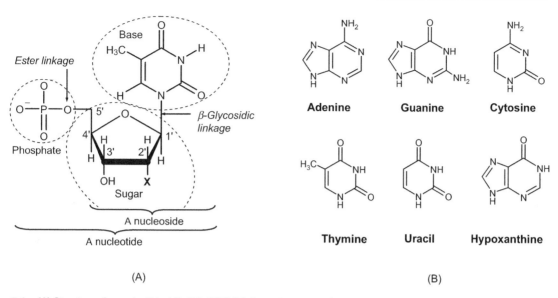

Figure 7.1 (A) Structure of a nucleotide. X is H in DNA (2′-deoxyribose sugar) and OH in RNA (ribose sugar). (B) Canonical nucleobases.

bases gives the *secondary structure* of a nucleic acid. Base pairing produces a *double strand* (ds) that folds into a double helix (Figure 7.3B), which is the typical conformation of DNA. An antiparallel double-strand sequence (*ds-nucleic acid*) can be symbolized as follows:

5′CGCGAATTCGCG
3′GCGCTTAAGCGC

Polynucleotides can adopt a large variety of conformations involving double-strand association. On the other hand, a single polynucleotide strand can fold into a specific conformation stabilized by intramolecular hydrogen bonds according to the Watson–Crick model. This is exemplified in Figure 7.4 that displays a *hairpin* conformation. The conformation of

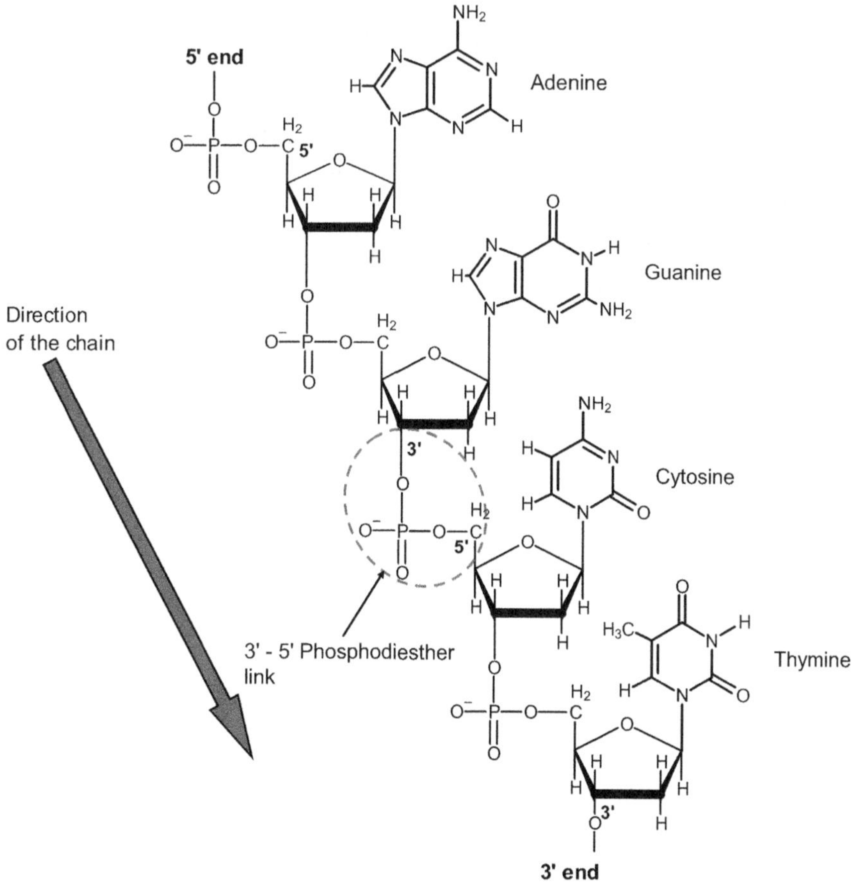

Figure 7.2 A polynucleotide chain (DNA primary structure).

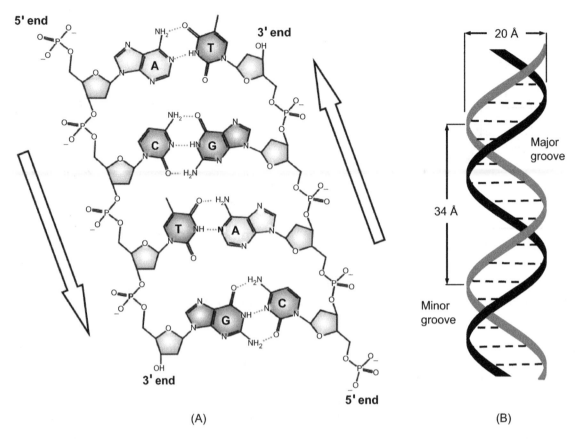

(A)   (B)

Figure 7.3 Structure of DNA. (A) Association of two antiparallel polynucleotide strands in DNA. Dotted lines represent hydrogen bonds between nucleobases. Note the opposite orientation of the strands. (B) Schematics of the DNA double helix with the sugar–phosphate backbones shown as ribbons. (A) Adapted from http://commons.wikimedia.org/wiki/File:DNA_chemical_structure.svg Last accessed 17/05/2012. Credit Madeleine Price Ball.

the nucleic acid molecule in three-dimensional space forms the *tertiary structure*. Particular features in the tertiary structure are minor and major *grooves* (Figure 7.3B) that play an essential role in supramolecular interactions of nucleic acids with proteins or small molecules leading to mixed adducts stabilized by shape fitting and noncovalent bonds.

Owing to the negative charge at the phosphate moieties in the nucleic acid chain, polynucleotides are polyanions. For this reason, cations such as Na^+ and Mg^{2+} are essential for stabilizing the structure.

When DNA duplex molecules are subjected to extreme conditions of temperature or pH, the hydrogen bonds of the double helix are cleaved and the two strands separate from each other in a process called *denaturation*. When heat is used to this end, the DNA is said to melt; this process occurs at a specific *melting temperature* T_m that is, by definition, the temperature at which only 50% of DNA is present in the duplex form. If the conditions change back to

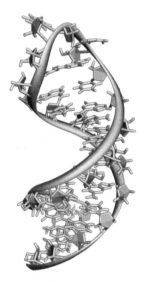

Figure 7.4 A hairpin loop from a pre-mRNA. Note that some nucleobases are not paired. Adapted from http://commons.wikimedia.org/wiki/File:Pre-mRNA-1ysv.png-tubes.png Last accessed 17/05/2012.

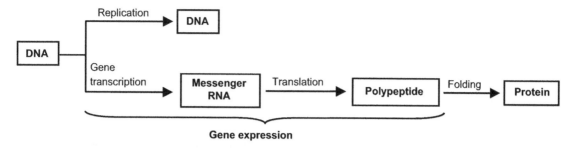

Figure 7.5　Pathways in the transfer of genetic information.

the natural parameters, two DNA strands may reassociate to restore the double helix. Denaturation–renaturation processes can also occur with single-strand tertiary structures.

Nucleic acids are the means by which genetic information is stored and transferred in living cells. The cell uses the sequence of bases in a nucleic acid to produce a sequence of amino acids in order to construct a particular protein that the cell needs. In other words, the primary structure of each protein is encoded by a nucleic acid fragment called *gene*. In a gene, each amino acid is indicated by a sequence of three nucleobases (a *codon*). The structure of a particular gene may vary slightly from individual to individual in a given species. Each particular form of a gene is called an *allele*. Hair color, for example, is encrypted in an allele that is specific to each individual and display only minute deviation from the similar allele of another individual.

Genetic information is stored in DNA molecules that are found in complex combination with proteins (chromatin) that make up *chromosomes*. Each gene is located at a particular site, termed the *locus,* along the chromosome. During cell reproduction, DNA molecules are *replicated* to produce identical DNA molecules in other cells. Protein synthesis, on the other hand, occurs in ribosomes under the control of messenger RNA (mRNA) that is synthesized as a replica of the relevant gene in DNA (*transcription*) before traveling to the ribosome. Conversion of genetic information from the codon language to the amino acid sequence during protein biosynthesis is called *translation*. The mechanism of transfer of genetic information is summarized in Figure 7.5. The process of protein biosynthesis under the control of genetic information is known as gene expression.

Polynucleotides are available for use in sensors from various natural sources. However, for many applications, custom-made oligonucleotides are required and are prepared automatically by *solid-phase synthesis* [7]. In this technique, a nucleotide is first covalently linked to a solid support (such as controlled-pore glass or macroporous polystyrene) and then chain elongation is carried out by assembling further nucleotides according to a predetermined sequence. Finally, the olygonucleotide is released by the cleavage of its bond to the support. This technique can be used to produce immobilized polynucleotides directly at the surface of a transducer.

7.2　Nucleic Acid Analogs

In addition to natural nucleic acids, medicine and molecular biology research make use of synthetic *nucleic acid analogs* [8]. Such an analog may include one or more of the natural components of a nucleic acid altered. Thus, in *peptide nucleic acids* (PNAs), the phosphate–sugar backbone is replaced by a polypeptide chain to which nucleobases are linked by methylene carbonyl bonds (Figure 7.6). Despite the different backbone structure, such

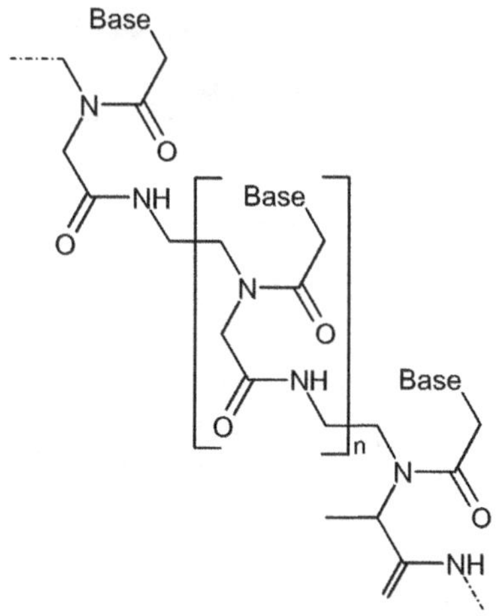

Figure 7.6　Structure of peptide nucleic acids (PNAs).

compounds can bind according to the Watson–Crick model. Since the backbone of PNAs contain no charged phosphate groups, the binding between PNA/DNA strands is stronger than between DNA/DNA ones owing to the lack of electrostatic repulsion. This brings about a higher melting temperature compared with that of the natural analog.

Hydrogen bonds in nucleobase pairs can be replaced by coordinated metal ion bridges such as T–Hg–T, A–Zn–T and G–Zn–C if the pH is sufficiently high to stimulate base deprotonation at the binding site. This allows the formation of additional linkages between bases that do not match the Watson–Crick model and led to development of metal ion sensors using nucleobases or oligonucleotides as receptors.

Nucleic acid analogs can be synthesized by inserting non-canonical nucleobases. This confers different base pairing and base stacking properties. Examples of this kind include common nucleobases, which can pair with all canonical bases.

7.3 Nucleic Acids as Receptors in Recognition Processes

7.3.1 Hybridization: Polynucleotide Recognition

An oligonucleotide can be used as the receptor for the recognition of a complementary strand by a process called *hybridization* (Figure 7.7) [9,10]. To this end, the recognition component (a hybridization *capture probe*) is attached to a solid support and incubated with a sample containing the *target* polynucleotide in the form of a single strand. A perfect match of the target–probe couple leads to the formation of a very stable duplex. Any mismatch results in a less-stable duplex. A properly selected probe of about 25 nucleotides length can recognize specifically very long polynucleotides. Compared with a very long probe, a short probe has the advantage of a much faster hybridization. Moreover, long olygonucleotide probes tend to fold over themselves, which impedes the hybridization process. At the same time, a longer probe is more selective as it binds to a longer sequence in the target. The introduction of four-to-six mismatches between the sensing interface and the target nucleic acid is sufficient to prevent hybridization and discriminates against a mutant. Ideally, one single mismatch should be sufficient for detecting a mutant.

For hybridization to occur it is essential to maintain the probe strand oriented perpendicularly to the support surface. Hydrophobic or positively charged surfaces promote a parallel orientation by interaction with the nucleobases or the backbone, respectively. Conversely, negatively charged surfaces compel the probe by electrostatic repulsion to adopt a perpendicular orientation to the support surface.

The hybridization event can be detected by means of a physical effect, such as mass change or modifications in the optical properties of the interface. Alternatively, indirect transduction can be carried out by means of a tag such as a fluorescent or electrochemically active compound that is brought by hybridization to be within the active region of the transducer (Figure 7.8).

The throughput of a hybridization assay depends on the reaction velocity. As hybridization at surface-attached probes involves prior target diffusion, this process dictates the overall reaction rate at low target concentrations.

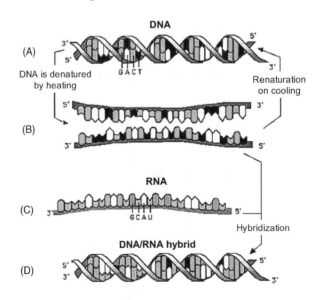

Figure 7.7 Thermal denaturation of a DNA duplex and hybridization with a complementary RNA sequence. (A) Double strand DNA; (B) two single-strand DNAs; (C) complementary RNA; (D) DNA–RNA hybrid. Adapted with permission from [11] by courtesy of Ewa Paszek. Last accessed 3/05/2012.

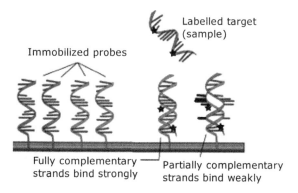

Figure 7.8 Hybridization of a target polynucleotide with a surface-immobilized probe and transduction by means of a reporter tag. Adapted from http://en.wikipedia.org/wiki/File:NA_hybrid.svg Last accessed 17/05/2012.

In this case, hybrid formation obeys the first-order Langmuir model:

$$\Gamma(t) = \Gamma_f[1 - \exp{(-t/\tau_D)}^{1/2}] \tag{7.1}$$

where t is the time, Γ is the surface concentration of the hybrid, Γ_f is the limiting Γ (at infinite time) and $\tau_D = \pi\Gamma_f^2/4Dc^2$, c being the target concentration and D its diffusion coefficient.

For hybridization to occur, the temperature should be below the melting temperature. The melting temperature depends on DNA sequence and length, cation concentration, target concentration, and the presence of denaturants (formamide or DMSO).

Any mismatch causes the melting temperature of the hybrid $(T_m{'})$ to be lower that that of a faultless double strand (T_m). In order to detect the latter, transduction should be performed at an intermediate temperature between $T_m{'}$ and T_m.

Hybridization can be accompanied by nonspecific adsorption of the target to the support material that gives rise to a background response. This problem can be alleviated by adding a mixture of high molecular weight polymers (Denhardt's solution) capable of saturating nonspecific binding sites at the solid support.

A high cation concentration is essential in hybridization assays in order to lessen the electrostatic repulsion between polynucleotide strands. When using peptide nucleic acids, the cation effect is less important.

Hybridization is a relatively slow process because it involves multiple site interactions of two oligonucleotides, accompanied by a radical change in molecular configuration. If the probe is immobilized at a solid support, diffusion of the target to the surface introduces an additional kinetic restriction. Moreover, the restricted mobility of the immobilized probe results in a marked sluggishness of the hybridization process compared with the process occurring in a homogeneous phase. In order to give the probe some freedom of movement, it is recommended to attach it to the surface through a long and flexible spacer arm. The hybridization rate is therefore the main limiting factor as far as the response time of the DNA sensor is concerned. The response time can be a few minutes under well-optimized conditions, but it can be a few hours in some cases, particularly at very low target concentrations.

If the sensor is exposed to crude cell extracts, care should be taken in order to avoid oligonucleotide decay by nuclease-catalyzed hydrolysis. A nuclease is an endogenous enzyme capable of cleaving the phosphodiester bonds between the nucleotide subunits of nucleic acids. RNA becomes nuclease resistant if the 2′ hydroxyl is methylated or substituted by fluoride. Another method for preventing nuclease degradation relies on substituting the phosphate in the backbone by *phosphorothioate* in which nonbridging oxygens are replaced by sulfur. This modification renders the internucleotide linkage resistant to nuclease degradation.

PNA probes are often more convenient that their DNA counterparts. As a PNA probe is neutral, it can function at a lower cation concentration. For the same reason, under similar conditions a PNA–DNA duplex is more stable than the DNA–DNA analog and, consequently a shorter PNA probe may give rise to a sufficiently stable hybrid. Moreover, the greater stability of the PNA–DNA duplex allows hybridization to be effected at higher temperatures, to induce a higher reaction rate. In addition, peptide nucleic acids that are missing phosphodiester bonds are immune to nuclease degradation.

Hybridization sensors can be used to reveal mutant genes associated with genetically determined diseases. This is because genetic mutations result in mismatching in the hybridization process. Another application of hybridization consists of identifying pathogen micro-organisms in water supplies, foodstuffs and biological samples. In such an assay, a pathogen can be detected by means of its specific nucleic acids. Similarly, genetically modified organisms can be detected by means of hybridization of a probe with a specific sequence in the target [12].

7.3.2 Recognition of Non-Nucleotide Compounds

Non-nucleotide molecules can interact with nucleic acids by noncovalent bonding to produce relatively stable adducts. This process may disturb the DNA function (that is, storage and transfer of genetic information) and is widely used in

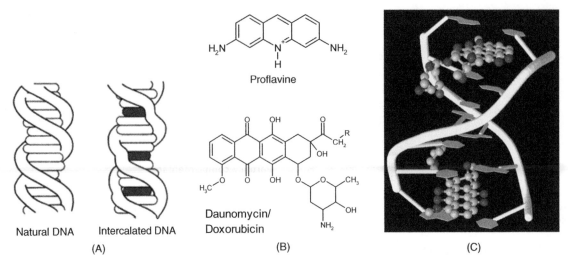

Figure 7.9 (A) Intercalative binding of a small molecule (ligand) into the DNA helix. (B) Intercalating organic molecules: proflavine, daunomycin (R = H) and doxorubicin (R = OH). (C) Two daunomycin molecules intercalated in a d(CGATCG) duplex. The upper part shows the planar, polycyclic moiety of the molecule inserted between two base pairs with the lateral ring accommodated in a minor groove. (A) Adapted from http://commons.wikimedia.org/wiki/File:DNA_intercalation.svg Last accessed 17/05/2012. (C) Adapted from http://www.rcsb.org/pdb/explore/explore.do?structureId=1da0 Last accessed 15/05/2012.

therapy of cancer or viral diseases or for disinfection purposes. Similar processes can be employed advantageously in order to perform recognition of non-nucleotide compounds in chemical sensors based on nucleic acid receptors.

Small molecules can form adducts with DNA by *intercalation*. Intercalation occurs when ligands of an appropriate size and chemical nature fit themselves in between base pairs of DNA (Figure 7.9A) [13]. Optimum intercalation occurs with planar, polycyclic molecules bearing a positive charge (for example, proflavine, Figure 7.9B). Intercalation proceeds as follows. In aqueous isotonic solution, the cationic intercalator is attracted electrostatically to the polyanionic DNA. The ligand displaces a cation (that is balancing the DNA charge) and forms a weak electrostatic bond with the outer surface of the DNA. From this position, the ligand may then slide into the hydrophobic environment found between the base pairs and away from the hydrophilic outer environment surrounding the DNA. The ensuing conformation is stabilized by hydrogen bonds and van der Waals interactions between the planar ligand and the base pairs around it.

As a result of intercalation, the helix expands because the distance between the bases pairs on either side of the ligand increases. This brings about a slight alteration in the helical twist (Figure 7.9A).

A small intercalator can be selected so as to be either electrochemically active or optically detectable. It can therefore be used as a label tag for detecting the DNA duplex.

More voluminous and asymmetrical molecules can be accommodated in DNA grooves where they are stabilized by van der Waals interactions and, in addition, by hydrogen bonds with adjacent bases, allowing site-selective interaction. Figure 7.9C demonstrates the binding of daunomycin (an anticancer drug shown in Figure 7.9B) that combines intercalation and minor groove interactions.

A number of metal chelates recognize specific sites in DNA by intercalation of the planar fragment, whilst the bulky nonplanar moiety of the complex fits closely against the walls and floor of a major groove [14]. Such complexes can act as optical or electrochemical indicators for double-strand nucleic acids.

Many drugs, as well as mutagenic pollutants perform as DNA intercalating compounds. Intercalation can therefore be employed for recognizing such analytes and allows the development of DNA sensors for these classes of substances [15]. In addition, intercalation has proved useful for attaching transduction tags to ds-nucleic acids.

7.3.3 Recognition by Nucleic Acid Aptamers

Nucleic acid *aptamers* are synthetic nucleic acids that manifest selective affinity for proteins or some small molecule compounds. The name of this class of compounds has been derived from the Latin *aptus* (meaning. "fitted") and the Greek *meros* (meaning "part"). Such compounds are being actively investigated for therapeutic applications [16,17]. At the same time, the high affinity of aptamers for specific analytes make them very useful as receptors in a series of analytical applications including biosensor development [18–21].

The design of aptamers is done by an automatic combinatorial procedure (systematic evolution of ligands by exponential enrichment (SELEX)), which is a trial and error method. This process begins with the synthesis of a very large library of oligonucleotides consisting of randomly generated sequences of fixed length flanked by constant 5′ and 3′ ends that serve as primers. For a randomly generated sequence including *n* nucleotides, the number of

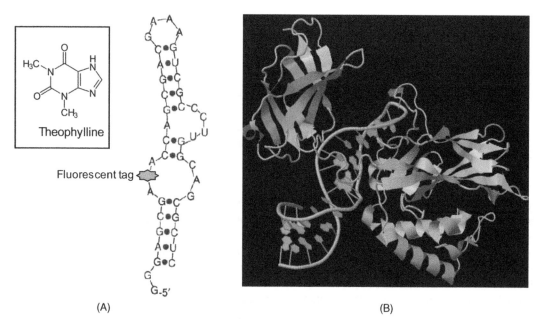

(A)   (B)

Figure 7.10 Examples of aptamers. (A) An antitheophylline aptamer. A site-specific fluorescent tag is inserted in the backbone by phosphodiesther links. (B) Structure of a protein–RNA aptamer complex. The aptamer in the hairpin configuration binds to the NF-kappaB p50 transcription factor protein. (A) Adapted with permission from [17]. Copyright 2006 Elsevier; (B) Adapted from pdb1ooa. (B) Adapted from http://www.rcsb.org/pdb/explore/explore.do?structureId=1OOA Last accessed 15/05/2012.

possible sequences in the library is 4^n. The sequences in the library are exposed to the target ligand and those that do not bind the target are removed. The bound sequences are separated and multiplied to prepare for subsequent rounds of selection in which the stringency of the separation conditions is increased in order to identify the tightest-binding sequences.

Aptamers are tailored so as to develop by folding a specific binding site for the target compound. As an example Figure 7.10A shows an aptamer with affinity to the small molecule of theophylline, a drug used in therapy of respiratory diseases. This figure demonstrates that a functional tag can be inserted readily in the aptamer molecule at the most convenient site. Small molecules are accommodated as a rule in a pocket, shaped by aptamer folding. Figure 7.10B displays the complex of an RNA aptamer with a protein (NF-kappaB p50 transcription factor). The aptamer molecule binds to the p50 N-terminal domain of the protein due to its particular secondary structure. Thus, the guanines, placed at the edge of the helix, together with the major groove, form the binding surface for the protein.

Aptamers are useful in affinity sensors in which they play a role similar to that of antibodies. However, when compared with antibodies, aptamers demonstrate a series of advantages. Aptamers can be obtained by chemical synthesis and are less expensive. They are more stable than antibodies and can be easily modified and immobilized on a solid support. After performing the recognition function, an aptamer can be efficiently regenerated without loss of either integrity or selectivity.

Aptamer selectivity for a selected target is expressed by the dissociation constant of the complex. This was found to be below 10^{-6} M for some thrombin aptamers, which demonstrates a huge stability of the aptamer–protein complex, comparable with that of an antigen–antibody conjugate.

Like monoclonal antibodies, aptamers can be prepared in epitope-selective form and by this means different aptamers for the same target molecules become available. For example, Figure 7.11 shows two aptamers partially hybridized and bearing on each arm an epitope-specific binding site. A protein molecule can be attached to both aptamer arms, which results in enhanced selectivity and better stability of the complex. The term *aptabody* was proposed to denote a duplex oligonucleotides that binds to two different sites of the target molecule [22].

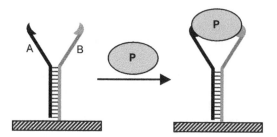

Figure 7.11 Schematic representation of protein binding by two partially hybridized epitope selective aptamers (A and B).

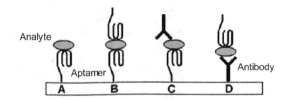

Figure 7.12 Possible configurations of aptamer-based sensors. (A) single-site format; (B) dual-site (sandwich) format with two aptamers; (C) and (D) mixed sandwich format with an a aptamer and an antibody. Adapted with permission from [24]. Copyright 2009 Wiley-VCH Verlag GmBH & Co. KGaA.

Aptamer-analyte recognition can be approached in several ways, as emphasized in Figure 7.12. Note that aptamers and antibodies can be used jointly in the sandwich format, as both of them act as affinity reagents. An overview of aptamer sensor design is presented in ref. [23].

Aptamers are therefore promising substitutes, or complements for antibodies in affinity sensors. Aptamers demonstrate some advantages in this respect. Thus, they can be readily synthesized in crude or modified form. Compared with antibodies, aptamers are more stable, are more resistant to multiple analysis–regeneration cycles, and show a lower propensity to nonspecific binding. RNA aptamers can decay in contact with biological samples by nuclease-catalyzed cleavage. Substitution of $-OH$ at the $2'$ position of the sugar by fluoride or amino groups stabilizes the oligonucleotide against this reaction.

7.4 Immobilization of Nucleic Acids

Probe immobilization is a crucial step in manufacturing DNA chips [25] and methods developed to this end are also applicable in the fabrication of DNA sensors [26,27]. Adsorption, covalent linking, chemisorption at surfaces or affinity reactions are the most common methods for attaching nucleic acid probes to a solid surface. Various materials such as polymers (polystyrene, polyacrylamide, nylon), inorganic materials (glass, carbon, gold) and bio-organic products (cellulose, chitosan) have been used as supports for probe immobilization. Immobilization of aptamers can be performed by similar methods [28].

An alternative to the immobilization of a preformed probe is the direct synthesis of the probe on the transducer surface. This approach is less time consuming but purification of the product as well as conventional purity assessment cannot be achieved in this case.

7.4.1 Adsorption

Adsorption is a facile method for attaching nucleic acids to surfaces since no reagents or modified polynucleotides are required. Although various noncovalent interactions could contribute to probe adsorption, electrostatic attraction between a positively charged surface and the negative adsorbate backbone is the most common procedure. Thus, a $-NH_3^+$ functionalized solid support binds oligonucleotides in an almost parallel orientation to the surface by electrostatic interactions involving multiple phosphate residues. If a positively charged gel is attached to the support, oligonucleotides bind by incorporation of a terminal phosphate residue in the gel and adopt an almost perpendicular orientation to the surface.

Much attention has been paid to adsorption at carbon materials for the purpose of fabricating electrochemical DNA sensors [27,29]. Various bare carbon materials (such as glassy carbon and pyrolytic graphite), composite carbon materials (such as carbon paste, rigid carbon composites [30] or screen-printed supports [31]) have been investigated to this end.

Adsorption of a single-strand sequence at glassy carbon is triggered by hydrophobic interaction of free bases with the surface. Conversely, a double-strand form is poorly adsorbed because its bases are hidden in the double-helix structure. As a rule, carbon electrodes are subject to energetic oxidation prior to adsorption in order to create oxygenated functionalities ($-OH$; $-COO^-$) that enhance the adsorption by hydrogen bonding. Performing adsorption at a positively polarized electrode also contributes to increasing the surface coverage by electrostatic interaction with the oligonucleotide backbone.

Adsorption yields a layer of sparsely bound molecules that do not resist thorough washing steps that are required to remove mismatched sequences. In addition, most of the immobilized probe lies in a position that is not favorable to hybridization causing this process to be relatively inefficient. Despite these inconveniencies, adsorption is considered as a promising immobilization method, particularly in conjunction with mass fabrication of disposable screen-printed sensors. The intrinsically poor sensitivity of such sensors can be ameliorated by prior polymerase chain reaction (Section 7.5.3).

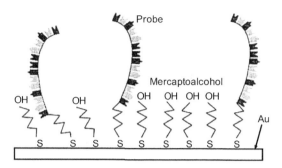

Figure 7.13 Immobilization of an oligonucleotide probe by self-assembly at a gold surface. Adapted with permission from [28]. Copyright 2000 Elsevier.

7.4.2 Immobilization by Self-Assembly

This method is based on the spontaneous reaction of the thiol group with gold to form a strong S−Au bond [32,33]; see also section 5.4.8). The sulfur-containing nucleotides are often supplied in the form of disulfides (for example, DNA-$(CH_2)_3$-S-S-$(CH_2)_3$-OH) to avoid thiol degradation before use. Prior to probe chemisorption, the disulfide linkage must be cleaved by reduction with dithiothreitol. As a rule, probe immobilization is followed by chemisorption of a mercapto-alcohol (for example, mercaptohexanol, Figure 7.13). The latter process prevents nonspecific adsorption of oligonucleotides and blocks the access of interferents to the transducer surface. Such sensors should be used with caution at elevated temperatures where the sulfur–gold bond becomes weak.

Sulfur–gold self-assembly enables a single-point attachment of the probe that gives it conformational mobility that, in turn, results in a relatively high hybridization velocity.

Probe immobilization by thiol self-assembly is a very popular immobilization method due to its simplicity, and because gold layers are widely used in various kinds of chemical sensors.

7.4.3 Immobilization by Polymerization

Oligonucleotides can be incorporated into a polypyrrole layer at a carbon electrode by electrochemical polymerization of pyrrole. As the polymer backbone has a positive charge, oligonucleotides bind electrostatically as a counter-anion. As an alternative to the electrostatic binding, covalent linking of the probe to the polymer backbone can be achieved by copolymerization of pyrrole with pyrrole-grafted probe molecules. This technique is outlined in Figure 7.14. Polymerization is conducted at high pyrrole monomer excess ($n + m = 20\,000$) in order to allow the

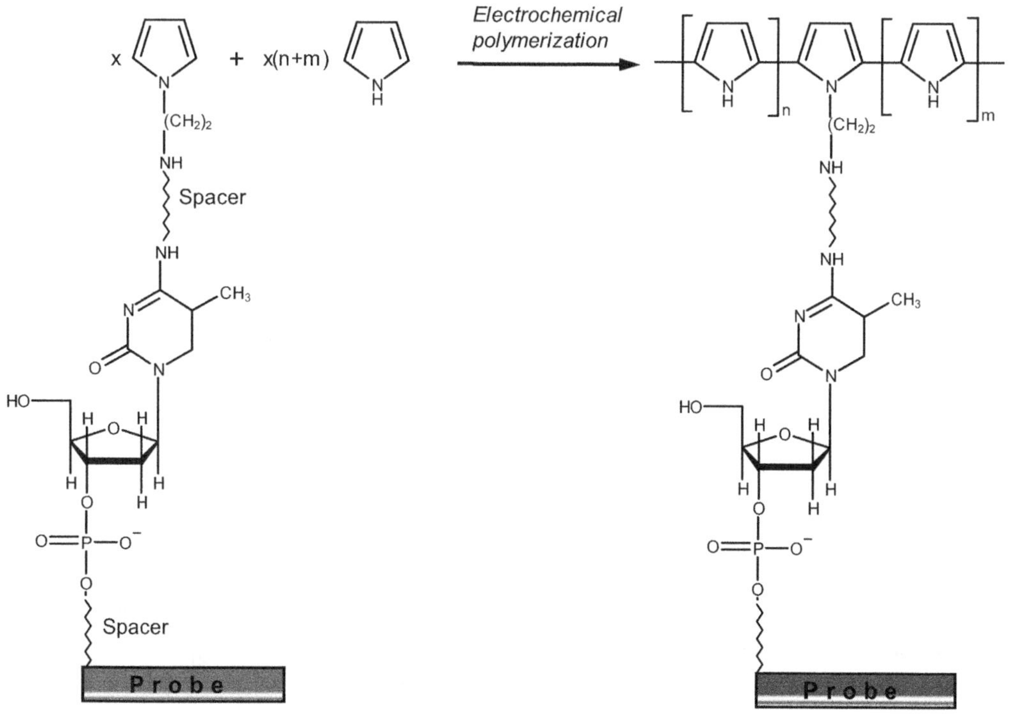

Figure 7.14 Grafting of a nucleic acid probe on a polypyrrole chain. Adapted with permission from [36]. Copyright 2004 Oxford University Press.

probe enough mobility. This method is suited for one-step spotting on continuous conducting surfaces (gold, silicon) or on individually addressable conducting spots [34].

Oligonucleotide probes have also been incorporated in redox hydrogels in order to allow an enzyme tag to exchange electrons with the underlying electrode. Care should be taken in this case to allow the probe sufficient mobility in order to react efficiently in the hybridization step.

7.4.4 Covalent Immobilization on Functionalized Surfaces

Covalent coupling of probes to a solid support may be time consuming and costly but it results in a rugged recognition layer that is suitable for multiple use. This method often requires prior formation of suitable functional groups on the support surface in order to perform further crosslinking of the probe. The most common groups for covalent immobilization of nucleic acids are carboxyl, amine and hydroxyl. Thus, carbon surfaces can be functionalized with carboxyl groups by energetic oxidation with a nitric acid and potassium dichromate mixture. Silicon, glass and carbon surfaces can be modified with organosilanes bearing various functionalities that are susceptible to crosslinking. So, materials with exposed hydroxyls at the surface (glass, cellulose, pretreated carbon) can be modified with amine groups by reaction with an aminosilane (for example, APTES, $H_2N(CH_2)_3Si(OC_2H_5)_3$). A similar treatment can be applied to silicon surfaces after preparing a silicon oxide layer by oxidation and hydrolysis to yield hydroxyl groups. Gold and indium-tin oxide can be modified by self-assembly of functionalized thiols containing a reactive group at the distal end. For example, gold surfaces can be modified with amine groups by self-assembly of an amino-thiol ($HS-(CH_2)_n-NH_2$). A functionalized polymer layer can be prepared by electrochemical copolymerization of pyrrole and a pyrrole derivative containing a linker (e.g., 3-carboxypyrrole).

As far as the binding site in the probe is concerned, it can be a pre-existing group, such as the $5'$ end phosphate residue or an exocyclic amine group in guanine. Often, the probe is synthesized so as to include a linking end, such as an amine or a short guanine sequence.

Figure 7.15A shows the coupling of the phosphate end to an amine-derivatized support through carbodiimide activation to yield a phosphoramide linking. Such a reaction is conducted in the presence of methylimidazole and N-hydroxy-succinimide (NHS) as stabilizers of intermediate products.

Covalent coupling to a carboxyl functionalized surfaces through guanine linkage is shown in Figure 7.15B. This method also makes use of carbodiimide activation and works equally well with amine-tethered probes.

Coupling of an amine-tethered probe to a hydroxyl-modified surface can be achieved in the presence of carbonyldiimidazole (CDI) as shown in Figure 7.15C.

7.4.5 Coupling by Affinity Reactions

The outstanding stability of the avidin–biotin linkage is the basis of a general procedure for assembling DNA sensors [36]. In this immobilization method, a biotinylated oligonucleotide probe is grafted to surface-linked avidin or strepavidin. However, the protein layer at the surface may induce nonspecific binding with negative effects on the sensitivity and the selectivity of some sensors. Biotin can be immobilized by any of the general methods for protein immobilization. It is also possible to synthesize biotin-tethered thioalkanes or pyrrole monomers for immobilization by chemisorption or gold or electropolymerization, respectively. Avidin–biotin interaction is also frequently used for obtaining oligonucleotide–enzyme conjugates in which the enzyme functions as a reactive hybridization reporter.

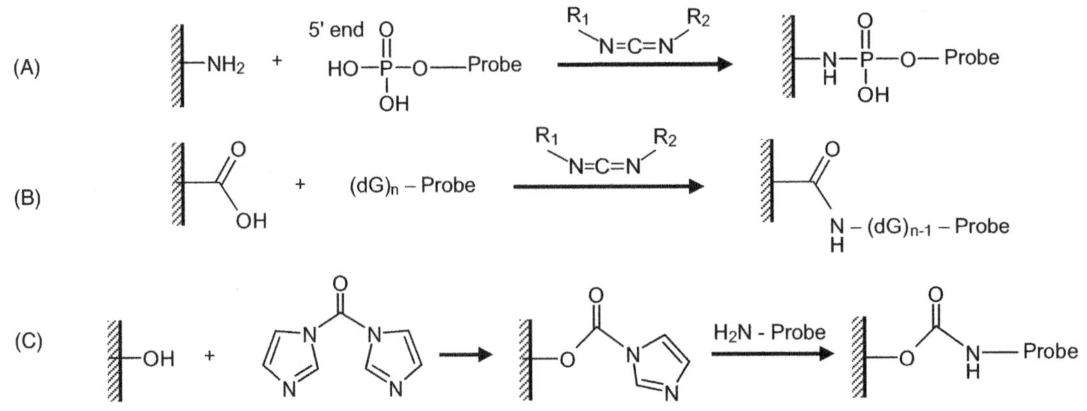

Figure 7.15 Covalent immobilization of a probe to a functionalized support. (A) Probe coupling to a $-NH_2$ functionalized support. (B) Coupling to a carboxyl-modified support. (C) Coupling to a hydroxyl-functionalized support.

7.4.6 Polynucleotides–Nanoparticles Hybrids

Nanoparticles are widely used in nucleic acid sensor research either as label tags, structural components, or for other purposes [37,38]. Conjugation of nucleic acids with properly functionalized nanoparticles can, as a rule, be achieved using the immobilization methods described above. Metal nanoparticles, semiconductor nanoparticles, carbon nanotubes and magnetic nanoparticle are among the nanomaterials currently investigated for applications in nucleic acid sensors.

Gold nanoparticles provide various possibilities for improving genetic assays either as supports for probe–target hybridization, as optical labels or as catalytic centers for chemical amplification [39,40]. Chemisorption of thiol-tethered probes allows straightforward immobilization on gold nanoparticles. As an alternative, gold nanoparticles can be functionalized first with biotin or an amine in view of further crosslinking of the probe by pertinent reactions.

Binding of amine-tethered nucleic acids to carbon nanotubes can be achieved by crosslinking to carboxyl groups at the nanotube terminus by means of carbodiimide activation. Vertically aligned carbon nanotubes allow the achievement of optimum spatial arrangement of nucleic acid probes for DNA arrays and sensors. A high density of probe molecules is achieved by immobilization at multiwalled carbon nanotubes that display multiple binding sites at the terminus.

Attaching nucleic acids to magnetic nanoparticles allows facile separation, concentration purification, and identification by the effect of a magnetic field. In order to attach nucleic acids to pure magnetite or silica-coated magnetite, nanoparticles will be first functionalized with an $-NH_2$ linker by reaction with an aminosilane. Next, glutaraldehyde is used for attaching an amine-derivatized nucleic acid. Alternatively, magnetic nanoparticles can be functionalized with avidin in view of the avidin–biotin linking of the nucleic acid.

7.5 Transduction Methods in Nucleic Acids Sensors

Recognition by nucleic acid receptors is essentially an affinity reaction. It is therefore not surprising that the transduction procedures used in conjunction with nucleic acid recognition are rather similar to those encountered in immunosensors. By the same reason, the response function of DNA sensors is similar to that of noncompetition immunosensors in that it displays a saturation trend at high analyte concentration.

7.5.1 Label-Free Transduction Methods

As in the case of standard affinity sensors, changes in the physical state of the recognition layer as a result of the recognition process can be used for direct, label-free transduction. Thus, a mass change can be detected by means of mass-sensitive transducers, whereas alteration in some optical properties (for example, the refractive index) can be detected by optical methods. A series of label-free transduction methods rely on the intrinsic electrochemical activity of nucleobases in nucleic acids.

Since both the nucleic acid probe and target are polyanions, their hybridization on an electrode enhances the surface negative charge that repels anionic redox probes such as $[Fe(CN)_6]^{3-/4-}$. Electrostatic repulsion inhibits the electrochemical reaction of the redox probe and enhances in this way the electron-transfer resistance, which can be monitored by electrochemical impedance spectrometry [41].

7.5.2 Label-Based Transduction

Transduction tags of various types are widely used in nucleic acids sensors for indirect detection by means of optical or electrochemical methods. Several label-based transduction strategies for hybridization sensors are given next.

Some procedures rely on selective linking of a signaling tag to the hybrid resulting from the recognition step. Thus, Figure 7.16 demonstrates a procedure based on the fact that the target could be longer than the probe and

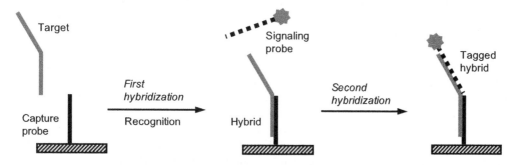

Figure 7.16 Labeling by hybridization of a tagged strand with the overhanging moiety of the target (sandwich hybridization).

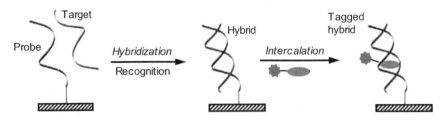

Figure 7.17 Posthybridization labeling using a label-tagged intercalator.

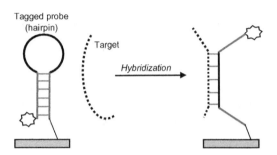

Figure 7.18 Configuration of a DNA hybridization sensor based on a hairpin-shaped probe (molecular beacon configuration).

preserves after recognition a single-strand fragment. A tagged oligonucleotide (signaling probe) that is complementary to this overhanging fragment can be attached to the initial hybrid in a second hybridization process leading to a tagged hybrid that can be detected in order to assess the extent of the recognition process. This technique is often termed *sandwich hybridization*.

Figure 7.17 illustrates an alternative postrecognition labeling method, which relies on the intercalation of a tagged intercalator. This method may be less specific than that based on sandwich hybridization, but it requires only some common chemicals and avoids the synthesis of a tagged oligonucleotide.

A frequently used transduction method relies on hairpin-shaped probes (Figure 7.18). The probe molecule includes complementary peripheral stems that form base pairs to stabilize the hairpin conformation. The label attached to the end of the chain shows a property that is decisively dependent on its position with respect to the transducer surface or the second end of the hairpin (spatially resolved transduction). Thus, an electrochemical label can accept electrons from the underlying electrode only if it is localized in the immediate proximity of the electrode. The recognition fragment of the probe is localized in the loop region such that the hybridization brings about the collapse of the hairpin conformation and shifts the label to a position where it is no longer active. So, in the case of electrochemical transduction, the label is moved too far from the electrode surface and current flow is interrupted ("signal-off"). Other strategies based on configuration change after hybridization have also been envisaged [42]. It is sometimes more advantageous to design a "signal-on" format that relies on a signal enhancement after hybridization. A much better signal:background ratio is obtained in this way. The term "*molecular beacon*" is often used to designate the DNA sensor configuration based on a hairpin probe.

7.5.3 DNA Amplification

Often, the amount of available DNA analyte is too low to allow reliable quantitation by a nucleic acid sensor and some amplification procedure has to be included in the protocol in order to enhance the sensitivity [43]. A straightforward amplification method relies on application of enzyme tags. High turnover of enzymes brings about an intrinsic signal enhancement as a single label can produce a large amount of product per time unit.

A general amplification method is based on multiplication of the original target by the polymerase chain reaction (PCB) [44–46]. This method aims at enhancing of the amount of analyte DNA prior to the assay by a sequence of enzyme-catalyzed reactions.

PCR involves a *polymerase* enzyme and two *primers*. Polymerases are enzymes that catalyze the polymerization of a polynucleotide as a complementary form to a template strand. The final product of this reaction is a double helix composed of the template and the newly synthesized complementary polynucleotide. The thermostable *Taq* polymerase is used in PCR because it allows the reaction to be carried out at a high temperature and consequently with a high reaction velocity. A primer is a short strand that is complementary to a specific sequence in the template chain. There are three basic steps in this method (Figure 7.19). First, the original ds-nucleic acid must be denatured, that is, the strands of its helix must be unwound and separated by heating to 90–96 °C. The second step (at about 50–55 °C) consists of primer hybridization in which the primers bind to their complementary sequences in the ss-DNA

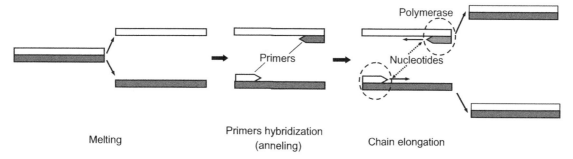

Figure 7.19 Schematics of the polymerase chain reaction (PCR). Nucleotides should be present in the triphospahate form.

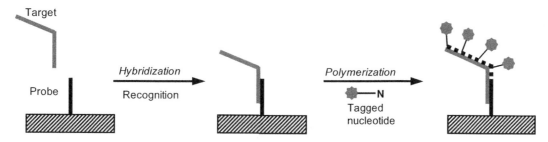

Figure 7.20 PCR application for adding multiple transduction tags after hybridization.

molecules. The third step is the synthesis of a complementary nucleic acid strand under the effect of the polymerase, at 72 °C. Starting from the primer, the polymerase can read the template strand and match it with complementary nucleotides. The products are two new double helices each composed of one of the original strands plus its newly assembled complement. This cycle can be repeated many times, the amount of final product molecules being 2^n, where n is the number of cycles. As each cycle takes only 1–3 min, millions of copies of the original DNA can be generated in about 1 h.

During the PCR process, label tags can be introduced into the newly synthesized sequence. This is achieved by including a tagged nucleotide in the reaction mixture.

Genetic assays are performed using *amplicons*, which are pieces of DNA formed as the products of natural or artificial amplification events. For example, amplicons can be formed via polymerase chain reactions or ligase chain reactions, but also by natural gene duplication.

Preliminary multiplication of the target is the most straightforward application of PCR. Sensitivity enhancement can also be achieved by posthybridization PCR. To this end, the PCR is effected after hybridization, using the probe as primer and the overhanging target moiety in the hybrid as template. For example, a complementary chain grown by PCR on the ss-fragment of the target will increase the mass variation in a mass-sensitive transduction method.

Moreover, template polymerization of a second chain provides the possibility of adding multiple transduction tags to the probe–target hybrid. (Figure 7.20). In order to include the label in the resulting structure, a tagged nucleotide is made available to the polymerization process, along with three normal ones. Consequently, an amplification of the response is achieved, each linked target being signaled by multiple tags.

7.6 DNA Microarrays

An essential process in cell growth is the transfer of genetic information from DNA to ribosomes via messenger RNAs (mRNAs). Anomalies in this process are responsible for many diseases. That is why a *genetic screen* is an essential tool in medicine and molecular biology. A genetic screen consists of seeking sample genes by means of a set of hybridization probes assembled in the form of a DNA *microarray*, also known as a DNA chip [47–50]. A microarray consists of a set of different probes each of them being specific to a particular gene and placed at a particular site on a small plate. A DNA microarray allows parallel and simultaneous detection of a great number of target genes. For example, more than ten thousands spots of the micrometer size can be fabricated on a plate a few centimeters in diameter. In standard microarrays, the probes are attached by covalent linking to a solid support (glass, polymers or silicon).

In a microarray assay (Figure 7.21), mRNA is first separated from the sample cells and then *reverse transcription* is applied in order to produce complementary DNA (cDNA) bearing similar genetic information. This step is necessary because RNA is decomposed by nuclease enzymes in real samples. Thereafter, PCR is applied in order to

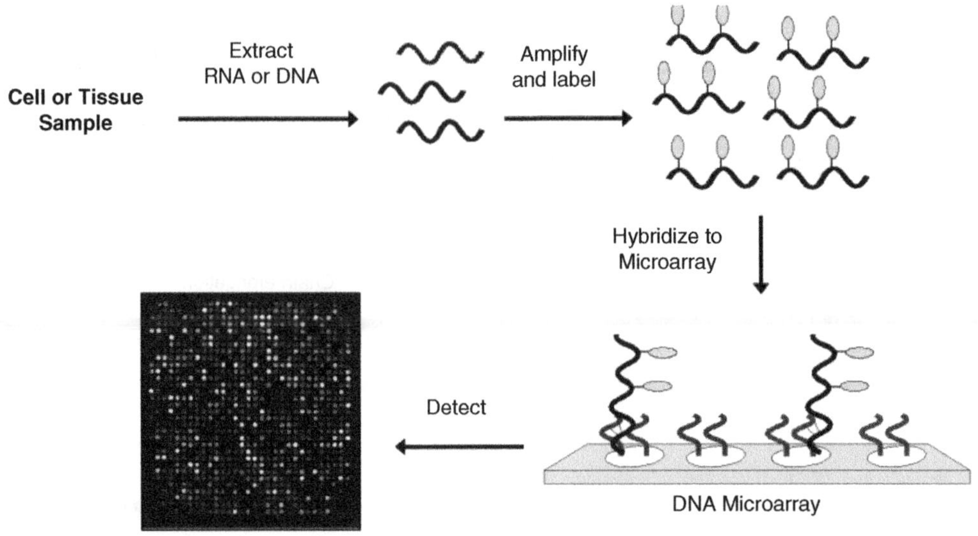

Figure 7.21 Main steps in a DNA microarray assay. Reproduced with permission from [53]. Copyright 2005 Springer Science + Business Media B.V.

increase the amount of cDNA and fluorescent (or other type) labels are affixed during this step by supplying labeled nucleotides. The labeled targets mixture is then incubated with the array and the hybridization is left to proceed. After washing out of the nonspecifically bound target, only strongly paired strands persist. Finally, the array is examined in order to measure the amount of label bound to each location. The identity of each target is indicated by its position, whereas the strength of the signal arising from a specific spot indicates the amount of each target in the sample. As a rule, a reference sample of healthy cells, labeled with a different color is run simultaneously. As a microarray generates a large amount of data, statistical algorithms are essential in data processing [51].

Although fluorescence is the standard method for examining a DNA microarray, other methods, (for example, electrochemical [53]), are actively investigated.

DNA microarrays are widely used for diagnostic purposes or for developing drugs against gene expression related diseases, such as cancer. The goal of such an assay is to assess the level or volume at which a certain gene is expressed (*microarray expression analysis*). Also, DNA microarrays are employed for detecting mutations or polymorphisms in a gene sequence. In this case, genomic DNA is targeted. A probe placed on a given spot will differ from other probes by only one or a few nucleotides.

The array throughput is determined by the spot density and spot miniaturization is the main way of improving the throughput. Much progress in this direction has been achieved by designing arrays of reaction chambers of nanoliter capacity or lower [54].

7.7 Outlook

Polynucleotides are used in affinity biosensors as recognition components for a series of applications. First, polynucleotide probes can recognize, by hybridization, complementary chains. By this means, key applications in medicine and molecular biology have been developed. Furthermore, double-strand nucleic acids can recognize by intercalation or covalent bonding some genotoxic substances with relevance to environment monitoring or the food industry. Intercalating analytes belong to various classes of organic substances, such as carcinogenic aromatic amines, polychlorinated biphenyls, aflatoxins, anthracene and acridine derivatives. Moreover, the ability of nucleic acid sensors to identify micro-organisms through their gene code render them valuable for detecting pathogens in foods, water and environment samples as well as for identification of biological warfare agents.

In addition, the advent of nucleic acid aptamers greatly expands the field of application of nucleic acid sensor. Aptamers perform recognition of proteins or certain small organic molecules with substantial advantages as compared with antibodies.

Questions and Exercises

1 What are the building blocks of nucleic acids?
2 What are the structural differences between DNA and RNA?
3 What are the biological functions of nucleic acids?
4 What is the mechanism of single-strand association? What are the structural restrictions in this process?

5 Comment on tertiary structures involving one single nucleic acid strand.

6 How can an oligonucleotide with a specific sequence be obtained?

7 Comment on the structure of nucleic acid analogs and their potential advantages with respect to natural nucleic acids.

8 Review the mechanism of nucleic acid hybridization.

9 What are the particular features of nucleic acid hybridization involving an immobilized partner?

10 What are the physicochemical parameters that determine the velocity of hybridization?

11 What is a nucleic acid capture probe? Discuss the effects of probe length and orientation with respect to the support surface.

12 Draw schematically possible probe orientations at positive, negative and hydrophobic surfaces.

13 How can double-strand nucleic acids interact with small molecules or proteins?

14 What are aptamers and what are their analytical application?

15 How can aptamers be used to develop chemical sensors? Discuss the advantages of aptamers in relation to their chemical structure.

16 What are the main methods for immobilization of nucleic acids to solid supports?

17 What are the advantages and drawbacks of adsorptive immobilization?

18 How can an oligonucleotide probe be immobilized by self-assembly and what are the main applications of this method?

19 Comment on the immobilization of nucleic acids by polymerization and its particular advantages.

20 What are the advantages of covalent immobilization?

21 Review the properties of solid surfaces used for covalent immobilization of nucleic acids.

22 What are the possible binding groups in nucleic acid probes?

23 How can affinity reactions be used for immobilization of nucleic acids?

24 How can nucleic acids be linked to nanoparticles?

25 Discuss the methods for label-free transduction in nucleic acid sensors.

26 What are the principles of the sandwich assay in nucleic acid sensors?

27 How can the intercalation of small molecules be exploited for detecting nucleic acid hybridization?

28 How can a hairpin-shaped probe be used in recognition–transduction processes?

29 Comment on possible methods for enhancing the sensitivity of nucleic acid sensors.

30 What are the principles of PCR? What is the name given to the product of such a process?

31 How can one make use of posthybridization PCR to enhance the sensitivity of nucleic acid sensors?

32 What is the structure of a DNA microarray and what are the applications of such devices?

33 Review the main fields of application of nucleic acid sensors.

References

1. Liu, J.W., Cao, Z.H., and Lu, Y. (2009) Functional nucleic acid sensors. *Chem. Rev.*, **109**, 1948–1998.
2. Yang, M.S., McGovern, M.E., and Thompson, M. (1997) Genosensor technology and the detection of interfacial nucleic acid chemistry. *Anal. Chim. Acta*, **346**, 259–275.
3. Wang, J. (2000) From DNA biosensors to gene chips. *Nucleic Acids Res.*, **28**, 3011–3016.
4. Sassolas, A., Leca-Bouvier, B.D., and Blum, L.J. (2008) DNA biosensors and microarrays. *Chem. Rev.*, **108**, 109–139.
5. Walker, J.M. and Wilson, K. (2010) *Principles and Techniques of Biochemistry and Molecular Biology*, Cambridge University Press, Cambridge.
6. Neidle, S. (2008) *Principles of Nucleic Acid Structure*, Elsevier, Amsterdam.
7. Reese, C.B. (2005) Oligo- and poly-nucleotides: 50 years of chemical synthesis. *Org. Biomol. Chem.*, **3**, 3851–3868.
8. Khudjakov, J.E. and Fields, H.A. (2003) *Artificial DNA: Methods and Applications*, CRC Press, Boca Raton.
9. Buzdin, A.A. and Lukyanov, S.A.(eds) (2007) *Nucleic Acids Hybridization: Modern Applications*, Springer, Dordrecht.
10. Caruana, D.J. (2004) Hybridization at oligonucleotide sensitive electrodes, in *Biosensors: A Practical Approach* (eds J.M. Cooper and A.E.G. Cass), Oxford University Press, New York, pp. 19–39.
11. Paszek, E. (2007) cDNA-Detailed Information (http://cnx.org/content/m12385/1.5/).
12. Minunni, M., Tombelli, S., Mariotti, E., *et al.* (2001) Biosensors as new analytical tool for detection of Genetically Modified Organisms (GMOs). *Fresenius J. Anal. Chem.*, **369**, 589–593.
13. Demeunynck, M., Bailly, C., and David Wilson, W. (eds) (2003) *Small Molecule DNA and RNA Binders: From Synthesis to Nucleic Acid Complexes*, Wiley-VCH, Weinheim.
14. Zeglis, B.M., Pierre, V.C., and Barton, J.K. (2007) Metallo-intercalators and metallo-insertors. *Chem. Commun.*, 4565–4579.
15. Mascini, M., Palchetti, I., and Marrazza, G. (2001) DNA electrochemical biosensors. *Fresenius J. Anal. Chem.*, **369**, 15–22.
16. Klussmann, S. (2006) *The Aptamer Handbook: Functional Oligonucleotides and their Applications*, Wiley-VCH, Weinheim.
17. Lee, J.F., Stovall, G.M., and Ellington, A.D. (2006) Aptamer therapeutics advance. *Curr. Opin. Chem. Biol.*, **10**, 282–289.
18. Jayasena, S.D. (1999) Aptamers: An emerging class of molecules that rival antibodies in diagnostics. *Clin. Chem.*, **45**, 1628–1650.

19. Mairal, T.T. (2008) Aptamers: molecular tools for analytical applications. *Anal. Bioanal. Chem.*, **390**, 989–1007.

20. Tombelli, S., Minunni, A., and Mascini, A. (2005) Analytical applications of aptamers. *Biosens. Bioelectron.*, **20**, 2424–2434.

21. Song, S.P., Wang, L.H., Li, J., *et al.* (2008) Aptamer-based biosensors. *TrAC-Trends Anal. Chem.*, **27**, 108–117.

22. Hianik, T., Porfireva, A., Grman, I. *et al.* (2008) Aptabodies - New type of artificial receptors for detection proteins. *Protein Pept. Lett.*, **15**, 799–805.

23. Cai, Z., Li, Z., Kang, Z., *et al.* (eds) (2009) Aptasensors design considerations, in *Computational Intelligence and Intelligent Systems*, Springer, Berlin, pp. 118–127.

24. Hianik, T. and Wang, J. (2009) Electrochemical aptasensors - recent achievements and perspectives. *Electroanalysis*, **21**, 1223–1235.

25. Heise, C. and Bier, F.F. (2005) Immobilization of DNA on microarrays, in *Immobilisation of DNA on Chips II* (ed. C. Wittmann), Springer, Berlin, pp. 1–25.

26. de-los-Santos-Alvarez, P., Lobo-Castanon, M.J., Miranda-Ordieres, A.J. *et al.* (2004) Current strategies for electrochemical detection of DNA with solid electrodes. *Anal. Bioanal. Chem.*, **378**, 104–118.

27. Pividori, M.I., Merkoci, A., and Alegret, S. (2000) Electrochemical genosensor design: Immobilisation of oligonucleotides onto transducer surfaces and detection methods. *Biosens. Bioelectron.*, **15**, 291–303.

28. Balamurugan, S., Obubuafo, A., Soper, S.A. *et al.* (2008) Surface immobilization methods for aptamer diagnostic applications. *Anal. Bioanal. Chem.*, **390**, 1009–1021.

29. Pividori, M.I. and Alegret, S. (2005) DNA adsorption on carbonaceous materials, in *Immobilisation of DNA on Chips I* (ed. C. Wittmann), Springer, Berlin, pp. 1–36.

30. Pividori, M.I. and Alegret, S. (2005) Electrochemical genosensing based on rigid carbon composites. A review. *Anal. Lett.*, **38**, 2541–2565.

31. Palchetti, I. and Mascini, M. (2005) Electrochemical adsorption technique for immobilization of single-stranded oligonucleotides onto carbon screen-printed electrodes, in *Immobilisation of DNA on Chips II* (ed. C. Wittmann), Springer, Berlin, pp. 27–43.

32. Luderer, F. and Walschus, U. (2005) Immobilization of oligonucleotides for biochemical sensing by self-assembled monolayers: thiol-organic bonding on gold and silanization on silica surfaces, in *Immobilisation of DNA on Chips I* (ed. C. Wittmann), Springer, Berlin, pp. 37–56.

33. Lucarelli, F., Marrazza, G., Turner, A.P.F. *et al.* (2004) Carbon and gold electrodes as electrochemical transducers for DNA hybridisation sensors. *Biosens. Bioelectron.*, **19**, 515–530.

34. Bidan, G., Billon, M., Calvo-Munoz, M.L. *et al.* (2004) Bio-assemblies onto conducting polymer support: Implementation of DNA-chips. *Mol. Cryst. Liquid Cryst.*, **418**, 983–998.

35. Livache, T., Roget, A., Dejean, E., *et al.* (1994) Preparation of a DNA matrix via an electrochemically directed copolymerization of pyrrole and oligonucleotides bearing a pyrrole group. *Nucleic Acids Res.*, **22**, 2915–2921.

36. Smith, C.L., Milea, J.S., and Nguyen, G.H. (2005) Immobilization of nucleic acids using biotin-strept(avidin) systems, in *Immobilisation of DNA on Chips II* (ed. C. Wittmann), Springer, Berlin, pp. 63–90.

37. de Dios, A.S. and Diaz-Garcia, M.E. (2010) Multifunctional nanoparticles: Analytical prospects. *Anal. Chim. Acta*, **666**, 1–22.

38. Chiu, T.C. and Huang, C.C. (2009) Aptamer-functionalized nano-biosensors. *Sensors*, **9**, 10356–10388.

39. Lin, Y.W., Liu, C.W., and Chang, H.T. (2009) DNA functionalized gold nanoparticles for bioanalysis. *Anal. Methods*, **1**, 14–24.

40. Wang, J., Katz, E., and Willner, I. (2005) Biomaterial-nanoparticle hybrid systems for sensing and electronic devices, in *Bioelectronics: From Theory to Applications* (eds I. Willner and E. Katz), Wiley-VCH, Weinheim, pp. 231–264.

41. Katz, E. (2003) Probing biomolecular interactions at conductive and semiconductive surfaces by impedance spectroscopy: Routes to impedimetric immunosensors, DNA-sensors, and enzyme biosensors. *Electroanalysis*, **15**, 913–947.

42. Miranda-Castro, R., de-los-Santos-Alvarez, N., Lobo-Castanon, M.J., *et al.* (2009) Structured nucleic acid probes for electrochemical devices. *Electroanalysis*, **21**, 2077–2090.

43. Willner, I., Shlyahovsky, B., Willner, B. *et al.* (2009) Amplified DNA biosensors, in *Functional Nucleic Acids for Analytical Applications* (eds Y. Li and Y. Lu), Springer, New York, pp. 199–252.

44. Mullis, K.B. (1990) The unusual origin of the polymerase chain-reaction. *Sci. Am.*, **262**, 56–65.

45. Arnheim, N. and Levenson, C.H. (1990) Polymerase chain-reaction. *Chem. Eng. News*, **68**, 36–47.

46. Erlich, H.A. (1989) Polymerase chain-reaction. *J. Clin. Immunol.*, **9**, 437–447.

47. Hegde, P., Qi, R., Abernathy, K., *et al.* (2000) A concise guide to cDNA microarray analysis. *Biotechniques*, **29**, 548–562.

48. Müller, H.J. and Röder, T. (2006) *Microarrays*, Elsevier, San Diego.

49. Bier, F.E., von Nickisch-Rosenegk, M., Ehrentreich-Forster, E. *et al.* (2008) DNA microarrays, in *Biosensing for the 21st Century*, Springer, Berlin, pp. 433–453.

50. Liljedahl, U., Fredriksson, M., and Syvänen, A.C. (2005) Analysis of DNA sequence variation in the microarray format, in *Microarray Technology and Its Applications* (eds R. Müller and V. Nicolau), Springer, Berlin, pp. 211–227.

51. Drăghici, S. (2003) *Data Analysis Tools for DNA Microarrays*, Chapman & Hall/CRC, Boca Raton, Fla.

52. Storhoff, J.J., Marla, S.S., Garimella, V. *et al.* (2005) Labels and detection methods, in *Microarray Technology and its Applications* (eds R. Müller and V. Nicolau), Springer, Berlin, pp. 147–179.

53. Dill, K. and Ghindilis, A. (2009) Electrochemical detection on microarrays, in *Microarrays* (eds K. Dill, R.H. Liu, and P. Grodzinski), Springer, New York, pp. 25–34.

54. Yamamura, S., Sathuluri, R.R., and Tamiya, E. (2007) Pico/nanoliter chamber array chip for single-cell, DNA and protein analysis, in *Nanomaterials for Biosensors* (ed. C. Kumar), Wiley-VCH, Weinheim, pp. 368–397.

8

Nanomaterial Applications in Chemical Sensors

8.1 Generals

We are currently witnessing a remarkable revolution in various fields of science and technology as a result of the development of nanomaterials [1,2]. This line of development has a considerable impact on the state-of-the-art of chemical sensors, and considerable progress in this area has already been achieved by the application of nano-materials [3–6].

A nano-object is characterized by the fact that at least one of its dimension is within the nanometer scale, that is, $< 10^{-6}$ m. From the standpoint of the macroscopic dimension, one can distinguish nanoparticles (zero-dimensional), nanowires (monodimensional) and nanosheets (two-dimensional) nanomaterials. Accordingly, none, one or two of these dimensions, respectively, are above the 10^{-6} m limit, the remaining dimensions falling within the nanometer range. Numerous biological entities, such as proteins, nucleic acids and viruses are within the nanosize scale and are often designated as biological nanomaterials. Synthetic nanomaterials of various chemical composition are currently available. Such nanomaterials may consist of pure chemical elements (metals, carbon or silicon), inorganic compounds (metal chalcogenides or metal oxides) or organic compounds (polymers).

What is characteristic of nanomaterials is that they normally display specific properties that are fundamentally different from the properties of the bulk material [7,8]. In the case of metal and semiconductor nanoparticles such differences arise first of all from the characteristic small size that has a major effect on the distribution of electronic levels in semiconductor nanoparticles. As a consequence, these materials exhibit specific and characteristic properties as far as light absorption and emission are concerned. In metal nanoparticles, for example, electron behavior in an electromagnetic field imparts characteristic light-scattering properties. Another significant characteristic is the high area/volume ratio. Due to the very small size, an important fraction of the component atoms are not completely involved in chemical bonding with vicinal atoms and this results in increased chemical reactivity and catalytic activity. Last but not least, the size of synthetic nanomaterials matches that of biopolymers such as antibodies, enzymes and nucleic acid. Combined with surface reactivity, the size compatibility allows to produce nanomaterials-biomolecule hybrids with wide applications in biomedicine for therapy or imaging purposes [9–11]. The size compatibility combined with the high specificity of the biocompounds mentioned led to a broad spectrum of applications of nanomaterials in the biosensors area [4,12–14].

As far as the production of synthetic nanomaterials is concerned, two main fabrication strategies can be distinguished, namely top-down and bottom-up techonologies.

The top-down approach is actually an extension of lithography [1]. In such a method, one starts with a macroscopic block of material that is processed by spatially controlled etching or layering in order to produce microscopic-sized patterns. Such technologies are widely used in the semiconductor industry and are characterized by a high throughput and flexibility. However, physical limitations do not allow the obtaining of structures smaller than 10 nm. In addition, conventional top-down methods may induce severe and uncontrollable structure defects during the etching steps. Nevertheless, such methods are very suitable for the fabrication of organized networks of nano-objects on solid surfaces.

The bottom-up approach is based on the assembling atoms, ions or molecules driven by specific physical or chemical interactions that determine the crystallographic structure of the product. Typical examples of this kind are crystal growth and the synthesis of large polymers. A critical issue is the control of the nano-object size. This can be achieved by adjusting specific physicochemical parameters. If this is not possible, a monodisperse nano-object product can be obtained by subsequent size-selective separation.

Bottom-up methods are based either on gas-phase processes (chemical vapor deposition) or solution-phase processes [8,15]. Chemical vapor deposition is a process where gaseous species react yielding a solid product on a solid support. Thus, carbon nanomaterials can be obtained by chemical decomposition of simple organic compounds in the presence of suitable catalyst.

A broad spectrum of solution-phase methods are based on colloid chemistry. Metallic, semiconductor and magnetic nanoparticles can be obtained in this way. Colloidal particles are produced by chemical reactions of suitable precursors, such as inorganic salts or organometallic compounds. The production of these colloidal particles involves initially a nucleation step that is followed by the subsequent growth of the preformed nuclei. The balance between the nucleation and growth velocities determines the size of the resulting particles. Thus, if the nucleation is much

Chemical Sensors and Biosensors: Fundamentals and Applications, First Edition. Florinel-Gabriel Bănică.
© 2012 John Wiley & Sons, Ltd. Published 2012 by John Wiley & Sons, Ltd.

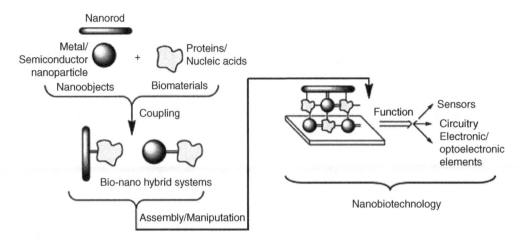

Figure 8.1 Integration of nanomaterials and biomolecules to yield functional devices. Adapted with permission from [14]. Copyright 2004 Wiley-VCH Verlag GmBH & Co. KGaA.

faster than the growth, relatively small particles result, whereas fast growth favors the growth of larger particles. These processes are conducted in the presence of organic surfactant molecules that adhere to the surface of the growing crystal. The surfactant layer prevents aggregation of particles and determines to a large extent both the size and size uniformity of the resulting particles.

Another class of solution-phase synthetic methods relies on the sol-gel technique that is well suited for obtaining oxide nanoparticles [15,16]. In this technique, a metal alkoxide (R-OM, where R is an alkane residue and M is a metal or semimetal) or metal chloride undergoes hydrolysis and polycondensation reactions to form a gel that consists of oxide- or alcohol-bridged networks. Subsequent aging, dehydration, drying and thermal treatment lead to solid particles.

A typical approach to using nanomaterials in biosensors is illustrated in Figure 8.1. In the first step, coupling of synthetic nanomaterials with biocompounds produces bionanohybrid systems. The biocomponent imparts specificity whereas the nanomaterial acts as a label, catalyst or signal-amplifying component [17]. In order to obtain a sensor, the hybrids are assembled on a solid support, which could be a transduction device. Nanomaterials in the resulting assembly can perform also as electron conductors or building blocks of an electronic device adapted for biosensing. Nanoparticle–biocompound hybrids can be obtained by various procedures based on either covalent or noncovalent interactions [18].

The following sections review the main types of nanomaterials as far as synthesis and main physicochemical properties as well as modification and bioconjugation are concerned. Possible applications in conjunction with various transduction methods are discussed in later chapters.

8.2 Metallic Nanomaterials

Metal nanoparticles arouse considerable interest mainly because of their particular electronic properties [19]. As is well known, valence electrons in metals are not bound but can move freely. This imparts good electrical conductivity to metals. However, size confinement in metal nanoparticles imparts to electrons the property of interacting in a specific way with an electromagnetic field. This makes metal nanoparticles valuable for imaging and optical sensing. Among various metallic materials, gold is particularly preferred because of its chemical inertness, which prevents surface oxidation [20].

8.2.1 Synthesis of Metal Nanoparticles

Gold nanoparticles with the size between 1 to 8 nm can be prepared by the reduction of the $AuCl_4^-$ anion with various reducing agents such as sodium borohydride ($NaBH_4$) or citric acid [21]. Stable colloidal gold systems are obtained if the reduction is performed in the presence of a protecting ligand that coats each particle with an adsorbed layer. Citric acid, for example, forms a negatively charged protecting layer that prevents coagulation of the nanoparticles by electrostatic repulsion. As is also well known, thiols form very stable surface layers by sulfur chemisorption on gold. Thiol-capped nanoparticles result from $AuCl_4^-$-reduction by $NaBH_4$ in the presence of a thiol that contains amine or carboxyl groups as the second terminal functionality. Such terminal groups lead to water-soluble

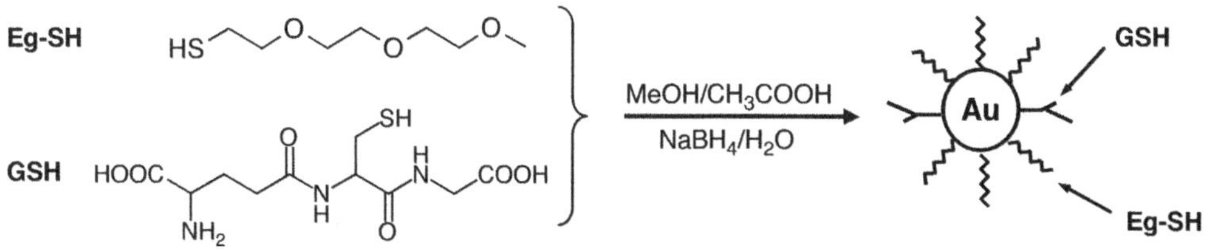

Eg-SH   HS⌒⌒O⌒⌒O⌒⌒O⌒

GSH   HOOC⌒⌒⌒NH₂ ... SH ... COOH

$\xrightarrow[\text{NaBH}_4/\text{H}_2\text{O}]{\text{MeOH/CH}_3\text{COOH}}$

GSH
Au
Eg-SH

Figure 8.2 Synthesis of gold nanoparticles covered by a mixed layer of glutathione (GSH) and thiolated ethylene glycol (Eg-SH). Redrawn with permission from [22]. Copyright 2005 Wiley-VCH GmbH.

nanoparticles and also provide linking sites for subsequent conjugation with biocompounds. Mixed layers on nanoparticles are particularly suited for biosensing. As shown in Figure 8.2, nanoparticles capped with a mixed layer can be obtained by the reduction of AuCl_4^- in the presence of two thiol derivatives: the glutathione (GSH) tripeptide and a thiolated (Eg-SH) triethylene glycol. The last component acts as a shielding component to minimize nonspecific interaction of the gold surface with biomolecules in the sample. At the same time, GSH can interact by hydrogen bonding or electrostatic attraction with the target compound. Moreover, GSH provides linking sites (such as $-\text{NH}_2$ and $-\text{COOH}$) that are useful for nanoparticle conjugation with biocompounds.

Metal nanoparticles can be produced not only as spherical granules but also with various geometries such as rods or cubes. Thus, metal nanorods can be prepared by template electrodeposition in a nanoporous host material such as alumina (Al_2O_3) or by chemical reduction within the nanopores of a polymer membrane. Such approaches are useful for preparing regular networks of nanosized disk electrodes. Metal nanorods can also be obtained as an aqueous suspension by the reduction of the metal ion in the presence of a surfactant that favors growth of an asymmetric shape. Either chemical or electrochemical reduction can be employed to this end [23]. An important feature of a nanorod is the *aspect ratio*, which represents the ratio between its length and diameter. Light interaction with metal nanorods depends to a large extent on this parameter.

Bimetallic nanoparticles can be prepared starting with silver or copper particles that are plated by more noble metal such as Au, Pt or Pd. Plating occurs by oxidation–reduction reactions of a compound of the plating metal. In platinum-metal core–shell nanoparticle, the platinum metal layer imparts specific catalytic activity.

8.2.2 Functionalization of Gold Nanoparticles

Various methods of functionalizing gold nanoparticles have been developed [10,14,22,24]. Some of them are summarized in Figure 8.3. In one method, nanoparticles capped with negatively charged carboxylic acids or phospholipids are modified by the adsorption of positively charged proteins by electrostatic attraction (Figure 8.3A). This approach is also suited for assembling multiple layers of oppositely charged components, such as proteins and polyelectrolytes. Multiple enzyme layers can be assembled in this way.

A common functionalization strategy relies on chemisorption of multifunctional thiol derivatives (see for example, Figure 8.2) that can function as bridges for tethering the nanoparticle to a biomolecule. Proteins including cysteine

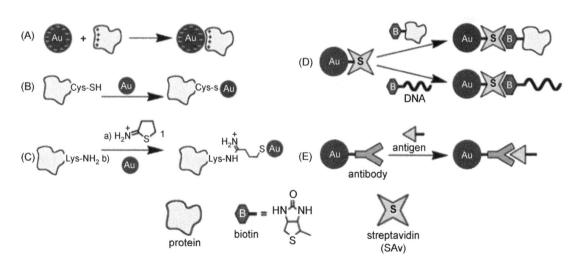

Figure 8.3 Methods for functionalization of gold nanoparticles with biological compounds. Adapted with permission from [14]. Copyright 2004 Wiley-VCH Verlag GmBH & Co. KGaA.

residues are suitable for direct chemisorption by thiol linkage to gold nanoparticles (Figure 8.3B). If no thiol residues are present in the native protein, thiol groups can be incorporated by chemical reaction of the amine groups with 2-iminothiolane (Traut's reagent) or by genetic engineering. A covalent method for the conjugation of proteins and gold nanoparticles is shown in Fig. 8.3C.

Affinity reactions are also widely used in order to link avidin (or streptavidin) tethered particles to biotin-derivatized proteins or nucleic acids (Figure 8.3D). Affinity reactions of carbohydrates with proteins are also useful in this respect.

Of great importance are the immunochemical reactions that make use of antibody-antigen reactions in order to attach proteins or living cells to nanoparticles (Figure 8.3E).

Gold nanoparticles can also be functionalized with metal-ion receptors such as macrocyclic ligands that allow specific detection of the ion of interest [25].

8.2.3 Applications of Metal Nanoparticles in Chemical Sensors

Assemblies of metal nanoparticles can be used for setting up electrical contact at a very small size scale. Moreover, including or growing metal nanoparticles within the sensing part can enhance dramatically the local electrical conductivity, which can be exploited for transduction. Metal nanoparticles can act as catalysts in chemical reactions involved with recognition or transduction processes. In mass-sensitive sensors (such as the quartz crystal microbalance) postrecognition attachment of metal nanoparticles to the receptor-bound analyte brings about a large mass increase with consequent enhancement of the response signal. Signal amplification can also be obtained by means of metal nanoparticles functionalized with multiple active components (for example, an enzyme).

A large number of applications are encountered in the field of optical sensors. Under the effect of the electromagnetic field, electrons at the surface of a metal particle oscillate collectively (localized surface plasmon resonance). This process can be noticed as a specific color, determined by the dispersion of light with the wavelength different from that of the resonance wavelength. At the same time electron oscillation results in a local enhancement of the electromagnetic field that enhances the intensity of Raman lines. On the other hand, metal nanoparticles can accept energy from excited fluorescent molecules and, in this way, reduce the fluorescence intensity (fluorescence quenching). The mentioned effects are useful for transduction in optical sensors and are discussed in more detail in Chapter 20.

Important applications rely on site-specific biocatalytic growth of metal nanoparticles in order to set up electrical contacts in nanostructured sensors [26].

Questions and Exercises (Section 8.2)

1 Draw a flow chart including the main steps in the synthesis of the gold nanoparticle shown in Figure 8.2 and sketch the relevant chemical reactions.
2 What are the possible shapes of metal nanoparticles? Outline in a flow chart a method for the synthesis of metal nanorods and point out the numerical parameter that characterizes the shape of nanorods.
3 What are core–shell metal nanoparticles and how can such particles be prepared?
4 Prepare a synoptic table including functionalization methods for gold nanoparticles, materials/reagents needed in each method, and relevant reactions.
5 Elaborate on metal nanoparticles applications to chemical sensors.

8.3 Carbon Nanomaterials

Natural carbon is found in two allotropic forms: diamond and graphite. Whereas the diamond structure consist of sp^3-hybridized carbons, graphite exhibits a honeycomb structure in which each atom is bound to three vicinal carbons in the sp^2-hybridization state. Such monoatomic layers can be obtained in the form of *graphene* (Figure 8.4A) [27,28]. Macroscopic graphite forms consist of assemblies of stacked layers of this type bound together by van der Waals forces.

Other carbon allotropic forms such as fullerenes have been obtained by synthetic procedures. A fullerene is any molecule composed entirely of carbon, in the form of a hollow sphere, ellipsoid, or tube (Figure 8.4B) [29] [30]. Fullerenes proved to be versatile materials for chemical sensors, for example being used as mediators in amperometric enzyme sensors [31].

Of particular importance in sensor applications are carbon nanotubes (CNTs) [32–36]. This section reviews the structure, synthesis and chemical properties of such materials. Various applications of CNTs as components of chemical sensors are emphasized in the following chapters.

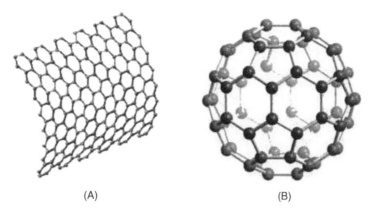

(A)   (B)

Figure 8.4   (A) Graphene structure. (B) Structure of C_{60} fullerene. In order to support the curving, both hexagon and pentagon carbon cycles are included in the structure. (A) Reproduced with permission from [32] Copyright 2007 Macmillan Publishers Ltd: Nature Materials. (B) Reproduced with permission from [30]. Copyright 2005 Wiley-VCH Verlag GmbH & Co. KGaA.

WALL

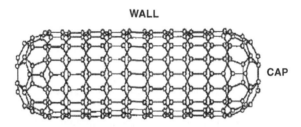

CAP

Figure 8.5   Structure of a single-walled carbon nanotube (SWCNT). Reproduced with permission from [37]. Copyright 2005 Wiley-VCH Verlag GmbH & Co. KGaA.

Carbon nanotubes (CNTs) are hollow cylinders made of graphite structured sheets (graphenes). Therefore, chemical bonds between carbons are of the π-type. Single-walled carbon nanotubes (SWCNTs) consist of a single carbon layer (Figure 8.5) whereas multiwalled species (MWCNTs) consist of a series of concentric nanotubes or a scrolled graphene ("papier mâchè" structure). A perfect SWCNT like that in Figure 8.5 is actually an extended fullerene. However, for practical applications, caps are removed so as to obtain a true tube.

CNTs are single-dimensional nanomaterials, with diameters of 1–2 nm and lengths ranging from 50 nm to about 1 cm. They exhibit outstanding electrical, mechanical and photophysical properties that are expected to revolutionize a series of scientific and technological fields including electronics, solid-state materials and biomedicine.

8.3.1 Structure of CNTs

Formally, an SWCNT can be imagined as graphene rolled at a certain chiral angle with a plane perpendicular to the tube long axis. Consequently, a SWCNT can be defined by its diameter and chiral angle, which can range from 0 to 30°.

However, it is more convenient to describe the tube geometry by means of a pair of integer indices (n, m). These indices refer to two equally long unit vectors at 60° angle to each other across a single 6-member carbon ring, as shown in the left-down corner in Figure 8.6. To build up a nanotube, draw a $n\mathbf{a_1}$ vector across the graphene sheet (OP), then draw a $m\mathbf{a_2}$ vector at 60° to the first one (PA) and add the two vectors. The length of the resultant vector $\mathbf{C}$ defines the circumference of the nanotube along a plane perpendicular to its axis. Perpendicular to $\mathbf{C}$, the vector $\mathbf{T}$ points to the long axis of the cylinder; its length is the shortest repeat distance along this axis. $\mathbf{C}$ and $\mathbf{T}$ are referred to as *chiral vectors* and *translational vectors*, respectively. Together, $\mathbf{C}$ and $\mathbf{T}$ define the unit cell of the nanotube as the rectangle OAB'B. The nanotube is obtained by rolling the graphene along the $\mathbf{C}$ direction making AB' and OB to coincide. Depending on the chiral angle θ, two particular nanotube types can be distinguished: the *zigzag* ($\theta = 0$, that is, $m = 0$) and the *armchair* ($\theta = 30°$, $m = n$) forms. For any intermediate θ value (that is, $n \neq m$, the tube is in the *chiral* configuration. Figure 8.6 displays the construction of a chiral (6,3) nanotube as well as the $\mathbf{C}$-axes for the zigzag and armchair particular forms. As this figure demonstrates, such designations derive from the path of the carbon atoms along the particular chiral axis.

The n and m indices determine whether the nanotube is a metal-type conductor or a semiconductor. The first case occurs when $|m - n| = 3k$ (k is an integer), whereas the second one corresponds to $|m - n| = 3k \pm 1$. The nanotube diameter d is related to n and m as [38]:

$$d = \frac{a}{\pi}\sqrt{(n^2 + m^2 + mn)} \qquad (8.1)$$

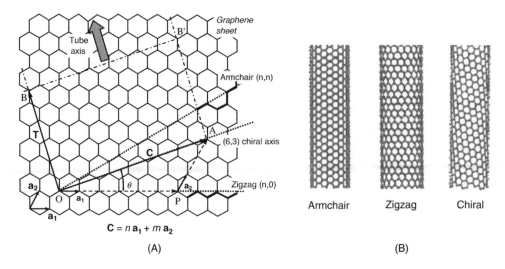

Figure 8.6 (A) Construction of the unit cell of a (6,3) SWCNT. Bold chemical bonds show the armchair and zigzag disposition of carbons along the cylinder circumference. (B) Armchair, zigzag, and chiral SWCNTs. Adapted with permission from [39]. Copyright 2006 Springer-Verlag.

where $a = 0.246$ nm is the magnitude of unit vectors. The chiral angle is also unambiguously determined by the n and m indices. Therefore, the n and m indices decide both crystallographic and geometrical characteristics as well as the metallic or semiconducting properties of a CNT.

As already mentioned, SWCNTs can be obtained as either semiconductor or metal-type conducting materials. In contrast, MWCNTs are always metal-type electric conductors.

8.3.2 Synthesis of CNTs

CNTs can be grown by sublimation of carbon vapors under arc discharge or laser ablation conditions [33,40,41]. However, the most promising method for CNT fabrication is based on chemical vapor deposition (CVD) [42]. In this process, CNTs form on metal catalyst particles (Co, Ni or Fe) supported on a refractory material (MgO or Al_2O_3) heated to about 700 °C, in the presence of a gas mixture that generates carbon vapors. This mixture consists of a carbon feedstock component (acetylene, ethylene or methane) and a process gas (nitrogen, hydrogen or ammonia). Finally, the metal catalyst is removed by acid treatment. As the product contains CNTs of various types and size and also other carbonaceous materials, CNTs of the expected type should be separated by a suitable method such as ultracentrifugation, electrophoresis or size-exclusion chromatography.

SWCNTs can be grown with a high yield by the HiPCO version of CVD. In this method, carbon monoxide disproportionation to carbon at high temperature and high pressures is catalyzed by iron particles that form by decomposition of gaseous iron pentacarbonyl, purposely added to the CO stream.

CNTs can be grown vertically aligned with respect to the support if the process is conducted in a plasma generated by an electric field, which dictates the growth direction. This method (known as plasma-enhanced CVD) allows regular networks of vertically aligned CNTs to be obtained by using support surfaces with a preformed array of catalyst nanoparticles [42]. Similarly, CNTs can be grown in the form of a nanobrush on a cylindrical support.

CNT assemblies with controlled geometry are very important in sensor development as they allow controlled tailoring of the sensing part. However, the growth of carbon nanotubes with a predefined microscopic structure still remains a major challenge.

CNT processing and application often relies on forming liquid suspensions. As CNTs are hydrophobic, water suspensions can be obtained by ultrasonic agitation in the presence of a surfactant such as sodium dodecylsulfate or sodium cholate, followed by centrifugation to remove bundles, ropes and residual catalyst. CNT wrapping with synthetic- or bio-polymers is equally successful.

8.3.3 Chemical Reactivity and Functionalization

Practical applications of CNTs requires some preliminary processing such as tube opening, changing of tube-wetting properties, tube filling, adsorption at the tube surface, charge transfer to or from the tube or tube doping [43].

Raw CNTs are closed by hemispherical fragments including carbon pentagons or metal catalyst inclusions. Opening is required for many applications; it can be accomplished by various methods including vapor-phase oxidation, plasma etching, or chemical reaction with strong oxidants (e.g., nitric acid, Figure 8.7). The opened end is terminated with different oxygen groups such as carboxyl, hydroxyl and quinones. Such groups are useful for covalently linking

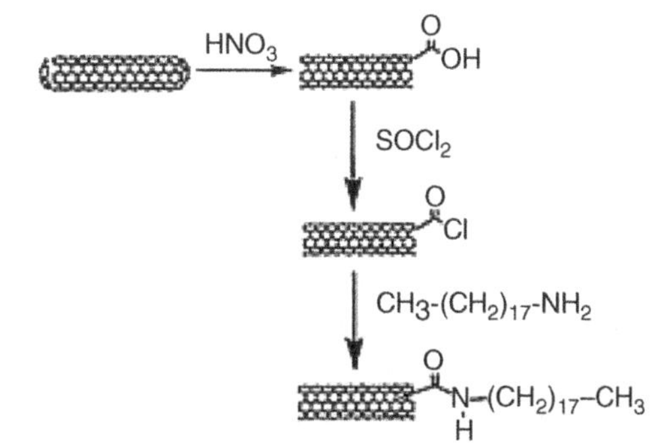

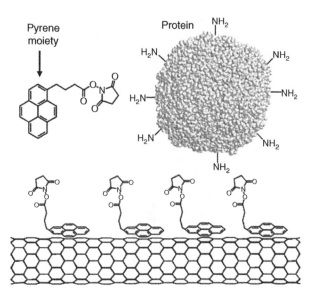

Figure 8.7 Cap opening and CNT functionalization by a peptide linkage. Reproduced with permission from [48]. Copyright 2005 Wiley-VCH.

CNTs to a support or to various biocompounds. Thus, the carboxylate group allows further nanotube modification via an ester or amide linkage (Figure 8.7). Oxidative treatment results as a rule in tube shortening and carboxylate formation on the side wall as well.

Covalent functionalization at the CNT side wall can be effected by various reactions producing reactive functionalities as linking sites [44–47]. However, it is important to note that covalent functionalization at the side wall disrupts the π orbitals system and affects the electronic properties of the nanotube.

Noncovalent functionalization can be achieved by adsorption on the side wall and has the advantage that the physical properties of the CNT are essentially preserved. Thus, polyaromatic molecules (such as pyrene) adsorb at the graphitic surface of CNTs via π-π - stacking. As shown in Figure 8.8, reactive groups can thus be attached to the nanotube surface to serve for further conjugation with proteins. Pyrene-derivatized receptors (e.g., cyclodextrin) or nanoparticles can be attached to CNTs in a similar way [49]. DNA can be attached to a pyrene methylammonium-covered CNT. DNA binds to the surface modifier by electrostatic interaction of the anionic phosphate backbone with the positive ammonium group in the modifier. π-π - stacking allows single-strand DNA molecules to wrap around a CNT.

Various amphiphilic compounds are currently used to stabilize CNT dispersions by adsorption. This approach can also be used in sensor development provided the amphiphile is biocompatible, nontoxic and sufficiently strongly adsorbed so as to resist desorption in biological media with high salt and protein content. Polyethylene glycol-derivatized phospholipids have proved particularly practical in this respect. The two hydrocarbon chains of the lipid strongly adsorb at the nanotube surface with the hydrophilic polyethylene glycol chain extending into the aqueous phase. Conjugation of the biological molecule can be achieved using a functional group (e.g., amine or carboxylate) purposely added at the PEG terminal.

CNTs can be covered by the adsorption of various biocompounds such as proteins, nucleic acids and polysaccharides (e.g., chitosan) that can act as anchors for attaching other compounds of interest.

Figure 8.8 Protein immobilization on a CNT via N-succinimidyl-1-pyrenebutanoate - π-π stacked at the nanotube surface. Adapted with permission from [50]. Copyright 2009 American Chemical Society.

Encapsulation into the nanotube cavity is another method for noncovalent modification. Various inorganic species (including pure elements, metallocenes, inorganic salts, metal oxides, and fullerenes) can be inserted into CNTs and bring about important changes in electronic properties. Metal nanowires can also be grown inside CNTs. Biomolecules (such as small proteins or single-stranded nucleic acids can be entrapped in CNTs by hydrophobic interactions due to the favorable tube diameter (2–10 nm).

Semiconducting CNTs are very promising materials for nanosized semiconductor devices including chemical sensors. Fabrication of CNTs with predefined semiconducting properties is therefore a matter of great interest [43]. Pristine CNTs are naturally p-doped, that is, display only hole conduction. Doping of CNTs shifts the Fermi level, while the band structure remains intact. This changes the population of electronic states near the Fermi level, which reduces the ohmic losses and facilitates carrier injection from contacts. Substitutional doping can be effected by the replacement of some carbon atoms with boron or nitrogen atoms. Boron induces holes and converts the tube into a p-semiconductor, whereas nitrogen, which imparts electrons, renders the tube an n-type semiconductor. Doping can also be achieved by encapsulating atoms, molecules or clusters exhibiting either electron-acceptor or electron-donor properties. Adsorption of some gases with electron-donor/acceptor properties (such as oxygen or ammonia) converts, in a reversible way, intrinsically semiconducting nanotubes into apparent metallic ones. Adsorption of an organic electron acceptor (tetracyanoquinodimethane, TCNQ) or donors (amines) also affects the electrical properties of SWCNTs.

8.3.4 CNT Applications in Chemical Sensors

The advent of CNTs has prompted a spectacular development in the research activity involving applications of such materials in analytical chemistry and, particularly, in chemical sensors and biosensors [51–54]. Integration of CNTs with sensor systems is facilitated by the large variety of available tube derivatization methods that allow the tuning of CNT properties and their conjugation with various biocompounds.

By physical adsorption, covalent binding or incorporation in a polymer, CNTs enable the structuring of chemical sensors at the nanometer scale. On the other hand, a CNT may serve as a support for multiple tags (e.g., fluorophores or enzymes). Once attached to a compound of interest, such a multiple label provides an appreciable enhancement of the response signal.

Excellent electrical conduction of metallic CNTs makes them suitable as molecular wires in electrochemical sensors [37,55,56]. Thus, in amperometric enzyme sensors A CNT can promote direct electron transfer between the active center of a redox enzyme and a macroscopic electrode. At the same time, catalytic properties, imparted by tube-end oxygen functionalities is useful in applications based on the electrochemical oxidation of the analyte. On the other hand, semiconducting CNTs integrated in field effect transistor structures allow chemical sensors to be developed for inorganic or biological species [49,55].

SWCNTs exhibit remarkable photophysical properties such as near-infrared fluorescence and very characteristic Raman scattering. Such properties render CNTs suitable for applications as optical labels in chemical sensors, as discussed in detail in Chapter 20.

8.3.5 Carbon Nanofibers (CNFs)

Carbon nanofibers (CNFs), also known as vapor-grown carbon nanofibers (VGCNFs), are cylindrical nanostructures with graphene layers arranged as stacked cones, cups or plates around the fiber axis [57].

CNFs are fabricated by chemical vapor deposition from hydrocarbon precursors using metal-particle catalysts [58]. The main steps of the synthesis are outlined in Figure 8.9. First, adsorption and decomposition of the hydrocarbon

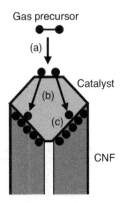

Figure 8.9 Grown of CNFs by chemical vapor deposition on a catalyst particle. (a) Decomposition of the precursor and carbon dissolution into the catalyst particle; (b) carbon diffusion through the catalyst particle; (c) nanofiber growth at the particle surface.

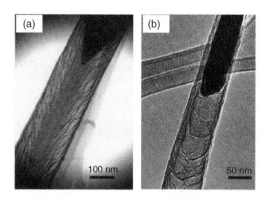

Figure 8.10 (A) CNF structure. (a) Herringbone structure formed by stacking of carbon nanocones grown on Ni catalyst particles by chemical vapor deposition; (b) bamboo-type CNF grown under the same conditions with an Fe catalyst. (B) Freestanding vertically aligned CNFs and forests of CNFs. Reproduced with permission from [59]. Copyright 2005 American Institute of Physics.

molecule at the catalyst particle takes place. Next, the carbon atoms thus formed are dissolved in the catalyst particle and undergo diffusion through it. Carbon reaching the opposite side of the particle precipitates to form a carbon network, thus promoting the growth of the fiber.

The diameter of CNFs can range between 5 to 500 nm; the diameter is determined by the size of the catalyst particle. The length of CNFs is in the micrometer region.

The structure of the nanofiber depends on the nature of the catalyst metal. Thus, Figure 8.10A(a) shows the so-called herringbone structure, obtained with nickel as catalyst. This structure is composed of stacked nanocones. Under similar conditions, but using iron as catalyst, one obtains a bamboo-type CNF (Figure 8.10A(b)).

CNFs can be obtained as bundles or as networks of randomly entangled fibers. For applications that require CNFs as individual elements, such as nanoelectronics or chemical sensing, ordered assemblies, such as vertically aligned CNFs, are more suitable. Vertically aligned CNFs can be grown in a nanoporous template (for example, alumina) or on catalyst particles immobilized by photolithography on a flat support in a predetermined pattern [59]. Figure 8.10B shows both individually freestanding structures of CNFs as well as dense mats ("forests") of fibers.

As far as the application of CNFs in sensor design is concerned, it is important to point out that the structure of CNFs contrasts strongly with that of CNTs. CNTs are well-defined three-dimensional macromolecular structures built up of hexagonal carbon cycles units while CNFs are assemblies of graphenes held together by van der Waals forces. While the functionalization of a CNT should be performed only at the tube end in order to prevent alteration of the conductive properties, CNFs display a high number of reactive sites at the edges of included graphenes. Oxidation of the fiber endows it with a high surface density of hydroxyl and carboxyl group that can interact physically or chemically with the surrounding environment. In addition, oxygen groups can function as binding sites for covalent attachment of biomolecules and other compounds of interest. Good electrical conductivity, combined with multiple possibilities of functionalization, render CNFs very attractive for applications in electrochemical sensors, particularly because CNFs allow direct electron transfer from the catalytic sites of redox enzymes to the electrode [60,61].

Questions and Exercises (Section 8.3)

1 What are the natural allotropic forms of carbon and what type of chemical bonding occurs in each of them?
2 What is the thinnest graphite-like material and what is its thickness?
3 What new allotropic carbon forms can be prepared by chemical synthesis?
4 What is the difference between fullerenes and carbon nanotubes?
5 What do a SWCNT and a MWCNT consist of?
6 Formally, a SWCNT can be obtained by scrolling a graphene sheet. How are the unit vectors defined? What is the chiral axis and how is it obtained by vector operation? What is the translational vector and what is its geometrical meaning?
7 What are the particular forms of carbon nanotubes as determined by the orientation of the chiral axis and how can they be symbolized by n and m indices?
8 SWCNTs can exhibit semiconductor- or metal-type electric conductivity. What are the conditions that determine such a particular behavior? How does a MWCNT behave from the standpoint of electrical conductivity?
9 Review the main methods for the synthesis of carbon nanotubes.
10 Is it possible to obtain regular assemblies of carbon nanotubes? If so, how?

11 Carbon nanotubes can be functionalized by an oxidative treatment. What chemical functionalities can be formed in this way? Where can they be located and what are their effects on the properties of the carbon nanotube?

12 From the structural standpoint, a carbon nanotube consists of a layer of π-bonded carbons. How can this property be used for noncovalent modification of carbon nanotubes?

13 How can biocompounds be attached directly to the surface of carbon nanotubes and what is the usefulness of this approach?

14 How can the electronic properties of semiconductor CNTs be tuned by doping and what are the main doping methods?

15 Review the main applications of CNTs as components of chemical sensors.

16 Comment on the structure of CNFs and the differences between CNFs and CNTs.

17 What are the particular features of CNFs that play a role in CNF modification and functionalization?

18 What methods can be used for immobilization of biomolecules on CNFs?

19 What are the benefits of the application of CNFs in sensor development?

8.4 Polymer and Inorganic Nanofibers

Nanofibers can be obtained by various methods such as template synthesis, phase separation or self-assembly. A versatile method for nanofiber fabrication is electrospinning that allows nanofibers to be formed from solutions or melts [62]. As shown in Figure 8.11A, in this process the liquid is pumped through a nozzle connected to a high-voltage supply. When a sufficiently high voltage is applied to a liquid droplet, the body of the liquid becomes charged, and electrostatic repulsion counteracts the surface tension. As a result, the droplet is stretched and at a critical point a stream of liquid erupts from the surface. This point of eruption is known as the Taylor cone. As the jet dries in flight, the charge migrates to the surface of the fiber. The jet is then elongated by a whipping process caused by electrostatic repulsion initiated at small bends in the fiber. Electrostatic repulsion also produces thinning of the fiber, which is finally deposited on a grounded support that collects the charge.

Electrospinning is typically employed to produce polymer fibers [63,64]. Inorganic fibers can be obtained in the same way, using a solution of precursors that yields by chemical reaction the expected material. Semiconductor metal oxide nanofibers produced by electrospinning followed by annealing, have proved to function as excellent sensing materials in certain gas sensors [65].

Polymer fibers can be obtained either from a single polymer or from polymer blends. In the second case, two different polymers are dissolved in the same solvent and the fiber is spun from this mixed solution. If the two solutions are incompatible, two syringes are used, each of them delivering one of the components to the spinning tip. The morphology of a nanospun polymer mesh of a polystyrene–polyaniline blend is shown in Figure 8.11B.

Owing to the very large specific area, electrospun polymer meshes are suitable as immobilization supports for enzymes since they allow high enzyme loading and facile mass transfer of the substrate. Enzyme immobilization can be performed by well-established methods such as entrapment, adsorption, crosslinking or covalent bonding [67]. In addition, a nanospun mesh can function as a size-selective separation membrane in enzyme sensors to discriminate between small substrate molecules and proteins so as to prevent fouling or interference by the proteins.

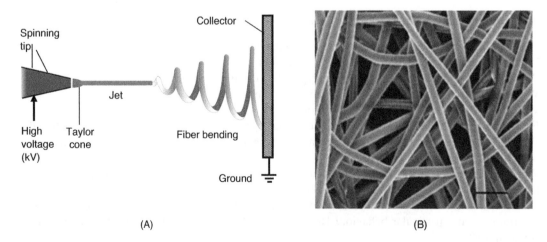

(A)   (B)

Figure 8.11 (A) Principle of nanofiber fabrication by electrospinning; (B) Scanning electron micrograph of polystyrene–polyaniline blend fibers. (A) Adapted from http://commons.wikimedia.org/wiki/File:Electrospinning_Diagram.jpg Last accessed 17/05/2012. (B) Reproduced with permission from [66] Copyright 2009 The Conference of Photopolymer Science and Technology.

Polymer blends that include conducting polymers provide an additional function, namely electrical conductivity, which makes them attractive for application in amperometric enzyme sensors. Nanofibers of polystyrene–polyaniline blends allow glucose oxidase to be immobilized by electrostatic adsorption. The conducting polymer component imparts electrical conductivity and provides direct electron transfer from the immobilized enzyme [66].

The two-syringe electrospinning method is also a handy technique for producing polymer–inorganic composite nanofibers such as carbon nanotube–polymer composites [68].

Metal nanofibers with excellent electrical conductivity can be obtained by electrospinning from a polymer solution containing a metal precursor, which provides nucleation seeds for subsequent growth of a metal coating over the polymer fiber. Thus, gold nanofibers have been fabricated by electrospinning of a polyacrylonitrile solution containing $HAuCl_4$. Nucleation seeds are formed by reduction of occluded $HAuCl_4$ to Au using an $NaBH_4$ solution. Finally, gold deposition has been produced from a solution containing $HAuCl_4$ with hydroxylamine as reducing agent [69]. Enzyme immobilization can be achieved by crosslinking to a layer of cysteamine chemisorbed at the gold surface. This design has been utilized in an amperometric fructose sensor based on fructose dehydrogenase, which performed well in the determination of fructose in blood and beverages.

Questions and Exercises (Section 8.4)

1 Describe a common method for the large-scale production of nanofibers.
2 What kinds of nanofiber material are useful in chemical sensors?
3 Comment on possible applications of polymer nanofibers in enzyme sensors.
4 How can metal nanofibers be produced and what are the applications of such nanofibers in sensor development?

8.5 Magnetic Micro- and Nanoparticles

Magnetic particles commonly consist of magnetic elements such as iron, nickel and cobalt or some of their chemical compounds. As will be shown later the magnetic behavior depends to a large extent on the particle size. That is why it is useful to distinguish two categories: magnetic beads (round particles of 100–400 μm size) and magnetic nanoparticles that are between 10 and 100 nm diameter. Both categories have proved extremely useful in biosciences owing to the possibility of remote manipulation of the magnetically tagged entities by means of magnetic fields. Remarkable achievements have been made using magnetic particles in diagnostic, therapy and bioimaging applications [70–72]. Also, magnetic particles have proved to be useful for separation, purification and identification of molecules, cells and micro-organisms tagged with such particles. As far as biosensors are concerned, magnetic particles are useful for remote manipulation of sample components and as transduction labels in magnetic sensors.

8.5.1 Magnetism and Magnetic Materials

The term magnetism denotes the property of some materials, molecules or elementary particles to be subject to forces under the effect of a magnetic field. Under the effect of a magnetic field, any magnetic entity experiences a force at each pole. These forces are equal to each other but are oriented in opposite directions and tend to rotate the magnet until it aligns with the magnetic field as a compass needle. Rigorously speaking, the effect of a magnetic field on a magnet results in a torque, a vector oriented perpendicularly to both forces and the pole axis and with the initial point at the mid-distance between poles. Its magnitude τ is:

$$\tau = dF \sin \theta \tag{8.2}$$

where d is the distance between poles, F is the force applied to each pole and θ is the angle between the force and the pole axis.

The strength of a magnet is given by its *magnetic dipole moment*, which is the ratio of the maximum torque (experienced at $\theta = 0°$) to the strength of the magnetic field. A specific magnetic property of a specimen is the *magnetization, M*, which is defined as the specimen magnetic moment divided by its volume. Magnetization depends on the strength of the magnetic field and a characteristic property of the specimen called magnetic susceptibility.

The source of magnetic properties is the electron orbital motion around the nucleus (*orbital magnetic dipole moment*) and the electron's intrinsic magnetic moment (*spin*). Vector addition of spin and orbital moments results in the *total magnetic dipole moment* that is responsible for the behavior of the electron in a magnetic field. Paired electrons are required by the Pauli exclusion principle to have their spin magnetic moments pointing to opposite directions, causing their magnetic fields to cancel out. However, an unpaired electron is free to align its magnetic moment with any direction and imparts to the atom a magnetic dipole moment. In the absence of a magnetic field, the atomic magnetic moments are oriented randomly and the overall magnetic moment of the specimen is zero. When an external magnetic field is applied, the atomic magnetic moments will tend to align themselves with the direction of the applied field. Materials displaying such a property are termed *paramagnetic* materials if the atoms return to random orientation after removal of the magnetic field. The oxygen molecule in the ground state displays typical paramagnetic behavior because it includes two unpaired valence electrons.

A *ferromagnetic* material is similar to a paramagnetic one in that it consists of atoms with a nonzero magnetic dipole moment. In addition, ferromagnetic materials are able to maintain the alignment of the atomic dipoles after removing the external magnetic field, which imparts to them permanent magnetism. This property is the result of the existence of *magnetic domains* in ferromagnetic materials. In each magnetic domain, atomic magnetic dipoles adopt, by mutual interaction, a parallel orientation. However, each domain adopts a random orientation and the overall magnetic moment of the specimen is zero. If a magnetic field is applied, these domains tend to line up along the direction of the magnetic field. However, due to restricted movement in the solid-state phase, after removal of the magnetic field the domains cannot return to perfectly random orientation and the specimen preserves residual magnetization.

8.5.2 Magnetic Nanoparticles

Colloidal particles of magnetic materials with sizes ranging from nanometer to micrometer sizes can be prepared by high-temperature liquid-phase reactions of suitable precursors in the presence of a surfactant [73–75]. As particle properties are size dependent, it is important to use monodisperse systems, which are characterized by a standard deviation of the diameter of less than 10%. Nanometer-sized magnetic particles are particularly suitable for biological applications because such particles consist of a single magnetic domain. Consequently, magnetic nanoparticles do not maintain residual magnetization after being exposed to a magnetic field, a property termed *superparamagnetism*. If residual magnetization persisted, particles would tend to form clusters by mutual attraction that would be detrimental to biological applications.

The most common materials for magnetic nanoparticles are α-Fe_2O_3 (hematite), γ-Fe_2O_3 (maghemite) and mixed oxides with the general formula MFe_2O_4, where M is a divalent ion, such as Fe, Co, Mg or Zn. Among these materials, magnetite (Fe_3O_4) is the most widely used.

Colloidal magnetic nanoparticles are covered by a hydrophobic layer (oleic acid or oleamine) that prevents self-aggregation. For biological applications it is essential to convert them into a hydrophilic form and attach functionalities or molecules that are able to interact specifically with the target compound or cell [76]. Conversion of magnetic particles into hydrophilic form can be effected either by surfactant addition or by surfactant exchange.

The surfactant-addition approach relies on forming an additional layer of an amphiphilic compound. The hydrophobic end of the amphiphilic compound interacts with the initial layer to form a double layer that displays a hydrophilic terminus in contact with the aqueous phase. As the surfactant layers are held together by hydrophobic interaction, this structure is not particularly stable. Alternatively, hydrophobic particles can be encapsulated in a polymer layer that contains hydrophilic groups oriented to the aqueous phase.

Surfactant exchange is achieved by direct replacement of the original layer by a compound that fulfills two conditions: it forms strong bonds with the particle surfaces and displays hydrophilic terminal groups at the opposite end. Small bifunctional molecules (such as cystamine or 2,3-dimercaptosuccinic acid) are practical for surfactant exchange either by forming strong metal–sulfur bonds or by chelate binding of carboxylate groups to the metal. Magnetic nanoparticles can also be coated with a silica shell by means of sol-gel chemistry methods [77].

A convenient alternative to the above strategies is based on particle synthesis in the presence of a hydrophilic multidentate ligand (such as citric acid) that covers, *in situ*, the particle with a hydrophilic layer.

Biological as well as sensor applications of magnetic nanoparticles require them to be able to bind target biocompounds or cells [76]. To this end, the surface layer should be engineered so as to include suitable anchoring groups to which receptors are attached by standard bioconjugation reactions. Thus, covalent linking can be effected if the particle envelope contains groups such as carboxyl, amine, thiol, or carbonyl. Antibodies, nucleic acids and affinity reagents (for example, streptavidin) can be conjugated thereafter with modified magnetic nanoparticles. Abiotic receptors such as macrocyclic ligands, cryptands or calixarenes can be appended in the same way.

Magnetic crystals can be extracted from certain bacteria and algae. Such particles are found as assemblies called magnetosomes that act together like a compass needle to orient the organism in the geomagnetic field. Each magnetite crystal within a magnetosome is surrounded by a lipid bilayer, and specific soluble proteins. Magnetosome crystals have high chemical purity, narrow size ranges and elicit species-specific crystal morphologies.

8.5.3 Magnetic Biosensors and Biochips

Magnetic biosensors rest on the detection of the stray magnetic field induced by magnetic nanoparticles accumulated near the transducer as a result of a suitable recognition process [78–82]. Various physical effects such as the Hall effect and the giant magnetoresistance effect form the basis of such transducers [78,80]. The Hall effect is the

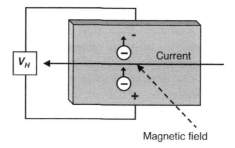

Figure 8.12   The Hall effect. A Hall potential difference (V_H) develops in a thin metal strip crossed by a current and subject to a magnetic field perpendicular to the surface.

consequence of the electron drift under the influence of a magnetic field (Figure 8.12). In a thin metal strip, this causes a potential difference to develop between the edges as a function of the strength of the magnetic field. The *anisotropic magnetoresistive effect* manifests as a change in material resistance when the magnetization changes from parallel to transverse with respect to the direction of current flow. The *giant magnetoresistance effect* is the basis of spin-valve transducers. Magnetoresistive transducers have been developed primarily for applications in magnetic random access memory devices but have also proved useful for magnetic biosensing purposes.

In order to detect magnetic nanoparticles, a conveniently oriented magnetic field is applied to the sensor. This magnetic field induces a magnetic dipole to each nanoparticle that results in an alteration of the local strength of the magnetic field.

Typical configuration of a magnetic affinity biosensor is shown in Figure 8.13. In the direct detection scheme (Figure 8.13A), the receptor interacts with the target analyte which was previously attached to magnetic nanoparticles. The unbound analyte molecules will then be washed away. The presence of the target in the sensing layer is indicated by the stray magnetic field induced by the nanoparticles. The sandwich assay format can also be implemented using magnetic nanoparticles. In this case, the sensing layer and nanoparticle tags consist of different ligands, each of them binding to a specific site of the analyte molecules.

A two-step DNA assay method is shown in Figure 8.13B. In the first step, the biotin-tagged DNA analyte binds by hybridization to the complementary probe attached to the transducer surface. Next, streptavidin-functionalized magnetic nanoparticles are added to the biotin tags of the hybridized probe. Any magnetic particles not bound to the receptor will then be driven away by the action of a magnetic field. Finally, the effect of the stray magnetic field within the sensing layers is recorded.

Current applications of magnetic biosensing are concerned chiefly with biochip development. Magnetic biochips consist of assemblies of minute magnetic sensors with a typical size in the micrometer region. This technology is suitable for developing hand-held biochip instrumentation [81,83,84].

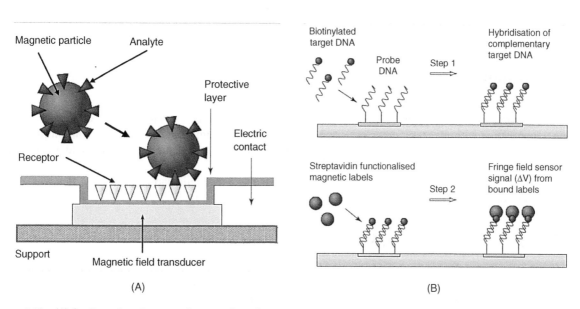

Figure 8.13   (A) Configuration of a magnetic sensor based on affinity recognition. (B) Sandwich DNA assay using magnetic nanoparticles as labels. Adapted with permission from [77]. Copyright 2004 Elsevier.

Magnetic biosensors exhibit a series of notable advantages. Magnetic labels are very stable and not affected by chemical reactions involved in the sensor operation. Even more important is the absence of interferences from biological media that exert almost no influence on magnetic nanoparticles. Consequently, magnetic transduction is practically free of interference from the sample matrix. It allows the development of low-cost components that provide compact, user-friendly and very sensitive devices. Remote manipulation by a magnetic field can be implemented in order to perform analyte separation or to remove unwanted magnetic particles. However, present magnetic biochips have a fairly low throughput owing to the relatively large area of each sensing site. Much progress is therefore anticipated from the reduction of transducer size.

8.5.4 Magnetic Nanoparticles as Auxiliary Components in Biosensors

Application of magnetic particles in biosensor design brings about a series of advantages as far as sensitivity, reliability, selectivity and fabrication methods are concerned [85–87].

First, magnetically tagged entities (proteins, DNA, living cells) can be easily separated from a crude sample in order to conduct the analysis under interference-free conditions. When necessary, the separated analyte can be subject to additional processing (such as DNA amplification) after separation. This approach works equally well with protein immunosensors and DNA hybridization sensors. It has also proved also useful for detecting living cells (such as pathogenic micro-organisms and tumor cells) by immunorecognition.

In standard affinity biosensors, the receptor is immobilized at the transducer surface in order to construct the sensing part. Under such circumstances, the recognition reaction is relatively sluggish owing to slow diffusion and the steric restriction imparted to the immobilized receptor. Magnetic particles allow the development of an alternative strategy based on receptor immobilization at the surface of magnetic particles thus letting the transducer interface free of additional components. Once mixed with the sample, the tagged receptor interacts with the analyte to give a tagged analyte–receptor–magnetic particle hybrid. This hybrid can be selectively accumulated at the transducer surface by means of a magnetic field in order to assess the response signal. As the recognition processes are much faster in the solution phase, this procedure results in a substantial reduction in the analysis time.

A series of promising applications of magnetic particles has been demonstrated in the field of enzyme sensors. Thus, in order to avoid the need for prior enzyme immobilization, the enzyme is tagged with magnetic particles and then accumulated at the transducer surface by means of an incorporated magnet. Owing to the size compatibility, enzyme tagging with magnetic nanoparticles does not affect enzyme activity to the same extent as the immobilization on macro-objects does. As a result, magnetic immobilization of the enzyme may result in improved sensitivity.

Signal enhancement in enzyme sensors can also be obtained by using a magnetically tagged cofactor. A rotating magnetic field drives the magnetic particles into motion, thus enhancing the diffusion rate by local convection. The simplest use of the magnetic-field effect consists of switching the enzyme reaction on/off alternately by the action of a magnet on magnetically tagged cofactor molecules that are present in the solution. The sensor turns on when a magnet is placed just under the sensor but it turns off if the magnet is moved to above the solution.

Magnetic switching has also been applied to control the state of a dual-enzyme sensor in which only one of the enzyme cofactors is magnetically tagged. This enzyme will be active only under the effect of a magnet placed under the sensor. Combined with a suitable selection of a specific working parameter (such as the electrode potential in the case of an amperometric sensor), the use of a magnetic field permits sequential determination of two different analytes using a sensing layer composed of a mixture of two enzymes. See ref. [86] for more details.

8.5.5 Outlook

Magnetic nanoparticles can be readily conjugated with biocompounds (such as proteins and nucleic acids) and thus allow remote manipulation by magnetic fields in order to perform separation or concentration of such entities. In addition, combination with magnetic particles is suited for triggering or amplifying certain recognition processes. Such applications can be implemented in conjunction with various biorecognition methods and transduction procedures.

Moreover, in combination with magnetic transducers, functionalized magnetic particles allow the development of magnetic biosensors and biochips characterized by high selectivity and sensitivity.

Questions and Exercises (Section 8.5)

1 What is the atomic basis of magnetism?
2 What is the effect of a magnetic field on paramagnetic and ferromagnetic materials?
3 What are the main materials that are used as magnetic nanoparticles?
4 Review the properties of magnetic nanoparticles in relation to biosensor applications.

5 How can one prevent self-aggregation of magnetic nanoparticles?

6 How can magnetic nanoparticles be conjugated with biocompounds?

7 What is the configuration of a typical magnetic biosensor? How can transduction be effected in such a sensor?

8 How can magnetic nanoparticles be used to enhance the response signal of biosensors?

9 What are the benefits of magnetic nanoparticle applications in biosensors?

8.6 Semiconductor Nanomaterials

Although doped silicon is currently the most widely used semiconductor material for applications in electronics, compound semiconductor nanomaterials are particularly applied in the field of chemical sensors and biosensors. Such nanocrystals consist of compounds of group II and VI elements (for example, CdSe and CdTe) or group III and V elements (for example, InP and InAs). Semiconductor nanocrystals are also known as *quantum dots*.

The main difference between bulk semiconductors and quantum dots arises from the distribution of the electronic energy levels. As shown in Figure 8.14, conduction and valence bands in a metal overlap partially allowing electron transition between them. This is the reason for the high electrical conductivity of metallic materials. In a bulk semiconductor, the valence and conduction bands are separated by an energy gap. By contrast, due to electron confinement a quantum dot displays a series of discrete energy levels in both the bonding and antibonding electronic states. These two groups of orbitals are separated by a relatively broad energy gap. The discrete level distribution is similar to that in atoms and imparts quantum dots with particular properties as far as light absorption or emission are concerned. Thus, light absorption occurs only if the photon energy matches the difference between a bonding and an antibonding level.

8.6.1 Synthesis and Functionalization of Quantum Dots

Quantum dots for use in analytical applications can be prepared from organometallic precursors by using one-step colloid chemistry methods [88–91]. In the initial stage, the synthesis is conducted in a high boiling point organic solvent whose molecules form coordination bonds with the particle surface (Figure 8.15A). Such a *capping group* performs two tasks: it saturates unoccupied metal orbitals at the surface and prevents, by steric hindrance, irreversible aggregation of the particles. Samples with the standard deviation of the particle diameter $\leq 5\%$ are referred to as *monodisperse*. Quantum dots thus prepared are coated with a hydrophobic layer and are therefore not suitable for analytical applications in aqueous solutions. In order to obtain hydrophilic quantum dots, the surface ligand layer needs to be exchanged with a hydrophilic one by prolonged contact with a $-COO^-$ or $-NH_2$ capped thiol, as shown in Figure 8.15B. The hydrophilic moiety in such ligands is exposed to the solution and renders the quantum dot compatible with water. In addition, this group allows biocompounds to be conjugated with the quantum dot by means of suitable linkers [92].

An alternative to previous the cap-exchange method relies on covering hydrophobic quantum dots with amphiphilic shells such as polymers or phospholipids. In such an arrangement, the hydrophobic moiety of the cap attaches to the original hydrophobic layer by hydrophobic interactions. In order to perform further functionalization of the quantum dot, the amphiphilic layer should include suitable reactive groups.

Another quantum-dot-encasing method involves the growth of a hydrophilic silica shell on the quantum dot by a sol-gel reaction [77]. Reactive functionalities attached to this layer provide convenient binding sites for conjugation with biocompounds.

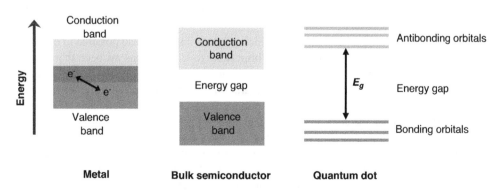

Figure 8.14 Distribution of electronic energy levels in metals, bulk semiconductors and semiconductor nanocrystals.

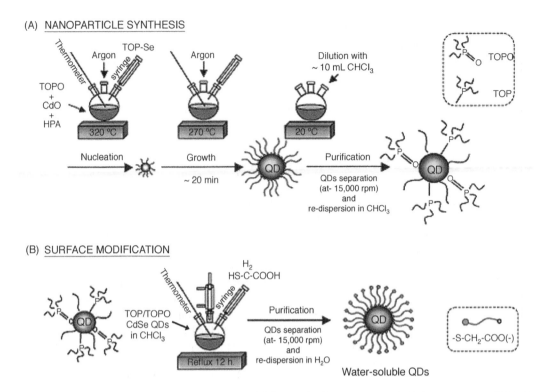

Figure 8.15 Synthesis of quantum dots as colloidal particles. (A) Synthesis of hydrophobic colloidal quantum dots. (B) Surface modification by cap exchange. TOP: trioctylphosphine; TOPO: tri-*n*-octylphosphine oxide; HPA: hexylphosphonic acid. Alternatively, the step (B) can be effected after chloroform removal by centrifugation followed by redispersion in methanol. Reproduced with permission from [93]. Copyright 2006 Elsevier.

As already mentioned, covalent bonding to functionalized capping ligands provides various possibilities for quantum dot bioconjugation [93]. In addition, affinity interactions (such as avidin–biotin) represent an alternative method for preparing quantum-dot–biocompound conjugates [94]. Host–guest chemistry provides another recognition method based on functionalized quantum dots. In this alternative, host molecules (cyclodextrin, crown ether or calixarene) are attached to the quantum dot to function analyte-recognizing receptor. Charged biocompounds (antibodies, DNA) can be appended by electrostatic attraction to quantum dots capped with charged ligands.

Hydrophilic quantum dots can also be obtained directly by conducting the synthesis in aqueous solutions in the presence of a thio-derivative (HS-R) as stabilizing and size-regulating agent [95]. For example, the synthesis of hydrophilic CdTe quantum dots proceeds according to reactions (8.3) and (8.4). In the first step, a precursor is formed by the reaction of Cd^{2+} with H_2Te in an alkaline aqueous solution under nitrogen. Quantum dots (QDs) grow subsequently under prolonged heating at $100\,°C$.

$$Cd^{2+} + H_2Te \xrightarrow{HS-R} Cd\text{-}(SR)_x Te_y + 2H^+ \tag{8.3}$$

$$Cd\text{-}(SR)_x Te_y \xrightarrow{100\,°C} CdTe\ (QD) \tag{8.4}$$

The capping ligand in this case is a short chain thiol containing one or more hydrophilic groups ($-NH_2$, $-COO^-$ or $-OH$) at the end. Such groups prevent aggregation by electrostatic repulsion and, in addition, permit facile linking to a biocompound using typical bioconjugation reactions. ZnSe quantum dots can be prepared in a similar way.

The previous methods refer to the preparation of colloidal quantum dots. It is also possible to obtain quantum dots embedded in solid materials such as polymers. It is important to note in this respect that various quantum-dot–polymer composites have been devised [96]. Thus, polymer quantum dot blends can be obtained simply by mixing these two materials taking care to avoid phase separation and interruption of polymer-chain entanglement. Nonaggregated nanoparticles can be grown straight in a polymer matrix with the advantage of obtaining controlled localization of nanoparticles within the composite. This technique can produce composites with a gradient of nano-crystals content along a selected direction. For the purpose of further functionalization of the composite, it is conve-nient to resort to thiol-terminated polymers or dendrimers as encapsulating materials. Dendritic encapsulation is advantageous in that it provides many chain ends for further covalent of suitable modifiers including biocompounds. Finally, polymer encapsulation can be achieved using quantum dots functionalized with molecules capable of initiat-ing polymerization reactions.

8.6.2 Applications of Quantum Dots

Application of quantum dots in biomedicine and chemical sensors relies on their luminescence properties, that is, light emission by excited species [97]. By excitation, electrons can be promoted from the ground bonding orbital to an antibonding one provided the excitation energy overcomes the energy gap. An excited quantum dot emits light at a characteristic wavelength that is determined by the energy gap. As the gap energy depends on the particle size, the emission wavelength can be tuned by adjusting the particle size. The optical properties and applications of quantum dots are discussed in detail in Chapter 20.

Questions and Exercises (Section 8.6)

1 Elaborate on the differences between bulk semiconductors and semiconductor nanoparticles as far as the distribution of the electronic energy levels is concerned.
2 What is the chemical composition of quantum dots? How can quantum dots be stabilized as colloidal solutions?
3 Outline by a flow chart the method of synthesis of colloidal hydrophobic quantum dots. List the reagents required and the role of each of them.
4 How can hydrophilic quantum dots be obtained?
5 Prepare a synoptic table including functionalization methods and receptor types that can be conjugated to quantum dots.

8.7 Silica Nanoparticles

8.7.1 Synthesis, Properties, and Applications

Silica (silicon dioxide) nanoparticles can be prepared by hydrolysis and polymerization of tetraethylorthosilicate (TEOS) in ethanol–water mixtures in the presence of ammonia (Ströber method [98]). The particle size is determined by the precursor concentration and the reaction time. Monodisperse spherical silica particles can be formed also by a similar reaction occurring within micelles in a nonpolar solvent (reverse microemulsion method [99]). The chemistry of such processes is typical of sol-gel chemistry, except that particular precautions are needed to prevent large-scale gelation.

Silica is chemically inert under normal conditions and its application rests on its porous structure that imparts a high surface area. Pore diameters can be adjusted between 2 and 10 nm. At the same time, the specific pore volume is very high (> 0.9 cm^3/g). In addition, silica particles exhibit good chemical stability and resistance to mechanical stress.

Such characteristics render silica particles well suited for immobilization of various compounds by pore entrapment or covalent linking [77,100,101]. Due to ionized Si-OH groups, the pore surface is negatively charged and positively charged compounds are therefore suitable for pore electrostatic immobilization (Figure 8.16A). Surface immobilization can be achieved if the particle is grown in the presence of a second precursor that is derivatized with

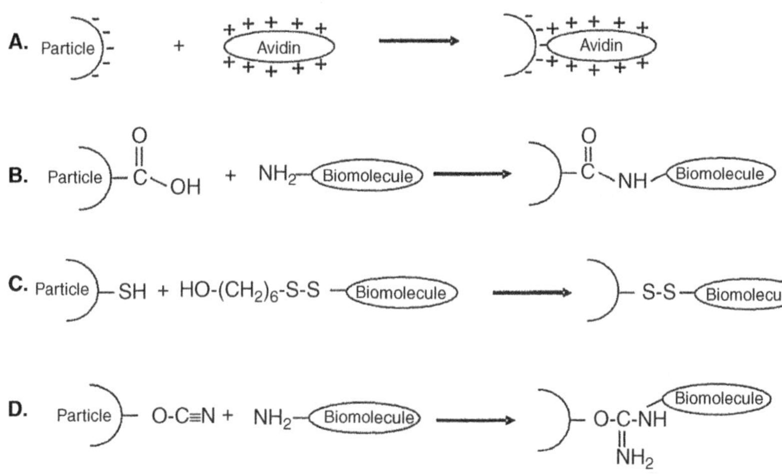

Figure 8.16 Various bioconjugation methods for silica nanoparticles. (A) avidin–biotin linking; (B) formation of peptide linkages by reaction with carboxyl-derivatized particles; (C) disulfide bonding; (D) formation of peptide linkages by cyanogen bromide modification. Adapted with permission from [101]. Copyright 2004 Wiley-VCH Verlag GmBH & Co. KGaA.

a reactive bridging group. In a further step, the target compound is linked to the functionalized surface by standard bioconjugation reactions (Figure 8.16).

Various functional nanomaterials (such as quantum dots, magnetic nanoparticles or metal nanoparticles) can be integrated with silica to yield nanohybrids. Growing a silica layer on functional nanoparticle prevents self-aggregation, reduces toxicity and brings about chemical groups useful for further functionalization.

Questions and Exercises (Section 8.7)

1 What properties of silica particles make them suitable for sensor applications?
2 Using the immobilization methods introduced in Chapter 5, write detailed mechanisms for bioconjugation methods in Figure 8.16.

8.8 Dendrimers

8.8.1 Properties and Applications

Dendrimers are repeatedly branched, roughly spherical, 2–7 nm diameter large polymer molecules [102]. The name comes from the Greek word dendron, which translates as "tree". Dendrimers have three major portions: a core, an inner shell, and an outer shell (Figure 8.17A). These compounds are grown as successive spherical layers assembled by covalent bonding, each layer being identified by its generation number.

A dendrimer can be synthesized so as to have different functionalities on each of these portions in order to control characteristic properties such as solubility, thermal stability, and its capability to be conjugated with various compounds. Dendrimer synthesis can be conducted so as to control precisely the size and number of branches of the dendrimer. It is thus possible to assemble a predetermined number of functional groups.

Related compounds are *dendrons* that contain a chemically addressable group called the focal point with an asymmetric branched structure grown asymmetrically (Figure 8.17B).

Dendrimer applications in chemical sensors rely on their outstanding property of displaying a high density of functional groups at the outer shell. This allows receptors and other molecules to be assembled so as to form multi-functional nanosized assemblies. To this end, one can resort to standard conjugation methods involving functionalized dendrimers [103]. In addition, properly designed dendrimers can encapsulate small molecules at certain sites of appropriate size within the dendritic structure.

Due to their multivalent nature, dendrimers can be used to add increased functionality to a solid surface. As an example, Figure 8.18 shows a dendrimer attached to a support via siloxane chemistry. Sulthydryl-focal point dendrons can be assembled on gold surfaces by sulfur chemisorption. Terminal groups in dendrimes and dendrons serve for subsequent conjugation with various compounds of interest.

It is not surprising that dendrimers have found various applications in chemical-sensor design [104]. Thus, dendrimers are used as nanosize scaffolds for assembling sensor active components via strong covalent bonds. On the other hand, dendrimers with a suitable structure elicit selective responses to various chemical species including gases, vapors inorganic ions and biomolecules that make them suitable as recognition elements.

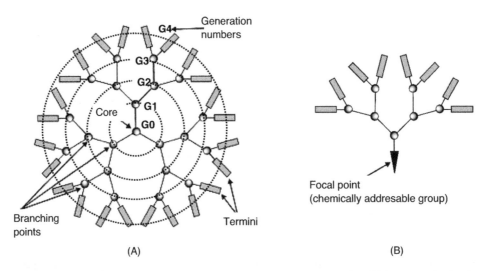

Figure 8.17 (A) Structure of a dendrimer. Branching points are located on equidistant circles. (B) Structure of a dendron. Adapted from http://commons.wikimedia.org/wiki/File:Graphs.jpg Last accessed 17/05/2012.

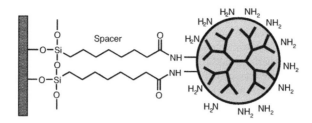

Figure 8.18 Surface functionalization by dendrimers.

Questions and Exercises (Section 8.9)

1 What are the particular features of the dendrimer chemical structure?
2 What characteristics of dendrimer structure are of interest to the development of chemical sensors?

8.9 Summary

Various nanomaterials have been applied to chemical-sensor design with extremely promising results. Nanomaterials can be used either as structural components of the sensing part or as functional components that perform a specific task in the recognition or transduction event.

Dendrimers, carbon nanotubes, metal nanoparticles and silica nanoparticles are useful for assembling recognition components such as enzymes or affinity receptors thus allowing to achieve of a high density of active component. This results, as a rule, in enhanced sensor sensitivity. On the other hand, nanostructuring of the sensing part enhances the rate of diffusion of the analyte to the receptor sites.

Conducting carbon nanotubes allow electrical wiring at the nanoscale and can also perform direct electron transfer from the active site of redox enzymes to the electrode in electrochemical sensors.

Nanomaterials are also useful as high-performance labels, particularly in optical sensors. Application of quantum dots, carbon nanotubes, and gold nanoparticles in such sensors have brought about considerable advantages over molecular label-based optical sensors.

In addition, nanomaterials can perform as amplification markers in certain types of sensors. Thus, postrecognition assembly of gold nanoparticles to the sensing element enhances considerably the total mass, which represents a valuable amplification method in mass-sensitive transduction. By postrecognition grafting of dendrimers or liposomes it is possible to enhance the local electric charge, which results in signal enhancement in certain types of electrochemical sensors.

No less interesting is the application of certain nanomaterials (such as semiconductor carbon nanotubes or dendrimers) as recognition materials. By interaction with specific analytes, such materials experience modifications in their physicochemical properties that can be translated into a response signal.

Application of nanomaterials is currently the most active research area in chemical-sensor science and there is no doubt that most of the traditional materials and manufacturing methods will be replaced by nanotechnology-derived ones.

References

1. Ozin, G.A., Arsenault, A.C., and Cademartiri, L. (2009) *Nanochemistry: A Chemical Approach to Nanomaterials*, RSC Publ., Cambridge.
2. Cademartiri, L. and Ozin, G.A. (2009) *Concepts of Nanochemistry*, Wiley-VCH, Weinheim.
3. Asefa, T., Duncan, C.T., and Sharma, K.K. (2009) Recent advances in nanostructured chemosensors and biosensors. *Analyst*, **134**, 1980–1990.
4. Kumar, C.S.S.R. (ed.) (2007) *Nanomaterials for Biosensors*, Wiley-VCH, Weinheim.
5. Pierce, D.T. and Zhao, J.X. (eds) (2010) *Trace Analysis with Nanomaterials*, Wiley-VCH, Weinheim.
6. Merkoçi, A. (ed.) (2009) *Biosensing Using Nanomaterials*, John Wiley & Sons, New York.
7. Vollath, D. (2008) *Nanomaterials: An Introduction to Synthesis, Properties and Applications*, Wiley-VCH, Weinheim.
8. Cao, G. (2004) *Nanostructures and Nanomaterials: Synthesis, Properties and Applications*, Imperial College Press, London.
9. Kumar, C.S.S.R. (ed.) (2005) *Biofunctionalization of Nanomaterials*, Wiley-VCH, Weinheim.
10. de Dios, A.S. and Diaz-Garcia, M.E. (2010) Multifunctional nanoparticles: Analytical prospects. *Anal. Chim. Acta*, **666**, 1–22.

11. Ruiz-Hitzky, E., Ariga, K., and Lvov, Y. (eds) (2008) *Bio-Inorganic Hybrid Nanomaterials: Strategies, Syntheses, Characterization and Applications*, Wiley-VCH, Weinheim.

12. Kalantar-zadeh, K. and Fry, B. (2008) *Nanotechnology-Enabled Sensors*, Springer, Boston.

13. Wang, J. and Katz, E. (2005) Biomaterial-nanoparticle hybrid systems for sensing and electronic devices, in *Bioelectronics: from Theory to Applications* (eds I. Willner and E. Katz), Wiley-VCH, Weinheim, pp. 231–264.

14. Katz, E. and Willner, I. (2004) Integrated nanoparticle-biomolecule hybrid systems: Synthesis, properties, and applications. *Angew. Chem. Int. Ed.*, **43**, 6042–6108.

15. Talapin, D.V., Lee, J.S., Kovalenko, M.V. *et al.* (2010) Prospects of colloidal nanocrystals for electronic and optoelectronic applications. *Chem. Rev.*, **110**, 389–458.

16. Shchipunov, Y. (2008) Entrapment of biopolymers into sol-gel-derived silica nanocomposites, in *Bio-inorganic Hybrid Nanomaterials: Strategies, Syntheses, Characterization and Applications* (eds E. Ruiz-Hitzky, K. Ariga, and Y. Lvov), Wiley-VCH, Weinheim, pp. 75–112.

17. Song, S., Qin, Y., Huang, Q., *et al.* (2010) Functional nanoprobes for ultrasensitive detection of biomolecules. *Chem. Soc. Rev.*, **39**, 4234–4243.

18. Arvizo, R.R., De, M., and Rotello, V.M. (2007) Proteins and nanoparticles: Covalent and noncovalent conjugates, in *Nanobiotechnology II: More Concepts and Applications* (eds C.A. Mirkin and C.M. Niemeyer), Wiley-VCH, Weinheim, pp. 65–97.

19. Kumar, C.S.S.R. (ed.) (2009) *Metallic Nanomaterials*, Wiley-VCH, Weinheim.

20. Daniel, M.C. and Astruc, D. (2004) Gold nanoparticles: Assembly, supramolecular chemistry, quantum-size-related properties, and applications toward biology, catalysis, and nanotechnology. *Chem. Rev.*, **104**, 293–346.

21. Schmid, G. (2010) Noble metal nanoparticles, in *Nanoparticles: From Theory to Application* (ed. G. Schmid), Wiley-VCH, Weinheim, pp. 214–239.

22. Zheng, M. and Huang, X. (2005) Biofunctionalization of gold nanoparticles, in *Biofunctionalization of Nanomaterials* (ed. C.S.S.R. Kumar), Wiley-VCH, Weinheim, pp. 99–124.

23. Feldheim, D.L. and Foss, C.A. (2001) Electrochemical synthesis and optical properties of gold nanorods, in *Metal nanoparticles: Synthesis, Characterization, and Applications* (eds D.L. Feldheim and C.A. Foss), Marcel Dekker, New York, pp. 163–182.

24. Shad Thaxton, C. and Mirkin, C.A. (2004) DNA-gold nanoparticles conjugates, in *Nanobiotechnology: Concepts, Applications and Perspectives* (eds C.M. Niemeyer and C.A. Mirkin), Wiley-VCH, Weinheim, pp. 288–307.

25. Drechsler, U., Erdogan, B., and Rotello, V.M. (2004) Nanoparticles: Scaffolds for molecular recognition. *Chem.-Eur. J.*, **10**, 5570–5579.

26. Baron, R., Willner, B., and Willner, I. (2007) Biocatalytic growth of nanoparticles for sensors and circuitry, in *Nanobiotechnology II: More Concepts and Applications* (eds C.A. Mirkin and C.M. Niemeyer), Wiley-VCH, Weinheim, pp. 99–121.

27. Allen, M.J., Tung, V.C., and Kaner, R.B. (2010) Honeycomb Carbon: A review of graphene. *Chem. Rev.*, **110**, 132–145.

28. Ratinac, K.R., Yang, W., Gooding, J.J. *et al.* (2011) Graphene and related materials in electrochemical sensing. *Electroanalysis*, **23**, 803–826.

29. Krueger, A. (2010) Fullerenes-cages made from carbon, in *Carbon Materials and Nanotechnology*, Wiley-VCH, Weinheim, pp. 33–122.

30. Hirsch, A. and Brettreich, M. (2005) *Fullerenes: Chemistry and Reactions*, Wiley-VCH, Weinheim.

31. Chaniotakis, N. (2007) Fullerene-based electrochemical detection methods for biosensing, in *Nanomaterials for Biosensors* (ed. C.S.S.R. Kumar), Wiley-VCH, Weinheim, pp. 101–122.

32. Geim, A.K. and Novoselov, K.S. (2007) The rise of graphene. *Nature Mater.*, **6**, 183–191.

33. Krueger, A. (2010) Carbon nanotubes, in *Carbon Materials and Nanotechnology*, Wiley-VCH, Weinheim, pp. 123–281.

34. Harris, P.J.F. (1999) *Carbon Nanotubes and Related Structures: New Materials for the Twenty-First Century*, Cambridge University Press, Cambridge.

35. Lukaszewicz, J.P. (2006) Carbon materials for chemical sensors: A review. *Sens. Lett.*, **4**, 53–98.

36. Strano, M.S., Boghossian, A.A., Kim, W.J., *et al.* (2009) The Chemistry of single-walled nanotubes. *MRS Bull.*, **34**, 950–961.

37. Wang, J. (2005) Carbon-nanotube based electrochemical biosensors: A review. *Electroanalysis.*, **17**, 7–14.

38. Terrones, M. (2004) Carbon nanotubes: Synthesis and properties, electronic devices and other emerging applications. *Int. Mater. Rev.*, **49**, 325–377.

39. Merkoci, A. (2006) Carbon nanotubes in analytical sciences. *Microchim. Acta*, **152**, 157–174.

40. Mann, D. (2006) Synthesis of carbon nanotubes, in *Carbon Nanotubes: Properties and Applications* (ed. M. O'Connell), Taylor & Francis, Boca Raton, pp. 19–49.

41. Joselevich, E., Dai, H., Liu, J. *et al.* (2008) Carbon nanotube synthesis and organization, in *Carbon Nanotubes: Advanced Topics in the Synthesis, Structure, Properties and Applications* (eds A. Jorio, G. Dresselhaus, and M.S. Dresselhaus), Springer, Berlin.

42. Vajtai, R., Wei, B., George, T.F. *et al.* (2007) Chemical vapor deposition of organized architectures of carbon nanotubes for applications, in *Molecular Building Blocks for Nanotechnology* (eds G.A. Mansoori, T.F. George, L. Assoufid, and G. Zhang), Springer, New York, pp. 188–209.

43. Sgobba, V. and Guldi, D.M. (2009) Carbon nanotubes-electronic/electrochemical properties and application for nanoelectronics and photonics. *Chem. Soc. Rev.*, **38**, 165–184.

44. Hermanson, G.T. (2008) Buckyballs, fullerenes, and carbon nanotubes, in *Bioconjugate Techniques*, Elsevier, Amsterdam, pp. 627–648.

45. Niyogi, S., Hamon, M.A., Hu, H., *et al.* (2002) Chemistry of single-walled carbon nanotubes. *Accounts Chem. Res.*, **35**, 1105–1113.

46. Tasis, D., Tagmatarchis, N., Bianco, A. *et al.* (2006) Chemistry of carbon nanotubes. *Chem. Rev.*, **106**, 1105–1136.

47. Yang, W.R., Thordarson, P., Gooding, J.J., *et al.* (2007) Carbon nanotubes for biological and biomedical applications. *Nanotechnology.*, **18**, 12.

48. Bekyarova, E., Haddon, R.C., and Parpura, V. (2005) Biofunctionalization of carbon nanotubes, in *Biofunctionalization of Nanomaterials* (ed. C.S.S.R. Kumar), Wiley-VCH, Weinheim, pp. 41–71.

49. Zhao, Y.L. and Stoddart, J.F. (2009) Noncovalent functionalization of single-walled carbon nanotubes. *Accounts Chem. Res.*, **42**, 1161–1171.

50. Chen, R.J., Zhang, Y.G., Wang, D.W. *et al.* (2001) Noncovalent sidewall functionalization of single-walled carbon nanotubes for protein immobilization. *J. Am. Chem. Soc.*, **123**, 3838–3839.

51. Le Goff, A., Holzinger, M., and Cosnier, S. (2011) Enzymatic biosensors based on SWCNT-conducting polymer electrodes. *Analyst*, **136**, 1279–1287.

52. Wang, J., Liu, G., and Lin, Y. (2007) Nanotubes, nanowires, and nanocantilevers in biosensor development, in *Nanomaterials for Biosensors* (ed. C.S.S.R. Kumar), Wiley-VCH, Weinheim, pp. 56–100.

53. Ye, J.S. and Sheu, F.S. (2007) Carbon nanotube-based sensor, in *Nanomaterials for Biosensors* (ed. C.S.S.R. Kumar), Wiley-VCH, Weinheim, pp. 27–55.

54. Balasubramanian, K. and Burghard, M. (2006) Biosensors based on carbon nanotubes. *Anal. Bioanal. Chem.*, **385**, 452–468.

55. Jacobs, C.B., Peairs, M.J., and Venton, B.J. (2010) Review: Carbon nanotube based electrochemical sensors for biomolecules. *Anal. Chim. Acta*, **662**, 105–127.

56. Gooding, J.J. (2005) Nanostructuring electrodes with carbon nanotubes: A review on electrochemistry and applications for sensing. *Electrochim. Acta*, **50**, 3049–3060.

57. Tibbetts, G.G., Lake, M.L., Strong, K.L. *et al.* (2007) A review of the fabrication and properties of vapor-grown carbon nanofiber/polymer composites. *Compos. Sci. Technol.*, **67**, 1709–1718.

58. De Jong, K.P. and Geus, J.W. (2000) Carbon nanofibers: Catalytic synthesis and applications. *Catal. Rev. -Sci. Eng.*, **42**, 481–510.

59. Melechko, A.V., Merkulov, V.I., McKnight, T.E., *et al.* (2005) Vertically aligned carbon nanofibers and related structures: Controlled synthesis and directed assembly. *J. Appl. Phys.*, **97**, 39.

60. Vamvakaki, V., Fouskaki, M., and Chaniotakis, N. (2007) Electrochemical biosensing systems based on carbon nanotubes and carbon nanofibers. *Anal. Lett.*, **40**, 2271–2287.

61. Huang, J.S., Liu, Y., and You, T.Y. (2011) Carbon nanofiber based electrochemical biosensors: A review. *Anal. Methods*, **2**, 202–211.

62. Ramakrishna, S. (2005) *An Introduction to Electrospinning and Nanofibers*, World Scientific, Hackensack, NJ.

63. Ramakrishna, S., Lala, N.L., Garudadhwaj, H., *et al.* (2007) Polymer nanofibers for biosensor applications. In *Molecular Building Blocks for Nanotechnology* (eds G.A. Mansoori, T.F. George, L. Assoufid, and G. Zhang), Springer, New York, pp. 377–392.

64. Potyrailo, R.A. (2006) Polymeric sensor materials: Toward an alliance of combinatorial and rational design tools? *Angew. Chem. Int. Ed.*, **45**, 702–723.

65. Ding, B., Wang, M.R., Yu, J.Y. *et al.* (2009) Gas sensors based on electrospun nanofibers. *Sensors*, **9**, 1609–1624.

66. Shin, Y.J., Wang, M., and Kameoka, J. (2009) Electrospun nanofiber biosensor for measuring glucose concentration. *J. Photopolym. Sci. Technol.*, **22**, 235–237.

67. Ramakrishna, S., Lala, N.L., Garudadhwaj, H., *et al.* (2007) Polymer nanofibers for biosensor applications, In *Molecular Building Blocks for Nanotechnology* (eds Mansoori, G.A., George, T.F., Assoufid, L., and Zhang, G.), Springer, Berlin, pp. 377–392.

68. Yeo, L.Y. and Friend, J.R. (2006) Electrospinning carbon nanotube polymer composite nanofibers. *J. Exp. Nanosci.*, **1**, 177–209.

69. Marx, S., Jose, M.V., Andersen, J.D. *et al.* (2011) Electrospun gold nanofiber electrodes for biosensors. *Biosens. Bioelectron.*, **26**, 2981–2986.

70. Kumar, C.S.S.R. (ed.) (2009) *Magnetic Nanomaterials*, Wiley-VCH, Weinheim.

71. Varadan, V.K., Chen, L., and Xie, J. (2008) *Nanomedicine: Design and Applications of Magnetic Nanomaterials, Nanosensors and Nanosystems*, John Wiley & Sons, Chichester.

72. Tamanaha, C.R., Mulvaney, S.P., Rife, J.C. *et al.* (2008) Magnetic labeling, detection, and system integration. *Biosens. Bioelectron.*, **24**, 1–13.

73. Varadan, V.K., Chen, L., and Xie, J. (2008) Magnetic nanoparticles, in *Nanomedicine: Design and Applications of Magnetic Nanomaterials, Nanosensors and Nanosystems*, John Wiley & Sons, Chichester, pp. 85–128.

74. Frey, N.A., Peng, S., Cheng, K. *et al.* (2009) Magnetic nanoparticles: Synthesis, functionalization, and applications in bioimaging and magnetic energy storage. *Chem. Soc. Rev.*, **38**, 2532–2542.

75. Krylova, G., Bodnarchuk, M.I., Tromsdorf, U., *et al.* (2010) Synthesis, properties and applications of magnetic nanoparticles, in *Nanoparticles: From Theory to Application* (ed. G. Schmid), Wiley-VCH, Weinheim, pp. 239–310.

76. Thode, C.J. and Williams, M.E. (2009) Approaches to the biofunctionalization of spherical and anisotropic iron oxide nanomaterials, in *Magnetic Nanomaterials* (ed. C.S.S.R. Kumar), Wiley-VCH, Weinheim, pp. 507–550.

77. Jin, Y.H., Li, A.Z., Hazelton, S.G., *et al.* (2009) Amorphous silica nanohybrids: Synthesis, properties and applications. *Coordin. Chem. Rev.*, **253**, 2998–3014.

78. Graham, D.L., Ferreira, H.A., and Freitas, P.P. (2004) Magnetoresistive-based biosensors and biochips. *Trends Biotechnol.*, **22**, 455–462.

79. Varadan, V.K., Chen, L., and Xie, J. (2008) Magnetic biosensors, in *Nanomedicine: Design and Applications of Magnetic Nanomaterials, Nanosensors and Nanosystems* (eds V.K. Varadan, L. Chen, and J. Xie), John Wiley & Sons, Chichester, pp. 329–454.

80. Wang, S.X. and Li, G. (2008) Advances in giant magnetoresistance biosensors with magnetic nanoparticle tags: Review and outlook. *IEEE Trans. Magn.*, **44**, 1687–1702.

81. Kasatkin, S.I., Vasil'eva, N.P., and Murav'ev, A.M. (2010) Biosensors based on the thin-film magnetoresistive sensors. *Autom. Remote Control*, **71**, 156–166.

82. Llandro, J., Palfreyman, J.J., Ionescu, A. *et al.* (2010) Magnetic biosensor technologies for medical applications: A review. *Med. Biol. Eng. Comput.*, **48**, 977–998.

83. Germano, J., Martins, V.C., Cardoso, F.A., *et al.* (2009) A portable and autonomous magnetic detection platform for biosensing. *Sensors*, **9**, 4119–4137.

84. Rife, J.C., Miller, M.M., Sheehan, P.E. *et al.* (2003) Design and performance of GMR sensors for the detection of magnetic microbeads in biosensors. *Sens. Actuators A-Phys.*, **107**, 209–218.

85. Sole, S., Merkoci, A., and Alegret, S. (2001) New materials for electrochemical sensing - III. Beads. *TrAC-Trends Anal. Chem.*, **20**, 102–110.

86. Willner, I. and Katz, E. (2003) Magnetic control of electrocatalytic and bioelectrocatalytic processes. *Angew. Chem. Int. Ed.*, **42**, 4576–4588.

87. Hsing, I.M., Xu, Y., and Zhao, W.T. (2007) Micro- and nano-magnetic particles for applications in biosensing. *Electroanalysis*, **19**, 755–768.

88. Rogach, A.L. (ed.) (2008) *Semiconductor Nanocrystal Quantum Dots: Synthesis, Assembly, Spectroscopy and Applications*, Springer, Vienna.

89. Murray, C.B., Kagan, C.R., and Bawendi, M.G. (2000) Synthesis and characterization of monodisperse nanocrystals and close-packed nanocrystal assemblies. *Annu. Rev. Mater. Sci.*, **30**, 545–610.

90. Trindade, T., O'Brien, P., and Pickett, N.L. (2001) Nanocrystalline semiconductors: Synthesis, properties, and perspectives. *Chem. Mater.*, **13**, 3843–3858.

91. Reiss, P. (2008) Synthesis of semiconductor nanocrystals in organic solvents, in *Semiconductor Nanocrystal Quantum Dots: Synthesis, Assembly, Spectroscopy and Applications* (ed. A.L. Rogach), Springer, Vienna, pp. 35–72.

92. Mazumder, S., Dey, R., Mitra, M.K., *et al.* (2009) Review: Biofunctionalized quantum dots in biology and medicine. *J. Nanomater.*, Article ID 815734.

93. Costa-Fernandez, J.M., Pereiro, R., and Sanz-Medel, A. (2006) The use of luminescent quantum dots for optical sensing. *TrAC-Trend Anal. Chem.*, **25**, 207–218.

94. Goldman, E.R., Balighian, E.D., Mattoussi, H., *et al.* (2002) Avidin: A natural bridge for quantum dot-antibody conjugates. *J. Am. Chem. Soc.*, **124**, 6378–6382.

95. Gaponik, N. and Rogach, A.L. (2008) Aqueous synthesis of semiconductor nanocrystals, in *Semiconductor Nanocrystal Quantum Dots: Synthesis, Assembly, Spectroscopy and Applications* (ed. A.L. Rogach), Springer, Vienna, pp. 73–99.

96. Skaff, H. and Emrick, T. (2004) Semiconductor nanoparticles, in *Nanoparticles: Building Blocks for Nanotechnology* (ed. V. Rotello), Kluwer Academic/Plenum Publishers, New York, pp. 29–82.

97. Zrazhevskıy, P., Sena, M., and Gao, X. (2010) Designing multifunctional quantum dots for bioimaging, detection, and drug delivery. *Chem. Soc. Rev.*, **39**, 4326–4354.

98. Stöber, W., Fink, A., and Bohn, E. (1968) Controlled growth of monodisperse silica spheres in micron size range. *J. Colloid Interface Sci.*, **26**, 62–69.

99. Arriagada, F.J. and Osseoasare, K. (1995) Synthesis of nanosize silica in aerosol OT reverse microemulsions. *J. Colloid Interface Sci.*, **170**, 8–17.

100. Wang, L., Zhao, W.J., and Tan, W.H. (2008) Bioconjugated silica nanoparticles: Development and applications. *Nano Res.*, **1**, 99–115.

101. Drake, T.J., Zao, J.X., and Tan, W.H. (2004) Bioconjugated silica nanoparticles for bioanalytical applications, in *Nanobiotechnology: Concepts, Applications and Perspectives* (eds C.M. Niemeyer and C.A. Mirkin), Wiley-VCH, Weinheim, pp. 444–457.

102. Newkome, G.R., Vögtle, F., and Moorefield, C.N. (2001) *Dendrimers and Dendrons: Concepts, Syntheses, Applications*, Wiley-VCH, Weinheim.

103. Hermanson, G.T. (2008) Dendrimers and dendrons, in *Bioconjugate Techniques*, Elsevier, Amsterdam, pp. 346–390.

104. Jayaraman, N. (2007) Dendrimers and their use as nanoscale sensors, in *Nanomaterials Chemistry: Recent Developments and New Directions* (eds C.N.R. Rao, A. Müller, and A.K. Cheetham), Wiley-VCH, Weinheim, pp. 249–297.

9

Thermochemical Sensors

Thermochemical transduction is intrinsically a universal method being based on a property common to a large number of chemical reactions, namely, the generation of heat by exothermal reactions. The thermal effect of a chemical reaction is indicated by the standard enthalpy of reaction (ΔH_r^0, in kJ mole^{-1}), which represents the change in the energy of the system at normal temperature and pressure when one mole of reactant is transformed by a chemical reaction.

Catalytic reactions are particularly suited for applications in chemical sensors since a large amount of reactant can be converted in a small catalytic reactor to produce a local change in temperature. This effect is quantified by means of a temperature transducer. Under steady-state conditions, the temperature variation with respect to a reference sensor indicates the concentration of the analyte in the sample. The steady state can be achieved by feeding the fluid sample to the reactor at a constant rate.

This chapter introduces first typical temperature transducers employed in thermochemical sensors and then presents two types of thermal sensors, namely enzymatic thermal sensors and thermocatalytic gas sensors.

9.1 Temperature Transducers

In order to convert temperature variations into an electric signal, two main kinds of transducer are used, namely resistive transducers and transducers based on the thermoelectric effect [1]. In the first case, the resistivity of the device varies with the temperature, whereas in the second case the device is a thermocouple that generates an electric voltage depending on the temperature difference between two points.

9.1.1 Resistive Temperature Transducers

Resistance thermometers, also called resistance temperature detectors or resistive thermal devices (RTDs), are temperature transducers that exploit the change in electrical resistance (R) of thin platinum wires whose resistivity increases with the temperature as:

$$R = R_0[1 + \alpha(T - T_0)] \tag{9.1}$$

where R_0 is the resistance at normal temperature T_0, and α is the temperature coefficient.

Another type of temperature transducer is the *thermistor*, which is a ceramic semiconductor device made of oxides of transition metals. Most thermistors have a negative temperature coefficient; that is, their resistance decreases with increasing temperature. Over a narrow temperature region the change in resistance is directly proportional to the temperature change.

In order to detect small resistance modifications the resistive temperature transducer is incorporated in a Wheatstone resistor bridge, which is a four resistor circuit as shown in Figure 9.1. A DC voltage is applied between the points C ad D and the output voltage is measured between A and B. It is easy to demonstrate that, if $R_1/R_2 = R_3/R_4$, the output voltage is zero. Very small deviations from the above condition give rise to a nonzero output voltage. In a thermochemical sensor setup, the actual sensor (R_4) is coupled with a reference sensor (R_3) whose temperature is not affected by the reaction. Measurement of the output bridge voltage allows detection of temperature variations down to 0.005 K.

9.1.2 Thermopiles

The basis of the thermopile is the thermoelectric effect, that is, conversion of the temperature difference directly into an electrical voltage in a device called *thermocouple* (Figure 9.2A). The thermocouple consists of a loop formed by two different materials (metals or semiconductors). The output voltage (ΔV) is proportional to the temperature difference between the two junctions, that is:

$$\Delta V = a_{XY}(T_1 - T_2) \tag{9.2}$$

Chemical Sensors and Biosensors: Fundamentals and Applications, First Edition. Florinel-Gabriel Bănică.
© 2012 John Wiley & Sons, Ltd. Published 2012 by John Wiley & Sons, Ltd.

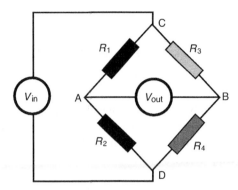

Figure 9.1 Wheatstone bridge circuit diagram. V_{in} is the applied DC voltage; V_{out} is the output bridge voltage; R_1 and R_2 are simple resistors. R_4 indicates the transducer resistance and R_3 is the resistance of a reference sensor.

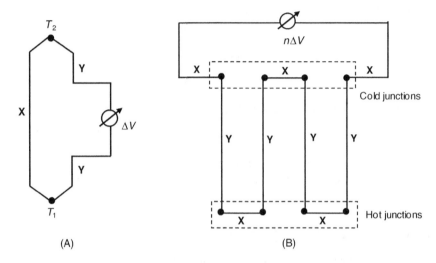

(A)   (B)

Figure 9.2 A thermocouple (A) and a thermopile (B). X and Y denote two different conducting materials; n is the number of thermocouples.

where a_{XY} (the relative Seebeck coefficient) depends upon the composition of the materials X and Y and upon the working temperature; it is approximately constant over a narrow temperature range. In order to increase the output voltage, thermocouples are connected in series to yield a thermopile (Figure 9.2B). The output voltage of this device is $n\Delta V$, where n is the number of thermocouples. When the cold junctions are maintained at constant temperature and the hot junctions are exposed to the heat evolved by a chemical reaction, the output voltage is related to the reactant concentration. Thermopiles can be fabricated easily on integrated circuits and display substantial advantages as compared with thermistors as far as sensitivity, noise cancellation and response time are concerned.

9.2 Enzymatic Thermal Sensors

9.2.1 Principles of Thermal Transduction in Enzymatic Sensors

Enzyme-catalyzed reactions are exothermic processes that allow for thermometric transduction in enzymatic sensors. Even when the substrate reaction is not thermally efficient, further heat can be generated by incorporating a subsequent enzymatic reaction. For example, if hydrogen peroxide is one of the primary products, catalase can be added in order to dissociate the H_2O_2 into water and oxygen to produce an appreciably enhanced thermal effect. A typical example of this kind is the determination of glucose using the glucose oxidase-catalyzed reaction. Additionally, proton exchange reactions involving a buffer component also contribute to an increase in the overall thermal effect; The Tris buffer (tris-(hydroxymethyl)aminomethane) is particularly effective in this respect.

A thermal enzymatic sensor can be looked upon as a microcalorimeter with the biological component closely integrated with a temperature transducer. This kind of analytical device was first developed in 1974 [2,3] and the technology has progressed steadily since then [4–6]. Manufacturing protocols for thermal enzymatic sensors can be found in ref. [7].

Changes in temperature caused by the enzymatic reaction can be probed by various temperature transducers, the thermistor being the most common.

The rate of heat production at the surface of the thermal transducer is limited by the diffusion of reactants but heat dissipation is a much faster process and this prevents substantial changes in temperature. In order to minimize heat

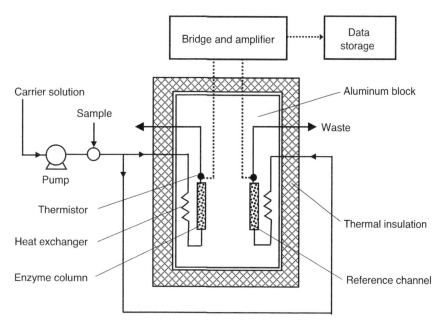

Figure 9.3 Configuration of a microcalorimetric enzyme analysis system.

dissipation, the overall sensor-solution system should be as small as possible. As organic solvents have lower thermal capacities than water, use of such solvents brings a significant improvement in sensitivity.

9.2.2 Thermistor-Based Enzymatic Sensors

The typical configuration of a thermal enzymatic analyzer is shown in Figure 9.3. The overall analysis system consists of two components, the enzyme thermistor calorimeter and the flow-injection system. A carrier buffer solution is pumped steadily to the calorimeter, and the stream is split so as to reach both the analytical and reference columns. The first column is filled with about 1 ml of biocatalytic material (for example, an enzyme immobilized by glutaraldehyde cross linking on controlled-pore glass beds). The second column is packed with inactive material. Before the buffer solution reaches the columns, the fluid temperature is stabilized by means of heat exchangers. Downstream from the columns, transducers are installed in order to detect the temperature changes in the effluent. The temperature changes are detected by a Wheatstone bridge and the output voltage is amplified and fed to the data storage and processing unit.

An important working parameter of this analyzer is the carrier flow rate. It should be selected so as to secure a sufficiently long contact time between the sample and the enzyme filling in order to achieve a sufficiently high substrate conversion, and, as a result, a suitably large heat output. The analysis system can be operated either in flow-injection mode or in flow-through mode. In the flow-injection mode, a limited volume of sample is carried to the column by a stream of buffer solution. The output is a peak-shaped ΔT vs. time curve. For quantitation purpose, either the peak height or the integrated signal (peak area) are correlated with the substrate concentration. Conversely, in the flow-through mode, a stream of substrate-containing solution is supplied continuously to the enzyme reactor that results in a greater ΔT value as compared with the flow-injection mode.

An alternative to the dual column system is presented in Figure 9.4. Here, the reference thermistor is installed before the catalytic column. The device in this Figure is a miniaturized one being suitable for 0.1 µl samples. The main unit is a stainless steel column packed with controlled-pore glass beds carrying an enzyme.

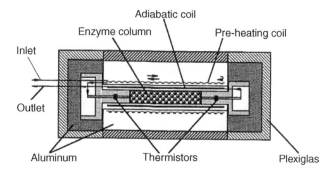

Figure 9.4 Miniaturized single-column thermal biosensor (54 mm in length and 24 mm in diameter). The enzyme column is 1.5/1.7 mm inner diameter/outer diameter and 15 mm in length. Arrows show the flow direction. Reproduce with permission from [6]. Copyright 2000 Elsevier.

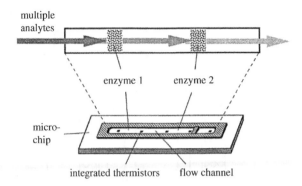

Figure 9.5 Schematics of an integrated thermal sensor array based on a combination of different enzyme columns. Reproduce with permission from [6]. Copyright 2000 Elsevier.

The most remarkable progress in this field has arisen from the application of integrated circuit technology to manufacture integrated microsystems. As demonstrated in Figure 9.5, this technology enables the construction of multianalyte sensors by integrating a series of enzyme columns and transducers in order to detect a range of substrates. The size of such a system can be as little as 10 mm, with the enzyme being immobilized directly onto the surface of microchannels etched within the wafer.

9.2.3 Thermopile-Based Enzymatic Sensors

An integrated thermal biosensor using a thermopile as transducer has been manufactured by microelectromechanical systems technology. It consists of a polymer microfluidic structure (chambers and channels) integrated with a silicon-based thermal transducer chip [8]. Enzyme-activated microbeads were loaded into the chambers using the inlet channels and were confined inside by a weir structure located at the junction between each chamber and its outlet channel. Chambers lie on a polymer diaphragm on which the thermopile is integrated in the form of thin-film chromium-nickel junctions. Metal resistive microheaters are included for device calibration and temperature control. Air gaps around each chamber as well as the plastic diaphragm provide a good thermal insulation.

Such a microfluidic device can be used in conjunction with a subcutaneous sampling method, such as microdialysis. The structure allows the handling of biological samples of very low volume (0.8 μl).

A gain in sample throughput is obtained if the microplate array format is used for the calorimetric detection of low molecular analytes. The transduction principle is based on the differential measurement of the heat generated between two wells, located at the cold and the hot junctions of a thermopile (Figure 9.6). The response is the difference in temperature between the active well the reference well (where no specific reaction takes place). Tens or hundreds of wells can be engraved on a wafer. The thermopiles are fabricated by integrated circuit technology as combinations of 64 aluminum-p-silicon thermocouples. As the required amount of enzyme is extremely small, dissolved enzyme can be dispensed to each measuring well instead of immobilizing it. Samples and reagents are dispensed by means of a software-controlled microdispenser. The feasibility of such a device was demonstrated by ascorbic acid determination using ascorbates oxidase-catalyzed oxidation by oxygen. In contrast to the sequential operation mode of a flow-injection analysis system, the microplate array allows simultaneous analysis of a great number of samples of volume in the nanoliter range.

9.2.4 Multienzyme Thermal Sensors

In general, the limit of detection of single-enzyme thermal biosensors is rather poor (10^{-4} M) and the response range ($10^{-4} - 10^{-2}$ M) is relatively narrow for the majority of applications. Better performances are obtained for the more

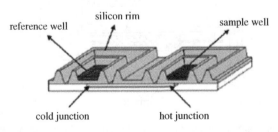

Figure 9.6 Structure of a microplate differential calorimeter. A full microplate consisted of 48 couples of wells. Reproduce with permission from [9]. Copyright 2007 American Chemical Society.

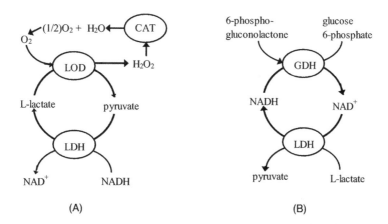

Figure 9.7 (A) Substrate recycling for thermal enzyme sensor applications. LOD = lactate oxidase; LDH = lactate dehydrogenase; CAT = catalase; (B) Coenzyme recycling for determining NAD(H). GDH = glucose-6-phosphate dehydrogenase; LDH = lactate dehydrogenase.

exothermic reactions (e.g., catalase). Sensitivity is substantially enhanced by increasing the heat output based on substrate or coenzyme recycling in multienzyme systems [4]. For example, a 1000-fold enhanced sensitivity was achieved for lactate or pyruvate by using coimmobilized lactate oxidase and lactate dehydrogenase (Figure 9.7A) [10]. The heat effect is enhanced by the decomposition of peroxide catalyzed by the catalase enzyme. This configuration allows the determination of lactate or pyruvate down to the 10 nM level.

An enlightening example of coenzyme recycling application refers to a device using the highly exothermic oxidation of NADH ($\Delta H = 225$ kJ/mole). The cycling process depicted in Figure 9.7B considerably enhances the sensitivity of NAD(H) determination by means of coimmobilized lactate dehydrogenase and glucose-6-phosphate dehydrogenase using glucose-6-phosphate and pyruvate, respectively, as substrates.

9.2.5 Outlook

In summary, what is usually termed as a thermal biosensor is in effect a kind of plug-flow enzyme reactor integrated with a temperature transducer incorporated within a microcalorimeter structure. The reactor can be configured in the packed-bed form, but tubular reactors, with the enzyme immobilized on the channel wall (e.g., nylon) can be a suitable alternative in some instances. The second configuration is advantageous when clogging of the packed beds may occur.

Thermal sensors have been designed for a large variety of applications including the determination of metabolites, bioprocess monitoring, and environmental control by means of enzyme inhibition [4,11]. The main advantage of the thermal sensor is its general applicability. As the temperature change induced by a simple enzymatic reaction is relatively low, an improvement in the limit of detection can be achieved by additional reactions that enhance the heat output. Further progress in this field will arise from miniaturization and increased automation, as well as from improved integration with microfluidic systems in order to increases the sample throughput and reduce the amount of reagents required.

Questions and Exercises (Sections 9.1–9.2)

1 What kind of temperature transducers are used in enzyme thermal sensors? Explain the physical principles of each of them.

2 Draw a simplified sketch of the experimental setup for thermal enzyme sensor, explain the role of each component and discuss on the alternative flow analysis methods that are applicable to thermal enzyme sensors.

3 Comment on methods for miniaturizing of thermal enzyme sensor and point out the advantages of miniaturization.

4 Review briefly the methods for multiple substrate determination by means of thermal enzyme sensors.

5 How can the limit of detection of thermal enzyme sensors be improved? For the examples in Figure 9.7, write the relevant chemical and biochemical reactions.

6 Assume that an enzyme layer is immobilized at the surface of a temperature transducer. Derive the steady-state difference in temperature (ΔT) between this one and a reference transducer that is dipped into the same solution, as a consequence of the catalytic conversion of the substrate. Calculate ΔT for the following parameter values: $s = 1.5 \times 10^{-3}$ M, $D_{S,m} = 10^{-6}$ cm^{2}s^{-1}, $\Delta H = 1.1 \times 10^4$ cal mole^{-1}, $k_t = 1.5 \cdot 10^{-3}$ cal/s cm K, $l_m = 0.005$ cm, $\alpha =$ (a) 10 and (b) 1 (alpha is the external substrate modulus).

Hint: Refer to the external diffusion model in Chapter 4. The reaction heat evolved per second and mole of substrate is $v_a\Delta H$, where ΔH is the heat of reaction. The heat flux to the solution is $k_t\Delta T/l_m$, where k_t is the thermal conductivity of the solution. Equate the evolved reaction heat to the heat flux, formulate the v_a term as a function of the external substrate modulus (α) and solve for ΔT. This model neglects the heat transfer to the enzyme layer and the result will therefore represent an overestimation.

Answer: $\Delta T = \Delta H (D_{S,m}/K_T)[\alpha/(1+\alpha)]s$; $\Delta T = 0.01$ (a) and 0.0055 K (b).

9.3 Thermocatalytic Sensors for Combustible Gases

Determination of various combustible gases is a stringent issue in mining and industry where accumulation of such gases brings about the risk of fire or explosion. Among the various types of combustible gas sensors, the catalytic type was the first to be produced on an industrial scale, and still arouses interest, despite the subsequent advent of more advanced gas sensors. This topic is well covered in refs. [12–14].

9.3.1 Structure and Functioning Principles

The typical configuration of a thermocatalytic gas sensor is shown in Figure 9.8A. It consists of a porous ceramic bead (typically alumina) into which is embedded a temperature-sensitive platinum coil resistor. The external layer of the bead contains a finely dispersed catalyst that is usually a platinum-group metal. Alternatively, the catalyst can be included through the whole ceramic bead. Commonly, this kind of gas sensor is termed a *pellistor*, the word being derived from "catalytic pelletized resistor". Other temperature probes, such as the thermistor or the thermocouple, can also be use in the thermocatalytic gas sensors.

In operation, the device is heated to 400–800 °C by passing a current through the Pt coil. In the presence of a combustible gas or vapor, the catalyst promotes the analyte combustion that causes the temperature of the device to increase. As a result, the resistance of the Pt wire also increases. The change in the resistance can be detected by a Wheatstone bridge circuit that permits the determination of the content of the gas of interest. Clearly the platinum wire serves as both heater and temperature transducer.

In order to enhance the thermal effect, the device should be fitted with effective thermal insulation. On the other hand, it should allow for efficient heat transfer from the reaction zone to the temperature probe.

A thermocatalytic gas sensor responds, in principle, to any flammable gas or vapor. However, some degree of selectivity can be achieved by proper selection of the operating temperature and the nature of the catalyst. Interfering gases can be retained by a sorbent filters installed between the sensor and the sample gas. A filter can also provide protection against damaging gases such as H_2S, SO_2, NO_2 or mercaptans that are catalyst poisons.

Fabrication of a standard pellistor, like that in Figure 9.8A involves much handwork. Progress in thermoelectric gas sensor technology came about with the introduction of thick-film technology that made possible the production of sensors in a planar configuration (see e.g., [15]).

Currently, research interest in thermocatalytic gas sensors is directed towards designing miniaturized devices and producing them by micromachining technology. As an example, Figure 9.8B shows a platform for

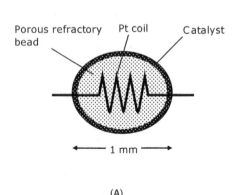
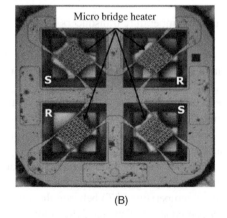

(A)   (B)

Figure 9.8 (A) The standard configuration of a thermocatalytic gas sensor (pellistor). Although this device is very small, the overall size is increased by the additional reference element and bridge resistors. (B) A platform for thermocatalytic gas sensors (2.4 × 2.4 mm) produced by micromachining. S means sensor; R means reference element. (B) Adapted with permission from [16]. Copyright (2011) Elsevier.

thermocatalytic gas sensors fabricated by micromachining on a silicon support coated with silicon dioxide and silicon nitride. Platinum has been deposited over cavities in the support so as to form four suspended resistors connected in a bridge circuit. Platinum layers provide heating and, at the same time, act as temperature probes. In order to obtain a sensor, alumina-supported catalyst films have been deposited over two opposite platinum layers, while the remaining two have been coated with plain alumina to form reference elements. This circuit deviates from the standard bridge configuration (Figure 9.1) that includes only one sensor and one reference element.

Suspended mounting of the platinum heater, as well as the small size of the device minimize heat loss to the support and the ambient atmosphere and hence provide good sensitivity.

The platform in Figure 9.8B has been employed to develop a hydrogen sensor using platinum as catalyst. Catalyzed combustion of hydrogen to water occurs with a very high reaction rate and, as a result, the sensor response to hydrogen is much greater than the response to various hydrocarbons, ethanol, and ether vapors.

Attempts at developing arrays of cross-selective thermocatalytic gas sensors have also been performed based on the effect of the catalyst nature upon the response to various gases. For example an array of four sensor has been designed to be used for the determination of methane, propane and ethanol vapors based on neural network data analysis [17]. The response of each individual sensor to a particular gas has been tuned by means of the catalyst nature (Pt or Pd), catalyst concentration and using mixtures of the above catalysts.

Despite the advent of other types of gas sensor, thermocatalytic gas sensors are still of interest in traditional applications such as detection of certain hazardous gases as well as in fire or explosion alarms. This is due to the extremely simple design and facile fabrication. As these devices should be operated at elevated temperature, the power consumption is a limiting parameter. Progress in this area is still possible by applications of advanced fabrication technologies that allow for inexpensive mass production of devices characterized by low power consumption, as well as for developing arrays of cross-selective sensors. The array approach can alleviate for the poor selectivity of thermocatalytic gas sensors.

Questions and Exercises (Section 9.3)

1 What kind of chemical reaction can be used in sensors for combustible gases and what physical effects of the reaction form the basis of the transduction principle in such sensors?
2 What are the components of a pellistor and what is the typical design of this kind of sensor?
3 What are the advantages and shortcoming of thermocatalytic gas sensors? How can some of the shortcomings be overcome?
4 Give a short overview of the contemporary trends in the technology of thermocatalytic gas sensors.

References

1. Sinclair, I.R. (2001) *Sensors and Transducers*, Newnes, Boston.
2. Mosbach, K. and Danielsson, B. (1974) Enzyme thermistor. *Biochim. Biophys. Acta*, **364**, 140–145.
3. Cooney, C.L., Weaver, J.C., Tannenbaum, S.R. *et al.* (1974) Thermal enzyme probe: A novel approach to chemical analysis, in *Enzyme Engineering* (eds E. Kendall Pye and L.B. Wingard), Plenum, New York, pp. 411–417.
4. Danielsson, B. and Windquist, F. (1990) Thermometric sensors, in *Biosensors: A Practical Approach* (ed. A.E.G. Cass), IRL Press, Oxford, pp. 191–209.
5. Ramanathan, K. and Danielsson, B. (2001) Principles and applications of thermal biosensors. *Biosens. Bioelectron.*, **16**, 417–423.
6. Xie, B., Ramanathan, K., and Danielsson, B. (2000) Mini/micro thermal biosensors and other related devices for biochemical/clinical analysis and monitoring. *TrAC-Trends Anal. Chem.*, **19**, 340–349.
7. Ramanathan, K., Khayayami, M., and Danielsson, B. (1998) Enzyme biosensors based on thermal transducer/thermistor, in *Enzyme and Microbial Sensors. Techniques and Protocols* (eds A. Mulchandani and K. R. Rogers), Humana Press, Totowa, pp. 175–186.
8. Wang, L., Sipe, D.M., Xu, Y. *et al.* (2008) A MEMS thermal biosensor for metabolic monitoring applications. *J. Microelectromech. Syst.*, **17**, 318–327.
9. Vermeir, S., Nicolai, B.M., Verboven, P. *et al.* (2007) Microplate differential calorimetric biosensor for ascorbic acid analysis in food and pharmaceuticals. *Anal. Chem.*, **79**, 6119–6127.
10. Scheller, F., Siegbahn, N., Danielsson, B. *et al.* (1985) High sensitivity enzyme thermistor determination of L-lactate by substrate recycling. *Anal. Chem.*, **57**, 1740–1743.
11. Danielsson, B. and Mosbach, K. (1988) Enzyme thermistors, in *Methods in Enzymology* vol. 137 (ed. K. Mosbach), New York, pp. 181–197.
12. Bársony, I., Dücsö, C., and Fürjes, P. (2009) Thermometric gas sensing, in *Solid State Gas Sensing* (eds E. Comini, G. Faglia, and G. Sberveglieri), Springer Science+Business Media, LLC, Boston, pp. 237–260.

13. Jones, G. (1987) The pellistor catalytic gas detector, in *Solid State Gas Sensors* (eds P.T. Moseley and B.C. Tofield), Adam Hilger, Bristol, pp. 17–31.

14. Symons, E.A. (1992) Catalytic gas sensors, in *Gas Sensors: Principles, Operation and Developments* (ed. G. Sberveglieri), Kluwer, Dordrecht, pp. 169–185.

15. Debeda, H., Massok, P., Lucat, C. *et al.* (1997) Methane sensing: From sensitive thick films to a reliable selective device. *Meas. Sci. Technol.*, **8**, 99–110.

16. Lee, E.B., Hwang, I.S., Cha, J.H. *et al.* (2011) Micromachined catalytic combustible hydrogen gas sensor. *Sens. Actuators B-Chem.*, **153**, 392–397.

17. Debeda, H., Rebiere, D., Pistre, J. *et al.* (1995) Thick-film pellistor array with a neural-network post-treatment. *Sens. Actuators B-Chem.*, **27**, 297–300.

10

Potentiometric Sensors

10.1 Introduction

Analytical determination of inorganic and organic ions is a matter of great interest to various areas such as biomedical sciences, environment monitoring and industry. Ion sensors respond to the demand of continuous ion monitoring, field analysis and *in vivo* ion determination.

Ion sensors rely on the general property of ions of bearing an electric charge. It is therefore not surprising that ion sensors were first developed based on electrochemical principles in the framework of potentiometry, that is, electrochemistry under equilibrium conditions. However, practical potentiometric ion sensors are not generally based on electrochemical oxidation or reduction processes because such processes display poor selectivity. Rather, potentiometric ion sensors are based on ion-transfer processes occurring at the interface of two immiscible phases. One phase is the test solution and the second phase is constructed so as to interact selectively with the analyte ion. The second phase, usually called an *ion-sensitive membrane,* can be in the solid state form or a liquid supported on an inert material. Due to the analyte interaction with the membrane, a nonuniform ion distribution appears at the membrane/solution interface. This results in a nonuniform electric charge distribution that gives rise to an electric potential difference. It is this potential difference that produces the response signal of potentiometric sensors. Such ion sensors are by tradition termed *ion-selective electrodes* (*ISEs*), although this name is not rigorously in agreement with the underlying physicochemical principles. That is why, the term potentiometric ion sensor will be used throughout this chapter to denote this kind of sensor.

Besides the determination of ions, potentiometric ion sensors can be used in an extensive series of applications involving the determination of nonionic compounds. Such applications have been developed by integrating an ion sensor with a component that reacts with the analyte to produce ions. In these devices, the ion sensor functions as a transducer that generates the response signal. Following this principle, sensors for gases that interact with aqueous solutions to form ions have been developed. Furthermore, if the sensing element contains an enzyme or living cells, biocatalytic potentiometric sensors can be developed providing that the biocatalytic reaction produces detectable ions. Enzymatic potentiometric sensors can be converted readily into immunosensors by including an immunoreactive component. Therefore, potentiometric ion sensors form the basis for the development of a broad range of chemical sensors.

One more point of interest regarding potentiometric ion sensors arises from the fact that sensing materials and reagents developed in the framework of this area are also useful in other types of ion sensors such as optical ion sensors (Section 19.2).

Good introductions to the field of ion-selective electrodes are available in various advanced analytical chemistry textbooks. Comprehensive presentations of this topic can be found in a series of monographs and book chapters [1–4]. Practical applications of ion-selective electrodes are amply discussed in refs. [5–7]. Specialized applications in the field of organic analysis [8,9], biology [10], and medicine [11] have also been comprehensively surveyed. IUPAC nomenclature recommendations for ion-selective electrodes can be found in ref. [12]. Recent advances in this field are presented in a series of review papers (see, e.g., refs. [13,14]). An historical perspective of ion-selective electrode development is available in ref. [15].

10.2 The Galvanic Cell at Equilibrium

Electrochemical processes involve electron- and ion-transfer processes. An electron transfer between two species produces modification in the oxidation state of the reactants in what is called a redox reaction in the case in which the involved reactants are ionic or molecular species in the solution. When the donor or acceptor species are macroscopic conductor materials, the electron-transfer process is termed an electrochemical reaction. Electrochemical reactions take place in electrochemical cells, that, in the simplest form, consist of two different electronic conductors (electrodes) immersed in an electrolyte solution. Each electrode may, in principle, exchange electrons with species in solutions either spontaneously or under the effect of an applied voltage. A cell in which the electron transfer at electrodes occurs spontaneously is termed a *galvanic cell*. It is important to keep in mind that the electrochemical cells used in potentiometric sensor applications are galvanic cell.

Before proceeding to the presentation of the galvanic cell characteristics, it is useful to give a short overview of the thermodynamics of electrolyte solutions.

10.2.1 Thermodynamics of Electrolyte Solutions

The thermodynamic state of any chemical species is defined by its *chemical potential,* μ_i, which indicates how much the free energy G of an open system changes in response to an infinitesimal variation in species amount (number of moles, (n_i)), when the temperature (T), the pressure (P) and the number of moles of other species (n_j) remain constant:

$$\mu_i = \left(\frac{\partial G}{\partial n_i}\right)_{T,P,n_j} \tag{10.1}$$

The chemical potential (in $J\,mol^{-1}$) depends on the thermodynamic activity *(a)* as follows:

$$\mu = \mu^0 + RT \ln a \tag{10.2}$$

where μ^0 is the chemical potential in the standard state. In turn, the activity is proportional to the concentration c, the proportionality coefficient γ being termed the activity coefficient ($\gamma \leq 1$):

$$a = \gamma c \tag{10.3}$$

In very dilute solutions, when γ is close to unity the activity is approximately equal to the concentration.

Only the *mean activity coefficient* of an ion can be determined experimentally. For a binary electrolyte, the mean activity coefficient $\gamma_{\pm}$ is, by definition:

$$\gamma_{\pm} = \sqrt{\gamma_+ \gamma_-} \tag{10.4}$$

where γ_+ and γ_- are the individual activity coefficients of the cation and the anion, respectively.

In electrolyte solutions, each ion exerts and, simultaneously, experiences electrostatic forces due to interaction with other ions. That is why the activity coefficient of an ion depends on the *ionic strength, I,* of the solution, which is determined by the activity (a_i) and the electric charge (z_i) of each ion present. So, the ionic strength is defined as:

$$I = \frac{1}{2}\sum_i a_i z_i^2 \tag{10.5}$$

where the summation involves the activity and the charge of each ion type in the system. At very low values of ionic strength, the mean activity coefficient can be calculated by the following equation:

$$\log \gamma_{\pm} = -A|z_+ z_-|\sqrt{I} \tag{10.6}$$

Here, A is parameter depending on the dielectric constant of water and the temperature. At a higher ionic strength $\gamma_{\pm}$ can be calculated by means of semiempirical formulas.

The chemical potential of a dissolved neutral species is determined mostly by the solvation process and short-range interactions. Conversely, in the case of an ion any change in the ion concentration brings about a variation in the electric charge density with important effects on the potential energy of the system. This effect can be accounted for by means of the concept of *electric potential.*

By definition, the electric potential difference between two points A and B (Figure 10.1) represents the change in the potential energy of the system when a unit positive charge arises from B to A in vacuum. The change in the potential energy equates the work W_{AB} associated with the charge movement in the electric field induced by the charge q_A. The charge probe experiences an electrostatic force F_x which is attractive if $q_A < 0$ (as in Figure 10.1) or repulsive when $q_A > 0$. Accordingly, the potential energy decreases by the term W_{AB} in the first case, as it occurs when an extended spring relaxes. Conversely, it increases by W_{AB} in the second case, similar to the case of a spring compression. By definition, the *electric potential difference* $\Delta\phi_{AB}$ between the two points is the quotient of the work and the local charge at the point A:

$$\Delta\phi_{AB} = \frac{W_{AB}}{q_A} \tag{10.7}$$

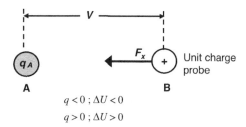

$q < 0 \; ; \Delta U < 0$

$q > 0 \; ; \Delta U > 0$

Figure 10.1 Definition of the electric potential difference between the points A and B. The unit charge at B moves to the point A where the charge q_A is localized. The electrostatic force F_x could be repulsive or attractive according to sign of the charge at the point A. ΔU is the change in the potential energy.

It is sometimes convenient to operate with the *absolute electric potential* that corresponds to the hypothetical situation in which the unit probe arrives at A from an infinite distance in vacuum. By means of this concept, the potential difference (or voltage) between A and B in volts (V) is:

$$\Delta\phi_{AB} = \phi_A - \phi_B \tag{10.8}$$

where ϕ_A and ϕ_B represents the absolute potential at the point A and B, respectively.

If the unit charge probe is replaced by a certain charge q, then the change in the potential energy due to the shift of this charge from infinity to the point A is:

$$\Delta U = \phi_A q \tag{10.9}$$

This is an important conclusion when assessing the effect of inserting an ion at a given point A. Letting z be the ion charge and F the Faraday constant (that is, the charge of a mole of a singly charged ion), it results that the change in potential energy produced by the transfer of one mole of ions is:

$$\Delta U = zF\phi_A \tag{10.10}$$

The state of an ion can now be described by taking into account both the chemical and electrostatic effects that are included in the *electrochemical potential* $\bar{\mu}$ of the ion. The change of the Gibbs free energy (in $J\,mol^{-1}$) when a mole of ions with charge z appears in the considered system is obtained by adding the chemical potential, as defined in Equation (10.2) and the electrical term given by Equation (10.10):

$$\bar{\mu} = \mu + zF\phi = \mu^0 + RT \ln a + zF\phi \tag{10.11}$$

Here, ϕ stands for the *inner potential* (also termed *the Galvani potential*) of a phase and the ion charge z should be included in this equation with the pertinent sign. The inner potential is the sum of two terms, the outer potential (ψ) and the surface potential (χ):

$$\phi = \psi + \chi \tag{10.12}$$

The surface potential is associated with the electrical charge localized at the surface of the considered phase. This charge is produced by oriented dipoles or ions accumulated at the surface. The outer potential is the potential difference between vacuum and a position close to the exterior of the surface where the electrostatic effect of the surface charge is absent.

10.2.2 Thermodynamics of the Galvanic Cell

A typical galvanic cell is the Daniel cell (Figure 10.2) which consists of copper and zinc electrodes immersed in solutions of their respective metal sulfates. The two solutions are separated by a porous diaphragm that prevents solutions from mixing but allows ions to migrate across it. Spontaneously, the zinc metal undergoes oxidation and zinc ions are released to the solution, whereas electrons are released to the electrode and that develops a negative charge. At the same time, copper ions in solution are reduced to copper metal by accepting electrons from the copper electrode that thus assumes a positive charge. Consequently, an electron flow from the zinc to the copper electrode takes place when the electrodes are connected via an external circuit. At the same time, anions from the right-hand part flow to the left in order to preserve the electric charge balance. In this way, an electric current is established across the systems, with electrons as charge carriers in the external circuit and ions as charge carriers in the solution.

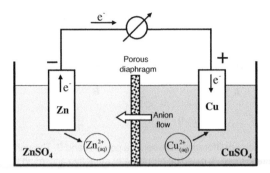

Figure 10.2 Schematic representation of a galvanic cell (the Daniel cell).

Transition from electronic to ionic conduction is secured by the electrochemical reactions occurring at each metal/solution interface. Schematically, this cell is represented as follows:

$$Zn(s)|ZnSO_4(aq)||CuSO_4(aq)|Cu(s) \tag{10.13}$$

Vertical bars in the above scheme represent phase interfaces; as a rule the activity or concentration of each species in the solution is also included in the scheme.

Potentiometry deals with electrochemical systems at equilibrium and it is hence of interest to discuss the characteristics of the electrochemical cell in this state. At equilibrium, the overall reaction rate is zero and no current flows through the system. The cell reaction involving $n = 2$ electrons is, in this case:

$$Zn(s) + Cu^{2+}(aq) \rightleftarrows Cu(s) + Zn^{2+}(aq) \tag{10.14}$$

The free energy change in this reaction represents the energy required to transfer nF electrons across a potential difference between the electrodes, which represents *the electromotive force* (EMF):

$$\Delta G = -nF(\text{EMF}) \tag{10.15}$$

The cell reaction (10.14) can be obtained by adding the two half-reactions that are zinc oxidation and copper ion reduction. A system composed of a metal electrode and the surrounding solution is termed a *half-cell* and the whole cell can be regarded as an assembly of two half-cells, each of them characterized by the pertinent half-reaction. In analytical applications, we are interested in only one of the half-cells, the second one acting as a *reference half-cell*. The standard half-cell consists of a platinum electrode immersed in a 1 M hydrogen ion activity solution saturated with hydrogen under a hydrogen gas pressure of 1 atm. Such a half-cell is termed the *standard hydrogen electrode* (SHE). Let us assume that a cell consists of a standard half-cell coupled with a half-cell containing a platinum electrode immersed in a solution containing a redox couple Ox and Red, where Ox is the oxidized form and Red is the reduced form of this couple. This cell is represented as follows:

$$Pt|Ox(a_O)Red(a_R)||H^+(a_{H^+} = 1\,M)|H_2(p_{H_2} = 1.00\,atm), Pt \tag{10.16}$$
$$\text{Indicator half-cell} \quad \text{Reference half-cell}$$

where a_O and a_R, and represent the activity of the oxidized and reduced form, respectively. a_{H^+} is the activity of the hydrogen ion in the solution and p_{H_2} is the partial pressure of hydrogen in the gas phase. The half-reaction at the left-hand half-cell is:

$$\nu_O Ox + ne^- \rightleftarrows \nu_R Red \tag{10.17}$$

where ν_O and ν_R are the stoichiometric coefficients of Ox and Red, respectively. The electrons needed in this reaction arise from the oxidation of the hydrogen molecule to hydrogen ions in the right hand half-cell. So the cell reaction is:

$$\nu_O Ox + (n/2)H_2 \rightleftarrows \nu_R Red + nH^+ \tag{10.18}$$

According to the general relationship between the Gibbs free energy and the equilibrium constant *(K)*, we have:

$$\Delta G = \Delta G^0 + RT \ln K = \Delta G^0 + RT \ln \frac{a_R^{\nu_R} a_{H^+}^n}{a_O^{\nu_O} p_{H_2}^{(n/2)}} \tag{10.19}$$

Here ΔG^0 is the standard Gibbs free energy change in reaction (10.18). Since both hydrogen ion activity a_{H^+} and hydrogen partial pressure p_{H_2} are equal to unity, they vanish from the above equations. The EMF of the cell in Scheme (10.16) will be denoted by E and, in view of Equation (10.15), is:

$$E = E^0 + \frac{RT}{nF} \ln \frac{a_O^{\nu_O}}{a_R^{\nu_R}} \tag{10.20}$$

The EMF of a cell like that in Equation (10.16) represents the *electrode potential* of the left-hand half-cell with respect to the standard hydrogen electrode. This implies that the electrode potential of the standard hydrogen electrode is taken by convention as being zero at any temperature and its value is the origin of the electrode potential scale. The constant term E^0 is the *standard electrode potential* of the considered half-cell; it depends on ΔG^0.

Equation (10.20) is the *Nernst equation* that relates the electrode potential of a particular redox couple to the activities of the reduced and oxidized forms. However, analytical chemistry operates with concentrations and not with thermodynamic activities. The Nernst equation in terms of concentrations is obtained by substituting each activity term by the γc product according to Equation (10.3). One obtains thus the following form:

$$E = E_f^0 + \frac{RT}{nF} \ln \frac{c_O^{\nu_O}}{c_R^{\nu_R}} \tag{10.21}$$

where the term $E_f^0 = (RT/nF)\ln(\gamma_O/\gamma_R)$ stands for the *formal potential*. In contrast to the standard potential, which is a thermodynamic constant, the formal potential depends on the activity coefficients and hence, on the solution properties.

The electrode potential of a selected half-cell can be measured by assembling it with a reference half-cell as shown in Scheme (10.16). However, preparation and manipulation of the standard hydrogen electrode is rather cumbersome and, as a rule, electrode potentials are measured against a *secondary reference electrode*. A common reference electrode of this type is the silver/silver chloride half-cell consisting of a silver wire immersed in an AgCl-saturated KCl solution. The half-cell reaction is in this case:

$$AgCl(s) + e^- \rightleftarrows Ag(s) + Cl^- \tag{10.22}$$

For this half-cell, the Nernst equation gives:

$$E_{Ag/AgCl} = E_{Ag/AgCl}^0 + \frac{RT}{F} \ln a_{Cl^-} \tag{10.23}$$

where a_{Cl^-} is the activity of the chloride ion and $E_{Ag/AgCl}^0$ is a constant. The activities of solid compounds do not appear in this equation because the activity of a pure solid is 1. If the chloride ion activity is kept constant, the electrode potential of the silver/silver chloride potential remains constant and serves as the origin of this electrode potential scale. As demonstrated by reaction (10.22), the metal Ag functions as a source of electrons, while Ag^+ in AgCl acts as a sink for electrons. Therefore, the Ag/AgCl system behaves as an electron buffer and any trend to potential shift is counteracted by a shift in the equilibrium (10.22).

A practical form of this reference electrode is shown in Figure 10.3A. It consists of a glass body containing a KCl solution and a silver wire coated with solid AgCl. At the lower end, it is closed by a porous plug that prevents the inner solution from leaking into the test solution but allows for ion transfer in order to set up electrical contact with the solution. According to Equation (10.23), the electrode potential depends on the KCl concentration and this concentration should be reported when mentioning the reference electrode (e.g., Ag|AgCl (KCl 3.5 M)).

Similarly functions the *calomel electrode* (Hg|Hg$_2$Cl$_2$ (KCl)). A common version of this electrode is the saturated calomel electrode (SCE) in which the internal solution is a saturated KCl solution.

Chloride ions can diffuse from the reference electrode into the test solution. If chloride ions interfere, the Ag/AgCl electrode can be coupled to the solution via an electrolyte bridge filled with an inert electrolyte solution (e.g., KNO$_3$). This combination is termed a *double-junction reference electrode* because the contact with the test solution is achieved via two liquid–liquid junctions. Alternatively, one can make use of the saturated mercury(I)/mercury sulfate electrode (Hg|Hg$_2$SO$_4$|K$_2$SO$_4$(saturated)). For the purpose of microanalysis or *in vivo* applications, miniaturized reference electrodes are needed. Such devices can be obtained by microfabrication technology [16].

It is often necessary to convert electrode potentials measured with respect to a secondary reference to corresponding values on the standard hydrogen electrode scale. Conversion can be effected according to the diagram in Figure 10.3B. As an aid to clarity, this figure shows an analogy with the rules for altitude determination with respect to sea level, which is conventionally set as the zero level. More details on reference electrodes can be found in ref. [17]

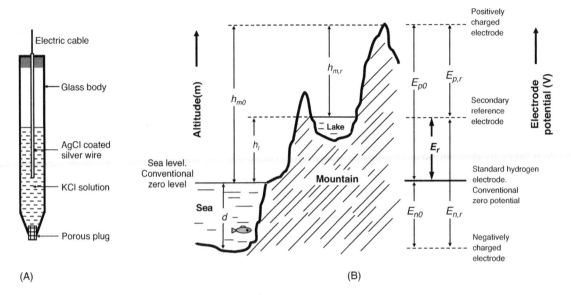

(A)   (B)

Figure 10.3   (A) A secondary reference electrode: the Ag/AgCl reference electrode; (B) An analogy between altitude measurement and the measurement of the electrode potential. The right-hand diagram shows the rules for electrode potential conversion from the secondary reference electrode scale to the standard hydrogen electrode scale.

By convention, the galvanic cell should be represented with the reference electrode at the left and the indicator electrode at the right as in Scheme (10.16), and the EMF is calculated as:

$$\text{EMF} = E_{\text{right}} - E_{\text{left}} \qquad (10.24)$$

where E_{right} and E_{left} are the pertinent half-cell potentials.

Questions and Exercises (Section 10.2)

1. Prepare a summary outlining the concepts of electric potential, chemical potential and electrochemical potential. What is the common feature of these concepts? What kind of interaction is involved in each case? Give some examples of systems in which each such concept is of relevance.

2. What is the meaning of thermodynamic activity and what solution properties determine the activity of an ion in solution?

3. What is the thermodynamic meaning of EMF?

4. What is a reference electrode and what is its role in a galvanic cell? Review the main types of reference electrode and formulate the corresponding half-cell for each of them. What parameter determines the electrode potential in each case?

5. What is the electrode potential at 25 °C of the half-cell below, with respect to the SHE and the Ag|AgCl (saturated),KCl(3.5 M)?

$$\text{H}^+(10^{-2}\,\text{M}), \text{V}^{3+}(10^{-1}\,\text{M}), \text{VO}^{2+}(10^{-3}\,\text{M})|\text{Pt}$$
$$E^0_{\text{VO}^{2+}/\text{V}^{3+}} = +0.337\,\text{V}; E_{\text{Ag/AgCl}} = 0.205\,\text{V}$$

Hint: the half-cell reaction is $\text{VO}^{2+} + 2\text{H}^+ + \text{e}^- \rightleftarrows \text{V}^{3+} + \text{H}_2\text{O}$.

Answer: E vs. SHE $= +0.692$ V.

10.3 Ion Distribution at the Interface of Two Electrolyte Solutions

10.3.1 Charge Distribution at the Junction of Two Electrolyte Solutions. The Diffusion Potential

Whenever different electrolyte solutions are put in contact, a nonuniform ion distribution develops at their interface. Two limiting situations can be envisaged: (a) a permeable interface is formed, which allows for unrestricted transfer

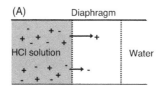
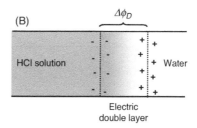

Figure 10.4 Development of the liquid junction potential at the junction of an HCl solution with water. (A) Initial state (no charge separation). Arrow lengths are proportional to ion mobility. (B) Steady-state conditions: a front of fast cations (H^+) precedes the front of slower anions (Cl^-) giving rise to a double electric layer. Ions in the intermediate region are not shown.

of any ion, and (b) a semipermeable interface is formed, which restricts the transfer of certain types of ion. The first case is addressed in this section, while the second case is the subject of the next section.

The case of a permeable interface is illustrated by the contact of an HCl solution with pure water (Figure 10.4A). Practically, such a system can be obtained by interposing a porous diaphragm between the two solutions. Due to the difference in concentrations, both H^+ and Cl^- do diffuse from the left to the right. However, the H^+ ion is much faster than Cl^-. As a result, a region rich in H^+ develops in front (Figure 10.4B), whereas a negatively charged region lags behind. Thus, a double electric layer develops and gives rise to an electric potential difference, $\Delta\phi_D$. Under the effect of $\Delta\phi_D$, the movement of H^+ is slowed down while the movement of Cl^- is accelerated. When the steady-state is attained, the electrical and diffusional driving forces balance each other and ion distribution across the interface remains invariable. $\Delta\phi_D$ is termed *the diffusion potential* or *the liquid junction potential*.

In order to calculate the junction potential, we have to consider the mass transport of ions across the interface. As already mentioned, ion transport occurs under the influence of both the activity gradient across the interface (*diffusion*) and the electric potential gradient (*migration*). The characteristic migration velocity of an ion is termed *ion mobility* and is defined as the increase in the ion velocity when crossing the distance between two points at 1 m distance from each other and subject to a potential difference of 1 V. The mobility unit is therefore the quotient of the velocity unit ($m\,s^{-1}$) and the electric field gradient unit ($V\,m^{-1}$) that is, $m^2\,s^{-1}\,V^{-1}$. The diffusion coefficient (D) of an ion is related to the ion mobility (u) as $D = RTu$, where R is the ideal gas constant and T is the absolute temperature. Selected ion mobility values are given in Table 10.1.

The general case is represented by the junction of two electrolyte solutions, each of them containing various ions. Under the simplifying assumption of a constant activity gradient across the double-layer zone, the diffusion potential is given by the *Henderson formula* [4]:

$$\Delta\phi_D = \phi_{\text{right}} - \phi_{\text{left}} = -\frac{RT}{F}\frac{\sum z_i u_i \Delta a_i}{\sum z_i^2 u_i \Delta a_i}\ln\frac{\sum z_i^2 u_i a_{i,\text{right}}}{\sum z_i^2 u_i a_{i,\text{left}}} \tag{10.25}$$

In this equation, z_i and u_i are the ion charge and mobility, respectively, $a_{i,\text{left}}$ and $a_{i,\text{right}}$ are ion activities at the left hand side and the right hand side of the double layer, respectively $\Delta a_i = a_{i,\text{right}} - a_{i,\text{left}}$. $a_{i,\text{right}}$ and $a_{i,\text{left}}$ could be different from the bulk solution activities. The summation is done over all the ions in the system.

Diffusion potential values predicted by the Henderson formula (Table 10.2) are in fair accord with the experimental ones. Data in this table can be interpreted by taking into account the ion mobility in Table 10.1. Thus, as K^+ and Cl^- have quite similar mobilities, the term in front of logarithm becomes very small. Therefore, a KCl solution on one side leads to a low junction potential and this potential decreases as the KCl concentration increases. That is why concentrated KCl solutions are customarily used to fill reference electrodes in order to secure as low a junction potential as possible. The large difference in mobility between H^+ and Cl^- produces a large junction potential when an HCl solution is put in contact with a diluted KCl solution, but this potential becomes much lower when the KCl solution is 3.5 M, that is, near saturation.

A junction potential develops always at the contact of the reference electrode with the test solution and adds to the EMF as an unknown and, possibly, randomly variable term. Therefore, although the liquid junction potential is, as a rule, very small, its effect cannot be neglected. Thus, for a potentiometric sensor with a sensitivity of 59 mV, an

Table 10.1 Ion mobility in water at 25 °C in $m^2\,s^{-1}\,V^{-1}$

Cation	Mobility	Anion	Mobility
H^+	36.25×10^{-8}	OH^-	20.5×10^{-8}
K^+	7.619×10^{-8}	$(1/2)SO_4^{2-}$	8.27×10^{-8}
NH_4^+	7.61×10^{-8}	Cl^-	7.912×10^{-8}
$(1/2)Ca^{2+}$	6.166×10^{-8}	NO_3^-	7.404×10^{-8}
Na^+	5.193×10^{-8}	ClO_4^-	7.05×10^{-8}
Li^+	4.010×10^{-8}	CH_3COO^-	4.24×10^{-8}

From Bard, A. J. and Faulkner, L. R. (2001) *Electrochemical Methods: Fundamentals and Applications*. Wiley, New York, p. 68.

Table 10.2   Liquid junction potentials at 25 °C estimated by Henderson equation

Junction	Potential/mV
0.1 M KCl ‖ 0.1 M NaCl	4.4
3.5 M KCl ‖ 0.1 M NaCl	−0.2
3.5 M KCl ‖ 1.0 M NaCl	1.9
0.1 M KCl ‖ 0.1 M HCl	−26.8
3.5 M KCl ‖ 0.1 M HCl	−4.2

From ref. [1], p 72.

error of 3 mV in the cell voltage brings about an error of 0.05 pX units, that is, a 12% concentration error. The effect of the junction potential can be minimized by using a double-junction reference electrode with the electrolyte bridge filled with a solution whose ions have similar mobilities. As the two junction potentials have opposite signs, the overall diffusion potential will be very small.

10.3.2 Ion Distribution at an Aqueous/Semipermeable Membrane Interface

The functioning of certain types of ion sensors is based on semipermeable membranes that restrict the mobility of some ions, as is the case, for example, with solid ion exchangers. A solid ion exchanger is a microporous material containing fixed ionic sites that prevent permeation by ions of a similar charge. Similar behavior is shown by liquid ion-exchanger membranes that include lipophilic charged molecules. The lipophilic ions cannot move into an aqueous solution and, at the same time, prevent the permeation of solution ions of a similar charge.

Figure 10.5 displays the situation at the contact of the solution 1 containing the MX salt with a cation-selective membrane. The membrane contains anionic sites R^- that cannot move into the aqueous solution. The anion X^- is confined to solution 1 due to electrostatic repulsion by the negative sites in the membrane. Conversely, the cations M^+ are free to diffuse between the two phases. If the cation activity in solution 1 (m_{aq}) is higher than that in the membrane (m_m), the cation diffusion to the right gives rise to a charge imbalance that counters cation diffusion (Figure 10.5). When the diffusional and electric driving forces balance each other, the potential difference between the two phases ($\Delta\phi_B$), called the *phase-boundary potential difference*, assumes a constant value. This quantity will be further denoted by the short-hand term boundary potential [1].

In order to derive an equation for the boundary potential, the partition of the cation between the two phases should be considered. At equilibrium, this process is characterized by the *partition coefficient* K_p defined as follows:

$$K_p = \frac{m_m}{m_{aq}} \tag{10.26}$$

where m_{aq} and m_m are the activities of the *free* metal ions in the solution and in the membrane, respectively. Free ions are present if there are no species able to bind the ion in the solution or in the membrane phase. As the partition coefficient is an equilibrium constant, it is correlated with the standard free energy change of transfer ΔG_p^0, and with the chemical potentials of the ion in each phase as follows:

$$\ln K_p = -\frac{\Delta G_p^0}{RT} = \frac{1}{RT}\left(\mu_{M^+,aq}^0 - \mu_{M^+,m}^0\right) \tag{10.27}$$

Ion transfer from phase 1 to phase 2 is favored by large ΔG_p^0 values, which implies a high partition coefficient.

The boundary potential equation can be derived from the equilibrium condition for the ion transfer between the two phases, which implies equality of electrochemical potentials (Equation 10.11):

$$\mu_{M^+,aq}^0 + RT\ln m_{aq} + zF\phi_{aq} = \mu_{M^+,m}^0 + RT\ln m_m + zF\phi_m \tag{10.28}$$

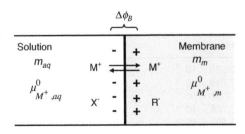

Figure 10.5   Ion distribution at a semipermeable phase boundary and development of the boundary potential. The left-hand phase is an aqueous solution, the right-hand one is a semipermeable phase.

Taking into account Equation (10.27), one obtains:

$$\Delta\phi_B = \phi_m - \phi_{aq} = \Delta\phi_0 + \frac{RT}{zF} \ln\left(\frac{m_{aq}}{m_m}\right) \tag{10.29}$$

where $\Delta\phi_0 = (RT/zF)\ln K_p$ represents the standard boundary potential. Equation (10.29) is formally analogous to the Nernst equation for the electrode potential.

If the metal ion can bind to a ligand to form a complex either in the solution or in the membrane phase, the free ion activity can be obtained from the stability constant of the complex.

Questions and Exercises (Section 10.3)

1 How does the diffusion potential develop at the contact of two solutions with different ionic compositions? Draw a sketch illustrating the formation of the diffusion potential at the contact of HCl and NaCl solutions of equal concentration.

2 Using the Henderson equation, estimate the diffusion potential for the following couples of solutions: (a) HCl (0.1M)||KCl (0.1M); (b) HCl (0.1M)||KCl (0.05M). *Hint*: assume that activity is equal to concentration.

   Answer: (a) 26.8; (b) 53 mV.

3 Point out the difference between the Nernst equation for the electrode potential and the equation for the boundary potential from the standpoint of the underlying physicochemical processes. What is the common physicochemical feature?

4 Derive the expression for the boundary potential at the interface of an aqueous solution with an anion-selective membrane assuming that the solution anion is singly charged.

   Answer: $\Delta\phi_B = \Delta\phi_0 - (RT/F)\ln(a_{aq}/a_m)$; a is the anion activity.

10.4 Potentiometric Ion Sensors – General

10.4.1 Sensor Configuration and the Response Function

The Nernst equation indicates that the electrode potential can be used to determine the concentration of redox species in solution. However, this approach is not selective at all, as any redox couple in solution will affect the electrode potential. A true potentiometric ion sensor is obtained by inserting an ion-selective membrane between two solutions containing the analyte ion as shown in Figure 10.6 for the particular case of a fluoride ion sensor. The galvanic cell is completed with a second reference electrode.

The membrane should interact in a specific way with the analyte ion. Thus, a lanthanum fluoride membrane presents selectivity to the fluoride ion. This is due to the fact that, in contact with water, the salt at the surface undergoes partial dissociation accompanied by the transfer of fluoride ions to the solution:

$$LaF_3(s) \rightleftarrows LaF_2^+(s) + F^-(aq) \tag{10.30}$$

Thus, a certain number of fluoride anions leave the surface and instead remain positive vacancies. If the fluoride ion is present in the solution, the equilibrium (10.30) will be shifted to the left, thus decreasing the density of positive charges at the membrane surface. Therefore, the positive charge density at the membrane surface depends on the analyte concentration in the solution. The same situation arises at the opposite side, where the membrane is in contact with the internal solution. However, the fluoride ion concentration in this solution is kept constant and the charge density at the interface remains constant as well.

It follows that the difference in electric charge density between the two sides of the membrane is dependent on the analyte concentration. Measuring charge densities on such a device is, however, impossible. Instead, one can resort to potential difference measurements. To this end, a reference electrode is immersed in each solution to act as probes that perform transition from ionic conductivity in solution and membrane, to electronic conductivity. A voltmeter connected to the two reference electrodes allows measuring the cell voltage that includes the potential difference across the membrane, termed the *membrane potential* ($\Delta\phi_m$).

For practical applications, the internal system and the membrane are assembled to form the actual sensor, as shown in Figure 10.7. In order to perform ion determinations, the sensor is dipped in the test solution along with the external reference electrode.

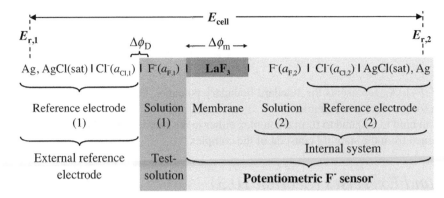

Figure 10.6 The structure of a galvanic cell for the determination of fluoride ion by means of a fluoride-selective membrane.

The response signal is the cell voltage that can be formulated using the symbols in Figure 10.6:

$$E_{cell} = E_{r,2} + \Delta\phi_m - E_{r,1} + \Delta\phi_D \tag{10.31}$$

The term $\Delta\phi_D$ is the liquid junction potential, which develops at the limit where the external reference electrode meets the test solution.

The key term in Equation (10.31) is the membrane potential that depends on the analyte activities as:

$$\Delta\phi_m = \frac{RT}{zF} \ln \frac{a_{X,1}}{a_{X,2}} \tag{10.32}$$

where $a_{X,1}$ and $a_{X,2}$ are analyte activities in the test solution and the internal solution, respectively; the charge term is positive for cations and negative for anions. This equation holds only under the assumption that no other ion, except the analyte, affects the membrane potential. As the analyte activity in the internal system is constant, Equation (10.31) can be formulated as:

$$E_{cell} = E_0 + \frac{RT}{zF} \ln a_{X,1} \tag{10.33}$$

The constant term E_0 in this equation includes the two reference electrode potentials as well as the junction potential that is assumed to be constant. Formally, E_0 represents the cell voltage at unit analyte activity; it should be not confused with the standard electrode potential.

In practical applications, the decimal logarithm is used instead of the natural one and the response function assumes the following form:

$$E_{cell} = E_0 + \frac{2.303\,RT}{zF} \log a_{X,1} \tag{10.34}$$

By analogy with the pH symbol, the pX symbol, defined in Equation (10.35), is often used instead of the log term (X stands for the symbol of the chemical element).

$$pX = -\log a_X \tag{10.35}$$

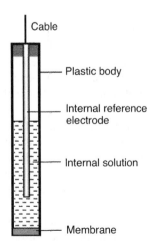

Figure 10.7 Typical configuration of a potentiometric ion sensor.

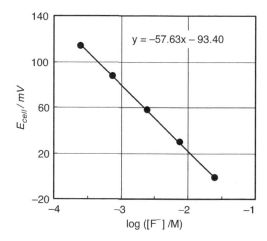

Figure 10.8 The response of a fluoride potentiometric sensor to the fluoride concentration at constant ionic strength. Temperature: 20 °C.

It follows therefore that the response of a potentiometric ion sensor is a linear function of the logarithm of the analyte activity with a theoretical sensitivity of $2.303\,RT/zF$, that is, $0.0591/z$ V per log unit at 25 °C. This ideal response is termed the *Nernstian response*. In practice, the sensitivity can be slightly smaller than the theoretical value. The E_{cell} response increases with the activity in the case of cations, but decreases in the case of anions. Note also that the sensitivity for divalent ions is only half of that for singly charged ions.

When the cell is coupled to the voltmeter, a current in the femtoampere range flows through the system. This tiny current does not disturb the equilibria. However, it is essential that the membrane elicits some minimal electrical conductivity. To this end, the LaF_3 membrane is doped with minute amounts of EuF_2. The conductivity issue should be kept in mind when designing any type of ion-sensing membrane.

A common problem arises from the fact that potentiometric sensors respond to ion activity and not to concentration (c), which is the required parameter in analytical applications. However, the activity is proportional to the concentration (Equation (10.25), and, further, the proportionality coefficient γ depends on the solution ionic strength (Equation (10.5)). Therefore, the activity term in Equation (10.34) can be substituted by concentration providing that the sample and the standard solutions have similar ionic strengths. The ionic strength can be adjusted by adding to each solution an excess of background electrolyte that does not interact with the membrane or the analyte (e.g., KNO_3). Thus, the contribution of sample ions to the total ionic strength becomes negligible and small variations in the analyte concentration have no effect on the activity coefficient. Under these conditions, E_{cell} is a linear function of log c. As an example, Figure 10.8 shows the calibration graph in terms of log c obtained with a series of fluoride solutions at constant ionic strength. The linear dependence of the signal on the logarithm of the analyte concentration is clear. The sensor sensitivity is given by the slope of the line; in this particular case, it is -57.6 V, that is, slightly lower than the theoretical value (-58.1 mV at 20 °C). This kind of less-than Nernstian response occurs often in practice.

10.4.2 Selectivity of Potentiometric Ion Sensors

The selectivity of the ion-recognition process can be achieved by complementarity of the analyte and receptor site as far as size, charge and chemical interactions are concerned. However, perfect selectivity cannot be achieved and ions related to the analyte can, in principle, also interact with the receptor. Thus, in the case of the fluoride sensor, interference is produced by hydroxyl ions whose size is quite similar to that of the fluoride ion. Fortunately, in this case, the concentration of the hydroxide ion can be made very low by adjusting the pH to around 5.

If interference by a secondary ion is possible, it can be accounted for by including an additional term in Equation (10.34) to obtain what is known as the *Nikolskii–Eisenman equation*:

$$E_{cell} = E_0 + \frac{2.303RT}{zF} \log\left(c_A + k_{A,B}^{pot} c_B\right) \tag{10.36}$$

where $k_{A,B}^{pot}$ is the *potentiometric selectivity coefficient* and c_A, c_B represent the concentrations of the analyte and the interfering ion, respectively, in the test solution. Clearly, good selectivity is obtained when the selectivity coefficient is sufficiently low in order to render the $k_{A,B}^{pot} c_B$ term negligible with respect to c_A. Simultaneous interference by a series of ions bearing the same charge as the analyte is accounted for by the generalized form of the Nikolskii–Eisenman equation:

$$E_{cell} = E_0 + \frac{2.303\,RT}{zF} \log\left(c_A + \sum_B k_{A,B}^{pot} c_B\right) \tag{10.37}$$

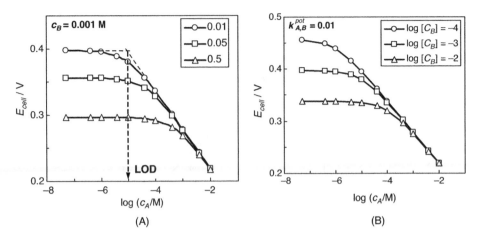

Figure 10.9 The effect of selectivity on an anion (A⁻) sensor response. (A) Effect of the selectivity coefficient (inset values) at a constant concentration of the interferent B⁻ (0.001 M). LOD is the lower limit of detection. (B) Effect of the interferent concentration (inset) at a constant selectivity coefficient of 0.01.

In order to account for the interference of an ion with a different charge, the following equation has been proposed:

$$E_{cell} = E_0 + \frac{2.303\,RT}{z_A F}\,\log\left(c_A + k_{A,B}^{pot}\,c_B^{z_A/z_B}\right) \tag{10.38}$$

where z_A and z_B represent the charges of the analyte and the interfering ion, respectively. However, due to the different units of the selectivity coefficients in Equations (10.37) and (10.38), this approach does not allow comparison of the effect of two interfering ions with different charges.

Equation (10.36) is based on the following assumptions: (a) both the analyte and the interfering ions have the same charge, and (b) the sensor response to each of the above ions is a Nernstian one. Each of the above assumptions can be inaccurate in many particular cases, which led to more careful approaches being made to the selectivity modeling problem [18,19].

There are several methods for the experimental determination of the selectivity coefficient based on Equation (10.36) [20,21]. Thus, in the *separate solution method*, the response to each ion is recorded in the absence of the second ion and Equation (10.36) is used to calculate the selectivity coefficient by means if ion activities producing a similar response to each ion. Other methods belong to the class of *mixed-solution methods* that rely on measurements in solutions containing both ions of interest. Thus, the *fixed interference method* is based on the determination of the response to increasing analyte concentration in the presence of a fixed interfering ion concentration, as shown in Figure 10.9A. According to Equation (10.36), the extended linear parts of the response line cross each other at a point where $c_A = k_{A,B}^{pot} c_B$. Hence, the ratio c_A/c_B at the intersection point is equal to the selectivity coefficient. The *fixed primary ion method* is based on the determination of the potential change in response to the addition of increasing interferent concentrations in a solution containing a fixed analyte concentration. The selectivity coefficient is calculated as in the previous method. The three methods above assume that the slope of the response line is the same for both the primary and secondary ion.

A method independent of Equation (10.36) is the *matched potential method* that overcomes the difficulty of obtaining accurate results when ions of unequal charge are involved [22]. The selectivity coefficient is defined as the $\Delta c_A/\Delta c_B$ ratio where Δc_A and Δc_B are concentration changes producing the same EMF change in a reference solution.

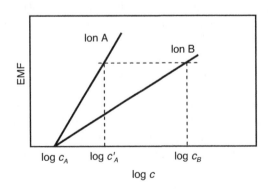

Figure 10.10 Determination of the selectivity coefficient by the matching potential method.

Accordingly, the sensor response to ions A and B is recorded independently in the reference solution (Figure 10.10). In the case of ion B, the solution should contain an initial background concentration of ion A. The ion concentrations yielding a similar response (c'_A and c_B) are recorded and the selectivity coefficient is determined using the following equation (See Figure 10.10 for symbol meanings):

$$k^{pot}_{A,B} = \frac{c'_A - c_A}{c_B} \tag{10.39}$$

The selectivity coefficient obtained by this method should agree with that obtained by methods based on Equation (10.36) if the ions have similar charge and the response to each of them is Nernstian. Otherwise, there could be large differences between the selectivity coefficients obtained using different methods.

Comprehensive compilations of potentiometric selectivity coefficients are available in refs. [21,23,24], whereas criteria and protocols for preparation and evaluation of potentiometric ion sensors can be found in ref. [25].

10.4.3 The Response Range of Potentiometric Ion Sensors

The response range of an ion sensor is the concentration interval within which the sensor response to the analyte activity is in accord with Equation (10.34). As demonstrated below, the limits of the response range are determined by the interaction of other sample components that may interact with the analyte in the solution or with receptor sites in the membrane.

The effect of the selectivity coefficient on the sensor response at a constant interferent concentration is represented in Figure 10.9A. The response is linear only as long as the analyte concentration is sufficiently high and the $k^{pot}_{A,B}$ term is negligible. At lower analyte concentrations, each ion contributes to some extent and the response function assumes a curved shape. Further, if $c_A \ll k^{pot}_{A,B}$, the contribution of the analyte to the total response becomes negligible and the response, which is now determined only by the interferent, assumes a constant value. The intersection of the two linear branches of the response curve defines the *lower limit of detection* of the ion A^- sensor. Curves in Figure 10.9A demonstrate that a low selectivity coefficient results in a lower limit of detection and a broader linear response range. Hence, a potentiometric ion sensor can be used even in the presence of an interferent, but both the limit of detection and the working range depend on interferent concentration. This conclusion is illustrated by the graph in Figure 10.9B which represents the response curve of a sensor with a selectivity coefficient of 0.01 in the presence of an interferent at various concentrations. This graph demonstrates clearly that both the limit of detection and the extent of the working rage depend on the concentration of the interferent.

It is clear from the above discussion that the lower detection limit is determined by ions bearing the same charge as the analyte itself. In turn, the *upper limit* of the response range is a consequence of the fact that ions of opposite charge penetrate into the membrane and produce a marked bending of the response line.

10.4.4 Interferences by Chemical Reactions Occurring in the Sample

As shown in the previous section, interference in the ion-recognition process can be avoided by designing ion-responsive membranes with very low potentiometric selectivity coefficients. Another source of interferences is connected to the chemistry of the analyte ion in the sample. This aspect can be illustrated for the particular case of the fluoride ion. A common feature of all potentiometric ion sensors is that they respond only to the analyte ion in the *free ion form*. Therefore, if the analyte can react with other sample components, it would be converted into a nondetectable form. Thus, in the case of the fluoride sensor, the fluoride ion is converted to HF and HF_2^- in acidic solutions. Also, metal ions in the sample (such as Fe^{3+} or Al^{3+}) form complexes with the fluoride ion. Both protonation and complex formation lead to the masking of the analyte. That is why the pH should be adjusted to about 5 by a buffer system in order to prevent the interference of both H^+ and OH^- ions. On the other hand, a strong complexing agent should be present in order to mask the metal ions. As the response depends on the ionic strength, an inert electrolyte at constant concentration should also be present in each sample and standard solution. A solution containing all these components is termed a total ionic strength adjustment buffer (TISAB). Thus, the data in Figure 10.8 have been obtained in the presence of a TISAB system that consists of a pH 5 acetate buffer, sodium chloride, and EDTA as a metal ion masking reagent.

So, in general, two kinds of interference can be expected when using a potentiometric ion sensor. These are (a) interaction of another ion with the receptor sites, and (b) analyte reactions in the solution leading to ion blocking. The first kind of interference is mostly decided by the characteristics of the receptor itself, that is, its selectivity. The second type of interference depends on the sample composition and can be dealt with by means of suitable chemical reactions.

10.4.5 The Response Time of Potentiometric Ion Sensors

The response time depends on the dynamics of all processes involved in the analyte–membrane interaction. Generally, it is defined as the time lapse between the moment of a sudden change in analyte concentration and the instant at which the response change (ΔE_{cell}) reaches its steady-state value within 1 mV or has reached 90% of the final value.

In the case of liquid membranes, the slowest step in the overall process is often the rate of analyte diffusion to the membrane surface and that is why is it is often recommended to provide gentle stirring during the measurement. With solid membranes, redistribution of ions within the membrane in response to the change in the membrane potential often determines the response time. The rate of ion redistribution depends on the membrane thickness (which determines the strength of the electric field) and the membrane conductivity.

10.4.6 Outlook

The key problem in developing a potentiometric ion sensor is the choice of preparation of the sensitive membrane that interacts in a characteristic way with the analyte in order to develop the membrane potential. The first potentiometric ion sensors (such as the pH electrode) were based on glass membranes that form a thin surface layer in contact with aqueous solutions that functions as a cation exchanger. Further, solid crystalline membranes formed of sparingly soluble salts (such as AgCl or metal sulfides) have been developed. Determination of ions that do not form sparingly soluble salts can be performed by means of ion exchangers dissolved in a water-immiscible solvent. The most advanced ion sensors are based on neutral receptors incorporated in liquid membranes. Each of the above alternatives will be reviewed in the next sections.

Questions and Exercises (Section 10.4)

1 Draw schematically the configuration of a galvanic cell that can be used for analytical determinations with potentiometric ion sensors. Comment on the role of each component and sketch a diagram showing the potential variation across the cell and the build up of the cell voltage.

2 What is the meaning of the Nernstian response concept in the case of potentiometric ion sensors?

3 What are the parameters that determine the theoretical sensitivity of a potentiometric ion sensor?

4 What precautions are necessary when performing analytical determinations with potentiometric ion sensors?

5 Calculate the relative error in the analyte concentration produced by a 3-mV change in the cell voltage. (This change could be due to a variation in the junction potential.) Assume that the sensitivity is the theoretical one and the ion charge is either 1 or 2.

   Hint: the relative error is defined as follows: error $= 100[(c_d - c_t)/c_t]\%$, where c_d is the concentration determined by neglecting the junction potential and c_t is the concentration determined by taking into account the junction potential.

   Answer: 12% for a monovalent ion.

6 For both the separate solution method and the mixed-solution method, draw schematically the response graph and derive equations for the calculation of the selectivity coefficient.

7 Use selected curves in Figure 10.9 to estimate the selectivity coefficient by the fixed interference method.

8 Demonstrate that in the case in which both the analyte and interfering ion have the same charge, the selectivity coefficients obtained by the matched potential method is similar to that obtained by a method based on the Nikolskii–Eisenman equation.

9 What determines the upper limit of detection of a potentiometric ion sensor?

10 Comment on the interplay between the selectivity of a sensor and its lower limit of detection.

11 What is the role of each component of the TISAB system used in the potentiometric determination of the fluoride ion?

10.5 Sparingly Soluble Solid Salts as Membrane Materials

10.5.1 Membrane Composition

A well-known crystalline membrane is the single-crystal lanthanum fluoride membrane discussed in Section 10.4.1. However, much more widely used are polycrystalline membranes composed of silver halides or sparingly soluble metal sulfides. Polycrystalline membranes can be obtained as pressed pellets or by incorporation of the salt in silicone rubber or in thermoplastic polymers. As already mentioned, the sensing mechanism of such membranes relies on the effect of the analyte on the dissociation equilibrium of the membrane salt. Therefore, such membranes would be expected to respond to both the composing cation and anion. However, due to the extremely low solubility of the membrane compound, only one of these ions can be present in solution at a significant concentration.

A simple example of this type is the silver chloride membrane that is responsive to both Ag^+ and Cl^-. However, it displays a series of drawbacks and, consequently, it is used as a $1:1$ molar ratio mixture with silver sulfide. Silver sulfide membranes are sensitive to both sulfide and silver ions. Mixtures of silver sulfide with a divalent ion sulfide respond to the divalent ion activity and form the basis for the development of potentiometric sensors for cations such as Cu^{2+}, Pb^{2+}, and Cd^{2+}. Determination of anions such as cyanide, iodide and thiocyanate are based on their reactions with the membrane cation to form sparingly soluble salts or stable metal complexes, as shown in next Section.

10.5.2 Response Function and Selectivity

In the following discussion it will be assumed that the membrane consists of a single salt, MX, which is in contact at both faces with solutions containing the ions M^+ and X^- (Figure 10.11). The behavior of such a membrane is determined by the dissolution of the solid compound MX into the aqueous phase:

$$MX(s) \rightleftarrows M^+(aq) + X^-(aq) \tag{10.40}$$

This equilibrium is characterized by the *solubility constant* K_s that depends on the ion activities in the saturated MX solution:

$$K_s = (mx)_{\text{saturation}} \tag{10.41}$$

where m and x represent the activity of the cation and anion, respectively. If either M^+ or X^- is present as an analyte, the activity of the second ion is related to that of the first by Equation (10.41). Figure 10.11 also shows the variation of the electric potential across the system. Note that the potential is constant across the membrane because the ions in the crystal lattice occupy fixed positions and no internal diffusion potential can develop.

The membrane potential difference $\Delta\phi_m$ can be derived from the equilibrium conditions for reaction (10.40), that, in the case of the first interface, is formulated as follows (the activity of the pure solid MX is taken as equal to one):

$$\mu_{M,1}^0 + RT \ln m_1 + F\phi_1 = \mu_{M,m}^0 + F\phi_m \tag{10.42}$$

$$\mu_{X,1}^0 + RT \ln x_1 - F\phi_1 = \mu_{X,m}^0 - F\phi_m \tag{10.43}$$

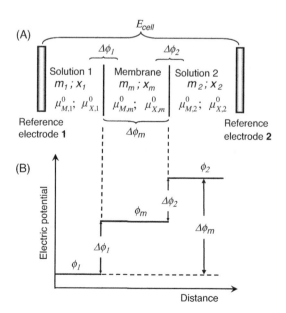

Figure 10.11 Variation of the electric potential across a membrane/solutions system in the case if a solid salt crystalline membrane. (A) Cell configuration; (B) potential variation across the system.

Equation (10.42) allows the potential difference at the first interface ($\Delta\phi_1$) to be calculated, as a function of the cation activity in solution 1:

$$\Delta\phi_1 = \phi_m - \phi_1 = \frac{1}{F}(\mu^0_{M,1} - \mu^0_{M,m} + RT \ln m_1) \tag{10.44}$$

Similarly, one obtains the following equation for the potential difference at the second interface:

$$\Delta\phi_2 = \phi_2 - \phi_m = -\frac{1}{F}(\mu^0_{M,2} - \mu^0_{M,m} + RT \ln m_2) \tag{10.45}$$

From the above equations, the *membrane potential* equation is obtained in the following form:

$$\Delta\phi_m = \Delta\phi_1 + \Delta\phi_2 = \frac{RT}{F} \ln \frac{m_1}{m_2} \tag{10.46}$$

Using the equilibrium conditions with respect to the anion (Equation (10.43)) and an analogous equation for the second interface, one obtains the membrane potential as a function of the anion activities (x_1; x_2):

$$\Delta\phi_m = -\frac{RT}{F} \ln \frac{x_1}{x_2} \tag{10.47}$$

It can be demonstrated in a similar way that a silver sulfide (Ag_2S) membrane responds to the sulfide ion (S^{2-}) with the prelogarithm factor of $-RT/2F$.

Ion activity in the solution 1 can be determined by keeping constant this ion activity in the reference solution 2. Under these conditions, Equation (10.46) gives:

$$\Delta\phi_m = \text{constant} + \frac{RT}{F} \ln m_1 \tag{10.48}$$

An analogous equation can be derived for the anion.

In order to assess the selectivity, it will be assumed that an X^- sensor is in contact with a solution containing both X^- and the an interfering anion, Y^-. Each anion is involved in a dissociation equilibrium and the activity of each anion is determined by the solubility constants of the pertinent salts:

$$\text{(a) } K_{s,MX} = m_1 x_1; \quad \text{(b) } K_{s,MY} = m_1 y_1 \tag{10.49}$$

x_1 and y_1 denotes activities of pertinent ions. From the above equations one obtains the activity of the second anion:

$$y_1 = \frac{K_{s,MY}}{K_{s,MX}} x_1 \tag{10.50}$$

As long as y_1 is lower than the value given by Equation (10.50), no solid MY forms in the solution and the membrane responds to the X^- activity only, according to Equation (10.47). Conversely, if y_1 exceeds the limit imposed by the Equation (10.21), solid MY does form in the solution and Equation (10.47) should be adapted by substituting the x_1 expression derived from Equation (10.50) into Equation (10.47). It results:

$$\Delta\phi_m = -\frac{RT}{F}\left(\ln \frac{K_{s,MX}}{K_{s,MY}} + \ln \frac{y_1}{x_2}\right) \tag{10.51}$$

Hence, if x_2 is kept constant, the above equation can be reduced to:

$$\Delta\phi_m = \text{constant} - \frac{RT}{F} \ln y_1 \tag{10.52}$$

The above considerations demonstrate that Y^- can interfere in the X^- determination when $K_{s,MY} \gg K_{s,MX}$. In the opposite case, the MX membrane turns into a Y^--sensitive membrane. For example, an AgCl membrane responds to the iodide concentration as the solubility of AgI ($\log K_{s,\ AgI} = -16.8$) is much less than that of AgCl ($\log K_{s,\ AgCl} = -9.75$). However, the limit of detection for the Y^- ion depends on the X^- concentration in the sample.

Quite a similar situation occurs if the membrane cation reacts with a solution component Z to form a soluble metal complex:

$$MX(s) + nZ(aq) \rightleftarrows [MZ_n](aq) + X^-(aq) \tag{10.53}$$

In such a situation, the sensor produces a mixed response to both X^- and Z species. However, if the X^- activity is negligible with respect to that of Z, the sensor responds in a Nernstian way to the Z activity. Thus, an AgCl membrane shows a Nernstian response to cyanide ion activity if this activity is much higher than that of the chloride ion.

A number of potentiometric ion sensors are based on mixed membranes made of a mixture of silver sulfide (M_2S) and a divalent cation sulfide (NS). This is because divalent cation sulfides are generally insulators and silver sulfide is needed in order to impart electric conductivity to the membrane. The dissolution of each sulfide is controlled by the pertinent solubility constant:

$$K_{M_2S} = m^2 s; \quad K_{NS} = ns \tag{10.54}$$

where s, m and n are the activities of S^{2-}, M^{2+} and N^{2+} ions, respectively, in the test solution. The membrane responds to the M^{2+} activity, but this, in turn, is determined by the activity of the N^{2+} ion, as is clear from Equations (10.54) which gives $m = \sqrt{K_{M_2S} n / K_{NS}}$. Upon inserting this expression into Equation (10.48) one obtains:

$$\Delta\phi_m = \text{constant} + \frac{RT}{2F} \ln \frac{K_{M_2S}}{K_{NS}} + \frac{RT}{2F} \ln n \tag{10.55}$$

The above principles have been used to develop potentiometric sensors for Cu^{2+}, Pb^{2+}, and Cd^{2+}.

Solid salt membrane sensors are robust devices characterized by a very long lifetime. The main drawbacks are the poor selectivity and a relatively high detection limit (in the 10^{-6} M region).

Questions and Exercises (Section 10.5)

1 What do solid salt membranes consist of and how can such membranes be prepared?
2 What is the mechanism of interaction with the solution ions in the case of solid salt membranes?
3 What physicochemical parameters determine the selectivity of a solid salt membrane?
4 Derive the response equation for an S^{2-} potentiometric sensor based on an Ag_2S membrane.
5 Review the possibilities of using solid salt membranes for the determination of ions that are not present in the membrane composition.
6 Assuming that the limit of detection for I^- and Br^- in determinations based on a AgCl membrane is given by y_1 in Equation (10.50), calculate the limit of detection for each of the above ions in a 0.001 M Cl^- solution. $\log K_{s, AgCl} = -9.75$; $\log K_{s, AgBr} = -12.28$; $\log K_{s, AgI} = -16.8$.

10.6 Glass Membrane Ion Sensors

10.6.1 Membrane Structure and Properties

In the case of membranes composed of sparingly soluble salts, both salt ions can undergo transfer between the solution and the membrane surface. A different kind of membrane is represented by solid ion exchangers in which one of the ions is covalently attached to a solid network. Typical materials of this kind are amorphous silicates (oxide glasses) such as the glass membrane forming the sensing element of the pH glass electrode. In contact with an aqueous solution, a layer of 50 to 100 nm thickness at the glass surface is impregnated with water and assumes a gel structure (Figure 10.12A). This layer displays cation-exchanger properties as it includes negative $Si-O^-$ groups attached to a random network which is typical of amorphous solids. Positive ions are free to cross the interface with a solution, but anion permeation is hindered by the fixed negative sites that exert electrostatic repulsion. The surface gel acts therefore as a *semipermeable membrane* and performs the actual sensing process in glass membranes.

Due to the cation-exchange process, the negative charge density within the sensing part can vary in response to the analyte ion activity in the test solution. The same happens at the opposite side of the glass membrane, which is in contact with the internal system of the sensor. The cell voltage of a glass electrode–reference electrode combination is a function of the hydrogen ion activity according to Equation (10.33).

The best-known application of the glass membrane is the potentiometric pH sensor commonly termed the pH glass electrode [26,27]. The membrane in this case is a sodium and calcium silicate glass. It displays an outstanding selectivity for the hydrogen ion as the single known interference arises from sodium ion at pH > 12. Nevertheless, the introduction of aluminum oxide into the glass imparts certain selectivity to alkali metal and ammonium ions and allows the development of sensors for these ions. However, such sensors should be operated at a high pH in order to

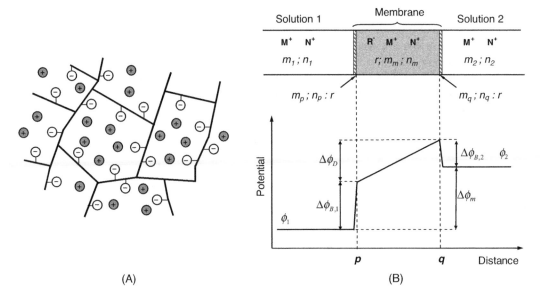

Figure 10.12 (A) Structure of a solid cation exchanger. The gel structure is impregnated with water. Thick lines represent the molecular network; empty circles are immobilized anions; full circles are mobile solvated cations in the embedded solution. (B) Configuration and potential distribution for a solid ion-exchanger membrane. R^- indicates immobilized anions within the membrane; M^+ and N^+ are mobile cations. Lower case symbols indicates the activities of ionic species indicated by pertinent upper case symbols.

avoid interference by the H^+ ion. Sufficient electrical conductivity within the compact part of the glass membrane is secured by mobile Na^+ ions.

10.6.2 Response Function and Selectivity

In order to derive the response function of a solid ion-exchanger membrane, it will be assumed that the whole membrane consists of a material like that in Figure 10.12A. In other words, it consists of a porous, solid ion-exchanger material containing fixed anionic sites R^- at the pore surface where a double electric layer develops by cation attraction. Complete exclusion of the anion X^- is achieved when the pore size is close to the double-layer thickness. This allows the repeling action of the R^- sites to be effective over the whole volume of the filling solution. If the pore size is slightly greater, moderate anion permeation occurs, whereas when the pore size is very large with respect to the double-layer thickness, the penetration of the anion is not hindered.

A solid ion-exchanger membrane can be approached by the three regions concept (also known as the Teorell, Meyer and Sievers model, more concisely as the TMS model). As the membrane is cation selective, a boundary potential ($\Delta\phi_{B,1}$ and $\Delta\phi_{B,2}$) develops at both faces of the membrane. At the same time, the difference in ion activity between the two sides of the membrane leads to the formation of an *internal diffusion potential* $\Delta\phi_D$ across the membrane (Section 10.3.1). Therefore, a sequence of three distinct regions form across the membrane, as shown in Figure 10.12B. According to this figure, the membrane potential can be obtained as follows:

$$\Delta\phi_m = \phi_2 - \phi_1 = \Delta\phi_{B,1} + \Delta\phi_D - \Delta\phi_{B,2} \tag{10.56}$$

If two cations M^+ and N^+ (e.g., H^+ and Na^+, respectively) are present in the solution, a competition between these cations will take place in the form of ion exchange between the solutions (s) and the surface layers of the membrane (p and q). At the left side of the membrane, the ion-exchange process is:

$$N^+(s) + M^+(p) \rightleftarrows N^+(p) + M^+(s) \tag{10.57}$$

An analogous equilibrium occurs at the right side of the membrane. The equilibrium constant of such a process is termed *the ion-exchange constant* K_{ex} and this can be formulated as follows using the activity symbols shown in Figure 10.12B

$$K_{ex} = \frac{m_1 n_p}{m_p n_1} = \frac{m_2 n_q}{m_q n_2} \tag{10.58}$$

In order to derive the internal diffusion potential be means of Equation (10.25) the activity differences (Δm and Δn) between p and q are needed. These differences can be found from the electrical neutrality conditions: ($m_p + n_p = r$ and $m_q + n_q = r$) that yield $\Delta m = \Delta n$. As the negative sites in the membrane are fixed, their mobility is zero ($u_R = 0$)

the diffusion potential results from the Henderson equation (10.25) as:

$$\Delta\phi_D = \phi_q - \phi_p = -\frac{RT}{F}\ln\frac{m_q + (u_N/u_M)n_q}{m_p + (u_N/u_M)n_p}\tag{10.59}$$

On the other hand, the boundary potential formulas are, according to Equation (10.29):

$$\Delta\phi_{B,1} = \Delta\phi_0 + \frac{RT}{F}\ln\frac{m_1}{m_p}; \quad \Delta\phi_{B,2} = \Delta\phi_0 + \frac{RT}{F}\ln\frac{m_2}{m_q}\tag{10.60}$$

The membrane potential can now be derived by substituting the above $\Delta\phi$ terms in Equation (10.56). One obtains thus:

$$\Delta\phi_m = \frac{RT}{F}\ln\left[\frac{m_1 m_q}{m_p m_2}\left(\frac{m_p + (u_N/u_M)n_p}{m_q + (u_N/u_M)n_q}\right)\right]\tag{10.61}$$

By taking into account the ion-exchange constant equation (10.58) and assuming that $n_2 = 0$ (hence $n_q = 0$), one obtains finally:

$$\Delta\phi_m = \frac{RT}{F}\ln\frac{m_1 + (u_N/u_M)K_{ex}n_1}{m_2}\tag{10.62}$$

This is the Nikolskii–Eisenman equation for a solid ion-exchanger membrane. Assuming that m_2 is a constant, Equation (10.62) can be converted into a form similar to Equation (10.38), with the potentiometric selectivity coefficient defined as:

$$k_{M,N}^{pot} = (u_N/u_M)K_{ex}\tag{10.63}$$

This equations accounts for the effect of both cations and demonstrates that the response to the M^+ activity is linear on a logarithm scale if $m_1 \gg k_{M,N}^{pot}n_1$. The ion-exchange constant, which is included in the $k_{M,N}^{pot}$, indicates in fact the difference between the ions M^+ and N^+ affinities for the membrane. The affinity depends on the ion size and the strength of its interaction with the negative sites within the membrane. Thus, the very small hydrogen ion interacts much more strongly than any other monopositive cation in which the charge is spread over a larger volume. The very high mobility of the H^+ ion ($u_M \gg u_N$) also imparts selectivity for this ion.

It was assumed in the approach above that the membrane behaves as a homogeneous ion exchanger. However, in a real glass membrane, ion-exchanger layers form only at the membrane surface, whereas the bulk of the membrane preserves a compact glass structure that hydrogen ions cannot permeate. Consequently, the application of the Henderson equation is a crude approximation. Nevertheless, Equation (10.62) provides a fair description of both sensor response and the effect of interferences by other ions. More advanced theoretical approaches to the glass membrane sensors can be found elsewhere [1,3].

10.6.3 Chalcogenide Glass Membranes

Chalcogenide glasses contain a large proportion of chalcogen elements, that is, sulfur, selenium and tellurium [28]. They are a kind of amorphous semiconductor with bandgap energies of 1–3 eV. There are marked differences between chalcogenide and oxide glasses due to the character of the metal–chalcogen bond. In oxide glasses, the oxygen–metal bond is purely ionic, whereas in chalcogenide glasses, the metal–chalcogen bond has a pronounced covalent character.

Chalcogenide glass materials have been used for preparing membranes that are selective for various cations and anions [29–31]. Chalcogenide glasses can be prepared from pure elements in evacuated quartz ampoules by heating at a defined temperature (up to 800–1200 K) for 6–20 h. The resulting amorphous material has to be quenched rapidly to room temperature. The solid material is cut into membranes 2–3 mm thick and 5–10 mm in diameter, which are then polished and sealed with epoxy resin into a plastic tube. A solid inner contact is obtained by metal deposition on the inner side of the membrane and attaching a metal wire by means of a conducting adhesive.

The membrane should include the analyte-element and, in addition, a group 13 (Ga), 14 (Ge) or 15 (As or Sb) element (Table 10.3).

The applications of chalcogenide glass sensors partially overlap those of solid salt membranes. However, better detection limit and selectivity have been reported for certain chalcogenide glass sensors. In addition, chalcogenide glasses allows the fabrication of solid membranes for ions such as Na^+, Fe^{3+} and Cr(VI), that cannot be obtained by the solid salt technology. Another advantage emerges from the easy fabrication of thin glass films by pulsed laser deposition [30].

Table 10.3 Chalcogenide glasses as recognition materials in potentiometric ion sensors [30]

Analyte ion	Glass composition	Sensitivity/mV	Detection limit/M
Na^+	$NaCl\text{-}Ga_2S_3\text{-}GeS_2$	50–55	10^{-5}
Cu^{2+}	$Cu\text{-}Ag\text{-}As\text{-}Se$	27–29	10^{-7}
Zn^{2+}	$ZnTe\text{-}ZnSe\text{-}GeSe_2$	22–41	10^{-6}
S^{2-}	$Ag_2S\text{-}As_2S_3$	45–50	10^{-7}
Br^-	$AgBr\text{-}Ag_2S\text{-}As_2S_3$	57–60	5×10^{-7}

Before being used, chalcogenide glass sensors should be conditioned in a solution containing the analyte ion. During the conditioning, a modified surface layer forms at the surface of the glass. This layer functions as a semipermeable membrane analogous to that in oxide glasses. The conditioning and response mechanisms can vary in accordance with the glass structure and the nature of the analyte ion [29]. Due to the semiconductor character of chalcogenide glasses, the development of the inner diffusion potential involves either electrons (in n-type glasses) or positive vacancies (holes) (in p-type glasses).

Questions and Exercises (Section 10.6)

1 What ions can be determined by means of glass membranes and what composition of membrane glass is suitable for such applications?
2 How is the ion-sensing layer at the surface of a glass membrane formed? What is the structure and the properties of the sensing part in glass membrane ion sensors?
3 What are the components of the membrane potential for a solid ion-exchanger membrane? What physicochemical parameters determine the magnitude of each component?
4 What physicochemical factors determine the selectivity of a solid ion-exchanger membrane?
5 What are the deviations of glass membrane behavior from the TMS model?
6 Comment on: (a) the characteristics of the chalcogenide glass membranes; (b) applications of such membranes; (c) advantages over other types of ion-selective membranes.

10.7 Ion Sensors Based on Molecular Receptors. General Aspects

Solid membranes used in potentiometric ion sensors are characterized by the presence of ion-binding sites that interact with ions in solutions. The selectivity of such membranes is mostly determined by their crystal structure and the possibility of adjusting the selectivity is quite limited. Moreover, determination of a series of important ions, such as Ca^{2+}, Mg^{2+} as well as organic ions cannot be approached by means of solid membranes. An alternative to solid membranes is based on molecular ion receptors that can interact selectively with the ion of interest. In order to obtain an ion-selective membrane, molecular receptors are embedded in a matrix that allows for ion transfer. Proper tailoring of the receptor structure, conformation and physical properties offers in principle an endless number of opportunities for obtaining receptors with good selectivity for particular ions.

Two different types of molecular ion receptors can be distinguished: *ionic receptors* (commonly known as liquid ion exchangers) and *neutral receptors* (also termed *ionophores*, that is, ion carriers). In the first case, the analyte–receptor interaction is based on electrostatic attraction, whereas in the case of neutral receptors a great variety of interactions come into play. Although certain natural ionophores are available, numerous synthetic neutral ion receptors have been developed in the framework of supramolecular chemistry [32].

In the initial stage of development, molecular receptors were used in the form of solutions in water-immiscible solvents. Such solutions can be supported in a porous plastic disc. Hence, the name *liquid membrane* sensor has been assigned to the resulting device. This approach is now obsolete as much better performances have been obtained by incorporating the receptors in a thin film of polymer material to form an ion-sensitive *polymeric membrane*. As a rule, a polymeric membrane behaves as a typical liquid membrane.

The outstanding importance of molecular receptors resides in their application in both potentiometric and optical ion sensors. The topic of molecular ion receptors has been amply reviewed (see, e.g., refs. [19,33,34]).

10.8 Liquid Ion Exchangers as Ion Receptors

10.8.1 Ion Recognition by Liquid Ion Exchangers

Customarily, charged ion receptors are known as ion exchangers because they allow ion exchange between two immiscible phases: an aqueous solution and a nonaqueous solvent in which the ion exchanger is dissolved. In a hydrophobic phase (e.g., a liquid membrane) the receptor is accompanied by a counterion that balance the receptor charge.

Several ionic receptors commonly used in potentiometric sensors are shown in Figure 10.13. The calcium receptor in A is a long-chain dialkyl phosphoric acid ester. Due to the difference in partition coefficients, a membrane based on this receptor discriminates between Ca^{2+} and Mg^{2+}. Another common ionic receptor for metal cations is dinonylnaphthylsulfonic acid whose anion is shown in Figure 10.13B.

Positively charge receptors are, of course, used in anion sensors. Thus, the nickel complex with *o*-phenanthroline in Figure 10.13C is suitable for use as a nitrate ion receptor. Fair selectivity arises mostly from the rather high lipophilicity of this ion but other anions in the sample can interfere and their removal is recommended prior to the analysis. Figure 10.13D shows a quaternary ammonium receptor for Cl^- and a phosphonium cation which is suitable as a receptor for either Cl^- or ReO_4^- ion. Quaternary ammonium ions are also suitable receptors for anionic surfactant [35]. A cationic receptor for the ClO_4^- is shown in Figure 10.13E. In this case, steric factors, in addition to the partition coefficient, contribute to the selectivity. In contrast to the previous receptors, in which the charge is localized at a single atom, the receptor in Figure 10.13F which is a guanidinium derivative, displays a positive charge distributed over three nitrogen atoms, and this compound has proved useful for sensing the oxygenated HSO_3^- anion.

Ionic receptors are in general inexpensive and ready available. However, the selectivity of ionic receptors is rather poor. In addition, certain cation receptors can be extracted into the aqueous solution in the protonated form. For this reason, low pH values should be avoided when using this kind of receptor.

10.8.2 Charged Receptor Membranes

One form of polymeric ion sensing membrane can be fabricated in a very simple way, by dissolving poly(vinyl chloride) PVC ($\approx$33%), a plasticizer ($\approx$66%) and the ion receptor ($\approx$1%) in a suitable solvent (e.g., tetrahydrofuran) and letting the solvent evaporate slowly from a thin solution layer on a glass plate [33,36]. The membrane is then cut and attached to a PVC tube holder. Other polymers, such as derivatized PVC, silicone rubber, polyurethanes, and polystyrene have also been tested as receptor matrices. However, in some applications, the use of a solvent causes problems. For this reason, photocurable, solvent-free polymers, such as methacrylates have been proposed as matrices.

The *plasticizer* (Figure 10.14) imparts plasticity to the membrane, but what is more important is that it functions as a solvent for the receptor and other membrane components. That is why the properties of the plasticizer determine to a large extent the performance of the ion sensor. Thus, in the case of an ionic receptor, which has no binding

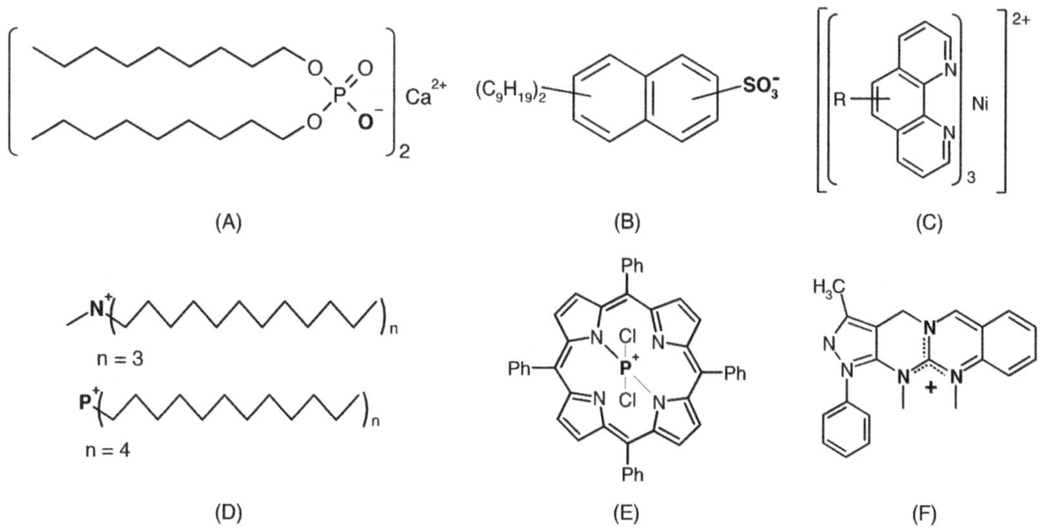

Figure 10.13 Liquid ion exchangers used as ion receptors in potentiometric ion sensors. (A) A dialkyl phosphoric acid ester as Ca^{2+} receptor; (B) the dinonylnaphthylsulfonic acid anion; a cation receptor; (C) a nickel complex used as nitrate ion receptor; (D) Quaternary ammonium and phosphonium ionic receptors; (E) a porphyrin derivative perchlorate receptor (Ph is a phenyl group); (F) a guanidinium derivative receptor for the HSO_3^- anion.

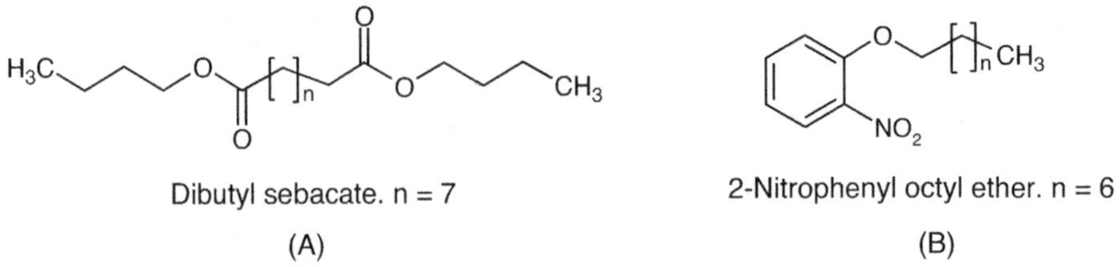

Dibutyl sebacate. n = 7

(A)

2-Nitrophenyl octyl ether. n = 6

(B)

Figure 10.14 Two common plasticizers used in ion-selective liquid membranes: A nonpolar (A) and a polar (B) plasticizer.

selectivity, the sensor selectivity arises from the partition of the analyte ion into the plasticizer solvent. A large number of plasticizers have been tested with a view to improving lipophilicity, solubility, and selectivity, and a strategy for the design of plasticizers has been advanced in ref. [37].

The useful lifetime of a polymeric ion-sensitive membrane is limited by the slow leakage of the membrane components into the aqueous solution.

10.8.3 Response Function and Selectivity

The functioning of a charged receptor-based membrane is based on the ion-exchange process involving an ion M^+ in solution (aq) and an ion N^+ in the membrane (m) which also contains the ionic receptor R^-:

$$M^+(aq) + \{N^+; R^-\}(m) \rightleftarrows \{M^+; R^-\}(m) + N^+(aq) \tag{10.64}$$
$$\underset{m_{aq}}{} \quad \underset{n_m}{} \quad \underset{m_m}{} \quad \underset{n_{aq}}{}$$

The symbols under the above equation indicate cation activities. This equilibrium is characterized by the ion-exchange constant:

$$K_{exch} = \frac{K_{p,M}}{K_{p,N}} = \frac{m_m n_{aq}}{m_{aq} n_m} \tag{10.65}$$

where $K_{p,M}$ and $K_{p,N}$ are the partition coefficients of M^+ and N^+, respectively (see Equation (10.26)). The extraction into the hydrophobic phase involves preliminary dehydration followed by solvation within the membrane. The hydration energy is much larger than the solvation energy, and, therefore, the partition coefficient depends mainly on the hydration energy of the ion. As a result, the partition coefficient increases with decreasing the degree of hydration. The degrees of hydration decrease with increasing ion size and decreasing ion charge. Thus, the partition coefficient for cation extraction into an anionic receptor-containing non-aqueous solvent increases in the following sequence (the Hofmeister series) [11]:

$$Mg^{2+} < Ca^{2+} < Li^+ < Na^+ < NH_4^+ < K^+ < Cs^+ < X^+ \tag{10.66}$$

where X^+ is a lipophilic non-polar cation. In the case of non-polar ions, the hydrophobic character, which is imparted by the organic substituents, leads to strong solvation within the membrane phase, which compensates for the hydration energy. For anion extraction the partition coefficient increases as follows:

$$HPO_4^{2-} < SO_4^{2-} < AcO^- \approx HCO_3^- < Cl^- < Br^- < NO_3^- < I^- < SCN^- < ClO_4^- < X^-, \tag{10.67}$$

where AcO^- is the acetate ion and X^- is a lipophilic organic anion.

In order to derive the response function of a liquid membrane sensor, consider a semipermeable liquid membrane (which contains the ionic receptor R^-) in contact with two solutions 1 and 2 containing the ions M^+ and N^+ with activities m_1 and n_1, respectively, in solution 1 and m_2 in solution 2 (Figure 10.15). At each side of the membrane, a boundary potential develops according to Equation (10.29), whereas the ion activity gradient across the membrane gives rise to a diffusion potential that can be expressed by the Henderson equation (10.25). Therefore, the potential difference between the two solutions is:

$$\Delta\phi_m = \phi_2 - \phi_1 = \Delta\phi_{B,1} + \Delta\phi_D - \Delta\phi_{B,2} \tag{10.68}$$

By making appropriate substitutions, the membrane potential equation is obtained in the following form which is valid when there is negligible ion association in the membrane [2]:

$$\Delta\phi_m = \frac{RT}{F} \ln\left(\frac{m_1 + (u_N/u_M)K_{exch}n_1}{m_2}\right) \tag{10.69}$$

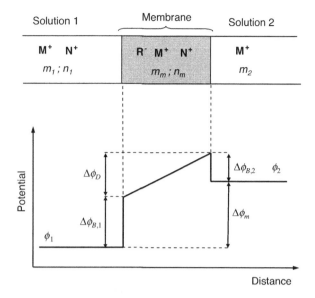

Figure 10.15   Variation of the electric potential across a liquid ion-exchanger membrane inserted between two solutions.

where u_M and u_N are the ion mobilities within the membrane. Hence, the potentiometric selectivity coefficient is:

$$k_{M,N}^{pot} = \frac{u_N}{u_M} K_{exch} \tag{10.70}$$

As a rule, there are no major differences between the mobilities of similar ions in the membrane phase and the potentiometric selectivity coefficient can be approximated by the ion-exchange constant, that is, the ratio of the partition coefficients.

10.8.4 Outlook

Because the selectivity of the liquid ion-exchanger membranes depends mostly on the ion hydration energy, no exceptional performances can be expected from this point of view. Despite this limitation, sensors based on liquid ion-exchanger of membrane have proved useful in a series of applications involving both inorganic and organic ions.

Questions and Exercises (Section 10.8)

1 What types of molecular ion receptors are employed in potentiometric ion sensors?
2 What are the structural and physicochemical features of liquid ion exchanger receptors?
3 What is the composition of a polymeric ion-sensitive membrane? What is the role of each component of a polymeric membrane formulation? Draw a flow chart for the preparation of such a membrane.
4 What are the factors that determine the selectivity of an ionic receptor membrane? What is the physicochemical constant that determines the selectivity and how does it depend on the properties of the analyte?
5 Why is NO_3^- preferred over Cl^- by the receptor in Figure 10.13C? What imparts selectivity for the specified analytes to the receptors in Figures 10.13E and F?
6 Why are organic ions preferred over inorganic ions in their interaction with an ionic receptor membrane?
7 What are the components of the membrane potential in the case of an ionic receptor membrane? Identify the physicochemical parameters that determine the magnitude of each component.
8 What factors influence the potentiometric selectivity coefficient of an ionic receptor membrane?

10.9 Neutral Ion Receptors (Ionophores)

10.9.1 General Principles

The first neutral receptors used in ion sensors were natural ionophores. Ionophores are lipophilic compounds that can bind ions and transport them across the hydrophobic lipid bilayer of cell membrane. Among ionophores, one can

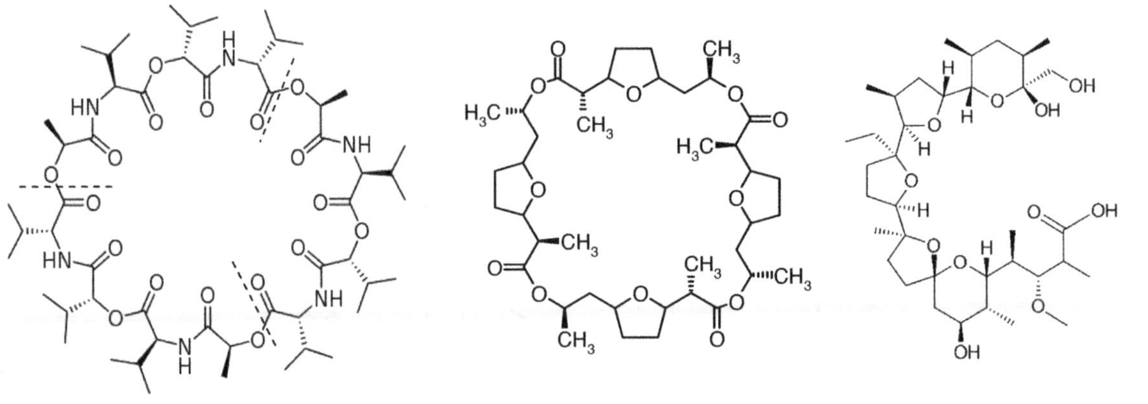

(A) Valinomycin   (B) Noanctin   (C) Monensin

Figure 10.16 Natural ionophores. Dotted lines in (A) indicate the limits of each of the three component peptides in valinomycin.

distinguish both cyclic and noncyclic molecule compounds (Figure 10.16). Thus, *valinomycin*, a cyclic polypeptide (Figure 10.16A) binds potassium ions with an exceptional selectivity over other alkali-metal ions. Ion binding is performed by multiple ion–dipole interactions involving the polar $-C{=}O$ groups in which the oxygen atom bears a partial negative charge. The selectivity for the K^+ ion is due to the fact that the ion diameter matches the internal diameter of the macrocycle, which implies that the ion can interact with all available polar oxygen atoms. Smaller ions (Li^+, Na^+) interact with a smaller number of oxygen atoms, whereas bulkier ions (e.g., Cs^+) do not fit inside the macrocycle.

Nonactin (Figure 10.16B) shows a slight preference for NH_4^+ over K^+ due to the formation of hydrogen bonds with the oxygen atoms directed towards the interior of the macrocycle.

An example of a noncyclic ionophore is *monensin* (Figure 10.16C), which binds various monovalent cations, particularly Na^+. As in the case of valinomycin, the cation is bound by ion–dipole interactions, but, due to the noncyclic conformation, ion wrapping is achieved by the receptor molecule folding around the ion.

The examples above indicate the main factors that determine the selectivity of neutral receptors, namely:

a. multiplicity of ion–receptor bonds;
b. the nature of the ion–receptor interactions;
c. steric factors, that is, size and shape complementarity of the analyte and the receptor, in analogy with the lock and key principle.

The effect of the bonding multiplicity is obvious as it determines to a large extent the strength of the analyte–receptor interaction. The interaction strength depends also on the nature of the chemical bonds involved. Finally, the shape and size fit is crucial for securing as strong as possible analyte–receptor bonding. This depends on the chemical structure and the conformation of the receptor molecule.

So, in contrast to liquid ion exchangers that bind the analyte by strong ion–ion interactions, neutral receptors form relatively weak bonds such as ion–dipole, dipole–dipole and hydrogen bonds. Bond multiplicity, however, enhances the stability of the analyte–receptor complex. In aqueous solutions, the stability of the complex is lessened by ion hydration, but ion solvation is absent in the lipophilic membrane environment, and this enhances the stability of the complex.

The effects of the chemical bonding nature and steric factors on ion recognition by neutral receptors are discussed in the next sections.

10.9.2 Chemistry of Ion Recognition by Neutral Receptors

The feasibility of ion recognition by molecular receptors depends on the type of ion–receptor interaction that takes place. The simplest type of interaction is that of an electrostatic nature which consists of electrostatic attraction between particles of opposite electric charge. Thus, an electrostatic interaction occurs when the receptor is a liquid ion exchanger, that is, a molecular ion. However, electrostatic interactions are also met in the case of neutral receptors provided that the receptor molecule includes strongly polarized chemical bonds. Thus, the carbon–oxygen bond in an organic compound is strongly polarized because of the large difference in electronegativity and the oxygen atom carries a fractional negative charge. A cation experiences attraction by the fractional charge on the oxygen in what is known as an ion–dipole interaction. This kind of interaction is weaker than an ion–ion interaction, but if the ion interacts with multiple dipoles at the receptor molecule, the strength of the ion-receptor bond can be quite large. This kind of interaction is a characteristic of alkali-metal ions.

Table 10.4 Lewis acids and bases classified according to the soft/hard character

	Hard	Borderline	Soft
Acids	H^+, Li^+, Na^+, K^+, Mg^{2+}, Ca^{2+} Al^{3+}, Cr^{3+}, Co^{3+}, Fe^{3+}	Fe^{2+}, Co^{2+}, Ni^{2+}, Cu^{2+}, Zn^{2+}, Pb^{2+}	Cu^+, Ag^+, Cd^{2+}, Hg^+, Hg^{2+}
Bases	F^-, $CH_3\text{-}COO^-$, PO_4^{3-}, SO_4^{2-}, Cl^-, CO_3^{2-}, ClO_4^-, NO_3^-, NH_3, RNH_2	Br^-, NO_2^-, SO_3^{2-} Aniline, pyridine, R-NH-R	R_2S, R-SH, R_3P $(H_3CO)_3P$ I^-, SCN^-, CN^-

Less-electronegative nonmetals substituents, such as nitrogen or sulfur, are less polarizable. However, they can donate a lone pairs of electrons to an empty valence orbital in transition-metal ions to form a coordinate bond.

Hydrogen bonding can be involved in the recognition of hydrated cations by means of water bridges between the ion and the negatively polarized atoms in the receptor molecule. In this case, the receptor size must be sufficiently large to incorporate the hydrated ion, which can be much larger than the nonhydrated form.

π-Electrons in aromatic rings can also take part in ion-recognition processes. Thus, transition-metal cations such as $Fe^{2+/3+}$ and $Co^{2+/3+}$ are known to form coordination compounds with aromatic hydrocarbons by accepting π electrons into an empty orbital. Aromatic molecules form weak bonds by $\pi - \pi$ stacking, which is due to van der Waals force interactions between face-to face oriented aromatic rings. $\pi - \pi$ Stacking is important in organic ion recognition where aromatic substituents in the receptor impart selectivity for aromatic over aliphatic analytes.

The above types of interaction can also operate in the case of anion receptors. In addition, coordination bonds involving anions can be formed when the receptor contains transition-metal sites.

Metal–ligand interactions have been rationalized within the framework of the *hard and soft acids and bases* concept (HSAB theory [38,39]) that refers to the interaction of Lewis acids and bases. A *Lewis acid* is a molecular entity that is an electron-pair acceptor, whereas a *Lewis base* is an electron-pair donor. A Lewis acid and a Lewis base are therefore able to form a *Lewis adduct* by sharing the electron pair provided by the Lewis base. In this context *hard* implies small and nonpolarizable entities and *soft* indicates larger entities that are more polarizable. Low polarizability is associated with small size and strong electronegative or electropositive character, whereas high polarizability occurs with bulky entities bearing a small charge. Borderline cases that show intermediate properties can also be identified. Several Lewis acids and bases are shown in Table 10.4.

The strongest Lewis acid–base interaction occurs when both reactants are either hard or soft, while the interaction of a hard and a soft reactant is much weaker. Thus, oxygen in ethers (R_1–O–R_2) is a hard base that interacts strongly with hard Lewis acids such as alkali-metal ions. Conversely, sulfur in thioethers (R_1–S–R_2) is a soft base and interacts preferentially with soft Lewis acids such as Ag^+ or Hg^{2+}. The limiting cases in Lewis acid–base interactions are the ionic bonding (hard–hard character) and the coordinate–covalent bonding (soft–soft character).

It is clear from the above discussion that the selectivity of a neutral ion receptor is determined largely by the hard/soft character of the binding groups. This suggests a broad range of possibilities for tuning receptor selectivity in conjunction with structural and conformational factors.

10.9.3 Effect of Bonding Multiplicity, Steric, and Conformational Factors

As already mentioned, the stability of the metal complex is enhanced if the receptor provides multiple binding sites to the analyte ion; such a molecule is called a *multidentate ligand*. But, in this case, steric factors also come into play. The conformation of the receptor molecule should be such as to allow the ion to make contact with a maximum number of binding sites at the receptor molecule.

The above principles are illustrated in Figure 10.17 that shows the effect of the receptor (host) shape on the stability of the resulting metal complex. The simplest receptor is a multidentate linear ligand (*podand*) that is able to interact with the ion by means of each binding site and that folds around the ion forming a *chelate complex*. Multiple bonding has a synergic effect as the strength of the resulting complex is greater than the sum of each ion-site interaction, an effect known as the *chelate effect*. Complex stability is impaired by any tension in the ligand molecule that would prompt it to unfold.

More stable complexes result when the receptor is a macrocyclic compound shaped such as to be able to wrap round the ion (*corand*). There is no cause for the ligand to unfold in this case, and this imparts greater stability to the complex. The chelate effect also operates in this case, but, in addition, the *macrocyclic effect* due to the stable cyclic conformation of the receptor comes into play.

Even greater stability is noticed when the receptor molecule has a three-dimensional configuration with more than one cyclic moiety (that is, a *cryptand*). Enhanced complex stability arises from the rigidity of the ligand molecule. While in the case of a corand the molecule can suffer slight deformation that reduces the stability of the complex, this is hardly possible with a cryptand.

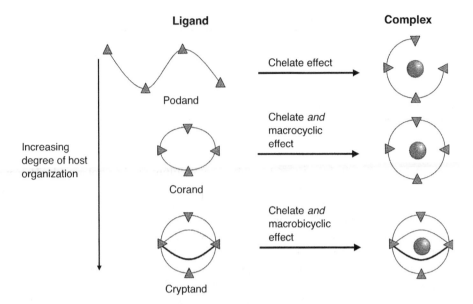

Figure 10.17 The chelate, macrocyclic, and macrobicyclic effect. Each triangle represents an ion binding site in the receptor molecule. Adapted with permission from [32]. Copyright 2009 John Wiley & Sons, Ltd.

In conclusion, the stability of the analyte–receptor complex increases with the degree of organization of the receptor molecule. Highly organized receptors display rigid conformation and only an ion matching the receptor size and shape can be accommodated.

Remembering that the receptor is intended to extract the ion from an aqueous solution into a lipophilic membrane, it is clear that the receptor should be a hydrophobic compound. Hydrophobicity is imparted by aromatic or long-chain aliphatic moieties.

10.9.4 Neutral Receptor Ion-Selective Membranes: Composition, Selectivity and Response Function

The fabrication of neutral receptor membranes is similar to that emphasized for the fabrication of liquid exchanger membranes [33]. The preparation consists of a mixture of matrix polymer (PVC), a plasticizer and the receptor. However, a neutral receptor membrane should contain, in addition, a *lipophilic ionic additive* with an electric charge opposite to that of the analyte. In practice, alkali salts of tetraphenylborate derivatives (Figure 10.18A) are used in cation-selective membranes and tetralkylammonium salts (Figure 10.18B) in anion-selective membranes. The role of the ionic additive is to prevent coextraction of solution counterions as ion pairs with the analyte.

Ion interaction with the membrane involves ion partitioning into the membrane followed by interaction with the neutral receptor R to yield a complex $[MR]^+$.

In a water/lipophilic solvent system, the partition equilibrium can be expressed as follows:

$$M^+(\text{solution}) \rightleftarrows M^+(\text{membrane}) \tag{10.71}$$

A partition coefficient ($K_{p,M}$) can be defined with respect to the activities of the free ion in the solution and membrane phase, (m_s and m_m, respectively):

$$K_{p,M} = \frac{m_m}{m_s} \tag{10.72}$$

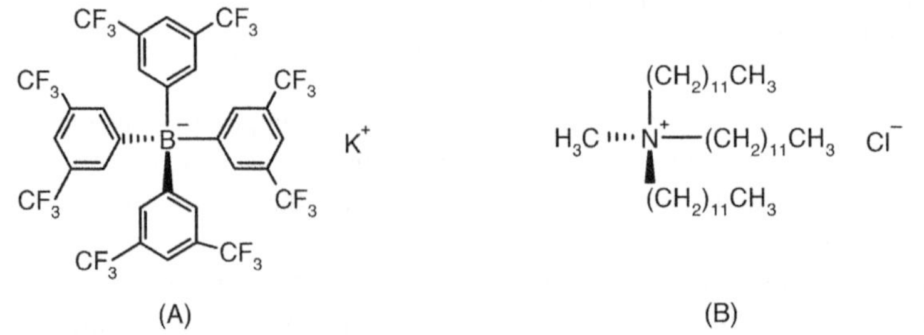

(A)

(B)

Figure 10.18 Ionic additives for neutral receptor liquid membranes. (A) Potassium tetrakis-[3,5,-bis(trifluoromethyl)phenyl]borate; (B) tridodecylmethyl ammonium chloride.

Formation of the ion complex with the receptor R is represented by the following reaction in which the activity symbol for each species is indicated below the species symbol:

$$\underset{m_m}{M^+(m)} + \underset{r}{R(m)} \rightleftarrows \underset{c_m}{[MR]^+(m)} \tag{10.73}$$

The driving force in the ion transfer from the solution to the membrane is the change in the free energy that accompanies the complex formation and is correlated with the stability constant of the ion–receptor complex given in the following equation:

$$K_{MR} = \frac{c_m}{r m_m} \tag{10.74}$$

The free metal ion activity in the membrane results from the partition coefficient Equation (10.72). On substituting this activity into Equation (10.74) one obtains the activity of the complex as a function of the ion activity in the solution, m_{aq}:

$$c_m = K_{p,M} K_{r,M} r m_{aq} \tag{10.75}$$

Assuming that an interfering cation N^+ is also present in the system, it follows that the analyte interaction with the membrane can be represented by the following reaction at equilibrium:

$$\underset{m_{aq}}{M^+(aq)} + \underset{c_N}{\{[R\,N]^+; X^-\}(m)} \rightleftarrows \underset{c_M}{\{[R\,M]^+; X^-\}(m)} + \underset{n_{aq}}{N^+(aq)} \tag{10.76}$$

(aq) and (m) denote the aqueous an membrane phase, respectively; X^- is the ionic additive in the membrane. As each cation forms a complex with the receptor, the selectivity coefficient of the recognition process can be expressed as $K_{M,N} = (c_M/c_N)_{m_{aq}=n_{aq}}$. By using Equation (10.75) and a similar equation for c_N, and assuming that the receptor is in high excess and its free-state concentration changes to a negligible extent, one obtains:

$$K_{M,N} = \frac{K_{p,M} K_{MR}}{K_{p,M} K_{NR}} \tag{10.77}$$

As in the case of the liquid exchanger membrane, the membrane potential can be expressed by an equation similar to Equation (10.68). However, the mobilities of the $[MR]^+$ and $[NR]^+$ complexes are similar, being determined by the properties of the receptor. Therefore, no diffusion potential forms across the neutral receptor membrane and the membrane potential is given by the difference in the boundary potentials. If the membrane is inserted between two solutions containing the ion M^+ with activities m_1 and m_2, respectively, the membrane potential is:

$$\Delta\phi_m = \frac{RT}{zF} \ln \frac{m_1}{m_2} \tag{10.78}$$

The effect of the interfering ion N^+ is accounted for as follows [2]:

$$E_{cell} = constant + \frac{RT}{zF} \ln \left(m_{aq} + k_{M,N}^{pot} n_{aq} \right) \tag{10.79}$$

where the potentiometric selectivity coefficient is $k_{M,N}^{pot} = K_{M,N}$. Therefore, the potentiometric selectivity coefficient is mostly determined by the stabilities of ion–receptor complexes involved. The partition coefficients ratio in Equation (10.77) is close to unity and has little effect on selectivity.

The ionic additive in the membrane plays an essential role in determining the upper limit of detection of the sensor. At a high anion concentration in the solution, the solution anion can be extracted into the membrane as an ion pair with a solution cation. Under these conditions, the sensor becomes anion responsive, which limits the cation response range. However, if an anionic additive is present, the permeation of the solution anions is prevented and the cation response range expands considerably.

The polarity of the plasticizer molecule can also exert a notable influence on the selectivity of neutral receptor-based membranes. When a more polar plasticizer is present, divalent ions are preferred over monovalent ions.

Although the above discussion has focused on cation sensors, similar conclusions apply to anion sensors as well.

In Section 10.9.1 several neutral ion receptors have been introduced. Such compounds are rather expensive and their applications are restricted to certain alkali-metal ions and ammonium. Synthetic chemistry has developed a large number of neutral receptors tailored so as to display selectivity for various cations and anions. By extension,

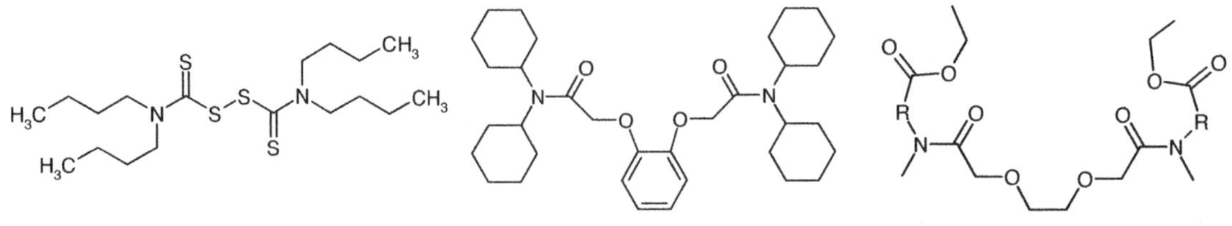

(A) A noncyclic Zn²⁺ receptor

(B) A noncyclic Na⁺ receptor

(C) An amide Ca²⁺ receptor.
R = (CH₂)₁₀

Figure 10.19 Noncyclic cation receptors.

the term ionophore is also used to designate synthetic neutral receptors. The following sections review a series of neutral receptors for both cations and anions.

10.9.5 Neutral Noncyclic Ion Receptors

Several podand-type *cation receptors* are shown in Figure 10.19. The molecule in (A) contains six soft base groups that interact preferentially with the Zn^{2+} ion forming a chelate complex. Other transition-metal ions (Cu^{2+}, Pb^{2+}, Hg^{2+}) interfere significantly, but alkali ions are well discriminated against. As the protonation of the nitrogen sites is possible, a good response to Zn^{2+} is obtained at pH 3.5–6.5. The receptor in Figure 10.19B displays selectivity to the Na^+ ion as it forms multiple ion–dipole bonds and folds optimally around this ion. A Ca^{2+} receptor is shown in Figure 10.19C. It contains six binding sites (ether oxygens) and has no protonable groups that would cause interference by the hydrogen ion, as occurs with ionic receptors.

Noncyclic, neutral *anion receptors* have also be designed [40–42]. Several examples are shown in Figure 10.20. Preorganized bis-thiourea derivatives provide anion recognition by hydrogen bonding. Thus, the compound in Figure 10.20A is selective for the sulfate ion, whereas a similar compound with thiourea moieties separated by a slightly longer distance demonstrates selectivity for Cl^-. Anion recognition can also be achieved by analyte coordination to metal centers in the receptor molecule as is the case in Figure 10.20B. The compound in (B) is an organo-metallic tin-fluorine derivative that can interact with the fluoride ion via coordinate bonding. A binding metal site can also be introduced into the receptor molecule by coordination, as shown in Figure 10.20C for a hydrogenphospahate receptor. In this case, the hydrogenphospahate ion can bind coordinatively to the uranyl center and also to the methoxy groups by hydrogen bonding.

It is of interest to note that neutral receptors for the hydrogen ion have also been developed and tested successfully. Such receptor molecules include a double-bonded tertiary nitrogen atom (−C=N−) which becomes positive by protonation. A liquid membrane pH sensor has been validated for pH determination in diary products [43]. An important advantage of the liquid membrane pH sensors arises from the facile miniaturization and the possibility of integration in sensor arrays.

Calixarenes provide a versatile scaffold for assembling binding sites, either at the upper or lower rim, in order to form podand-type receptors. Thus, exceptional Na^+ selectivity has been achieved with simple ester substituents such as $-OCH_2COOCH_3$. Anion recognition can be performed by calixarenes substituted with coordinated metal ions and amidic or urea functions [44]. For example, functionalization with upper rim sulfonamide substituents leads to a three-dimensional cavity that binds the tetrahedral HSO_4^- ion stronger than the spherical Cl^- or the planar NO_3^- ions.

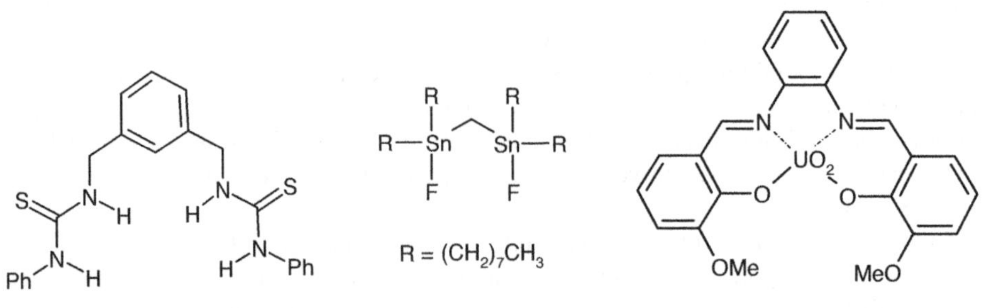

(A) A SO_4^{2-} receptor

(B) A F^- receptor

(C) A HPO_4^{2-} receptor

Figure 10.20 Noncyclic neutral anion receptors.

10.9.6 Macrocyclic Cation Receptors

Macrocyclic cation receptors are synthetic analogs of natural cyclic ionophores. However, while natural ionophores display selectivity to only a couple of ions, synthetic macrocyclic receptors have been developed for a very broad range of both cations and anions.

Macrocyclic ethers (also termed crown ethers, Figure 10.21) were the first compounds of this type to be synthesized and tested. The name of a crown compound includes the term "crown" and two numbers; the first one indicates the total number of atoms in the cycle and the second one, the number of heteroatoms. Besides oxygen in oxa groups $(-O-)$, aza groups $(-N-)$ or thia groups $(-S-)$ can be included in the macrocycle. The selectivity of such receptors is determined by size fitting and the nature and multiplicity of the analyte–binding-site interactions.

The importance of the steric factor is illustrates in Figure 10.21 that shows that a good match of the ion and receptor size allows the maximum possible ion–dipole interactions to occur, thus imparting maximum stability to the complex.

Beside the size matching, the nature of the heteroatoms in the crown receptor contributes substantially to the selectivity in accordance with the HSAB principle. Thus, the macrocyclic ether in Figure 10.22A includes strong Lewis bases and displays selectivity to the K^+ strong Lewis acid. Introduction of tertiary nitrogen groups that are much softer bases (Figure 10.22B) imparts selectivity for the Hg^{2+} ion. The Lewis base hardness of the nitrogen site depends on the electronic effect applied by the substituent R that allows variation in the polarity of the $C-N$ group.

The soft base thia group in Figure 10.22C imparts very good selectivity for the Ag^+ soft acid, although its ionic diameter (2.00 Å) is somewhat larger than the cavity diameter (1.1–1.4 Å). Binding of more voluminous soft acid ions (e.g., Hg^{2+}) is prevented by the bulky substituent at the upper part of the molecule. The combined effect of size fitting and steric hindrance leads to a selectivity coefficient $k^{pot}_{Ag,Hg}$ of about 0.03 for an Ag^+ sensor based on this receptor [45].

A higher degree of organization of the receptor molecule is expected to impart more selectivity. This is demonstrated by examining the binding energy of alkali ions to the receptors in Figure 10.23. The compound in A is a *spherand*, that is, a macrocyclic molecule in which binding groups are not constituents of the macrocycle itself but are appended to the cycle. The binding energy sequence demonstrates a clear preference for small ions, particularly for Li^+. Compounds in Figures 10.23B–D are cryptands (macrobicycles). In such compounds, one or more additional chains link distant atoms in the macrocycle. As a result, the conformation of the molecule is very rigid, which prevents the molecule from adapting the cavity size to that of the ion upon binding. Consequently, the compound in Figure 10.23B displays exceptional selectivity for the Li^+ ion due to the size-exclusion effect. The larger molecule in Figure 10.23C can bind all alkali ions (except Cs^+) but with a clear preference for Na^+. An even larger bicyclic molecule like that in Figure 10.23D prefers more voluminous ions but does not bind to the small Li^+ ion at all.

An enhancement of the binding strength is obtained by adding to the macrocycle a side arm that contains an ion binding site to form a *lariat crown ether*. The side arm could position itself over the cavity of the crown compound and contribute to ion encapsulation (Figure 10.24A). Alternatively, the macrocycle can function only as a scaffold for assembling binding sites in a suitable spatial arrangement (Figure 10.24B). Proper selection of the side arm binding group leads to enhanced selectivity in accordance with the HSAB principle.

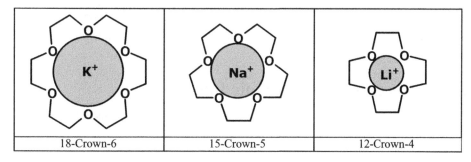

| 18-Crown-6 | 15-Crown-5 | 12-Crown-4 |

Figure 10.21 Macrocyclic ethers as alkali ion receptors.

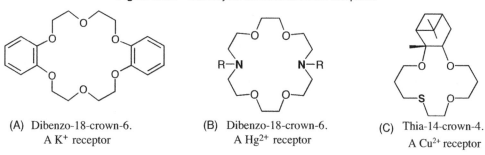

(A) Dibenzo-18-crown-6.
    A K^+ receptor

(B) Dibenzo-18-crown-6.
    A Hg^{2+} receptor

(C) Thia-14-crown-4.
    A Cu^{2+} receptor

Figure 10.22 Effect of the Lewis base hardness of the binding sites in macrocyclic ion receptors.

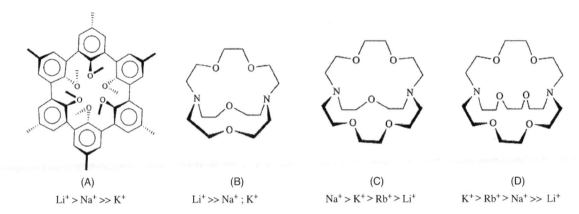

(A)	(B)	(C)	(D)
$Li^+ > Na^+ \gg K^+$	$Li^+ \gg Na^+ ; K^+$	$Na^+ > K^+ > Rb^+ > Li^+$	$K^+ > Rb^+ > Na^+ \gg Li^+$

Figure 10.23 Highly organized alkali ion receptors. (A) A spherand; (B–D) cryptands. The series of binding energy for alkali ions is indicated for each compound. Adapted with permission from [46]. Copyright 1986 Wiley-VCH Verlag GmBH & Co. KGaA.

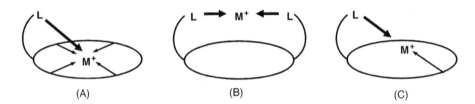

(A)	(B)	(C)

Figure 10.24 Metal ion binding to lariat crown compounds. (A) Binding to both the macrocycle and the side arm; (B) binding to two or more side arms; (C) binding to the side arm and a site at the macrocycle. Thin arrows indicate binding to the macrocycle heteroatoms; thick arrows indicate the displacement of the binding group (L) at the side arm in order to bind to the ion.

10.9.7 Macrocyclic Anion Receptors

Macrocyclic receptors for anions are of great interest because, with the notable exception of the solid membrane fluoride sensor, other anion sensors based on solid-state membranes are generally plagued by poor selectivity. Anion receptors and potentiometric anion sensors are comprehensively reviewed in refs. [34,40,47]. Several macrocyclic anion receptors are shown in Figure 10.25. Thus, the aza crown compounds in Figure 10.25A can form hydrogen bonds with an oxygenated anion such as NO_3^-. Hydrogen bonding with anionic receptors can also take place with imidazolium moieties [48]. For example, Figure 10.25B shows a calixpyrrole compound that can adopt either a cone or 1,3-alternate conformation like calyx[4]arenes. Upon F^- binding by hydrogen bonding, it is fixed in a cone conformation and preferentially discriminates well against the Cl^- ion. The compound in Figure 10.25C includes an organometallic aluminum substituent. As aluminum is a hard Lewis acid, it interacts strongly with the hard Lewis base F^-. Metal centers can also be included by coordination as shown in Figure 10.25C for a cyanide ion receptor. Each Cu^{2+} ion, which is a soft Lewis acid, can form coordinate bonds with a cyanide ion (soft Lewis base) which imparts selectivity for this ion.

10.9.8 Neutral Receptors for Organic Ions

The interest in determining organic ions arises from the great importance of such ions in physiological processes as well as from the great number of pharmaceutical products of this type. Cationic receptors have been extensively employed for such applications, but, in a search for better selectivity, the attention has shifted to cyclic or noncyclic neutral receptors [34].

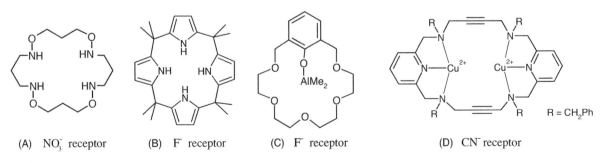

(A) NO_3^- receptor	(B) F^- receptor	(C) F^- receptor	(D) CN^- receptor

Figure 10.25 Macrocyclic anion receptors. Me in (C) stands for methyl.

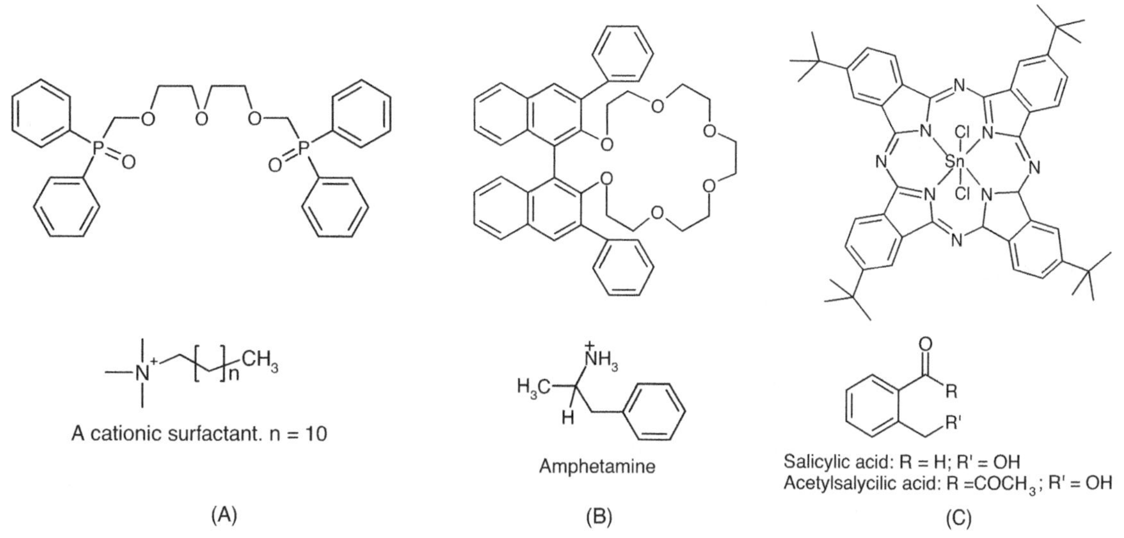

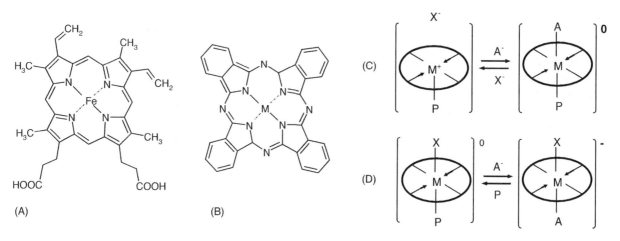

Figure 10.26 Receptor for organic ions (up) and corresponding analytes (down). (A) A podand-type receptor for cationic surfactants; (B) a macrocyclic receptor for a primary ammonium aromatic compound; (C) a phthalocyanine-type receptor for salicylic acid and its derivatives.

Of great interest is the determination of organic ammonium ions. It was found that aliphatic quaternary ammonium ions interact by ion–dipole interactions with strongly polarized oxa groups such as that in the crown ether in Figure 10.22A but much better discrimination of alkali-metal ions has been found with the podand-type receptor shown in Figure 10.26A. Primary ammonium ions can form hydrogen bonds with oxa sites in macrocycles such as that shown in Figure 10.26B. This compound, which is rich in phenyl residues displays particular selectivity for aromatic derivatives such as amphetamine due to the $\pi - \pi$ stacking interactions.

Determination of carboxylates and their esters is of particular interest due to the broad applications of salicylic acid and its derivatives in therapy. Figure 10.26C shows a receptor for compounds of this type. This receptor is a tin phthalocyanine and the analyte interacts with the metal center according to the HSAB principle.

Calixarenes present particular advantages because such receptors can form inclusion compounds with organic ions, thus providing an additional selectivity factor in addition to the ion–dipole interactions.

Of particular interest is the chiral recognition of ions [49]. Chiral recognition requires at least three simultaneous interactions between the receptor and one of the enantiomers, with at least one of these interactions being stereochemically dependent.

10.9.9 Porphyrins and Phthalocyanines as Anion Receptors

Both porphyrins (Figure 10.27A) and phthalocyanines (Figure 10.27B) feature four pyrrole-like subunits linked to form a 16-membered ring which can bind a transition-metal ion at the center. Closely related to porphyrins are vitamin B_{12} derivatives. The metal ion in such compounds can be coordinated axially with either one or two ligands P and X^-. In the first case (Figure 10.27C) the receptor is positively charged and the interaction with the analyte anion A^- involves an ion-exchange process. This process is formally like that in scheme (10.64) except for the fact

Figure 10.27 (A) A porphyrin (hemoglobin); (B) a phthalocyanine; (C) anion recognition by the ion-exchange mechanism; (D) anion recognition by ion-neutral ligand exchange.

that the analyte in the membrane phase is not included in an ion pair but is bound to the metal center in the receptor. In the second case (Figure 10.27D), the receptor is doubly coordinated, hence it is a neutral species. Therefore, the exchange of the neutral ligand P with the ionic analyte results in a charge modification. Therefore, the membrane should contain an ionic additive to preserve the charge balance.

The strength of the analyte–receptor interaction depends on the nature of the metal center, its oxidation state, and the number and the nature of axial ligands. There are, therefore, various possibilities for tuning the receptor selectivity for various anion sensing applications [47].

10.9.10 Outlook

Neutral ion receptors offer endless possibilities for developing ion sensors with suitable selectivity for both inorganic and organic ions by rational design of the molecular structure and conformation. Selectivity arises from the multiple bonding character of the analyte–receptor interaction and also from the careful selection of the most suitable type of chemical bonding.

Currently, neutral receptors are dominating the field of ion sensor science and practical applications are restricted only by their availability. The single drawback of these devices is the relatively short lifetime of the membrane that is often limited to several months due to the leaking of the membrane components. The major advantage of neutral receptor-based ion sensors is their exceptional selectivity, which, in general, is not matched by any other type of ion-selective membrane.

Questions and Exercises (Section 10.9)

1 What are the structural and conformational features that provide selectivity in the interaction of natural ionophores with certain metal ions? Why does monensin display selectivity to NH_4^+ over K^+?

2 What are the conformational factors that impart selectivity in the interaction of synthetic ionophores with ions? Discuss in detail why the selectivity increases in the following sequence: podand < corand < cryptand.

3 What kinds of interaction are involved in the recognition of an ion by neutral receptors? What are the rationales of the strength of interaction?

4 What are the structural factors that influence the selectivity of a neutral ion receptor?

5 How can the selectivity of the neutral ion receptor be quantified?

6 What is the typical composition of a neutral receptor-based ion-sensitive membrane? Comment on the function of each component within the membrane. What factors determine the selectivity of such a membrane?

7 What are the particular features of neutral receptor-based membrane as far as ion transport within the membrane is concerned. What are the consequences of these features as far as the potentiometric selectivity coefficient is concerned?

8 For each cation receptor in Figure 10.19: (a) give a detailed overview of the molecular features that provide selectivity for a particular ion; (b) for each receptor in this figure, identify the molecule moiety that imparts lipophilicity; (c) why are long aliphatic chains included in the structure of the Ca^{2+} receptor in Figure 10.19C?

9 For each anion receptor in Figure 10.20: (a) give a detailed overview of the molecular features that provide selectivity for a particular ion; (b) for each receptor in this figure, identify the molecule moiety that impart lipophilicity.

10 Inorganic cations have a spherical shape. What is the effect of this shape on cation recognition by crown ethers?

11 What kind of heteroatoms can be included in the macrocycle of a crown compound and how may the heteroatom influence the selectivity?

12 How can substituents to the crown compound influence the selectivity in interactions with ions? Comment on the substituent effect for the receptors in Figure 10.22, Figure 10.23A, and Figure 10.24.

13 Comment on the factors that provide selectivity to the macrocyclic anion receptors in Figure 10.25.

14 What kind of chemical and conformational factors make possible the recognition of certain organic ions?

15 What are the rationales of the $\pi - \pi$ stacking? Identify among the neutral receptors in the text certain receptors that interact with the analyte by $\pi - \pi$ stacking.

16 What structural factors render porphyrins and phthalocyanines suitable as anion receptors?

17 Give a short overview of the applications of calixarenes as ion receptors.

18 Classify neutral receptors that are mentioned in the text according to the type of interaction in view of the HSAB principle.

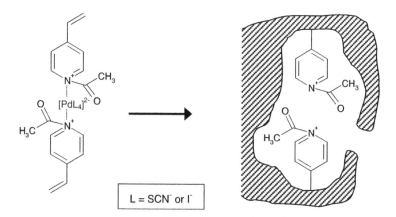

Figure 10.28   Molecular imprinting of a Pd^{2+} ion receptor. Left: template ion complex including monomers; right: imprinted receptor site. Adapted with permission from [50]. Copyright 2006 Elsevier.

10.10 Molecularly Imprinted Polymers as Ion-Sensing Materials

Molecularly imprinted polymers (introduced in Section 6.7.3) provide excellent opportunities for ion sensing and are of particular interest when no suitable molecular receptor is available [50–52]. In order to obtain an ion-selective polymer, specific ion ligands, appended with monomer tails are subjected to copolymerization with another monomer that produces the polymer matrix (Figure 10.28). The polymerization is conducted in the presence of the ion of interest which acts as a template and obliges the ligand to adopt the spatial orientation specific to the metal-ion complex (Figure 10.28). After leaching out the ion from the resulting polymer, the matrix includes small pores that provide specific binding sites for the ion. In the case shown in Figure 10.8, two binding sites covalently bonded to the polymer backbone are present. However, binding sites can also be included by trapping receptor molecules in the matrix along with covalently bound ligands.

Once formed, the polymer is ground down to some 60 mesh and cast in PVC membranes. Alternatively, the polymerization can be conducted at a solid support surface. Electrochemical polymerization at a conducting material surface represents yet a further method of preparing molecularly imprinted polymer membranes. Ion binding sites can be obtained using well-established metal ion reagents developed in classical analytical chemistry. Molecular receptors (ionophores) can also be appended to the polymer backbone; this results in enhanced selectivity compared with that of the free receptor.

Molecularly imprinted polymers have been used to produce sensors for more exotic ions such as uranyl, or dysprosium as well as for common ions such as Ca^{2+}. A nitrate ion sensor with good selectivity has been obtained by electrochemical polymerization of pyrrole onto carbon electrodes in a $NaNO_3$ solution [53]. This appears as a remarkable achievement if we remember that molecular receptor-based membranes for the nitrate ion are not particularly selective.

Questions and Exercises (Section 10.10)

1 Outline the detailed steps in the synthesis of a La^{3+}-selective imprinted polymer using the reagents below. The binding site should include 2 molecules of (1) covalently attached to the polymer backbone and 3 molecules of (2) included by trapping them in the matrix.
   Hint: see ref. [50].

(1)   (2)

2 Outline the main steps in the fabrication of a potentiometric ion sensor using this polymer.
3 What are the advantages of imprinted polymers as ion-recognition materials?

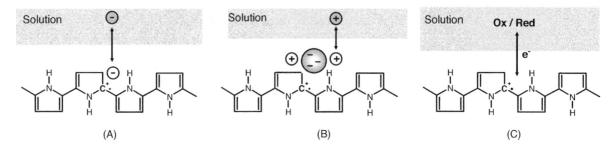

Figure 10.29 Charge-exchange processes involving polypyrrole. (A) Anion exchange involving the polymer counterion; (B) cation exchange involving a bulky anion embedded in the polymer. (C) electron exchange with a redox species.

10.11 Conducting Polymers as Ion-Sensing Materials

The general characteristics of conducting polymers have been emphasized already in Section 5.9. In summary, a conducting polymer molecule includes an extended π-orbital system. By oxidation, positive charges (in the form of bound radical ions) can be imparted to the polymer backbone. Charge balance is secured by mobile counterions (dopants) associated with the polymer in a gel-like structure. Various substituents can be appended to the polymer backbone in order to enhance the selectivity.

In potentiometric ion sensors, conducting polymers can be used as (a) ion-exchanger recognition materials, and (b) immobilization matrices for molecular ion receptors [13,54,55].

As a conducting polymer molecule contains fixed positive charges, it can act as an anion exchanger, as shown in Figure 10.29A. In order to perform cation exchange, the dopant should be a lipophilic anion (e.g., tetraphenylborate or dodecylsulfate) that is entrapped within the polymer matrix and acts as a fixed anionic site (Figure 10.29B). Sensors for various cations (including the hydrogen ion) have been developed in this way. Moreover, as conducting polymers can perform electron exchange with redox species in solution (see Figure 10.29C), this allows ionic and neutral redox species to be determined. However, redox reactions involving a conducting polymer can be a source of interference in ion determinations based on ion-exchange processes.

Improved selectivity over the ion-exchange recognition method is obtained by including a molecular ion receptor within the polymer matrix by entrapment or covalent binding. This alternative is characterized by enhanced stability of the sensing material.

Questions and Exercises (Section 10.11)

1 What properties of conducting polymers are of interest in the design of potentiometric ion sensor?
2 Review the possible applications of the conducting polymers in potentiometric ion sensors as recognition materials and receptor matrix.

10.12 Solid Contact Potentiometric Ion Sensors

Direct electrical contact of the sensing membrane with the external circuit can be achieved using a solid contact, thus avoiding the use of an internal reference electrode and internal solution. This simplifies the fabrication process, allows for miniaturization and imparts enhanced robustness. The solid contact should be designed such as to allow for facile transition from ionic conduction in the membrane to electronic conduction in the connecting wire. As a potential difference develops at the contact, it is included as an additional term in the cell voltage and, therefore, it should be very stable.

In the case of solid membranes, solid contact can be achieved by means of conductive epoxy resins. Better contact is achieved by depositing a thin metal layer over the top side of the membrane.

Direct contact to polymeric sensing membranes can be achieved simply by casting the membrane over the surface of a metal wire in order to obtain a so-called *coated wire potentiometric sensor*. However, the contact potential difference is not sufficiently stable. Nevertheless, such devices have proved useful as indicator electrodes in potentiometric titrations [56,57]. It has been proved recently that a very stable response can be obtained by including highly porous materials (such as nanostructured macroporous carbon, porous silicon or carbon nanotubes) between the membrane and the metal wire [58].

Another approach is based on the inclusion of the receptor into a carbon paste mixture that functions as the sensing membrane [59–61]. Thus, perchlorate and tetrafluoroborate ion-selective carbon paste sensors demonstrated similar characteristics to those obtained with commercial polymer membranes [62]. However, the carbon pastes provide a faster response and have very low ohmic resistance.

Screen-printing technology is very convenient for developing direct contact potentiometric sensors [63]. In this design, the sensing membrane is formed as a thin layer over the surface of the working electrode. A molecular ion receptor can be incorporated into the conventional insulating paste. Alternatively, a carbon paste containing metal sulfide can act as a metal ion-sensing element. Ag/AgCl reference electrodes can also be fabricated by screen printing. Arrays of millimeter-sized sensors assembled on the same strip have been developed.

A good membrane-metal contact is obtained by means of materials that elicit both ionic and electronic conductivity, such as conducting polymers [13,55]. In such materials, ionic conductance is provided by counterion displacement, whereas electronic conductance arises from electrons jumping from filled to empty orbitals in the polymer (hole conduction mechanism). The transition from ionic to electronic conductivity is achieved according to reaction (10.80) in which P^+ is a positive site in the polymer backbone and A^- is the counteranion:

$$(P^+ A^-)_{Polymer} + (e^-)_{Metal} \rightleftarrows (P^0)_{Polymer} + (A^-)_{Solution} \tag{10.80}$$

It is easy to realize that this process is analogous to that occurring in a standard reference electrode (reaction (10.22)).

Direct contact with a metal conductor can be achieved if the sensing membrane is made of a conducting polymer that is deposited by electrochemical polymerization onto a metal or carbon electrode surface. When using a sensing polymer membrane, a layer of conducting polymer between the top side of the membrane and a metal contact piece secures the contact. Direct contact between the membrane and a metal conductor can also be obtained by including a small amount of conducting polymer in the membrane material. In order to secure a stable potential, it is important to avoid the formation of a solution layer between the sensing membrane and the conducting polymer plug.

Questions and Exercises (Section 10.12)

1 What are the advantages of solid contact potentiometric ion sensors?
2 Review the methods for achieving internal solid contact for (a) solid membranes; (b) liquid membranes.
3 What are coated-wire potentiometric sensors? What are the possible problems that might arise in using such sensors and how can these problems be overcome?

10.13 Miniaturization of Potentiometric Ion Sensors

Potentiometric microsensors are used in the biosciences for the measurement of ion activity in individual cells or in intracellular liquid and also in environmental science for ion determination in pore water samples or for profiling the ion concentration in sediments. Such sensors have also proved useful as detectors in liquid chromatography and capillary zone electrophoresis and are promising for applications in micro-total analysis systems. Principles and applications of potentiometric microsensors are amply reviewed in refs. [64,65] and the applications in plant-cell biology are surveyed in ref. [66].

Microsensors for common ions (such as H^+, K^+, Ca^{2+}, Cl^-) have been developed using suitable recognition materials or reagents. They are manufactured from glass micropipettes with a tip diameter down in the micrometer range. pH microsensors of this type rely on glass membranes, whereas metal cation sensors are based on molecular receptors dissolved in a nonaqueous solvent to form a liquid membrane, as shown in Figure 10.30. In order to prevent water penetration through the micropipette tip, the inner surface should be made hydrophobic by silanization. The internal filling solution can be replaced by a hydrogel layer.

When a single-barreled sensor (Figure 10.30A) is used in a single cell, the reference electrode is placed outside the cell. This procedure needs additional corrections to the voltage read-out and is not suitable for electrically excitable cells. Such problems can be circumvented by using double-barreled microsensors (Figure 10.30B). Although micropipette-type sensors are physically large devices, they can be used with sample volumes down to the picoliter range. However, such microelectrodes are fragile, difficult to prepare, and the lifetime is only in the order of days.

Recent research work has been focused on improving the ruggedness and stability of microsensors by using new materials and design principles [13,58]. A more robust design relies on direct internal contact achieved by means of conducting polymers, which exclude the need for reference electrodes. The fragile glass micropipette can be replaced by monolithic capillaries as holders for the ion-sensing membrane [67].

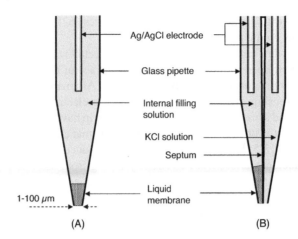

Figure 10.30 Cross section of micropipette potentiometric ion sensors. (A) Single-barreled sensor; (B) double-barreled sensor. A septum separates the reference (right) and the measuring (left) compartments.

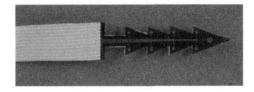

Figure 10.31 Microfabricated nine-sensor array (12 × 4 mm) for ionic distribution measurements in porcine beating hearts. Adapted with permission from [15]. Copyright 2001 American Chemical Society.

An alternative to the micropipette design is based on sensing membranes attached over a recessed gold microelectrode of some 25 μm diameter [68]. Advanced miniaturization has been achieved by using conducting polymers as sensing materials. Thus, a sub-μm-sized pH sensor with excellent selectivity has been developed by coating polyaniline over the surface of a carbon microdisk [69].

Microfabrication technology has been successfully employed to fabricate sensor microarrays for application in physiological research (Figure 10.31) [70]. Also, silicon nitride micropipette arrays have been obtained by microfabrication.

Miniaturized reference electrodes that are needed in such applications can be obtained from an ion microsensor that responds to an ion with constant concentration (e.g., Na^+ in blood serum analysis). Ion-independent miniaturized reference electrodes have also been obtained on the basis of hydrophobic membrane materials [71].

As *in vivo* determination of pH is a matter of great concern, much attention has been paid to the development of miniature pH sensors. A comprehensive overview of this topic is give in ref. [72].

Questions and Exercises (Section 10.13)

1 Why miniature potentiometric sensors are of interest
2 Review the applications of polymer membranes in miniaturized potentiometric ion sensors.
3 Give an overview of the methods of fabrication of potentiometric ion microsensors focusing on alternatives to micropipette technology.

10.14 Analysis with Potentiometric Ion Sensors

Direct determinations by means of potentiometric ion sensors can be performed by means of a calibration graph like that shown in Figure 10.8. Care should be taken to secure identical pH and ionic strength in all the standard and sample solutions. IUPAC recommended procedures for calibration of potentiometric sensors are presented in ref. [73]

Ion determination can also be performed by resorting to controlled modification of the analyte concentration in what is known as the Gran method [74,75]. In this method, the analyte concentration is varied by adding a known volume V_a of an analyte solution with known concentration c_a (standard solution). If c_0 and V_0 are the sample concentration and volume, respectively, then, the total concentration after performing the standard addition is:

$$c_t = \frac{c_0 V_0 + c_a V_a}{V_t}$$

(10.81)

where the total volume is $V_t = V_0 + V_a$. On substituting c_t for the analyte concentration in Equation (10.34) and rearranging one obtains:

$$y = V_t 10^{E_{cell}/s} = k(c_0 V_0 + c_a V_a) \tag{10.82}$$

where s is the empirical sensitivity of the sensor and $k = 10^{E_0/s}$. By plotting y vs. V_a a straight line is obtained whose intersection with the horizontal axis ($y = 0$) allows the calculation of the unknown concentration c_0. This concentration can also be obtained numerically from the quotient of the slope and the intercept of the regression equation.

A simple alternative to the above method is the single standard addition method. After several algebraic manipulations of Equation (10.82) one obtains the following equation that allows the unknown concentration to be calculated:

$$c_0 = \frac{V_a c_a}{V_t 10^{-\Delta E_{cell}/s} - V_0} \tag{10.83}$$

where ΔE_{cell} is the difference between the cell voltage before and after performing a single standard addition.

The above methods are advantageous when the sample matrix cannot be reproduced in the standard solutions. If $V_a \ll V_0$, only negligible modification in the matrix component concentration occurs. Thus, the effects of variations in the activity coefficient ant the junction potential are minimized. Moreover, the analyte concentration can be changed by adding a ligand that forms a very stable complex with the analyte. In this case, the $c_a V_a$ term in the Equation (10.82) is negative.

Potentiometric ion sensors are also commonly used for detecting endpoints in titrimetric analysis. More details on analytical methods based on potentiometric sensors can be found in refs. [5,7,76].

Questions and Exercises (Section 10.14)

1 In order to perform pH determinations with a glass electrode, the cell potential was measured for three standard solutions with the following pH values at 25 °C: 2.04, 7.05, and 9.20. The cell voltage readout (in mV) for each of the above solutions was 238.0, −37.5 and, −164.5, respectively. Calculate: (a) the sensitivity of the pH sensor; (b) the pH of an unknown sample yielding a cell voltage of 20.5 mV; (c) the pH deviation from the actual value if the sample temperature is 35 °C.

   Answer: (b) pH = 6.07.

2 In order to determine the fluoride concentration in a mouthwash by means of a F⁻ sensor, 50 ml of the original sample was mixed with 50 ml of TISAB solution. The cell voltage for this solution was 54.8 mV. Using the calibration data in Figure 10.8, calculate the fluoride concentration in the original sample.

   Answer: 0.010 g/100 mL

3 The content of F⁻ in a contaminated water sample was measured by means of a fluoride potentiometric sensor by the standard addition method. A cell voltage of 66.2 mV was recorded with the original sample. Then, 1 ml of 0.0207 M standard F⁻ solution was added to 10 ml of this sample and the cell voltage for the resulting solution was 28.6 mV. Calculate the concentration of F⁻ in the sample by taking into account that the sensitivity is 59.1 mV per unit pF.

   Answer: 5.5×10^{-4} M.

4 In order to determine the sulfide ion concentration by means of a S^{2-} potentiometric sensor, the cell voltage for the sample (100 ml volume) was first measured yielding $E_{cell} = -845.0$ mV. After adding 1 ml of 0.100 M $AgNO_3$ the voltage became −839 mV. What is the sulfide concentration in the sample? Assume that the sensitivity has the ideal value and the temperature was 25 °C.

   Hint: the following reaction takes place:

   $$S^{2-}(aq) + 2Ag^+(aq) \rightleftharpoons Ag_2S(s). \quad \textit{Answer} : 1.35 \times 10^{-3} \text{ M}.$$

10.15 Recent Advances in Potentiometric Ion Sensors

For a long time, it has been assumed that the lower limit of detection of potentiometric sensors is around 10^{-6} M, which makes such sensors unsuitable for use in most environmental monitoring applications. However, recent theoretical and experimental research efforts led to the conclusion that the lower detection limit can be lessened to about 10^{-10} M by means of straightforward modifications in the sensor design and the operational procedure [77–81]. Such performances emerged from a careful analysis of the factors that determine the limit of detection.

The standard definition of the lower limit of detection assumes that this limit is determined by analyte competition with interfering ions for the recognition sites. However, for highly selective membranes, the calculated limit of detection may be in the femtomolar region. Such a low limit has been approached in the presence of a strong complexing agent that produces only a very low free ion activity in the solution even if the total ion concentration is much larger.

What actually restricts the limit of detection is the contamination of the sample by analyte ions originating from the internal solution of the sensor. At first glance, this is not possible because the sensor is operated at zero current and the ion flux through the membrane should also be zero. However, even under zero-current conditions, ion diffusion across the membrane is possible to some extent, either by codiffusion with a counterion or by letting an ion with the same charge sign diffuse in the opposite direction from the test-solution.

Therefore, the limit of detection can be improved by minimizing the analyte ion flux across the membrane. The ion flux can be minimized by adjusting the analyte activity in the internal solution to a value close to that in the sample. By this means, a low concentration gradient across the membrane is achieved. A very low ion activity in the internal solution is obtained by adding a strong complexing agent (such as EDTA) along with the ion in the normal concentration range. As a result, most of the ions in the internal solution are bound in a hydrophilic complexe and only the extremely low free ion activity contributes to the activity gradient across the membrane. The same effect can be obtained by including a solid ion-exchanger in the internal system of the sensor.

Alternatively, the analyte flux can be reduced by decreasing the diffusion coefficient in the membrane. This can be achieved by decreasing the plasticizer content, by using a polymer other than PVC as the membrane matrix, or by using covalently bound receptors that are not free to diffuse and carry the ion across the membrane.

Another method for minimizing the detection limit is based on operating the sensor under strong convection conditions. In this way, accumulation of analyte arising from the internal solution at the sample/membrane interface is prevented. Strong convection can be secured with a rotating sensor setup in which the membrane is placed off-center from the rotation axis. Similarly, good results have been obtained in wall-jet systems, in which a sample stream impinges with a high velocity on the sensor surface. This approach has proved to be effective not only with polymeric membrane sensors, but also with solid salt membranes. Thus, impressive detection limits for Cu^{2+} have been obtained with a jalpaite (Ag_3CuS_2) polycrystalline membrane operated under rotating sensor conditions.

Diffusion of the analyte from the internal solution is completely eliminated when an internal solid contact is set up. However, possible instability in the contact potential will result in random potential fluctuations that represent a noise signal and can drastically degrade the limit of detection. As already emphasized, conducting polymers represent very convenient materials for contacting the sensing membrane with a metallic conductor. However, if a solution layer forms between the metal and the polymer, this layer leads to a cell voltage instability. Such a problem has been mitigated by using solvent casting to form the conducting polymer layer. For a Pb^{2+} sensor this led to a detection limit in the nanomolar range.

As far as glass membranes are concerned, exceptionally low detection limits for Hg^{2+} in sea water have been obtained with a chalcogenide glass sensor [82].

Potentiometric sensor arrays combined with multivariate data analysis appears to be a promising approach to trace analysis by potentiometric ion sensors. Thus, an array of chalcogenide glass and polymeric membrane sensors enabled the determination of Cu^{2+}, Zn^{2+}, Pb^{2+}, and Cd^{2+} activity at the nanomolar level in artificial seawater [83].

Clearly, recent progress in potentiometric ion-sensor technology has led to devices and methods that show extremely low limits of detection that compare favorably with those characteristic of advanced spectrometric methods. This opens up new opportunities for ion sensor applications to environment monitoring, particularly in speciation analysis.

New applications of potentiometric ion sensors have arisen from operations under nonequilibrium conditions [84]. This operation method provides a means of solving analytical problems that cannot be approached by equilibrium potentiometry. Such problems are the determination of the total ion concentration in a sample or determination of polyanions. In the case of polyanion determination, the high charge of the analyte renders the Nernstian response sensitivity extremely low. Operation under nonequilibrium conditions alters substantially the response function and makes possible the potentiometric determination of such ions.

Questions and Exercises (Section 10.15)

1 What are the causes of the relatively poor limit of detection of potentiometric ion sensors?
2 What methods can be used to improve the limit of detection of polymer membrane ion sensors?
3 Indicate possible methods for improving the limit of detection of solid membrane ion sensors.

10.16 Potentiometric Gas Sensors

Potentiometric ion sensors can function as transducers in neutral-molecule sensors provided that the recognition of the neutral species results in ions being produced or consumed. The best-known application of this kind is the potentiometric carbon dioxide sensor (the Severinghaus sensor) which was initially developed in 1958 for the determination of the partial pressure of carbon dioxide in blood [85,86]. Determination of carbon dioxide is also of interest to food industry [87]. This sensor is based on the property of carbon dioxide to dissolve in water forming carbonic acid. By dissociation, carbonic acid gives rise to hydrogen cations and hydrogen carbonate anions. The activity of the hydrogen ion can be assessed by a suitable pH sensor (e.g., a pH glass electrode) and correlated with the carbon dioxide concentration in the sample. The pH sensor is immersed in the internal solution (int) separated from the external, test solution (ext), by a CO_2-permeable membrane (Figure 10.32). In order to form a galvanic cell, a reference electrode is immersed in the test-solution.

For a liquid sample, the overall process is described by the following sequence:

$$CO_2(aq)_{ext} \underset{(1)}{\rightleftharpoons} CO_2(g) \xrightarrow[(2)]{\text{Diffusion}} CO_2(aq)_{int} \xrightarrow[(3)]{H_2O} H^+ + HCO_3^- \tag{10.84}$$

The membrane can be made of a microporous hydrophobic polymer (e.g., polypropylene) that allows the gas molecule to cross it by effusion through the pores. Alternatively, the membrane can be formed of a compact polymer (such as silicone rubber) that dissolves the gas and lets it diffuse across from the sample to the internal solution. Carbon dioxide dissolution and carbonic acid dissociation occurs to a large extent within a thin solution layer located between the membrane and the pH sensor surface.

The response signal of this sensor is the potential difference between the internal reference electrode of the pH sensor glass electrode and the external reference electrode (E_{cell}). The response time depends on the time needed to attain the equilibrium and is hence determined by the rate of gas transfer through the membrane. In order to obtain a convenient response time, the membrane should be very thin. It should be about 0.1 mm thick if the transfer proceeds by effusion, but it can be thicker (0.1 to 0.3 mm) when the transfer is achieved by diffusion.

As is typical for the pH glass electrode, the cell voltage depends on the hydrogen ion activity in the internal solution as follows:

$$E_{cell} = E_0 - b \log a_{H^+} \tag{10.85}$$

where b is the response slope. In turn, the pH depends on the carbon dioxide activity that can be expressed by taking into account the following equilibrium constant of the overall process in (10.84):

$$K = \frac{a_{H^+} \cdot a_{HCO_3^-}}{a_{CO_2}} \tag{10.86}$$

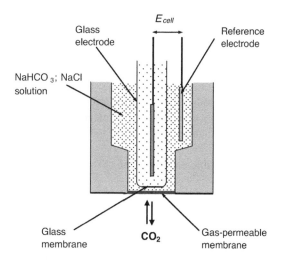

Figure 10.32   The lower section of a potentiometric CO_2 sensor. Adapted with permission from [88]. Copyright 2002 A. Bănică and F.G. Bănică.

Here, a_{H^+} and $a_{HCO_3^-}$ are ion activities in the internal solution and a_{CO_2} is the gas activity in the sample. On substituting into Equation (10.85) the a_{H^+} expression derived from (10.86), one obtains the response function as:

$$E_{cell} = \text{constant} + b \log \frac{a_{CO_2}}{a_{HCO_3^-}} \tag{10.87}$$

In order to obtain a nonequivocal response, the internal solution contains $NaHCO_3$ at a sufficiently high concentration so that the concentration of hydrogen carbonate ion remains practically constant. Under these conditions the response is a linear function of $\log a_{CO_2}$ with a slope similar to the glass electrode sensitivity. This gas sensor works equally well with both liquid and gas samples.

Carbon dioxide determination is of great interest to food industry and attempts at improving the design and developing new sensing procedures is a subject of current research interest [87,89]. In order to perform miniaturization and impart enhanced robustness, the pH glass electrode can be replaced by a more convenient sensor. By using an iridium/iridium dioxide pH sensor and a solid, gas-permeable film of nonaqueous polymer-gel electrolyte, a planar CO_2 sensor has been fabricated by the screen-print method [90].

A CO_2 sensor has been obtained by using a polypyrrole polymer matrix impregnated with hydrogen carbonate as the counterion [91]. Miniaturized design has allowed integration of multiple sensors (for oxygen, pH and carbon dioxide) within a catheter-type sensor [92].

Similar sensors for other reactive gases have been developed using suitable potentiometric ion sensors as transducers [3]. Thus, pH monitoring can be used in sulfur dioxide or ammonia sensors. A more selective ammonia sensor can be obtained by using an ammonium sensor as transducer. Hydrofluoric acid, hydrogen cyanide and hydrogen sulfur sensor have been obtained using fluoride, cyanide or sulfide ion sensors, respectively, as transducers. Nitrogen dioxide, which yields nitrate and nitrite on dissolution, can be detected by potentiometric monitoring of nitrate ions.

Due to the slow diffusion of the gas analyte through the membrane, potentiometric gas sensors can display a slow response and long recovery times at low gas concentrations. When using a pH sensor as the transduction element, the selectivity over other acidic or basic gases is poor.

Questions and Exercises (Section 10.16)

1 Write the chemical reactions occurring when the following gases are dissolved in water; NH_3; HCN, HF, H_2S, SO_2, NO_2. For each of the above gases, select an ion sensor as transducer and comment on the composition of the internal solution. Review the possible interferences.
2 Why should NaCl be present in the internal solution of the sensor shown in Figure 10.32?

10.17 Solid Electrolyte Potentiometric Gas Sensors

10.17.1 General Principles

Gas determination at elevated temperatures is a critical problem in combustion control and various industrial processes. Potentiometric gas sensors for such applications have been developed using solid electrolytes instead of aqueous electrolyte solutions [93].

Solid electrolytes are ionic crystalline compounds containing mobile ions that impart electrical conductivity. In a gas sensor, the solid electrolyte surface is coated with a thin, porous layer of a metal with catalytic properties, which also forms the electrode. In this way, triple phase boundaries are formed, as shown in Figure 10.33. At elevated temperatures, neutral gas molecules X, can undergo an electrochemical reaction leading to an ionic product $Y^{z+/-}$ that appears in the solid phase. This process is accompanied by electron exchange with the metal phase. At equilibrium, the potential of the metal electrode indicates the extent of the gas reaction and, hence, the analyte concentration in the gas phase.

A simple reaction of this type is the reversible electrochemical reaction of hydrogen gas that results in hydrogen ions in a proton-conducting solid:

$$H_2(g) + 2e^-(Pt) \rightleftarrows 2H^+(\text{protonic solid}) \tag{10.88}$$

This reaction occurs at the contact point of three phases: the gas, the platinum electrode and the solid electrolyte (Figure 10.33). At equilibrium, the electrical potential of the Pt electrode is determined by the ratio of H_2 activity in the gas phase and H^+ activity in the solid electrolyte.

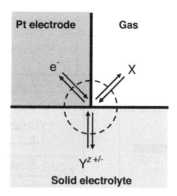

Figure 10.33 Electrochemical reaction of a gas X at the triple phase boundary gas-metal-solid electrolyte. $Y^{z+/-}$ is an ion produced by the electrochemical reaction of the gas molecule X.

Determination of oxygen in a gas mixture can be performed by means of a solid electrolyte containing mobile O^{2-} ions, such as zirconia (ZrO_2). In the pure state, the crystal lattice of this compound is formed of Zr^{4+} and O^{2-} ions located at fixed positions so that no ionic conductivity is present. However, if ZrO_2 is doped with a divalent (e.g., Ca^{2+}) or a trivalent (e.g., Y^{3+}) ion, O^{2-} vacancies appear in the lattice in order to preserve the charge balance. Consequently, O^{2-} ions can jump from their position in the lattice to an adjacent vacancy, thus providing electrical conductivity. If yttria is used as dopant, the material thus obtained is called yttrium-stabilized zirconia (in short, YSZ). The term "stabilized" indicates the fact that yttrium stabilizes the tetragonal crystal structure of ZrO_2.

Oxygen determination by means of a YSZ-based sensor is made possible by the following electrochemical reaction:

$$O_2(\text{gas}) + 4e^-(\text{Pt}) \underset{\longleftarrow}{\overset{T>800\,°C}{\longrightarrow}} 2O^{2-}(\text{YSZ}) \tag{10.89}$$

A high reaction rate in each direction is important in order to provide fast equilibration of the chemical system, which secures a short response time. This condition is fulfilled at temperatures over 800 °C. This is not a drawback because the sensor is designed for applications involving high-temperature samples, such as exhaust gases or molten metals and alloys.

In order to obtain a gas sensor, an indicator half-cell, such as that in Figure 10.33 should be combined with a reference half-cell to form a galvanic cell. The reference half-cell can be similar to the indicator one, except for the fact that the content of the analyte gas is held constant in the adjacent gas phase. The electromotive force of this cell depends on the gas concentration according to the Nernst equation.

10.17.2 Solid Electrolyte Potentiometric Oxygen Sensors

An oxygen sensor can be produced by combining two half-cells in order to obtain the galvanic cell below, which consists of a reference and an indicator half-cell:

$$\begin{array}{cc} O_2(p_{O_2})_{\text{ref}}, \text{Pt}|O^{2-}(\text{YSZ})|\text{Pt}, O_2(p_{O_2}) \\ \text{Reference half-cell} \qquad \text{Indicator half-cell} \end{array} \tag{10.90}$$

p_{O_2} indicates the partial pressure of oxygen in the gas phase. In electrochemistry, such a cell is called a concentration cell because the sole difference between the half-cells resides in the concentration of the electrochemically active species. The electromotive force of this cell can be derived from thermodynamic considerations and is given by the following Nernst-type equation [94,95]:

$$\text{EMF} = \frac{RT}{4F} \ln \frac{p_{O_2}}{\left(p_{O_2}\right)_{\text{ref}}} \tag{10.91}$$

As the partial pressure of oxygen in the reference half-cell is constant, the actual response function is:

$$\text{EMF} = E_0 + \frac{RT}{4F} \ln p_{O_2} \tag{10.92}$$

where the constant term is $E_0 = -(RT/4F)\ln\left(p_{O_2}\right)_{\text{ref}}$. However, the actual cell voltage differs from the EMF due to additional terms arising from the secondary effects. Therefore, the sensor response function in terms of measured cell voltage (V) is:

$$V = V_0 + \frac{RT}{4F} \ln p_{O_2} = V_0 + 0.576 \frac{RT}{F} \log p_{O_2} \tag{10.93}$$

where the V_0 term includes the E_0 constant as well as contributions from secondary effects.

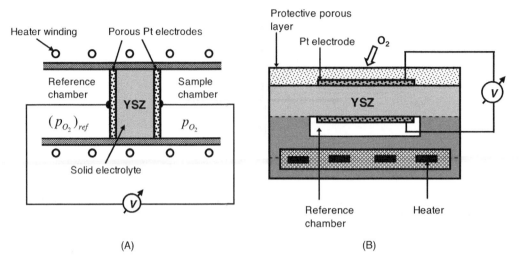

Figure 10.34 (A) Basic structure of a solid electrolyte oxygen sensor. A temperature probe (not shown) is also included in the sensor design. (B) Cross section of a potentiometric oxygen sensor in the planar configuration

Commonly used reference gases are pure oxygen and air. A rigorous control of the oxygen partial pressure in the reference chamber can be achieved by means of a sealed-in mixture of metal and metal oxide that forms a reference electrode [96].

Equation (10.93) indicates that a decade change in the partial pressure of oxygen results in a change in EMF of approximately 50 mV at 1000 K. This kind of response occurs over a very broad range of oxygen partial pressure. As the response depends on temperature, this parameter should be rigorously controlled; to this end, a temperature probe is included in the sensor design. Temperature can also affect other parameters included in the V_0 term in Equation (10.93).

The limits of the response range are determined by possible electronic conduction at extreme oxygen contents in the sample. At very low oxygen content, electron conductivity becomes significant, while at high oxygen content, hole conduction plays an important role. In both cases, the response deviates from Equation (10.93). In addition, interferences can be produced by other gases that can react with either the oxygen gas (thus depleting it at the sensor surface) or with O^{2-} ions in the solid electrolyte. Thus, oxygen depletion can be caused by catalytic oxidation of NO, CO, or hydrocarbon. The same compounds can undergo electrochemical reactions that modify the O^{2-} concentration in the solid electrolyte, such as:

$$CO + O^{2-} \rightleftarrows CO_2 + 2e^-$$ (10.94)

$$NO + O^{2-} \rightleftarrows NO_2 + 2e^-$$ (10.95)

Under suitable conditions, the error in oxygen partial pressure is not greater than 2%.

A possible configuration of potentiometric oxygen sensors is shown in Figure 10.34A. It includes a YSZ membrane coated on each side with a porous platinum layer or a platinum net. The operating temperature is controlled by means of an electrical heater winding and measured by a temperature probe (not shown).

In addition to the sensor design in Figure 10.34, other configurations have been implemented [97]. A planar configuration (Figure 10.34B) is advantageous due to its compact form, small size, and compatibility with thick-film microfabrication technology [97].

10.17.3 Applications of Potentiometric Oxygen Sensors

The most common application of the potentiometric oxygen sensor is combustion control in motor vehicles in order to optimize fuel consumption and avoid air pollution by carbon monoxide [98–100]. The oxygen content in the exhaust gas indicates the degree of completion of the combustion reaction. From the chemical standpoint, the combustion process can be characterized by the nondimensional combustion parameter λ defined in Equation (10.96), which indicates to what extent the combustion of an air–fuel mixture approaches stoichiometric conditions. It is defined as:

$$\lambda = \frac{\text{Actual(air/fuel)volume ratio}}{\text{Soichiometric(air/fuel)volume ratio}}$$ (10.96)

A stoichiometric mixture is characterized by $\lambda = 1$; for a fuel-lean mixture, $\lambda > 1$, whereas for a fuel-rich mixture, $\lambda < 1$. Figure 10.35 demonstrates the typical response of the potentiometric oxygen sensor in terms of the combustion

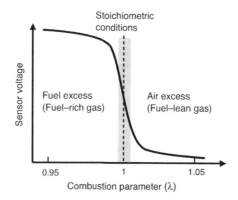

Figure 10.35   Typical response of a lambda sensor.

parameter. Near the stoichiometric condition there is an abrupt drop in the sensor response due to a several orders of magnitude change in the oxygen concentration. Therefore, in practice, this kind of sensor works with a binary response, assuming only two possible states that correspond to either air excess or fuel excess. In order to achieve complete combustion, a slight excess of air should be maintained. In the case of complete combustion, no carbon monoxide is present in the exhaust gas.

Also known as the *lambda sensor*, this device was first introduced in 1961 by Peters and Möbius [101] and Weissbart and Ruka [102] and became commercially available by 1965. It is currently produced on an industrial scale and is widely used due to the strict regulations concerning traffic pollution.

10.17.4   Types of Solid Electrolyte Potentiometric Gas Sensors

Solid electrolyte potentiometric gas sensors can be classified according to the relation between the gas analyte and the ions present in the structure of the solid electrolyte. Three types of sensors can be distinguished, as summarized in Figure 10.36.

i. *Type I sensors* (Figure 10.36A) are characterized by the fact that the mobile ion in the solid electrolyte is in equilibrium with the analyte via an electrochemical reaction. The oxygen sensor shown in Figure 10.34A is a typical sensor in this class.

   Another type I gas sensor is the hydrogen sensor based upon reaction (10.88) and making use of a proton-conducting solid electrolyte such as hydrogen uranyl phosphate tetrahydrate ($UO_2HPO_4 \cdot 4H_2O$). In this case, reduction of the hydrogen gas results in hydrogen ions (protons), which are the charge carriers in the electrolyte. Remarkably, this solid electrolyte displays its highest protonic conductivity at temperatures near the normal one (282–323 K) [103].

   Type I gas sensors for water vapor have been developed using protonic conductor solid electrolytes such as $SrCeO_3$ or $BaCeO_3$ [104]. When using platinum electrodes, the electrochemical reaction proceeds as follows:

$$H_2O(g) \rightleftarrows 2H^+(s) + \frac{1}{2}O_2(g) + 2e^-(Pt) \tag{10.97}$$

   Determination of dissolved hydrogen in molten aluminum can be performed by sensors based on proton-conducting perovskite oxides such as indium-doped $CaZrO_3$ [103].

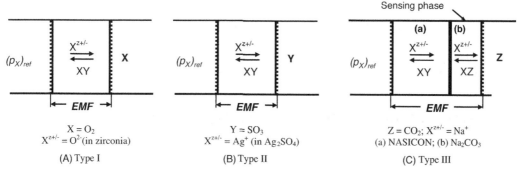

Figure 10.36   Various types of potentiometric gas sensor based on solid electrolytes. XY = solid electrolyte; $X^{z+/-}$ = mobile ion (charge carrier). Adapted with permission from [106]. Copyright 1992 Kluwer Academic Publishers.

ii. *Type II sensors* are based on electrochemical reactions that convert the gas analyte into an ion similar to the fixed ion in the solid electrolyte (Figure 10.36B). For example, in the presence of oxygen, sulfur dioxide is converted into sulfate ion, which is the fixed ion in the Ag^+-conducting silver sulfate electrolyte. This process occurs at the indicator half-cell. The reference half-cell can be based on oxidation of a pure silver electrode to form the mobile Ag^+ ion:

$$Ag(electrode) \rightleftarrows Ag^+(electrolyte) + e^-(electrode) \tag{10.98}$$

Similarly, an alkali-metal carbonate solid electrolyte responds to CO_2. Another example is the type II chlorine sensor that makes use of Ag^+-conducting silver chloride as solid electrolyte and is based on the following galvanic cell which includes an all-solid reference electrode [105]:

$$\underset{\text{Reference}}{Ag|Ag^+} \underset{\text{Indicator}}{(AgCl)|Cl_2, Pt} \tag{10.99}$$

This sensor functions at room temperature. Analogously, the silver nitrate solid electrolyte responds to nitrogen dioxide.

Application of type II sensors is restricted by the limited number of suitable solid ionic conductors.

iii. *Type III sensors* include two solid electrolyte phases in contact, each phase containing the same type of mobile ion but different fixed ions (Figure 10.36C). The sensing phase (b) interacts with the gas analyte to produce the fixed ion in this phase. The other phase (a) forms, along with a metal electrode and a reference gas phase, the reference half-cell. An example of a type III gas sensor is the CO_2 sensor obtained by means of sodium carbonate (a Na^+-ionic conductor) as the sensing phase (b). The phase (a) can be made of NASICON ($Na_3Zr_2Si_2PO_{12}$) or sodium beta alumina, which is a nonstoichiometric sodium aluminate [107]. Both NASICON and β-alumina are sodium ion conductors. The sensing and reference reactions occur as follows at about 900 °C:

$$\text{Sensing(phase(b))}: \quad 2Na^+ + CO_2 + \frac{1}{2}O_2 + 2e^- \rightleftarrows Na_2CO_3 \tag{10.100}$$

$$\text{Reference(phase(a))}: \quad 2Na^+ + \frac{1}{2}O_2 + 2e^- \rightleftarrows Na_2O(\text{in NASICON}) \tag{10.101}$$

In this arrangement, Na^+ is the common, mobile ion. The reaction of carbon dioxide (10.100) alters the concentration of Na^+ in phase (b) and hence, this ion concentration in phase (a). As a result, the cell voltage changes owing to the implication of the Na^+ ion in reaction (10.101) that occurs at the reference half-cell. Consequently, the EMF is dependent on the carbon dioxide concentration according to the Nernst equation:

$$EMF = E_C + \frac{RT}{2F} \ln p_{CO_2} \tag{10.102}$$

where E_C is a constant term and p_{CO_2} is the partial pressure of CO_2 in the sample gas.

Similarly, a solid metal nitrate or nitrite salt can be used as the sensing phase for NO_2 in combination with NASICON as reference phase. However, much better performances have been obtained with a sensor composed of nickel oxide as sensing phase, in conjunction with YSZ as reference phase. Both electrolytes share the same mobile ion, namely O^{2-}, and nitrogen dioxide reacts as follows:

$$NO_2 + NiO + 4e^- \xrightleftharpoons{800 \,°C} NO + Ni + 2O^{2-} \tag{10.103}$$

Sulfur oxides can be determined by means of sodium sulfate as sensing phase in conjunction with a proton-conducting reference phase [103].

It is clear that type III potentiometric gas sensors offer a large variety of applications using a given solid electrolyte as the phase (a) in combination with various metal salts as sensing phases.

10.17.5 Mixed Potential Potentiometric Gas Sensors

It has already been mentioned that certain gases that can undergo electrochemical reactions at the triple phase boundary affect the response of the potentiometric oxygen sensor and cause the response to deviate from that predicted by the Nernst equation. This arises because, in addition to the oxygen electrochemical reaction, additional reactions such as (10.94) or (10.95) contribute to the establishment of a certain electron density at the metal electrode and thus affect its electrical potential. Thus, oxygen reduction depletes the electrode of electrons, while carbon monoxide oxidation supplies electrons to the metal. At equilibrium, the electrode potential assumes a value determined by the extent of each of the above reactions and the electrode potential thus developed is a *mixed potential* [94,108].

The effect of each reaction on the mixed potential is determined by its reaction rate. In turn, the reaction rate depends on three factors: gas concentration, temperature, and the nature of the metal catalyst. Optimal selection of these factors can render a given solid electrolyte suitable for detection of a particular gas with minor interferences from other gases present in the sample. For example, if gold is used as electrode catalyst, carbon monoxide oxidation (reaction (10.94)) is much faster than oxygen reduction (reaction (10.89)) and the sensor responds reliably to carbon monoxide at constant oxygen content in the gas mixture. The mixed potential is in this case a logarithmic function of the partial pressure of carbon monoxide. Such a response has been noted when the difference between equilibrium potentials of the pertinent reactions is small. Conversely, a linear dependence on the carbon monoxide content has been found when the electrode was made of platinum, which allows oxygen reduction to occur faster.

The effect of the operating temperature can be illustrated by the case of the mixed potential sensor for nitrogen oxide based on the Pt/YSZ system. At temperatures above 800 °C, oxygen reaction is very fast and the mixed potential is determined mainly by this gas. Conversely, at temperatures below 600 °C, the oxygen reaction is much slower and the mixed potential depends on the nitrogen oxide concentration.

10.17.6 Outlook

Solid-state potentiometric gas sensors have been developed as a consequence of the demand for monitoring the content of hazardous compounds in hot gases. These sensors make use of a solid electrolytes containing ions that are in equilibrium with the gas phase through an electrochemical reaction that occurs at the triple contact phase gas–solid electrolyte–metal electrode. In this reaction, the metal electrode functions as an electron sink. In order to obtain a sensor, two systems of this kind can be assembled to form a galvanic cell. The above-mentioned reactions give rise to a nonuniform distribution of mobile ions across the solid electrolyte, which determine the electromotive force of the cell. The electromotive force is related to the gas content by a Nernst-type equation.

The best-known sensor of this type is the oxygen sensor used to monitor the combustion process in car engines. It is the prototype of the type I sensor. Other detection schemes have been designed in order to develop type II and III solid electrolyte potentiometric gas sensors.

If a reducing gas is present along with oxygen, this gas undergoes oxidation with electrons being released to the electrode. In this case, the electrode reaction depends on two processes: electron donation by the reducing gas and electron withdrawing by oxygen. Proper selection of the catalyst and the operating conditions allows the reducing gas to be determined. Owing to the involvement of two gases in the potential-determining processes, the name mixed potential sensor is used to denote this kind of sensor. Very often, the response of mixed potential sensors deviates from the Nernst equation.

Application of solid electrolytes allowed the development of sensors of particular relevance to the monitoring of air quality as well as for industrial applications. Because these sensors operate at high temperatures, they can be designed so as to allow dissolved gases to be determined in molten metals or alloys.

The principles and applications of solid electrolyte gas sensors are amply reviewed in the literature (see, e.g., see above references and also [109–111]).

Questions and Exercises (Section 10.17)

1 What are solid electrolytes and what properties of these materials are applied in gas sensing?
2 What is the mechanism of gas sensing by means of solid electrolytes?
3 Write several gas reactions that can be employed in type I gas sensors.
4 Draw schematically the configuration of a solid electrolyte oxygen sensor in the flow-through and the thimble-type configuration (see ref. [96] for orientation).
5 Give an account of the principles of solid electrolyte oxygen sensor application in combustion control.
6 Write the reactions involved in nitrogen oxide determination by means of a type II solid electrolyte sensor.
7 Sketch the mechanism of sulfur dioxide determination by means of a type III solid electrolyte sensor using sodium sulfate as sensing phase.
8 Write the reaction involved in a mixed potential sensor for nitrogen monoxide and comment on the possibilities of adjusting the selectivity to this gas.

10.18 Potentiometric Biocatalytic Sensors

Many enzyme-catalyzed reactions involve an ionic species either as reactant or product. Therefore, a potentiometric enzyme sensor can be obtained by integrating a biocatalytic layer with a potentiometric ion sensor. In such a device, the ion sensor functions as a transducer that monitors the concentration of the relevant ion. The enzyme can be immobilized over the ion sensor surface by crosslinking within nylon netting followed by entrapment under a dialysis membrane. Alternatively, the enzyme can be attached by covalent immobilization.

Further, certain enzymatic reaction yield gaseous products, such as CO_2 or NH_3 that makes possible the application of potentiometric gas probes for performing the transduction.

Due to the simple configuration and wide availability of potentiometric sensors, a great number of potentiometric transduction schemes for enzymatic sensors have been developed [112–114]. A good example is the application of urease in potentiometric urea sensors. The enzymatic hydrolysis of urea leads to carbon dioxide and ammonia (Figure 3.11). At pH values around 7, both these products undergo protonation to HCO_3^- and NH_4^+, respectively. Therefore, the transduction can be performed either by ion sensors (H^+ or NH_4^+) or by gas sensors (CO_2 or NH_3).

The configuration of a gas probe-based enzyme sensor is shown in Figure 10.37. The potentiometric gas probe consists of a potentiometric ion sensor 5 (a pH glass electrode in this case) in contact with a thin layer of internal solution 4 confined by means of a gas-permeable membrane 3. A biocatalytic layer is immobilized below and is separated from the solution by a semipermeable membrane 1. The substrate crosses the external membrane 1 and undergoes enzymatic conversion within the biocatalytic layer 2. The gas product thus formed diffuses across membrane 3 and, by reaction within the internal solution, modifies the pH in the thin layer adjacent to the pH sensor surface and, therefore, the response of the pH sensor. Consequently, the voltage developed between the two reference electrodes is dependent on the substrate concentration in the sample.

As many enzyme reactions involve hydrogen ions, the use of a pH glass electrode as transducer is probably the most straightforward alternative. However, in order to detect the pH change produced by enzymatic reaction, a buffer system with a low buffer capacity should be used (e.g., 1 mM phosphate buffer). This may bring about some difficulties in sensor operation as possible pH variation may affect enzyme activity. However, this method has proved useful, particularly with both oxidase- and hydrolase-type enzymes. An example of this kind is the potentiometric penicillin sensor based on penicillase (beta-lactamase) that catalyzes the hydrolysis of the substrate to penicilloic acid. Hence, the pH increases in proportion to the substrate concentration. Analogously, acetylcholine sensors have been developed using acetylcholine esterase that gives rise to acetic acid by substrate hydrolysis.

Oxidase-type enzymes that give rise to hydrogen peroxide as a side product can be used in combination with an iodide ion sensor provided that the iodide ion is present in the solution. In the presence of horseradish peroxidase, hydrogen peroxide oxidizes iodide ion to iodine and the transducer monitors the decrease of the iodide concentration at the surface. Sensors for glucose and amino acids based on this principle have been developed. An alternative approach relies on the reaction of hydrogen peroxide with 4–fluorophenol that releases fluoride ions. A potentiometric fluoride sensor can therefore be used as the transducer in this case.

The response function of a potentiometric enzyme sensor can be derived using the results of the approach in Chapter 4. The response function will be derived next for the case of external diffusion control and first-order kinetics. In this case, the rate-determining step is the substrate diffusion across the external membrane and the product is uniformly distributed within the biocatalytic layer. If the product does not undergo further conversion or is totally converted to another detectable species, the response of the potentiometric transducer is determined by the product

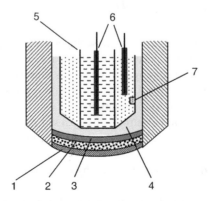

Figure 10.37 A biocatalytic potentiometric sensor based on a gas sensor as transducer element. (1) External membrane; (2) biocatalytic layer; (3) gas-permeable membrane; (4) gas sensor internal solution; (5) combination pH glass electrode; (6) reference electrodes; (7) liquid junction. Adapted with permission from [88]. Copyright 2002 A. Bănică and F.G. Bănică.

concentration p_e which is given by Equation (4.17). Assuming that the potentiometric transducer provides a Nernstian response to the product concentration with the sensitivity b:

$$E_{\text{cell}} = E_0 + b \log p_{\text{e}} \tag{10.104}$$

the response to the substrate concentration is:

$$E_{\text{cell}} = E_0 + b \log \left(\frac{k_{\text{m,S}}}{k_{\text{m,P}}} \frac{\alpha}{\alpha + 1} \right) + b \log s \tag{10.105}$$

Here, $k_{\text{m,S}}$ and $k_{\text{m,P}}$ are the mass-transfer coefficients of the substrate and product, respectively, in the external membrane, s is the substrate concentration in the test solution and α is the substrate modulus for external diffusion. The α parameter is directly proportional to the enzyme loading (that is, the product of the enzyme concentration and the thickness of the biocatalytic film). At the same time, α is inversely proportional to the mass transfer coefficient of the substrate in the external membrane. Therefore, the α value can be adjusted by the selection of the enzyme loading and the characteristics of the external membrane (thickness and permeability to the substrate).

According to Equation (10.105), the sensor sensitivity relative to substrate concentration is determined by the sensitivity of the ion sensor used as transducer. However, the second term in this equation is a function of the α parameter and, therefore, is not sufficiently reliable. The response function assumes a more convenient form if $\alpha \gg 1$; in this case, the term in α vanishes and the response becomes independent of the parameters of the enzyme layer.

If the enzyme sensor is devised for applications in inhibitor determination, it is convenient to design it such as to function under the zero-order kinetic regime in order to secure maximum sensitivity.

Potentiometric enzyme sensors are simple and inexpensive devices that can be fabricated readily even in the disposable form [115]. The range of applications encompasses the determination of various substrate compounds for diagnostic purposes as well as monitoring of toxic species in the environment by enzyme inhibition.

Questions and Exercises (Section 10.18)

1 Review the underlying principles of potentiometric ion and gas sensors application as transducers in enzymatic sensors.
2 Comment on the function of each component of the biocatalytic sensor in Figure 10.37.
3 Derive the response function of a potentiometric enzymatic sensor for the case of internal diffusion and comment on the effect of relevant physicochemical and geometrical parameters.
   Hint: see Section 4.3.

10.19 Potentiometric Affinity Sensors

Potentiometric transduction in affinity sensors can be achieved either by means of redox electrodes or by means of potentiometric ion or gas sensors [116–118].

A very simple approach to potentiometric transduction in immunosensors rests on redox electrodes combined with a redox enzyme as label. A platinum electrode in contact with a solution containing a redox couple develops an electrode potential according to the Nernst equation (10.21). Substrate oxidation leads to a change in the molar ratio of the oxidized/reduced forms of the redox couple that induces a change in the electrode potential. The sensor can be assembled by immobilizing the receptor at the surface of a platinum electrode. In a competition assay, the enzyme label is brought to the electrode surface by the reaction of the receptor with the enzyme-labeled analyte analog, while in the sandwich assay the enzyme is held by the formation of a ternary complex involving a labeled entity.

Ion-selective membranes can be used as transducer in conjunction with enzyme labels that catalyze ion-releasing reactions. Gaseous products, such as ammonia or carbon dioxide can also function in the transduction process when the sensing layer is build up at the surface of a potentiometric gas sensor.

Immunorecognition, as well as other affinity interactions, involves charged particles and the recognition event results in a partial neutralization of the charges on the reactants. The change in the electric charge density can be detected by means of a PVC-coated electrode to which the receptor is appended [119]. This is an example of a potentiometric label-free immunoassay based on the polyelectrolyte character of the involved reactants.

An interesting approach to potentiometric immunosensors is based on ion channels formed by the interaction of the immunocomplex with a complementary species supplied by a blood serum sample [120]. In this approach, a marker-ion solution is entrapped in liposomes that act as a label integrated with the receptor antigen. The analyte–receptor complex forms at the liposome surface and interacts further with the complementary species. The ternary complex thus formed gives rise to ion channels that allow the marker ion to leak out and makes it available for detection by an ion sensor. This method has been designed for assays with a very small sample volume (down to 1 μL) in the form of a thin solution layer. A protocol for manufacturing and assay applications of such a sensor is available in ref. [121].

Questions and Exercises (Section 10.19)

1 Draw schematically the main steps in a potentiometric sandwich immunoassay based on a redox electrode and write the relevant electrochemical reactions.
2 Review the principle of signal amplification in an immunosensor using an enzyme label and a polyanion sensor as transducer.
3 Draw a sketch showing the principle of a potentiometric immunosensor based on ion-channel formation in liposome labels. Indicate the mechanism of signal amplification in this kind of sensor.
    Hint: see ref. [120].

10.20 Summary

Potentiometric ion sensors are based on specific interactions between a selective membrane and the analyte ion that leads to the alteration of the charge density at the membrane surface and, consequently, a change in the membrane potential. The membrane potential is monitored by means of two reference electrodes placed in the solution on each side of the membrane. By keeping the ion concentration constant in one of the two solutions, the measured potential difference depends only on the ion concentration in the second solution. The potential difference is a logarithmic function of the analyte concentration, this being called a Nernstian response.

Ion-responsive membranes can be prepared in the solid-state form (e.g., sparingly soluble salts or glass materials) or can consist of a polymeric matrix embedding ionic or neutral ion receptors. Solid membranes are very rugged, whereas polymeric membranes are less resilient but allow achieving excellent selectivity and limit of detection.

Potentiometric ion sensors are simple, robust, and inexpensive analytical devices, well suited for either laboratory of field applications as well as for continuous concentration monitoring and *in vivo* chemical analysis. The accuracy of potentiometric ion sensors, which is commonly of ±2 to 3%, compares favorably with analytical techniques that require more complex and expensive instrumentation.

It should be made clear that, as far as the field of applications is concerned, potentiometric ion sensors have competition from other widely used analytical methods such as atomic spectrometry, mass spectrometry and ion chromatography. Such methods are superior to potentiometry since they allow for simultaneous determination of more than one analyte. In addition, the limits of detection of the above methods are generally better than that of standard potentiometry. That is why potentiometric ion sensors appeared early in their development not to be suitable for monitoring toxic ions in biological and environment samples. However, recent progress in this field led to developments that allowed potentiometric sensor to achieve detection limits that compare favorably with those characteristic of mass spectrometry.

Further, the problem of multianalyte potentiometric determination can be addressed by means of potentiometric sensor arrays. Although the degree of generality achieved by contending methods cannot be achieved in this way, potentiometric sensor arrays represent the method of choice in certain specialized applications.

Potentiometric sensor arrays can provide a relief of the problems of interferences arising from crossreactivity. Indeed, data from an array of multiresponsive sensors can be processed so as to extract the unbiased concentration of each ion of interest.

As far as ruggedness is concerned, best performances are achieved by solid-state membranes. However, the selectivity of such sensors is not of the best. Polymer membranes that can be made very selective by using suitable ion receptors are less resilient. Nevertheless, this apparent drawback can be offset by including replaceable membranes in the sensor design.

As certain biochemical and nonbiochemical reactions produce ions, potentiometric ion sensors can function as transduction elements in various kinds of sensor based on ion-generating reactions. This principle is used in the development of gas sensors, biocatalytic sensors, and immunosensors.

As ions are found in both environmental and biological samples, as well as in certain industrial products, the potential field of application of ion sensors is very broad. Comprehensive surveys of potentiometric ion sensor application in various areas can be found in the literature as follows: monitoring of environment pollution [83,122]; ion determination in marine chemistry [123]; biomedical sciences [11,124,125]; pharmaceutical analysis [126];

petroleum industry [127]. A spectacular example of field analysis application is the *in situ* determination of ions leached from Mars planet soil by an automatic wet laboratory fitted with an array of potentiometric ion sensors [128]

Last but not least, the large volume of expertise developed in the framework of potentiometric ion sensor research has proved invaluable in the development of other types of ion sensors such as field effect transistor ion sensors (Chapter 11) and optical ion sensors (Chapter 19).

In addition to the applications outlined above, potentiometry principles have allowed development of high-temperature gas sensors based on solid electrolytes as gas-sensitive materials. Sensors of this kind are widely used in monitoring burning processes as well as for the determination of various hazardous gases in the atmosphere.

References

1. Morf, W.E. (1981) *The Principles of Ion-Selective Electrodes and of Membrane Transport*, Elsevier, Amsterdam.
2. Koryta, J. and Štulík, K. (1983) *Ion-Selective Electrodes*, Cambridge University Press, Cambridge.
3. Lakshminarayanaiah, N. (1976) *Membrane Electrodes*, Academic Press, New York.
4. Fawcett, W.R. (2004) *Liquids, Solutions, and Interfaces: From Classical Macroscopic Descriptions to Modern Microscopic Details*, Oxford University Press, New York, pp. 383–505.
5. Cammann, K. (1979) *Working with Ion-Selective Electrodes: Chemical Laboratory Practice*, Springer, Berlin.
6. Covington, A.K. (1979) *Ion-Selective Electrode Methodology*, CRC Press, Boca Raton, Fla.
7. Cammann, K. and Galster, H. (1996) *Working with Ion-Selective Electrodes (in German)*, Springer, Berlin.
8. Ma, T.S. and Hassan, S.S.M. (1982) *Organic Analysis using Ion-Selective Electrodes*, Academic Press, London.
9. Cosofret, V.V. (1977) *Applications of Ion-Selective Membrane Electrodes in Organic Analysis*, Ellis Horwood, Chichester.
10. Havas, J. (1985) *Ion- and Molecule-Selective Electrodes in Biological Systems*, Springer, Berlin.
11. Spichiger-Keller, U.E. (1998) *Chemical Sensors and Biosensors for Medical and Biological Applications*, Wiley-VCH, Weinheim.
12. Buck, R.P. and Lindner, E. (1994) Recommendations for nomenclature of ion-selective electrodes – (IUPAC Recommendations 1994). *Pure Appl. Chem.*, **66**, 2527–2536.
13. Bobacka, J., Ivaska, A., and Lewenstam, A. (2008) Potentiometric ion sensors. *Chem. Rev.*, **108**, 329–351.
14. Bakker, E. and Pretsch, E. (2007) Modern potentiometry. *Angew. Chem. Int. Edit.*, **46**, 5660–5668.
15. Buck, R.P. and Lindner, E. (2001) Tracing the history of selective ion sensors. *Anal. Chem.*, **73**, 88A–97A.
16. Shinwari, M.W., Zhitomirsky, D., Deen, I.A. *et al.* (2010) Microfabricated reference electrodes and their biosensing applications. *Sensors*, **10**, 1679–1715.
17. Kahlert, H. (2010) Reference electrodes, in *Electroanalytical Methods: Guide to Experiments and Applications* (ed. F. Scholz), Springer-Verlag, Berlin, pp. 291–308.
18. Bakker, E., Pretsch, E., and Buhlmann, P. (2000) Selectivity of potentiometric ion sensors. *Anal. Chem.*, **72**, 1127–1133.
19. Bakker, E., Buhlmann, P., and Pretsch, E. (1997) Carrier-based ion-selective electrodes and bulk optodes. 1. General characteristics. *Chem. Rev.*, **97**, 3083–3132.
20. Umezawa, Y., Umezawa, K., and Sato, H. (1995) Selectivity coefficients for ion-selective electrodes – recommended methods for reporting $K_{A,B}^{Pot}$ values. *Pure Appl. Chem.*, **67**, 507–518.
21. Umezawa, Y., Buhlmann, P., Umezawa, K. *et al.* (2000) Potentiometric selectivity coefficients of ion-selective electrodes Part I. Inorganic cations. *Pure Appl. Chem.*, **72**, 1851–2082.
22. Gadzekpo, V.P.Y. and Christian, G.D. (1984) Determination of selectivity coefficients of ion-selective electrodes by a matched-potential method. *Anal. Chim. Acta*, **164**, 279–282.
23. Umezawa, Y., Umezawa, K., Buhlmann, P. *et al.* (2002) Potentiometric selectivity coefficients of ion-selective electrodes. Part II. Inorganic anions (IUPAC technical report). *Pure Appl. Chem.*, **74**, 923–994.
24. Umezawa, Y., Buhlmann, P., Umezawa, K. *et al.* (2002) Potentiometric selectivity coefficients of ion-selective electrodes Part III. Organic ions – (IUPAC technical report). *Pure Appl. Chem.*, **74**, 995–1099.
25. Lindner, E. and Umezawa, Y. (2008) Performance evaluation criteria for preparation and measurement of macro- and microfabricated ion-selective electrodes (IUPAC technical report). *Pure Appl. Chem.*, **80**, 85–104.
26. Bates, R.G. (1973) *Determination of pH: Theory and Practice*, Wiley, New York.
27. Galster, H. (1991) *pH Measurement: Fundamentals, Methods, Applications, Instrumentation*, VCH, Weinheim.
28. Tanaka, K. (2001) Chalcogenide glasses, in *The Encyclopedia of Materials: Science and Technology* (eds K.H.J. Buschow, R.W. Cahn, M.C. Flemings, B. Ilschner, E.J. Kramer and S. Mahajan), Elsevier, Amsterdam.
29. Vassilev, V.S. and Boycheva, S.V. (2005) Chemical sensors with chalcogenide glassy membranes. *Talanta*, **67**, 20–27.
30. Schoning, M.J. and Kloock, J.P. (2007) About 20 years of silicon-based thin-film sensors with chalcogenide glass materials for heavy metal analysis: Technological aspects of fabrication and miniaturization. *Electroanalysis*, **19**, 2029–2038.
31. Vlasov, Y.G., Bychkov, E.A., and Legin, A.V. (1994) Chalcogenide glass chemical sensors – research and analytical applications. *Talanta*, **41**, 1059–1063.
32. Steed, J.W. and Atwood, J.L. (2009) *Supramolecular Chemistry*, Wiley, Chichester.
33. Johnson, R.D. and Bachas, L.G. (2003) Ionophore-based ion-selective potentiometric and optical sensors. *Anal. Bioanal. Chem.*, **376**, 328–341.
34. Buhlmann, P., Bakker, E., and Pretsch, E. (1998) Carrier-based ion-selective electrodes and bulk optodes. 2. Ionophores for potentiometric and optical sensors. *Chem. Rev.*, **98**, 1593–1687.

35. Sanchez, J. and del Valle, M. (2005) Determination of anionic surfactants employing potentiometric sensors – A review. *Crit. Rev. Anal. Chem.*, **35**, 15–29.

36. Craggs, A., Moody, G.J., and Thomas, J.D.R. (1974) PVC matrix membrane ion-selective electrodes – construction and laboratory experiments. *J. Chem. Educ.*, **51**, 541–544.

37. Eugster, R., Rosatzin, T., Rusterholz, B. *et al.* (1994) Plasticizers for liquid polymeric membranes of ion-selective chemical sensors. *Anal. Chim. Acta*, **289**, 1–13.

38. Pearson, R.G. (1967) Hard and soft acids and bases. *Chem. Br.*, **3**, 103–107.

39. Pearson, R.G. (1997) *Chemical Hardness*, Wiley-VCH, New York.

40. Antonisse, M.M.G. and Reinhoudt, D.N. (1998) Neutral anion receptors: Design and application. *Chem. Commun.*, 443–448.

41. Snowden, T.S. and Anslyn, E.V. (1999) Anion recognition: synthetic receptors for anions and their application in sensors. *Curr. Opin. Chem. Biol.*, **3**, 740–746.

42. Bianchi, A., Bowman-James, K., and García-España, E. (1997) *Supramolecular Chemistry of Anions*, Wiley-VCH, New York.

43. Upreti, P., Metzger, L.E., and Bühlmann, P. (2004) Glass and polymeric membrane electrodes for the measurement of pH in milk and cheese. *Talanta*, **63**, 139–148.

44. Lhotak, P. (2005) Anion receptors based on calixarenes, in *Anion Sensing*, Springer-Verlag Berlin, Berlin, pp. 65–95.

45. Siswanta, D., Nagatsuka, K., Yamada, N. *et al.* (1996) Structural ion selectivity of thia crown ether compounds with a bulky block subunit and their application as an ion-sensing component for an ion-selective electrode. *Anal. Chem.*, **68**, 4166–4172.

46. Cram, D.J. (1986) Preorganization – from solvents to spherands. *Angew. Chem. Int. Edit. Engl.*, **25**, 1039–1057.

47. Antonisse, M.M.G. and Reinhoudt, D.N. (1999) Potentiometric anion selective sensors. *Electroanalysis*, **11**, 1035–1048.

48. Xu, Z., Kim, S.K., and Yoon, J. (2010) Revisit to imidazolium receptors for the recognition of anions: highlighted research during 2006–2009. *Chem. Soc. Rev.*, **39**, 1457–1466.

49. Stibor, I. and Zlatuskova, P. (2005) Chiral recognition of anions, in *Anion Sensing* (ed. P. Stibor), Springer-Verlag, Berlin, pp. 31–63.

50. Rao, T.P., Kala, R., and Daniel, S. (2006) Metal ion-imprinted polymers – novel materials for selective recognition of inorganics. *Anal. Chim. Acta*, **578**, 105–116.

51. Rao, T.P. and Kala, R. (2008) Potentiometric transducer based biomimetic sensors for priority envirotoxic markers – an overview. *Talanta*, **76**, 485–496.

52. Suryanarayanan, V., Wu, C.T., and Ho, K.C. (2010) Molecularly imprinted electrochemical sensors. *Electroanalysis*, **22**, 1795–1811.

53. Hutchins, R.S. and Bachas, L.G. (1995) Nitrate-selective electrode developed by electrochemically mediated imprinting/doping of polypyrrole. *Anal. Chem.*, **67**, 1654–1660.

54. Bobacka, J., Ivaska, A., and Lewenstam, A. (2003) Potentiometric ion sensors based on conducting polymers. *Electroanalysis*, **15**, 366–374.

55. Bobacka, J. (2006) Conducting polymer-based solid-state ion-selective electrodes. *Electroanalysis*, **18**, 7–18.

56. Vytras, K. (1991) Coated wire electrodes in the analysis of surfactants of various types – an overview. *Electroanalysis*, **3**, 343–347.

57. Vytras, K. (1984) Determination of some pharmaceuticals using simple potentiometric sensors of coated-wire type. *Mikrochim. Acta*, **3**, 139–148.

58. Bakker, E. and Pretsch, E. (2008) Nanoscale potentiometry. *TrAc-Trends Anal. Chem.*, **27**, 612–618.

59. Kalcher, K., Švancara, I., Metelka, R. *et al.* (2006) Heterogeneous carbon electrochemical sensors, in *Encyclopedia of Sensors* (eds C.A. Grimes, E.C. Dickey, and M.V. Pishko,), American Scientific Publishers, Stevenson Ranch, CA, pp. 283–429.

60. Švancara, I., Kalcher, K., Walcarius, A. *et al.* (2012) *Electroanalysis with Carbon Paste Electrodes*, CRC Press, New York.

61. Kalcher, K., Švancara, I., Buzuk, M. *et al.* (2009) Electrochemical sensors and biosensors based on heterogeneous carbon materials. *Monatshefte für Chemie*, **140**, 861–889.

62. Jezkova, J., Musilova, J., and Vytras, K. (1997) Potentiometry with perchlorate and fluoroborate ion-selective carbon paste electrodes. *Electroanalysis*, **9**, 1433–1436.

63. Tymecki, L., Glab, S., and Koncki, R. (2006) Miniaturized, planar ion-selective electrodes fabricated by means of thick-film technology. *Sensors*, **6**, 390–396.

64. Ammann, D. (1986) *Ion-Selective Microelectrodes: Principles, Design and Application*, Springer, Berlin.

65. Dierkes, P.W., Neumann, S., Müller, A. *et al.* (2002) Multi-barrelled ion-selective microelectrodes. Measurements of cell volume, membrane potential, and intracellular ion concentrations in invertebrate nerve cells, in *Electrochemical Microsystem Technologies* (eds J.W. Schultze, T. Osaka, and M. Datta), CRC Press, Boca Raton.

66. Felle, H.H. (1993) Ion-selective microelectrodes – their use and importance in modern plant-cell biology. *Botanica Acta*, **106**, 5–12.

67. Vigassy, T., Huber, C.G., Wintringer, R. *et al.* (2005) Monolithic capillary-based ion-selective electrodes. *Anal. Chem.*, **77**, 3966–3970.

68. Sundfors, F., Bereczki, R., Bobacka, J. *et al.* (2006) Microcavity based solid-contact ion-selective microelectrodes. *Electroanalysis*, **18**, 1372–1378.

69. Zhang, X.J., Ogorevc, B., and Wang, J. (2002) Solid-state pH nanoelectrode based on polyaniline thin film electrodeposited onto ion-beam etched carbon fiber. *Anal. Chim. Acta*, **452**, 1–10.

70. Cosofret, V.V., Erdosy, M., Johnson, T.A. *et al.* (1995) Microfabricated sensor arrays sensitive to pH and K^+ for ionic distribution measurements in the beating heart. *Anal. Chem.*, **67**, 1647–1653.

71. Bakker, E. (1999) Hydrophobic membranes as liquid junction-free reference electrodes. *Electroanalysis*, **11**, 788–792.

72. Zhang, X., Ju, H., and Wang, J. (2008) Microelectrodes for *in-vivo* determination of pH, in *Electrochemical Sensors, Biosensors, and their Biomedical Applications* (eds X. Zhang, H. Ju, and J. Wang), Academic Press, Amsterdam, pp. 261–305.

73. Buck, R.P. and Cosofret, V.V. (1993) Recommended procedures for calibration of ion-selective electrodes. *Pure Appl. Chem.*, **65**, 1849–1858.

74. Gran, G. (1988) Equivalence volumes in potentiometric titrations. *Anal. Chim. Acta*, **206**, 111–123.

75. Michalowski, T., Toporek, M., and Rymanowski, M. (2005) Overview on the Gran and other linearisation methods applied in titrimetric analyses. *Talanta*, **65**, 1241–1253.

76. Bailey, P.L. (1980) *Analysis with Ion-Selective Electrodes*, Heyden, London.

77. Bakker, E. and Pretsch, E. (2002) The new wave of ion-selective electrodes. *Anal. Chem.*, **74**, 420A–426A.

78. Pretsch, E. (2007) The new wave of ion-selective electrodes. *TrAc-Trends Anal. Chem.*, **26**, 46–51.

79. Bakker, E. and Pretsch, E. (2005) Potentiometric sensors for trace-level analysis. *TrAc-Trends Anal. Chem.*, **24**, 199–207.

80. Bakker, E., Bühlmann, P., and Pretsch, E. (2004) The phase-boundary potential model. *Talanta*, **63**, 3–20.

81. Bakker, E., Buhlmann, P., and Pretsch, E. (1999) Polymer membrane ion-selective electrodes – What are the limits? *Electroanalysis*, **11**, 1088–1088.

82. De Marco, R. and Shackleton, J. (1999) Calibration of the Hg chalcogenide glass membrane ion-selective electrode in seawater media. *Talanta*, **49**, 385–391.

83. Rudnitskaya, A., Legin, A., Seleznev, B. *et al.* (2008) Detection of ultra-low activities of heavy metal ions by an array of potentiometric chemical sensors. *Microchim. Acta*, **163**, 71–80.

84. Bakker, E. and Meyerhoff, M.E. (2000) Ionophore-based membrane electrodes: new analytical concepts and non-classical response mechanisms. *Anal. Chim. Acta*, **416**, 121–137.

85. Severinghaus, J.W. and Bradley, A.F. (1958) Electrodes for blood pO_2 and pCO_2 determination. *J. Appl. Physiol.*, **13**, 515–520.

86. Severinghaus, J.W., Astrup, P., and Murray, J.F. (1998) Blood gas analysis and critical care medicine. *Am. J. Respir. Crit. Care Med.*, **157**, S114–S122.

87. Neethirajan, S., Jayas, D.S., and Sadistap, S. (2009) Carbon dioxide (CO_2) sensors for the agri-food industry-a review. *Food Bioprocess Technol.*, **2**, 115–121.

88. Ion, A. and Bănică, F.G. (2002) *Electrochemical Methods in Analytical Chemistry*, Ars Docendi, Bucharest.

89. Senorans, F.J., Ibanez, E. and Cifuentes, A. (2003) New trends in food processing. *Crit. Rev. Food Sci. Nutr.* **43**, 507–526.

90. McMurray, H.N., Lewis, M.J., and Brinz, T. (2003) Planar solid-state potentiometric carbon dioxide sensors incorporating polymer-gel electrolyte. *Electrochem. Solid State Lett.*, **6**, H5–H7.

91. Tongol, B.J.V., Binag, C.A., and Sevilla, F.B. (2003) Surface and electrochemical studies of a carbon dioxide probe based on conducting polypyrrole. *Sens. Actuators B-Chem.*, **93**, 187–196.

92. Meruva, R.K. and Meyerhoff, M.E. (1998) Catheter-type sensor for potentiometric monitoring of oxygen, pH and carbon dioxide. *Biosens. Bioelectron.*, **13**, 201–212.

93. Zhuiykov, S. (2008) *Electrochemistry of Zirconia Gas Sensors*, CRC Press, Boca Raton, FL.

94. Park, C.O., Akbar, S.A., and Weppner, W. (2003) Ceramic electrolytes and electrochemical sensors. *J. Mater. Sci.*, **38**, 4639–4660.

95. Weppner, W. (1987) Solid-state electrochemical gas sensors. *Sens. Actuators*, **12**, 107–119.

96. Maskell, W.C. and Steele, B.C.H. (1986) Solid-state potentiometric oxygen gas sensors. *J. Appl. Electrochem.*, **16**, 475–489.

97. Riegel, J., Neumann, H., and Wiedenmann, H.M. (2002) Exhaust gas sensors for automotive emission control. *Solid State Ion.*, **152**, 783–800.

98. Docquier, N. and Candel, S. (2002) Combustion control and sensors: a review. *Prog. Energy Combus. Sci.*, **28**, 107–150.

99. Lopez-Gandara, C., Ramos, F.M., and Cirera, A. (2009) YSZ-based oxygen sensors and the use of nanomaterials: A review from classical models to current trends. *J. Sensors*, Article ID 258489.

100. Lee, J.H. (2003) Review on zirconia air-fuel ratio sensors for automotive applications. *J. Mater. Sci.*, **38**, 4247–4257.

101. Peters, H. and Möbius, H.H. (1961) Procedure for the gas analysis at elevated temperatures using galvanic solid electrolyte elements. DDR-Patent 21673. 20.5.1958.

102. Weissbart, J. and Ruka, R. (1961) Oxygen gauge. *Rev. Sci. Instrum.*, **32**, 593–&.

103. Jacob, K.T. and Mathews, T. (1990) Solid-state electrochemical sensors in process control. *Indian J. Technol.*, **28**, 413–427.

104. Traversa, E. (1995) Ceramic sensors for humidity detection – the state-of-the-art and future-developments. *Sens. Actuators B-Chem.*, **23**, 135–156.

105. Hotzel, G. and Weppner, W. (1986) Application of fast ionic conductors in solid-state galvanic cells for gas sensors. *Solid State Ion.*, **18–9**, 1223–1227.

106. Yamazoe, N. and Miura, N. (1992) New approaches in the design of gas sensors, in *Gas Sensors: Principles, Operation and Developments* (ed. G. Sberveglieri), Kluwer, Dordrecht, pp. 1–42.

107. Maruyama, T., Sasaki, S., and Saito, Y. (1987) Potentiometric gas sensor for carbon-dioxide using solid electrolytes. *Solid State Ion.*, **23**, 107–112.

108. Park, C.O., Fergus, J.W., Miura, N. *et al.* (2009) Solid-state electrochemical gas sensors. *Ionics*, **15**, 261–284.

109. Miura, N., Elumalai, P., Plashnitsa, V.V. *et al.* (2009) Solid-state electrochemical gas sensing, in *Solid State Gas Sensing* (eds E. Comini, G. Faglia, and G. Sberveglieri), Springer Science+Business Media, LLC, Boston, MA, pp. 181–207.

110. Zhuiykov, S. and Miura, N. (2007) Development of zirconia-based potentiometric NO$_x$ sensors for automotive and energy industries in the early 21st century: What are the prospects for sensors? *Sens. Actuators B: Chem.*, **121**, 639–651.

111. Garzon, F.H., Mukundan, R., Lujan, R. *et al.* (2004) Solid state ionic devices for combustion gas sensing. *Solid State Ion.*, **175**, 487–490.

112. Koncki, R. and Glab, S. (1991) Potentiometric enzymatic sensors. *Chem. Anal.*, **36**, 423–446.

113. Kauffmann, J.M. and Guilbault, G.G. (1992) Enzyme electrode biosensors – theory and applications. *Method Biochem. Anal.*, **36**, 63–113.

114. Guilbault, G.G. (1988) Enzyme electrode probes. *Method Enzymol.*, **137**, 14–29.

115. Gaberlein, S., Knoll, M., Spener, F. *et al.* (2000) Disposable potentiometric enzyme sensor for direct determination of organophosphorus insecticides. *Analyst*, **125**, 2274–2279.

116. Janata, J. and Blackburn, G.F. (1984) Immunochemical potentiometric sensors. *Ann. NY Acad. Sci.*, **428**, 286–292.

117. Yakovleva, J. and Emneus, J. (2008) Electrochemical immunoassay, in *Bioelectrochemistry: Fundamentals, Experimental Techniques and Applications* (ed. P.N. Bartlett) Wiley, Chichester, pp. 377–410.

118. Stefan, R.I., van Staden, J.F., and Aboul-Enein, H.Y. (2000) Immunosensors in clinical analysis. *Fresenius J. Anal. Chem.*, **366**, 659–668.

119. Janata, J. (1975) Immunoelectrode. *J. Am. Chem. Soc.*, **97**, 2914–2916.

120. Umezawa, Y., Sofue, S., and Takamoto, Y. (1984) Thin-layer ion-selective electrode detection of anticardiolipin antibodies in syphilis serology. *Talanta*, **31**, 375–378.

121. Radecka, H. and Umezawa, Y. (1998) Immunosensors based on ion-selective electrodes, in *Affinity Biosensors: Techniques and Protocols* (eds K.R. Rogers and A. Mulchandani), Humana Press, Totowa, pp. 149–160.

122. De Marco, R., Clarke, G., and Pejcic, B. (2007) Ion-selective electrode potentiometry in environmental analysis. *Electroanalysis*, **19**, 1987–2001.

123. Denuault, G. (2009) Electrochemical techniques and sensors for ocean research. *Ocean Sci.*, **5**, 697–710.

124. Solsky, R.L. (1982) Ion-selective electrodes in biomedical analysis. *Crit. Rev. Anal. Chem.*, **14**, 1–52.

125. Lunte, C.E. and Heineman, W.R. (1988) Electrochemical techniques in bioanalysis. *Top. Curr. Chem.*, **143**, 1–48.

126. Campanella, L. and Tomassetti, M. (1989) Sensors in pharmaceutical analysis. *Sel. Electrode Rev.*, **11**, 69–110.

127. Badoni, R.P. and Jayaraman, A. (1988) Use of ion-selective membrane electrodes in the petroleum-industry. *Erdol & Kohle Erdgas Petrochemie*, **41**, 23–30.

128. Kounaves, S.P., Lukow, S.R., Comeau, B.P. *et al.* (2003) Mars surveyor program '01 mars environmental compatibility assessment wet chemistry lab: A sensor array for chemical analysis of the martian soil. *J. Geophys. Res. -Planets*, **108**, Article Number: 5077.

11

Chemical Sensors Based on Semiconductor Electronic Devices

A potentiometric ion sensor cell develop a potential difference (voltage) that depends on the activity of the target ion. This voltage is applied to a semiconductor electronic device called field-effect transistor (FET) that is included in the measuring instrument. This FET imparts a very high input impedance that prevents current flow through the cell and allows the measurement to be performed under equilibrium conditions. The current flowing through the FET depends on the applied voltage and additional electronic circuitry allows the voltage to be measured.

A further step in the development of potentiometric sensors was achieved by integrating a ion-responsive membrane with a FET-type structure (P. Bergveld, 1970). In this way, the potential difference developed at the membrane-solution interface directly affects the electrical parameters of the device and the response signal is represented by an electrical parameter of the FET structure.

As in the case of standard potentiometric sensors, FET-based ion sensors can be utilized not only for ion determination but also as transducers in sensors based on recognition processes that give rise to ions, such as enzymatic reactions and gas dissolution that alter the ionic composition of the solution.

A standard FET includes a metal layer separated by a thin insulator film from the semiconductor component of the device. If the metal film is able to interact with a gas, the device functions as a gas sensor. The first gas sensor of this type was the hydrogen sensor introduced by I. Lundström *et al.* in 1975.

This chapter gives first a brief overview of semiconductor properties and introduces several electronic semiconductor devices commonly emplyed in the development of chemical sensors, such as FETs and metal-insulator-semiconductor capacitors. The next section addresses ion sensors based on field effect devices. This section also presents gas sensors and enzymatic sensor using semiconductor device ion sensors as transducers. The final section is devoted to gas sensors based on field effect semiconductor devices.

Doped silicon is the typical semiconductor material used in field effect device chemical sensors. As will be shown in this chapter, new semiconductor materials (such as certain nanomaterials and organic semiconductors) have been successfully applied to the development of chemical sensors.

11.1 Electronic Semiconductor Devices

Electronic semiconductor devices, such as diodes and transistors, are essential components of any kind of electronic equipment. Such a device allows the current flow in electrical circuits to be controlled in order to perform information conveying and processing [1].

The functioning of semiconductor devices depends in an essential way on the density of the electric charge at the semiconductor surface. The charge density can build up simply under the effect of applied electric field. However, the charge density can also be modulated upon chemical interaction of the device with ions or molecules. This is the basis for the application of semiconductor devices as chemical sensors [2–4].

11.1.1 Semiconductor Materials

Currently, the standard semiconductor material in electronic devices is silicon, a group IV element that has four valence electrons available for chemical bonding. In a pure silicon crystal each valence electron is paired with another electron in a covalent bond with an adjacent atom. Apparently, no free electrons exist in pure silicon and the material is expected to behave as an insulator. However, the thermal agitation can release sporadically free electrons into the crystal lattice leaving a partially filled valence orbital (Figure 11.1A). So, charge carriers appear in the form of free electrons and positive vacancies (*holes*). Free electrons can move in an ordered way under the action of an electrical field as they do in a metal. At the same time, a bound electron can jump to a hole at an adjacent atom and this effect gives rise to a hole drift in a direction opposite to that of the free electrons. However, at normal temperatures, the free electrons and holes density are extremely low and pure silicon has therefore a very low conductivity. Such a material is an *intrinsic semiconductor* and is characterized by the occurrence of both free electrons and holes with equal concentrations.

Chemical Sensors and Biosensors: Fundamentals and Applications, First Edition. Florinel-Gabriel Bănică.
© 2012 John Wiley & Sons, Ltd. Published 2012 by John Wiley & Sons, Ltd.

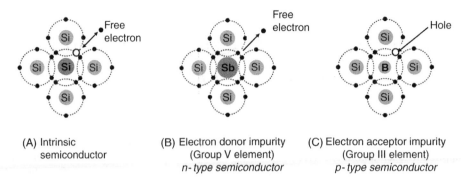

Figure 11.1   Effect of silicon doping with group V or group III elements. Adapted from http://hyperphysics.phy-astr.gsu.edu/hbase/solids/dope.html#c2. Last accessed 17/05/2012.

Enhanced silicon conductivity can be obtained by controlled addition of tiny amounts of impurities in a process termed semiconductor doping. Silicon doping can be performed by either group V or group III elements as shown in Figure 11.1. If a group V element serves as dopant (antimony in Figure 11.1B), its atoms enters the crystal lattice with five valence electrons. Four of these electrons form covalent bonds with silicon atoms, whereas the fifth one cannot find an available bonding orbital and remains in the unbound state. At the same time, the antimony atom becomes positively charged. As a result of the antimony doping, a large density of free electrons appears in the crystal lattice that leads to an enhanced conductivity. Such a doped crystal is termed an *n-type semiconductor* because the majority charge carriers are negatively charged electrons.

If the doping element belongs to the group III (boron in Figure 11.1C), its atom brings only three valence electrons. The boron atom forms four valence bonds with silicon atoms, but one of the bonding orbitals is partially filled with one electron contributed by a silicon atom. This orbital is actually a positive hole; its charge is balanced by the negative charge of the boron atom assumed after accepting an electron from an adjacent silicon atom. A silicon crystal doped with a group III element is termed a *p-type semiconductor* because the majority charge carriers are positively charged holes.

11.1.2   Band Theory of Semiconductors

The properties of semiconductors can be formulated more accurately in terms of the electronic band structure that accounts for the electrical resistivity of metal and semiconductors. An electronic band is formed of a large number of closely spaced molecular orbitals. Metals contain a band that is partly empty and partly filled regardless of temperature. Consequently, electrons can move freely in a lattice of positively charged atoms, which explains the high electrical conductivity of metals.

In semiconductors, one can distinguish a valence band and a conduction band. The *valence band* is the highest range of electron energies in which electrons are normally present at absolute zero temperature. Electrons with the energy in this rage are bound to individual atoms. The uppermost, almost unoccupied band is called the *conduction band* because only when electrons are excited to the conduction band can current flow in these materials. In semiconductors, the valence and conduction bands are separated by an energy bandgap (of about 1.1 eV for silicon) where no electron state exists (Figure 11.2A).

The probability that a given available electron energy state E will be occupied at a given temperature is indicated by the Fermi–Dirac function $f(E)$:

$$f(E) = \frac{1}{e^{(E-E_\mathrm{F})/k_\mathrm{B}T} + 1} \qquad (11.1)$$

where k_B is the Boltzmann constant and T is the absolute temperature. The constant E_F represents the *Fermi level* that is the energy at which the probability of occupation by an electron is one-half. The Fermi–Dirac function

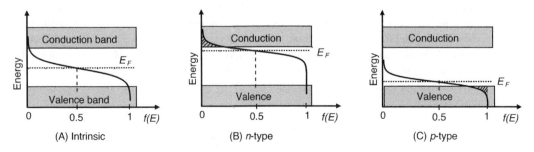

Figure 11.2   Band structure of semiconductors. The curve represents the Fermi–Dirac distribution.

demonstrates that at ordinary temperatures most of the allowed levels up to the Fermi level are filled, and relatively few electrons have energies above the Fermi level.

In an intrinsic semiconductor, the Fermi level lies close to midbandgap and, even if an electron is released from the valence band, it is not allowed to occupy the Fermi level. The single alternative for the electron is to be promoted to the conduction band by excitation. However, at normal temperatures this is a much less probable event because the bandgap is much larger than the average thermal energy. Consequently, at normal temperatures the electron population of the conduction band is extremely sparse; thereby the very low conductivity of intrinsic semiconductors. This intuitive conclusion is confirmed by the Fermi–Dirac function that predicts an extremely low probability that levels in the conduction band are occupied by electrons.

Doping of the semiconductor affects significantly the position of the Femi level due to the charges located at doping atoms. In n-semiconductors (Figure 11.2B) the Fermi level shifts close to the conduction band. In this situation, the Fermi distribution indicates a small but still significant probability that electrons occupy energy levels at the bottom of the conduction band (hatched area) and contribute to electrical conductivity.

In the case of a p-semiconductor (Figure 11.2C), the Fermi level is shifted near the conduction band. The Fermi–Dirac function indicates that the degree of occupation of energy states by electrons is smaller than unity at the top of the valence band (hachured area). The unoccupied states represent hole levels. Therefore, valence electrons can jump from a neutral atoms to an adjacent hole under the action of an electric field. This process is equivalent to hole drift in the opposite direction and accounts for hole conduction in p-semiconductors.

Therefore, by doping a pure silicon crystal one obtains either an n-semiconductor or a p-semiconductor, according to the number of valence electrons of the doping element. In the first case, the current flow through the material is due to the movement of free electrons in the conduction band, whereas in the second case the electric conduction is ensured by the holes drift in the valence band.

11.1.3 Metal-Insulator-Semiconductor (MIS) Capacitors

Certain electronic semiconductor devices are based on a sequence of well-matched layers of metal, insulator and semiconductor (MIS). The thickness of the metal and insulator layer is about 50–100 nm. Such a simple structure forms a MIS capacitor (Figure 11.3A), whilst more advanced devices, such as the MIS transistor include additional elements (Section 11.1.4).

In a MIS capacitor the semiconductor body is connected to the ground through an ohmic contact and a bias voltage (V_B) is applied between the metal and the semiconductor (Figure 11.3A). As a result, the same amount of charge, but with opposite sign, appears in the metal film and at the upper limit of the semiconductor. As will be shown later, the particular feature of this capacitor is that its specific capacitance depends on the *electric field* created by the applied voltage. Hence, the MIS capacitor is a *field effect device* (*FED*), that is, a semiconductor device whose properties are determined largely by the effect of an electric field upon a certain region within the semiconductor.

The capacitance can be measured by superimposing a small-amplitude AC voltage over the DC bias voltage. The AC current produced in this way is proportional to the capacitance.

There are three possible regimes of operation for a MIS capacitor, as shown in Figures 11.3B–D for the case of a p-semiconductor. In the *accumulation* regime (B), a negative bias is applied on the metal electrode. In response, positive holes accumulate at the semiconductor surface and the device behaves as a normal plate capacitor (C_I) with

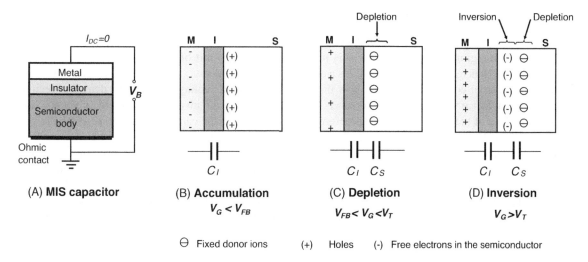

Figure 11.3   (A) Cross section of a *p*-semiconductor based MIS capacitor; (B–D) operation regimes of a p-semiconductor MIS capacitor. M, I, and S stands for metal, insulator and semiconductor, respectively.

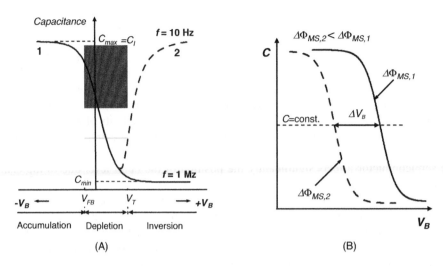

Figure 11.4 (A) Capacitance–voltage curves of a MIS capacitor as measured at high-frequency (curve1) and low-frequency (curve 2) applied AC voltage; (B) effect of the work-function difference at a high-frequency AC voltage.

the insulator as intermediate dielectric. Under these conditions, the device capacitance is at its maximum level ($C_I = C_{max}$, Figure 11.4A).

The *depletion* regime occurs when a slight positive bias applied to the metal repels the holes away, leaving at the surface only the fixed doping atoms that bear negative charges. As a result, two different dielectric layers are present, namely the actual insulator and the depleted region. Under these conditions, the MIS capacitor behaves as two capacitors in series (C_I and C_S, Figure 11.3C) and the overall capacitance is:

$$C = \left(\frac{1}{C_I} + \frac{1}{C_S}\right) \tag{11.2}$$

Hence, in this case, the overall capacitance is smaller than C_I.

On increasing the positive bias voltage, the thickness of the depletion layer increases and, consequently, the C_S capacitance decreases while the surface is being depleted. Therefore, the total capacitance also decreases (Figure 11.4A, curve 1). The voltage at which the depletion layer sets on is termed *flat-band voltage* (V_{FB}). This name derives from the fact that under flat-band conditions there is no excess charge in the semiconductor and the energy band is flat throughout the semiconductor up to the surface.

Over a particular positive bias voltage level called the *threshold voltage* (V_T), free electrons, which are present as minority carriers, are driven to the silicon surface thus turning locally the p-semiconductor into an n-semiconductor (Figure 11.3D). Inversion of the carrier type gives rise in this way to an *inversion layer*. Its position in the depletion layer depends on the relation between the speed of the charge carrier generation/recombination and the frequency of the excitation AC voltage.

The measured capacitance of a MIS capacitor depends on both the bias voltage and the frequency of the applied AC voltage, as shown in Figure 11.4A.

The curve 1 in Figure 11.4A represents the capacitance measured at a high-frequency AC voltage (for example, 1 MHz). The effect of frequency is due to the fact that this parameter determines the position of the inversion layer. At very high frequencies the measured capacitance decreases within the depletion region (Figure 11.4A, curve 1). Within the inversion region, the inversion layer is formed at the inner limit of the depletion layer that is, in the semiconductor bulk. Hence, the capacitance attains a constant, minimal value C_{min}. At low frequencies (for example, 10 Hz) the charge-carrier concentration can follow the variation in the AC voltage. Within the depletion region, the low-frequency capacitance varies according to curve 1. However, within the inversion region, the inversion layer is built up at the surface of the semiconductor. Therefore, within this region the capacitance increases with the voltage (Figure 11.4A, curve 2) and, in strong inversion, the capacitance recovers the maximal value.

The threshold voltage is the key parameters in the functioning of FED sensors. Its physical meaning can be derived by considering the charge balance under inversion conditions. To this end, the charge per unit area (surface charge density) for each charged layer in the system will be considered. It should be noted that due to the characteristics of the fabrication process, there are always fixed and mobile charged sites within the insulator layer and the overall charge in the insulator should also be taken into account. Therefore, the charge balance condition of the MIS structure under inversion conditions can be formulated as follows:

$$\underset{\text{Metal}}{(Q_M)_{V_B > V_T}} = -(\underset{\text{Insulator}}{Q_I} + \underset{\text{Depletion layer}}{Q_{D,max}} + \underset{\text{Inversion layer}}{Q_{Inv}}) \tag{11.3}$$

Notice that $Q_{D,max}$ is the maximum achievable charge in the depletion layer. At the threshold voltage ($V_B = V_T$), the inversion layer is not yet set in, so, the charge-balance equation in this particular case is:

$$(Q_M)_{V_B=V_T} = -(Q_I + Q_{D,max}) \tag{11.4}$$

Equation (11.4) represents the physical condition for the onset of the inversion regime. An analysis of the potential distribution across the MIS system demonstrates that the threshold voltage depends essentially on the work-function difference between the metal and the semiconductor.

The work function is the minimum energy needed to remove an electron from the Fermi level of a neutral metal or semiconductor to vacuum. The emitted electron should be located at a point immediately outside the solid surface such that the electron does not experience electrostatic effects induced by the solid. The energy level occupied by this electron is the first nonbounded state and is usually termed the *vacuum level*. According to this definition, the work function is determined by both the bulk properties of the solid and the state of its surface (which determines the vacuum level).

The functioning parameter that depends on the work function is the threshold voltage of the MIS device. More precisely, the threshold voltage depends on the work-function difference between the metal (Φ_M) and the semiconductor (Φ_S), as follows:

$$V_T = \frac{\Delta\Phi_{MS}}{e} - \Delta\phi_{IS} \tag{11.5}$$

where $\Delta\Phi_{MS} = \Phi_M - \Phi_S$, e is the elementary charge and $\Delta\phi_{IS}$ is a constant term.

Equation (11.5) is the basis of the chemical sensing by means of MIS devices. MIS sensors are designed such as to undergo a variation in $\Delta\Phi_{MS}$ in response to the interaction with the analyte. Details of the analyte action on $\Delta\Phi_{MS}$ are given in further sections dealing with specific MIS sensors. In the particular case of a MIS capacitor, a change in $\Delta\Phi_{MS}$ results in the current–voltage curve being shifted in accordance with Equation (11.5) as shown in Figure 11.4B. The sensor response in this case is the bias voltage difference at a constant capacitance (ΔV_B).

Figure 11.4B displays schematically capacitance–voltage curves recorded at a high frequency (1 MHz). The bias voltage shift can also be inferred from low-frequency curves that are similar to curve 2 in Figure 11.4A. In this case, the voltage shift is indicated by the shift of the minimum point on the curve.

It should be kept in mind that modulation of the work function by the interaction of the analyte with the sensor is the key functioning principle of semiconductor-based chemical sensors. A detailed approach to this topic is available in refs. [5,6].

11.1.4 Metal-Insulator-Semiconductor Field Effect Transistors (MISFETs)

MISFETs are electronic devices with tunable electric resistance. In order to understand the functioning of MISFETs it is useful to consider first at a simpler device, namely the p-n junction diode. As its name suggests, this device consists of two semiconductor zones of different type (Figure 11.5A). When a p-n junction is build up, some of the electrons from the n-region are free to diffuse across the junction and recombine with holes. Filling a hole makes a negative ion and leaves behind a positive ion on the n-side. Thus, the junction region turns depleted of charge carriers.

A current can cross the diode only if a voltage source is connected with the negative pole to the n-region and the positive pole to the p-region (forward bias, Figure 11.5B). In this case, the electrons from the n-type semiconductor are injected in the p-type one and also, holes from the p-type semiconductor are injected in the n-type one. Hence, the depletion region collapses, becoming thinner. In this state, the diode shows less resistance to current passing through it. In order for a sustained current to go through the diode, though, the forward voltage should be above a certain limit such that the depletion region becomes fully collapsed. In practice, the current has to be limited by a resistor, otherwise the diode will be thermally destroyed.

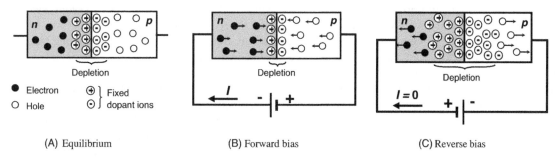

(A) Equilibrium   (B) Forward bias   (C) Reverse bias

Figure 11.5 The semiconductor diode.

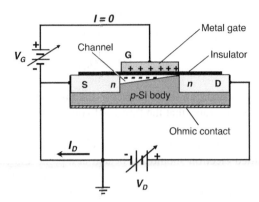

Figure 11.6 An n-channel enhancement mode MISFET. S: source; D: drain; G; gate.

If the source polarity is reversed (*reverse bias*, Figure 11.5C) charge carriers in each zone are attracted to the periphery leaving a wide depletion zone in the junction region. Under these conditions, the diode displays a very large resistance and the current through the diode is insignificant. Diodes are used to convert alternating current to direct current, a process known as *rectification*.

A diode displays only two states, on or off, in accordance to the polarity of the applied voltage. In contrast, the transistor is a device that allows tuning the current between zero and a certain limit by proper adjustment of some electrical parameters. The structure of a field-effect transistor is shown schematically in Figure 11.6A. It consists of a small piece of p-type silicon (*substrate*) with two n-type regions formed at the upper left and right extremities. The two n-type regions are called *source* and *drain*, respectively. The surface is covered by an insulating layer (for example, SiO$_2$) and a thin metal layer is deposited above it to form the *gate*. If a voltage V_D is applied between the source and the drain, no current flow is possible because reverse biasing occurs at one of the two p-n junctions no matter what the drain source polarity is. The situation changes if a positive electric charge is loaded on the gate by means of the V_G voltage source. By electrostatic interaction, this positive charge drives free electrons from the source to the substrate surface layer. In this way, the p-semiconductor turns into an n-type one within a very thin layer below the gate; hence the name *inversion layer* ascribed to this zone (see Section 11.1.3 for details). Now, there are no more p-n junctions, the source and the drain are shortcut and current flow is allowed. In other words, the electric resistance between the source and drain assumes a much lower value. Because the current flow is confined within the inversion layer, this zone is called a *channel*.

As the current flow is controlled by the electrical field produced by the gate, such a device is termed a *field-effect transistor* (FET). In accordance with the sequence of layers, this a device is also called a metal-oxide-semiconductor field-effect transistor, a MOSFET. As the insulator can be made of a nonoxide material (such as silicon nitride) such devices are more generally termed metal-insulator-semiconductor field-effect transistors (MISFETs).

It is important to note that the inversion layer forms only after the gate voltage overcomes the threshed voltage V_T. If $V_G > V_T$, the volume density of electrons in the inversion layer is higher than the volume density of holes in the semiconductor body. The gate voltage therefore determines the electron density in the channel and hence, its electric resistance. The FET operation regime depends crucially on the quotient of V_D and $(V_G - V_T)$, If the $V_D < (V_G - V_T)$ condition is fulfilled, the drain current is given by the following equation in which β is a parameter depending on the physical and geometrical characteristics of the transistor:

$$I_D = \beta \left[(V_G - V_T) - \frac{V_D}{2} \right] V_D \tag{11.6}$$

Hence, the current varies nonlinearly with the drain voltage, which is characteristic of the *subthreshold region* on the characteristic curve (Figure 11.7A). At very small drain voltages, the $I_D - V_D$ relationship is linear.

However, the source–drain potential difference produces a variation in the electron density along the channel. This is negligible at the onset of the subthreshold region, but increases considerably at $V_D > (V_G - V_T)$. On increasing the drain voltage, a situation will arise when the free electrons are missing near the drain, which is connected to the positive pole of the drain source. In this case, the channel does not extend across the whole distance between the source and the drain. The onset of the depletion region is also known as *pinch-off* to indicate the lack of channel region near the drain. Nevertheless, the drain current can flow due to the high electron velocity caused by the source–drain potential difference. Any increase in the drain voltage speeds up the electrons, but, at the same time, increases the length of the depleted part of the channel and thereby moves the pinch-off point towards the source. As a result, the current displays only a slight increase with increasing drain

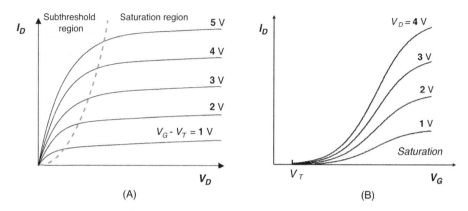

Figure 11.7 Characteristic curves of a MOSFET. (A) Drain current vs. drain voltage at constant gate voltage. The broken line indicates the limit of the saturation region. (B) Drain current vs. gate voltage at constant drain voltage.

voltage. This situation is characteristic of the *saturation region* in Figure 11.7A. In the saturation region, the drain current varies as:

$$I_D = (\beta/2)(V_G - V_T)^2 \tag{11.7}$$

As will be shown later, the dependence of the drain current on the gate voltage (Figure 11.7B) plays an essential role in chemical-sensor applications. In accordance with the previous considerations, the current is set up at $V_G = V_T$ and increases as a parabola with increasing gate voltage within the subthreshold region according to Equation (11.6). In the saturation region, the drain current increases proportionally with $(V_G > V_T)^2$, as shown in Equation (11.7).

A suggestive analogy for the FET is the behavior of a water tap. The gate is analogous to the knob that controls the flow rate. If the knob is screwed up just a little, probably no water will flow at all. Further screwing up causes the flow to increase proportionally (subthreshold region). However, over a certain limit, the flow does not increase any longer (saturation). Under these circumstances, the flow can be increased only by increasing the water reservoir pressure (drain voltage).

p-Channel transistors are also available; such a transistor is the mirror twin of the n-channel one. Hence, it is formed of an n-type semiconductor body and the gate is negatively polarized. Therefore, at a gate voltage over the threshold voltage, holes accumulate at the semiconductor surface to form the channel. Current flow through the channel is secured by hole drift under the action of the source–drain voltage.

The above types of transistor operate in the *enhancement mode* because the drain current is zero at zero gate bias, but increases with increasing gate voltage due to the increase in charge-carrier density within the channel. Common symbols for an enhancement-mode MOSFET are shown in Figure 11.8.

Another possibility is to form, by doping, a channel during the fabrication process. Hence, the drain current can flow at zero gate voltage. In order to control the drain current, the gate is biased so as to decrease the charge carrier density within the channel; hence a decrease in the drain current results. Such a device is a *depletion mode* transistor.

In summary, one can distinguish two operational regions of a MOSFET. For values of $V_D < (V_G - V_T)$, the drain current is given by Equation (11.6). The device operates in the nonsaturated mode and behaves as a nonlinear, gate-voltage-controlled resistor. For values of $V_D > (V_G - V_T)$, the drain current is given by Equation (11.7). In this case, the device operates in the saturation mode and the current is almost independent of the drain voltage.

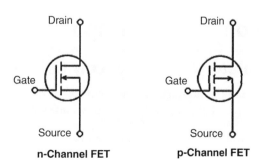

Figure 11.8 Symbols for enhancement mode MOSFET devices.

The field-effect transistor is the key component in contemporary electronic equipment. It is available as an independent unit (in the mm size range), but, as a rule, a great number of submicrometer-size transistors are produced on a single silicon chip, along with other passive components to form integrated circuits. Transistor applications are based on the fact that the drain current can be modulated by means of the gate voltage at a constant drain voltage.

11.1.5 Outlook

Pure semiconductor materials (intrinsic semiconductors) have a very low electrical conductivity because most of the valence electrons are localized in the valence band, that is, are bound to the lattice atoms. The conductivity of a semiconductor material can be enhanced considerably by doping. If the doping element is an electron donor, one obtains an n-semiconductor in which the electric current is carried by free electrons. Doping by an electron acceptor element leads to p-semiconductors, which are characterized by hole conduction.

Field effect electronic devices are based on the control of charge distribution at the surface of a semiconductor by means on an electric field. Typical devices of this type are the MIS capacitor and the MISFET. These devices are formed of a sequence of semiconductor-insulator-metal (gate) layers in which the metal gate is used to create the electric field that controls the device behavior.

A key parameter of FEDs is the threshold voltage, which is the gate voltage limit over which the inversion layer forms. The threshold voltage depends on the work-function difference between the semiconductor and the metal layer. As will be shown in next sections, the work potential difference plays an essential role in transduction by FEDs.

Questions and Exercises (Section 11.1)

1 What is an intrinsic semiconductor? Comment on the electrical conductivity of an intrinsic semiconductor in relation to its band structure.
2 What is semiconductor doping and how does doping affect the band structure and the electrical conductivity of a semiconductor?
3 What kinds of charge carrier can be found in a semiconductor and what is the mechanism of charge transport in each case?
4 What parameter determines the charge distribution in a MIS structure? What is the charge distribution under accumulation, depletion and inversion conditions?
5 What is the particular charge distribution at the flat-band voltage and at the threshold voltage?
6 Why and how does the capacitance of a MIS capacitor vary with the bias voltage?
7 What are the parameters that determine the threshold voltage?
8 What is the effect of the metal nature on the MIS capacitance?
9 What kind of semiconductor device is based on a p-n semiconductor junction? How does the resistance of this device vary with the applied voltage?
10 Draw schematically the structure of a MISFET and indicate on the sketch the name of each component part.
11 Why does the source-drain resistance of a MISFET vary with the gate voltage?
12 What is the state of the MISFET channel under unsaturation and saturation conditions?
13 Draw schematically a graph showing the variation of the source–drain resistance as a function of the drain voltage at a constant gate voltage.
14 Sketch a graph representing the drain current–gate voltage curve at a constant drain voltage. Indicate what region of the curve is suitable for sensor applications and why.
15 What makes the difference between enhancement- and depletion-mode MISFETs?

11.2 FED Ion Sensors and Their Applications

11.2.1 Electrolyte-Insulator-Semiconductor (EIS) Devices

A field effect device can be converted into ion sensor if the metal gate is omitted, and, instead, an ion-responsive layer is build up over the insulator layer and put in contact with the test solution as shown in Figure 11.9A. The ion-responsive layers are generally similar to those developed in the field of potentiometric ion sensors. As the system includes an electrolyte solution, it is termed an electrolyte-insulator-semiconductor (EIS) that differs from the MIS system by the nature of the layer over the insulator. In order to control the charge distribution at the semiconductor surface, a bias voltage is applied through a reference electrode. In an EIS system, the metal in the

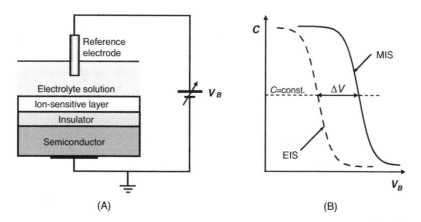

Figure 11.9 (A) Schematics of an electrolyte-insulator-semiconductor ion sensor. The solution should be insulated from the solid device. (B) Capacitance–voltage curves for an MIS system and an EIS ion sensor.

reference electrode is an analogue of the metal gate in the corresponding MIS structure. By ion exchange with the solution, a characteristic potential develops at the interface between the solution and the ion-responsive layer. This potential alters a particular parameter of the device depending on the ion concentration in the solution. Comprehensive surveys of FED ion sensors are available in refs. [2,4,6–8].

In order to understand the transduction mechanism in an EIS device, the device will be represented as follows, for the case in which an Ag/AgCl reference electrode is used:

$$\overset{\phi_S}{Si} \mid \underset{\Delta\phi_{SI}}{Insulator} \mid \underset{\Delta\phi_{I-m}}{Membrane} \mid \underset{\Delta\phi_m}{Solution} \parallel \underset{\phi_{M-Sol}}{KCl(aq)|AgCl,} \overset{\phi_M}{Ag} \tag{11.8}$$

As shown in Scheme (11.8), a series of potential differences develops across this system, Thus, $\Delta\phi_{SI}$ is the potential difference between the semiconductor and the insulator and is similar to that found in an MIS device. $\Delta\phi_{I-m}$ is the potential difference between the insulator and the membrane and $\Delta\phi_m$ is the membrane potential that develops across the ion-sensitive layer. Finally, the potential difference between the reference electrode metal and the solution is written as ϕ_{M-Sol}.

In the case of a true MISFET, the threshold voltage is a constant parameter that is determined by the charge in the depletion layer, the work-function difference between the gate metal and the semiconductor and insulator charges. As shown before, the threshold voltage represents the particular value of the gate voltage needed for setting up the inversion layer. In the case of an EIS structure, additional potential difference terms come into play as shown in Scheme (11.8). Therefore, an electron that hypothetically passes from the semiconductor to vacuum has to cross several charged regions that affect its energy. The same happens with an electron transferred from the reference electrode metal to vacuum. Consequently, the threshold voltage equation (Equation (11.5)) should be altered in order to account for the additional potential terms in the EIS system, and the threshold voltage for the EIS system (V_T^*) assumes the following form [2]:

$$V_T^* = V_T - (\Delta\phi_{M-Sol} - \Delta\phi_m - \Delta\phi_{I-m}) - \Delta\phi_{SI} \tag{11.9}$$

Here $\Delta\phi_{M-Sol}$ represent the reference electrode potential plus the junction potential and can be assumed to be constant. If $\Delta\phi_{I-m}$ is held constant, the threshold voltage difference between the EIS and the MIS systems is represented by the membrane potential corrected by a constant term:

$$\Delta V_T = V_T^* - V_T = constant + \Delta\phi_m \tag{11.10}$$

Hence, the capacitance–voltage curve of an EIS ion sensor shifts with the ion concentration in the solution as shown in Figure 11.9B. The change in the threshold voltage can be detected as the change in the bias voltage at a constant capacitance:

$$(\Delta V)_{C=const} = constant + \Delta\phi_m \tag{11.11}$$

As the membrane potential is a logarithmic function of the ion concentration in the solution, a similar variation in the bias voltage will occur when varying the ion concentration.

By suitable modification of the device in Figure 11.9A, it can be turned into an ion-sensitive field-effect transistor (ISFET). In this case, the alteration of the threshold voltage produces a change in the drain current at a constant gate voltage. As will be shown in the next section, the response can be translated into a variation in the applied gate voltage at a constant drain current, which is formally similar to the capacitor response shown in Figure 11.9B.

Hence, the mechanism by which FED ion sensor devices respond to the analyte ion concentration is similar to that involved with potentiometric ion sensors. The response can be obtained as the variation in the gate voltage, which is analogous to the voltage response of a potentiometric sensor. Moreover, ion-recognition methods and materials developed in the framework of standard potentiometric sensors can be easily transposed to the design of FED ion sensors. That is why FED sensors are categorized as a particular kind of potentiometric sensor.

As shown before, FED ion sensors can in principle be designed in either the capacitor or the FET configuration. As it is simpler the capacitor design is preferable in the fabrication of ion sensor arrays [9,10]. On the other hand, the capacitor version is more demanding as far as the measuring instrumentation and the data-processing procedures are concerned.

11.2.2 FED pH Sensors

The first application of field effect devices in ion chemical sensing was the FED-based pH sensor initially developed by Bergveld [11–14].

In a pH ISFET, the sensing element is the surface of the gate insulator (for example, silicon dioxide) that is put in direct contact with the solution. Under these conditions silanol sites (formulated as A-OH) develop at the oxide surface. The pH effect on such sites can be rationalized by the site-dissociation model, which assumes that a silanol site is able to undergo both acid dissociation and protonation as follows:

$$A - OH \xrightleftharpoons{K_a} A - O^- + H_s^+; \quad K_a = \frac{[A - O^-][H^+]_s}{[A - OH]} \tag{11.12}$$

$$A - OH + H_s^+ \xrightleftharpoons{K_b} A - OH_2^+; \quad K_b = \frac{[AOH_2^+]}{[A - OH][H^+]_s} \tag{11.13}$$

Here, $[H^+]_s$ represent the hydrogen ion concentration near the interface. As a result of the reaction of $A - OH$ sites with the hydrogen ions, a double electric layer forms at the insulator/solution interface (Figure 11.10). It consists on the one side of the ionic sites at the oxide surface and, on the other side, of solution anions attracted by the positive charges at the surface. Consequently, a potential difference $\Delta\phi_s$ builds up at the interface of the oxide and the solution. This determines the surface concentration of the hydrogen ion at the surface as a function of the bulk concentration ($[H^+]_b$) according to the Boltzmann equation:

$$[H^+]_s = [H^+]_b \exp\left(-\frac{e\Delta\phi_s}{k_B T}\right) \tag{11.14}$$

where e is the elementary charge, k_B is the Boltzmann constant, and T is the absolute temperature.

The $A - OH$ site can behave as either an acid (reaction (11.12)) or base (reaction (11.13)), which implies an amphoteric character. Consequently, the assembly of ionogen sites forms a surface-confined pH buffer system. The buffer capacity of this system (β_s) plays an essential role in determining the pH response. The buffer capacity measures the ability of the buffer to counteract a pH change when hydrogen ions are added or removed. This parameter depends on the above equilibrium constants and the total number of active sites per unit area ($N_s = [A - OH] + [A - OH_2^+] + [A - O^-]$), as follows:

$$\beta_s = kN_s\sqrt{K_a K_b} \tag{11.15}$$

where k is a specific parameter of the insulator.

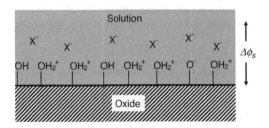

Figure 11.10 Structure of the double electric layer at the oxide/solution interface. The double layer consists of fixed charges at the oxide surface and mobile anions (X^-) driven from the solution by electrostatic attraction.

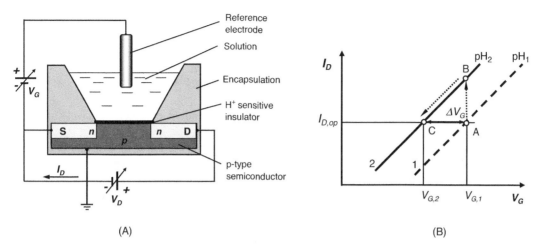

Figure 11.11 (A) The layout of a cell for pH determination by means of a pH ISFET; (B) the principle of constant drain current operation of an ISFET sensor.

If $\beta_s \gg F\Delta\phi_s/RT$, the following linear equation relates the surface potential to the solution pH:

$$\Delta\phi_s = 2.303\frac{RT}{F}\frac{\beta_s}{\beta_s + 1}(pH_0 - pH) \tag{11.16}$$

Here, pH_0 is the pH value at which $\Delta\phi_s = 0$; it is determined by the equilibrium constants ($pH_0 = -\log\sqrt{K_a/K_b}$). A crucial issue is that the response sensitivity is determined by β_s, and a Nernstian response is obtained only if $\beta_s \gg 1$, that is, if there is a high surface buffer capacity. This condition is fulfilled when the surface density of ionogen sites is very high. In addition, the insulator layer has to be stable in contact with the solution so as to prevent solution from reaching the semiconductor surface. The most suitable insulator materials for use in pH detection are Al_2O_3, Ta_2O_5 (which can be formed by pulsed laser deposition) and Si_3N_4 (which can be formed by plasma-enhanced chemical vapor deposition). In contrast to oxide materials, the Si_3N_4 surface displays two kinds of active site, namely silanol and primary silamine ($Si-NH_2$) sites. The characteristics of various materials used in pH-sensing layers are surveyed in ref. [15].

Therefore, an insulator layer that develops a high ionogen site density by conditioning in water forms the hydrogen-responsive element of this pH sensor. By ion exchange with the solution it gives rise to an interface potential that is a linear function of the solution pH.

The layout of a pH FET sensor is shown in Figure 11.11A. In contrast to the standard FET, the metal gate is missing in this device. Instead, the H^+-sensitive insulator layer is set in contact with the test solution and a gate voltage (V_G) is applied via a reference electrode. The sequence of contacting phases can be formulated as follows:

$$\overset{\phi_S}{Si} \mid \underset{\Delta\phi_{SI}}{Insulator} \mid \underset{(\Delta\phi_s+\chi_{sol})}{Solution} \underset{E_r}{\parallel KCl(aq)\mid AgCl,} \overset{\phi_M}{Ag} \tag{11.17}$$

The relevant potentials are also shown in Scheme (11.17). Here, E_r is the reference electrode potential including the junction potential, $\Delta\phi_s$ is the potential difference produced by the proton-transfer reactions (11.12) and (11.13), χ_{sol} is a constant potential difference due to the oriented solvent dipoles at the solution/insulator interface, and $\Delta\phi_{SI}$ is a constant potential difference contributed by the solid-state part. Hence, the $\Delta\phi_s$, χ_{sol} and E_r terms would alter the threshold voltage in the drain current equation of the FET device and the pertinent term in Equation (11.9) should be substituted so as to give Equation (11.18).

$$V_T^* = V_T - (E_r - \Delta\phi_s - \chi_{sol}) - \Delta\phi_{SI} \tag{11.18}$$

In this equation, each term except $\Delta\phi_s$ is pH independent and their sum will be further substituted by the constant k_1. Under these circumstances, the drain current equation in the subthreshold regime assumes the following form:

$$I_D = \beta(V_G + \Delta\phi_s + k_1 - V_D/2)V_D \tag{11.19}$$

Taking into account Equation (11.16) it results that at constant gate voltage and drain voltage the drain current depends on the pH via the $\Delta\phi_s$ term.

The response in terms of voltage (potentiometric response) can be obtained under constant drain-current conditions. Thus, if an explicit relationship for the gate voltage is derived from Equations (11.16) and (11.19), one obtains upon assuming that $\beta_s \gg 1$ the following expression:

$$(V_G)_{I_D=const} = \text{constant} + 2.303\frac{RT}{F}\text{pH} \qquad (11.20)$$

Here, the constant term includes contributions from any pH-independent parameter as well as the constant drain current.

Therefore, the sensor response in terms of gate voltage at constant drain current is a Nernstian one with the ideal sensitivity determined by the prelogarithm factor (59 mV/pH unit at 25 °C). Actual sensors can display a slightly lower sensitivity. That is why both the constant term and the sensitivity in Equation (11.20) should be treated as empirical parameters and their values are obtained by calibration with standard pH buffer solutions, as in the case of the pH glass electrode.

The principles of constant drain-current operation can be made clear by means of the diagram in Figure 11.11B. Thus, according to Equation (11.19), at a given pH, the drain current is a function of the gate voltage. Assume that at pH_1, the operating point is A with the coordinates $V_{G,1} - I_{D,op}$, where $I_{D,op}$ is the operational drain current. If the pH changes to pH_2 at a constant gate voltage, the operating point shifts to B. However, if the gate voltage is lessened, the operating point moves downhill along the line 2. When the gate voltage value becomes $V_{G,2}$, the $I_{D,op}$ drain current will be restored to the initial value (point C). The $V_{G,2}$ value is related to pH_2 by an equation derived from Equation (11.20).

The change in the gate voltage compensates for the variation in the interface potential in response to the pH modification. In practice, the gate-voltage adjustment is performed automatically by suitable electronics incorporated in the pH meter.

pH ISFET devices compare favorably with the pH glass electrode. Thus, the working range spans from pH 2 to 12 and the response time is less than 1 min. pH ISFET sensors with a remarkable ruggedness, stability, and accuracy are currently commercially available. In contrast to the pH glass electrode, a pH ISFET resists dry storage and does not need prolonged rehydration before use.

In an analogous way, capacitive pH sensors can be fabricated on the basis of electrolyte-insulator-semiconductor structures. In this case, the response is represented by the shift of the capacitance–voltage curve with the change in pH.

A great advantage of the pH ISFET sensor is the possibility of fabrication by microelectronics technology, which facilitates the mass production. Such a sensor is of about one millimeter size and is therefore suitable for integration in sensor microarrays along with additional electronics for parameter control and signal processing. Due to the simple structure of the device, the fabrication of capacitive sensor arrays is more facile.

The small size makes the pH FED sensor suitable for *in vivo* applications [16,17]. Thus, a catheter-type pH FED sensor of about 1 mm diameter has been developed [18].

11.2.3 pH ISFET-Based Gas Probes

A Severinghaus-type CO_2 sensor (analogous to that introduced in Section 10.16) can be obtained by using a pH ISFET sensor as transduction element. The small size of the FED sensor allows for miniaturization in view of *in vivo* applications, as shown in Figure 11.12. The included pH sensor has been fabricated on an elongated semiconductor substrate that has the sensitive region at one end and the bonding-pad region for lead wires at the opposite end. The sensing and bonding-pad regions have been isolated from each other by a SiO_2-N_3O_4 double layer that extends also to front and back surfaces as well as to the gate wall. This provides good electrical insulation and makes the device waterproof. An Ag/AgCl reference electrode has been formed by vapor deposition of silver followed by chlorination. After covering the electrical bonding wires with epoxy resin, the chip is mounted in a nylon tube and fastened with silicone rubber. The pH-sensitive gate region has been coated with a hydrogel. In this configuration, the device functions as a pH sensor. In order to convert it into a CO_2 probe, a thicker gel impregnated with a NaCl-NaHCO$_3$ solution was placed over the gate region and the whole assembly was coated with a gas-permeable silicone membrane by dip coating.

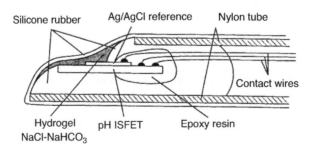

Figure 11.12 A catheter tip FET device for *in vivo* measurement of CO_2 partial pressure in blood. Adapted with permission from [18]. Copyright 1980 Springer Science+Business Media B.V.

Long-term functioning of such a CO_2 probe can be impaired by the drift in the sensor response. This problem can be overcome by including an additional metal electrode that is used to produce by electrolysis OH^- ions that neutralize the extra acidity caused by CO_2 dissolution. So, the device performs a coulometric titration with the pH ISFET acting as indicator electrode [19]. As a titration lasts for about 10 s, this device is insensitive to ISFET drift and shows very good long-term stability.

11.2.4 Membrane-Covered ISFETs

In the pH ISFET, it is the insulator surface that serves as ion-responsive element. As already shown, sensors for other ions can be obtained by coating the insulator surface with an ion-responsive membrane. The responsive membranes used in ISFETs are similar to those introduced in Chapter 10. Hence, the ion-responsive membrane can be either an ionophore-based polymeric membrane or can be made of a solid ion-responsive material. IUPAC terminology and conventions for ISFED sensors can be found in ref. [20].

A polymer membrane ISFET was first developed by attaching an ion-sensitive polymeric membrane over the insulator layer on the MIS structure [21]. As in the case of pH ISFET, the membrane–solution potential affects the channel current. The trouble with this configuration is that an ill-defined potential difference develops at the membrane–insulator surface, similar to the situation arising in the case of coated-wire potentiometric ion sensors (Section 10.12). This problem can be overcome by inserting in-between these two layers an aqueous hydrogel in which an aqueous pH buffer is incorporated (Figure 11.13A) [22]. In this structure, the silanol groups at the oxide surface are in equilibrium with the buffered solution and the surface potential will be stable as long as the pH of the aqueous hydrogel remains constant. Alternatively, a conducting polymer can be inserted between the membrane and the insulator.

The structure of a neutral ionophore sensing membrane is shown in Figure 11.13B. As is typical of potentiometric ion sensors, the analyte ion first undergoes partition between the solution and the membrane, followed by recognition by the receptor, with the characteristic constants K_p and K_r, respectively. Potential distribution across the system is shown at the lower part of this figure, where $\Delta\phi_{B,1}$ and $\Delta\phi_{B,2}$ are boundary potentials, $\Delta\phi_D$ is the diffusion potential across the membrane, $\Delta\phi_m$ is the membrane potential, and $\Delta\phi_s$ is the surface potential at the semiconductor/hydrogel interface. The membrane potential is, therefore:

$$\Delta\phi_m = \Delta\phi_{B,2} - \Delta\phi_D - \Delta\phi_{B,1} \tag{11.21}$$

where $\Delta\phi_{B,1}$ is dependent on the analyte concentration while $\Delta\phi_{B,2}$ is held constant by the buffer solution in the hydrogel layer. By analogy with Equation (11.18) the threshold voltage of the ISFET can be formulated as follows:

$$V_T^* = V_T - (E_r - \Delta\phi_m - \Delta\phi_s) - \Delta\phi_{SI} \tag{11.22}$$

where E_r is the reference electrode potential including the junction potential. Each term in this equation is a constant, except the membrane potential, that depends on the analyte ion activity as shown in section 10.8.3. Upon making appropriate substitutions and rearrangements, one obtains the following response equation in terms of gate voltage at constant drain current:

$$(V_G)_{I_D=const} = \text{constant} + \frac{2.303RT}{zF} \log a_i \tag{11.23}$$

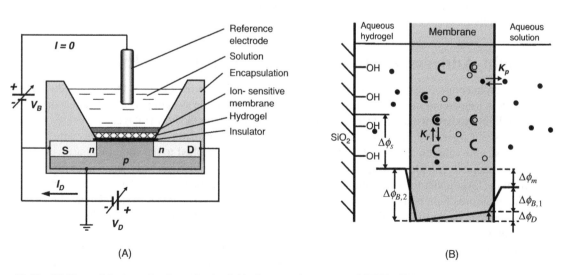

(A)   (B)

Figure 11.13 (A) The architecture of an ion-selective field-effect transistor sensor (ISFET); (B) Layer sequence for a polymeric membrane ISFET. (B) was adapted with permission from [22]. Copyright 1996 John Wiley & Sons.

where a_i is the activity of the analyte ion and z is its charge. The effect of interference by other ions (j) on the response to the analyte ion is accounted for by a Nikolskii–Eisenman-type equation in which $k_{i,j}^{pot}$ is the potentiometric selectivity coefficient:

$$(V_G)_{I_D=const} = constant + \frac{2.303RT}{zF} \log \left(a_i + k_{i,j}^{pot} a_j \right) \tag{11.24}$$

Hence, the potentiometric ISFET response to ion activity is similar to that of a standard potentiometric ion sensor. In practice, the sensitivity factor can deviate from the theoretical value and should be determined by calibration along with the constant term.

Polymeric ion-sensitive layers can be formed by solvent casting of PVC membranes (with incorporated ionophore and a lipophilic ion) on top of the gate insulator layer. However, cast membranes display poor adhesion and some kind of covalent grafting to the insulator layer imparts much better adhesion. In order for the membrane potential to affect the drain current, the sensing membrane should be extremely thin. For a PVC membrane, this limits drastically the lifetime because of fast leaching of the membrane components into the aqueous solution. Better results have been obtained with polysiloxane membranes prepared *in situ* by condensation, hydrosilylation or photopolymerization. The membrane is covalently attached to the gate oxide by silanisation of the silanol groups at the gate oxide surface, to introduces methacryl functionalities.

The best sensor stability has been obtained by covalent grafting of both the ion receptor and the lipophilic ion to the polymer network. In this way, the lifetime of the sensor can be extended up to half a year or more.

An alternative to ionophore-based polymeric membranes is the use of *molecularly imprinted polymers* [23,24]. Polymerization can be conducted *in situ* so as to allow covalent bonds to form between the gate insulator oxide and the growing polymer layer. Both organic and inorganic polymers can be imprinted for the purpose of ion sensing. Thus, organic anion FET sensors have been obtained by imprinting the analyte in an inorganic TiO_2 polymer formed by sol-gel chemistry over the SiO_2 gate insulator [25].

Solid, ion-sensitive membranes (for example, a AgCl–AgBr mixture) can be produced over the gate insulator by vapor deposition to obtain an ISFET responsive to Ag^+, Cl^- and Br^-. In the search for better selectivity, chalcogenide glass materials have been found to perform very well as sensing membranes in ISFETs [26]. A chalcogenide glass film can be formed over the gate insulator by physical vapor deposition in which preformed glass serves as a source of vapors. However, this method yields a glass film with a stoichiometry different from that of the original material. This disadvantage is circumvented by using pulsed laser vapor deposition. In this method, a pulsed laser beam is focused on the glass target in order to produce vapors that are deposited onto the gate insulator. The main advantages of this method are the well-defined reproduction of the stoichiometry of the glass coating and the short process times, as well as the possibility of carrying out the process in a reactive atmosphere.

Chalcogenide glass ISFETs for determining Ag^+, Cd^{2+}, Cu^{2+}, Hg^{2+}, Pb^{2+} with a near-Nernstian response (according to Equation (11.24)), and a detection limit in the μM region, have been developed.

The main advantage of the chalcogenide glass approach lies in the possibility of manufacturing ion sensors by using microfabrication technology, with no need for additional chemical processing. Compared with polymeric membranes, chalcogenide glass membranes display good chemical durability and excellent lifetimes. Their small size allows development of sensor arrays for multicomponent analysis of complex samples. The main drawback of chalcogenide glass membranes is their relatively poor limit of detection that cannot match that of the ionophore-based polymer membranes.

11.2.5 Light-Addressable Potentiometric Sensors (LAPS)

A typical ISFET architecture is derived from that of a FET device and no current crosses the semiconductor body. In contrast, a LAPS is formed out of a simple semiconductor-insulator-sample system [27–29]. The response is generated by infrared light irradiation that produces a photocurrent controlled by the bias voltage.

A LAPS consists of a piece of doped semiconductor coated with an insulator and a chemically sensitive layer in contact with the sample solution (Figure 11.14A). A bias voltage V_B is applied to the solution phase via a reference electrode and the circuit is completed with an AC amperometer. If the bias voltage V_B is set so as to repel the majority carriers from the layer under the insulator, a depletion layer forms and the DC current across the circuit is negligible. When silicon absorbs light of appropriate wavelength, electron–hole pairs are generated. Under the effect of the electric field, electrons move to the silicon/insulator surface and holes accumulate towards the silicon bulk. This charge separation causes a capacitively coupled displacement of charges in the external circuit. As a result, a transient photocurrent I_{ph} appears that the amperometer can detect. In practice, the light source (a light-emitting diode) is modulated as shown in Figure 11.14B to induce an AC photocurrent. If the bias voltage is too low, the depletion layer cannot form and the photocurrent is zero. However, with increasing bias voltage, the photocurrent increases until it attains a limiting value (Figure 11.14C).

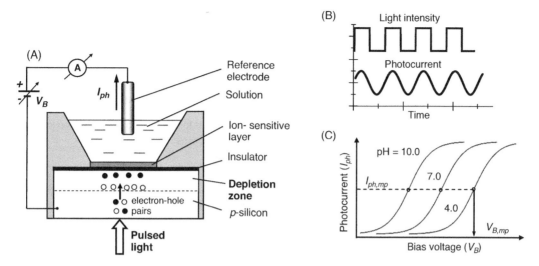

Figure 11.14 (A) The configuration of a light-addressable potentiometric sensor (LAPS); (B) time variation of light intensity and photocurrent; (C) the alternating photocurrent as a function of the bias voltage for a pH LAPS.

As the bias voltage is in series with the membrane potential, this causes the I_{ph}–V_B curve to shift in response to a change in the analyte concentration. This is exemplified in Figure 11.14C for the case of a pH sensor. The mid-point on the I_{ph}–V_B curve yields a particular $V_{B,mp}$ value that is related to the logarithm of analyte concentration by means of a linear equation [30].

The light beam can be focussed so as to excite only a limited portion of the silicon body. Therefore, the response is specific to a limited area over the insulator. This allows for facile multiplexing by installing a series of different sensing areas, each of them being specific to a particular analyte. Sequential interrogation of each sensing area by a light beam provides the response values necessary for multiple analyte determination.

The redox potential can be measured by coating metallic gold over the insulator layer. In this case, the photo-current is modulated by the redox potential that builds up at the metal layer according to the Nernst equation.

The volume of sample-solution needed in LAPS operation can be 100 nL or less, which makes the LAP suitable for microchemical analysis.

LAPS devices can be coupled with enzymes that produce either pH or redox-potential change. Combined with the very low sample volume and the multiplexing capability, this makes LAP devices attractive for applications in enzyme-linked immunoassay.

11.2.6 Reference Electrodes for ISFET Sensors

In order to take advantage of the small size of the ISFET sensor, equally small reference electrodes are needed. More-over, it is desirable to make available a FET-type reference element that can be integrated with the sensor on a single-chip device. As shown previously (Section 11.2.3) an Ag/AgCl reference electrode can be readily integrated with the sensor. However, its proper functioning is dependent on the presence of a chloride salt dissolved in a hydrogel coating. Hence, the lifetime of such electrodes is limited by chloride leakage and contamination with sample ions.

An ideal way to overcome this problem would be to develop a reference FET device (REFET) that, in contact with the solution, does not interact with the ions at all. The REFET and the ISFET can be integrated and connected as shown in Figure 11.15. In this arrangement, grounding of the test solution is performed by a pseudoreference electrode that can be a platinum layer integrated with the ISFET. The potential of the pseudoreference electrode manifests itself as a common signal in the differential system and its possible fluctuations are not disturbing. Each of

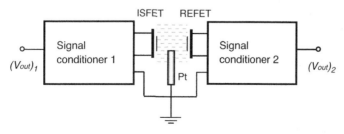

Figure 11.15 Block diagram of a differential ISFET/RFET measuring system.

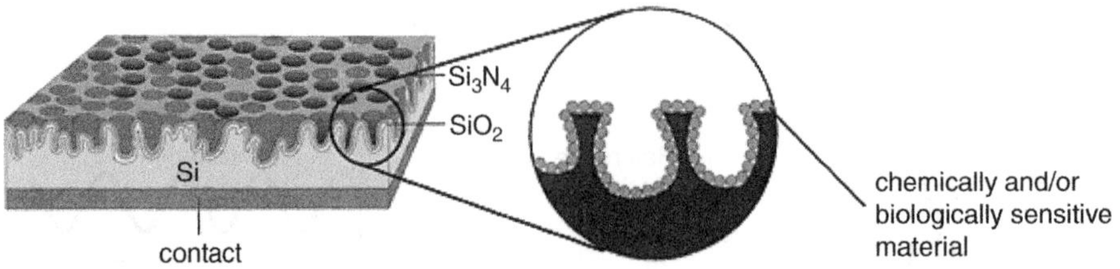

Figure 11.16 A porous EIS structure for FED sensors. The porous silicon surface is formed by anodic electrochemical etching. Adapted with permission from [10]. Copyright 2001 John Wiley & Sons Ltd.

the ISFETs and REFETs is connected to a signal conditioner that provides an output voltage. The output voltages are fed to the inputs of a differential amplifier that provides at the output a voltage difference that depends on the analyte concentration c:

$$(V_{out})_1 - (V_{out})_2 = f(\log c) \tag{11.25}$$

In this way, a totally solid-state probe, including both the sensor and the reference element, can be constructed.

As the most common ion in aqueous samples is the hydrogen ion, an ideal REFET should not be responsive to this ion. So, in principle, a REFET can be obtained by coating a compact hydrophobic polymer membrane (for example, Teflon) over the gate insulator. Rather than their hydrophobicity, it is its very low surface buffer capacity that makes such membranes suitable for REFET applications.

If the sample pH is controlled by a buffer system, a pH FET sensor can function as the reference element. Thus, in enzymatic FET sensors, the reference element can be a pH ISFET sensor coated with a membrane with no enzymatic activity.

11.2.7 Enzymatic FET Sensors (EnFETs)

An enzymatic ISFET sensor can be obtained by coating an enzyme layer over the insulator in a pH ISFET [31]. Clearly, this approach works well with enzymes that produce a change in pH on conversion of the substrate. Principally, this approach is similar to that already introduced in the Section 10.18. In the first approach of this type, penicillinase was used to develop a penicillin sensor [32]. However, the change in the pH can affect the enzyme activity and, in addition, the products of the reaction can be poorly dissociated acids or bases. In such a case, the amount of H^+ or OH^- ions produced per mole of converted substrate also depends on pH. Such factors, which are also present in the case of enzymatic potentiometric sensors, cause the response to be nonlinear.

Such problems can be alleviated by operating the sensor under pH-static conditions [33]. As in the case of the CO_2 probe introduced previously, a noble-metal electrode is included closely around the ISFET gate area. This electrode produces either hydrogen or hydroxyl ions by water electrolysis. By a feedback from the pH sensor, the electrolysis current is regulated so as to keep the pH at a constant level. This current is therefore proportional to the rate of product formation, independently of the buffer capacity of the system. The current response varies linearly with substrate concentration, as is typical of a enzymatic sensor with the response proportional to the product concentration.

In addition to the EnFET approach, an enzymatic FED sensor can be designed as a LAPS or capacitor-type sensor. For the particular case of penicillin sensors, it was shown that all three of the above devices give similar performances [34]. Selection of a particular approach should be made by taking into account the complexity of the readout equipment the sensor configuration and the available manufacturing technology with regard to the particular application to be addressed.

The performances of capacitive FED sensors are considerably improved by using a porous surface silicon body upon which the insulator layer is formed by chemical vapor deposition, as shown in Figure 11.16. Deposition of the enzyme layer over the porous insulator layer allows for higher enzyme loading and, in this way, enhances the durability of the device. At the same time, the higher surface area brings about an increased capacitance, which is advantageous from the standpoint of response measurement [35]. Nonmodified porous insulator surfaces proved to give superior performances in pH detection.

Further advances in EnFET technology aim at expanding the response range, lowering the limit of detection and improving the design of multiplexed EnFET sensors [36,37].

11.2.8 Outlook

FED ion sensors are developed from semiconductor-insulator structures by adding a responsive layer over the insulator instead of the metal gate, which is replaced with a reference electrode. A potential difference formed in response to the analyte interaction with the responsive layer determines a modification in a characteristic electrical parameter of the device. This modification arises from the alteration of the device threshold voltage.

There are three possible approaches to the design of FED ion sensors, namely the EIS ion-sensitive capacitor, the ISFET and the LAPS. EIS capacitor sensors rely on the shift of the capacitance–voltage curve due to the ion interaction with the device. In the case of ISFETs, it is the shift of the drain current–gate voltage curve that provides the sensor response. In contrast to the above two designs, which involve only electrical devices, the LAPS resorts to light in order to excite a photocurrent that depends on the electric potential at the device/solution interface. The feasibility of each of the above type of sensor is well documented and an option for a particular approach is determined mostly by the complexity of the measurement setup. Although the LAPS seems to be more intricate, it offers an important advantage in the case of sensor multiplexing. This is due to the fact that multiple LAPSs can be build up on a single EIS structure on which a series of different sensitive layers can be formed. The response of each layer is obtained by scanning the device with a light beam.

The most straightforward application of FED device sensors is represented by the pH sensors in which the sensing layer is formed by ionogen groups at the insulator surface. FED pH sensors can function as transducers in gas probes for detecting gases that modify by dissolution the local pH. On the other hand, upon adding an enzymatic layer to a FED pH sensor, one can obtain a FED biocatalytic sensor. Ion-sensitive FED devices are fabricated by including an ion-responsive membrane, either in the solid-state form or as a polymeric membrane. Ion-responsive membranes used in FED ion sensors are, in general, similar to those developed in the frame of potentiometric ion sensors. When using polymer membranes, care should be taken to secure a stable potential difference between the membrane and the insulator.

The most attractive feature of FED sensors is the easy manufacturing by microfabrication technology. Taking also into account the small size of the device, it results that FED ion sensors and biosensors are suitable for multiplexing in order to obtain multisensor arrays, which, in addition to chemical sensors, can also include sensors for physical parameters.

An instructive survey of various kinds of FED ion sensors and biosensors is presented in ref. [34].

Questions and Exercises (Section 11.2)

1 Sketch the structure of an EIS device and note down on it each relevant potential difference.
2 What is the role of the reference electrode in a EIS device?
3 What physical parameter of the device is dependent on the ion interaction with the sensing element?
4 How does the hydrogen ion interact with the sensing part of a pH ISFET?
5 What parameter of the sensing part determines the sensitivity of a pH ISFET? How can the sensor sensitivity be made close to the ideal one?
6 What materials are suitable as insulators in pH ISFETs?
7 How can the potentiometric response of a pH ISFET be derived?
8 What alternative semiconductor device can be used in pH measurements? Comment on advantages and disadvantage of this device as compared with the pH ISFET.
9 How can a pH ISFET be used for the determination of acid(base)-generating gases? What are the advantages of such gas probes?
10 Sketch the general configuration of an ISFET for other ions than H^+ and indicate on the sketch the relevant potential differences.
11 Try to derive the potentiometric response function of an ISFET using the characteristic equations of the ISFET and the membrane potential.
12 How can a stable response of a polymer membrane ISFET be secured?
13 Give a short overview of polymeric materials used in the sensing component of ISFETs.
14 What kind of solid materials can be used in solid-state ion-responsive membranes? Give an account of the advantages and disadvantages of such materials.
15 Discuss the configuration and response mechanism of a LAPS. Briefly review the advantages and disadvantages of LAPSs.
16 Discuss the problem of the reference electrode for ISFETs.
17 Prepare a list of enzymes and enzymatic reactions that are suitable for application in EnFET sensors (Hint: see Chapter 3).
18 Give a brief overview of enzyme immobilization methods for EnFETs (Hint: see Chapter 5)
19 Comment on methods for improving the performance of EnFETs.
20 Elaborate on the application of coulometric titration in EnFETs and FET gas probes.

11.3 FED Gas Sensors

As pointed out before, a key parameter in the functioning of field effect devices is the threshold voltage, which depends on the difference between the work function of the metal and that of the semiconductor. Upon interaction of a gas with either the metal or the semiconductor the work function of the sensing part of the device can be altered [5,38]. The resulting change in the threshold voltage can be measured and used to calculate the gas-analyte content in a gas mixture.

11.3.1 FED Hydrogen Sensors

Metal-oxide-semiconductor layered structures respond to the hydrogen concentration in a gas if the metal layer consists of palladium or platinum. The general structure of such a sensor is shown in Figure 11.17. The response mechanism is based on the catalytic properties of the platinum metal which can absorb molecular hydrogen from the gas sample and promotes molecule dissociation to hydrogen atoms dissolved in the bulk of the metal. Further, hydrogen atoms undergo fast diffusion through the metal and reach the metal/oxide interface. Hydrogen atoms adsorb at the interface and turn polarized, forming a dipoles layer. This layer generates a potential difference across the interface that affects the charge distribution at the semiconductor surface. As a consequence, the electrical parameters of the MIS system are altered in relation to hydrogen concentration in the gas sample [4,39,40]. The surface dipoles modify the work function of the metal with an amount that is the electrical work required to transfer an electron throughout the dipolar layer on its way out of the metal.

Therefore, the response to hydrogen can be interpreted by taking into account the effect of the work function on the threshold voltage. The first hydrogen sensor based on this principle was developed by Lundström in 1975 [41]. A retrospective overview of the development of MIS gas sensors is presented in refs. [42,43].

Two alternative types of FED gas sensor are possible, namely the MIS capacitor type and the MISFET type (Figure 11.18). The MISFET-type sensor is often called ChemFET, an acronym for chemically sensitive field-effect transistor. In both devices, the 100–200-nm thick palladium layer functions as analyte-sensing element. In the case of the capacitor device, the response signal is the change in the bias voltage needed to keep the capacitance at a constant value (Figure 11.18A). As is typical of MIS capacitors, the capacitance–voltage curves are obtained by AC current measurements.

The structure of a MISFET sensor is more intricate, but the operation of this device involves only DC signals. As shown in Figure 11.18B, the change in the threshold voltage produced by hydrogen absorption induces a shift of the drain current–gate voltage curve at a constant drain current. The operation of this sensor is therefore similar to that of the ion-sensitive FET devices.

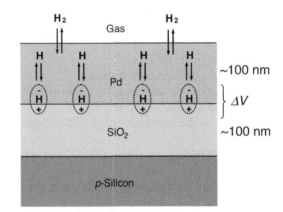

Figure 11.17   Structure of a Pd-SiO₂-p-silicon MIS hydrogen sensor.

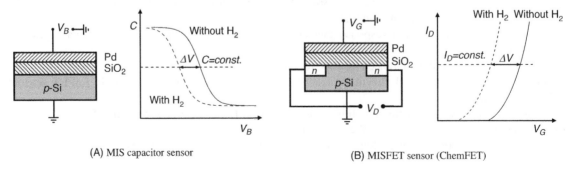

(A) MIS capacitor sensor   (B) MISFET sensor (ChemFET)

Figure 11.18   FED hydrogen sensors and corresponding characteristic curves.

With both MIS capacitors and MISFET sensors the response is determined by the alteration of the threshold voltage. In turn, the change in the threshold voltage depends of the amount of gas absorbed in the metal phase at equilibrium with the gas phase. Further, the concentration of the gas in the metal is determined by the partition coefficient in the gas–metal system, which is defined as follows:

$$\text{Partition coefficient} = \frac{\text{Analyte concentration in metal}}{\text{Analyte concentration in the gas}} \tag{11.26}$$

The response function in terms of voltage change (ΔV in Figures 11.18A and B, respectively) is the same for both the MIS capacitor and the MISFET sensor. Denoting by $V_{T,H}$ and $V_{T,0}$ the threshold voltage in the presence of or in the absence of hydrogen, respectively, the measured voltage change is:

$$\Delta V = V_{T,H} - V_{T,0} \tag{11.27}$$

The response function that relates the voltage change to the hydrogen content can be approximated by the Langmuir isotherm equation as follows [44]:

$$\Delta V = \Delta V_{\max} \frac{\alpha \sqrt{p_{H_2}}}{1 + \alpha \sqrt{p_{H_2}}} \tag{11.28}$$

where p_{H_2} is the partial pressure of hydrogen while $\Delta V_{\max}$ and α are empirical parameters. At very low partial pressure, Equation (11.28) can be reduced to the following linear function:

$$(\Delta V)_{\text{Low partial pressure}} = \alpha \sqrt{p_{H_2}} \tag{11.29}$$

Here, the proportionality constant α includes the effect of oxygen, if oxygen is present in the sample. Therefore, the calibration and the measurement should be performed at a similar oxygen content in the standard gas and the unknown sample.

The response time of the sensor is determined by the rate of hydrogen diffusion within the palladium layer. As this process is very fast even at the room temperature, the response time under these conditions is of about 0.15 ms. The response time can be dropped to about 10 μs by operating the device at 150 °C.

The reported limit of detection of MIS hydrogen sensors is close to 3×10^{-5} Pa hydrogen partial pressure in an inert atmosphere and 5×10^{-4} Pa in air. The higher detection limit in air is due to oxygen interference. Oxygen and hydrogen react at the metal surface to form water and, in this way, hydrogen atoms are depleted in the metal surface layer. Consequently, the hydrogen atom concentration in the metal is reduced.

More details about the response mechanism of MIS hydrogen sensors can be found in ref. [44].

11.3.2 Metal Gate FED Sensors for Other Gases

Ammonia can also be detected by a palladium-layer MIS device. Ammonia molecules dissociate under the catalytic effect of the metal producing hydrogen atoms that are detected according to the previously emphasized mechanism. In order to achieve a reasonable sensitivity, the sensor should be operated at a temperature over 100 °C in order to promote the dissociation. Sensitivity to ammonia can be enhanced by adding an about 3-nm layer of a catalytic metal (Pt or Ir) over the Pd layer. In this version, the operation temperatures can be only slightly above room temperature and the sensor can be used to monitor enzymatic reactions that evolve ammonia.

Palladium MIS devices are also responsive to other gases such as carbon monoxide, hydrogen sulfide, alcohols, amines and unsaturated hydrocarbons. However, as such gases are poorly soluble in palladium, relatively few molecules reach the metal/semiconductor interface to form a dipole layer. Hence, the sensitivity of a sensor in a configuration like that in Figure 11.17 is poor when above gases are concerned. A reasonable sensitivity can be achieved by means of a discontinuous palladium gate (Figure 11.19A). In this case, gas molecules adsorb at the triple phase

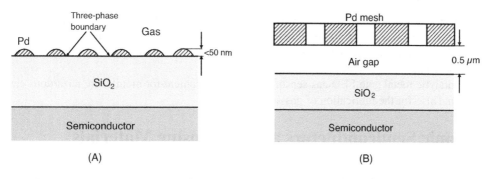

Figure 11.19 Alternative configurations of MIS gas sensors. (A) Discontinuous metal gate; (B) suspended-gate MIS sensor.

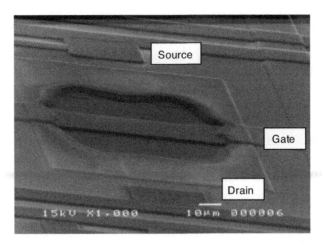

Figure 11.20 Scanning electron microscope image of a suspended gate FET device. The device area is 130 μm × 70 μm. Reproduced with permission from [47]. Copyright Elsevier 2010.

boundary formed by the gas, the insulator and the metal spots. As a result, the effect of the gas on the work function of the metal is much greater than in the case of a compact metal layer.

The detection limit for the above-mentioned gases is generally of the order of 1 ppm in air. The response function of the sensor is given in Equation (11.30) where p_G is the partial pressure of the analyte whilst α and a are empirical parameters.

$$\Delta V = \alpha p_G^a \tag{11.30}$$

An alternative approach to sensing palladium-insoluble gases relies on devices including a *suspended metal gate* that is not in direct contact with the insulator surface. The suspended gate can be a metal mesh (Figure 11.19B), a perforated metal plate or, in general, a metal plate mounted so as to allow communication between the air gap and the gas sample. The air gap between the gate and the insulator functions as an additional insulator layer. When polar gas molecules adsorb either at the metal or the insulator surface, the molecular dipoles give rise to a surface potential that is proportional to the density of adsorbed molecules. This potential difference is in series with the gate voltage and, therefore, alters the drain current. A suspended-gate FED sensor can be produced by microfabrication, with the gate made of heavily doped silicon, which is an electronic conductor [45]. In an alternative design, the solid insulator is absent and the single present insulator is the air gap. In this instance, the analyte gas can adsorb on the semiconductor surface and affect directly the charge transport through the channel.

Not only metals, but also other conductive materials can be used to fabricate the suspended gate. The gate material should be selected so as to be able to interact with the analyte. For example, Figure 11.20 shows a humidity sensor including a polycrystalline silicon suspended gate. The gate has been obtained by chemical vapor deposition over a preformed siloxane layer that has been removed after completing the deposition.

As shown before, a catalytic metal gate FED sensor can respond to a broad variety of different gases. That is why it is important to underline several methods for controlling the selectivity of these sensors. A straightforward method relies on the control of the operation temperature that affects both the rate of chemical reactions and the partition coefficient of the gas. On the other hand, one can resort to a particular catalytic metals or alloy as the gate materials in order to obtain convenient selectivity.

Due to the very small size, MIS gas sensors are suitable for the fabrication of sensor arrays. An array can be formed of several sensors, each of them responding selectively to a certain analyte. As advanced selectivity is hardly achievable, it is more convenient to make use of an assembly of cross-selective sensor and extract the concentration of each analyte by multivariate data analysis (Section 1.7). This approach is also suitable in the design of electronic nose devices [47].

In many industrial applications the gas sensor should operate at a high temperature. Under such conditions the semiconductor properties of silicon are altered due to the relatively low bandgap of this material. High-temperature MIS gas sensors rely on semiconductor with a wide bandgap, such as silicon carbide [48], group III element nitrides (AlN, GaN) and semiconductor diamond.

In conclusion, catalytic metal gate FED gas sensors are very convenient for monitoring hazardous gases in industrial environment and also for the fabrication of gas-sensor arrays.

11.3.3 Organic Semiconductors as Gas-Sensing Materials

The application of metal gate FED gas sensors is mostly limited to the detection of hydrogen and certain hydrogen-containing gases and vapors. This limitation arises from the catalytic nature of the gas interaction with the metal gate. Conversely, organic semiconductors provide a broader variety of possible interactions with gases and vapors and are well suited for detecting polar molecule compounds. Both the sensitivity and the selectivity of organic semiconductor sensors can be tuned by doping or by proper tailoring of certain components of the device. As in the case of the metal gate ChemFETs, the transduction mechanism relies on the modification of the work function of the sensing material.

Two kinds of materials have been used for sensing purposes in organic semiconductor gas sensors, namely molecular semiconductors (phthalocyanines and porphyrins) and conducting polymers [49].

Phthalocyanines and porphyrins have already been introduced in Section 10.9.9. Both kinds of compound consist of macrocycles including four pyrrole units and features a conjugated π-bond system that is responsible for the semiconducting properties. At the same time, the four pyrrole nitrogens can form coordination bonds with various metal ions. In the solid-state form, such planar molecules form molecular stacks and can accommodate dopant molecules in channels adjacent to the stacks. Metal phthalocyanines can be obtained either as insulators (for example, NiPc; Pc stands for phthalocyanine) or intrinsic p-semiconductors (for example, $LuPc_2$). Insulator phthalocyanines are easily converted into p-semiconductors by doping.

Depending on the properties of the analyte gas, the phthalocyanine–analyte interaction can proceed according to various mechanisms. Thus, by interaction with the metal center, an oxidizing species generates extrinsic positive charges, while a reducing species gives rise to negative charges. That is why phthalocyanine films are suitable for the detection of oxidizing gases such as ozone and nitrogen dioxide. Weak charge-transfer interactions with the organic region of the molecule are relevant in the case of analytes that do no interact with the metal ion.

Phthalocyanine sensing films can be obtained by vapor deposition or by the Langmuir–Blodgett technique. Sensors based on phthalocyanine can operate at temperatures up to 200 °C. Phthalocyanine applications in FED gas sensors has been reviewed in ref. [50].

Conducting polymers, (introduced in Section 5.9) can be deposited by spin coating, drop casting or electrochemical polymerization. The cast layer incorporates an important proportion of solvent. Hence, by solvent evaporation, the properties of the film are altered. A better alternative relies on gel-like films formed of a mixture of polymer and an ionic liquid. Ionic liquids are salts of certain organic cations and are present in the liquid state at ambient temperature. Low volatility, good thermal stability and good solubility of gases in ionic liquids result in superior performances over the solid polymer coatings.

Sensitivity improvement can be achieved by incorporating binding sites into the polymer coating. Binding sites can be formed of dielectric polymers, carbon fibers, metal nanoparticles or metal oxides.

Conducting polymer gas sensors can be operated at temperatures up to 90 °C.

11.3.4 Organic Semiconductors FED Gas Sensors

Organic semiconductors are suitable as sensing materials in gas sensors and allow potentiometric sensors for neutral molecules to be devised based on the modulation of the work-function. In order to perform this task, an organic semiconductor should be integrated within a field effect device. There are two possible integration alternatives. The first one relies on inorganic semiconductor FED devices in which the gate is made of an organic semiconductor, as shown in Figure 11.21A. This is actually a modified ChemFET device. Upon gas interaction with the organic sensing layer, the work-function difference between the semiconductor and the gate is altered and the threshold voltage changes in relation to the gas concentration.

Another approach is based on organic field-effect transistors (OFET, Figure 11.21B). An OFET has the typical configuration of MISFETs but are fabricated as thin-film transistors, which implies that the thickness of the semiconductor body is very low [51–53]. Gas interaction with the organic semiconductor does modify the threshold voltage of the device, which makes OFET suited for gas-sensing applications [54].

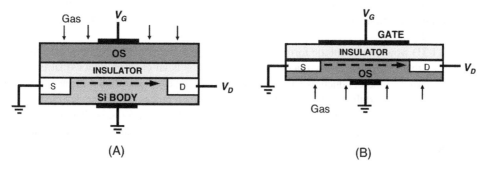

Figure 11.21 Two types of gas sensors including an organic semiconductor (OS) sensing layer. (A) Modified gate silicon FET (ChemFET); (B) organic semiconductor transistor (OFET). Dashed arrows indicate the current flow through the device.

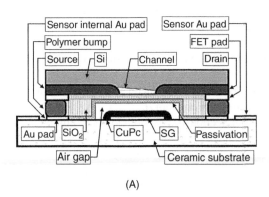 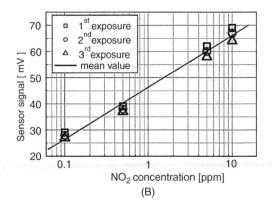

(A)   (B)

Figure 11.22 (A) Structure of a suspended-gate copper phthalocyanine (CuPc) based FET nitrogen dioxide sensor SG indicates the suspended gate on which the phthalocyanine layer is loaded. (B) The calibration graph of a sensor like that in (A) at 25 °C. Reproduced with permission from [57]. Copyright 2006 Elsevier.

There is, however, an essential difference between the organic layer ChemFET sensor and the OFET-type one. In the first case, no current flows through the organic layer, whereas in the second case, the current passes from the source to the drain throughout the organic semiconductor and encounters three resistances: two at the contacts and one in the organic semiconductor. The change in the work function can affect each of these three resistances. If a voltage is applied to the gate (Figure 11.21B) this will affect not only the drain current, but also all these resistances. Hence, the response is nonlinear. It was therefore asserted that the OFET device acts actually as a chemically modulated resistor and the field effect modulation of the current is of a minor importance [49].

OFET application in gas sensors is still in its infancy. Conversely, modified gate ChemFET sensors are well-established devices. In addition to the organic semiconductor approach underlined in Figure 11.21A, modified suspended-gate devices are also of interest. In this case, the sensing layer is supported on a metal gate separated from the insulator by an air gap.

Thus, an alcohol vapor sensor has been obtained by depositing a polypyrrole sensing layer over a platinum mesh suspended gate [55]. Figure 11.22A shows the structure of and NO$_2$ sensor based on a 150–200-nm thick phthalocyanine sensing layer. This layer has been produced by spraying from polyacrylic acid solutions onto gold-patterned alumina through a mechanical mask. In addition to the measuring unit, the device includes reference units for humidity and drift compensation. Humidity compensation is important because organic semiconductors respond in general to the content of water vapors in the gas sample. The graph in Figure 11.22B demonstrates a linear response in a semilogarithmic representation and also proves the very good reproducibility of the results.

11.3.5 Response Mechanism of FED Gas Sensors

In the case of potentiometric ion sensors, the key parameter is the surface potential that forms in response to the ion transfer between a solution and a semipermeable membrane. The transfer process involves the transfer of an *integer number of elementary charges* that are borne by each ion.

The situation is radically different when a polar molecule A enters a metal or semiconductor material. This becomes polarized, which is similar to the transfer of a *partial charge* δe, where e is the elementary charge and δ is a fractional number:

$$A \rightleftharpoons A^\delta + \delta e; \quad K_{ct} = \frac{[e]^{2\delta}}{k_H p_A} \tag{11.31}$$

where K_{ct} is the equilibrium constant for the gas dissolution in the solid material, p_A is the partial pressure of the gas A and k_H is the absorption coefficient in the Henry isotherm. Hence, the $k_H p_A$ product is the concentration of the dissolved gas in the solid phase.

If $\delta < 1$, reaction (11.31) would represent the formation of a charge-transfer complex between the solid matrix and the dissolved gas. Assuming that the matrix is a semiconductor and the gas is an electron donor (that is, a Lewis base), a negative charge is transferred to the conduction band. For an electron-acceptor gas (that is, Lewis acid), electrons are withdrawn from the conduction band thus increasing the hole density. Such processes can be viewed as doping processes that shift the Fermi level. Consequently, the work function of the semiconductor material is altered.

Based on these premises, an equation relating the Fermi level of an n-semiconductor to the partial pressure of the donor gas has been derived in the following form [38]:

$$E_F = E_D^* + \frac{k_B T}{2\delta} \ln p_A \tag{11.32}$$

where k_B is the Boltzmann constant and T is the absolute temperature. The constant term E_D^* is the standard donor-dopant energy level at $p_A = 1$, that is, in the pure analyte gas. For a p-type semiconductor, the term E_D^* should be replaced by the standard acceptor-dopant energy level E_A^*. Equation (11.32) has been derived under the assumption that the equilibrium constant K_{ct} is not affected by gas absorption, which is only an approximation. Despite this limitation, Equation (11.32) represents a sound basis for the rationalization of the response of semiconductor-based sensors to neutral molecules.

There is a striking analogy between Equation (11.32) and the Nernst equation that applies to electron or ion transfer between two phases. The analogy is more evident if one remember that $R = k_B N_A$ and $F = e N_A$, where N_A is the Avogadro number and F is the Faraday constant. The main difference originates from the charge term in the denominator of the prelogarithm term. While the charge z in the standard Nernst equation is an integer number, the term δ in Equation (11.32) is a fractional number.

The change in the Fermi level of the sensing material can be assessed by means of the change in the threshold voltage of a FED including a semiconductor-based sensing layer. An important point is that the change in the threshold voltage can be observed only if the gas-sensitive coating interface is capacitive, that is, if no interphase electron transfer is possible [38]. This implies that the alteration of the work function originates only in the sensing layer adjacent to the insulator. Thus, in the case of the hydrogen sensitive Pd-ChemFET, the formation of a dipole layer at the Pd/insulator interface is responsible for the local modulation of the work function.

Sensor operation in the saturation regime ($V_D \gg (V_G - V_T)$) is preferable because it implies an explicit relationship between the work function and the operating gate voltage. Using Equation (11.7) one obtains the following relationship between the gate voltage and the work function at constant drain current:

$$V_G = \sqrt{2 I_D / \beta} + V_T \tag{11.33}$$

On the other hand, the dependence of the gate voltage on the partial pressure of the analyte gas (p_A) can be formulated as follows [49]:

$$V_G = V_G^* + \frac{k_B T}{2 \delta_A} \ln \left(p_A + \sum_i k_i^{\text{pot}} p_i \right) \tag{11.34}$$

where V_G^* includes all constant terms related to the transistor and the sensing layer as well as $\sqrt{I_D}$. p_i is the partial pressure of any other gas in the sample and k_i^{pot} is the potentiometric selectivity coefficient. The above equation is alike to the Eisenman-Nikolskii equation, with the notable difference that instead of the integer ion charge z, the fractional charge δ_A appears in the prelogarithm factor. The background response V_G^0, is obtained with a reference gas in which $p_A = 0$: Therefore, the gate voltage change with respect to the background is:

$$\Delta V_G = V_G - V_G^o = \frac{k_B T}{2 \delta_A} \ln \left[p_A \left(\sum_i k_i^{\text{pot}} p_i \right)^{-1} + 1 \right] \tag{11.35}$$

If the interference is negligible, ΔV_G is proportional to the logarithm of the analyte partial pressure. This kind of response is illustrated in Figure 11.22B.

11.3.6 Outlook

FET gas sensors represent an application of FET devices to sensing of neutral molecules. In contrast to FED ion sensors, which are obtained by forming an ion-responsive layer over the insulator film, FET gas sensors preserve the standard structure of MIS devices. The sensing element in FET gas sensors is a structural component of the MIS devices that is typically the gate.

In order to perform gas recognition, the gate is formed of a metal that can absorb or adsorb the target analyte. In the case of hydrogen sensors, sorption leads to the dissociation of the hydrogen molecules yielding hydrogen atoms. Hydrogen atoms become polarized at the metal/insulator surface and in this way alter the work function of the metal. As a result, the threshold voltage of the MIS device is modified, leading to the modification of a characteristic functioning parameter. In the case of MIS capacitors, this parameter is the capacitance, whilst in the case of a MISFET device, the drain current is modified. This principle has also been applied to FED sensors for gases that give rise to hydrogen atoms upon dissolution in a the gate metal, such as ammonia.

Gas sensing by FED devices is also feasible for gases that are not soluble in the gate metal but undergo adsorption at the gate surface accompanied by molecule polarization. Best performances are obtained in this case by means of suspended-gate devices. Further, modification of the suspended gate with an organic semiconductor film opened the way to the development of FED sensors gas for various gases and vapors that cannot be sensed by a metal gate.

FED gas sensor relies on the formation of polar atoms or molecules upon interaction between the sensing element and the analyte. This process is similar to the transfer of a partial charge between the analyte and the sensing material. Consequently, the sensor response can be described by a generalized form of the Nernst equation, in which a fractional charge replaces the integer charge included in the standard form of this equation.

Questions and Exercises (Section 11.3)

1 What is the mechanism of hydrogen sensing by means of a MIS device?
2 What parameter of a MIS device is affected by hydrogen action upon the sensing element?
3 What are the measurement methods in the cases of a MIS capacitor and a ChemFET hydrogen sensor?
4 How do hydrogen-containing gas heteromolecules interact with a catalytic metal-based MIS device?
5 Why a compact Pd layer is not suitable for the detection of hydrogen containing gas heteromolecules? What are the alternatives to this approach and why do such alternatives perform better?
6 What kind of organic materials are appropriate for gas detection by means of FEDs? Briefly review the relevant structural and physical characteristics of such materials.
7 What is the response mechanism of organic semiconductor FED gas sensors? Comment on the differences and similarities between ISFETs and gas FET sensors.
8 Review the two types of FET gas sensors based on organic semiconductors and comment on the transduction mechanism and the response function in each case.

11.4 Schottky-Diode-Based Gas Sensors

A Schottky diode is formed of a metal film coated over a semiconductor surface to form a heterogeneous metal–semiconductor junction. The Schottky diode is a rectifier, which implies very low resistance under forward bias and very high resistance under reverse-bias conditions.

Gas sensors based on the Schottky diode can be obtained by coating a thin layer of a catalytic metal (for example, palladium) over a semiconductor oxide such as titanium dioxide. The response of this sensor is based on the variation of the work function of the metal upon interaction with the gas [40,57].

In order to understand the response mechanism of the diode gas sensor it is necessary to examine first the band diagram of the diode in the absence and then in the presence of an analyte like hydrogen. The band diagram of a palladium Schottky diode in an inert atmosphere is shown in Figure 11.23A. The direct contact between the metal and semiconductor leads to equalization of the Femi levels in each component under equilibrium conditions. Due to surface charges, both the conduction and the valence bands in the semiconductor bend upwards. When an electron moves from the metal to the semiconductor, it encounters an energy barrier Φ_b that is the difference between the Fermi level and the edge of the conduction band at the interface. The energy barrier is determined by two material properties, namely the work function of the metal (Φ_M) and the electron affinity of the semiconductor (χ_s). By definition, the electron affinity is the energy required to bring an electron

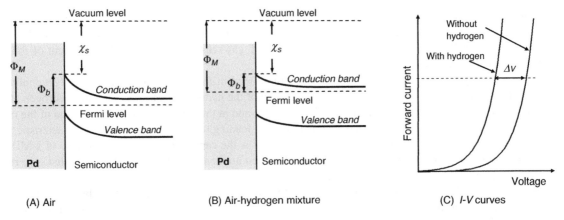

Figure 11.23   Band energy diagram of a palladium–semiconductor system in pure air (A) and in hydrogen-containing air (B). (C) Current–voltage curves of a palladium-semiconductor Schottky diode in the absence or in the presence of hydrogen.

from the vacuum level to the bottom of the conduction band at the surface. Therefore, according to the diagram in Figure 11.23A the energy barrier is given by:

$$\Phi_b = \Phi_M - \chi_s \tag{11.36}$$

An electron from the metal can move to the semiconductor only if its thermal energy overcomes the energy barrier Φ_b. On the other hand, an applied voltage affects the position of the Fermi level in the semiconductor relative to the Fermi level in the metal. As a consequence, the electron flow across the diode is determined by the applied voltage and, under forward-bias conditions the current is given by the following equation [1]:

$$I = A_b \left[\exp\left(\frac{eV}{k_B T} \right) - 1 \right] \tag{11.37}$$

Here, V is the applied voltage, k_B is the Boltzmann constant, T is the absolute temperature e is the elementary charge and A_b is a potential barrier-dependent term:

$$A_b = k_S \left[\exp\left(-\frac{\Phi_b}{k_B T} \right) \right] \tag{11.38}$$

where k_S is constant parameter of the semiconductor. Equation (11.36) indicates that the A_b term depends on the work function of the metal and hence can vary upon hydrogen absorption. Therefore, the forward current increases exponentially with the voltage (Equation (11.37)) and the pre-exponential term depends on the hydrogen partial pressure. As a result, hydrogen absorption causes the current–voltage curve to be translated with respect to the reference curve obtained in an inert atmosphere, as shown in Figure 11.23C. The voltage shift ΔV at a constant current is a function of the partial pressure of hydrogen in the gas sample.

The structure of a Schottky-diode hydrogen sensor is shown in Figure 11.24A, whilst the response of Schottky diode to the hydrogen content in air is shown in Figure 11.24B. Data in Figure 11.24B have been obtained with a Pd-TiO$_2$ diode operated at the room temperature. A clear voltage shift at constant forward current can be noted at hydrogen concentrations as low as 10 ppm and the response range extends over a broad interval.

Schottky-diode sensors are very simple and inexpensive devices. As is typical of semiconductor gas sensors, the response is generated by modulation of the work function under the action of the gas analyte. If silicon is used as the semiconductor, Schottky-diode sensors are compatible with monolithic silicon integrated circuit technology. The current–voltage curve obeys an exponential low and exhibit excellent hydrogen detection sensitivity. However, the operation temperature of silicon diodes is limited to 250 °C, due to the small energy gap of silicon. Sensors operating at higher temperature can be obtained by using semiconductors with a larger energy bandgap, such as silicon carbide or semiconductor diamond.

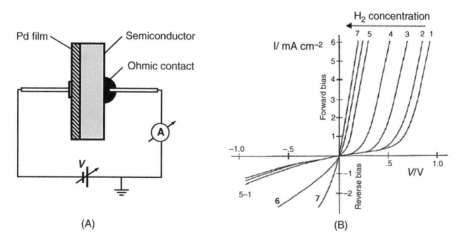

Figure 11.24 (A) A Pd Schottky-diode hydrogen sensor; (B) current–voltage characteristic curves of a Pd-TiO$_2$ Schottky diode at different concentrations of hydrogen in air at 25 °C. Hydrogen concentration: (1) 0 ppm; (2) 14 ppm; (3) 140 ppm; (4) 1400 ppm; (5) 7150 ppm; (6) 1%; (7) 1.5%. (B) Adapted with permission from [59]. Copyright 1980 Elsevier.

11.5 Carbon-Nanotube-Based Field-Effect Transistors

Semiconductor nanomaterials, such as graphene, carbon nanotubes and silicon nanowires are currently being intensively investigated for possible applications in microelectronics. It is therefore not surprising that nanomaterial-based semiconductor devices have raised considerable interest from the standpoint of chemical sensing. This section introduces the carbon nanotube field-effect transistor (CNFET) and some of its possible applications in chemical sensors.

As shown in Figure 11.25A, a nanosize field-effect transistor can be constructed by connecting a single-walled semiconductor carbon nanotube to source and drain contacts over an insulating film coated over a conducting material, such as a very highly doped semiconductor. The underlying material functions in this case as a gate, whereas the nanotube forms the channel. The device characteristic curve (that is, the $I_D - V_G$ curve) is similar to that of a standard FET. If a sufficiently negative gate potential is applied, holes are created in the nanotube resulting in increased conductance. However, the manufacturing reproducibility of single-nanotube FETs is poor and, from this standpoint, it is more convenient to use instead a network of randomly oriented nanotubes. Another CNFET device geometry involves suspending the nanotube over a trench to prevent the contact with the substrate and gate oxide. This technique has the advantage of less charge-carrier scattering at the CNT/substrate interface, which improves the device performance.

The current across the nanotube channel is very sensitive to the state of the nanotube surface. As a result, adsorption of simple gases such as oxygen or ammonia on the nanotube have a great effect on the source–drain current at constant gate voltage. A CNFET can therefore act as a chemical sensor and this property is manifested not only in the gas phase but also in liquids. In the case of liquids (Figure 11.25B) a second gate (a "liquid gate") can be created by inserting a reference electrode into the sample solution.

Two different mechanisms can account for the change in the channel conduction when a CNFET interacts with an adsorbed compound. In the first case (Figure 11.26A), electrons from the adsorbate molecules are transferred to the π orbitals of the nanotube leading to partial hole depletion. In order to achieve the same current, the gate voltage needs to be shifted to more negative values compared with the case of a plain surface. Consequently, the characteristic curve is displaced as shown in Figure 11.26A. In the second mechanism (Figure 11.26B), a charged adsorbate gives rise to an electric potential that scatters the charge carriers in the nanotubes and causes the current to become lesser at any particular applied gate voltage.

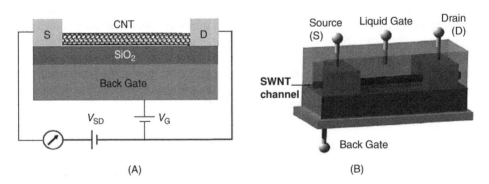

Figure 11.25 (A) Structure and wiring of a CNFET. (B) A CNFET-based chemical sensor for liquid samples. SWNT stands for single-walled carbon nanotube. Reproduced with permission from [60]. Copyright 2007 Wiley-VCH Verlag GmBH & Co. KGaA.

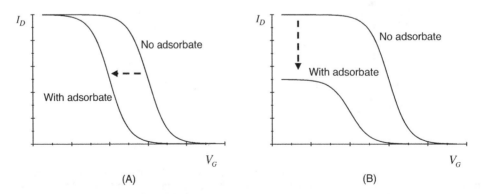

Figure 11.26 Effect of surface-adsorbed species on the characteristics of a CNFET sensor. (A) The effect of electron transfer from adsorbate to the nanotube; (B) the effect of charge carriers scattering by the electric potential created by the adsorbate.

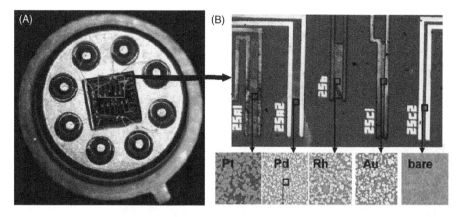

Figure 11.27 A CNFET gas-sensor array. (A) A silicon chip (2 mm × 2 mm) containing multiple carbon-nanotube devices. (B) Enlarged view of five pairs of metal electrodes connecting networks of CNTs. The first four networks are plated with the metals indicated in figure; the bare device functions as reference element. Reproduced with permission from [64]. Copyright 2006 American Chemical Society.

In order to develop a CNFET-based biosensor, it is best to prevent nonspecific interactions by coating the nanotubes with a polymer layer such as polyethylene glycol/polyethyleneimine. A suitable receptor can be attached to this layer in order to achieve analyte binding. A charged analyte molecule (protein or DNA) brings about a decrease in the channel current (Figure 11.26B), which is in relation to the analyte concentration and the incubation time.

Various types of CNFET biosensors based on immunoreceptors, DNA hybridization or enzyme reactions have been reported [59–62]. Because of their submicrometer size CNFET sensors show promising prospects for application in implantable devices and for the development of sensor microarrays.

Semiconducting CNTs are very sensitive to gas adsorption, which provides the basis for applications in gas sensors. However, bare CNTs are poorly selective. Selectivity to a particular gas can be secured by suitable modification of the CNT. For example, deposition of palladium nanoparticles causes the conductance of the CNT to vary in the presence of hydrogen. Selectivity to other gases can be obtained using other metal nanoparticles, such as gold, platinum and rhodium. As full selectivity cannot be achieved, it is best to build up an array of gas sensors such as each of them responds to a gas component to a different extent. So, each sensor provides a mixed response but data processing by partial least square regression yields the magnitude of the response to each gas (see Section 1.7). A device of this kind is shown in Figure 11.27. It consists of an assembly of five CNFET formed on a silicon chip. Each device consists of a network of semiconducting CNTs gathered between two metal electrodes (source and gate). CNTs coverage with metal nanoparticles is best achieved by electrochemical deposition such that metal nanoparticles are grown at each predefined site. The response is recorded as the drain current at zero gate voltage. This sensor network allowed selective quantitation of H_2, H_2S, NH_3 and NO_2.

In conclusion, carbon nanomaterial semiconductor devices provide new possibilities in the design of chemical sensor for both liquid and gaseous samples. Besides CNTs, graphenes are also of interest in the design of FET-type chemical sensors [64].

Questions and Exercises (Section 11.5)

1 What makes the difference between a standard FET and a CNFET?
2 How does a CNFET respond to the interaction with an analyte?
3 What method is used to impart selectivity to a CNFET gas sensor?
4 Using ref. [62] for orientation, prepare a short overview of CNFET applications in biosensors

11.6 Concluding Remarks

The basic principles of semiconductor electronic devices have been successfully applied to develop chemical sensors for both ionic species an neutral molecule compounds. Two main strategies can be distinguished in these applications.

The first strategy relies on a FED structure integrated with a distinct chemically sensitive element, as it occurs in the case of FED ion sensors. Sensing reagents and materials developed in the frame of potentiometric ion sensors area are, in general, suitable for the development of FED ion sensors.

The second strategy relies on field effect devices configured in the standard structure. In this case the analyte interacts with one of the device components, such as the gate or the semiconductor body material. This approach is typical of FED gas sensors. Notwithstanding, certain types of FED gas sensors include a distinct organic sensing layer coated over the gate.

In general, sensing by means of FED devices relies on the alteration of the work function of a metal or semiconductor part of the device.

An important advantage of FED sensors is their compatibility with microelectronics technology that allows mass production in miniaturized format and facile fabrication of sensor arrays.

References

1. Sze, S.M. and Ng, K.K. (2007) *Physics of Semiconductor Devices*, Wiley-Interscience, Hoboken, N.J.
2. Blackburn, G.F. (1987) Chemically sensitive field effect transistors, in *Biosensors: Fundamentals and Applications* (eds A.P. F. Turner, I. Karube, and G.S. Wilson), Oxford University Press, Oxford, pp. 481–530.
3. Huber, R.J. and Janata, J. (eds) (1985) *Solid State Chemical Sensors*, Academic Press, Orlando.
4. Lundström, I., van den Berg, A., van der Schoot, B.H., and van den Vlekkert, H.H. (1991) Field effect chemical sensors, in *Sensors: A Comprehensive Survey* (eds W. Göpel, J. Hesse, and J.N. Zemel), Wiley-VCH, Weinheim, pp. 467–528.
5. Oprea A., Bârsan, N., and Weimar, U. (2009) Work function changes in gas sensitive materials: Fundamentals and applications. *Sens. Actuators B*, **142**, 470–493.
6. Bergveld, P., Hendrikse, J., and Olthuis, W. (1998) Theory and application of the material work function for chemical sensors based on the field effect principle. *Meas. Sci. Technol.*, **9**, 1801–1808.
7. Zemel, J.N. (1975) Ion-sensitive field effect transistors and related devices. *Anal. Chem.*, **47**, 255A–268a.
8. Janata, J. (1985) Chemically sensitive field effect transistors, in *Solid State Chemical Sensors* (eds R.J. Huber and J. Janata), Academic Press, Orlando, pp. 66–118.
9. Schöning, M.J., Thust, M., Müller-Veggian, M. *et al.* (1998) A novel silicon-based sensor array with capacitive EIS structures. *Sens. Actuators B-Chem.*, **47**, 225–230.
10. Schöning, M.J. and Luth, H. (2001) Novel concepts for silicon-based biosensors. *Phys. Status Solidi A-Appl. Res.*, **185**, 65–77.
11. Bergveld, P. (1970) Development of an ion-sensitive solid-state device for neurophysiological measurements. *IEEE Trans. Biomed. Eng.*, **BM17**, 70–71.
12. Bergveld, P. (1972) Development, operation, and application of ion-sensitive field-effect transistor as a tool for electrophysiology. *IEEE Trans. Biomed. Eng.*, **BM19**, 342–351.
13. Bergveld, P. (1988) Analytical and bioanalytical applications of ion-selective field-effect transistors, in *Comprehensive Analytical Chemistry*, vol. 23 (ed. G. Svehla), Elsevier, Amsterdam, pp. 1–172.
14. Bergveld F P. (2003) Thirty years of ISFETOLOGY - What happened in the past 30 years and what may happen in the next 30 years. *Sens. Actuators B-Chem.*, **88**, 1–20.
15. Vlasov, Y.G., Tarantov, Y.A., and Bobrov, P.V. (2003) Analytical characteristics and sensitivity mechanisms of electrolyte-insulator-semiconductor system-based chemical sensors - a critical review. *Anal. Bioanal. Chem.*, **376**, 788–796.
16. Zhou, D. (2008) Microelectrodes for in-vivo determination of pH, in *Electrochemical Sensors, Biosensors, and Their Biomedical Applications* (eds X. Zhang, H. Ju, and J. Wang), Academic Press, Amsterdam, pp. 261–305.
17. Korostynska, O., Arshak, K., Gill, E. *et al.* (2008) Review paper: Materials and techniques for in vivo pH monitoring. *IEEE Sens. J.*, **8**, 20–28.
18. Shimada, K., Yano, M., Shibatani, K. *et al.* (1980) Application of catheter-tip ISFET for continuous in vivo measurement. *Med. Biol. Eng. Comput.*, **18**, 741–745.
19. Van der Schoot, B. and Bergveld, P. (1988) Coulometric sensors, the application of a sensor actuator-system for long-term stability in chemical sensing. *Sens. Actuators*, **13**, 251–262.
20. Covington, A.K. (1994) Terminology and conventions for microelectronic Ion-Selective Field-Effect Transistor devices in electrochemistry. *Pure Appl. Chem.*, **66**, 565–569.
21. Moss, S.D., Janata, J., and Johnson, C.C. (1975) Potassium ion-sensitive field-effect transistor. *Anal. Chem.*, **47**, 2238–2243.
22. Reinhoudt, D.N. (1996) Transduction of molecular recognition into electronic signals. *Rec. Trav. Chim. Pays-Bas-J. Roy. Neth. Chem. Soc.*, **115**, 109–118.
23. Suryanarayanan, V., Wu, C.T., and Ho, K.C. (2010) Molecularly imprinted electrochemical sensors. *Electroanalysis*, **22**, 1795–1811.
24. Rao, T.P., Kala, R., and Daniel, S. (2006) Metal ion-imprinted polymers - Novel materials for selective recognition of inorganics. *Anal. Chim. Acta*, **578**, 105–116.
25. Zayats, M., Lahav, M., Kharitonov, A.B. *et al.* (2002) Imprinting of specific molecular recognition sites in inorganic and organic thin layer membranes associated with ion-sensitive field-effect transistors. *Tetrahedron*, **58**, 815–824.
26. Schöning, M.J. and Kloock, J.P. (2007) About 20 years of silicon-based thin-film sensors with chalcogenide glass materials for heavy metal analysis: Technological aspects of fabrication and miniaturization. *Electroanalysis*, **19**, 2029–2038.
27. Hafeman, D.G., Parce, J.W., and McConnell, H.M. (1988) Light-addressable potentiometric sensor for biochemical systems. *Science*, **240**, 1182–1185.
28. Owicki, J.C., Bousse, L.J., Hafeman, D.G. *et al.* (1994) The light-addressable potentiometric sensor - principles and biological applications. *Annu. Rev. Biophys. Biomolec. Struct.*, **23**, 87–113.
29. Wagner, T. and Schöning, M.J. (2007) Light addressable potentiometric sensors (LAPS): Recent trends and applications, in *Electrochemical Sensor Analysis* (eds S. Alegret and A. Merkoçi), Elsevier, Amsterdam, pp. 87–128.
30. Adami, M., Sartore, M., Baldini, E. *et al.* (1992) New measuring principle for LAPS devices. *Sens. Actuators B-Chem.*, **9**, 25–31.

31. Shiono, S., Hanazato, Y., and Nakako, M. (1992) Advances in enzymatically coupled field-effect transistors. *Methods Biochem. Anal.*, **36**, 151–178.
32. Caras, S. and Janata, J. (1980) Field-effect transistor sensitive to penicillin. *Anal. Chem.*, **52**, 1935–1937.
33. Van der Schoot, B.H. and Bergveld, P. (1987) ISFET based enzyme sensors. *Biosensors*, **3**, 161–186.
34. Schöning, M.J. (2005) "Playing around" with field-effect sensors on the basis of EIS structures, LAPS and ISFETs. *Sensors*, **5**, 126–138.
35. Schöning, M.J., Kurowski, A., Thust, M. *et al.* (2000) Capacitive microsensors for biochemical sensing based on porous silicon technology. *Sens. Actuators B-Chem.*, **64**, 59–64.
36. Schöning, M.J. and Poghossian, A. (2002) Recent advances in biologically sensitive field-effect transistors (BioFETs). *Analyst*, **127**, 1137–1151.
37. Schöning, M.J. and Poghossian, A. (2006) Bio FEDs (field-effect devices): State-of-the-art and new directions. *Electroanalysis*, **18**, 1893–1900.
38. Janata, J. and Josowicz, M. (1998) Chemical modulation of work function as a transduction mechanism for chemical sensors. *Accounts Chem. Res.*, **31**, 241–248.
39. Spetz, A., Winquist, F., Sundgren, H. *et al.* (1992) Field effect gas sensors, in *Gas Sensors: Principles, Operation and Developments* (ed. G. Sberveglieri), Kluwer, Dordrecht, pp. 219–279.
40. Mandelis, A. and Christofides, C. (1993) Gas-sensitive solid state semiconductor sensors, in *Physics, Chemistry, and Technology of Solid State Gas Sensor Devices* (eds A. Mandelis and C. Christofides), John Wiley & Sons, New York, pp. 19–132.
41. Lundström, K.I., Shivaraman, M.S., and Svensson, C.M. (1975) Hydrogen-sensitive Pd-gate MOS-transistor. *J. Appl. Phys.*, **46**, 3876–3881.
42. Lundström, I., Svensson, C., Spetz, A. *et al.* (1993) From hydrogen sensors to olfactory images - 20 years with catalytic field-effect devices. *Sens. Actuators B-Chem.*, **13**, 16–23.
43. Lundström, I., Sundgren, H., Winquist, F. *et al.* (2007) Twenty-five years of field effect gas sensor research in Linköping. *Sens. Actuators B-Chem.*, **121**, 247–262.
44. Ekedahl, L.-G., Eriksson, M., and Lundström, I. (1998) Hydrogen sensing mechanisms of metal-insulator interfaces. *Accounts Chem. Res.*, **31**, 249–256.
45. Mahfoz-Kotb, H., Salaün, A.C., Bendriaa, F. *et al.* (2006) Sensing sensibility of surface micromachined suspended gate polysilicon thin film transistors. *Sens. Actuators B-Chem.*, **118**, 243–248.
46. De Sagazan, O., da Silva Rodrigues, B., Crand, S. *et al.* (2010) Investigation on suspended gate field effect transistor as humidity sensor. *Proc. Eng.*, **5**, 1434–1437.
47. James, D., Scott, S.M., Ali, Z. *et al.* (2005) Chemical sensors for electronic nose systems. *Microchim. Acta*, **149**, 1–17.
48. Soo, M.T., Cheong, K.Y., and Noor, A.F.M. (2010) Advances of SiC-based MOS capacitor hydrogen sensors for harsh environment applications. *Sens. Actuators B-Chem.*, **151**, 39–55.
49. Janata, J. and Josowicz, M. (2009) Organic semiconductors in potentiometric gas sensors. *J. Solid State Electrochem.*, **13**, 41–49.
50. Bouvet, M. (2006) Phthalocyanine-based field-effect transistors as gas sensors. *Anal. Bioanal. Chem.*, **384**, 366–373.
51. Katz, H.E. (1997) Organic molecular solids as thin film transistor semiconductors. *J. Mater. Chem.*, **7**, 369–376.
52. Horowitz, G. (2011) The organic transistor: state-of-the-art and outlook. *Eur. Phys. J.-Appl. Phys.*, **53**, Paper 33602.
53. Mabeck, J.T. and Malliaras, G.G. (2006) Chemical and biological sensors based on organic thin-film transistors. *Anal. Bioanal. Chem.*, **384**, 343–353.
54. Torsi, L. and Dodabalapur, A. (2005) Organic thin-film transistors as plastic analytical sensors. *Anal. Chem.*, **77**, 380A–387A.
55. Josowicz, M. and Janata, J. (1986) Suspended gate field effect transistors modified with polypyrrole as alcohol sensor. *Anal. Chem.*, **58**, 514–517.
56. Oprea, A., Weimar, U., Simon, E. *et al.* (2006) Copper phthalocyanine suspended gate field effect transistors for NO$_2$ detection. *Sens. Actuators B-Chem.*, **118**, 249–254.
57. Potje-Kamloth, K. (2008) Semiconductor junction gas sensors. *Chem. Rev.*, **108**, 367–399.
58. Yamamoto, N., Tonomura, S., Matsuoka, T. *et al.* (1980) Study on a palladium-titanium oxide Schottky diode as a detector for gaseous components. *Surf. Sci.*, **92**, 400–406.
59. Allen, B.L., Kichambare, P.D., and Star, A. (2007) Carbon nanotube field-effect-transistor-based biosensors. *Adv. Mater.*, **19**, 1439–1451.
60. Kichambare, P.D. and Star, A. (2007) Biosensing using carbon nanotube field-effect transistors, in *Nanomaterials for Biosensors* (ed. C. Kumar), Wiley-VCH, Weinheim, pp. 1–26.
61. Gruner, G. (2005) Carbon nanotube transistors for biosensing applications. *Anal. Bioanal. Chem.*, **384**, 322–335.
62. Tarakanov, A.O., Goncharova, L.B., and Tarakanov, Y.A. (2010) Carbon nanotubes towards medicinal biochips. *Wiley Interdiscip. Rev.-Nanomed. Nanobiotechnol.*, **2**, 1–10.
63. Star, A., Joshi, V., Skarupo, S. *et al.* (2006) Gas sensor array based on metal-decorated carbon nanotubes. *J. Phys. Chem. B*, **110**, 21014–21020.
64. Hu, P. A., Zhang, J., Li, L. *et al.* (2010) Carbon nanostructure-based field-effect transistors for label-free chemical/biological sensors, *Sensors*, **10**, 5133–5159.

12

Resistive Gas Sensors (Chemiresistors)

Certain kinds of conducting material experience a change in their electrical resistance in response to an interaction with gases and vapors. Among these materials, the most typical are semiconductor metal oxides and organic semiconductors. Such materials are inexpensive and resistance measurements require simple DC electronics. This prompted the development of simple and inexpensive gas sensors based on the measurement of the device resistance. The name *chemiresistor* is ascribed to this kind of sensor. A recent trend in this field is based on the application of various nanomaterials in chemiresistor manufacturing.

This chapter introduces the main kinds of gas-sensitive materials used in chemiresistors, the configuration of these sensors, the design principles, manufacturing methods, and applications of resistive gas sensors.

12.1 Semiconductor Metal Oxide Gas Sensors

12.1.1 Introduction

Polycrystalline semiconductor metal oxides interact with oxygen to form active oxygen species that alter the electrical charge at the grain surface. By reaction with combustible gases, oxygen active species are depleted, which results in an alteration of the resistance of the device.

Gas sensors based on semiconductor metal oxides are widely used for monitoring flammable gases (such as hydrocarbons) in industrial environment and also for detecting hazardous gases (such as carbon monoxide and nitrogen oxides) in ambient air. Such sensors are small, inexpensive, reliable and rugged. As the demand for such sensors is very high, mass production has been developed by various companies around the world.

The physical background, design principles, and fabrication methods of semiconductor oxide gas sensor are amply surveyed in the literature (see, for example, [1–4]).

12.1.2 Gas-Response Mechanism

The most common sensing material in resistive gas sensing is SnO_2, but other metal oxides such as ZnO, TiO_2, WO_3 and In_2O_3 are also used [5]. As a rule, the oxide material is formed of nanosized grains that adhere partially to each other (Figure 12.1). This porous structure allows gas molecules to penetrate into the bulk of the material.

Nonstoichiometric tin oxide (SnO_{2-x}; $x \ll 1$) behaves as an n-type semiconductor due to oxygen vacancies in the crystal structure. The presence of a vacancy causes two electrons at a tin atom to become free. If tin dioxide is exposed to ambient oxygen, oxygen molecules adsorb at the surface and each oxygen molecule captures one electron, forming adsorbed anionic oxygen species such as O_2^-, O^- and O^{2-} (Figure 12.1A). Electron trapping by oxygen depletes the surface of charge carriers, the depletion degree being maximal at the neck between two particles. Hence, electron transfer between adjacent grains is hindered by a potential barrier. The electrical resistance of the sensor (R_{Air}) is determined by this barrier according to the following approximate equation:

$$R_{Air} = R_0 \exp \frac{eV_{Air}}{k_B T} \tag{12.1}$$

where R_0 is a constant, e is the elementary charge, V_{Air} is the barrier height in pure air, k_B is the Boltzmann constant, and T is the absolute temperature. Based on this principle, oxygen sensors have been developed using TiO_2 or $BaTiO_3$ as sensing materials.

If a reducing gas (for example, carbon monoxide) is present in the air sample, it reacts with adsorbed oxygen species as follows:

$$2CO + O_2^- \rightarrow CO_2 + e^- \tag{12.2}$$

In this way, free electrons are released into the conduction band (Figure 12.1B). As a result, the intergrain energy barrier decreases and the sensor resistance becomes:

$$R_{Gas} = R_0 \exp \frac{eV_{Gas}}{k_B T} \tag{12.3}$$

Chemical Sensors and Biosensors: Fundamentals and Applications, First Edition. Florinel-Gabriel Bănică.
© 2012 John Wiley & Sons, Ltd. Published 2012 by John Wiley & Sons, Ltd.

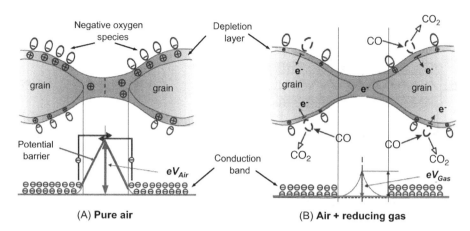

Figure 12.1 A porous semiconductor metal oxide. Formation of a charged layer by oxygen chemisorption (A) and the effect of a reducing gas (B). V_{air} and V_{gas} are the potential differences between the conduction band and the barrier top in pure air and air with reducing gas, respectively. Adapted with permission from [6]. Copyright 2004 Polish Academy of Science.

where V_{Gas} is the barrier height in the presence of the reducing gas. As $V_{Gas} < V_{Air}$, the resistance becomes smaller under the action of a reducing gas. The material sensitivity to the reducing gas (S_{Mat}) can be expressed as:

$$S_{Mat} = \frac{R_{Gas}}{R_{Air}} = \exp\frac{eV_{Air} - eV_{Gas}}{k_B T} \tag{12.4}$$

Oxidizing gases (such as NO_2 or chlorine) can also be detected by certain metal-oxide sensors. In this case, the analyte causes the electron depletion to increase, which results in a greater resistance. More details on the response mechanism of metal-oxide chemiresistors can be found in the literature [3,4,7–9].

The relationship between sensor resistance (R_S) and the concentration of the analyte gas (c) can be derived from on a theoretical basis [9,10]. In many cases, the following function represents fairly the sensor response:

$$\log\frac{R_S}{R_{ref}} = A + \alpha \log c \tag{12.5}$$

where A and α are empirical parameters and R_{ref} is the resistance measured with a reference gas. Therefore, the response function is linear on a logarithmic scale. This kind of response can extend over a broad concentration range, from several ppb to several thousand ppm. Other empirical correlation functions have also been advanced and are summarized in [3].

Metal-oxide chemiresistors should by operated at elevated temperatures (200–800 °C), the temperature being regulated by means of an electrical heater included in the sensor layout.

The detection principle of semiconductor oxide gas sensors is based on adsorption and chemical reactions at oxide grain surfaces. Consequently, the operating temperature will affect the sensitivity by determining the rate of chemical reactions. That is why, a circuit for compensation of the temperature variations should be included in the measurement setup.

12.1.3 Response to Humidity

Water vapors adsorb at the surface and cause a certain change in sensor resistance. This property can be advantageously used in resistive humidity sensors for warm gases, the best materials in this respect being perovskite-type mixed oxides containing an alkali-earth metal, such as $SrSnO_3$ and $CaTiO_3$ [11,12]. The marked response of such materials to water vapors has been ascribed to the strong affinity of alkali-earth ions for water molecules. Very good sensitivity to water vapors has been obtained with $Sr_{0.95}La_{0.05}SnO_3$, that is, a mixed oxide in which the divalent alkali-earth metal ion is partially substituted by the trivalent lanthanum ion. The above materials are n-type semiconductors and the response to humidity has been assigned to ionized oxygen sites (O^{2-}) that promote water dissociation at elevated temperature accompanied by release of a trapped electron from a vacancy (V_o^-):

$$H_2O(g) + O^{2-} + V_o^- \rightleftarrows 2OH^-(ads) + e^- \tag{12.6}$$

This mechanism, which involves one electron donation, produces a conductivity modification proportional to $p_w^{1/3}$ (p_w is the partial pressure of water vapor) and has been observed with $Sr_{0.95}La_{0.05}SnO_3$. Materials with doubly ionized vacancies produce a response proportional to $p_w^{1/4}$, which implies a lower sensitivity.

A similar response has been noted with other hydroxyl-containing compounds, such as ethanol.

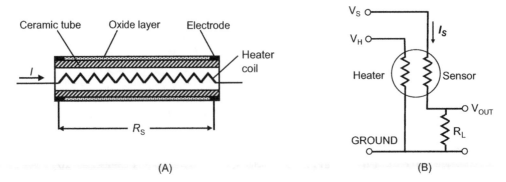

(A)

(B)

Figure 12.2 (A) Design of a tin oxide tubular gas sensor; (B) electrical wiring of a resistive gas sensor. V_S: sensor voltage; V_H: heating voltage; R_L: load resistance; V_{OUT}: output voltage.

Certain oxides allow humidity to be gaged by the ionic conduction mechanism (Section 17.8.2). Although many oxides used in such applications are semiconductors, semiconductor properties are not relevant in this case. However, this property can be exploited in the design of multifunctional sensors that respond to humidity at low temperatures and to reducing gases at high temperatures. In the second case, the mechanism presented in Section 12.1.2 is operating.

12.1.4 Sensor Configuration

Initially, gas chemiresistors were fabricated by depositing a thin oxide layer over a ceramic tube as shown schematically in Figure 12.2A. This kind of sensor, which is known as a Taguchi gas sensor, has been commercially available since 1968. The electrical contact to the oxide layer is provided by two metal electrodes placed at the tube ends. Sensor heating is provided by a resistor coil inside the tube. The electrical wiring of the sensor is shown in Figure 12.2B. In order to perform resistance measurements, a constant DC voltage (V_S) is applied to the series circuit composed of the sensor and the load resistance R_L. According to Ohm's law, the voltage drop across the load resistance (V_{OUT}) is inversely proportional to the sensor resistance. This voltage is the output signal response of the sensor. As can be seen, the measurement principle is very simple and involves only DC circuits.

Although the tubular configuration in Figure 12.2 has been in use for long time, it is currently being replaced by a planar configuration, which is more advantageous from the standpoint of fabrication technology. In the planar design (Figure 12.3), the metal oxide layer is deposited over a flat dielectric support and the heater is placed underneath. The electrodes are formed as an interdigitated structure, which makes it possible to use gas-sensitive materials with a very high offset resistivity. Such materials are characterized by a high sensitivity to gas interaction. The planar configuration that allows for miniaturization is compatible with microfabrication technologies such as screen printing and micromachining. Micromachining allows manufacturing very small sized devices characterized by low power consumption and easy multiplexing [13].

The configurations described above involve two distinct electrical circuits: the heating circuit and the measuring circuit. A simpler configuration that is based on a single electrical circuit is obtained by embedding a platinum coil in a semiconductor oxide bead as shown in Figure 12.4A [15,16]. This coil functions as a heater but, at the same time, the metal oxide partially short circuits the turns of the coil, as shown in Figure 12.4B. The overall resistance of the device (R_S) depends on both the constant coil resistance and the gas-modulated oxide resistance. Such a sensor is called a *semistor*, a word that is a blend of semiconductor and resistor. The semistor is a very simple device, but since the oxide is shunted by platinum turns, its overall resistance cannot be lower that that of the coil. Hence, small

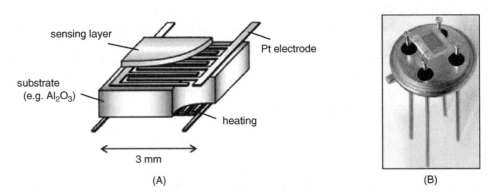

(A)

(B)

Figure 12.3 (A) Schematics of a planar resistive gas sensor; (B) photograph of a planar gas sensor. Reproduced with permission from [14]. Copyright 2007 Wiley-VCH Verlag GmBH & Co. KGaA.

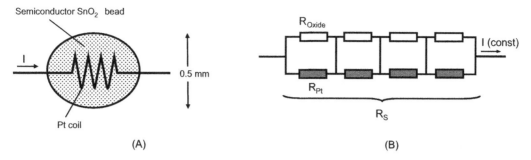

Figure 12.4 The semistor. (A) Configuration; (B) equivalent circuit. R_{Pt} represents the resistance of a platinum turn; R_{Oxide} is the resistance of the oxide contained between two platinum turns.

variations of the oxide resistance cannot be detected, which limits the sensor sensitivity. However, this device is useful in applications that are not demanding as far as sensitivity is concerned.

12.1.5 Synthesis and Deposition of Metal Oxides

Metal oxides for chemiresistor applications can be obtained by either wet or dry synthesis methods [17].

Sol-gel chemistry was introduced in Section 5.7 and it represents the usual wet synthesis method. Depending on reaction conditions, sol-gel synthesis produces either amorphous or crystalline particles. Their size depends on synthesis parameters such as the precursor concentration or the reaction temperature. Subsequent crosslinking of the sol particles leads to a gel, that is, a porous network impregnated with a liquid phase. To obtain a supported film, the sol is applied to the substrate before gelation. Finally, the solvent is removed from the gel to yield a porous solid, which is subject to calcinations, in order to improve its stability. Sintering of the particles, which occurs during the calcination step, increases the grain connectivity but decreases the specific area. Alternatively, sol-gel synthesis of the oxide can be conducted directly on the support. Nonconventional sol-gel methods have recently been advanced [18].

Dry synthesis offers broad possibilities for rapid production of metal-oxide films of various morphologies, degree of porosity, and thickness. Notably, dry synthesis is compatible with microfabrication technology. A series of dry synthetic methods are based on chemical vapor deposition. This is performed by supplying a gaseous precursor to the high-temperature reaction zone so that the synthesis of the metal oxide takes places directly on the target substrate. In this way, SnO_2 can be formed by high-temperature hydrolysis of a $SnCl_4$ solution. An alternative deposition method relies on aerosol synthesis in a flame. In this case, a solution containing the precursor is fed as an aerosol to the flame.

Currently, a broad spectrum of dry synthesis methods are available, each of them providing some particular advantages as far as the control of the grain size, film morphology, and film thickness are concerned [14,17].

12.1.6 Fabrication of Metal-Oxide Chemiresistors

Planar gas-sensitive layers can be deposited as a thin film between two interdigitated electrodes. Sensors of this type can be manufactured by screen printing on a thin ceramic support.

A more advanced fabrication method is based on micromachining that allows production of small-size devices characterized by low power consumption and easy multiplexing [13,19,20]. An example of a gas sensor manufactured by micromachining is presented in Figure 12.5. The gas-sensitive layer is formed over a thin dielectric

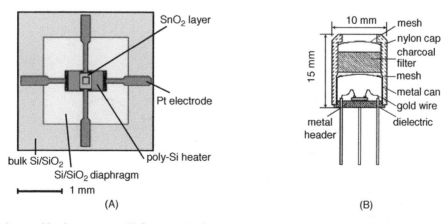

Figure 12.5 A micromachined gas sensor. (A) Sensor unit; (B) sensor housing. Reproduced with permission from [13]. Copyright 2001 Elsevier.

membrane (SiO$_2$ or Si$_3$N$_4$) deposited on a silicon support. The low thermal conductivity of this membrane secures good thermal insulation between the support and the sensitive region. In this way, power consumption can be kept very low, typically between 30 and 150 mW and the distance between electrodes can be decreased to the μm range, which permits using interdigitated electrodes integrated with highly resistive sensing materials. In addition, sensor arrays can be easily implemented in this technology.

12.1.7 Selectivity and Sensitivity

A crucial parameter that determines the sensitivity is the size of the oxide grains [21,22]. The effect of the grain size can be rationalized if the sensitive layer is viewed as a three-dimensional network of grains and necks, each particle being composed of a couple of adherent grains (Figure 12.6). The total resistance of the sensing layer is controlled by the resistance of the grains, necks, and the contact region between particles. In general, a strong increase in the sensor sensitivity is obtained by decreasing the grain size (D) to the nanometer region. The critical parameter in this respect is the *Debye length* (δ), which is the distance over which significant charge separation can occur in the semiconductor. As shown in Figure 12.6A, if the particle size is very large with respect to the Debye length, the depletion layer occupies a very thin region and affects, to a small extent, the current flow along the grain and across the necks. The material resistivity is in this case determined by discontinuities at interparticle zones of contact. If the particle size is just slightly greater that the Debye length (Figure 12.6B), the depletion layer expands over an appreciable fraction of the grain and the conduction is limited to a channel whose thickness is mostly determined by the thickness (L_C) of the nondepleted zone within the neck regions. This results in a higher resistivity compared with the previous case. If the particle size is close to twice the Debye length (Figure 12.6C) nearly all the electrons are trapped at the particle surface and the maximum resistivity is obtained. Under these conditions, very small changes in the trapped electron density in response to the gas reaction induce a noticeable variation in the resistance. In general, particle sizes below 100 nm provide a high sensitivity.

The particulate film morphology allows for gas diffusion into the bulk of the film. Nevertheless, compact oxide films can also be used in gas sensing. In this case, the interaction with the gas is limited to the surface of the sensitive layer [9]. Compact layers can be formed as thick or as thin layers. Film thickness is expressed in relative units as the film thickness/Debye length ratio. For a *thick film*, the film is much thicker than the Debye length and the current

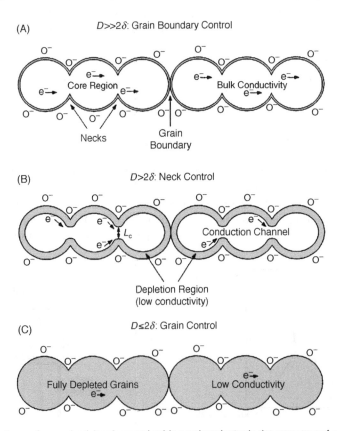

Figure 12.6 Effect of grain size on the conductivity of a metal oxide semiconductor in the presence of oxygen. D is the grain size; δ is the Debye length. (A) For $D \gg 2\delta$ the extension of the depletion layer in the neck region is so small that the necks have a minor effect on the resistivity. Resistivity is mostly determined by grain boundaries (bulk conductivity). (B) For D values moderately greater than δ, the thickness of the nondepleted zone in the neck region (L_C), is very small and determines the resistivity (neck control); (C) for $D \leq \delta$ the depletion layer expands over the whole grain thickness. The oxide layer has the lowest conductivity and responds with maximum sensitivity to the analyte gas. Reproduced with permission from [17]. Copyright 2010 Wiley-VCH Verlag GmBH & Co. KGaA.

flow is limited to the surface conducting channel. Gas interaction modulates the thickness of this channel and hence the film resistivity. Conversely, the thickness of a *thin film* is comparable with the Debye length and the influence of gas interaction is extended over the whole thickness of the layer.

Sensitivity to a particular gas results from the interplay of a great number of geometrical factors (grain size, porosity, area of intergrain contact and orientation of nanocrystals) and physicochemical parameters (chemical and phase composition, bulk and surface density of oxygen vacancies, and properties of the additives) [23,24]. Such parameters can be tuned by proper selection of the synthesis method and by post-synthesis treatment of the oxide material.

Generally, metal-oxide resistive gas sensors are not selective and those developed for the determination of reducing gases respond to many compounds of this kind. In order to enhance the response to a particular gas or a class of gases particular additives are added to the semiconductor material. Noble metals that display catalytic properties are common additives that can be applied as a surface layer or can be incorporated through the body of the sensitive layer. Other kinds of additive have also been used.

Selectivity can be enhanced by the addition of a component with which the target analyte will react. Thus, basic oxide additives increase the response to H_2S, while acidic oxide additives enhance the sensitivity to NH_3.

The effect of noble-metal additives can be rationalized by taking into account the partial charge transfer between adsorbed oxygen and the metal clusters embedded in the oxide [25]. Oxygen adsorption at the metal surface leads to electron depletion in the metal, which raises the work function of the metal and, consequently, leads to an increase in the barrier height and the appearance of an inversion layer at the surface of oxide grains. Maximum resistance is obtained when the density of metal clusters is such as to allow inversion layers in the oxide grains to overlap. Reducible gases, that can perform partial electron transfer to metal clusters, lower the barrier height and thus enhance the conductivity.

In some instances, a filter that removes interfering gas components provides sufficient selectivity. For example, interference by hydrocarbons in the determination of carbon monoxide can be avoided by using a charcoal filter.

12.1.8 Outlook

Metal-oxide chemiresistors are currently well established devices in industrial and environmental monitoring of atmosphere quality. Industry makes extensive use of such sensors for the detection of hazardous gases, whereas the control of air quality in areas exposed to high traffic or in garages is another important application. Metal-oxide chemiresistors are rugged and inexpensive devices. From the standpoint of functioning characteristics, they are reasonably accurate, reversible, and have a short response time.

Although the gas-recognition selectivity of metal oxides is poor, improvements can be achieved by including various additives, particularly catalytic metals. If high selectivity is needed, one can resort to sensor arrays that allow for accurate determination of the sample components by suitable data-processing methods (see Section 1.7).

The science and technology of metal-oxide chemiresistors is currently at an advanced level and the sound theoretical background of this field provides reliable premises to rational sensor development. Further technological advances are expected from the improvement in the degree of control of the structure and morphology of the oxide layer by microfabrication methods. Much progress in the field of the synthesis of oxidic sensing material is expected from the application of the methods of combinatorial chemistry. Other materials, such as porous silicon, are currently being investigated for application in resistive gas sensors [26,27].

Questions and Exercises (Section 12.1)

1 What kind of metal oxides are suitable for resistive gas sensing? What are the structural and electrical characteristics that allow such applications?

2 What is the effect of oxygen and other oxidizing gases on the charge distribution at the oxide surface? How does the resistivity respond to the adsorption of oxidizing gases?

3 What are the effects of reducing gases on the electrical properties of a metal-oxide semiconductor in the presence of oxygen? How varies the resistivity in response to a reducing gas?

4 Make a comparison between the tubular and planar configurations of chemiresistors and outline the advantages of the latter.

5 Review the structure and the functioning principles of semistors. Point out their advantages and drawbacks.

6 Review the methods for the synthesis of metal oxides and the fabrication of chemiresistors based on metal oxides.

7 Give an overview of the grain-size effect on the sensitivity of the chemiresistor and outline what factors can limit the sensitivity.

8 Comment on the selectivity of metal-oxide chemiresistors and on methods of controlling the selectivity.

12.2 Organic-Material-Based Chemiresistors

Easily available *polymer-carbon black composites* have demonstrated good performances as resistive gas-sensing materials that respond to organic vapors. Carbon particles provide electric conductivity in such a composite film. Vapor absorption in the polymer matrix causes the composite to swell and, as a result, some of the conductive pathways are broken, which leads to increased resistance.

Thin films of carbon-loaded polymer are well suited for manufacturing inexpensive sensor arrays that can discriminate between various compounds in an organic vapor mixture [28,29]. Each array unit includes a different polymer type, such as to endow it with particular sensitivity to each component in a mixture. A rational selection of the composite can be made using correlations of sensor response with molecular descriptors [30]. Combinatorial chemistry represent an alternative route to the design of gas polymer composites sensitive to a species of interest [31].

Organic semiconductors, such as phthalocyanines and conducting polymers, are commonly used in chemiresistors [32]. In order to obtain the sensor, organic semiconductors are deposited in the gap between two electrodes over an insulator substrate.

Phthalocyanines are able to form adducts with electron-donor molecules (such as O_2, NO and NO_2) by electron transfer to an orbital at the central metal ion. This process injects electrons into the conduction band and thus enhances the conductivity of the phthalocyanine film.

Conducting polymers (introduced in Section 5.9) can respond to a wide range of polar or nonpolar gases and vapors depending on various interaction mechanisms [33–35]. Possible interactions between gas molecules and a conducting polymer are represented in Figure 12.7 and are reviewed below in accordance with the site labels in this figure.

1. The analyte molecule X can affect the charge transfer between the polymer and the electrode contact.
2. The analyte causes oxidation or reduction of the polymer chain, thus changing the density of charge carriers. This mechanism is relevant to reactive gases such as ammonia or hydrogen sulfide.
3. The analyte can interact with the mobile charge carriers at the polymer backbone thus changing the carrier mobility along the chain.
4. The analyte can interact with the counterion C^- within the polymer film and thus modulates the mobility of the charge carriers along the chain.
5. The analyte can alter the probability of carrier hopping between chains and can thus affect the resistivity of the film.

One more possible mechanism is based on polymer swelling under the effect of an organic vapor analyte, which alters the density of counterions and thus modifies the polymer resistivity.

No matter what recognition mechanism operates, the response is generally rapid and reversible at room temperature, in contrast to metal-oxide sensors that should be operated at elevated temperature. The response function is of the Langmuir type. Accordingly, the resistance increases nonlinearly with the analyte concentration at concentrations below the saturation level. The sensor response can be represented as follows [33]:

$$\frac{1}{\Delta R_g} = \frac{K + c}{K(R_{sat} - R_0)} \tag{12.7}$$

where ΔR_g is the change in resistance in the presence of the analyte ($\Delta R_g = R_g - R_0$), c is the concentration of the analyte, K is the binding constant, R_0 is the resistance in the absence of the analyte, and R_{sat} is the resistance at saturation.

Although conducting polymers are also useful in other types of gas sensor, the chemiresistor configuration possesses the great advantages of simplicity and easy fabrication. Selectivity can be readily modulated by post-synthesis treatments that alter the chemical structure. In addition, these devices are small and function with very low power

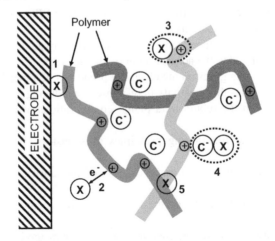

Figure 12.7   Possible interactions of a gas molecule with a conducting polymer. X is the analyte; C^- is the doping counterion.

consumption. Conducting polymers are suitable for applications in sensor arrays for multianalyte determination and artificial nose applications.

Questions and Exercises (Section 12.2)

1 Elaborate on the application of dielectric semiconductors in the development of sensors for volatile organic compounds.
2 What kind of gases can be determined by phthalocyanine-based chemiresistors? How does this material respond to the interactions with certain gases?
3 Comment on the possible interaction of conducting polymers with (a) oxidizing and reducing gases; (b) polar gases; (c) nonpolar gases.
4 What are the expected advantages and drawbacks of polymer-based chemiresistors?

12.3 Nanomaterial Applications in Resistive Gas Sensors

Semiconductor metal oxides, that are well-established gas-sensitive materials, have been recently shown to be suitable for applications in gas sensors in the form of *nanofibers* of about 100 nm diameter. Tin dioxide nanofibers can be obtained by electrospinning of a $SnCl_4$-polyvinyl alcohol solution with subsequent stepwise annealing at 300, 500 and 700 °C. Such nanofibers are woven to obtain a gas-sensing membrane. Due to the small diameter, the depletion layer produced by oxygen extends over a large fraction of the fiber section which gives high sensitivity, short response times, and fast recovery [36].

Conducting polymers, that display excellent qualities as gas-sensing materials can also be used in the form of nanofibers of 100–500 nm diameter. Either a single fiber or a nanowoven membrane can be employed as the sensing element [36].

New opportunities in gas sensing arise from the application of *metal nanoparticle–organic material composites*, such as gold nanoparticles with alkanethiol ligand shells or gold nanoparticle–dendrimer composite films [37]. The response mechanism in this case is based on the absorption of the analyte gas into the organic matrix. Selective recognition can be obtained by introducing specific chemical functionalities into the organic ligand shell. Current flow through nanoparticle–organic composites proceeds by electron hopping between adjacent particles. As a result of analyte interaction with the organic matrix, the interparticle spacing is altered, which can affect either the interparticle energy barrier or the transmission probability for electron tunneling through the barrier. A change in the work function of the metal in response to gas adsorption is another possible mechanism that may operate in the case of gases that display Lewis acid or base properties. By partial charge transfer between such an analyte and the nanoparticle, the Fermi level of the metal is altered. The resulting modification of the metal work function brings about a change in the potential barrier to the electron hopping between particles.

Details of the structure of a resistive gas sensor based on gold nanoparticles–dendrimer composite material are shown in Figure 12.8A. This dendrimer consists of poly(propyleneimine) and the sensing layer is deposited on a

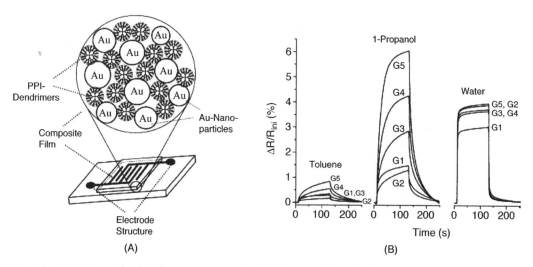

Figure 12.8 A chemiresistor with a sensitive layer composed of gold nanoparticles–dendrimer composite material. (A) Structure of the sensing layer and the sensor layout; (B) Sensor response to several vapors as a function of dendrimer generation (G). PPI stands for poly(propyleneimine). Reproduced with permission from [38]. Copyright 2003 Elsevier.

couple of interdigitated electrodes. Sensitivity to various analytes (toluene, propanol, and water vapor) depends to a large extent on dendrimer generation, as demonstrated in Figure 12.8B. Clearly, the best sensitivity to 1-propanol is obtained with the generation 5 dendrimer, whereas the sensitivity to toluene is relatively low, although it increases with increasing dendrimer generation. It was assumed in this work that the effect of the dendrimer generation is determined by the capture of organic molecules within the interior structure of dendrimers. The response to water vapor is affected to a very small extent by the dendrimer generation, which suggests that water vapor interacts with primary amino groups at the surface of the dendrimers. Therefore, the sensitivity of this kind of sensor can be tuned by adjusting the structure of the organic component in the composite material.

Metal nanoparticle–organic matrix composites are very versatile sensing materials in gas and vapor sensors. The main limitation of such materials arises from their relatively poor thermal stability.

Single-walled carbon nanotubes represent another kind of advanced material that is suitable for gas detection [39]. By charge-transfer interaction with electron-donor or -acceptor gases, the density of charge carrier in the nanotube is altered and, consequently, the conductivity is modified. For example, ammonia (an electron donor) lessens the conductivity of p-type semiconducting nanotubes, while NO_2 (an electron acceptor) causes an increase in conductivity.

Adsorption of such gases to the nanotube surface is irreversible, which causes the nanotube device to act like a dosimeter which reports on the levels of exposure to the analyte-gas. Sensor regeneration can be performed by exposing the device to heat or ultraviolet light.

Carbon-nanotube gas sensors can be produced by depositing a network of nanotubes between two interdigitated electrodes. Some degree of selectivity can be imparted by special sorbent polymers coated as thin films (<100 nm) over the nanotube layer. Resistive carbon-nanotube gas sensors show promising perspectives for applications in sensor arrays.

Questions and Exercises (Section 12.3)

1 How can metal-oxide semiconductors be obtained in the form of nanofibers and what are the advantages of using this particular form in gas sensors?
2 What are the possibilities of using metal nanoparticles in resistive gas sensing? What is the structure of composite materials for such applications and how do they interact with nonpolar vapors and with polar gases.
3 What are the basic principles of resistive gas sensing with carbon nanotubes and how can the selectivity be controlled?

12.4 Resistive Gas Sensor Arrays

Due to the simple fabrication technology, resistive gas sensors are very attractive for application in the array format. A sensor array compensates for the poor selectivity of sensing materials and can be operated as an electronic nose device.

A gas sensor array can be obtained simply by assembling a number of discrete sensors based on different sensing materials. However, the size of the array increases with the number of individual sensors and, in the case of high-temperature-operated sensors, the power consumption also increases.

A more versatile alternative is based on an array of different sensing materials deposited over a common substrate, as shown in Figure 12.9. The sequence of sensing layers is crossed by a constant current (I_{DC}) supplied by a common voltage source. In order to asses the resistance of each sensing layer, the voltage drop (ΔV_i) over each on them is measured individually and the particular resistance is obtained as $R_i = \Delta V_i / I_{DC}$.

A sequence of different sensing layers as that shown in Figure 12.9 can be produced by depositing first the same type of sensing material (e.g., a metal oxide) over the whole sensing area. A varying degree of selectivity is imparted then to each section by individual modification with suitable additives, such as platinum metals. A similar structure can be built up using various carbon-black–polymer composites or organic semiconductors.

Advanced design of gas sensor arrays is based on integration of a number of discrete sensors produced by microfabrication technology, such as complementary metal-oxide-semiconductor (CMOS) technology that is widely used in manufacturing integrated electronic microcircuits. As demonstrated in Figure 12.10A, each individual sensor can be formed on a platform suspended over a cavity etched in the silicon chip and supported by four legs. This arrangement minimizes the heat transfer to the ambient atmosphere. Details of this sensor structure are illustrated by the cross section in Figure 12.10B. A Pt/Ti bilayer pattern deposited over a SiO_2 insulating layer (TO) functions as both heater and resistive thermometer. Next, the SiO_2-Si_3N_4-SiO_2 multilayer (ML) provides insulation of the sensing layer from the Pt/Ti electrodes. These electrodes are coated with a passivation SiO_2 film (PO). Finally, the SnO_2 sensing layer is deposited by oxygen-radical-assisted electron beam evaporation of a metal target. The oxide formed in this

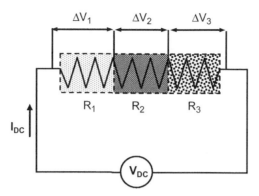

Figure 12.9 Multipoint probe measurement in a resistive sensor array. Each section of the device is coated with a layer with particular responsive characteristics. R_1, R_2, and R_3 indicates the resistance of each layer. Probe contacts are formed at the boundary of each section to measure the pertinent resistance.

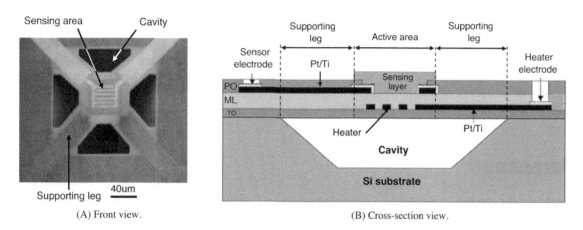

(A) Front view.   (B) Cross-section view.

Figure 12.10 Structure of a resistive gas sensor fabricated by micromachining technology. TO: SiO_2 insulating layer; ML: SiO_2-Si_3N_4-SiO_2 multilayer insulator; PO: passivation SiO_2 film. Reproduced with permission from [40]. Copyright 2001 Elsevier.

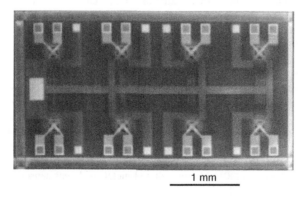

Figure 12.11 Topography of a micromachined 2 × 4 resistive gas sensor array produced by micromachining. Reproduced with permission from [40]. Copyright 2001 Elsevier.

way proved to respond with high sensitivity to ethanol vapors with negligible interference of other gases such as carbon monoxide, hydrogen, methane and isobutene vapors.

Figure 12.11 displays a magnified view of a sensor array composed of eight distinct sensors similar to that shown in Figure 12.10. This array is characterized by low power consumption and can be produced by standard microelectronics technology, which allows for a high degree of integration.

12.5 Summary

Gas and vapor determination is a current analytical problem in various areas such as industry and in air-quality monitoring, which prompted the development of gas sensors, particularly for the detection of flammable or toxic gases and vapors. There are various types of gas sensor that can, in principle, fulfill such functions. Among them, resistive gas sensors based on semiconductor sensing materials occupy an important position. Well-established sensors of this type make use of metal-oxide semiconductors and are currently produced on a large-scale basis.

What is important with regard to metal-oxide resistive gas sensors is the simple constructive principle and facile measurement procedure, combined with excellent ruggedness and reliability.

Besides metal oxides, organic semiconductors have emerged as a new class of gas sensing material with applications in resistive gas sensors. Organic semiconductors provide a large number of possibilities of adjusting the sensing parameters either directly, by synthesis, or by postsynthesis modification. As organic semiconductors can interact with gases and vapors by a multitude of different mechanisms, their field application is potentially very broad. In addition, organic-semiconductor-based sensor can be operated at ambient temperature, which has the advantage of very low power consumption.

Nanomaterial application in the field of resistive gas sensors represents another line of development. Actually, well-established metal-oxide semiconductors function best when the material is structured at the nanoscale. Distinct nanoparticles, such as carbon nanotubes and metal nanoparticles are newcomers in this field, but have aroused much interest and a number of applications have already been developed.

A new trend in the fabrication of gas sensors is based on printing on flexible, plastic foils [41]. This technology is suitable for manufacturing both resistive and capacitive sensors and allows integration of multiple sensors into sensor arrays. Microelectronic circuits can also be integrated with the sensor for signal processing and interfacing

Typically, resistive gas sensors are characterized by moderate selectivity. This is not a major drawback in gas-leakage monitoring in industry, where no complex mixtures of gases have to be tested. As usually, the selectivity problem can be conveniently addressed by resorting to arrays of cross-selective sensors. In this respect, resistive gas sensors have the advantage of simple fabrication technology and resistive gas sensor arrays are already well established in the development of electronic nose devices.

References

1. Dinesh, K. and Gupta, S.K. (eds) (2007) *Science and Technology of Chemiresistor Gas Sensors*, Nova, New York.
2. Hoffheins, B. (1996) Solid state, resistive gas sensors, in *Handbook of Chemical and Biological Sensors* (eds R.F. Taylor and J.S. Schultz), Institute of Physics Publishing, Bristol, pp. 371–397.
3. Bârsan, N., Schweizer-Berberich, M., and Göpel, W. (1999) Fundamental and practical aspects in the design of nanoscaled SnO_2 gas sensors: a status report. *Fresenius J. Anal. Chem.*, **365**, 287–304.
4. Bârsan, N., Koziej, D., and Weimar, U. (2007) Metal oxide-based gas sensor research: How to? *Sens. Actuators B-Chem.*, **121**, 18–35.
5. Korotcenkov, G. (2007) Metal oxides for solid-state gas sensors: What determines our choice? *Mater. Sci. Eng. B-Solid State Mater. Adv. Technol.*, **139**, 1–23.
6. Licznerski, B.W. (2004) Thick-film gas microsensors based on tin dioxide. *Bull. Polish Acad. Sci.-Technical Sci.*, **52**, 37–42.
7. Kohl, D. (2001) Function and applications of gas sensors. *J. Phys. D-Appl. Phys.*, **34**, R125–R149.
8. Park, C.O. and Akbar, S.A. (2003) Ceramics for chemical sensing. *J. Mater. Sci.*, **38**, 4611–4637.
9. Bârsan, N. and Weimar, U. (2001) Conduction model of metal oxide gas sensors. *J. Electroceram.*, **7**, 143–167.
10. Williams, D.E. (1999) Semiconducting oxides as gas-sensitive resistors. *Sens. Actuators B-Chem.*, **57**, 1–16.
11. Seiyama, T., Yamazoe, N., and Arai, H. (1983) Ceramic humidity sensors. *Sens. Actuators*, **4**, 85–96.
12. Traversa, E. (1995) Ceramic sensors for humidity detection – the state-of-the-art and future-developments. *Sens. Actuators B-Chem.*, **23**, 135–156.
13. Simon, T., Bârsan, N., Bauer, M. *et al.* (2001) Micromachined metal oxide gas sensors: opportunities to improve sensor performance. *Sens. Actuators B-Chem.*, **73**, 1–26.
14. Tiemann, M. (2007) Porous metal oxides as gas sensors. *Chem.-Eur. J.*, **13**, 8376–8388.
15. Williams, G. and Coles, G.S.V. (1999) The semistor: A new concept in selective methane detection. *Sens. Actuators B-Chem.*, **57**, 108–114.
16. Korotcenkov, G. (2007) Practical aspects in design of one-electrode semiconductor gas sensors: Status report. *Sens. Actuators B-Chem.*, **121**, 664–678.
17. Tricoli, A., Righettoni, M., and Teleki, A. (2010) Semiconductor gas sensors: Dry synthesis and application. *Angew. Chem.-Int. Edit.*, **49**, 7632–7659.
18. Neri, G. (2010) Non-conventional sol-gel routes to nanosized metal oxides for gas sensing: From materials to applications. *Sci. Adv. Mater.*, **2**, 3–15.
19. Spannhake, J., Helwig, A., Schulz, O. *et al.* (2009) Micro-fabrication of gas sensors, in *Solid State Gas Sensing* (eds E. Comini, G. Faglia, and G. Sberveglieri), Springer Science+Business Media, LLC, Boston, MA, pp. 1–46.
20. Yamazoe, N. (2005) Toward innovations of gas sensor technology. *Sens. Actuators B-Chem.*, **108**, 2–14.
21. Rothschild, A. and Komem, Y. (2004) The effect of grain size on the sensitivity of nanocrystalline metal-oxide gas sensors. *J. Appl. Phys.*, **95**, 6374–6380.
22. Korotcenkov, G., Han, S.D., Cho, B.K. *et al.* (2009) Grain size effects in sensor response of nanostructured SnO_2-and In_2O_3-based conductometric thin film gas sensor. *Crit. Rev. Solid State Mater. Sci.*, **34**, 1–17.

23. Korotcenkov, G. (2005) Gas response control through structural and chemical modification of metal oxide films: state of the art and approaches. *Sens. Actuators B-Chem.*, **107**, 209–232.

24. Korotcenkov, G. (2008) The role of morphology and crystallographic structure of metal oxides in response of conductometric-type gas sensors. *Mater. Sci. Eng. R-Rep.*, **61**, 1–39.

25. Licznerski, B.W., Nitsch, K., Teterycz, H. *et al.* (2001) The influence of Rh surface doping on anomalous properties of thick-film SnO_2 gas sensors. *Sens. Actuators B-Chem.*, **79**, 157–162.

26. Ozdemir, S. and Gole, J.L. (2007) The potential of porous silicon gas sensors. *Curr. Opin. Solid State Mater. Sci.*, **11**, 92–100.

27. Korotcenkov, G. and Cho, B.K. (2010) Porous semiconductors: Advanced material for gas sensor applications. *Crit. Rev. Solid State Mater. Sci.*, **35**, 1–37.

28. Lonergan, M.C., Severin, E.J., Doleman, B.J. *et al.* (1996) Array-based vapor sensing using chemically sensitive, carbon black-polymer resistors. *Chem. Mater.*, **8**, 2298–2312.

29. Patel, S.V., Jenkins, M.W., Hughes, R.C. *et al.* (2000) Differentiation of chemical components in a binary solvent vapor mixture using carbon/polymer composite-based chemiresistors. *Anal. Chem.*, **72**, 1532–1542.

30. Shevade, A.V., Homer, M.L., Taylor, C.J. *et al.* (2006) Correlating polymer-carbon composite sensor response with molecular descriptors. *J. Electrochem. Soc.*, **153**, H209–H216.

31. Potyrailo, R.A. (2006) Polymeric sensor materials: Toward an alliance of combinatorial and rational design tools? *Angew. Chem. Int. Edit.*, **45**, 702–723.

32. Sadaoka, Y. (1992) Organic semiconductor gas sensors, in *Gas Sensors: Principles, Operation and Developments* (ed. G. Sberveglieri), Kluwer, Dordrecht, pp. 187–218.

33. Gardner, J.W. and Bartlett, P.N. (1999) *Electronic Noses: Principles and Applications*, Oxford University Press, Oxford.

34. Lange, U., Roznyatouskaya, N.V., and Mirsky, V.M. (2008) Conducting polymers in chemical sensors and arrays. *Analyt. Chim. Acta*, **614**, 1–26.

35. Bai, H. and Shi, G.Q. (2007) Gas sensors based on conducting polymers. *Sensors*, **7**, 267–307.

36. Ding, B., Wang, M.R., Yu, J.Y. *et al.* (2009) Gas sensors based on electrospun nanofibers. *Sensors*, **9**, 1609–1624.

37. Franke, M.E., Koplin, T.J., and Simon, U. (2006) Metal and metal oxide nanoparticles in chemiresistors: Does the nanoscale matter? *Small*, **2**, 36–50.

38. Krasteva, N., Guse, B., Besnard, I. *et al.* (2003) Gold nanoparticle/PPI-dendrimer based chemiresistors Vapor-sensing properties as a function of the dendrimer size. *Sens. Actuators B-Chem.*, **92**, 137–143.

39. Snow, E.S., Perkins, F.K., and Robinson, J.A. (2006) Chemical vapor detection using single-walled carbon nanotubes. *Chem. Soc. Rev.*, **35**, 790–798.

40. Mo, Y.W., Okawa, Y., Tajima, M. *et al.* (2001) Micro-machined gas sensor array based on metal film micro-heater. *Sens. Actuators B-Chem.*, **79**, 175–181.

41. Briand, D., Oprea, A., Courbat, J. *et al.* (2011) Making environmental sensors on plastic foil. *Mater. Today*, **14**, 416–423.

13

Dynamic Electrochemistry Transduction Methods

13.1 Introduction

Electrochemical reactions can be monitored under either equilibrium or non-equilibrium conditions. Equilibrium conditions are characteristic of potentiometric sensors in which the response is represented by the electromotive force of the cell. The overall reaction rate under equilibrium conditions is zero.

Under non-equilibrium conditions, the electrochemical reaction proceeds in a definite direction as either oxidation or reduction. Consequently, reactants are consumed and products form by electron transfer reactions that give rise to an electrical current flowing through the cell. In many electrochemical methods, the current measured at a given electrode potential represents the analytical response as the current is dependent on the reactant concentration. These methods are known as *amperometric methods*. In certain methods the potential is varied over a particular region in order to record a current-potential curve on which the current is measured at a particular potential so as to obtain good sensitivity and minimal interference. A method in which the current-potential curve is recorded for analytical purposes is termed a *voltammetric method*.

A series of dynamic electrochemistry methods require rigorous control of the electrode potential, which function as the independent variable (the excitation signal) whilst the current is the dependent variable. That is the reason why such methods are classified as potentiostatic methods.

Other dynamic electrochemistry methods (termed *galvanostatic methods*) involve experiments carried out at controlled current. In these methods, the response is obtained in the form of a potential-time curve.

This current chapter addresses electrochemical transduction methods under non-equilibrium conditions. The first section outlines the design principles of electrolytic cells for analytical applications. The next section reviews the theoretical principles that govern electrochemical reactions. Mastering of these principles allows for rational selection of optimum operating parameters of an electrochemical sensor. The following section presents electrochemical measurement methods that have been developed in order to achieve good sensitivity and selectivity in electroanalytical determinations. As catalyzed electrochemical reactions are of great importance in the electrochemical sensor area, a special section is devoted to this topic. Next, an overview of materials used as electrodes in electroanalytical applications is presented. Finally, amperometric gas sensors are introduced.

Theoretical principles of electrochemical methods are amply reviewed in various monographs (e.g. [1–4]). Practical aspects of experimental electrochemistry are comprehensively covered in refs. [5–8]. Advances in electrochemical sensors are surveyed in ref [9].

13.2 Electrochemical Cells in Amperometric Analysis

Electrolysis involves a reduction and an oxidation reaction, each of them occurring at a particular electrode called the *cathode* and *anode*, respectively. Each half-cell reaction can involve different compounds. Hence, the cell is designed and operated such that the analyte reacts only at one of the two electrodes, called the *working electrode*. As will be demonstrated in the following sections, the electrolytic current depends on the electrical potential applied to the working electrode. That is why it is essential to control rigorously the potential of this electrode and to design the cell so that the current is determined by the electrochemical reaction occurring at the working electrode.

There are two alternative methods for controlling the potential of the working electrode, namely the two-electrode cell and the three-electrode cell method (Figure 13.1).

In the two-electrode configuration (Figure 13.1A), the working electrode is combined with a reference electrode that is expected to keep constant its potential (E_r) during the run. If no voltage is applied to the cell, each electrode assumes a specific, equilibrium potential. However, if a voltage V is applied by means of an external voltage source, the potential of the working electrode shifts from the equilibrium value whilst the potential of the reference electrode remains unchanged. In addition, current flow through the electrolyte gives rise to a potential difference across the solution in agreement with Ohm's law. This voltage drop is given by the $R_s i$ term, where R_s is the electrical resistance of the solution and i is the electrolytic current. Therefore, the relationship between the applied voltage and

Chemical Sensors and Biosensors: Fundamentals and Applications, First Edition. Florinel-Gabriel Bǎnicǎ.
© 2012 John Wiley & Sons, Ltd. Published 2012 by John Wiley & Sons, Ltd.

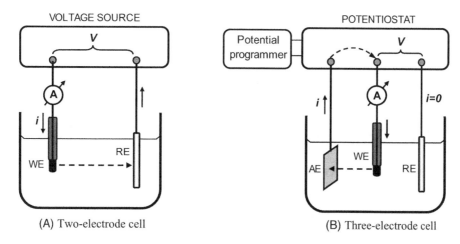

Figure 13.1 Electrochemical cells for amperometric determinations. WE: working electrode; RE: reference electrode; AE: auxiliary electrode (counterelectrode). Arrows indicate the current flow. (A) Two-electrode cell; (B) Three-electrode cell.

the working electrode potential is:

$$V = E_w - E_r + R_s i \tag{13.1}$$

Therefore the applied voltage determines the working electrode potential as follows:

$$E_w = V + E_r - R_s i \tag{13.2}$$

So, the working electrode potential can be tuned simply by adjusting the applied voltage if the ohmic drop is negligible. This condition is fulfilled at currents in the microampere region, provided that the solution resistance is reasonably small. Due to its simplicity the two-electrode configuration is widely used in electrochemical sensors. If the current is measured at a constant potential, the current is also constant and the ohmic drop is a constant correction term.

The ohmic drop becomes a critical issue if the measurement involves recording of a current–potential curve. In this case, the ohmic drop varies during the run and leads to unpredictable variations in the working electrode potential.

A more rigorous control of the working electrode potential is achieved by means of a three-electrode cell (Figure 13.1B). In this case, an auxiliary electrode is added to the cell and the voltage is applied by means of an electronic instrument called a *potentiostat*. This instrument allows the electrolytic current to flow between the working and auxiliary electrodes, whilst no current flow through the reference electrode. Therefore, the equilibrium at the reference electrode is not disturbed. At the same time, the working electrode potential is adjusted with no interference from the ohmic voltage drop in the solution. This last conclusion is only partially valid. Actually, due to the nonuniform distribution of the current in the cell, some interference of the ohmic drop can be experienced, particularly in solutions of high resistivity. Nevertheless, the effect of the ohmic drop is much less that in the case of the two-electrode cell. However, if the ohmic drop is significant, the potentiostat can be fitted with a device for automatic compensation of the ohmic drop.

Ohmic-drop correction is important in methods based on fast variation of the electrode potential, such as in pulse potential methods, AC methods and fast scan rate cyclic voltammetry. In such methods, the ohmic drop causes the time variation of the applied potential to deviate from that expected. Automatic ohmic-drop correction allows to alleviate this problem.

In order for the three-electrode cell to function correctly, the auxiliary electrode should be designed so as to prevent the reaction occurring at this electrode to affect the current. To this end, the area of the auxiliary electrode should be much larger than that of the working electrode. This brings about a very low current density at the auxiliary electrode, and hence only small deviations from the equilibrium state. Platinum and glassy carbon are typical auxiliary electrode materials. As a rule, the reaction at these electrodes is oxygen or hydrogen evolution, depending on the electrode polarity.

In some applications it is necessary to control the potential of two working electrodes with respect to a reference electrode. The device performing such a task is called a *bipotentiostat*.

Amperometric methods have been developed under the assumption that the transport of reactants and products occur by diffusion and possibly by convection. However, under the effect of the electric field in the cell, ions undergo displacement by migration. Migration of ionic reactants and products should be prevented by adding to the solution a *supporting electrolyte* (also called a *background electrolyte*). This is an electrolyte whose ions do not interfere with the electrochemical reaction under investigation. A simple salt (such as KNO_3, $NaClO_4$ or KCl) can act as supporting electrolyte at concentrations at least 100 times higher than that of the analyte. In the presence of a large excess of supporting electrolyte, migration of the analyte ion becomes negligible since its transport number assumes a very low value. In addition, the supporting electrolyte imparts good electrical conductivity to the solution and lessens the ohmic potential drop. If the analyte has acidic or basic properties, its protonation state is controlled by a pH buffer system which can also function as the supporting electrolyte.

13.3 The Electrolytic Current and its Analytical Significance

13.3.1 Current–Concentration Relationships

Current flow within an electrode occurs in the form of an electron stream, but free electrons cannot exist in the solution. In order for the current to cross the electrode/solution interface, an electrochemical reaction needs to occur. This reaction involves species in the solution and can be formulated in a general form as follows:

$$\text{Ox} + ne^- \underset{\text{Oxidation}}{\overset{\text{Reduction}}{\rightleftharpoons}} \text{Red} \tag{13.3}$$

where Ox and Red are the oxidized and reduced forms of a redox couple, respectively, both of them being assumed to be soluble species. So, the current intensity is limited by the velocity of the electrochemical reaction, which determines the amount of electrons crossing the interface in unit time.

The reaction rate of an electrochemical reaction (v_e) is defined in accord with its heterogeneous character as follows:

$$v_e = -\frac{1}{A}\frac{dN_O}{dt} = \frac{1}{A}\frac{dN_R}{dt} \tag{13.4}$$

where N_O and N_R represent the number of moles of oxidized and reduced form, respectively. In this equation, each derivative term represents the variation of the number of moles of either Red or Ox per unit time. As the Ox concentration decreases when the reaction (13.3) proceeds to the right, the minus sign is introduced in Equation (13.4) in order to obtain a positive quantity. A represents the electrode surface area.

According to Faraday's laws, reduction of 1 mole Ox (and formation of 1 mole Red) involves the exchange of nF coulombs of electricity (F is the Faraday constant). Hence, the current i is proportional to the reaction velocity:

$$i = nFAv_e = -nF\frac{dN_O}{dt} = nF\frac{dN_R}{dt} \tag{13.5}$$

The occurrence of the electrochemical reaction has a marked effect on the solution concentration of the compounds involved. In the equilibrium state (that is, $i = 0$), concentrations are uniform across the solution; these are bulk solution concentration, denoted $c_{O,b}$ and $c_{R,b}$ for Ox and Red, respectively. However, if the electrochemical reaction proceeds to the right, the Ox species is consumed, whilst the Red species is produced. This alters each of the above concentrations near the electrode surface. As shown in Figure 13.2, under electrolysis conditions each concentration

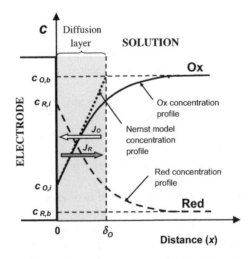

Figure 13.2 Concentration profiles at the electrode/solution interface and diffusion fluxes for the Ox (J_O) and Red (J_R) species. This figure emphasizes the concentration profiles in the case in which the electrochemical reaction (13.3) occurs to the right as a reduction (cathodic) reaction. For simplicity, only the Ox diffusion layer is represented.

varies with the distance from the electrode and tends asymptotically towards the bulk concentration value. The rate of concentration change with the distance is indicated by the *concentration gradient*, which is the concentration derivative with respect to the distance (dc/dx). Nonuniform distribution of the concentration induces diffusion. This is a spontaneous displacement of molecules or ions from a region of a higher concentration to a region of a lower concentration. The rate of the diffusion process is expressed by the diffusion flux (J) which represents the number of moles crossing a section of unit area in unit time. According to Fick's first law, the flux of a species of concentration c is proportional to the concentration gradient, the proportionality constant being the diffusion coefficient (D).

$$J = -D\frac{dc}{dx} \tag{13.6}$$

Under steady-state conditions, concentration profiles are invariable and the amount of substance transported by diffusion equates to the velocity of the electrochemical reaction. However, as this reaction occurs at the electrode/solution interface (that is, at $x = 0$ in Figure 13.2), the reaction velocity is equal to the flux occurring at this interface.

It will be assumed next that the electrode potential is lower than the equilibrium potential of the redox couple. In this case, the current corresponds to a reduction (cathodic) process and is designed as a *cathodic current* (i_c). The cathodic current relationship with the Ox and Red fluxes is as follows:

$$i_c = -nFAD_O\left(\frac{dc_O}{dx}\right)_{x=0} = nFAD_R\left(\frac{dc_R}{dx}\right)_{x=0} \tag{13.7}$$

The last term in the above equation is positive because the Red flux is opposite to the direction to the Ox flux. The above relationships are applicable in the presence of an excess of supporting electrolyte, which makes ion migration negligible.

In order to derive a relationship between current and concentration, the differential Equation (13.7) has to be solved using appropriate limiting conditions.

The above problem can be solved in a simpler way using the *Nernst diffusion layer* concept. This approach is based on the assumption that in a well-stirred solution, concentration gradients are localized within a thin solution layer, called the *diffusion layer* (Figure 13.2), near the electrode surface. Within this layer, the concentration gradient is constant and equal to the gradient at $x = 0$. Hence, the concentration profile in this model is represented by a straight line that is tangential to the actual concentration profile at $x = 0$ (dotted line in Figure 13.2). The thickness of the diffusion layer (δ) is defined by the intercept of the $c_{O,b}$ level with this hypothetical concentration profile. Under these conditions, concentration gradients can be expressed as follows:

$$\frac{dc_O}{dx} = \frac{c_{O,b} - c_{O,i}}{\delta_O} \quad \text{and} \quad \frac{dc_R}{dx} = \frac{c_{R,b} - c_{R,i}}{\delta_R} \tag{13.8}$$

where δ_O and δ_R represents the thickness of the diffusion layer for Ox and Red, respectively. These expressions can be substituted in Equation (13.7) to derive current–concentration relationships that are obtained in the following form:

$$i_c = -nFAk_{m,O}\left(c_{O,b} - c_{O,i}\right) = nFAk_{m,R}\left(c_{R,b} - c_{R,i}\right) \tag{13.9}$$

The terms $k_{m,O}$ and $k_{m,R}$ in this equation represent the *mass-transfer coefficients* of the Ox and Red species, respectively. The mass-transfer coefficient is the quotient of the diffusion coefficient and the thickness of the diffusion layer:

$$k_{m,O} = \frac{D_O}{\delta_O} \quad \text{and} \quad k_{m,R} = \frac{D_R}{\delta_R} \tag{13.10}$$

For simplicity, the following proportionality constants will be introduced:

$$\kappa_O = nFAk_{m,O}; \quad \kappa_R = nFAk_{m,R} \tag{13.11}$$

Upon substitution of these constants in the current–concentration relationships (13.9) it results:

$$i_c = -\kappa_O\left(c_{O,b} - c_{O,i}\right) \tag{13.12}$$

$$i_c = \kappa_R\left(c_{R,b} - c_{R,i}\right) \tag{13.13}$$

It should be borne in mind that the mass-transfer coefficient is not a true constant because it depends on the convection conditions in the solution, which determine the thickness of the diffusion layer. However, the mass-transfer coefficient is essentially constant under well-defined convection conditions.

In electroanalytical transduction methods, electrolysis is conducted so as to hold the bulk concentrations constant. This is achieved by keeping both the current and the electrolysis time sufficiently low.

Equation 13.12 includes the analytical concentration ($c_{O,b}$), but it is of no analytical use because it contains also the concentration at the interface ($c_{O,i}$), which is unknown. In order to derive a current expression of analytical relevance, it is necessary to look for some particular conditions under which the interface concentration is negligible. Such conditions can be found by considering the effect of the electrode potential on the surface concentrations, which can be expressed by the Nernst equation written in the following particular form:

$$E = E_f^0 + \frac{RT}{nF} \ln \frac{c_{O,i}}{c_{R,i}} \tag{13.14}$$

where E_f^0 is the formal potential of the redox system. This equation can be interpreted as follows. When a particular potential is imposed on the electrode, concentrations at the interface are altered by the electrochemical reaction so as to satisfy Equation (13.14). This equation allows a direct relationship between concentration and potential to be derived in the following form:

$$\frac{c_{O,i}}{c_{R,i}} = \exp\left[\frac{nF}{RT}\left(E - E_f^0\right)\right] \tag{13.15}$$

This equation demonstrates that the $c_{O,i}/c_{R,i}$ ratio becomes negligible when the electrode potential is much lower than the formal potential. In other words, this limiting situation can be describes as follows:

$$(c_{O,i})_{E \ll E_f^0} = 0 \tag{13.16}$$

Under the above conditions, Equation (13.12) gives a particular value of the current called the *limiting cathodic diffusion current*, $i_{c,d}$:

$$i_{c,d} = (i_c)_{E \ll E_f^0} = -\kappa_O c_{O,b} \tag{13.17}$$

Equation 13.17 indicates that the limiting current is directly proportional to the analytical concentration of the reactant and can serve as response signal in analytical determinations. The proportionality constant depends on the diffusion conditions, which are included in the mass-transfer coefficient.

If the Red concentration is of interest, reaction (13.3) should be made to proceed to the left as an oxidation (anodic) reaction. This can be achieved by making the electrode potential lower than the formal potential. Using the same reasoning as above, the anodic current (i_a) observed in this case depends on the Red concentration as:

$$i_a = nFAk_{m,R}(c_{R,b} - c_{R,i}) \tag{13.18}$$

Note that, in accordance with the IUPAC sign convention, a positive sign is assigned to the anodic current. If the electrode potential is set at a value much higher than the formal potential, it follows from Equation (13.14) that the $c_{R,s}$ term becomes negligible and the current assumes a particular value termed the *limiting anodic diffusion current* ($i_{a,d}$):

$$i_{a,d} = (i_a)_{E \gg E_f^0} = \kappa_R c_{R,b} \tag{13.19}$$

It remains an ambiguity as to how far from the formal potential the electrode potential should be in order to obtain the limiting current. This problem is addressed in the next section.

13.3.2 The Current–Potential Curve: Selecting the Working Potential

In order to derive a criterion for the selection of the electrode potential, a mathematical expression for the current–potential relationship is needed. This relationship can be derived by means of Equation (13.14) upon substituting the concentration terms by current–concentration relationships. The current–potential equation will be derived next under the assumption that the Red species is not initially present in the solution ($c_{R,s} = 0$). Combining Equations (13.9) and (13.17), one obtains:

$$c_{O,i} = \frac{i_c - i_{c,d}}{\kappa_O} \tag{13.20}$$

On the other hand, taking into account that the Red species is absent in the bulk of the solution, one obtains its interface concentration from Equation (13.13) as:

$$c_{R,i} = -\frac{i_c}{\kappa_R} \tag{13.21}$$

Upon substituting the above expressions for concentrations in Equation (13.14) and some rearrangements, it follows that:

$$E = E_f^0 + \frac{RT}{nF} \ln \frac{k_{m,R}}{k_{m,O}} + \frac{RT}{nF} \ln \frac{i_{c,d} - i_c}{i_c} \tag{13.22}$$

It is easy to prove that if $i_c = i_{c,d}/2$, the potential assumes a characteristic value called the *half-wave potential* ($E_{1/2}$):

$$E_{1/2} = (E)_{i_c = i_{c,d}/2} = E_f^0 + \frac{RT}{nF} \ln \frac{k_{m,R}}{k_{m,O}} \tag{13.23}$$

Substituting this constant in Equation (13.22), it follows that:

$$E = E_{1/2} + \frac{RT}{nF} \ln \frac{i_{c,d} - i_c}{i_c} \tag{13.24}$$

A simple rearrangement of the above equation gives the current–potential relationship as follows:

$$i_c = \frac{i_{c,d}}{1 + \exp\left[\dfrac{nF}{RT}\left(E - E_{1/2}\right)\right]} \tag{13.25}$$

A plot of the current–potential curve according to Equation (13.25) is shown in Figure 13.3A, for particular values of the equation parameters, as indicated in the figure. Owing to its characteristic shape, such a curve is called a *voltammetric wave* and is characteristic of electrochemical reactions in the *steady state*. As can be seen, the cathodic current is zero at $E \gg E_{1/2}$. As the potential is made more negative, the current increases abruptly over a potential region of about 0.2 V centered around the half-wave potential and reaches the limiting current value at $E \ll E_{1/2}$. The symbols "$\gg$" and "$\ll$" used above imply a potential shift of only about 0.15 V with respect to the half-wave potential. So, for $n = 1$, the cathodic limiting current is obtained if the electrode potential is about $(E_{1/2} - 0.2)$ V.

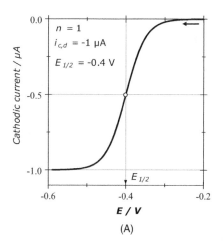
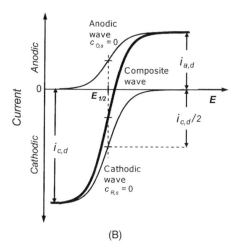

Figure 13.3 Current–potential curves (voltammograms) for electrochemical reactions under steady-state conditions. (A) A cathodic wave; (B) a composite, cathodic–anodic wave. A simple cathodic or anodic wave is recorded if only the Ox or Red species, respectively, is present in the solution. The composite wave results when both species are present. Adapted with permission from [10]. Copyright 2002 A. Bănică and F.G. Bănică.

For $n > 1$ the steepness of the current variation around the half-wave potential is even greater, in accordance with Equation (13.25).

The shape of the wave and its position on the potential axis depends on two parameters: the number of electrons involved in the electrochemical reaction and the half-wave potential. This characteristic potential, which is defined in Equation (13.23), is independent of the reactant concentration and is useful for identification purpose. The second term in Equation (13.23) is very small because the mass-transfer coefficients of Ox and Red are almost similar and their ratio appears under logarithm. Therefore, the half-wave potential is a good approximation of the formal potential of the redox couple.

If only the Red species is present in the solution, the recorded wave corresponds to an anodic process and is described by the following equation:

$$i_a = \frac{i_{a,d}}{1 + \exp\left[-\dfrac{nF}{RT}\left(E - E_{1/2}\right)\right]} \qquad (13.26)$$

This equation indicates that the anodic current is zero when the potential is much lower than the half-wave potential but reaches the limiting current value at potentials much higher than the half-wave potential. An anodic wave is shown in Figure 13.3B.

If both Ox and Red species are present in the solution one obtains a composite wave which displays both kinds of limiting current at the extreme limits of the potential region, as shown in Figure 13.3B. This figure includes also the cathodic and anodic waves which are obtained when only one of the redox couple forms is present. The half-wave potential of the composite wave is the potential at which the current is half the difference of the limiting currents. This value is similar to the half-wave potentials of individual waves obtained with only one species present in the solution.

The above treatment applies to so-called *reversible (or Nernstian) electrochemical reactions*. For reversible reactions, the reactant and product concentrations at the interface satisfy the Nernst Equation (13.14).

Reaction (13.3) assumes that neither the reactant nor the product is involved in some chemical reaction. This mechanism is characteristic of metal complex redox mediators used in amperometric enzyme sensors.

If either the reactant or the product is involved in a chemical reaction at equilibrium, then the Nernst equation is satisfied by the concentrations of species directly involved in the electrochemical reaction. Thus, a hypothetical reaction involving both electrochemical and chemical reactions can be formulated as follows:

$$Ox + e^- \rightleftarrows Red^- \qquad (13.27)$$

$$Red^- + H^+ \overset{K_p}{\rightleftarrows} HRed \qquad (13.28)$$

The electrochemical process occurs in a reversible way if the Nernst equation with respect to the interfacial concentrations of Ox and Red^- is satisfied. However, the Red^- concentration depends on the pH through the step (13.28) which is characterized by the equilibrium constant K_p. If this step proceeds reversibly (that is, it can occur in both directions with very high rate constants), then a pH-dependent term should be included in the equation. This additional term includes the equilibrium constant K_p. In order to obtain reliable results, such electrochemical processes must be conducted in solutions containing a pH buffer system in order to keep the pH constant.

13.3.3 Irreversible Electrochemical Reactions

Many electrochemical reactions are *irreversible*, which cause deviation from the Nernst equation with respect to the interfacial concentrations. This deviation arises from the sluggishness of the reaction in both directions. That is why it is common (and more informative) to term an irreversible reaction a *slow electrochemical reaction*. In contrast, a reversible (Nernstian) electrochemical reaction is often termed a *fast electrochemical reaction*.

If the reaction (13.3) proceeds as a slow (irreversible) electrochemical reaction, the current potential relationship is given by the following equation, which applies to a cathodic current:

$$i_c = \frac{i_{c,d}}{1 + \exp\left[\dfrac{\alpha nF}{RT}\left(E - E_{1/2}\right)\right]} \qquad (13.29)$$

This equation differs from Equation (13.25) by the presence of the subunit coefficient α called the *transfer coefficient* ($\alpha < 1$) in the argument of the exponential. Owing to this coefficient, the half-wave potential is very different from the standard potential and the current variation in response to the potential change is less steep than is the case for a reversible process. However, the limiting current is independent of the reversibility of the electrochemical reaction.

In general, fast (reversible) electrochemical reactions are preferred in the design of amperometric sensor. However, by the force of circumstances, slow (irreversible) reactions are also used.

More details on the reversibility of electrochemical reactions are presented later in Section 13.6.

13.3.4 Sign Convention

When plotting a current–potential curve, a distinction is made between anodic and cathodic current by assigning a particular sign to each of them. In accordance with the IUPAC recommendation, it was assumed before that an anodic current is positive whilst a cathodic current is negative. Accordingly, axes orientation is selected according to the scheme in Figure 13.4A.

However, in the literature, voltammetric curves are often reported in a format different from the IUPAC recommendation, as shown in Figure 13.4B. In this American (or polarographic) convention, an anodic current is negative and a cathodic one is positive. In addition, on the potential axis, increasing negative values are oriented to the right, whilst increasing positive values are directed to the left.

In order to avoid confusion, it is best to indicate on the graph the cathodic or anodic nature of the current as well as the sign of the potential on the horizontal axis.

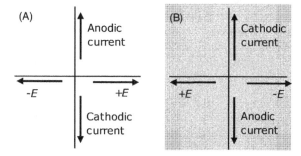

Figure 13.4   Sign conventions in the representation of current–potential curves. (A) IUPAC convention; (B) American (polarographic) convention.

13.3.5 Geometry of the Diffusion Process

It has been implicitly assumed in the above approach that diffusion occurs at a planar electrode. Under these circumstances, diffusion paths are parallel and perpendicularly oriented to the electrode surface. It was also assumed that the solution expands over a distance much greater than the thickness of the diffusion layer. These conditions are typical of *planar, semi-infinite diffusion*.

However, electrodes with other shapes, such as cylindrical and spherical electrodes are often used in amperometric determinations. For such electrodes, diffusion paths are not parallel but are convergent. However, if the thickness of the diffusion layer is negligible with respect to the electrode diameter, the degree of convergence is also negligible. Consequently, the planar diffusion model is a good approximation in such cases, despite the nonplanar shape of the electrode. In general, electrodes with a diameter above 1 mm fulfill this condition.

13.3.6 Outlook

Many electroanalytical methods rely on the measurement of diffusion currents that are directly proportional to analyte concentration. The proportionality constant depends on the diffusion coefficient and the thickness of the diffusion layer.

In the steady state, the plot of current vs. potential gives a typical curve called a voltammetric wave. For reversible processes, at any particular potential, the reactant and product concentrations at the electrode surface are in accord with the Nernst equation. If this condition is not satisfied, the electrochemical reaction is to a greater or lesser extent irreversible.

A reversible wave is completely defined by three parameters: the diffusion current, the half-wave potential and the number of electrons in the electrochemical reaction. A limiting current is obtained when the electrode potential is sufficiently far from the standard potential in order to achieve complete depletion of the reactant at the electrode surface.

The half-wave potential is very close to the formal potential of the redox couple if no chemical reaction is coupled with the electrochemical one.

If a very fast and reversible chemical reaction is coupled with the electrochemical reaction, the half-wave potential depends also on the equilibrium constant of the chemical reaction and the concentration of any reactant involved in this reaction.

Questions and Exercises (Sections 13.2; 13.3)

1 How do Faraday's laws determine the strength of the electrolytic current?

2 What effect has an electrochemical reaction on the concentration of the reactant and the product in the solution?

3 What process in the solution accompanies the electrochemical reaction? What quantity indicates the rate of this process and what is the unit of this quantity?

4 How can an electrochemical reaction affect the rate of the associated diffusion process?

5 What is the diffusion layer and what parameters determine its thickness?

6 What do the terms cathodic and anodic current mean? What are limiting diffusion currents and how do they depend on reactant concentration?

7 What are the characteristic parameters of a steady-state voltammetric wave in the case in which, a) only the Ox or Red species is present in the solution; b) both Ox and Red are present? What factors determine each of these parameters?

8 Prove that at the half-wave potential the interface concentration of the reactant is half the bulk concentration.

9 The table below includes current–potential data for a reversible cathodic steady-state wave with the limiting diffusion current $-25\,\mu A$. (a) Using these data, plot the logarithm term in Equation (13.24) as a function of potential. (b) Use the plotted straight line to determine the half-wave potential and the number of electrons in the electrochemical reaction. (c) Assuming that the potential was measured with respect to the saturated calomel electrode (SCE), calculate the half-wave potential with respect to the standard hydrogen electrode (SHE). $E_{SCE} = +0.244\,V$ vs. SHE. This exercise demonstrates the logarithmic analysis of a steady-state voltammetric wave.

E/V	-0.27	-0.26	-0.25	-0.24	-0.23
$i/\mu A$	-20.7	-17.1	-12.5	-7.9	-4.4

Answer: $E_{1/2} = -0.25\,V$ vs. SCE; $n = 2$.

10 (a) Draw schematically a reversible anodic wave assuming that the formal potential is 0.0 V and using both the IUPAC and the American sign convention. (b) Draw an anodic irreversible wave assuming that the formal potential, reactant concentration and diffusion coefficients are similar to those of the reaction in (a).

13.4 Membrane-Covered Electrodes

According to the conclusions in Section 13.3.1, the limiting diffusion current depends on the thickness of the diffusion layer, which, in turn, is determined by the convection conditions (e.g., stirring rate and stirring procedure). Reliable amperometric determinations can be performed only under stable and reproducible convection conditions. This limitation is not compatible with sensor design requirements because a sensor should be as independent as possible of the experimental conditions. This problem is more severe when the sensor is operated in standing solutions. Even if no stirring is provided, slight spontaneous convection occurs always in a random way and produce current fluctuations.

A solution to this problem is an electrode covered by a membrane permeable to the analyte such that about a 10-μm solution layer is sandwiched between them, as shown in Figure 13.5A. Adhesion to the electrode surface prevents convection within this solution layer. A potential gradient develops due to the electrochemical reaction and the analyte has to cross the membrane before reaching the electrode surface. However, the diffusion coefficient within the membrane is much lower compared to that in the solution. Consequently, for a given flux, the concentration gradient across the membrane will be much greater than that in the solution phase ($\Delta c_2 \gg \Delta c_3$). In the steady state, the analyte flux (J_A) should be the same across each layer in Figure 13.5A, that is:

$$J_A = \frac{D_1}{\delta_1}\Delta c_1 = \frac{D_2}{\delta_2}\Delta c_2 = \frac{D_3}{\delta_3}\Delta c_3 \tag{13.30}$$

We shall now introduce for convenience the *mass-transfer resistance* (r) which is the reciprocal of the mass transfer coefficient:

$$r = \frac{\delta}{D_A} = \frac{1}{k_{m,A}} \tag{13.31}$$

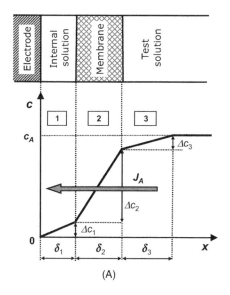
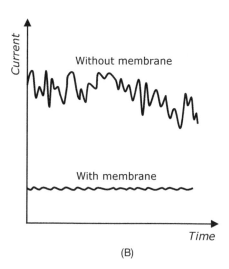

(A)   (B)

Figure 13.5   (A) Concentration profile at a membrane-covered metal electrode under the conditions of the limiting current (that is, $c_{A,i} = 0$); (B) Membrane effect on the limiting current.

Taking into account Equation (13.31), Equation (13.30) becomes:

$$J_A = \frac{\Delta c_1}{r_1} = \frac{\Delta c_2}{r_2} = \frac{\Delta c_3}{r_3} \tag{13.32}$$

The subscript to r indicates the pertinent layer in Figure 13.5A. Simple algebraic manipulation gives:

$$J_A = \frac{\Delta c_1 + \Delta c_2 + \Delta c_3}{r_1 + r_2 + r_3} = \frac{c_A}{r_1 + r_2 + r_3} \tag{13.33}$$

As a rule, r_2 is much greater that r_1 and r_3 because the diffusion coefficient in the membrane is very small. Additionally, both r_1 and r_2 are constants because the relevant δ values are invariable.

If we inspect now the effect of convection in the test solution, we realize that only r_3 depends on this factor. But, because $r_3 \ll r_2$, fluctuations in the r_3 values (due to some factors that are not under control) have a negligible effect on J_A. Consequently, the membrane causes the current to be almost independent of the convection in the test solution, which is very convenient for practical applications. Figure 13.5B displays in a schematic way the effect of the membrane on the current. In the absence of the membrane, the current shows large fluctuations owing to spontaneous convection in the solution that affects randomly the thickness of the diffusion layer. If the electrode is covered by a membrane permeable to the analyte, the current is lower with respect to the free-surface electrode owing to the high value of r_2. At the same time, current fluctuations are very small.

Besides removing the current fluctuations, a semipermeable membrane protects the metal surface against contact with damaging components of the sample, such as, for example, large organic molecules. Such molecules can form by adsorption surface layers that block the access of the analyte to the electrode surface.

The membrane can also enhance the selectivity by preventing the access of interfering compounds to the electrode surface. For example, glucose determination can be perturbed by anionic species such as the ascorbate anion. A cation-selective membrane prevents the transport of ascorbate but does not hinder the diffusion of glucose and oxygen, which are neutral molecules. Membranes with selective permeability are widely used in amperometric sensors [11].

A more comprehensive theoretical approach to membrane-covered electrodes is available in ref. [12].

13.5 Non-Faradaic Processes

13.5.1 Origin of Non-Faradaic Currents

The current used as the analytical signal in amperometric methods originates from an electrochemical reaction and is related to the reaction rate through Faraday's laws. Hence the name *Faradaic current* is assigned to such a current in order to distinguish it from that types of current that are not associated with an electrochemical reaction. Non-Faradaic currents arise from the displacement of electric charges (electrons and ions) in order to achieve charge balance at the electrode solution interface. In order for the electrode potential to be modified, electrons are supplied

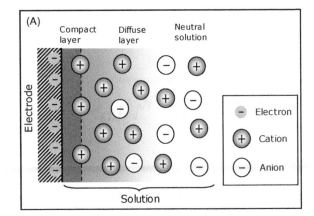
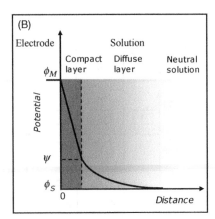

Figure 13.6 Double electric layer at the electrode/solution interface in the case of a negatively polarized electrode. (A) Charge distribution at the interface; (B) variation of the electric potential at the interface. Adapted with permission from [10]. Copyright 2002 A. Bănică and F.G. Bănică.

or withdrawn thus changing the electron density at the electrode surface. Consequently, ions in solution migrate to or from the interface in order to compensate for the change in the electrode charge and impart to the interface an overall zero charge. This process gives rise to a non-Faradaic current that is usually designated a *charging* (or *capacitive*) *current* because of the similarity of the electrode/solution interface to an electrical capacitor. If an electrochemical reaction also occur the capacitive current adds to the Faradaic one to give the overall current, which is the measurable quantity. This effect determines the limit of detection in amperometric methods. The analyte can be determined only if its concentration is sufficiently large to give a Faradaic current that is significantly higher that the capacitive current. That is why it is important to design an amperometric method so as to render the capacitive current as low as possible with respect to the Faradaic current.

13.5.2 The Electrical Double Layer at the Electrode/Solution Interface

In order to understand the origin of the capacitive current it is necessary to consider the distribution of the electric charge at the electrode/solution interface. This is shown schematically in Figure 13.6A for the case of a negatively charged electrode. The negative charge at the electrode is balanced by cations attracted to the electrode surface to form an *electrical double layer* in which electrons are localized at the electrode surface, whilst the counterions in solution are distributed over a certain distance from the electrode. Within the solution charge layer, at least two regions can be distinguished if no adsorption at the electrode surface occurs. A *compact ion layer* forms in immediate contact with the electrode surface. It consists of hydrated ions strongly attracted by the electric charge at the electrode. Water dipole molecules are also present in a fixed orientation determined by the electric field. Thus, at a negative electrode, water molecules are oriented with the hydrogen atoms towards the electrode. Beyond the compact layer, the electrostatic forces are weaker and the ion distribution is affected by random thermal agitation. This second layer is called the *diffuse layer*. It can include both cations and anions but cations prevail in order to balance the excess charge at the electrode. The diffuse layer has no well-defined boundary at the solution side. The excess charge in this layer decreases gradually with the distance from the electrode until the neutral solution zone is reached.

This charge distribution gives rise to a particular distribution of the electric potential in the solution phase (Figure 13.6B). Within the compact layer, the potential varies linearly with the distance, from the potential level in the metal (ϕ_M) to a characteristic potential ψ. Further, in the diffuse layer, the potential decreases asymptotically to the potential level in the solution (ϕ_s). The thickness of the compact layer is close to the size of hydrated cations (that is, in the angstrom range) whilst the diffuse layer can expand over a distance in the nanometer region. With increasing electrolyte concentration, more cations are included in the compact layer. For charge-balance reasons fewer cations will be found in the diffusion layer. Hence, the thickness of the diffusion layer decreases with increasing ion concentration and the ψ potential varies accordingly.

If the electrode is positively charged, a similar charge distribution occurs, except for the fact that anions function in this case as counterion in the double layer.

The transition point between the cation and anion layer occurs at a particular potential at which the charge density at the electrode is zero. Hence, this potential is called *the potential of zero charge* (E_{pzc}). This potential depends on the nature of the electrode material as well as the nature and concentration of ions in solution. Therefore, the charge density at the electrode is determined by the difference between the actual electrode potential (E) and the potential of zero charge:

$$\Delta E_e = E - E_{pzc} \tag{13.34}$$

13.5.3 The Charging Current

In order to derive a mathematical expression for the charging current, it will be assumed that the double layer behaves as a planar capacitor composed of two metal plates with an area (A) similar to that of the electrode surface area. The dielectric of the equivalent capacitor has a dielectric constant (ε_d) similar to that of the solution near the electrode surface. This capacitor is loaded with a charge Q_M similar to that present at the electrode surface. According to the general definition of the current, the capacitive current (i_C) is given by the time derivative of the capacitor charge:

$$i_C = \frac{dQ_M}{dt} \tag{13.35}$$

The electrical charge at the metal surface depends on the applied potential; more precisely, it is proportional to ΔE_e (Equation (13.34)), the proportionality constant being the integral capacitance (C_i) of the double layer:

$$Q_M = C_i \Delta E_e \tag{13.36}$$

The capacitance of a planar capacitor depends on the plate surface area (A), the distance between plates (l) and the dielectric constant (ε_d) as follows:

$$C_i = \varepsilon_d \frac{A}{l} \tag{13.37}$$

Upon suitable substitutions, one obtains from the above equations the following relationship between charge, on the one hand, and the geometrical and physical parameters of the capacitor on the other hand:

$$Q_M = A\varepsilon_d \frac{\Delta E_e}{l} \tag{13.38}$$

Assuming that the distance between plates is constant, the time variation of the charge can be caused by variations in either the electrode surface area, the applied potential or the dielectric constant. Hence, the time derivative of the charge (that is, the current) can be expressed as a sum of partial derivatives as follows:

$$i_C = \left[\varepsilon_d \Delta E_e \left(\frac{\partial A}{\partial t} \right)_{\varepsilon_d, \Delta E_e} + A\varepsilon_d \left(\frac{\partial (\Delta E_e)}{\partial t} \right)_{A, \varepsilon_d} + A\Delta E_e \left(\frac{\partial \varepsilon_d}{\partial t} \right)_{A, \Delta E_e} \right] \frac{1}{l} \tag{13.39}$$

When using solid electrodes, the surface area is a constant, but it varies in the case of the dropping mercury electrode used in polarography. Hence, a capacitive current is produced at this electrode even if the electrode potential is held constant.

Many voltammetric methods involve fast variation of the electrode potential that generate a capacitive current according to the second term in the right-hand side of Equation (13.39). In order to minimize the contribution of the capacitive current to the overall current, the difference between the time variation of the Faradaic and capacitive current is used (Section 13.7).

The dielectric constant can also vary if the solution contains an organic substance that adsorbs at the electrode surface. As the dipole moment of an organic compound is much lower than that of water, adsorption causes a decrease in the capacitance according to Equation (13.37). On the other hand, adsorption occurs within a limited potential range depending on the charge of the adsorbate. Neutral compounds adsorb within a potential region around the potential of the zero charge. In contrast, adsorption of charged molecules is determined by the electrostatic interaction between adsorbate and the charged electrode. Thus, cations adsorb when the electrode is negative of the potential of zero charge ($E < E_{pzc}$) whereas anion adsorption occurs under the opposite conditions ($E > E_{pzc}$). At the limits of the adsorption potential range there are narrow regions where the degree of coverage varies strongly with the potential. If a small-amplitude periodic potential is superimposed on a DC polarization potential, potential oscillations give rise to periodic changes in the degree of coverage. As a consequence, the dielectric constant also varies periodically and gives rise to a periodic current in accordance with the third term in the right-hand side of Equation (13.39). However, this is not a purely capacitive current because it involves periodical modifications in the properties of the insulator. This kind of non-Faradaic current is called a pseudocapacitive current. It is used in an electroanalytical method called *tensammetry* that allows the determination of electrochemically inactive compounds based on their adsorption at the electrode surface [13,14].

13.5.4 Applications of Capacitance Measurement in Chemical Sensors

As mentioned above, adsorption of organic compounds at the electrode surface causes the surface capacitance to decrease for at least two reasons. First, the adsorbed layer modifies the dielectric constant at the interface. Secondly, if the surface layer is formed of bulky molecules, the distance between the charged electrode surface and the ionic part of the double layer increases.

Such changes are also expected to occur when the electrode surface is coated with a recognition layer that can bind by affinity interactions an analyte from the solution. As a consequence of the recognition event, the thickness of the surface layer increases. In addition, if charged particles (such as proteins or nucleic acids) are involved, the recognition event produces a redistribution of electrical charges in the solution part of the double layer. Consequently, the recognition event can produce a significant change in the surface capacitance and allows for capacitive transduction in sensors based on affinity interactions. The measurement of the capacitance at the electrode/solution interface is performed by *electrochemical impedance spectrometry* methods that are presented in detail in Chapter 17.

Questions and Exercises (Section 13.5)

1 What is the difference between a Faradaic and a non-Faradic electrochemical process?
2 How are electric charges distributed at the electrode/solution interface?
3 What is the origin of the charging current? What parameters determine the strength of the charging current?
4 How can the charging current affect the limit of detection of an electroanalytical method of analysis?
5 What is the definition of the potential of zero charge and how is the adsorption of neutral or charged molecules at the electrode surface affected by this parameter?
6 What are the applications of electrode capacitance measurements in chemical sensing?

13.6 Kinetics of Electrochemical Reactions

13.6.1 The Reaction Rate of an Electrochemical Reaction

It was proved in the early history of electrochemistry that the electrolytic current depends essentially on two parameters: the reactant concentration and the electrode potential. The rationale of this finding is provided by a close inspection of the kinetics of electrochemical reaction. An electrochemical reaction involving two species in solution can be formulated as follows:

$$\text{Ox(aq)} + \text{e}^- \underset{\text{Anodic}(k_a)}{\overset{\text{Cathodic}(k_c)}{\rightleftarrows}} \text{Red(aq)} \tag{13.40}$$

The reaction rate of an electrochemical reaction is defined in Equation (13.4). According to this definition, the reaction rate represents the number of moles converted in the unit time per unit surface area and is measured in $\text{mol m}^{-2}\text{ s}^{-1}$. For a first-order reaction (13.40), the reaction rate is proportional to the reactant concentration, the proportionality constant being the *rate constant* (k_c and k_a). Denoting by $v_{e,c}$ and $v_{e,a}$ the reaction rate for the cathodic and anodic reaction, respectively, we have:

$$v_{e,c} = k_c c_{O,i}; \quad v_{e,a} = k_a c_{R,i} \tag{13.41}$$

Here, $c_{O,i}$ and $c_{R,i}$ represent concentrations at the solution/electrode interface, which can be different from the bulk concentrations (see Figure 13.2). The overall reaction velocity is the difference of the partial reaction velocities:

$$v_e = v_{e,a} - v_{e,c} \tag{13.42}$$

At equilibrium, the partial reaction velocities are equal ($v_{e,a} = v_{e,c}$) and the overall reaction rate is zero.

In general, the rate constant k of a chemical reaction is an exponential function of temperature according to the Arrhenius equation:

$$k = A \exp\left(-\frac{\Delta G^*}{RT}\right) \tag{13.43}$$

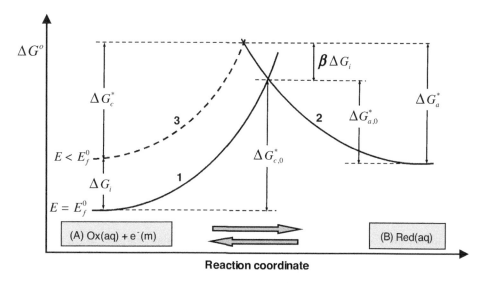

Figure 13.7 Evolution of the free energy during an electrochemical reaction. Curves 1 and 2 pertain to the reference state ($E = E_f^0$); curve 3 refers to a cathodic process. Adapted with permission from [10]. Copyright 2002 A. Bănică and F.G. Bănică.

Here, $\mathcal{A}$ is a temperature-independent parameter, ΔG^* is the *activation energy* in J mol^{-1}, R is the gas constant and T is the absolute temperature. This equation implies that, in order for the initial state to turn into the final state, the free energy of the system should first reach a maximum value, which is ΔG^*. After reaching this energy, the system shifts immediately to the final state. As the additional energy originates from thermal motion, the higher the temperature, the more frequent the transitions occur. In other words, a higher temperature implies a greater rate constant.

In contrast to typical chemical reactions, an electrochemical reaction involves the electron as a reactant. Electrons are carried by the metal electrode and hence the electron energy depends on the electric potential applied to the electrode. That is why the activation energy of an electrochemical reaction depends not only on the temperature but also on the electrode potential.

The effect of the electrode potential is demonstrated in Figure 13.7 that shows the change in the energy of the system during the transition from the initial to the final state. Curves 1 and 2 indicate the change in energy for the direct and reverse reactions, respectively, at the reference potential, which in this approach is assumed to be the formal potential of the redox system under consideration. Both the transition from Ox to Red and the reverse transition involve an increase in energy up to the intersection of curves 1 and 2. Once this maximum point is reached, the system turns instantly into the final state. The activation energy is represented by the energy difference between the initial state and the state corresponding to the intercept of curves 1 and 2. As can be seen, the activation energy of the cathodic reaction ($\Delta G_{c,0}^*$) and that of the anodic reaction ($\Delta G_{a,0}^*$) are different owing to the different energy level of each stable state.

If the electrode potential (E) is modified (for example, $E < E_f^0$), the energy of the state (1) (which includes electrons) is modified by:

$$\Delta G_i = -F\Delta E \tag{13.44}$$

where

$$\Delta E = E - E_f^0 \tag{13.45}$$

As a result, the energy curve of the state (A) will be translated to the curve 3, while the energy curve 2 remains unchanged because the state (B) does not involve electrons. However, the activation energy of both the forward and reverse process is altered in response to the change in the electrode potential. These modifications are given by the $\beta\Delta G_i = -\beta nF\Delta E$ where β is a subunity constant ($\beta < 1$). According to Figure 13.7, the activation energy of the anodic reaction becomes:

$$\Delta G_a^* = \Delta G_{a,0}^* - \beta F\Delta E \tag{13.46}$$

In the case of the cathodic reaction, the activation energy changes to:

$$\Delta G_c^* = \Delta G_{c,0}^* + \beta\Delta G_i - \Delta G_i \tag{13.47}$$

Upon setting $\alpha = 1 - \beta$, one obtains:

$$\Delta G_c^* = \Delta G_{c,0}^* + \alpha F\Delta E \tag{13.48}$$

The constants α and β are the *transfer coefficients* for the cathodic and anodic reactions, respectively. In many cases, $\alpha = \beta = 0.5$. The $\alpha + \beta = 1$ equality is rigorous only for single-electron reactions.

It is possible now to derive the rate-constant equations upon substituting Equations (13.46) and (13.48) for the activation energy in the Arrhenius equation. It is convenient to introduce first the f parameter defined as:

$$f = \frac{F}{RT} \tag{13.49}$$

Upon substituting Equations (13.46) and (13.48) in the Arrhenius equation (Equation 13.43) one obtains:

$$k_c = k'_c \exp(-\alpha f \Delta E) \tag{13.50}$$

$$k_a = k'_a \exp(\beta f \Delta E) \tag{13.51}$$

In the above equations, the constants k'_c and k'_a include the $\mathcal{A}$ constant and the $\Delta G^*_{c,0}$ and $\Delta G^*_{a,0}$ parameters, respectively.

The meaning of the above rate constants can be further elucidated upon considering the standard state of the system. In this state the redox system is at equilibrium ($k_c c_O = k_a c_R$), the concentrations of both Ox and Red are 1 mol m^{-3}, and the electrode potential is $E = E^0_f$. It follows that in the standard state the constants k'_c and k'_a have the same value, which is the *standard rate constant* of the electrochemical reaction (k_s, in m s^{-1}):

$$(k'_c)_{\text{standard state}} = (k'_a)_{\text{standard state}} = k_s \tag{13.52}$$

Therefore, using the k_s symbol, the reaction rate of the cathodic and anodic reaction become, respectively:

$$v_{e,c} = k_s c_{O,i} \exp(-\alpha f \Delta E) \tag{13.53}$$

$$v_{e,a} = k_s c_{R,i} \exp(\beta f \Delta E) \tag{13.54}$$

and the overall reaction rate is:

$$v_e = v_{e,a} - v_{e,c} = k_s \left[\underset{\text{Anodic}}{c_{R,i} \exp(\beta f \Delta E)} - \underset{\text{Cathodic}}{c_{O,i} \exp(-\alpha f \Delta E)} \right] \tag{13.55}$$

Therefore, at negative potentials ($\Delta E \ll 0$) the cathodic reaction prevails and the reaction (13.40) proceeds to the right as a reduction process. In the opposite case ($\Delta E \gg 0$), the anodic reaction is much faster and the overall reaction occurs to the left. At equilibrium, the partial reaction rates equal each other and the overall reaction rate is zero.

13.6.2 Current–Potential Relationships

According to Faraday's laws, conversion of one mole of reactant through a single-electron reaction involves the transfer of one farad of electricity. Therefore, the current is proportional to the reaction velocity as follows:

$$i = FA v_e \tag{13.56}$$

It is convenient to operate further with the current normalized to the electrode surface area, which represents the *current density* (j). In this way, Equation (13.56) gives:

$$j = \frac{i}{A} = F v_e \tag{13.57}$$

The current–potential equation can now be derived upon combining equations (13.55) and (13.57). It results thus

$$j = F k_s \left[\underset{\text{Anodic}}{c_{R,i} \exp(\beta f \Delta E)} - \underset{\text{Cathodic}}{c_{O,i} \exp(-\alpha f \Delta E)} \right] \tag{13.58}$$

According to this equation, the total current results from the algebraic summation of two components: the cathodic current (j_c) and the anodic current (j_a):

$$j = j_c + j_a \tag{13.59}$$

where:

$$j_a = Fk_s c_{R,i} \exp(\beta f \Delta E) \tag{13.60}$$

$$j_c = -Fk_s c_{O,i} \exp(-\alpha f \Delta E) \tag{13.61}$$

Both the cathodic and anodic component vary exponentially with the electrode potential. While the cathodic current increases with $-\Delta E$ value, the anodic current increases with ΔE.

13.6.3 Mass-Transfer Effect on the Kinetics of Electrochemical Reactions

Using the same reasoning as in Section 13.3.1, interface concentrations can be expressed as functions of current densities by the following equations in which $j_{c,d}$ and $j_{a,d}$ are the limiting diffusion current densities for the cathodic and anodic reaction, respectively:

$$c_{O,i} = \frac{j - j_{c,d}}{Fk_{m,O}}; \quad c_{R,i} = \frac{j_{a,d} - j}{Fk_{m,R}} \tag{13.62}$$

Upon substituting the above expressions for concentration terms in Equation (13.58) and assuming for simplicity that $k_{m,O} = k_{m,R} = k_m$ one obtains:

$$j = \frac{k_s}{k_m} \frac{j_{c,d}\exp(-\alpha f \Delta E) + j_{a,d}\exp(\beta f \Delta E)}{1 + \frac{k_s}{k_m}[\exp(-\alpha f \Delta E) + \exp(\beta f \Delta E)]} \tag{13.63}$$

In this equation, reactant and product concentrations are expressed in an indirect way, through the relevant limiting current that is proportional to the concentration. The assumption of equal mass-transfer coefficients is valid when the electron transfer does not produce modifications in the reactant size and shape but affects only the charge of the redox center.

According to Equation (13.63), the shape of the current potential curve is determined to a large extent by the quotient of the standard rate constant and the mass-transfer coefficient. This dimensionless quotient represents the Damköhler number for an electrochemical reaction $(Da)_e$ and its meaning can be formulated as follows:

$$(Da)_e = \frac{k_s}{k_m} = \frac{\text{Rapidity of the electrochemical reaction}}{\text{Rapidity of the mass transfer process}} \tag{13.64}$$

In other words, the k_s/k_m ratio indicates how fast an electrochemical reaction could be with respect to the associated mass-transfer process.

A plot of current potential curves according to Equation (13.63) is shown in Figure 13.8 for the particular case in which Ox and Red species have similar concentrations (that is, $j_{a,d} = -j_{c,d}$) and the transfer coefficient is 0.5. Each curve corresponds to a different k_s/k_m ratio that varies from 100 to 0.001.

At a very high k_s/k_m ratio (curve 1) the current varies abruptly over a narrow potential region and one single composite wave, similar to that in Figure 13.3B, is obtained. Its half-wave potential is similar to the formal potential of the redox system, as is typical of a reversible process. Curve 1 is characteristic of a *fast* electrochemical reaction

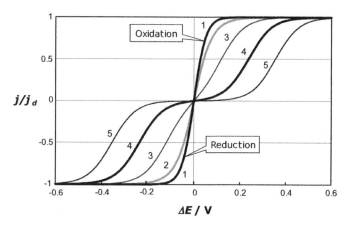

Figure 13.8   Current–potential curves under steady-state conditions for various values of the k_s/k_m ratio: (1) 100; (2) 1; (3) 0.1; (4) 0.01; (5) 0.001. $\alpha = \beta = 0.5$; $j_{a,d} = -j_{c,d} = j_d$. Reproduced with permission from [10]. Copyright 2002 A. Bănică and F.G. Bănică.

and this condition is equivalent to the reversibility condition introduced in Section 13.3.2. Therefore, if $k_s/k_m \gg 1$, an electrochemical reaction like reaction (13.40) is reversible. It is easy to demonstrate using Equation 13.63 that the half-wave potential for a reversible electrochemical reaction is independent of the transfer coefficients α and β and is in agreement with Equation (13.22).

As curve 2 in Figure 13.8 demonstrates, if $k_s/k_m = 1$, a single, composite wave is still obtained, but the steepness of the current variation decreases with respect to curve 1.

At $k_s/k_m = 0.1$ (curve 3), a more prominent change in the curve shape is noticed. In this case, instead of a single composite wave, two distinct waves, a cathodic and an anodic one are profiled. Each individual wave is characterized by a particular value of the half-wave potential. The above trend becomes more evident on curves 4 and 5 that correspond to even lower values of the k_s/k_m ratio. Curves 4 and 5 are typical of a *slow* electrochemical reaction and such reactions are also termed as *irreversible* electrochemical reactions.

The rationale of the reversible or irreversible character assigned to an electrochemical reaction is now evident from Figure 13.8 if we focus on the variable-current regions. For the reversible case (curve 1) a slight change in ΔE from positive to negative values is sufficient to reverse the current direction. Conversely, in the irreversible case (e.g., curves 4 and 5) a large variation of ΔE is needed in order to obtain the same effect.

As demonstrated in Figure 13.8, the half-wave potential of an irreversible wave is very different from the formal potential of the redox system and depends on the k_s/k_m ratio. At the same time, current variation with the potential is much less steep than in the case of a reversible wave. This is due to the fact that the subunit transfer coefficients α and β are included in the current–potential equation.

In conclusion, the reversible or irreversible character of an electrochemical reaction is determined by the value of the k_s/k_m ratio. A reversible character is noted when $k_s/k_m \gg 1$ whereas the irreversible character becomes evident when $k_s/k_m \ll 1$. Electrochemical reactions occurring under intermediate value of the k_s/k_m ratio are termed *quasireversible* reactions.

The degree of reversibility depends on the experimental conditions through the mass-transfer coefficient, which is inversely proportional to the thickness of the diffusion layer. Therefore, a reversible (fast) electrochemical reaction can turn into an irreversible (slow) one if the convection in the solution is adjusted so as to decrease sufficiently the thickness of the diffusion layer. On the other hand, the standard rate constant of the electrochemical reaction (k_s) can be enhanced by means of catalysis.

As will be shown in detail in Section 13.7, the best analytical conditions are obtained when the electrochemical reaction is reversible (fast).

13.6.4 Equilibrium Conditions

It is useful now to consider the electrochemical reaction at equilibrium. In this particular case, the reaction occurs with the similar velocities to each direction, which means that the anodic and cathodic currents at equilibrium fulfil the following condition:

$$j_{c,eq} = -j_{a,eq} = j_0 \tag{13.65}$$

The quantity j_0 thus defined is the *exchange-current density*, while $i_0 = Aj_0$ is the *exchange current*.

The characteristics of the exchange current can be inferred upon adapting Equation (13.58) to the particular case of the equilibrium state. In this case, the interface concentrations are the same as the bulk concentrations ($c_{O,i} = c_{O,b}$ and $c_{R,i} = c_{R,b}$). Denoting the equilibrium potential by E_{eq}, one obtains from Equations (13.65) and (13.61):

$$j_0 = j_{c,eq} = Fk_s c_{O,b} \exp\left[-\alpha f\left(E_{eq} - E_f^0\right)\right] \tag{13.66}$$

In order to derive the effect of Ox and Red concentrations, one makes use of the Nernst equation written in the following form:

$$\frac{c_{O,b}}{c_{R,b}} = \exp\left[f\left(E_{eq} - E_f^0\right)\right] \tag{13.67}$$

Upon substituting the exponential term in Equation (13.67) into Equation (13.66) and after some rearrangements the following equation is obtained:

$$j_0 = Fk_s c_{O,b}^{(1-\alpha)} c_{R,b}^{\alpha} \tag{13.68}$$

The above equation demonstrates that the exchange-current density depends on the standard rate constant, reactant concentrations and the transfer coefficient. If $\alpha = 0.5$ and $c_{O,b} = c_{R,b} = c$, then the exchange-current density becomes:

$$(j_0)_{c_{O,b}=c_{R,b}=c} = Fk_s c \tag{13.69}$$

Therefore, the exchange-current density, which is proportional to the standard rate constant, indicate the rapidity of an electrochemical reaction.

13.6.5 The Electrochemical Reaction in the Absence of Mass-Transfer Restrictions

A mass-transfer process always accompanies an electrochemical reaction that involves soluble species. However, in order to highlight some particular features, it is useful to consider the limiting situation in which the electrochemical reaction is not coupled with a mass-transfer process. Such a situation is approached when the reactant concentrations are sufficiently high and at the same time vigorous convection conditions are provided. In fact, the mass-transfer restriction is negligible when the current is below 10% of the diffusion current. Under these conditions, the concentration at the interface can be considered to be essentially the same as the bulk concentration and Equation (13.58) assumes the following particular form:

$$j = Fk_s \left[c_{R,b} \exp(\beta f \Delta E) - c_{O,b} \exp(-\alpha f \Delta E) \right] \tag{13.70}$$

Dividing this equation by Equation (13.68), one obtains:

$$\frac{j}{j_0} = \left(\frac{c_{R,b}}{c_{O,b}} \right)^{\beta} \exp(\beta f \Delta E) - \left(\frac{c_{O,b}}{c_{R,b}} \right)^{\alpha} \exp(-\alpha f \Delta E) \tag{13.71}$$

In this equation, the potential is referred to the formal potential of the redox system. For the purposes of further discussion, it is more convenient to select the equilibrium potential as the origin of the potential axis. To this end, the concept of *overvoltage* (η) will be introduced. By definition, the overvoltage is the difference between the actual potential applied to the electrode and the equilibrium potential:

$$\eta = E - E_{eq} \tag{13.72}$$

According to this definition, the current will be zero at zero overvoltage. Taking into account the expression for the equilibrium potential given by the Nernst equation and also the definition of ΔE (Equation (13.45), one obtains:

$$\Delta E = \frac{1}{f} \ln \frac{c_{O,b}}{c_{R,b}} \tag{13.73}$$

Substituting the above expression of ΔE in Equation (13.71), one obtains the following equation which is known as the *Butler–Volmer equation*:

$$j = j_0 [\exp(\beta f \eta) - \exp(-\alpha f \eta)] \tag{13.74}$$

This equation includes two exponential terms in η; the first one corresponds to the anodic component, whilst the second represents the cathodic component of the total current. The effect of the concentrations is expressed in an indirect way, by means of the j_0 term. The Butler–Volmer equation is plotted as curve 1 in Figure 13.9 for a slow electrochemical reaction. As can be seen in this figure, at overvoltage values far from zero, one of the exponentials in Equation (13.74) becomes negligible and the current has a pure anodic or cathodic character. A plot of $\ln j$ vs. η at high overvoltage (Tafel plot) gives a straight line that allows the exchange current and the transfer coefficient to be calculated from the line intercept and slope, respectively.

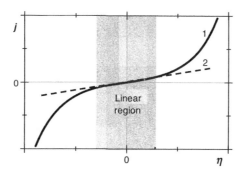

Figure 13.9 Current–potential curves in the absence of mass-transfer restriction for a slow electrochemical reaction. (1) curve generated with the Butler–Volmer Equation (13.74); (2) approximate linear relationship corresponding to the limit of very low overvoltage (Equation (13.75)).

The behavior at a very low overvoltage is also of interest. In this situation, the argument of each exponential is much smaller than unity and the exponential can be approximated as $\exp x \approx 1 + x$ if $x \ll 1$. Using this approximation, Equation (13.74) gives:

$$j = f j_0 \eta \tag{13.75}$$

Accordingly, at very low overvoltage, the current approximates well to a linear function of the overvoltage, as indicated by line 2 in Figure 13.9. In general, this approximation holds at overvoltage values less than 0.01 V in absolute value.

Taking into account the definition of the electrical resistance as the derivative of the of the voltage with respect to the current, the *electron transfer-resistance* (R_{et}), (also termed the *polarization resistance* or *charge transfer resistance*) can be introduced according to the following definition, in which $i = Aj$:

$$R_{et} = \left(\frac{\partial \eta}{\partial i} \right)_{\beta f \eta \ll 1} \tag{13.76}$$

Upon derivation of Equation (13.75), one obtains:

$$R_{et} = \frac{RT}{FAj_0} \tag{13.77}$$

In the particular case in which $c_{O,b} = c_{R,b} = c$, upon substituting Equation (13.69) for j_0, it results at very low overvoltage:

$$(R_{et})_{c_{O,b}=c_{R,b}=c} = \frac{RT}{F^2 A k_s c} \tag{13.78}$$

Therefore, the charge-transfer resistance is inversely proportional to the standard rate constant and indicates the opposition exerted by the electrochemical reaction to the current flow at the electrode/solution interface.

Taking into account the fact that the exchange current density is directly proportional to the standard rate constant (Equation 13.68), it follows that the charge-transfer resistance is inversely proportional to the standard rate constant. This property is used to determine the standard rate constant by a method called electrochemical impedance spectrometry, which is dealt with in detail in Chapter 17. In addition, electrochemical impedance spectrometry is useful in monitoring affinity reactions that occur at the electrode surface using a redox system as a probe. The recognition event alters the interface properties and, therefore, affects the charge-transfer resistance, which can function as the response signal of the sensor.

13.6.6 Polarizable and Nonpolarizable Electrodes

An electrode becomes *polarized* when its potential is shifted from the equilibrium value. The degree of polarization is indicated by the applied overpotential. Depending on the electrode material and any possible electrochemical reaction, an electrode can behave as a polarizable or nonpolarizable electrode. Current–potential curves are often termed *polarization curves*.

The potential of a *nonpolarizable* electrode remains close to the equilibrium potential even if the current is flowing through the cell but $j \ll j_d$. In other words, a minute overpotential gives rise to a high current density. This behavior is characteristic of very fast electrochemical reactions and the corresponding current–potential curve looks like curve 1 in Figure 13.10. This condition is essential in the functioning of a reference electrode, which should keep the

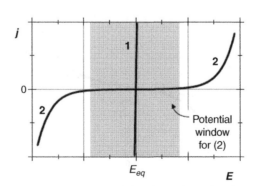

Figure 13.10 Current–potential curves for a nonpolarizable (1) and a polarizable (2) electrode. The gray area indicates the potential window of the polarizable electrode 2.

potential constant during the electrochemical run. The reason why silver is widely used in the construction of reference electrodes is the very large standard rate constant of the pertinent electrochemical reaction ($Ag^+ + e^- \rightleftharpoons Ag$). At high overvoltages, a nonpolarizable electrode becomes polarizable.

By contrast, at a polarizable electrode, a high overvoltage is needed in order to generate a measurable current (Figure 13.10, curve 2). This situation arises when the electrochemical reaction occurring at this electrode is extremely slow.

In order for an electrode to function as a working electrode in a cell, it should behave as a polarizable electrode in the supporting electrolyte solution. The single possible cathodic reaction in this system is reduction of hydrogen ions or of alkali metal ions present in the supporting electrolyte. Possible anodic reactions are the oxidation of the metal electrode itself to form ions that can react with solution components to form soluble or insoluble products (see, for example, the anodic reaction of silver in the Ag/AgCl reference electrode). Another possible anodic reaction is oxygen evolution. Such cathodic and anodic reactions determine the potential region within which the residual (background) current is negligible in the supporting electrolyte solution. This potential region forms the *potential window* of the electrode (Figure 13.10). As the anodic and the cathodic limits can be determined by different electrochemical reactions, the potential window is often asymmetrical, in contrast to the situation depicted in Figure 13.10. In order to investigate the electrochemistry of a given compound at a given electrode, the half-wave potential of this compound should be located within the potential window.

13.6.7 Achieving Steady-State Conditions in Electrochemical Measurements

The treatment of the electrochemical reaction in the above sections has been carried out under the assumption that the process occurs in the steady state. Such a condition supposes that, at a fixed electrode potential, the concentration profiles are time independent and, therefore, the current is also constant. As the current depends on the thickness of the diffusion layer, the steady state can be achieved under rigorously controlled convection conditions. Controlled convection can be obtained with particular electrode configurations such as the rotating-disk electrode, the channel electrode and the wall-jet electrode [15,16].

The rotating-disk electrode (Figure 13.11A) is a disk electrode embedded in a Teflon holder and rotated at a controlled rotation speed, w (in revolutions per second, that is, s^{-1} or Hz). As a result, the solution is caused to flow along the surface, which gives rise to a pressure difference between the bulk of the solution and the solution layer adjacent to the electrode. Therefore, a solution stream moves constantly towards the electrode surface and the rotating electrode acts as a centrifugal pump.

For the case of a rotating disk, in the absence of any chemical reaction, the thickness of the diffusion layer is:

$$\delta = 0.643 D^{1/3} \nu^{1/6} w^{-1/2} \tag{13.79}$$

where the diffusion coefficient D is in $cm^2 s^{-1}$ and the kinematic viscosity of the solution ν is in $cm^2 s^{-1}$.

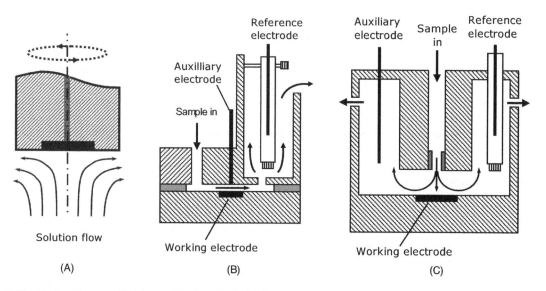

Figure 13.11 Devices for controlled convection electrolysis. (A) Rotating-disk electrode; (B) channel electrode cell; (C) wall-jet electrode cell. Adapted with permission from [10]. Copyright 2002 A. Bănică and F.G. Bănică.

By inserting this expression in the limiting current equation derived above, one obtains the following equation for the limiting current at the rotating-disk electrode:

$$i_d = \pm 1.55 nFAD^{2/3} v^{-1/6} w^{1/2} c \tag{13.80}$$

where c is the reactant concentration, n is the electron number and the plus or minus sign indicates the anodic or cathodic character. As is typical of a limiting diffusion current, the limiting diffusion current at a rotating-disk electrode is proportional to the reactant concentration. A test for a diffusion-controlled electrochemical reaction is the direct proportionality relationship between the limiting current and the square root of the rotation speed. This kind of dependency is the result of the fact that with increasing the rotation rate, the flow rate along the surface increases and the thickness of the diffusion layer decreases, as indicated in Equation (13.79).

The rotating-disk electrode is used in kinetic investigations of electrochemical reactions, as well as in certain electroanalytical methods.

Flow-through cells with controlled convection are suitable for application in flow-analysis systems. Two flow-through cell configurations are presented here.

The *channel electrode* (Figure 13.11B) consists of a flat electrode embedded in the wall of an insulator tube down which solution flows under laminar conditions. As the solution moves along the electrode surface, the reactant is gradually depleted, which causes the thickness of the diffusion layer to vary along the electrode. In practice, the length of the electrode is made sufficiently small so that this variation is negligible. The limiting current is directly proportional to the reactant concentration and $v_f^{1/3}$, where v_f is the volume flow rate of the solution (in $cm^3 \, s^{-1}$).

The wall-jet electrode is a standing-disk electrode. As shown in Figure 13.11C, a jet of solution is directed to the disk surface. The thickness of the diffusion layer increases from the disk center to the periphery owing to the flow direction at the surface. Provided that the jet diameter is very small compared with the disk radius, the limiting current is proportional to the bulk concentration of the analyte and to $v_f^{1/4}$.

In the previous examples, a steady state is imposed owing to the controlled convection conditions. The steady state can also be obtained with membrane-covered electrodes (Section 13.4). In this approach, diffusion is mostly confined within the membrane, which has a fixed thickness.

As will be shown in Section 13.7.9, it is possible to achieve steady-state electrolysis in the absence of convection using ultramicroelectrodes, which are electrodes with the size in the micrometer region.

13.6.8 Outlook

The velocity of an electrochemical reaction is proportional to the reactant concentration at the electrode/solution interface, the proportionality constant being the rate constant of the electrochemical reaction. The interface concentration depends on the diffusion process which supplies the reactant to the interface.

A particular feature of electrochemical reactions is that the rate constant depends exponentially on the electrode potential. That is why electrochemical kinetics operate with the rate constant defined at the standard potential of the redox couple.

The dependence of the rate constant on the electrode potential suggests that the current can increase exponentially with the potential. However, current increase above a certain limit is restricted by the limited capacity of reactant supply by diffusion. The current obtained when the maximum supply capacity is attained is the limiting diffusion current. This corresponds to the situation in which the reactant is completely depleted at the interface.

As the diffusion and the electrochemical reaction occur sequentially, the behavior of the electrochemical system is critically determined by the rate constants of each of the above steps. If the electrochemical reaction is very fast with respect to the diffusion step, the electrochemical reaction has a fast (reversible) character. In this case, concentrations at the interface satisfy the Nernst equation. However, if the electrochemical reaction is very sluggish compared with the diffusion process, the electrochemical reaction is said to be slow (irreversible). The above definition of reversibility and irreversibility assumes that the reaction product does not undergo irreversible chemical conversion, in other words that the electrochemical reaction is *kinetically reversible*. This implies that the backconversion of the product is thermodynamically possible in a reversible reaction, but it faces a very high activation barrier in the case of an irreversible electrochemical reaction.

A reversible electrochemical reaction is much more convenient in analytical applications. A reversible character can be imparted by means of catalysis that increases the rate constant of the electrochemical reaction.

Under equilibrium conditions, an electrochemical reaction proceeds with similar velocities in both directions. Therefore, the overall current is zero at equilibrium. However, even at equilibrium, electrons cross the interface in both directions. The electron flux at equilibrium is indicated by the exchange current, which depends on both the standard rate constant and reactant concentrations.

The deviation of an electrochemical system from the equilibrium state is measured by the overvoltage. The overvoltage is zero at equilibrium but its increase in absolute value determines a shift from the equilibrium, which gives rise to a current. A nonzero overvoltage implies that the electrode is polarized and the magnitude of the overvoltage is a measure of the degree of polarization.

From the standpoint of electrode polarizability, one can distinguish two limiting situations, namely nonpolarizable electrodes and polarizable electrodes. Nonpolarizable electrodes correspond to the situation in which the electrochemical reaction is extremely fast. If a nonpolarizable electrode is crossed by a moderate current density, it undergoes negligible change in the electrode potential. Nonpolarizable electrodes are used to fabricate reference electrodes. In the absence of the analyte, polarizable electrodes are characterized by a very sluggish electrochemical reaction, which can occur only at a high overvoltage and involves either the components of the supporting electrolyte, the solvent or the material of the electrode. Therefore, it remains a potential region where no electrochemical reaction takes place in the absence of the analyte. This region forms the potential window within which electrochemical reactions of interest can be conducted.

The very low overvoltage region is of interest in the determination of the standard rate constant of the electrochemical reaction. Electrical quantities related to the standard rate constant are the exchange current and the electron-transfer resistance. Determination of the electron-transfer resistance is of interest in sensors based on recognition films formed at the electrode surface, as changes in film properties affect the electron-transfer resistance.

Questions and Exercises (Section 13.6)

1 What constants and parameters determine the velocity of a chemical reaction?
2 How does the temperature affect the velocity of a chemical reaction?
3 Explain why the velocity of an electrochemical reaction depends not only on the temperature but also on the electrode potential.
4 What are the components of the overall reaction rate of an electrochemical reaction? How does each component vary with the electrode potential?
5 What is the dimensionless parameter that determines the reversible or irreversible character of an electrochemical reaction? Discuss the meaning of this parameter in terms of reactant supply capacity and reactant conversion capacity.
6 Prove that the Damköhler number for an electrochemical reaction is indeed a dimensionless parameter.
7 Comments on the differences between a reversible and an irreversible current–potential curve recorded under steady-state conditions.
8 Adapt Equation (13.63) to the case in which the Red species is absent from the solution and $k_s/k_m \gg 1$. Demonstrate that the resulting equation is similar to Equation (13.26) that was derived under the assumption that interface concentrations satisfy the Nernst equation. Remember that Equation (13.63) has been derived under the assumption that $k_{m,O} = k_{m,R}$. Comment on the reasons for the similarity of Equation 13.26 and the equation derived in this exercise.
9 What is the exchange current and what is its correlation with the standard rate constant of the electrochemical reaction?
10 How can the irreversible character of an electrochemical reaction be changed into a reversible one?
11 How can the thickness of the diffusion layer determine the reversibility of an electrochemical reaction?
12 What is the overvoltage and how does it influence the electrochemical current?
13 (a) Write approximate forms of Equation (11.74) for $\eta \gg 0$ and $\eta \ll 0$; (b). Using data in the table A and B below, plot the logarithm of the current density (in absolute value) vs. the overvoltage (Tafel plot) according to these equations; (c) use this plot to determine the exchange current density, the transfer coefficients α and β and the electron-transfer resistance for a 3×10^{-4} m^2 electrode area (hint: omit points on the graph that deviate from the linear trend). (d) Calculate the standard rate constant for 1 mM concentrations of both Ox and Red. (e) Determine the electron-transfer resistance from the η–j curve at very low overpotentials.

η/V	−0.20	−0.14	−0.1	−0.06	−0.04	−0.02	−0.01	−0.005
j/A m^{-2}	−49.13	−15.21	−6.87	−2.91	−1.72	−0.81	−0.4	−0.19

η/V	0.005	0.01	0.02	0.04	0.06	0.1	0.14	0.20
j/A m^{-2}	0.21	0.38	0.79	1.72	2.91	6.87	15.21	49.13

Answer: $j_0 = 0.81$ A m^{-2}; $\alpha \approx \beta \approx 0.5$; $R_{et} = 85.6$ Ohm; $k_s = 8.4 \times 10^{-6}$ m s^{-1}.

14 Comment on the meaning of reversible and irreversible electrochemical reactions in terms of (a) the Nernst equation; (b) the rapidity of involved steps; (c) the propensity of the reaction to change direction at slight modifications of the electrode potential; (d) the chemical stability of the product.

15 What are the differences between a polarizable and a nonpolarizable electrode? What are the applications of each of the above types of electrode?

16 What factors determine the limits of the potential window of a working electrode?

13.7 Electrochemical Methods

Electrochemical methods can be classified as steady-state and nonsteady-state methods. The steady-state approach needs simple electronic equipment, but is prone to interferences and may not provide satisfactory sensitivity. That is why, a series of nonsteady-state electroanalytical methods have been developed in a search for better sensitivity and selectivity.

This section gives first a short overview of the application of steady-state electrochemical methods in chemical sensors. The next part introduces nonsteady-state methods that are useful both as transduction methods and as investigative tools in electrochemical kinetics.

Most of the nonsteady-state methods presently available rely on the control of the potential electrode and the measurement of the current produced in response to the applied potential. Such methods are included in the class of *voltammetric methods*.

Other methods rely on imposing a constant or variable current. The response is obtained from the variation of the electrode potential as a function of time. In this case, the method is termed a *chronopotentiometric* method.

Good introductions to electrochemical methods of investigation and analysis can be found in dedicated texts (for example [6,8,17]).

13.7.1 Steady-State Methods

In a steady-state method, the stable limiting current is measured at a fixed potential in order for it to serve as the sensor response. Steady-state methods are also called simply amperometric methods and are widely used in electrochemical enzyme sensors or in sensors using enzymes as labels. This approach is very simple and needs minimal electronic equipment. However, these method are prone to interference from other electrochemically active compounds which are present in the sample. An interference occurs when the half-wave potential of the interferent is close to that of the analyte. Concomitant electrochemical reactions affect the limit of detection owing to the additional residual current that overlaps the response current.

Interference in steady-state methods can be alleviated by coating the electrode with a membrane that is permeable to the analyte but not to the interferent. If this approach is not feasible, the sensor can be fabricated so as to include two working electrodes, one of which is generating the background current. For example, in the case of an enzymatic sensor, one of the two electrodes is coated with an enzyme layer, whilst the second is coated with an enzyme-free layer. At the first electrode, both the analyte and the concomitant contribute to the total current. On the other hand, the second electrode generates only the background current produced by the concomitant. The current difference between the two electrodes represents the actual sensor response. This method can be easily implemented by screen-printing technology.

13.7.2 Constant-Potential Chronoamperometry

This simple amperometric method is based on a sudden change in the applied potential from a value E_i, where no electrochemical reaction occurs, to a value E_S that promotes reduction or oxidation of an electrochemically active species in an unstirred solution (Figure 13.12A). Immediately on applying the potential step, the current–time variation is recorded; hence the name *chronoamperometry* assigned to this method.

In the absence of convection, the electrochemical reaction depletes the reactant in the solution layer adjacent to the electrode and the thickness of the diffusion layer increases with time. Consequently, the Faradaic current decays with time (t) according to the *Cottrell equation*:

$$i_f = nFA\sqrt{\frac{D}{\pi t}}c_b \tag{13.81}$$

where c_b is the bulk concentration of the reactant and D is its diffusion coefficient. This equation holds when the step potential E_s is set in the limiting-current region. A current decay curve for a reversible process is shown in Figure 13.12B by the curve labeled i_f.

In the absence of any electrochemical reaction, the potential step produces only a change in the electron density at the electrode surface and, simultaneously, a modification of the ion distribution in the solution in order to maintain the charge balance at the interface. This variation in the total charge at the electrode surface give rise to a current that

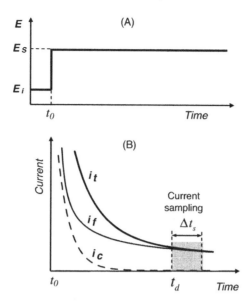

Figure 13.12 Principle of the chronoamperometric method. (A) Potential step variation; (B) time variation of the Faradaic current (i_f), capacitive current (i_c) and total current (i_t). Current sampling after a time delay (t_d) lessens the contribution of the capacitive current to the total current. Δt_s is the width of the sampling interval.

is not associated with an electrochemical reaction. The metal/solution interface behaves in this case as a capacitor and the transient current thus generated is a *capacitive current* (i_C) (see Section 13.5). As typical of a capacitor, the capacitive current decays according to an exponential law in response to the potential step:

$$i_C = \frac{E_s - E_i}{R_s} \exp\left(-\frac{t}{R_s C}\right) \tag{13.82}$$

where R_s is the electrical resistance of the solution and C is the capacitance of the double electric layer at the electrode/solution interface.

If an electrochemical reaction occurs in the system, the recorded current (i_t) is the sum of the Faradaic and capacitive currents (Figure 13.12B). Hence, at very low reactant concentrations, the Faradaic current, which is providing analytical information, is overshadowed by the capacitive current that is irrelevant from the analytical point of view. In other words, it is the capacitive current that determines the limit of detection.

The limit of detection can be improved by taking into account the different rates of decay of the capacitive and Faradaic currents (Figure 13.12B). As the capacitive current decays much faster, after a sufficiently long delay its value becomes negligible and the total current becomes practically equal to the Faradaic current. Thus, delayed current sampling provides a response that is less affected by the capacitive current and produces a better limit of detection. Low solution resistance secures fast decay of the capacitive current and efficient elimination of this component.

Current sampling consists of integration of the current over the sampling interval (Δt_s) for a certain time delay (t_d) after the application of the potential step (Figure 13.12B). The integral represents the amount of electricity that has flowed during the sampling interval. On dividing by the sampling interval (Δt_s) one obtains the mean current over the sampling interval (i_m):

$$i_m = \frac{1}{\Delta t_s} \int_{t_d}^{t_d + \Delta t_s} i(t) \, \mathrm{d}t \tag{13.83}$$

According to the Cottrell equation, the sampled current is directly proportional to the analyte concentration. As will be shown later, a series of voltammetric methods characterized by low limits of detection are based on the application of a sequence of small potential steps. After each step, the current is sampled so as to reduce the effect of the capacitive current according to the principle presented earlier.

13.7.3 Polarography

The polarographic method, which was the first widely used voltammetric analytical method was invented in 1922 by Jaroslav Heyrovský who was awarded the Nobel Prize for Chemistry in 1959 for this discovery. The working

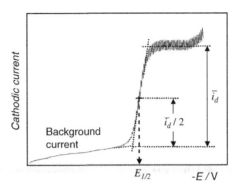

Figure 13.13 Polarographic wave for a cathodic (reduction) reaction. Current oscillation is due to the periodical variation of the mercury drop area. The oscillation is damped during current recording and the average current is measured for analytical purposes. This curve is plotted according to the polarographic sign convention. Adapted with permission from [10]. Copyright 2002 A. Bănică and F.G. Bănică.

electrode in this method is the *dropping mercury electrode*, a glass capillary tube through which mercury flows at a constant rate [18,19]. A sequence of mercury drops form at the end of the capillary and during its lifetime (1–5 s) each drop functions as the working electrode in the electrochemical cell. The applied potential is varied slowly (scan rate, 50–200 mV min^{-1}) so that the potential variation over the drop lifetime is negligible. As the drop surface increases with time, the current also varies and the average current for each drop $(\bar{i})$ is recorded as a function of the potential to give a potential–current curve called a *polarogram* (Figure 13.13). An electrochemical reaction yields a polarographic wave, which is similar to a steady-state voltammogram. Hence, the mean limiting current $(\bar{i}_d)$ is directly proportional to the analyte concentration and the half-wave potential is close to the formal potential of the redox system in the case of a reversible electrochemical reaction.

The background current in polarography is represented by the charging current produced by the periodical variation of the electrode surface area (see the first term in Equation (13.39)).

In an advanced version of polarography (called *current-sampled polarography*), the applied potential is increased in small steps of 1–5 mV. Step application is synchronized with drop formation so that the potential is rigorously constant during the drop life time. The current is sampled shortly before the drop is dislodged in order to reduce the contribution of the capacitive current.

Drop dislodgement produces local stirring that homogenizes the solution near the electrode so that each drop functions in a fresh solution, with almost no effect from the electrolysis that has occurred at the previous drop. This is why the polarographic wave is shaped like a steady-state voltammogram and is described by a similar equation.

Continuous renewal of the electrode surface prevents surface contamination with reaction products or solution impurities. In addition, the mercury electrode displays a very broad potential window in the negative-potential region, which makes it particularly suitable for investigating cathodic reactions. However, electrochemical reactions of dissolved oxygen produce a large current within this potential region and oxygen must be removed by flushing the sample with pure nitrogen for 5–10 min prior to the run. The limit of detection in polarographic analysis is about 10^{-5} M.

13.7.4 Linear-Scan Voltammetry (LSV) and Cyclic Voltammetry (CV)

The effect of the electrode potential on the rate of an electrochemical reaction can be investigated by steady-state voltammetry and polarography. However, in the case of steady-state voltammetry, a special cell with rigorously controlled convection is required. On the other hand, application of polarography in the investigation of anodic reactions is limited by the anodic reaction of mercury. A simple alternative is *linear-scan voltammetry* at a stationary electrode in quiescent solutions [16,20]. As the name of the method implies, the potential is varied linearly with time starting from a selected initial potential E_i and going towards the equilibrium potential and beyond it:

$$E = E_i \pm vt \tag{13.84}$$

where v is the potential scan rate in V s^{-1}. The initial potential is selected so that no electrochemical reaction occurs at the beginning. A typical LSV curve is presented in Figure 13.14. At the beginning of the scan, only the capacitive current is recorded, but as the potential approaches the equilibrium value, the Faradaic current increases abruptly until a maximum point is reached. At this point, the reactant concentration becomes zero at the electrode surface. Further, the increase in the thickness of diffusion layer causes the reactant flux to decrease gradually. Hence, after the maximum point, the current decreases asymptotically.

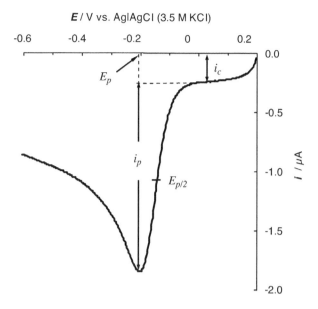

Figure 13.14 Linear scan voltammogram for the one-electron reduction of $[Ru(NH_3)_6]^{3+}$ (1 mM) at a 2 mm diameter gold disk electrode in phosphate buffer (pH 7). Scan rate, 100 mVs^{-1}.

For a reversible process under semi-infinite diffusion conditions, the peak current is given by the Randles–Sevčik equation that has the following form at 25 °C, with A in cm^2, D in cm^{2}s^{-1}, c in mol/cm^{-3} and v in Vs^{-1}, and the peak current in amperes:

$$i_p = \left(2.69 \times 10^5 n^{3/2} A D^{1/2}\right) v^{1/2} c \tag{13.85}$$

This equation demonstrates that the peak current is directly proportional to the reactant concentration and the square root of the scan rate. The direct proportionality between the peak current and $v^{1/2}$ is a test for diffusion control over the electrode process. The peak potential for a reversible process is:

$$E_p = E_{1/2} \pm 1.109 \frac{RT}{nF} = E_{1/2} \pm \frac{28.5}{nF} \text{ mV at 25 °C} \tag{13.86}$$

As the potential at $i = i_p/2$ (called *half-peak potential, $E_{p/2}$*) can be measured more accurately, it is often reported instead of the peak potential. It is given by the following relationship:

$$E_{p/2} = E_{1/2} \pm 1.09 \frac{RT}{nF} = E_{1/2} \pm \frac{28.0}{n} \text{ mV at 25 °C} \tag{13.87}$$

In the above two equations, $E_{1/2}$ is the half-wave potential of the steady-state wave for the same reaction (Equation 13.23). The positive and negative signs apply to anodic and cathodic reactions, respectively. Hence, both E_p and $E_{p/2}$ values are determined by the formal potential of the redox system and can be used to determine this parameter. Characteristic potentials of the reversible LSV peak are independent of concentration and scan rate.

In the case of an irreversible process, the peak current depends also on the transfer coefficient α and has a lower value compared to that of a reversible peak current for the same reactant concentration. At the same time, the peak potential is different from that indicated in Equation (13.86) and depends on α, the standard rate constant, and the scan rate.

The Faradaic current is proportional to $v^{1/2}$ whilst the capacitive current increases proportionally with v (see Equation (13.39), in which the time derivative of the potential represents the potential scan rate, v). Therefore, interference from the capacitive current increases with the scan rate. This interference can be avoided using a method called *staircase voltammetry* in which the potential increases in small steps and the current is sampled for each step as shown in Figure 13.12B. Owing to the sampling procedure, the measured current is lower than that recorded in LSV under similar conditions. Elimination of the capacitive current is effective if the step duration is sufficiently long. At very short step durations (that is, high scan rates) the staircase voltammogram becomes similar to the LSV curve.

More information about a redox system is provided by *cyclic voltammetry* (CV) that is presented in Figure 13.15. The time evolution of the potential and current are shown in A. In this method, a linear potential scan is applied in order to record the current for the reaction of the investigated compound, as in LSV. After reaching a point S beyond the maximum current, the direction of the potential scan is reversed. During this reverse scan, the product of the forward scan reaction undergoes electrochemical conversion back to the initial reactant giving rise to a second peak.

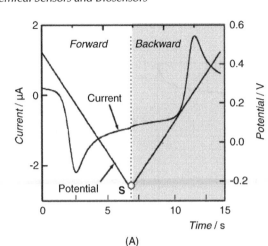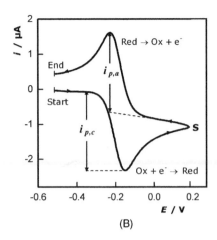

Figure 13.15   Cyclic voltammetry for 1 mM $[Fe(CN)_6]^{3+}$ in 0.1 M KCl at a 3 mm diameter gold disk electrode at 100 mVs^{-1} scan rate. (A) Time evolution of potential and current. S represents the switching point. (B) current–potential curve (cyclic voltammogram) for the same system. Arrows on the curve indicate the scan direction. Reference electrode: Ag|AgCl (3.5 M KCl). Adapted with permission from [10]. Copyright 2002 A. Bănică and F.G. Bănică.

As a rule, data are plotted as a current–potential curve (cyclic voltammogram) as shown in Figure 13.15B on which one can distinguish the forward and reverse peaks. If the process is reversible and both forms of the redox couple are soluble, the cathodic and anodic peak currents are similar in magnitude. The reversibility of the electrochemical reaction is indicated by the following relationship:

$$\Delta E_p = E_{p,a} - E_{p,c} = 2.303RT/nF = 59/n \text{ mV at } 25\,^\circ\text{C} \tag{13.88}$$

At the same time, peak potentials allow the determination of the half-wave potential of the steady-state voltammogram and, by this mean, the formal potential of the redox couple. The half-wave potential of the steady-state wave for the same system is situated at the midpoint between the peak potentials.

For slow electrochemical reaction, the peak separation is greater than that indicated in Equation (13.88) and peak currents are lower. If the product of the forward scan reaction undergoes an irreversible chemical conversion, the reverse scan peak can be completely absent. Such a process is kinetically irreversible.

Cyclic voltammetry is a versatile diagnostic tool that is used to assess the reversibility of the electrochemical reaction and its diffusion-controlled character. For reversible reactions, this method provides an estimation of the formal potential of the redox system and the number of electrons in the electrochemical reaction (Equation 13.88). Data fitting and simulation by means of dedicated software packages allow the determination of the reaction mechanism of complex electrochemical reactions and provide the values of relevant constants.

In sensor development, CV is used to check the integrity of molecular layers formed at the electrode surface. The ferricyanide ion can be used as a redox probe to this end. As an example, Figure 13.16 shows an application of CV to monitoring the state of a gold electrode modified by means of a lipoic acid self-assembled monolayer that served for

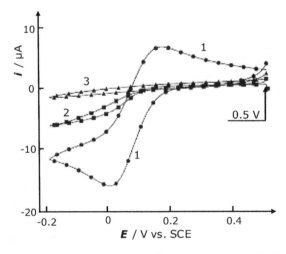

Figure 13.16   Effect of the state of the electrode surface on the cyclic voltammograms recorded in 5 mM $[Fe(CN)_6]^{3-}$ at 10 mVs^{-1} scan rate with a gold electrode (3 mm diameter) in a citrate buffer (pH 7.4). (1) Lipoic acid-modified electrode (2) after antibody immobilization (3) after 1-dodecanethiol adsorption at the electrode in (2). Adapted with permission from [21]. Copyright 1997 American Chemical Society.

covalent linking of an antibody. The aim of this experiment was to obtain an insulated gold surface after antibody immobilization. Curve 1, recorded with a lipoic acid-modified electrode, deviates to some extent from the typical shape of the reversible voltammogram due to partial coverage by the adsorbed layer. Curve 2 was recorded after the antibody was linked to the lipoic acid monolayer. This curve indicates that the penetration of the redox probe to the electrode surface is prevented to a large extent, but not completely. If 1-dodecanethiol is then adsorbed over the exposed gold surface, one obtains the curve 3 which displays only the capacitive current and proves that access of the redox probe to the metal surface is completely blocked. Therefore, curve 3 conditions secure good protection against the penetration of ions and the electrode behaves mostly as a pure capacitor.

13.7.5 Pulse Voltammetry

In order to improve the limit of detection of the polarographic method, G. C. Barker developed between 1950 and 1960 several methods based on the application of potential pulse waveforms.

Normal pulse voltammetry (NPV) is based on the application of a sequence of potential pulses of increasing amplitude superimposed onto a constant potential, E_i (Figure 13.17A). After each pulse, the potential is held at the initial value, where no electrochemical reaction occurs, so that the concentration gradient created previously vanishes. Over each pulse, the current varies as in Figure 13.17B and, in order to remove the effect of the capacitive current, the current is sampled during a short period near the pulse end. The sampled current (i_s) is plotted as a function of the pulse amplitude to yield the NPV voltammogram, as shown in Figure 13.17C for the case of Pb^{2+} and Cd^{2+} reduction at a dropping mercury electrode. Although the current is measured under nonsteady-state conditions, the curve shape is similar to that of a steady-state voltammogram. Each reducible or oxidizable species in solution gives rise to a wave characterized by its half-wave potential and limiting current. The limiting current is proportional to the analyte concentration.

As the height of the potential pulse increases steadily, the capacitive current also increases, in accord with Equation (13.82). Therefore, current sampling is not particularly efficient in removing the effect of the capacitive current over the whole range of the voltammogram. The residual capacitive current is responsible for the slight tilting of the limiting current region of the Cd^{2+} wave in Figure 13.17C. The limit of detection of the NPV method is about 10^{-6} M.

Much lower detection limits are obtained by *differential pulse voltammetry* (DPV). As shown in Figure 13.17B, in this method a sequence of equally high pulses is superimposed on a stepwise increasing potential. For each pulse, the current is sampled twice: first just before the pulse application ($i_{s,1}$) and then near the pulse end ($i_{s,2}$). The voltammogram is obtained by plotting the current difference ($\Delta i_s = i_{s,2} - i_{s,1}$) as a function of the potential at the foot of the pulse. As demonstrated in Figure 13.17C, DPV curves display a symmetrical peak-shaped pattern, with the maximum potential similar to the half-wave potential on the NPV curve. The peak height of the DPV curve is proportional to the analyte concentration.

As each point on the DPV curve is a current differential, the shape of this curve is fairly represented by the derivative of the equation of the steady-state voltammetric wave (Equation 13.25). That is why the DPV curve displays a

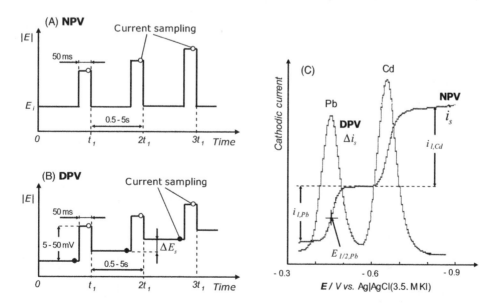

Figure 13.17 Pulse voltammetry. Potential waveform in normal pulse polarography (A) and differential pulse polarography (B). t_1 is the pulse application period. (C) NPV and DPV curves for a Pb^{2+} and Cd^{2+} solution (10^{-4} M in each) in 0.1 M HNO_3 recorded with a dropping mercury electrode. Each pulse is applied to a new drop. Curves are plotted according to the polarographic convention. Note that the DPV peak potentials are similar to half-wave potentials on the NPV curve, which is typical of reversible electrochemical reactions. Adapted with permission from [10]. Copyright 2002 A. Bănică and F.G. Bănică.

maximum at the half-wave potential of the steady-state voltammogram, where di/dE has a maximum value. In turn, the derivative is close to zero at the foot of the wave and in the region of the limiting current, where di/dE assumes negligible values.

DPV makes use of small-amplitude pulses. Hence, according to Equation (13.82), the capacitive current is very small. As a result, current sampling removes to a large extent the effect of the capacitive current. That is why the limit of detection of DPV is close to 10^{-7} M. The sensitivity increases with the pulse height but, at the same time, the DPV peak broadens, which brings about a loss of resolution. An additional advantage results from the peak-shaped form of the response. After reaching a maximum, the signal lessens close to the zero line and if another peak follows (as in Figure 13.17C) its current does not add to that of the previous peak. In addition, the peak-shaped response brings about a much better resolution compared with steady-state voltammograms and allows for simultaneous monitoring of electrochemical reactions characterized by close half-wave potentials. In such cases, steady-state voltammetry produces overlapped waves that do not permit accurate estimation of the individual limiting currents.

The above advantages apply to fast electrochemical reactions that display a sharp current variation in response to the potential change. If the electrochemical reaction is slow, the peak Δi_s value is lower for the same analyte concentration. Moreover, the peaks are broader and sometimes asymmetrical.

13.7.6 Square-Wave Voltammetry (SWV)

An even better detection limit is achieved in square-wave voltammetry also pioneered by G. C. Barker. The fundamental principles of this method are presented in ref [22]. In SWV, a stepwise varying base potential (dotted line in Figure 13.18A) is applied. On each potential step (E_s), a symmetrical double pulse is superimposed. The first pulse increases the potential over the step level, while the second shifts the potential below this level. The current is sampled near the end of each pulse. The time variation of the current profile is given in Figure 13.18B. It will be assumed further that an analyte A undergoes a reversible cathodic reaction and the product B is not present in the solution:

$$A + ne^- \rightleftarrows B \tag{13.89}$$

Consider first the case in which the electrode potential is close to the half-wave potential. When the forward pulse is applied, the current increases suddenly and then decays with $t^{-1/2}$. When the reverse pulse is applied, the product B undergoes oxidation and gives rise to an anodic current that also varies with $t^{-1/2}$. For each pulse, the current is sampled within the intervals 1 and 2 yielding the current values i_1 (cathodic) and i_2 (anodic). Each sampled current is mostly composed of a Faradaic component, with a negligibly small capacitive contribution.

Typical SWV voltammograms obtained with the ferricyanide ion are shown in Figure 13.19. Both i_1 and i_2 currents are plotted vs. the step potential value. Both curves are shaped as asymmetrical peaks, with the maximum values slightly shifted with respect to the half-wave potential. In absolute value, the reverse peak current is less than the forward peak current because a fraction of the B product formed during the forward pulse diffuses away from the interface. A signal greater that each of the above currents is obtained by taking the algebraic difference of the above currents. ($\Delta i = i_2 - i_1$). This signal, which is also plotted in Figure 13.19, appears as a symmetrical, peak-shaped curve, with the peak potential nearly identical with the half-wave potential. The Δi value is very close to zero at potentials far from the half-wave potential for the same reasons as in the case of NPV. It is this kind of curve that is used in analytical applications. Its peak height is proportional to the analyte concentration.

The main operating parameters in SWV are the step height (ΔE_s), the pulse height (ΔE_p), the potential scan rate ($v = \Delta E_s/t_1$, in V s^{-1}), and the period of pulse application (t_1). Typical values for the first two parameters are given

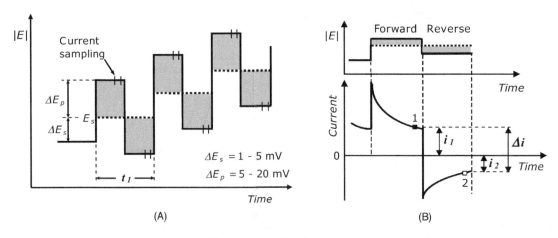

Figure 13.18 Square-wave voltammetry. (A) Waveform of the applied potential in SWV. (B) Measurement principle in SWV. Reproduced with permission from [10]. Copyright 2002 A. Bănică and F.G. Bănică.

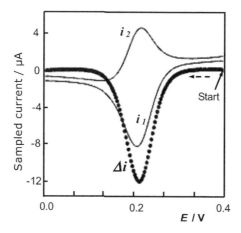

Figure 13.19 SWV curves for the single-electron reversible reaction. The brocken arrow indicates the direction of the potential scan. Adapted with permission from [10]. Copyright 2002 A. Bănică and F.G. Bănică.

in Figure 13.18A. As a rule, instead of the time period, the pulse application frequency is reported ($f = 1/t_1$, in Hz). As in the case of DPV, the height of the SWV peak increases with pulse height, but, at the same time, the capacitive current also increases. In order to preserve the effectiveness of capacitive current elimination, the frequency is adjusted between 20 and 100 Hz. On the other hand, the scan rate can be varied from 0.01 to 5 V s^{-1}. This means that the voltammogram can be recorded very quickly, which is an advantage over DPV.

For fast and reversible electrode reactions, SWV provides a limit of detection close to 10^{-8} M, which makes SWV the most sensitive voltammetric method. As in the case of DPV, the sensitivity is much lower in the case of slow electrode reactions.

13.7.7 Alternating-Current Voltammetry

Differences between time variation for the Faradaic and capacitive currents are used in DPV and SWV to improve the limit of detection. A similar principle is used in alternating-current voltammetry (ACV). As in DPV and SWV, the electrode is polarized with a stepwise varying potential. Over each constant step potential (E_s), a small-amplitude AC potential (E_{AC}) is superimposed so that the overall applied potential is:

$$E = E_s + E_{AC} = E_s + E_{AC,max} \sin(2\pi f t) \tag{13.90}$$

where f is the frequency and $E_{AC,max}$ is the amplitude of the AC potential.

In response to this excitation, a composite current results:

$$i = i_{DC} + (i_f + i_C)_{AC} \tag{13.91}$$

where i_{DC} is the current response to the potential step, while i_f and i_C are the Faradaic and capacitive components, respectively of the AC current.

The slowly varying DC component can be separated from the AC components by means of suitable electronics and only the AC current is recorded. However, as shown in Equation (13.91), the AC current consists of two components. Only the Faradaic current is relevant, whereas the capacitive component worsens the limit of detection. In order to extract the Faradaic component, one makes use of the different time evolution of these currents. Both the Faradaic and the capacitive AC currents are shifted on the time axis with respect to the AC potential as shown in Figure 13.20A. This is indicated by the phase shift ϕ in the following equation:

$$i_{AC} = i_{AC,max} \sin(2\pi f t + \phi) \tag{13.92}$$

The phase shift is $-\pi/2$ radians for the capacitive current and $-\pi/4$ radians for the Faradaic current. According to Figure 13.20A, when the AC capacitive current is zero, the Faradaic current at this moment ($i_{f,s}$) is about $0.7 i_{f,max}$. Therefore, if the total AC current is measured under such particular conditions, it will consist mostly of a Faradaic component. The electronic device performing this type of measurement is called a lock-in amplifier (also known as a phase-sensitive detector).

In ACV, experimental data are collected in the form of AC current vs. the step potential. In the fundamental frequency mode (AC1), the current is measured at the frequency of the AC potential and the voltammogram is peak shaped, as shown in Figure 13.20B, curve 1. For a reversible electrochemical reaction, the peak potential is equal to

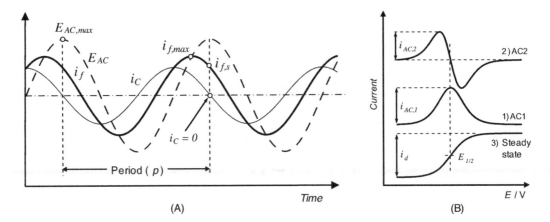

Figure 13.20   Alternating-current voltammetry. (A) Time evolution of AC potential (E_{AC}) and AC currents (i_f: Faradaic AC current; i_C: capacitive AC current). (B) AC voltammograms for a reversible electrochemical reaction. (1) fundamental harmonic current; (2) second-harmonic current; (3) steady-state voltammogram.

the half-wave potential of the steady-state voltammogram, represented by curve 3. There is one more possibility to exploit ACV data, that is, recording of the second harmonic of the AC current. This is the current measured at a frequency twice that of the AC potential frequency and its dependence on the step potential gives the AC2 voltammogram (Figure 13.20B, curve 2). The AC2 current is much smaller than the AC1 current, but the AC2 mode permits a more advanced removal of the capacitive current. In both AC1 and AC2 mode, the peak current is proportional to the analyte concentration.

The main operating parameters in ACV are the frequency ($f = 1/p$, where p is the period) and the potential scan rate ($v = \Delta E_s / t_1$; ΔE_s is the step height and t_1 is the step duration). The scan rate should be selected so that t_1 is at least $20p$ in order to provide good sampling conditions. The AC peak current increases with the square root of the frequency whereas the capacitive current increases proportionally with frequency. Typically, the frequency is set between 50 and 100 Hz. The AC peak current increases with frequency, but at higher frequencies the removal of the capacitive current is less effective. The limit of detection in ACV is near 10^{-7} M for fast and reversible processes but it is less satisfactory in the case of slow processes.

13.7.8   Chronopotentiometric Methods

Chronopotentiometric methods belong to the class of *galvanostatic methods*, in which the current is controlled and another variable (e.g., the electrode potential) is recorded as the system response. A very simple method of this type involves a sudden variation of the current from zero to a constant value in a quiescent solution, as shown in Figure 13.21A. Before the application of the current step, the electrode potential is at its equilibrium value, E_{eq}.

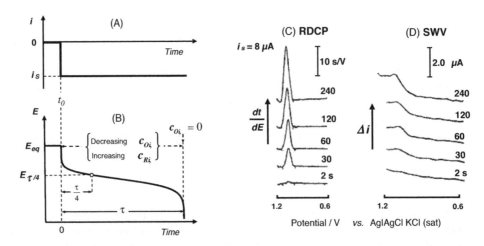

Figure 13.21   The chronopotentiometric method. (A) The current step applied to the electrode; (B) time variation of the electrode potential in response to the current step: (C) RDCP curves for a ribonucleic acid solution in acetate buffer (pH 5.0) at a carbon paste electrode (3.5. mm diameter) after adsorptive accumulation at the electrode surface under stirring at +0.5 V. Accumulation time (in seconds) is shown on each curve. Applied current: 8 μA. (D) SWV curves recorded after adsorptive accumulation with the same solution as in (C). Pulse amplitude: 10 mV; frequency: 40 Hz. (C) and (D) were adapted with permission from [24]. Copyright 1995 American Chemical Society.

If the current direction is selected so as to promote the reduction of the Ox species, current application causes the $c_{O,i}/c_{R,i}$ ratio in the Nernst Equation (13.14) to decrease. This results in a decrease in the electrode potential, as indicated in Figure 13.21B. As long as the Ox species is still present at the electrode surface, the potential varies relatively slowly. However, when $c_{O,i}$ becomes zero, the current can no longer be supported by the Ox reaction and the potential shifts suddenly to a value determined by another redox couple present in the solution. In the absence of a second redox couple, the second reaction will involve solvent molecules.

The time required for $c_{O,i}$ to become zero is the *transition time*, τ. It decreases with increasing the current because the depletion of Ox proceeds faster at a higher current. On the other hand, τ increases with the analyte concentration. The dependence of τ on both current and concentration is expressed by the Sand equation:

$$i\sqrt{\tau} = \frac{1}{2}nFA\sqrt{\pi D_O}\, c_{O,b} \tag{13.93}$$

A characteristic point on the potential–time curve is located at $\tau/4$ s after the application of the current step. For a reversible electrochemical reaction, the potential at this point is:

$$E_{\tau/4} = E_f^0 - \frac{RT}{nF}\ln\sqrt{\frac{D_O}{D_R}} \tag{13.94}$$

The constant-current chronopotentiometric method is useful in the investigation of the mechanism of electrochemical reactions but it is of minor analytical relevance. However, an alternative form of this method, called *reciprocal derivative chronopotentiometry* (RDCP) has found promising applications in sensor development. In this method, the reciprocal of the time derivative of the potential $(dE/dt)^{-1} \equiv dt/dE$ is plotted as a function of the measured potential [23]. An electrochemical reaction gives rise to a characteristic peak on such a plot. Either the peak height or the peak area can be correlated with the analyte concentration to perform quantitative analysis.

As an example, RDCP curves recorded with a solution containing a nucleic acid are shown in Figure 13.21C. These curves have been obtained at a carbon paste electrode after prior adsorptive accumulation of the analyte at the electrode surface for various time intervals shown on each curve. The response is due to the electrochemical reaction of guanine residues in the nucleic acid. As demonstrated in this figure, well-shaped peaks are recorded. At the same time, the background current is almost constant, which allows for accurate background correction. As expected, the peak height increases with the accumulation time. The calibration graph in terms of peak area vs. the analyte concentration is linear and the reported limit of detection is 4×10^{-16} mol ribonucleic acid [24].

For comparison, Figure 13.21D presents SWV curves recorded with the same solution as in C and under similar accumulation conditions. SWV curves display a large and potential-dependent background that distorts the SWV peak and damages the accuracy of the peak current. Despite its reputed sensitivity, SWV does not perform well in this case owing to the irreversible character of the electrochemical reaction. RDCP appears as a suitable alternative in such situations.

13.7.9 Electrochemistry at Ultramicroelectrodes

Traditional electroanalytical chemistry makes use of working electrodes with sizes in the millimeter region. The very small area of these electrodes leads to very small currents (in the μA region) which do not alter the bulk concentration of the analyte during the run. As shown schematically in Figure 13.22A, the diffusion layer developed at such an electrode is much thinner than the electrode size. Consequently, even if the diffusion paths converge to the center, the effect of convergence can be neglected as the diffusion paths within the diffusion layer are nearly parallel. A good analogy is the fall of two objects from a height much lower than the Earth radius. Although the fall paths converge to the Earth center, the convergence effect is negligible and the objects' paths are practically parallel.

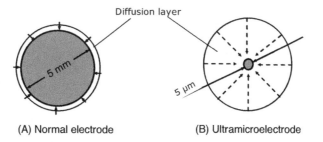

Figure 13.22 Diffusion geometry at a millimeter-sized electrode (A) and at an ultramicroelectrode (B). Adapted with permission from [10]. Copyright 2002 A. Bănică and F.G. Bănică.

As a consequence of the diffusion conditions discussed above, the linear diffusion model is a very good approximation to the diffusion process at nonplanar electrodes, such as spherical or cylindrical electrodes.

The situation changes fundamentally when the electrode size is in the micrometer region or lower. Such electrodes are termed *ultramicroelectrodes* or, more simply, *microelectrodes* [25,26]. In this case, the thickness of the diffusion layer becomes much greater then the electrode size and the convergence of the diffusion paths cannot be neglected (Figure 13.22B). Ultramicroelectrodes are widely used in the design of very small sensors for applications in the analysis of microsamples or in *in vivo* sensing.

The particular features of ultramicroelectrode electrolysis will be first illustrated for the case of a spherical ultramicroelectrode. A theoretical approach to diffusion at a spherical electrode demonstrates that the current produced in response to a potential step (under the limiting-current conditions) is given by the following equation:

$$i = \frac{nFDAc}{r}\left[1 + \frac{r}{(\pi Dt)^{1/2}}\right] \tag{13.95}$$

Here, r is the electrode radius, c is the reactant concentration and the other symbols have their usual meaning. The term $(\pi Dt)^{1/2}$ actually represents the time-dependent thickness of the diffusion layer. Accordingly, the thickness of the diffusion layer expands and can reach a value much higher that the electrode radius. Under these conditions the term in square parenthesis in Equation (13.95) assumes a near-unity value and the current becomes time independent, that is, a steady-state current, i_{st} which is given by the following equation:

$$i_{st} = \frac{nFDAc}{r} \tag{13.96}$$

As an illustration of the previous conclusions, Figure 13.23A presents the response of a spherical ultramicroelectrode to a potential step in the case of an anodic reaction, for two values of the electrode radius. In both cases the current decreases asymptotically to the steady-state value, but the current variation is faster in the case of a smaller electrode radius. Assuming that the steady state is attained when the thickness of the diffusion layer is about 10 times the electrode diameter, it follows that this thickness is about 50 μm for a 10- μm diameter ultramicroelectrode. Such a thin layer is hardly disturbed even in stirred solutions. Therefore, the current recorded with an ultramicroelectrode is almost independent of the stirring conditions.

A stable diffusion layer brings about important consequences for the shapes of cyclic voltammograms recorded with an ultramicroelectrode, as demonstrated in Figure 13.23B. At any point on the voltammogram, the Faradaic current is at its steady-state value and a limiting current develops at potentials sufficiently far from the formal potential of the redox system. At the same time, the forward and backward currents should be identical at any potential. However, on an actual curve like that in Figure 13.23B, there is some difference between the two scans owing to the additional capacitive current that changes orientation according to the scan direction. The capacitive current is also the reason why the limiting current region of the curve is tilted.

Owing to the very small surface area, the current at an ultramicroelectrode is less than 1 nA. Consequently, the ohmic voltage drop is very small and, therefore, ultramicroelectrodes can be operated in two-electrode cells and also function well in highly resistive solutions.

Ultramicroelectrodes can be shaped as hemispheres, disks, cylinders, or flat bands. The previously emphasized characteristics of the diffusion process are preserved provided that at least one of the electrode dimensions is within the micrometer region.

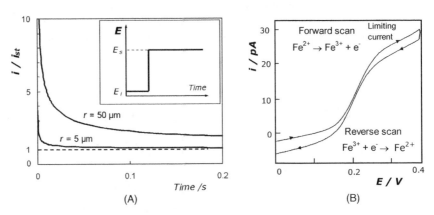

Figure 13.23 (A) Response of an ultramicroelectrode to a potential step (chronoamperometric experiment); (B) cyclic voltammogram recorded with a10 μm diameter gold UME for the reaction of the ferrocyanide ion (10 μM) in 0.05 M KNO₃. Adapted with permission from [10]. Copyright 2002 A. Bănică and F.G. Bănică.

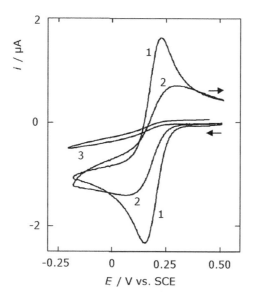

Figure 13.24 Effect of electrode coverage on the diffusion conditions. Cyclic voltammograms of the ferricyanide ion (5 mM) in 0.1 M KNO$_3$. The electrode surface was covered by self-assembly of a thione carotenoid derivative. (1) free surface; (2) partial coverage; (3) high coverage degree. Adapted with permission from [27]. Copyright 2002 Elsevier.

Ultramicroelectrodes can be made of noble metals or carbon materials such as carbon microfibers embedded in epoxy resin. Flat ultramicroelectrodes are obtained by vapor deposition over an insulator support through a mask, which is a method suited for mass production.

Ultramicroelectrodes can be assembled in arrays of connected electrodes or of individually addressable electrodes. Arrays of ultramicroelectrodes with preselected size and spacing are obtained by vapor deposition through a mask (photolithography) or by soft lithography. Diffusion at an array of connected electrodes depends on the electrode size and spacing, as well as on the duration of the electrochemical run. Thus, for a short run time, each electrode develops an individual diffusion layer and the total current is a multiple of the current produced by each electrode. However, if the run time is longer (and electrode spacing is sufficiently low), individual diffusion layers will overlap and the array behaves as a macroscopic electrode, the current being proportional to the overall area of the array.

An array of ultramicroelectrodes can be obtained by partially coating the surface of a normal electrode with an insulating layer. Figure 13.24 displays the behavior of such an electrode in the cyclic voltammetry of the ferricyanide ion. The uncoated electrode yields the curve 1 that is typical of linear diffusion conditions. Coverage by a self-assembled monolayer of a long, hydrophobic molecule leads to the formation of a network of pinholes where the reactant is free to access the metal surface, the pinholes being incorporated within an insulating layer. As indicated by curve 3, which corresponds to a high degree of coverage, this electrode behaves as an array of ultramicroelectrodes, the cyclic voltammograms being shaped as the curve in Figure 13.23B. At an intermediate degree of coverage (curve 2), the curve has an intermediate shape, between that typical of linear diffusion (curve 1) and that of spherical diffusion (curve 3).

In recent years, nanosized electrodes have been intensively explored for applications in electroanalytical chemistry and in electrochemical sensors [28]. Carbon-nanotube or nanofiber electrodes can be grown individually or as regular assemblies ("forests"). The behavior of nanoelectrodes is of the steady-state type, similar to that of ultramicroelectrodes.

13.7.10 Current Amplification by Reactant Recycling

Of particular importance in sensor applications are interdigitated electrode couples (Figure 13.25A) that can be used to perform electrochemical recycling of the reactant in order to enhance the sensitivity. Recycling is achieved by controlling each electrode potential such that one of them supports the reaction of the analyte (e.g., oxidation of a Red species) whilst the second electrode is biased so as to promote the opposite reaction (Figure 13.25B). If the electrode spacing is lower than the thickness of the diffusion layer, the Ox species formed at the generator electrode is converted back to the Red form at the adjacent, collector electrode. As this sequence is repeated many times, the recorded current is much higher than that for a similar reaction occurring at a single electrode.

Operation of twin interdigitated electrodes needs a bipotentiostat to control independently the potential of each component. However, interdigitated electrodes can be operated by means of a normal potentiostat if one of the components is electrically connected to a large-area electrode. Owing to the large area, the two electrodes assume a

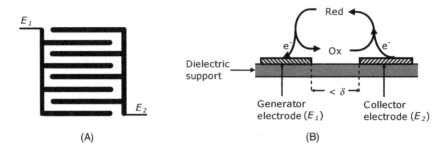

Figure 13.25 (A) Schematics of a twin interdigitated electrodes assembly. (B) Electrochemical cycling and at a pair of interdigitated electrodes. δ is the thickness of the diffusion layer.

potential close to the equilibrium potential. Therefore, only the potential of the second component of the interdigitated system needs to be controlled by means of the potentiostat.

The limiting current recorded with interdigitated electrodes is proportional to the analyte concentration. However, the above-mentioned current enhancement can be achieved only with reversible electrochemical reactions.

13.7.11 Scanning Electrochemical Microscopy

The advent of ultramicroelectrodes paved the way to performing local electrochemistry at selected microscopic spots on large-area electrodes [29,30]. The instrument dedicated to this task, called the *scanning electrochemical microscope* is presented in Figure 13.26A. The system includes an ultramicroelectrode that functions as a probe. A micromanipulator adjusts the position of the probe with respect to the specimen. Both the probe and specimen potentials are controlled by means of a bipotentiostat. These electrodes are parts of an electrochemical cell containing the solution of a redox probe, Ox. The current measured at the probe electrode depends to a high extent on the l/δ ratio, where l is the probe–specimen spacing and δ is the thickness of the diffusion layer developed at the probe electrode (Figure 13.26B). If $l/\delta \gg 1$ (B.1), diffusion at the probe electrode is not disturbed and the recorded current represent the limiting diffusion current at this electrode (i_l). However, if $l/\delta < 1$ the presence of the specimen affects the electrochemical reaction at the probe. If the probe is placed over a dielectric spot, diffusion of the redox probe is restricted and the recorded current is lower than the limiting current (B.2). However, if the probe is situated over a conducting area, the probe–specimen system performs redox recycling and the current is enhanced with respect to the limiting current obtained in case B.1.

Scanning electrochemical microscopy can be used as an imaging tool in which the probe or current potential reflects both topography and chemical reactivity across the explored surface. It is also a practical method for investigating metal surfaces covered with thin molecular layers, such as conducting polymers. The activity of immobilized oxidase enzymes can be investigated by monitoring the current due to a redox mediator involved in the enzyme reaction. Therefore, scanning electrochemical microscopy is a valuable method in the research and development of biosensors [31].

In microfabrication, scanning electrochemical microscopy is used to alter in a controlled way the chemical properties of selected spots of micrometer size. This alteration is obtained by electrochemical reactions promoted at the probe-support electrode couple.

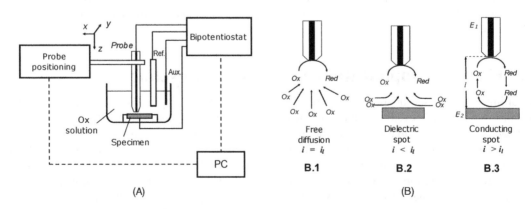

Figure 13.26 Scanning electrochemical microscopy. (A) Principle of the method. Ref and Aux. denote the reference and auxiliary electrodes, respectively. (B) Electrochemical reactions at the probe electrode under various conditions. Adapted with permission from [10]. Copyright 2002 A. Bănică and F.G. Bănică.

13.7.12 Outlook

Electrochemical methods have been developed for two kinds of application: investigation of electrochemical reactions and analytical determinations.

Cyclic voltammetry is a valuable tool in the investigation of electrochemical reactions in order to assess the reaction mechanism and characteristic parameters of the reaction. This provides essential information that is needed in sensor design. Graphical processing of voltammograms provides information in regard to the reversibility of the electrochemical reaction, the standard redox potential, the reaction stoichiometry, and the effect of reactant diffusion. Much more accurate and reliable information is obtained by data fitting and simulation using dedicated software packages.

Electrochemical transduction can be achieved under steady-state conditions of the overall recognition–transduction process. Steady-state methods are convenient because they require only simple electronic equipment and data processing methods and are widely used in electrochemical sensors. However, steady-state methods can be impaired by poor detection limits and interference from electrochemically active concomitants.

Much better limits of detection and selectivity, and resolution are obtained by methods such as SWV, DPV, and ACV. Very low limits of detection are achieved in these methods owing to the advanced elimination of the contribution of the capacitive current to the total recorded current. Good selectivity is due to the peak-shaped response voltammograms, which reduces considerably the interference by electrochemical reactions that occur at potentials close to that of the analyte reaction. However, the above methods require more elaborate electronics and additional steps in the data-processing method.

Questions and Exercises (Section 13.7)

Steady-state methods

1 What are the advantages and drawbacks of steady-state methods as electrochemical transduction methods?
2 Explain how the steady state can be achieved by proper design of the cell, the electrode and the sensing layer?
3 How can interferences be avoided in steady-state methods?

Chronoamperometry

4 How does an electrochemical reaction respond to a potential step?
5 How does the capacitive current vary in response to this excitation?
6 How can a response current be obtained free of interference from the capacitive current? What conditions should the solution composition fulfill in order to achieve effective elimination of the capacitive current?

Polarography

7 What are the characteristics of the working electrode used in polarography?
8 This electrode has enjoyed a huge popularity in fundamental electrochemistry and electroanalytical chemistry. Why?

Linear-scan voltammetry and cyclic voltammetry

9 For an anodic reaction investigated by linear-scan voltammetry draw schematically a) the time variation of the applied potential, and b) the current–potential curve. Mark on this curve characteristic points and discuss their dependence on system parameters.
10 Using the curve in Figure 13.14, determine the peak potential, the half-peak potential, and estimate the formal potential of the $[Ru(NH_3)_6]^{3+}/[Ru(NH_3)_6]^{2+}$ couple with respect to the reference electrode indicated in this figure and the standard hydrogen electrode ($E_{Ag|AgCl(3.5MKCl)} = +0.205$ V vs. SHE).
11 Data in the table below indicate the dependence of the peak current of a linear-scan voltammogram on the scan rate. Plot the logarithm of the current vs. the logarithm of the scan rate and decide as to whether or not the current has a diffusional character. How can a reversible reaction be distinguished from an irreversible one using the effect of the scan rate?

$v/V\,s^{-1}$	0.01	0.05	0.1	0.2	0.5
$i_p/\mu A$	20	39	59	84	130

12 How can the effect of the capacitive current be suppressed in potential-scan voltammetry?

13 Draw schematically the time variation of potential and current in the cyclic voltammetry of a species in the solution than can undergo electrochemical oxidation. Draw also the cyclic voltammogram for this species.

14 Peak potentials on a cyclic voltammogram are $E_{p,a} = 0.243$ and $E_{p,c} = 0.213$ V at 25 °C. (a) Is the solution reactant reduced or oxidized? (b) Calculate the number of electrons if the electrochemical reaction is reversible; (c) calculate the half-wave potential; (d) estimate the formal potential of the redox couple.

Answer: (b) $n = 2$; (c) 228 mV.

Pulse voltammetry and square-wave voltammetry

15 Why is the elimination of the capacitive current not particularly effective in normal pulse voltammetry?

16 What are the improvements brought about by differential pulse voltammetry as far as the limit of detection and the selectivity are concerned?

17 What is the effect of pulse duration on the limit of detection in differential pulse voltammetry?

18 What types of response are provided by SWV? Make a comparison between SWV and cyclic voltammetry from the standpoint of the provided information.

19 Why is it recommended to use low frequencies in SWV?

Alternating-current voltammetry

20 What are the principles of capacitive current elimination in ACV.

21 What are the common features of NPV, SWV and ACV from the standpoint of the measuring principles?

22 Why do these methods produce peak-shaped voltammograms? What are the advantages of this particular shape?

23 What parameters determine the peak potential and the peak current in the above methods?

Chronopotentiometric methods

24 What are the excitation signal and the response signal in chronopotentiometry? What parameter is related to the reactant concentration and what parameter depends on the formal potential of the redox system?

25 Why does the potential vary sharply at the transition time in chronopotentiometry? What factors limit the potential variation?

26 What version of chronopotentiometry is useful for application in nucleic acid sensors? Comment on the shape of the response. What response parameters are related to the analyte concentration.

Electrochemistry at ultramicroelectrodes and scanning electrochemical microscopy

27 What are the particular features of diffusion at ultramicroelectrodes?

28 What are the consequences of the diffusion geometry on the current response at ultramicroelectrodes in the case of (a) chronoamperometry; (b) linear-scan voltammetry?

29 Draw schematically the shape of various types of ultramicroelectrodes and indicate on each of them the micrometer-sized dimension.

30 How can ultramicroelectrode arrays be obtained? Sketch the profile of the diffusion layer at an array of disk ultramicroelectrodes and discuss the particular features of the diffusion process and current response at this array.

31 Discuss the application of ultramicroelectrodes in current amplification by reactant recycling.

32 What are the applications of ultramicroelectrodes and reactant recycling in surface investigation?

13.8 Electrode Materials

Selection of the electrode material is an important step in the development of electrochemical sensors. First, the electrode should provide a suitable potential window for the relevant electrochemical reaction. Secondly, it is preferable to select an electrode material at which the relevant electrochemical reaction occurs with as low an overpotential as possible. If a high overpotential is needed, the electrode must be polarized at an extreme potential and this can promote concomitant electrochemical reactions with disturbing effects. A low overpotential can be secured by means of electrocatalysis (Section 13.9). Thirdly, the electrode material should provide possibilities for facile surface modification in order to build up the sensing layer. Last but not least, the electrode material should be readily available and suitable for processing by large-scale fabrication technologies such as microfabrication.

This section reviews three main kinds of electrode materials, namely carbon materials, metals, and metals oxides. The properties of such materials, as well as practical aspects in regard to their application in electrochemistry are covered in refs. [6,8,32]. As nanomaterial applications in electrochemistry bring about new perspectives, this topic is addressed in the final section.

13.8.1 Carbon Electrodes

Various forms of graphite are popular in electrochemistry owing to their chemical inertness, wide potential windows (especially, in the anodic region) and easy availability [33–36]. Undisturbed graphite structures consist of parallel sheets formed of sp^2-hybridized carbons atoms grouped in a polyhexagon structure (Section 5.4.7). Two different kinds of planes can be distinguished in the graphite structure, namely basal planes and edge planes (Figure 13.27A). The *basal plane* is a peripheral graphite sheet in the graphite crystallite. As all bonding orbitals on this plane are filled, the basal plane is chemically inert and hydrophobic. The edges of the graphite sheets form the *edge plane* that is perpendicular to the basal plane. Various oxygen functionalities ($(-COO^-, -OH, -C=O)$) that can be present at the edge impart hydrophilicity and the capacity to interact with solution species through noncovalent interactions.

A common graphite material is *pyrolytic graphite* that is produced by carbon crystallization from light hydrocarbons near the decomposition temperature. Pyrolytic graphite includes a high density of graphite crystallites. Pressure annealing of pyrolytic graphite at 3000 °C removes much of the structure defects and produces a material called *highly ordered pyrolytic graphite* (HOPG) which contains graphite crystallites of at least 1 μm size and structured as shown in Figure 13.27A.

Pyrolytic graphite electrodes are cut so as to expose preponderantly at the surface either edge or basal planes. Certain electrochemical reactions are critically dependent on the orientation of the crystallites at the electrode surface. As a result, the standard reaction rate of an electrochemical reaction occurring at the edge plane can be much higher than that at the basal plane due to the electrocatalytic effect exerted by oxygen groups and structural defects present at the edge. The catalytic edge effect is demonstrated in Figures 13.27B and C for the electrochemical reaction of the Co^{2+}–phenanthroline complex. If the edge density at the surface is very low (curve B), this reaction occurs as a slow (irreversible) process that is characterized by a very low standard rate constant. Conversely, a high density of edges imparts to this reaction a fast (reversible) character that implies a very high rate constant (curve C).

Glassy carbon, also known as *vitreous carbon*, has a disordered structure including a tangled mass of very small graphitic ribbons. It has been proved recently that commercially available glassy carbon also contains a high proportion of fullerene-related structures [39]. Glassy carbon displays an isotropic distribution of edges, is impermeable to liquids and gases, and can be easily mounted and polished. The potential window of glassy carbon electrodes is very broad, with the negative limit close to -1 V vs. the Ag/AgCl reference electrode and the positive limit at about 1.5 V.

Carbon fibers have a 5–15 μm diameter and exhibit at the surface a high fraction of edge planes. They are used to produce ultramicroelectrodes for *in vivo* and microanalysis applications.

Boron-doped diamond is a p-type semiconductor obtained as thin films by vapor deposition methods. It is characterized by very large overpotential for both oxygen and hydrogen evolution, which results in a wide potential window ranging from -0.75 to $+2.35$ V in 0.5 M H_2SO_4. This wide range is unparalleled by any other electrode material. The negative limit of the potential window is shifted by about 1 V by fluorination of the diamond surface. Diamond electrodes can be obtained not only as compact films, but also as nanostructured layers that can be formed by oxygen plasma etching. Ion implantation or immobilization of metal nanoparticles are used to impart electrocatalytic activity

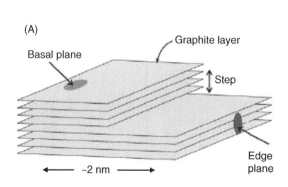
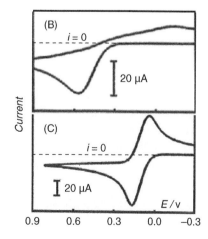

Figure 13.27   (A) Structure of a graphite crystallite in highly oriented pyrolytic graphite (HOPG). (B) and (C) Effect of the carbon surface structure on the cyclic voltammetric response of the phenanthroline complex of Co^{2+} (2 mM) in 1 M KCl recorded at 0.2 V/s scan rate. Quasireference electrode: silver wire dipped in the test solution. (B) Low edge plane defect density HOPG; (C) high defect density HOPG. (A) was reprinted from [37] Reproduced with permission of the Royal Society of Chemistry. (B) and (C) were reprinted with permission from [38]. Copyright 1992, American Chemical Society.

to the diamond electrode. In order to achieve surface functionalization by covalent immobilization, the diamond surface is activated by forming oxygen, hydrogen or amine terminal groups to serve as binding sites.

Owing to the outstanding properties of doped diamond, this material is currently used extensively as the electrode of choice in fundamental research and electroanalytical applications [40–42] and shows promising perspectives for application in electrochemical sensors [43].

Carbon composite electrodes are obtained by mixing graphite powder with an insulator binding material so as to obtain a conductive mixture. The most common electrode of this type is the *carbon paste electrode* [44–46], which is made of a mixture of graphite powder and a pasting oily liquid packed in an insulator holder. The electrode surface can be easily renewed by gentle polishing. Suitable modifiers (such as solid or molecular catalysts or enzymes) can be incorporated in the paste in order to impart specificity. The fabrication of carbon paste electrodes is facile but not suited for mass production. However, carbon paste electrodes are useful in the early stage of the development of an electrochemical sensor. For mass-production purposes, the paste composition should be adapted for fabrication based on screen printing technology (Section 13.8.4).

Composite carbon electrodes can also be obtained by incorporating graphite powder in epoxy resin [47].

Pretreatment of carbon electrodes determines in a critical way the standard rate constant of certain electrochemical reactions [35,48]. Pretreatment removes adsorbed impurities at the surface and also determines the density of edge defects and oxygen groups, with an appreciable effect on the electrocatalytic properties of the electrode.

Carbon electrodes are first polished with alumina slurries of successively smaller particle size, typically finishing with 0.05 μm. Then, the surface is carefully cleaned by ultrasound treatment and then with an organic solvent containing activated carbon to remove surface-adsorbed material. Further treatment is needed in order to control the surface density of edges and oxygen groups. When necessary, oxygen groups can be removed by heat treatment in vacuum. Laser beam treatment can remove adsorbed impurities and increases the density of exposed edge planes.

Electrochemical oxidation at 1–2 V vs. SCE in acidic solutions has a beneficial effect on the electrocatalytic activity of carbon electrodes as this treatment enhances the density of oxygen groups at the surface. A sufficiently long anodic treatment can disrupt significantly the surface to the point of forming a film that contains a large amount of oxygen, particularly in the form of anionic sites ("electrogenerated graphitic oxide"). This film is permeable to solvent and small molecules and displays a much higher active surface. Owing to the anionic sites, energetic oxidation brings about selectivity to cationic reactants (such as dopamine) over anionic reactants (such as the ascorbate ion).

Short anodization of carbon electrodes in alkaline solution removes the polishing residues but does not give rise to oxygen groups because these groups are hydrolyzed at pH > 10.

13.8.2 Noble-Metal Electrodes

Owing to their chemical stability, gold and platinum are popular materials for electrode fabrication.

Gold is particularly convenient because it allows facile surface modification by chemisorption of thiol derivatives to form self-assembled monolayers (Section 5.4.8). Under anodic polarization, gold undergoes oxidation by reaction with water to form a chemisorbed oxygen layer, which can be seen as a gold oxide monolayer (Figure 13.28). This layer undergoes reductive dissolution on the reverse scan of the voltammogram yielding a sharp cathodic peak. However, formation of the oxide layer is prevented by coating the surface with a long-chain thioalkane layer that prevents

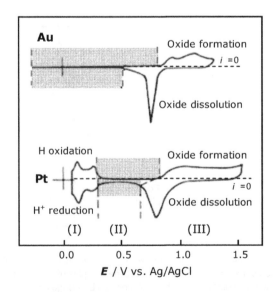

Figure 13.28 Potential windows of gold and platinum electrodes in 0.5 M H_2SO_4. Gray area indicates the potential window. Adapted with permission from [49]. Copyright 1976 American Chemical Society.

water access to the surface. Chloride ion should be absent as it limit the potential window by forming a soluble gold complex under anodic polarization. After mirror polishing and sonication, the gold surface is subject to an electrochemical treatment in an H_2SO_4 solution by prolonged potential cycling between the limits of the oxide dissolution shown in Figure 13.28. In this way, successive oxidation and oxide dissolution bring about a redistribution of surface gold atoms and improves the uniformity of the surface.

Platinum shows a more intricate behavior owing to its property of adsorbing hydrogen. As the reduction of the hydrogen ion proceeds with an extremely low overvoltage, this reaction occurs at an electrode potential very close to the equilibrium potential. Hydrogen atoms thus formed are adsorbed by platinum and can undergo oxidation to H^+ under cathodic polarization. These processes give rise to the peaks in the region (I) of the Pt voltammogram in Figure 13.28. In addition, formation and reduction of a chemisorbed oxygen surface layer is also possible (region III). Therefore, the potential window of a Pt electrode is restricted to region (II).

13.8.3 Metal-Oxide Films

Ruthenium dioxide (RuO_2) electrodes can be obtained by screen printing a paste containing this compound. This electrode displays catalytic activity in many anodic reactions that are difficult at metal electrodes. The potential window extends from -0.4 to $0.7\,V$ vs. the Ag/AgCl reference in neutral solutions.

Tin oxide (SnO_2) and indium oxide (In_2O_3) can be obtained as thin films by vapor deposition. Both these oxides are n-type semiconductors with reasonable electrical conductivity at normal temperatures. In addition, they are transparent and are therefore suitable as electrodes in photoelectrochemical experiments. Surface modification can be performed by common methods for oxide surfaces, such as the reaction of alkoxysilanes with surface hydroxyl groups. Mixed indium-tin oxide electrodes are also commonly employed in photoelectrochemistry.

13.8.4 Electrode Fabrication

Mass production of planar electrodes for sensor application can be achieved either by thick-film technology (screen printing) or thin-film technology [50,51]. An overview of these fabrication methods is presented in Section 5.13). Briefly, screen printing is based on the application of a paste on a flat support through openings in a screen. Adhesion is imparted by subsequent thermal treatment. Conducting layers for electrodes are obtained using pastes including conducting microparticles of carbon or metal precursors (e.g., a metal oxide). During the thermal treatment, the metal oxide gives, by reduction, a metal film.

A typical sensor platform produced by screen printing is shown in Figure 13.29. On a small size strip, the working, reference and auxiliary electrodes can be formed in a concentric configuration. Reference Ag/AgCl electrodes can be obtained by the anodic reaction of silver in HCl or by the chemical oxidation of silver in a $FeCl_3/KCrO_3Cl$ solution. Connection to the output contacts is made by insulated silver tracks. Being inexpensive, screen-printed electrodes are suitable for fabricating disposable sensors.

Owing to the versatility of this technique, screen-printed devices are widely used as platforms in electrochemical enzymatic sensors [52,53].

Thin-film electrodes (thickness $< 1\,\mu m$) can be obtained from various conductive materials (Au, Pt) by sputtering or vapor deposition through a mask over a dielectric glass or oxidized silicon substrate [50,55]. Alternatively, polycrystalline silicon (polysilicon) can serve as a conducting substrate. A few monolayers of Ti or Cr coated on glass prior the deposition improve considerably the adhesion of gold and platinum layers. Metal oxide thin films (SnO_2 and In_2O_3) on glass or quartz are commercially available.

Patterned carbon thin layers can be obtained by pyrolysis of hydrocarbons onto metal or quartz surfaces. Another method is based on pyrolysis of photoresist films made of phenolic resins [48].

← 1 cm →

Figure 13.29 A three-electrode screen-printed platform for amperometric sensors. (a) Gold working electrode; (b) Ag/AgCl reference electrode; (c) gold auxiliary electrode; (d) silver output contacts. Reproduced with permission from [54]. Copyright 2008 Faculty of Health and Social Studies, University of South Bohemia, Czech R.

Shaping of the deposited layer as well as construction of integrated sensor platforms is achieved by photolithography. Microfabrication is the method of choice for producing electrochemical microsensors and sensor arrays.

13.8.5 Carbon Nanomaterial Applications in Electrochemistry

Carbon nanomaterials have been introduced in Section 8.3. The main types of carbon nanomaterials with applications in electrode preparation are carbon nanotubes, carbon nanofibers and graphenes.

Good electrical conductivity, catalytic properties and size compatibility with organic and biological molecules has stimulated a great interest in application of carbon nanomaterials in electroanalytical chemistry and electrochemical sensors [28,37,56,57].

Carbon nanotubes (CNTs) are available as either single-walled (SWCNT) or multiwalled (MWCNT) forms. *Carbon nanofibers* are formed of more or less regularly stacked graphite nanocones producing more edge sites at the outer wall than in CNTs, which enhances electrocatalytic activity [58].

CNTs can be used as modifiers of metal or carbon electrodes. Prior to the integration with an electrode, CNTs are subject to an acid purification process which also opens tube ends.

Carbon-nanotube-modified electrodes can be obtained by drop casting nanotube suspensions into organic solvents onto underlying Pt, Au or glassy carbon electrodes. This method produces an uneven distribution of nanotube over the surface and the layer thus obtained include large nanotube bundles and also amorphous carbon. Alternatively, CNTs can be dispersed at the electrode surface as composites with different binders or can be included in a screen-printed film.

When using CNT dispersions on a conducting substrate, an electrochemical reaction can occur at both nanotubes and the substrate surface which complicates the interpretation of experimental data. This drawback is alleviated by growing single CNTs on a dielectric support such as oxidized silicon. Ordered arrays of vertically aligned CNTs can be obtained by vapor deposition over a preformed metal catalyst nanoparticle pattern.

The main benefit of CNTs is the fast character of many electrochemical reactions in contrast to the slow character observed at macroelectrodes. Generally, this feature is ascribed to catalytic effects of CNTs.

CNT behavior is strongly dependent on the single-walled or multiwalled configuration, and also on the synthesis method. The acid purification process may also influence in a critical way the behavior of the CNTs owing to oxidative damage at the sidewall.

A closer inspection of the reason why CNT-modified electrodes allow for more facile electron transfer revealed that this property may be determined by the occurrence of metal nanoparticles and structural defects. Metal nanoparticles included in CNTs originate from the catalyst nanoparticles used in CNT synthesis. For example, iron residual nanoparticles release Fe^{2+} in a neutral solution containing a ligand that stabilizes this ion and the Fe^{2+}/Fe^{3+} couple can act as a redox mediator system.

Structural defects at the end of CNTs are supposed to be responsible for the catalytic effect, by analogy with the behavior of edge planes in HOPGs. Defects can be present also at the side wall in the form of rotated bonds, vacancies and sp^3 defects (that is, carbon atoms in the sp^3 configuration and terminated by –H or various functionalities). In general, sidewall defects reduce the conductance of SWCNTs.

Structural defects at the side wall are of relevance in SWCNTs, but the behavior of MWCNTs is dominated by the tube end that behaves as edge plane graphite crystallites [37]. This conclusion applies also to bamboo or herringbone-like carbon nanofibers that display a high density of edge-like sites at the fiber side wall.

Graphene is a single atomic planar sheet of sp^2-bonded carbon atoms forming a two-dimensional honeycomb lattice structure (Figure 8.4A). Stacks of 2–10 graphene sheets are known as graphene nanoplatelets. Although until recently CNTs have dominated the fields of carbon nanomaterials, graphenes are expected to bring about new opportunities [59–61].

As in the case of CNTs, the most striking feature of graphenes is their electrocatalytic property, which are ascribed to oxygen groups and structural defects at the sheet edge. Owing to its shape, graphene can support a higher density of defects than CNTs. Heavily oxidized graphene is known as *graphene oxide*, which is obtained by treating graphite with strong oxidizers to form graphite oxide. During this treatment, graphite bulk material is dispersed in basic solutions to yield oxidized graphene sheets.

In contrast to CNTs, graphene can be obtained free of metal nanoparticle impurities, which allows unambiguous assignment of the catalytic effect to the carbon structure itself.

13.8.6 Outlook

Carbon materials, noble metals and certain metal oxides are the main classes of materials used in electrodes for electrochemical sensors. Each of them displays specific advantages and drawbacks that determine the selection of the electrode material for a specific application. Platinum and carbon electrodes are suitable when electrocatalytic activity is needed. Gold displays a broader potential window than platinum and its surface can be prepared more easily in a pure state. Gold surfaces are prone to facile modification by chemisorption of thiol derivatives.

Mass production of electrochemical sensors is based on high-throughput technologies such as screen printing and vapor deposition.

Questions and Exercises (Section 13.8)

1 What structural properties of graphite are of relevance to graphite electrodes?
2 Give a comparison between the structure of pyrolytic graphite and vitreous carbon and point out some advantages of each of them as electrode materials.
3 What particular properties make doped diamond attractive as an electrode material?
4 How can the properties of a carbon electrode surface be modified in order to impart specific properties?
5 Comment on the advantages of composite carbon electrodes, their applications and the possibility of mass production.
6 Discuss the processes that limit the potential window of gold and platinum electrodes. Why does the width of the potential window depend on the previous anodic polarization of the electrode?
7 Draw a flow chart of the fabrication of a screen-printed device including two graphite electrodes and an Ag/AgCl reference electrode.
8 Draw a summary flow chart representing the fabrication of gold interdigitated electrodes.
9 What kind of nanomaterials are of interest in the fabrication of electrodes?
10 How can nanomaterials be used in the fabrication of electrodes?
11 What properties can nanomaterials impart to the electrodes?

13.9 Catalysis in Electrochemical Reactions

Many electrochemical reactions of interest in sensor applications occur with a high activation energy and need a high overvoltage in order for them to proceed. Therefore, such reactions may not occur within the potential window of the electrode. Even if such a reaction occurs within the potential window of the electrode it will display a marked irreversible character that results in poor sensitivity and an increased risk of interference. A high activation energy is associated with breaking or formation of covalent bonds. Another example of difficult reaction is the electron transfer from the active site of a redox enzyme to the electrode. Such a reaction is hindered by the fact that the active site is buried within the bulky enzyme molecule and cannot approach the electrode surface.

In order to enable such difficult electrochemical reactions, catalysis is brought into play [62]. According to the IUPAC definition a catalyst is a substance that increases the rate of a reaction without modifying the overall standard Gibbs energy change in the reaction. The increase in the reaction rate is caused by a decrease in the activation energy whilst the absence of a change in the Gibbs energy implies that the equilibrium constant is not affected by catalysis. In electrochemical terms, the catalysis decreases the overvoltage and thus allows the electrochemical reaction to proceed as a fast (reversible) reaction.

Distinction should be made between homogeneous catalysis and heterogeneous catalysis. In *homogeneous catalysis*, both the reactant and the catalyst are present in the same phase, for example, in an aqueous solution. Conversely, in *heterogeneous catalysis*, two distinct phases, for example a solid catalyst (which might be the electrode itself) and a solution containing the reactants, are involved.

13.9.1 Homogeneous Redox Catalysis

Homogeneous redox catalysis involves indirect electron transfer from the species of interest to the electrode via an intermediate redox system called the *electron-transfer mediator* (in short, the *mediator*) which is free to diffuse in the solution. The mechanism of a mediated electrochemical oxidation is introduced in Figure 13.30A. Accordingly, the reactant X undergoes oxidation in the solution by electron transfer to the oxidized form of the mediator (M_O) to yield the reduced mediator form (M_R) and a product Y (reaction I). By diffusion, the reduced form M_R reaches the electrode surface and transfers the electron to the electrode through an electrochemical reaction (reaction II). The oxidized form M_O resulting from this reaction diffuses back into the solution and will oxidize another reactant molecule. Therefore, the overall reaction consists of an electron transfer from the reactant X to the electrode, accompanied by mediator recycling.

The mediator should be selected so as to fulfill two conditions. First, it should be able to perform electron exchange with the reactant. For an oxidation reaction, this condition can be expressed by means of the standard electrode potentials as $(E_f^0)_X > (E_f^0)_M$. The opposite relationship is required in the case of a reduction reaction. Secondly, the mediator should undergo a reversible electrochemical reaction in order to achieve good sensitivity and selectivity.

In a mediated electrochemical reaction the chemical step (I) proceeds in a thin solution layer called the reaction layer, where mediator recycling is possible. The reactant X is supplied to this layer by diffusion from the bulk of the

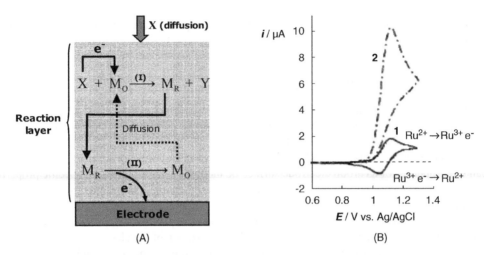

Figure 13.30 (A) Principle of mediated electron transfer. Oxidation of the species X to Y proceeds by mediation of the redox couple M_O/M_R which is freely diffusing in the solution. The thick, solid line indicates the path of the electron from the donor species X to the electrode. (B) Cyclic voltammograms demonstrating the mediated electrochemical oxidation of guanosine-5'-monophosphate (0.3 mM) by $[Ru(bypy)_3]^{2+}$ (25 μM) at an indium-tin oxide electrode (area 0.32 cm^2) in phosphate buffer (pH 7). Scan rate 25 mV/s^{-1}. (1) No guanosine in the solution; (2) with guanosine in the solution. (B) was reprinted with permission from [64]. Copyright 1997 American Chemical Society.

solution. Therefore, the current may depend on both the rate of reactant diffusion and the reaction rate of the step (I). However, if the reactant concentration is sufficiently high, its depletion is negligible and the velocity of the overall process is determined by the chemical step (I). The current becomes in this case a *kinetic current*, that is, a current determined by the reaction rate of an intermediate chemical reaction.

Cyclic voltammograms in Figure 13.30B demonstrate the electrochemical oxidation of the guanidine residue in guanosine-5'-monophosphate (GMP) by means of a $Ru^{2+/3+}$ complex with 2,2'-bipyridine (bypy) as mediator. In the absence of GMP (curve 1), only the $Ru^{2+/3+}$ electrochemical reactions take place, producing the curve 1 that is characteristic of a reversible process. If GMP is present, it becomes the primary electron donor, according to the scheme in Figure 13.30A. In this case, the anodic current becomes much larger (curve 2 in Figure 13.30B) and depends on the GMP concentration. As a consequence of redox mediation, GMP oxidation occurs within the potential region of the standard potential of the redox mediator.

A common application of mediated electron transfer is found in certain electrochemical nucleic acid sensors in which the transduction is based on the electrochemical reaction of the guanine residue discussed above [63].

13.9.2 Homogeneous Mediation in Electrochemical Enzymatic Reactions

Of great interest in the development of electrochemical enzyme sensors is the electron transfer from a substrate (S) to the electrode under the catalytic action of a redox enzyme ($E_{red/ox}$), which is mediated by a redox couple according to the reaction sequence in Figure 13.31A. The first step consists of substrate oxidation by interaction with the oxidized form of the enzyme (E_O), leading to the product (P) and the reduced enzyme (E_R). The active form of the enzyme is regenerated in the next step (II) by electron transfer to the oxidized form of the mediator (M_O) to give the reduced mediator form (M_R). Finally, the M_R form conveys by diffusion the electron to the electrode (step (III)) giving rise to the electrolytic current. This process can be viewed as a doubly mediated electron transfer from the substrate to the electrode involving the redox enzyme in step (I) and the redox mediator in step (II). The current will be dependent on the velocity of one of the steps in the sequence: these are the substrate diffusion and reactions (I) and (II). If the enzyme concentration is very small, the rate-determining step could be the step (I) and the current in this case will be a kinetic current. Conversely, if the enzyme concentration is sufficiently high, steps (I) and (II) could be very fast and the current would be determined by the rate of substrate diffusion.

Typical cyclic voltammograms for the process outlined above are shown in Figure 13.31B for the oxidation of glucose catalyzed by glucose oxidase. The mediator is in this case a ferrocene derivative that contains iron (II). Curve 1 in this figure is recorded in the absence of the enzyme, when only the diffusion-controlled reversible electrochemical reaction of the mediator (step (III)) is possible and the substrate is inert. In the presence of the enzyme (curve 2) the whole reaction sequence in Figure 13.31A occurs. As a consequence, the current becomes dependent on the concentration of the substrate that functions as the primary electron donor. In the case shown in Figure 13.31B, the slowest step is reaction (I) and the current is not influenced by substrate diffusion. Therefore, the current has a kinetic character and a clear limiting current is visible on curve 2. As no significant substrate depletion occurs, the forward and reverse scans are nearly identical, the small difference between the two branches scans of

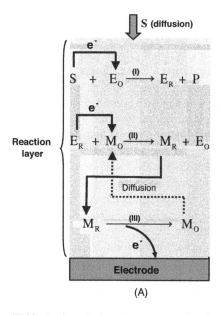 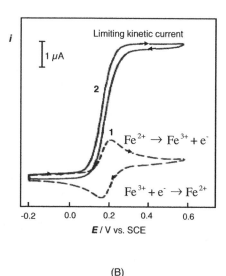

Figure 13.31 (A) Mechanism of a homogeneous mediated oxidation catalyzed by a redox enzyme. (B) Cyclic voltammograms for a mediated enzymatic electrochemical reaction. Curve 1: 0.1 mM ferrocene methanol (Fc-CH$_2$OH) and 0.5 M glucose in phosphate buffer (pH 7.5). Curve 2: 2.7 μM glucose oxidase added to the previous solution. Electrode: 3 mm diameter glassy carbon. Scan rate, 80 mVs^{-1}. (B) was reprinted with permission from [65]. Copyright 1993 American Chemical Society.

curve 2 being due to the capacitive current. A conclusive test for a kinetic current is the absence of any effect of the scan rate, which is due to the absence of any influence of substrate diffusion on the overall reaction rate.

If the enzyme concentration is sufficiently high, the reaction (II) becomes very fast and substrate diffusion becomes the rate-determining step. In this case, the current assumes the characteristics of a diffusion current.

Cyclic voltammetry with substrate–enzyme–mediator solutions is a valuable tool in kinetic investigations of enzyme-catalyzed reactions [66].

13.9.3 Catalysis by Immobilized Enzymes

Mediated electrochemical enzyme sensors make use of enzyme layers immobilized at the electrode surface along with a mediator. Transduction in this case is based on the electrochemical reaction of the mediator that exchanges electrons with the enzyme and conveys them to the electrode.

In order to understand the functioning principle of this type of sensor, it is necessary to inspect first the electrochemistry of a redox system confined in a thin film at the electrode surface. Typically, the cyclic voltammogram of a surface-confined reversible redox system is shaped like the curve in Figure 13.32A. During the forward scan, the

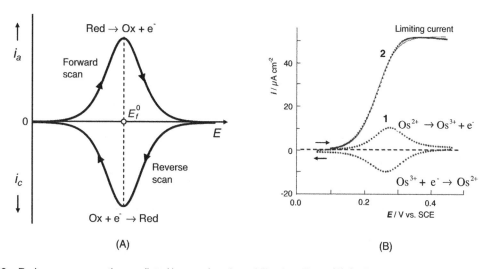

Figure 13.32 Redox enzyme reaction mediated by a surface-immobilized mediator. (A) Cyclic voltammogram for the reversible electrochemical reaction of an immobilized redox couple. (B) An immobilized Os$^{2+/3+}$ complex as mediator in an amperometric glucose sensor. The surface film includes the mediator and glucose oxidase. Curve 1: no substrate in the solution; curve 2: 50 mM glucose added. Conditions: 10 mM HEPES buffer (pH 7.5), 0.2 M KNO$_3$. (B) was reprinted with permission from [67]. Copyright 2001 American Chemical Society.

reactant is totally converted to its redox conjugate; then, the reaction proceeds in the opposite direction during the reverse scan. Both peak potentials are similar to the formal potential of the redox system, which imparts to the curve a symmetrical shape. Each peak current is proportional to the concentration of reactant in the immobilized layer, whilst the charge obtained by time integration of the current is proportional to the amount of immobilized reactant. The peak current is directly proportional to the potential scan rate, which is characteristic of reactions occurring in immobilized layers. Deviation from reversibility causes the curve to assume an asymmetrical shape, with a different peak potential on the forward and reverse scans, peak potentials being different from the formal potential.

The electrochemical behavior of a surface layer containing both a redox enzyme and a mediator is illustrated in Figure 13.32B. In this example, the surface film contains glucose oxidase and an osmium complex as mediator. In the absence of the glucose substrate (curve 1), the only possible electrode process is the oxidation/reduction of the mediator, which gives rise to the curve 1 that is similar to that in Figure 13.32A. The slight asymmetry of this curve is due to the contribution of the capacitive current to the total recorded current. If the glucose substrate is present in the solution, electrons are transferred sequentially from the glucose molecule to the electrode via the enzyme and the mediator as successive conveyors. As indicated by curve 2 in Figure 13.32A, the voltammogram is in this case shaped as a steady-state voltammetric wave, with a clear limiting current. In addition, the forward and reverse tracks are almost similar. Both the above features demonstrate that no substrate depletion occurs because the velocity of the enzymatic reaction is much lower than the rate of substrate supply by diffusion. Therefore, the limiting current is determined by the rate of the enzymatic reaction and is a kinetic current. Owing to the absence of any diffusional restriction, the kinetic current is independent of the substrate concentration but depends on the enzyme concentration in the surface film.

If the enzyme concentration in the surface layer is very high or a diffusion barrier in the form of a semipermeable membrane is present over the surface film, diffusion can become more sluggish than the enzymatic reaction. In this case, the current is determined by the rate of substrate diffusion and becomes a limiting diffusion current. Under these conditions, the current is proportional to the substrate concentration in solution. More details on reaction kinetics at enzyme-modified electrodes are available in Chapter 15.

13.9.4 Heterogeneous Redox Catalysis

Heterogeneous catalysis (or *electrocatalysis*) implies specific interactions between the reactant or reaction intermediates and the electrode material or a solid modifier coated over the electrode surface. When this interaction is an electron-transfer process, one has to deal with a heterogeneous redox catalysis process. On the other hand, catalytic effects can be caused by a noncovalent interaction of the reactant with the electrode surface with no accompanying electron transfer. This topic is addressed in the next section.

Heterogeneous redox catalysis occurs when the electrode is coated with a modifier film that contains metal ion sites that can swing between two oxidation states, Ox and Red. The modifier can be made of a metal oxide or a metal complex compound.

The mechanism of heterogeneous redox catalysis is outlined in Figure 13.33A. The reactant X in the solution undergoes oxidation by electron transfer to the Ox site in the modifier film, which leads to the conversion of this site into the reduced form Red. Further, the Red site transfers an electron to the electrode and is converted back to the Ox state. In this way, the electron acceptor site Ox is recycled and can react with another X molecule.

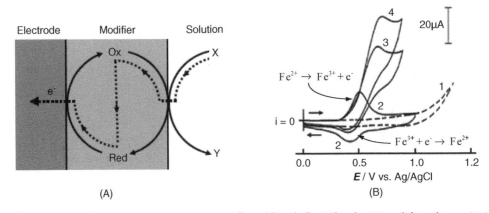

Figure 13.33 (A) Mechanism of heterogeneous redox catalysis. Dotted lines indicate the electron path from the reactant X to the electrode. (B) Heterogeneous redox catalysis in the hydroxylamine anodic reaction at a glassy carbon electrode modified with a cobalt(II) hexacyanoferrate film. (1) Bare carbon electrode; 9 mM NH_2OH; (2) modified electrode, no NH_2OH (the CV peaks are due to the electrochemical processes involving the immobilized modifier); (3 and 4) as (2) with 6 and 9 mM NH_2OH, respectively. (B) was adapted with permission from [68]. Copyright 1998 Elsevier.

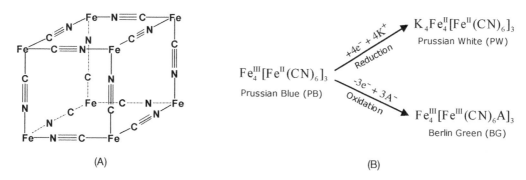

Figure 13.34 Prussian Blue structure **(A)** and electrochemical reactions **(B)**. A^- is an anion supplied by the electrolyte.

The behavior of such a system is illustrated by cyclic voltammograms in Figure 13.33B that correspond to the electrochemical oxidation of hydroxylamine catalyzed by a cobalt(II) hexacyanoferrate film. Curve 1 in this figure was recorded at the bare electrode and shows no electrochemical reaction, although hydroxylamine was present in the solution. Curve 2 is the voltammogram of the modified electrode in the absence of hydroxylamine. It displays characteristic peaks for Fe^{2+} oxidation and Fe^{3+} reduction within the modifier film. If hydroxylamine is added (curves 3 and 4), Fe^{3+} sites formed during the anodic scan prompts electron transfer from hydroxylamine to the electrode according to the mechanism in Figure 13.33A. The peak current obtained in this case is a linear function of the hydroxylamine concentration.

A widely used solid redox catalyst is *Prussian Blue* (PB), a solid compound with the formula $Fe_4^{III}[Fe^{II}(CN)_6]_3$ [69]. It consists of a network of iron ions crosslinked by cyanide (CN^-) anions, as shown in Figure 13.34A. This compound can be deposited on various electrode materials from a 0.1 M HCl aqueous solution containing both ferric (Fe^{3+}) and ferricyanide ($[Fe^{III}(CN)_6]^{3-}$) ions. The reaction can occur spontaneously or can be driven by an applied electrode potential that promotes the reduction of $[Fe^{III}(CN)_6]^{3-}$ to $[Fe^{II}(CN)_6]^{3-}$ [70,71].

The chemical formula of Prussian Blue indicates that this material contains iron in two oxidation states. Therefore, a Prussian Blue film coated over an electrode can undergo either electrochemical reduction or oxidation, depending on the electrode potential as shown in Figure 13.34B. Reduction of Prussian Blue gives the Prussian White (PW) compound whilst oxidation leads to Berlin Green (BG).

The cyclic voltammogram in Figure 13.35 was recorded with a glassy carbon electrode coated with Prussian Blue and shows that both reactions in Figure 13.34B are kinetically reversible. Both reactions involve ion entrapment within the solid polymer lattice for charge-balance reasons. Thus, reduction involves K^+ inclusion whilst oxidation is accompanied by inclusion of an anion. As the included ion should fit available sites in the crystal lattice, the reduction is a cation-selective process. In addition to K^+, Cs^+, Rb^+, Tl^+ and NH_4^+, can also be included, but the reaction cannot proceed in the presence of bulky cations.

As Prussian Blue can undergo both electrochemical oxidation and reduction, it can function as a redox catalyst in both anodic and cathodic reactions.

The main application of Prussian Blue derives from its catalytic function in the electrochemical reduction of hydrogen peroxide. This property can be exploited in the amperometric determination of this compound with no interference from dissolved oxygen.

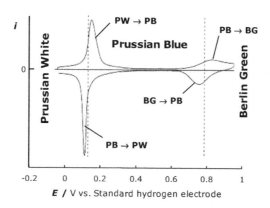

Figure 13.35 Cyclic voltammetry of a Prussian Blue layer at a glassy carbon electrode in 0.1 M KCl (scan rate, $40\,mV s^{-1}$). Adapted with permission from [71]. Copyright 2001 Wiley-VCH Verlag GmBH & Co. KGaA.

On the other hand, hydrogen peroxide results as a byproduct in reactions catalyzed by oxidase enzymes. This allow the transduction in oxidase-based sensor to be performed by monitoring the hydrogen peroxide by its anodic reaction. Owing to the catalytic properties of platinum, platinum electrodes were used in such sensors in the early stage of development. Inexpensive carbon electrodes covered with Prussian Blue represent a more convenient alternative to platinum electrodes in such amperometric sensors [70–72]. However, electrochemically synthesized Prussian Blue is not particularly stable in neutral media that are typical of enzyme sensors. Improved stability is achieved by coating the Prussian Blue layer with a Nafion film or by using tetrabutylammonium toluene 4-sulfonate as an additive in the electrochemical synthesis. Chemically prepared Prussian Blue is more stable and can be applied to the support electrode by screen printing [73].

As Prussian Blue synthesis is carried out in strongly acidic solution, it is not practical to include an enzyme into the film at the same time. Instead, gel entrapment is suitable for preparing enzyme layers over preformed Prussian Blue films.

Other metal hexacyanoferrates have also been investigated as possible catalysts in the electrochemical oxidation of various organic, inorganic, and biological species [71,74]. Furthermore, the equilibrium potential of a metal ferrocyanate is dependent on the concentration of the available counterion, such as potassium or ammonium ions. This property has been exploited in the development of potentiometric sensors for the above ions. Thus, a potassium sensor can be developed by coating a nickel or copper heacyanoferrate layer over a glassy carbon electrode. For a copper hexacyanoferrate layer, the potential-determining electrochemical reaction can be formulated as follows:

$$Cu_3^{II}[Fe^{III}(CN_6)]_2 + 2K^+ + 2e^- \rightleftarrows K_2\{Cu_3^{II}[Fe^{II}(CN_6)]_2\} \tag{13.97}$$

The equilibrium potential of such electrodes is a Nernstian function of the potassium ion concentration.

Other materials of interest in redox catalysis are porphyrins and phthalocyanines as thin layers at the electrode surface [75]. Complexed metal ions in these compounds can act as electron conveyors between the solution reactant and the electrode.

13.9.5 Surface Activation of Electrochemical Reactions

Another type of heterogeneous catalysis involves some kind of interaction between the reactant and the electrode surface that is not accompanied by electron transfer. Traditionally, this kind of catalytic effect is termed an electrode-activation process. Well known in this respect is the property of platinum metals to catalyze electrochemical reactions involving hydrogen, which is due to the fact that such metals promote the dissociation of the hydrogen molecule to give adsorbed hydrogen atoms. The hydrogen atom is much more reactive than the hydrogen molecule itself. Efficient electrocatalysis is promoted by a high specific area. That is why metal nanoparticles are being actively investigated as catalysts in electrochemical reactions [76].

Carbon electrodes represent another typical example of catalysis by surface interaction with the reactant. As discussed in Section 5.4.7, the edge of graphite sheets contains various oxygen functional groups such as the anionic carboxyl group and the neutral hydroxyl and carbonyl groups. Depending on the reactant nature, such groups can interact with the reactant by electrostatic interaction, coordinate bonding, or hydrogen bonding. As a result, the reactant molecule can be oriented in a favorable position at the electrode surface. On the other hand, the electron acceptor or donor level in the reactant molecule can be affected by interface interactions. This can lead to a considerable decrease in the activation energy of an electrochemical reaction. A well-known example is the $Fe(H_2O)_6^{3+/2+}$ redox system that displays a strong dependence of the activation energy on the presence of oxygen at the carbon electrode surface. Catalytic activity can be imparted to carbon electrodes by particular pretreatment methods (Section 13.8.1). In addition, catalytic properties of carbon materials can be exploited by using carbon nanomaterials that display a high density of sites similar to that at the edge plane in graphite. For example, Figure 13.36 demonstrates the catalytic effect of a graphene layer coated over the glassy carbon surface. Cyclic voltammograms in this figure have been obtained in the presence of paracetamol as reactant. The reaction at the plain glassy carbon surface (curve 1) has an irreversible character, which is demonstrated by the high difference between the anodic and cathodic peak potentials. However, the reaction shows a reversible character when it occurs at the graphene-modified electrode (curve 2).

13.9.6 Outlook

Catalysis offers a broad range of possibilities for increasing the reaction rate of electrochemical reactions and converting slow (irreversible) reactions into fast (reversible) processes. Therefore, catalysis contributes to the improvement of sensitivity and allows the implementation of electrochemical reactions that cannot occur at all in the absence of a catalyst. Moreover, as catalysis is often a selective process, it can alleviate interference problems.

Redox catalysis implies mediation of the electron transfer from the reactant to the electrode by a redox mediator. This can be achieved in a solution phase that contains both the reactant and the mediator so as to promote homogeneous catalysis.

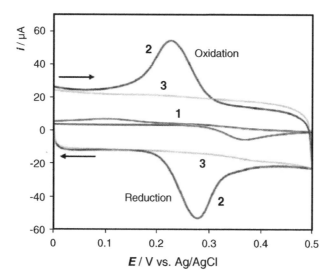

Figure 13.36 Graphene electrocatalysis in the electrochemical reaction of paracetamol (100 μm) in ammonia buffer (pH 9.3) at 50 mV/s^{-1} scan rate. (1) Plain glassy carbon electrode (3 mm diameter); (2) graphene-modified glassy carbon electrode; (3) same electrode with buffer alone (residual current). Adapted with permission from [77]. Copyright 2010 Elsevier.

Catalysis application in electrochemical sensors relies often on surface films with catalytic properties. This film can be a redox-type catalyst that promotes the electrochemical reaction of the reactant with no need of a catalyst in the solution.

Amperometric enzyme sensors make use of gel-type surface films which contain both the enzyme and a mediator. As the mediator displays some mobility within the film, this case is in some respects similar to the case of typical homogeneous catalysis.

Catalytic effects can be noted when using electrodes that interact with the reactant at the surface so as to reduce the activation energy of the electrochemical reaction. Typical electrode materials that behave in this way are platinum metals and carbon electrodes rich in oxygen functional groups. At the same time, metal or carbon nanomaterials display significant catalytic properties and can be used as catalysts in the form of coated layers over metal or carbon electrodes.

Questions and Exercises (Section 13.9)

1 What reaction parameters are affected by catalysis in general and what electrochemical parameter changes in response to catalysis?

2 What are the differences between homogeneous catalysis and heterogeneous catalysis?

3 Using Figure 13.30A as a template, sketch the reaction mechanism for a cathodic reaction involving homogeneous redox catalysis. Write the necessary conditions for this kind of reaction to occur. Draw schematically the cyclic voltammograms for this reaction in the absence and in the presence of the reducible reactant.

4 What steps in the overall process can determine the current in a homogeneous mediated electrochemical reaction involving a redox enzyme?

5 Figure 13.31B shows a cyclic voltammogram for a mediated enzymatic electrochemical reaction controlled by the substrate conversion step. Draw schematically a cyclic voltammogram for a diffusion-controlled reaction of the same kind.

6 List the response of a kinetic current and that of a diffusion current to the change in the following parameters: (a) scan rate; (b) substrate concentration; (c) mediator concentration.

7 For a surface-immobilized redox mediator in the absence of the redox enzyme draw schematically the cyclic voltammogram for (a) a reversible electrochemical reaction, and (b) an irreversible electrochemical reaction.

8 What are the differences between a homogeneous mediated enzymatic electrochemical reaction and a reaction involving a surface layer containing both an enzyme and a mediator, as far as the mobility of involved species is concerned?

9 What conditions determine the kinetic or diffusional character of the current produced by such a sensor? What analytical applications are possible in each case?

10 Using Figure 13.31A for orientation draw a reaction scheme for a cathodic reaction catalyzed by a surface layer of a solid redox catalyst. Draw schematically the relevant cyclic voltammograms.

11 Write the reactions involved in the chemical and electrochemical synthesis of Prussian Blue.
12 What are the applications of metal ferrocyanates in amperometric sensors? Draw schematically cyclic voltammograms for the cathodic reaction of Prussian Blue in the absence and in the presence of increasing concentrations of hydrogen peroxide.
13 What are the effects of cations on the chemical equilibrium between Prussian Blue and Prussian White at an electrode surface? How can this effect be exploited to develop potentiometric cation sensors? Write the Nernst equation for the electrochemical reaction (13.97) and determine the theoretical sensitivity of a potentiometric potassium sensor based on this reaction.
14 Why can carbon electrodes behave as catalysts in certain electrochemical reactions? What kinds of interactions are involved in the catalysis of electrochemical reactions by a carbon surface?
15 Why can carbon nanomaterials and certain metal nanoparticles display catalytic activity?

13.10 Amperometric Gas Sensors

Certain gases in the dissolved state undergo typical electrochemical reactions that can be exploited in the design of amperometric gas sensors [78]. In order for the sensor to function in the steady state, the electrode is coated with a semipermeable membrane that provides necessary diffusion resistance and, in addition, prevents electrode clogging by organic impurities. As no recognition element is present in these devices, a better term to define them would be amperometric gas probes. However, by tradition, the term amperometric gas sensor is widely used.

In a more elaborated design, gas diffusion electrodes are employed in order to construct amperometric sensors for gaseous samples.

The electrochemical reaction in a gas sensor can be prompted by a voltage supplied by an external voltage source. In another approach, the counterelectrode reaction can be selected so as to provide the needed polarization potential to the working electrode. In this case, the sensor functions as a Galvanic cell and needs no external voltage supply.

13.10.1 The Clark Oxygen Sensor

The first amperometric sensor was the oxygen sensor developed by L.L Clark. in 1956 for the determination of dissolved oxygen in blood [79–81]. It is based on the electrochemical reduction of dissolved oxygen that can occur in two steps as follows:

$$O_2(aq) + 2H^+ + 2e^- \rightleftharpoons H_2O_2 \tag{13.98}$$

$$H_2O_2 + 2H^+ + 2e^- \rightarrow 2H_2O \tag{13.99}$$

The first step is a reversible process whilst the second is irreversible. In accordance with this reaction sequence, the current–potential curve for O_2 reduction displays two waves, as shown in Figure 13.37A. If the potential is located in the region of the first wave, O_2 is reduced to H_2O_2, whereas in the region of the second wave, step (13.99) occurs in addition and the overall process consists of the $4e^-$ reduction of oxygen to water. As hydrogen ions are consumed the electrochemical reaction should be carried out in pH-buffered solutions. The curves in Figure 13.37A have been

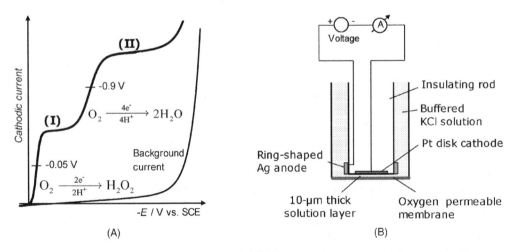

Figure 13.37   The amperometric oxygen sensor (Clark sensor). (A) Polarogram for oxygen reduction at a dropping mercury electrode at pH 7. SCE stands for saturated calomel reference electrode. (B) Schematics of the Clark oxygen sensor. (B) Adapted with permission from [10]. Copyright 2002 A. Bănică and F.G. Bănică.

obtained with a dropping mercury electrode. However, actual oxygen sensors use a noble metal (Au, Pt) as working electrode. At solid electrodes the second wave is obscured by the large background current due to hydrogen evolution and only the first wave is useful for analytical purposes. The electrode should be polarized in the region of the limiting current of the wave (I) in Figure 13.37A.

The Clark oxygen sensor (Figure 13.37B) is actually an electrochemical cell with the base consisting of a thin oxygen-permeable membrane made of Teflon. This material is hydrophobic and is hence not wetted by the aqueous solution but allows the transport of oxygen molecules. A platinum disk cathode is placed inside the cell at a very small distance from the membrane. The cell is completed with a silver anode and the internal solution is a buffered KCl solution. Oxygen diffuses across the membrane from the sample to the internal solution and undergoes reduction at the surface of the polarized platinum electrode. If the potential of the working electrode is fixed within the range of the limiting current of wave (I), the recorded current is proportional to the oxygen concentration in the test solution. The membrane in this sensor plays a double role. It protects the electrode surface against clogging and, in addition, provides a diffusion barrier that leads to a stable current.

The solution layer between the membrane and the Pt electrode should be very thin in order to obtain good sensitivity and a fast response time.

Commercial instruments provide the response in the form of oxygen partial pressure, which is the oxygen partial pressure in air at equilibrium with the liquid sample. This parameter depends on the salt content in the sample and salinity correction should be applied when necessary. The linear calibration graph is plotted by means of two points. The first is recorded with an air-saturated solution, whilst the second (background current) is obtained with an oxygen-free solution [82].

Since reaction (13.98) is reversible, the Clark sensor can also be used to determining hydrogen peroxide through its anodic reaction. In this case the electrode potential is adjusted to about +0.6 V.

Practical and theoretical aspects of the Clark oxygen sensor are amply presented in refs. [12,83].

A similar arrangement can be used for determining other electrochemically active gases, such as chlorine in swimming pools. Sulfide ions in aquatic systems can be determined by using indirect oxidation with the ferrocyanide/ferricyanide couple acting as mediator [84].

The Clark oxygen sensor is of remarkable importance to industry, biomedical and environmental research, and clinical medicine. Typical applications are the determination of dissolved oxygen in various samples such as blood, sea water, sewage, and effluents from chemical plants [85]. Physiological applications are based on microsensors that are suited for *in vivo* measurements [83]. Oxygen determination in gas samples is also feasible but is limited by poor response stability due to water evaporation from the thin solution film between the membrane and the electrode.

Moreover, the amperometric oxygen sensor can function as a transducer in biosensors based on oxygen consumption in biocatalytic reactions (Chapters 14 and 23). Either oxygen or hydrogen peroxide (which results by biochemical reduction of oxygen) can be monitored in this way, depending on the potential applied to the working electrode.

13.10.2 Nitric Oxide Sensors

Nitric oxide (NO) is an important regulator and mediator of numerous processes in the nervous, immune and cardiovascular systems. Besides mediating normal functions, NO is involved in various pathophysiological states such as septic shock, hypertension, stroke, and neurodegenerative diseases. Therefore, *in vivo* determination of nitric oxide is a matter of great interest and prompted the development of electrochemical sensors for this compound. The first reported nitric oxide sensor was based on the Clark oxygen sensor design [86]. Further progress in this field is well covered in refs. [87,88].

Nitric oxide can undergo either electrochemical reduction or oxidation. As dissolved oxygen interferes with the reduction reaction, the oxidation reaction is preferred in sensor applications. It occurs in three steps, as follows:

$$NO \rightarrow NO^+ + e^- \tag{13.100}$$

$$NO^+ + OH^- \rightarrow HNO_2 \tag{13.101}$$

$$HNO_2 + H_2O \rightarrow NO_3^- + 3H^+ + 2e^- \tag{13.102}$$

Biological samples contain as a rule a series of anionic species (such as ascorbic acid, uric acid, nitrite and others) that can also undergo anodic reactions. Interferences from such compounds is avoided by coating the electrode with a Nafion membrane. Being a cation exchanger, Nafion prevents the access of the anionic species to the electrode but is permeable to the neutral nitric oxide molecule. However, Nafion is not effective in eliminating interference from cationic species (such as dopamine and serotonin) or neutral compounds (for example, hydrogen peroxide and acetaminophen). A second, hydrophobic membrane prevents the access of such species to the electrode surface but is not an obstacle to nitric oxide, which is a hydrophobic compound.

Figure 13.38 displays an amperometric nitric oxide sensor designed according to the above principles. The working electrode consists of a carbon microfiber electrode (7 μm diameter) with a conical active tip coated with Nafion

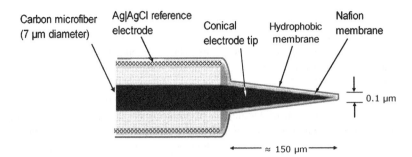

Figure 13.38 A nitrogen oxide microsensor. The surface is coated with Nafion and WPI membranes. Adapted with permission from [89]. Copyright 2002 Elsevier.

and a hydrophobic WPI membrane. Integrated with this is an Ag/AgCl reference electrode coated as a thin layer over the carbon fiber holder. This sensor has a linear response to NO in a stirred buffer solution over the concentration range from 10 to 1000 nM. It is suitable for *in vivo* and real-time detection of nitric oxide in single cells or micro-samples [89].

A substantial improvement in the sensitivity is obtained by coating the electrode with a catalytic layer formed of metal porphyrins. Again, a cation exchanger membrane is formed over the catalyst layer in order to prevent interferences from anionic compounds.

Nitric oxide microsensors are widely used for determining this compound and investigating its spatial and temporal distribution in living tissues.

13.10.3 Other Types of Amperometric Gas Sensors

There are various gases and vapors that can undergo electrochemical reactions. Thus, hydrogen, carbon monoxide, ethanol, ammonia and hydrocarbons can be oxidized, whilst carbon dioxide and chlorine can be reduced, provided that the electrode displays catalytic activity. However, the catalytic activity requires a high specific surface area, which is not available with compact electrodes.

Liquid electrolytes used in the Clark design pose some problems related to stability and maintenance. More rugged devices are obtained using a solid polymer electrolyte such as Nafion (see Section 5.12.2). As Nafion is a cation exchanger, water embedded in Nafion displays electrical conductivity owing to the mobility of hydrogen ions within its pores. Therefore, electrochemical reactions involving hydrogen ions can proceed with no impediment. In addition to oxygen sensors, Nafion-based sensors for a series of other gases have been developed [78]. Such sensors are free from problems related to water evaporation and can function as gas-phase sensors as well.

Substantial progress in the development of amperometric gas sensors has been achieved through the introduction of *gas diffusion electrodes* (Figure 13.39A). Such an electrode is made of a porous composite material including

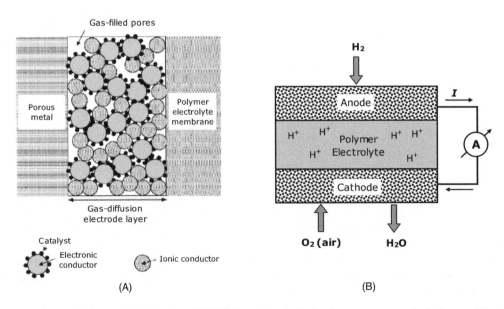

Figure 13.39 (A) Structure of a gas diffusion electrode. (B) Schematics of a fuel-cell-type gas sensor for hydrogen. (A) was reproduced with permission from [90]. Copyright 2005 Elsevier.

particles of polymer electrolyte, metal and catalyst. The rear-side of the membrane is coated with a thin porous metal layer to provide electrical contact. The electrochemical reaction occurs in this case at the three-phase boundary of the electrolyte, metal and Teflon. The mass transfer of the analyte from the test solution to the working electrode is much faster, resulting in shorter response times and greater sensitivity.

13.10.4 Galvanic Cell-Type Gas Sensors

An amperometric sensor is an electrolytic device in which the analyte reaction is prompted by a suitable electrode potential provided by an external voltage source. An alternative to this design is represented by the Galvanic-cell-type sensor [83]. Its design is mostly similar to that of an amperometric gas sensor, but the operating conditions are selected so as to allow the cell reaction to proceed spontaneously. Consequently, an electromotive force develops between the electrodes and the current flows across the external circuit in the absence of an applied voltage. As the oxygen reaction is a cathodic one, the anode should develop a negative potential with respect to that of the cathode. Thus, a combination of a lead anode and a silver cathode in an alkaline solution was used in the early stage of development as well in subsequent developments. If a high boiling point solvent is used instead of water, this sensor can be operated at higher temperatures. In addition, it withstands thermal sterilization, which is essential in oxygen monitoring in industrial bioreactor fluids.

Galvanic-cell-type oxygen sensors have a simple configuration and no potentiostat is needed in operation. The design is more rugged and more suitable for industrial applications.

Another type of Galvanic cell, namely the *fuel cell* has been applied to electrochemical sensing of various reducing gases. A fuel cell is a galvanic cell that produces electrical energy by electrochemical reactions of certain gases supplied to electrodes. Fuel cells electrodes are manufactured as gas diffusion electrodes that include a suitable catalyst. The cell electrolyte is a polymer electrolyte, for example, Nafion in normal-temperature-operated cells. Solid electrolytes are used in high-temperature fuel cells.

This design can be used for developing gas sensors for various reducing gases such as hydrogen [91], carbon monoxide, hydrocarbons, and volatile alcohols. The working principle of a fuel-cell hydrogen sensor is outlined in Figure 13.39B. It is based on the reaction of hydrogen with oxygen to give water. Electrochemical reactions proceed at the gas-metal-electrolyte triple phase boundary. Under the effect of the catalysts, the electrochemical reactions occur as follows:

$$
\begin{aligned}
\text{Anode:} & \quad H_2(\text{gas}) \rightarrow 2H^+(\text{electrolyte}) + 2e^-(\text{metal}) \\
\text{Cathode:} & \quad (1/2)O_2(\text{gas}) + 2H^+(\text{electrolyte}) + 2e^-(\text{metal}) \rightarrow H_2O(\text{vapor})
\end{aligned}
\tag{13.103}
$$

Current flow within the cell is provided by migration of hydrogen ions in the polymer electrolyte. At the same time, electron flow from the anode to the cathode through the external circuit produces a measurable current that depends on the hydrogen content in the gas sample supplied to the anode.

Fuel-cell-type gas sensors have the common advantages of galvanic cell gas sensors, that is, simple configuration, rugged design and operation without a potentiostat. However, fabrication of gas diffusion electrodes is rather intricate.

13.10.5 Solid Electrolyte Amperometric Gas Sensors

Solid electrolytes are ionic compounds that display electrical conduction by ion displacement (see Section 10.17). It was shown there that galvanic cells formed of a solid electrolyte inserted between two thin, porous metal electrodes develop an electromotive force that is correlated with the partial pressure of a reacting gas by a Nernst-type equation. An oxygen sensor based on these principles is used to determine the oxygen/fuel ratio in the combustion mixture in the form of the λ parameter.

A similar configuration is used in amperometric gas sensors functioning at elevated temperatures, as shown in Figure 13.40, which refers to the particular case of the oxygen sensor [92,93]. In this case, a voltage is applied to the cell in order to prompt the ion transfer across the solid electrolyte membrane. As a result, gas–solid equilibria at the triple-phase boundaries are disturbed. As the potential of the reference electrode is kept constant by a constant oxygen concentration at this side, the applied voltage affects only the potential of the working electrode that is in contact with the gas sample. The working electrode is negatively biased so as to promote the reduction of gaseous oxygen to O^{2-} ions that migrate through the solid electrolyte towards the anode. At this electrode, O^{2-} ions are oxidized to gaseous oxygen. So, this device functions as an oxygen pump, which transfers oxygen from the sample to the reference gas under the effect of the pumping voltage (V_p). The current–voltage curve shows the typical shape of a steady-state voltammetric curve and displays a plateau corresponding to the limiting diffusion current. As expected, the diffusion current is proportional to the partial pressure of oxygen in the sample gas.

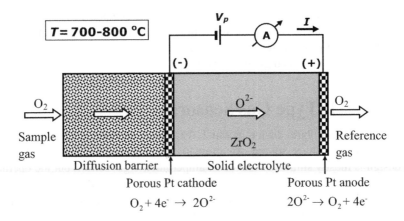

Figure 13.40 Principle of the amperometric oxygen sensor based on a solid electrolyte.

Proper functioning of this sensor requires a diffusion barrier to be interposed between the sample and the sensing surface in order to stabilize the mass-transfer conditions near the surface and secure the development of the limiting current. Such a barrier can be provided by a porous layer or a gap in the front of the cathode.

The setup in Figure 13.40 is used to control the combustion in the case of a lean exhaust gas ($\lambda < 1$), which contains nonreacted residual oxygen. In the case of a fuel-rich gas ($\lambda > 1$), which does not contain residual oxygen, the polarity of the pumping voltage is reversed and the electrode reactions occur in the opposite direction.

A sensor covering the whole range of λ, from fuel-rich to fuel-lean region is based on a dual-cell design. This is a combination of a pumping cell and a galvanic cell separated by a 50- μm gap. Depending on the polarity of the pumping voltage, oxygen can be pumped into or out of the gap. The response of the galvanic cell is used as a feedback signal that controls the limiting current so as to keep the galvanic cell EMF at a constant level. The limiting current thus regulated is proportional to the oxygen content in the sample.

Compared with the potentiometric λ sensor, the amperometric version allows determining this parameter over a wider range. In addition, the response is less sensitive to temperature variations.

References

1. Bard, A.J. and Faulkner, L.R. (2001) *Electrochemical Methods: Fundamentals and Applications*, John Wiley & Sons, New York.
2. Girault, H.H. (2004) *Analytical and Physical Electrochemistry*, EPFL Press, Lausanne.
3. Galus, Z. (1994) *Fundamentals of Electrochemical Analysis*, Ellis Horwood, New York.
4. Brett, C.M.A. and Brett, A.M.O. (1993) *Electrochemistry: Principles, Methods, and Applications*, Oxford University Press, Oxford.
5. Sawyer, D.T., Sobkowiak, A., and Roberts, J.L. (1995) *Electrochemistry for Chemists*, John Wiley & Sons, New York.
6. Zoski, C.G. (ed.) (2007) *Handbook of Electrochemistry*, Elsevier, Amsterdam.
7. Holze, R. (2009) *Experimental Electrochemistry: A Laboratory Textbook*, Wiley-VCH, Weinheim.
8. Kissinger, P.T. and Heineman, W.R. (eds) (1996) *Laboratory Techniques in Electroanalytical Chemistry*, Marcel Dekker, New York.
9. Zhang, X.J., Ju, H.X., and Wang, J. (eds) (2007) *Electrochemical Sensors, Biosensors and their Biomedical Applications*, Academic Press, New York.
10. Ion, A. and Bănică, F.G. (2002) *Electrochemical Methods in Analytical Chemistry*, Ars Docendi, Bucharest.
11. Reddy, S.M. and Vadgama, P.M. (1997) Membranes to improve amperometric sensor characteristics, in *Handbook of Biosensors and Electronic Noses: Medicine, Food, and the Environment* (ed. E. Kress-Rogers), CRC Press, Boca Raton, Fla, pp. 111–135.
12. Hitchman, M.L. (1978) *Measurement of Dissolved Oxygen*, John Wiley & Sons, New York.
13. Breyer, B. and Bauer, H.H. (1963) *Alternating Current Polarography and Tensammetry*, Interscience, New York.
14. Bersier, P.M. and Bersier, J. (1988) Polarographic adsorption analysis and tensammetry – toys or tools for day-to-day routine analysis. *Analyst*, **113**, 3–14.
15. Mount, A.R. (2003) Hydrodynamic electrodes, in *Instrumentation in Electroanalytical Chemistry* (ed. P.R. Unwin), Wiley-VCH, Weinheim, pp. 134–159.
16. Compton, R.G. and Banks, C.E. (2011) *Understanding Voltammetry*, Imperial College Press, London.
17. Scholz, F. and Bond, A.M. (eds) (2002) *Electroanalytical Methods: Guide to Experiments and Applications*, Springer-Verlag, Berlin.

18. Heyrovský, J. and Kuta, J. (1965) *Principles of Polarography*, Publishing House of the Czechoslovak Academy of Sciences, Prague.
19. Zuman, P. (2001) Electrolysis with a dropping mercury electrode: J. Heyrovky's contribution to electrochemistry. *Crit. Rev. Anal. Chem.*, **31**, 281–289.
20. Marken, F., Neudeck, A., and Bond, A.M. (2010) Cyclic voltammetry, in *Electroanalytical Methods: Guide to Experiments and Applications* (ed. F. Scholz), Springer-Verlag, Berlin, pp. 57–106.
21. Berggren, C. and Johansson, G. (1997) Capacitance measurements of antibody-antigen interactions in a flow system. *Anal. Chem.*, **69**, 3651–3657.
22. Mirceski, V., Komorsky-Lovrić, S., and Lovrić, M. (2007) *Square-Wave Voltammetry: Theory and Application*, Springer-Verlag, Berlin.
23. Britz, D. (2006) Setting the record straight on reciprocal derivative chronopotentiometry. *Int. J. Electrochem. Sci.*, **1**, 379–382.
24. Wang, J., Cai, X.H., Wang, J.Y. *et al.* (1995) Trace measurements of RNA by potentiometric stripping analysis at carbon-paste electrodes. *Anal. Chem.*, **67**, 4065–4070.
25. Zoski, C.G. (1996) Steady-state voltammetry at microelectrodes, in *Modern Techniques in Electroanalysis* (ed. P. Vanýsek), John Wiley & Sons, New York, pp. 241–312.
26. Amatore, C. (1995) Electrochemistry at ultramicroelectrodes, in *Physical Electrochemistry: Principles, Methods, and Applications* (ed. I. Rubinstein), M. Dekker, New York, pp. 131–208.
27. Ion, A., Partali, V., Sliwka, H.R., and Bănică, F.G. (2002) Electrochemistry of a carotenoid self-assembled monolayer. *Electrochem. Commun.*, **4**, 674–678.
28. Dumitrescu, I., Unwin, P.R., and Macpherson, J.V. (2009) Electrochemistry at carbon nanotubes: perspective and issues. *Chem. Commun.*, 6886–6901.
29. Nagy, G. and Nagy, L. (2000) Scanning electrochemical microscopy: a new way of making electrochemical experiments. *Fresenius J. Anal. Chem.*, **366**, 735–744.
30. Bard, A.J. and Mirkin, M.V. (2001) *Scanning Electrochemical Microscopy*, Marcel Dekker, New York.
31. Stoica, L., Neugebauer, S., and Schuhmann, W. (2008) Scanning electrochemical microscopy (SECM) as a tool in biosensor research. *Adv. Biochem. Engin./Biotechnol.*, **109**, 455–492.
32. Komorsky-Lovrić, S. (2002) Working electrodes, in *Electroanalytical Methods: Guide to Experiments and Applications* (eds F. Scholz and A.M. Bond), Springer-Verlag, Berlin, pp. 273–290.
33. McCreery, R.L. and Cline, K.K. (1996) Carbon electrodes, in *Laboratory Techniques in Electroanalytical Chemistry* (eds P.T. Kissinger and W.R. Heineman), Marcel Dekker, New York, pp. 293–333.
34. Kinoshita, K. (1988) *Carbon: Electrochemical and Physicochemical Properties*, John Wiley & Sons, New York.
35. McCreery, R.L. (1991) Carbon electrodes, in *Electroanalytical Chemistry* (ed. A.J. Bard), Marcel Dekker, New York, Vol. 17, pp. 221–374.
36. McCreery, R.L. (1999) Electrochemical properties of carbon surfaces, in *Interfacial Electrochemistry: Theory, Experiment, and Applications* (ed. A. Wieckowski), Marcel Dekker, New York, pp. 631–647.
37. Banks, C.E. and Compton, R.G. (2006) New electrodes for old: from carbon nanotubes to edge plane pyrolytic graphite. *Analyst*, **131**, 15–21.
38. Kneten, K.R. and McCreery, R.L. (1992) Effects of redox system structure on electron-transfer kinetics at ordered graphite and glassy-carbon electrodes. *Anal. Chem.*, **64**, 2518–2524.
39. Harris, P.J.F. (2004) Fullerene-related structure of commercial glassy carbons. *Philos. Mag.*, **84**, 3159–3167.
40. Compton, R.G., Foord, J.S., and Marken, F. (2003) Electroanalysis at diamond-like and doped-diamond electrodes. *Electroanalysis*, **15**, 1349–1363.
41. Peckova, K., Musilova, J., and Barek, J. (2009) Boron-doped diamond film electrodes-new tool for voltammetric determination of organic substances. *Crit. Rev. Anal. Chem.*, **39**, 148–172.
42. Luong, J.H.T., Male, K.B., and Glennon, J.D. (2009) Boron-doped diamond electrode: synthesis, characterization, functionalization and analytical applications. *Analyst*, **134**, 1965–1979.
43. Zhou, Y.L. and Zhi, J.F. (2009) The application of boron-doped diamond electrodes in amperometric biosensors. *Talanta*, **79**, 1189–1196.
44. Švancara, I., Kalcher, K., Walcarius, A., and Vytřas, K. (2012) *Electroanalysis with Carbon Paste Electrodes*, CRC Press, New York.
45. Kalcher, K., Švancara, I., Metelka, R. *et al.* (2006) Heterogeneous carbon electrochemical sensors, in *Encyclopedia of Sensors* (eds C.A. Grimes, E.C. Dickey, and M.V. Pishko), American Scientific Publishers, Stevenson Ranch, CA, pp. 283–429.
46. Kalcher, K., Švancara, I., Buzuk, M. *et al.* (2009) Electrochemical sensors and biosensors based on heterogeneous carbon materials. *Monatshefte für Chemie*, **140**, 861–889.
47. Pividori, M.I., Merkoçi, A., and Alegret, S. (2003) Graphite-epoxy composites as a new transducing material for electrochemical genosensing. *Biosens. Bioelectron.*, **19**, 473–484.
48. McCreery, R.L. (2008) Advanced carbon electrode materials for molecular electrochemistry. *Chem. Rev.*, **108**, 2646–2687.
49. Armstrong, N.R., Lin, A.W.C., Fujihira, M., and Kuwana, T. (1976) Electrochemical and surface characteristics of tin oxide and indium oxide electrodes. *Anal. Chem.*, **48**, 741–750.
50. Lambrechts, M. and Sansen, W. (1992) *Biosensors: Microelectrochemical Devices*, Institute of Physics Publ., Bristol.
51. Suzuki, H. (2000) Advances in the microfabrication of electrochemical sensors and systems. *Electroanalysis*, **12**, 703–715.
52. Hart, J.P., Crew, A., Crouch, E. *et al.* (2007) Screen-printed (bio) sensors in biomedical, environmental and industrial applications, in *Electrochemical Sensor Analysis* (eds S. Alegret and A. Merkoçi), Elsevier, Amsterdam, pp. 497–557.

53. Albareda-Sirvent, M., Merkoçi, A., and Alegret, S. (2000) Configurations used in the design of screen-printed enzymatic biosensors. A review. *Sens. Actuators B-Chem.*, **69**, 153–163.

54. Pohanka, M. and Skladal, P. (2008) Electrochemical biosensors – principles and applications. *J. Appl. Biomed.*, **6**, 57–64.

55. Anderson, J.L. and Winograd, N. (1996) Film electrodes, in *Laboratory Techniques in Electroanalytical Chemistry* (eds P.T. Kissinger and W.R. Heineman), Marcel Dekker, New York, pp. 333–365.

56. Gooding, J.J. (2005) Nanostructuring electrodes with carbon nanotubes: A review on electrochemistry and applications for sensing. *Electrochim. Acta*, **50**, 3049–3060.

57. Wang, J. (2005) Carbon-nanotube based electrochemical biosensors: A review. *Electroanalysis*, **17**, 7–14.

58. Huang, J.S., Liu, Y., and You, T.Y. (2011) Carbon nanofiber based electrochemical biosensors: A review. *Anal. Methods*, **2**, 202–211.

59. Brownson, D.A.C. and Banks, C.E. (2010), Graphene electrochemistry: an overview of potential applications. *Analyst*, **135**, 2768–2778.

60. Pumera, M. (2010) Graphene-based nanomaterials and their electrochemistry. *Chem. Soc. Rev.*, **39**, 4146–4157.

61. Kauffman, D.R. and Star, A. (2010) Graphene versus carbon nanotubes for chemical sensor and fuel cell applications. *Analyst*, **135**, 2790–2797.

62. Bănică, F.G. and Ion, A. (2000) Electrocatalysis-based kinetic determination, in *Encyclopedia of Analytical Chemistry: Instrumentation and Applications* (ed. R.A. Meyers), J. Wiley & Sons, New York, pp. 11115–11144.

63. Thorp, H.H. (1998) Cutting out the middleman: DNA biosensors based on electrochemical oxidation. *Trends Biotechnol.*, **16**, 117–121.

64. Napier, M.E., Loomis, C.R., Sistare, M.F. *et al.* (1997) Probing biomolecule recognition with electron transfer: Electrochemical sensors for DNA hybridization. *Bioconjugate Chem.*, **8**, 906–913.

65. Bourdillon, C., Demaille, C., Moiroux, J., and Savéant, J.M. (1993) New insights into the enzymic catalysis of the oxidation of glucose by native and recombinant glucose oxidase mediated by electrochemically generated one-electron redox cosubtrates. *J. Am. Chem. Soc.*, **115**, 1–10.

66. Savéant, J.-M. (2006) *Elements of Molecular and Biomolecular Electrochemistry: An Electrochemical Approach to Electron Transfer Chemistry*, Wiley-Interscience, Hoboken, N.J.

67. Calvo, E.J., Etchenique, R., Pietrasanta, L. *et al.* (2001) Layer-by-layer self-assembly of glucose oxidase and Os(Bpy)(2) ClPyCH$_2$NH-poly(allylamine) bioelectrode. *Anal. Chem.*, **73**, 1161–1168.

68. Chen, S.M. (1998) Characterization and electrocatalytic properties of cobalt hexacyanoferrate films. *Electrochim. Acta*, **43**, 3359–3369.

69. Karyakin, A. (2008) Chemical and biological sensors based on electroactive inorganic polycrystals, in *Electrochemical Sensors, Biosensors, and Their Biomedical Applications* (eds X. Zhang, H. Ju, and J. Wang), Academic Press, Amsterdam, pp. 411–439.

70. Ricci, F. and Palleschi, G. (2005) Sensor and biosensor preparation, optimisation and applications of Prussian Blue modified electrodes. *Biosens. Bioelectron.*, **21**, 389–407.

71. Karyakin, A.A. (2001) Prussian Blue and its analogues: electrochemistry and analytical applications. *Electroanalysis*, **13**, 813–819.

72. Koncki, R. (2002) Chemical sensors and biosensors based on Prussian blues. *Crit. Rev. Anal. Chem.*, **32**, 79–96.

73. Ricci, F., Moscone, D., and Palleschi, G. (2007) Mediated enzyme screen-printed electrode probes for clinical, environmental and food analysis, in *Electrochemical Sensor Analysis* (eds S. Alegret and A. Merkoçi), Elsevier, Amsterdam, pp. 559–584.

74. de Mattos, I.L. and Gorton, L. (2001) Metal-hexacyanoferrate films: A tool in analytical chemistry (in Portuguese). *Quim. Nova*, **24**, 200–205.

75. Nyokong, T. and Bedioui, F. (2006) Self-assembled monolayers and electropolymerized thin films of phthalocyanines as molecular materials for electroanalysis. *J. Porphyr. Phthalocyanines*, **10**, 1101–1115.

76. Gileadi, E. (2011) *Physical Electrochemistry: Fundamentals, Techniques and Applications*, Wiley-VCH, Weinheim.

77. Kang, X.H., Wang, J., Wu, H. *et al.* (2010) A graphene-based electrochemical sensor for sensitive detection of paracetamol. *Talanta*, **81**, 754–759.

78. Stetter, J.R. and Li, J. (2008) Amperometric gas sensors – A review. *Chem. Rev.*, **108**, 352–366.

79. Clark, L.C. (1956) Monitor and control of blood and tissue oxygen tensions. *Trans. Am. Soc. Artif. Intern. Organs*, **2**, 41–48.

80. Clark, L.C., and Lyons, C. (1962) Electrode systems for continuous monitoring in cardiovascular surgery. *Ann. N.Y. Acad. Sci.*, **102**, 29–45.

81. Severinghaus, J.W. (2002) The invention and development of blood gas analysis apparatus. *Anesthesiology*, **97**, 253–256.

82. Falck, D. (1997) Amperometric oxygen electrodes *Curr. Separ.*, **16**, 19–22.

83. Fatt, I. (ed.) (1976) *The Polarographic Oxygen Sensor: Its Theory of Operation and its Application in Biology, Medicine, and Technology*, CRC Press, Cleveland, Ohio.

84. Kühl, M. and Seuckart, C. (2000) Sensors for in situ analysis of sulfide in aquatic systems, in *In Situ Monitoring of Aquatic Systems: Chemical Analysis and Speciation* (eds J. Buffle and G. Horvai), John Wiley & Sons, New York, pp. 121–159.

85. Gnaiger, E. and Forstner, H. (eds) (1983) *Polarographic Oxygen Sensors: Aquatic and Physiological Applications*, Springer-Verlag, Berlin.

86. Shibuki, K. (1990) An electrochemical microprobe for detecting nitric oxide release in brain tissue. *Neurosci. Res.*, **9**, 69–76.

87. Bedioui, F. and Villeneuve, N. (2003) Electrochemical nitric oxide sensors for biological samples – Principle, selected examples and applications. *Electroanalysis*, **15**, 5–18.

88. Privett, B.J., Shin, J.H., and Schoenfisch, M.H. (2010) Electrochemical nitric oxide sensors for physiological measurements. *Chem. Soc. Rev.*, **39**, 1925–1935.

89. Zhang, X.J., Kislyak, Y., Lin, H. *et al.* (2002) Nanometer size electrode for nitric oxide and S-nitrosothiols measurement. *Electrochem. Commun.*, **4**, 11–16.

90. Sundmacher, K., Rihko-Struckmann, L.K., and Galvita, V. (2005) Solid electrolyte membrane reactors: Status and trends. *Catal. Today*, **104**, 185–199.

91. Korotcenkov, G., Do Han, S., and Stetter, J.R. (2009) Review of electrochemical hydrogen sensors. *Chem. Rev.*, **109**, 1402–1433.

92. Riegel, J., Neumann, H., and Wiedenmann, H.M. (2002) Exhaust gas sensors for automotive emission control. *Solid State Ion.*, **152**, 783–800.

93. Ivers-Tiffee, E., Hardtl, K.H., Menesklou, W., and Riegel, J. (2001) Principles of solid state oxygen sensors for lean combustion gas control. *Electrochim. Acta*, **47**, 807–814.

14
Amperometric Enzyme Sensors

Amperometry is a suitable transduction method when coupled with an enzymatic reaction involving oxidation/reduction steps, such as those catalyzed by oxidase and dehydrogenase enzymes. The prototype is the amperometric glucose sensor, which was first introduced by Clark and Lyons in 1962 [1], and that was the starting point for further impressive advances and improvements.

During the evolution of the technology of oxidase-based amperometric sensors three main transduction methods were developed. These are summarized below with reference to oxidase-type enzymes. The general scheme for an oxidase-catalyzed reaction is:

$$S + O_2 + 2H^+ \xrightarrow{\text{Oxidase}} P + H_2O_2 \tag{14.1}$$

where S an P are the substrate and the product of the enzymatic reaction.

Hence, *first-generation* amperometric sensors relied on electrochemical monitoring of either oxygen depletion or hydrogen peroxide formation. In s*econd-generation sensors* oxygen is replaced by an artificial electron acceptor (that is, a redox mediator) that could be incorporated along with the enzyme in the biocatalytic layer [2]. In this way, the sensor becomes independent of any additional reagent in solution. Attempts at performing direct electron transfer from the enzyme to the electrode led to the *third-generation*, mediatorless sensors [2]. Current research is directed at improving sensor performance by rational use of new materials, particularly nanomaterials [3].

14.1 First-Generation Amperometric Enzyme Sensors

The most straightforward amperometric transduction relies on the detection of natural cosubstrates, such as oxygen in the case of oxidase catalyzed reactions. In its simplest form, such a sensor can be obtaining by entrapping an enzyme solution between the membrane of an oxygen probe and an additional semipermeable membrane (Figure 14.1A). Glucose, along with dissolved oxygen diffuses into the enzyme layer where the catalyzed substrate conversion results in a decrease in the oxygen concentration. Residual oxygen diffuses further to the electrode surface where its reduction generates the response current. The glucose concentration is indicated by the proportional decrease in the oxygen probe response relative to the response in the absence of the substrate. In more advanced versions, the enzyme was incorporated in structured materials using various immobilization methods [4,5]. Although successful, this transduction scheme suffers shortcomings because the oxygen probe is sensitive to uncontrolled variations in the oxygen concentration within the sample, particularly in the case of *in vivo* applications. In addition, solution pH should be controlled by a pH buffer system.

A second transduction alternative is based on hydrogen peroxide detection through its electrochemical anodic reaction (Figure 14.1B). Hydrogen peroxide, which is the product of the enzymatic conversion, travels to the anodically polarized electrode where it is oxidized to water. However, at the peroxide oxidation potential, endogenous interferents (such as urate, ascorbate and acetaminophen) also undergo anodic reactions resulting in a bias current. As some interferents are anions, their involvement can be avoided if the external membrane is made of a cation exchanger (such as Nafion). This membrane is permeable to small neutral molecules, such as glucose and oxygen, but rejects anionic species.

Another method for alleviating interference from electroactive concomitants relies on the lessening of the peroxide oxidation potential by means of a catalyst such as Prussian Blue [7]. This catalyst is amenable to integration into screen-print technology. However, Prussian Blue is not stable at pH > 7, which limits its application in conjunction with enzymes that display maximum activity in the alkaline region. Chemical, instead of electrochemical preparation of the Prussian Blue layer was reported to produce enhanced stability in this pH region.

An alternative catalyst is a peroxidase enzyme that can be integrated with an oxidase in the sensing layer. Hydrogen peroxide resulting from substrate conversion oxidizes the peroxidase and the resulting oxidized peroxidase then accepts electrons from the electrode either by direct transfer or via a mediator [8]. Thus, a glucose bienzyme sensor was obtained by integrating glucose oxidase and horseradish peroxidase immobilized on carbon nanotubes [9]. As usual, interference by anionic interferents was prevented by means of a Nafion external membrane. This sensor showed a high sensitivity for glucose detection at zero applied potential, which prevents interfences.

Chemical Sensors and Biosensors: Fundamentals and Applications, First Edition. Florinel-Gabriel Bănică.
© 2012 John Wiley & Sons, Ltd. Published 2012 by John Wiley & Sons, Ltd.

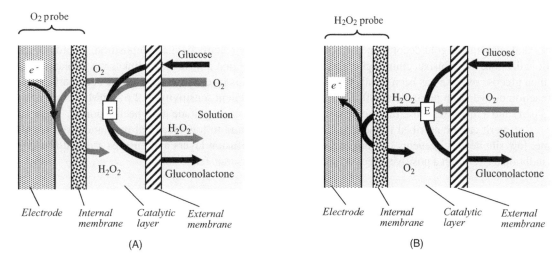

Figure 14.1 Configurations of oxygen-linked **(A)** and hydrogen peroxide-linked **(B)** amperometric enzyme sensors. Adapted with permission from [6]. Copyright 2002, A. Bănică and F.G. Bănică.

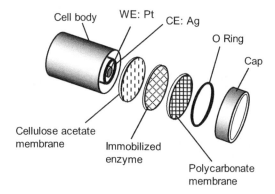

Figure 14.2 Typical arrangement of an amperometric sensor based on O_2 or H_2O_2 detection. WE: working electrode; CE: counter (reference) electrode. Adapted with permission from [6]. Copyright 2002, A. Bănică and F.G. Bănică.

Similar transduction methods have been applied with a series of oxidase enzymes, enabling the determination of a large variety of compounds such as glucose, lactose, amino acids, cholesterol and phenols [4,5]. The standard configuration of an amperometric sensor of this type is shown in Figure 14.2.

Despite the dependence on oxygen supply, the simplicity of this transduction method makes it still attractive for applications relying on new enzyme immobilization methods [10] or the use of nanomaterials [11,12]. An example of nanomaterial application is shown in Figure 14.3, which represents a network of carbon nanotubes grown in a

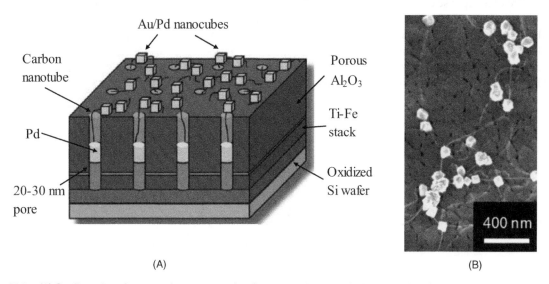

Figure 14.3 **(A)** Configuration of a nanocube-augmented carbon-nanotube network. The palladium deposit in each pore secures the electric contact of the carbon nanotube to the underlying titanium layer. **(B)** A field emission scanning electron microscopy picture showing carbon nanotubes protruding from the porous Al_2O_3 layer and attached Au/Pd nanocubes. Adapted with permission from [13]. Copyright 2009 American Chemical Society.

porous Al_2O_3 layer and connected to gold-covered palladium nanocubes [13]. Carbon nanotubes have been individually grown within the alumina pores by microwave plasma-enhanced chemical deposition. Formation of palladium nanocubes and subsequent gold decoration has been carried out by electrochemical deposition. In order to perform biofunctionalization, dithiobis(succinimidyl undecanoate) has been first attached as a linker via thiolate–gold bonding and then glucose oxidase has been covalently immobilized to this layer. Transduction has been performed by the anodic reaction of hydrogen peroxide at +0.5 V vs. Ag/AgCl. Excellent sensitivity (5.2 μA mM^{-1} cm^{-2}), detection limit (1.3 μM) and a linear response range extending over 4 orders of magnitude have been reported. These satisfactory figures of merit can be ascribed to the high enzyme loading and to low electrical resistance pathways. At the same time, low site density prevents overlapping of individual diffusion layers and ensures radial diffusion taking place at individual sites with a positive effect on the rate of the mass transfer.

Questions and Exercises (Section 14.1)

1 Give a short account of the main amperometric transduction methods used in amperometric enzyme sensors.
2 Write the chemical and electrochemical reactions involved in amperometric enzyme sensors of the first generation.
3 Comment on the application of nanomaterials in first-generation amperometric enzyme sensors.

14.2 Second-Generation Amperometric Enzyme Sensors

14.2.1 Principles

In order to alleviate the sensor's dependence on oxygen supply, an artificial electron acceptor (the mediator M_O) was introduced as an oxygen substitute [14,15]. The reaction scheme for such a sensor is:

$$S + E_O \underset{k_{-1}}{\overset{k_1}{\rightleftharpoons}} ES \xrightarrow{k_2} P + E_R \tag{14.2}$$

$$E_R + mM_O \xrightarrow{k_M} E_O + mM_R \tag{14.3}$$

$$M_R \xrightarrow{E^0} M_O + ne^- \tag{14.4}$$

So, in the first step, electrons are transferred from the enzyme active site (E_O) to the substrate (S) that is converted into a product (P), whereas the active site of the enzyme is converted into the reduced form (E_R). Next, the oxidized enzyme form is regenerated by electron transfer to the electron acceptor M_O. The resulting reduced form M_R undergoes further electrochemical reaction at the electrode, which regenerates the M_O species. The overall process consists therefore of a stepwise electron transfer from the substrate to the electrode with the enzyme and the mediator as intermediate relays (see also Sections 13.9.2 and 13.9.3). Transduction is performed by the reaction (14.4) that yields the response current. The process is independent of any additional reactant except the hydrogen ion that may be involved in substrate conversion.

The process in a mediated amperometric sensor is schematically outlined in Figure 14.4. Charge transport to the electrode is a key process that can occur by various mechanisms depending on the architecture of the biocatalytic

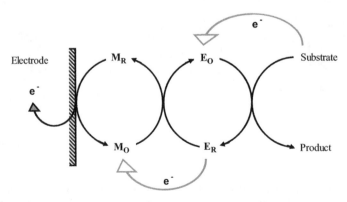

Figure 14.4 Schematics of the reactions in a mediated amperometric enzyme sensor. Gray arrows indicate intermolecular electron transfer; dark arrows show actual chemical reactions.

layer [2]. The mediator can, for example, be added to the sample solution and allowed to reach the biocatalytic layer by diffusion. Electrons are in this case conveyed to the electrode by M_R diffusion. However, such a sensor is not a true reagentless one. It is best to get the mediator entrapped along with the enzyme within the biocatalytic layer. Often, mediated sensors are prepared by incorporating the enzyme and the mediator within a composite conductive material, such as carbon paste or screen-printing carbon ink. In such a case, electron transport will be achieved by sequential electron transfer between mediator molecules and conductive particles. If the sensing part consists of only one or a few molecular layers, direct electron transfer from mediator to the electrode can be achieved.

The key problem in developing a mediated amperometric sensor is the selection of the mediator. The first condition a mediator should fulfill is the thermodynamic possibility of performing electron uptake from the enzyme. To this end, the standard electrode potential of the mediator must be positive relative to that of the enzyme. Investigation of redox processes in biological systems led to the discovery of a large variety of mediators characterized by various standard electrode potentials [16]. In addition, a suitable mediator should be very stable, compatible with the immobilization technique and, preferably, independent of hydrogen ions. Moreover, a mediator should exhibit as high a reaction rate as possible in both enzyme regeneration (reaction (14.3)) and mediator regeneration (reaction (14.4)) steps. Preferably, the electrochemical reaction (14.4) should be of reversible (fast) type. In order to drive the process, the electrode should be polarized at a potential close to the standard potential of the mediator redox couple. A mediator with a lower standard potential performs best because the interference of electrochemical active concomitants can be avoided. If the sensor is designed for *in vivo* applications, care should be exercised to prevent a toxic mediator from leaking out into the tissue under investigation.

A common interference in mediated amperometric sensors arises from dissolved oxygen, which competes with the mediator in reaction (14.3) and gives rise to a bias in the response current [17]. This interference can be alleviated if the mediator concentration is very high and the enzyme reoxidation (14.3) as well as the electron transport is very fast. Oxygen interference can be prevented by immobilizing the enzyme in a permselective material that hampers oxygen access to the enzyme [18]. In some instances, an oxygen-independent enzyme can be used instead of an oxidase. Thus, PQQ-dependent glucose dehydrogenase can be a convenient alternative to glucose oxidase in amperometric enzyme sensors.

The next sections introduce the main classes of mediators for amperometric enzyme sensors.

14.2.2 Inorganic Mediators

Ferricyanide ion ($[Fe(CN)]_6^{3-}$) was among the first mediators that were tested. It undergoes a one-electron reduction to ferrocyanide with no hydrogen ion participation. Such an anion-mediator can be entrapped by electrostatic attraction into a positively charged matrix (e.g., oxidized polypyrrole). Ferricyanide has also been included in screen-printed structures along with carbon nanotubes that enhance the rate of the electron transport [19]. Ferricyanide is suitable for disposable sensors because mediator leakage is not a problem when a very short operation time is expected.

Metal oxides, such as Fe_3O_4, Fe_2O_3, MnO_2, and SnO_2 behave as efficient mediators when incorporated in carbon paste or screen-printing ink. Best results have been obtained, however, with platinum-group metal oxides, particularly IrO_2, which allows the sensor to be operated at quite a low potential [20]. A tyrosinase-based catechol sensor was constructed by entrapping the enzyme and Fe_3O_4 nanoparticles in a chitosan matrix [21]. A high enzyme loading was achieved in this case as a result of the porous structure of chitosan. At the same time, the large specific area of the mediator nanoparticle secure a high reaction rate of the enzyme-mediator reaction.

14.2.3 Organic Mediators

Well-known organic redox couples can function as expedient mediators in redox enzyme-based biosensors. Thus, benzoquinone (Figure 14.5A) and its derivatives, as well as other diquinones (such as 1,2-naphthoquinone-4-sulfonate) have been successfully employed in mediated sensors. Many organic mediators have been selected from among the well-known class of redox indicators (redox dyes) that are used in redox titrimetry. To this class belong various derivatives of phenazine (Figure 14.5B) phenoxazine (Figure 14.5C) and phenothiazine (Figure 14.5D).

The redox reactions of such mediators involve hydrogen-ion transfer and, therefore, the standard electrode potential is pH dependent. Quinones undergo $2e^-$, $2H^+$ reactions, whereas the remaining compounds in Figure 14.5 experience $2e^-$, $1H^+$ reactions. Consequently, the sample pH should be controlled by means of a pH buffer. When selecting such a mediator, care should be exercised with regard to its possible instability under the effect of light or oxygen.

Another group of organic mediators are heteroatomic nonsaturated compounds whose ionic or radical forms are stabilized by charge delocalization over a conjugated π orbitals system. As a result of this structure, such compounds undergo H^+-independent redox reactions. Typical examples are tetracyanoquinodimethane (TCNQ) and tetrathiafulvalene (TTF) shown in Figure 14.6. The first compound is an electron acceptor, whereas the second compound acts as an electron donor. Both of them undergo a sequence of two single-electron-transfer reactions leading to a radical-ion first and a di-ion finally. The standard potential of each redox step depends on the nature of the solvent and counterions in solution. As a rule, only the first electron transfer step is involved in biosensor applications. The properties of TTF and its derivatives have been extensively reviewed recently [22].

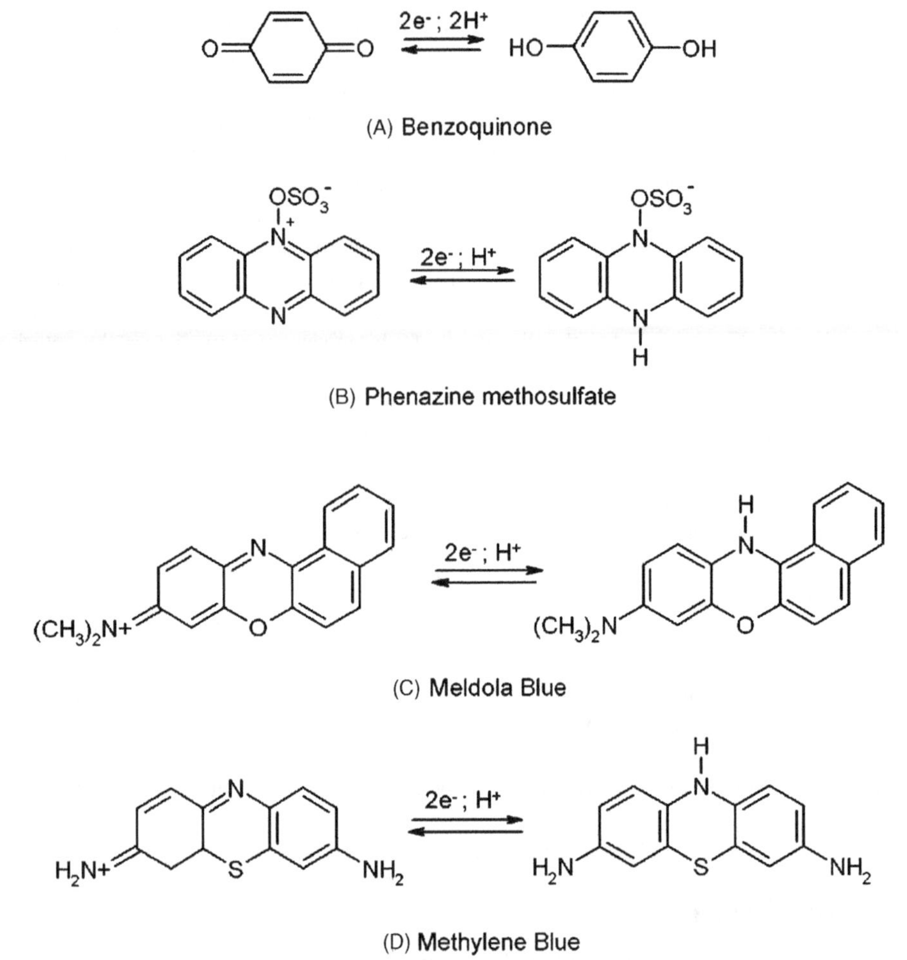

(A) Benzoquinone

(B) Phenazine methosulfate

(C) Meldola Blue

(D) Methylene Blue

Figure 14.5 Examples of H^+-dependent organic mediators used in amperometric enzyme sensors.

TTF-modified multiwalled carbon nanotubes are capable of facilitating electron transfers between the active sites of a redox enzyme and the electrode surface [23].

A TCNQ-modified graphite electrode was used for detecting the product of the acetylcholine esterase reaction in order to determine a series of insecticides by enzyme inhibition [24]. This method allows operation at an electrode potential lower than that required by the cobalt-phthalocyanine-mediated reaction [25].

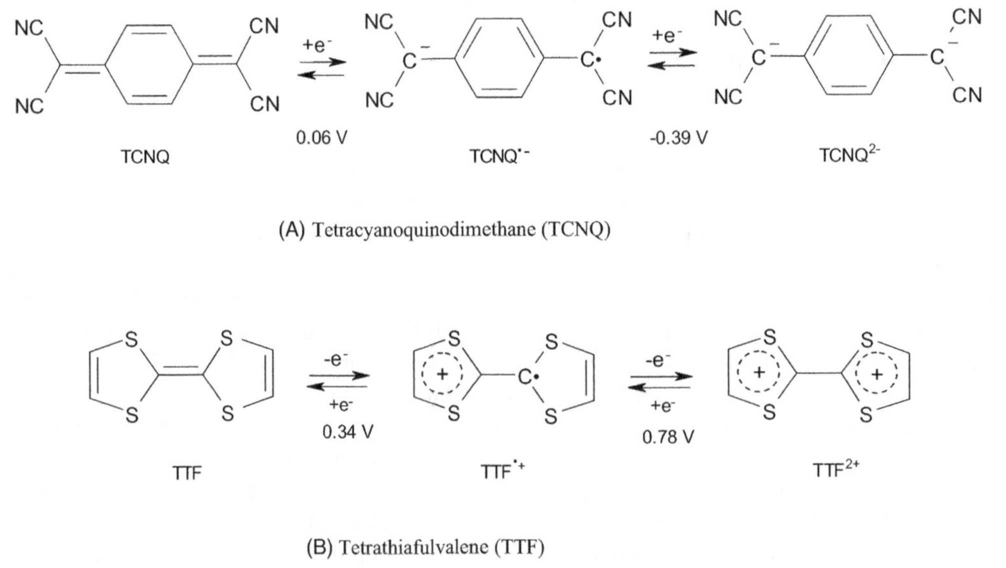

(A) Tetracyanoquinodimethane (TCNQ)

(B) Tetrathiafulvalene (TTF)

Figure 14.6 Structure and electrochemical reactions of tetrathiafulvalene (TTF) and tetracyanoquinodimethane (TCNQ). Specific potentials (vs. Ag/AgCl) have been determined in acetonitrile.

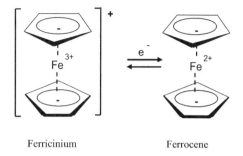

Ferricinium   Ferrocene

Figure 14.7 Structure and redox reaction of ferrocene. Note that each cyclopentadienyl ligand bears a negative charge (π electrons) delocalized over the whole ring.

14.2.4   Ferrocene Derivatives as Mediators

Among the large variety of investigated mediators, ferrocene derivatives, first introduced by Hill *et al.* [26], performs almost ideally. Ferrocene (Figure 14.7) is a charge-transfer complex of iron with the cyclopentadienyl aromatic anion. The chemical bonding in this compound results from the partial electron transfer from the ligand π orbital to the empty orbitals at the iron ion. Electrochemical interconversion of the reduced and oxidized form proceeds reversibly with a half-wave potential of about +165 mV vs. SCE. Both solubility and the standard potential can be adjusted by adding a substituent to the ligand. Thus, electron-attracting groups, (such as carboxylate or aromatic groups) increase the standard potential, whereas electron-repelling groups (such as alkyls or amines) shift it to more negative values. Although ferrocene itself is soluble in water, solubility can be lessened by hydrophobic substituents. Ferrocene mediation was used in the first glucose sensor designed for home use [27].

Construction of a ferrocene-mediated sensor is quite straightforward. A water-insoluble ferrocene (such as 1,1′-dimethyl ferrocene) is first deposited on the surface of a carbon electrode by evaporation of its solution in toluene. Next, the enzyme is immobilized by a suitable method, such as covalent binding or adsorption. Finally, the sensing layer can be coated with a perm-selective membrane. Alternatively, both the enzyme and mediator can be incorporated in a carbon paste or in a screen-printed structure [28]. Screen-printing technology is particularly appropriate for mass production of amperometric enzyme sensors. For example, the configuration in Figure 14.8 has been employed in glucose or ethanol sensors using pertinent PQQ-dependent dehydrogenases with 4-ferrocenylphenol as the electron-transfer mediator. The carbon ink layer is first covered with a perm-selective film formed by electrochemical polymerization of 2,4,7-trinitro-9-fluorenone (TNF). The mediator is adsorbed from acetone onto the carbon ink and, thereafter, the enzyme is immobilized by crosslinking with glutaraldehyde. The poly(TNF) coating prevents electroactive concomitants from reaching the electrode which improves the detection limit.

Recent research work in this field focuses on ferrocene integration with nanostructured materials [30] or nanoparticles such as gold [31] or carbon nanotubes [32]. As an example, Figure 14.9 shows a biocatalytic layer that includes glucose oxidase covalently linked to a gold electrode and to a gold nanoparticle. This device functions as a platform for appending ferrocene via a thiol bond in the close proximity of the enzyme molecule. In this configuration, the rate of electron transfer from the reduced enzyme to the electrode via the mediator is enhanced [33].

A carbon-nanotube-based glucose sensor was obtained by means of carbodiimide coupling of ferrocene monocarboxylic acid to carbon nanotubes previously derivatized by an amino-silane. The biocatalytic layer was then prepared by casting onto a graphite electrode a suspension of derivatized nanotubes in a chitosan and glucose oxidase solution [32]. The response graph of such a sensor is shown in Figure 14.10. Note the negative bias induced by dissolved oxygen that, as mentioned above, can occur with mediated sensors.

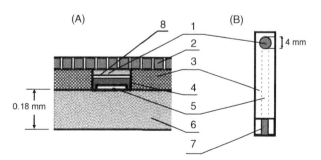

Figure 14.8 A screen-printed amperometric enzyme sensor. (A) cross section; (B) general view. 1) enzyme layer; 2) semipermeable terylene film; 3) insulating film; 4) carbon ink layer; 5) silver track; 6) polyethylene terephthalate film; 7) contact zone; 8) Perm-selective poly (TNF) film. Adapted with permission from [29]. Copyright 2001 Elsevier.

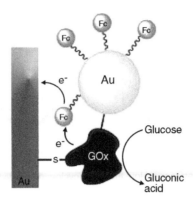

Figure 14.9   An electrically contacted Au nanoparticle–glucose oxidase composite electrode. 6-Ferrocenyl hexanethiol was attached to the nanoparticle in order to provide electron-transfer mediation. Reproduced with permission from [33]. Copyright 2008 John Wiley and Sons.

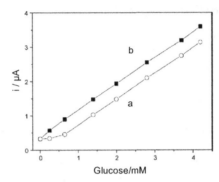

Figure 14.10   Calibration graph for a glucose sensor mediated by carbon-nanotube-attached ferrocene in the presence (a) and in the absence of oxygen (b). Adapted with permission from [32]. Copyright 2008 Elsevier.

14.2.5   Electron-Transfer Mediation by Redox Polymers

Initial applications of mediated electron transfer relayed on mediators as independent molecules randomly distributed within the biocatalytic layer. More efficient mediating was subsequently demonstrated by anchoring mediator moieties to a polymer backbone that allows, in addition, the enzyme to be immobilized by covalent bonding [34–36]. A typical structure of this type is shown in Figure 14.11A. It can be built up starting from poly(vinylpyridine). First, a fraction of the pyridine residues in the polymer is used for attaching by coordination an osmium complex (Figure 14.11B) to yield a *redox polymer*. Further, remaining pyridines are derivatized so as to allow enzyme immobilization by covalent bonding. The resulting macromolecular structure (Figure 14.11A) includes the enzyme and the osmium complex mediator firmly attached to the backbone. In addition, it contains charged moieties that impart hydrophilicity and allow hydration of the polymer network thus facilitating the access of substrate to the enzyme. The resulting material is termed a *redox hydrogel*.

In such a structure, electron transfer from enzyme to mediator is facilitated by the close proximity of the latter to the active site of the enzyme, although direct contact is prevented by steric factors. Further, electrons can be transferred directly from the mediator to the electrode within a few polymer monolayers. However, in a thicker system, some kind of electron diffusion must occur in order to achieve electrical communication between the redox centers and the electrode. This can be achieved by percolative electron hopping, long-range chain motion and short-range chain motion. Hence, electron diffusion is to a great extent dependent on the flexibility of the spacer chain between

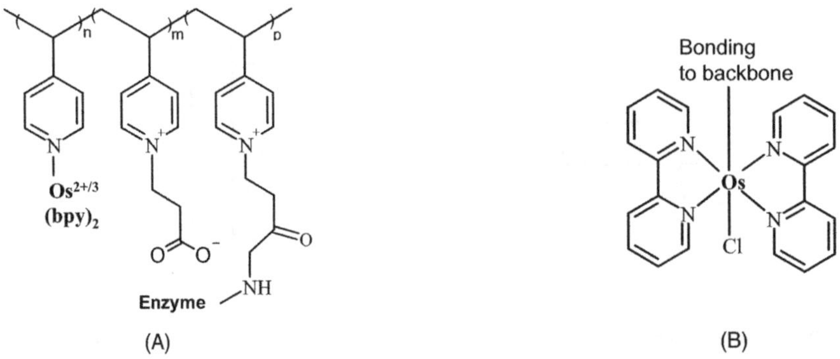

(A)   (B)

Figure 14.11   General representation of enzyme wiring. (A) Polymer backbone and functional side groups. (B) A typical redox center/mediator, the [Os(bipirydine)Cl]$^{2+/3+}$ complex.

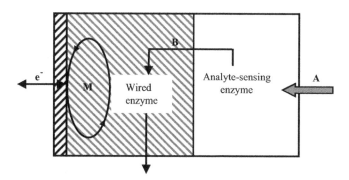

Figure 14.12 A bienzyme sensor based on enzyme wiring. The enzyme in the first layer converts the analyte (A) into the intermediate product B that is further converted by the wired enzyme in order to generate the response current.

the mediator and the backbone. Electron diffusion can be limited by the diffusion velocity of counterions required for preserving electroneutrality in the system. The redox polymer accomplishes therefore a kind of electrical wiring between enzyme and electrode. Hence, it is said that an enzyme integrated with a redox polymer is a *wired enzyme*.

Other polymers, such as polysiloxanes and polyethylene oxide have been considered in order to construct more flexible structures with enhanced hydrophilicity. Also, various mediators such as ferrocene [35], tetrathiafulvalene, benzoquinone and viologen [36] have been tested in order to exploit their particular standard electrode potentials and their compatibility with particular enzymes.

Various applications of redox hydrogel enzyme sensors for determining oxidase and dehydrogenase substrates have been reported [36,37]. It appears that redox hydrogels represent a versatile alternative for preparing biocatalytic layers for amperometric enzyme sensors. Their use guarantees a high mediator concentration and very fast electron transport, although this is still much slower than substrate diffusion.

Redox hydrogels are also suitable for manufacturing multienzyme sensors, as demonstrated in Figure 14.12. In this design, the first enzyme layer converts the analyte A into a product B that acts as the substrate to a second enzyme that is wired to the electrode through a redox polymer. This approach allows the determination of an analyte that is not involved in an enzyme-catalyzed redox reaction.

Interestingly, osmium redox polymers can also be applied to achieve efficient electrical communication between the gram-positive micro-organism *Bacillus subtilis* and an electrode in order to develop an amperometric succinate sensor [38].

14.2.6 Sensing by Organized Molecular Multilayer Structures

Preparation methods highlighted above lead as a rule to three-dimensional structures with the enzyme and mediator randomly distributed. Close control over the properties of the sensing layer can be accomplished by assembling successive molecular layers in order to integrate the component into a well-organized structure [10,39–41] (see also Section 5.6.3). A sensing layer prepared by electrostatic, sequential self-assembly is depicted in Figure 14.13. In the first step, a

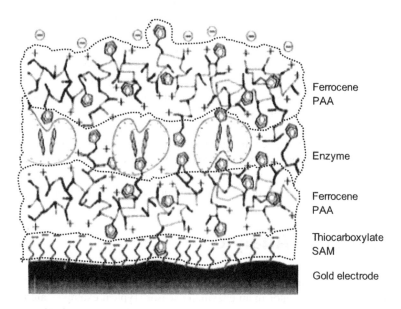

Figure 14.13 A wired enzyme biocatalytic layer built up by successive electrostatic assembly of various monolayers. Double pentagons represent ferrocene. Note the protrusion of ferrocene mediator moieties into the enzyme layer. PAA stands for poly(allylamine). Adapted with permission from [42]. Copyright 1997 Copyright American Chemical Society.

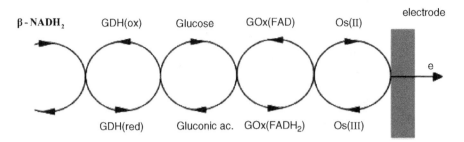

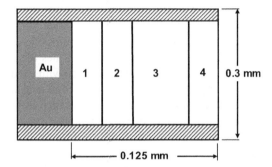

Figure 14.14 Configuration of a glucose sensor with substrate amplification assembled by the layer-by-layer method. GDH-glucose dehydrogenase; GOx-glucose oxidase. Reproduced with permission from [40]. Copyright 2000 The Royal Society of Chemistry.

Figure 14.15 An implantable, multienzyme glucose electrode. Adapted with permission from [44]. Copyright 1994, American Chemical Society.

negatively charged thiocarboxylate self-assembled monolayer was appended to a gold electrode surface. Next, the mediator, in the form of a cationic ferrocene polymer (poly(allylamine)) was attached to the first layer by electrostatic assembly. Finally, the negatively charged enzyme was allowed to form a monolayer over the second film. Such a bilayer film acts as the sensing element as it includes both the enzyme and a mediator. In order to increase the enzyme loading, successive, alternating mediator and enzyme layers can be assembled by repeating the previous procedure.

In the previous example, adhesion of successive layers was imparted by electrostatic interaction. The preparation method is very simple, but the resulting structure is somewhat sensitive to changes in pH and ionic strength that may causes electric charges to be altered. More robust catalytic films can be obtained by forming covalent bonds between consecutive layers. The usual methods and reagents for covalent immobilization of enzymes can be employed to this end. As an alternative, affinity reactions can be used for performing self-assembly of enzyme and redox polymer layers.

The layer-by-layer method outlined above allows a rigorous control of enzyme loading and also makes possible straightforward preparation of multienzyme layers. An example of a multienzyme sensor performing amplification by substrate recycling is presented in Figure 14.14. In this system, glucose oxidase converts the substrate (glucose) to gluconic acid that is recycled to glucose by a β-NADH$_2$-assisted glucose dehydrogenase reaction. Wiring of glucose oxidase was achieved by an Os$^{II//III}$ functionalized polymer. As usually, substrate recycling results in greatly enhanced sensitivity.

Incorporation of carbon nanotube–enzyme conjugates in redox hydrogels through a layer-by-layer self-assembly process leads to enhanced current density that gives better sensitivity relative to the film containing no nanotubes [43].

A redox polymer has been used in a flexible, implantable multienzyme sensor for the determination of glucose in blood. As shown in Figure 14.15, this sensor consists of several distinct layers at the surface of a recessed gold electrode. The first layer forms the actual sensing part and includes glucose oxidase entrapped in a redox polymer. The third layer contains a peroxidase coimmobilized with lactose oxidase and is intended to destroy common interferents in blood, such as urate, ascorbate and acetaminophen. In this layer, lactate (typically present in blood at about 1 mM) reacts with oxygen to form pyruvate and hydrogen peroxide that oxidizes the interferents (by peroxidase catalysis) and is itself reduced to water. The intermediate layer 2 consists of a polycationic polymer that secures insulation of the enzyme in the first layer. The outermost layer 4, which consists of tetraacrylated poly (ethylene oxide), provides biocompatibility.

14.3 The Mediator as Analyte

The key role played by the mediator in an amperometric enzyme sensor suggests a straightforward method for detecting a compound that can function as mediator in an enzymatic sensor. Typical examples are represented by phenolic compound sensors based on redox enzymes such as tyrosinase, laccase and peroxidase [45]. As shown in Figure 14.16, the analyte (a phenolic compound), along with an oxidizing agent, diffuses from the solution to the enzyme layer

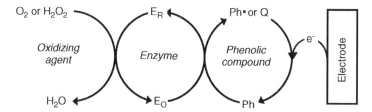

Figure 14.16   Mechanism of an amperometric phenol sensor. E_R and E_O are the reduced and oxidized forms, respectively, of the enzyme. Ph, Ph* and Q are the phenolic molecule, its phenoxy radical and its quinone form, respectively. Adapted with permission from [45]. Copyright 1995, Elsevier.

where the enzyme reaction, in conjunction with the electrode reaction recycles the analyte between two oxidation states. The oxidizing agent is H_2O_2 in the case of peroxidase or O_2 in the cases of laccase or tyrosinase enzyme.

In contrast to this configuration, which is based on enzymatic oxidation reactions, dehydrogenases (such as cellobiose dehydrogenase or PQQ-dependent glucose dehydrogenase) allow the development of reduction–based sensors [46]. In this case, the phenolic analyte is first oxidized at the electrode and the product then accepts electrons via the enzyme from a suitable substrate. Whereas the oxidation-based sensor operates at a relatively low potential (-0.1 to 0 V vs. Ag/AgCl), the reduction-based version needs a relatively high anodic polarization (0.3–0.4 mV vs. Ag/AgCl) in order to perform electrochemical oxidation of the analyte. Consequently, the second version is subjected to the risks of interference by electroactive concomitants.

Substrate recycling can in some cases be assisted by a reducing agent, such as ascorbic acid. This task can also be performed by a second enzyme that converts the product of the electrochemical reaction into the substrate of the main enzyme. Thus, a phenol sensor was constructed by coimmobilization of poly-phenol oxidase and horse radish peroxidase by electrochemical polymerization [47]. The first enzyme converts phenol to *o*-quinone that can enter an amplification cycle involving electrochemical reduction to catechol, followed by peroxidase-catalyzed conversion of catechol to *o*-quinone. A theoretical model for such a bienzyme sensor has recently been worked out [48]. As some phenolic compounds are insoluble in water, they can be extracted into an organic solvent and then determined in the extract by means of a properly designed sensor [49,50].

Questions and Exercises (Sections 14.2–14.3)

1 What is a redox mediator and how does it function in an amperometric enzyme sensor?
2 What are the advantages of mediator-based amperometric enzyme sensors?
3 What kind of interferences can occur in this type of enzyme sensor and how can such interferences be alleviated?
4 What kind of inorganic mediators are suitable for applications in amperometric enzyme sensors?
5 Review the main kinds of organic compounds used as mediators in amperometric enzyme sensors. What precautions are needed when using organic mediators?
6 Outline some differences between TCNQ and TTF on the one hand and the mediators shown in Figure 14.5 on the other hand, as far as their redox behavior is concerned.
7 Give an overview of ferrocenes as mediators in amperometric enzyme sensors and outline their advantages when compared with other types of mediator.
8 Give an account of the configuration of screen-printed amperometric enzyme sensors of the second generation.
9 What benefits can be obtained by including conducting nanomaterials in the structure of mediated amperometric enzyme sensors?
10 What are redox polymers and how can they be used in the construction of amperometric enzyme sensors? What are the advantages of redox polymer-based sensing layers?
11 Prepare a short overview of molecular layer structures used in amperometric enzyme sensors.
12 What kinds of analytical application result from the utilization of the analyte itself as a mediator in amperometric enzyme sensors?

14.4 Conducting Polymers in Amperometric Enzyme Sensors

Conducting polymers can be used for various purposes in the design of enzyme amperometric sensors [51–54]. A simple application involves conducting polymers (such as polypyrrole or polyaniline) as entrapment matrix for

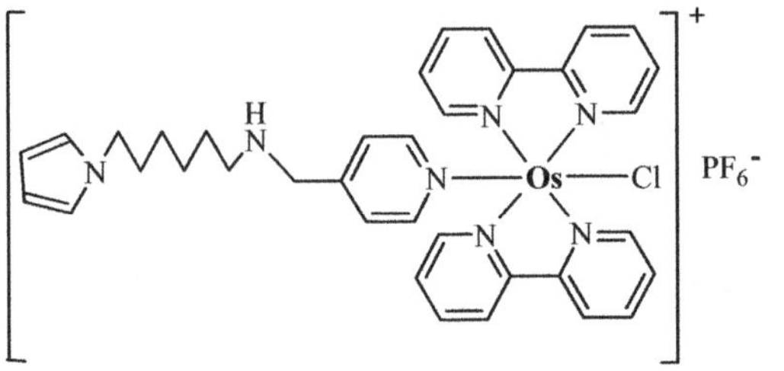

Figure 14.17 An osmium-bipyridine-derivatized pyrrole monomer used in the synthesis of a redox hydrogel by copolymerization with pyrrole.

enzyme immobilization. If amino groups are attached to the polymer backbone, they can act as a binding site in enzyme immobilization by covalent linking.

A more advanced application resorts to conducting polymers to form the backbone in the structure of a redox hydrogel. For example, PQQ-dependent glucose dehydrogenase was included in a three-dimensional polymeric structure along with osmium bipyridine by electrochemical copolymerization of pyrrole and the monomer in Figure 14.17 in the presence of the enzyme [55]. Enzyme immobilization was achieved by the reaction of the radical cations formed during the polymerization reaction with nucleophilic functions on the enzyme surface.

Certain multifactor enzymes (e.g., PPQ and heme-c-based alcohol dehydrogenase) are able to perform direct electron transfer to the polypyrrole backbone. Therefore, enzyme sensors using a polypyrrole coated electrode can be designed with no need for a redox mediator. However, the number of enzymes that function in this way is limited.

Questions and Exercises (Section 14.4)

1 Review the properties of conducting polymers used in amperometric enzyme sensors.
2 What functions can a conducting polymer perform in amperometric enzyme sensors?

14.5 Direct Electron Transfer: 3rd-Generation Amperometric Enzyme Sensors

Achieving direct electron transfer between the prosthetic group of an oxidoreductase and the electrode brings about a major simplification in the sensor configuration by precluding sensor dependence on a cosubstrate or a mediator. The next sections present several methods that proved successful in this respect

14.5.1 Conducting Organic Salt Electrodes

Direct electron transfer from redox enzymes can be achieved with electrodes consisting of conducting organic salts. Such compounds are formed by the combination of an electron donor and an electron acceptor such as tetrathiafulvalene (TTF) and tetracyanoquinodimethane (TCNQ), respectively (Figure 14.6). The TTF·TCNQ salt can be prepared by direct reaction of the above compounds in acetonitrile. The product is a planar molecule with delocalized π orbitals extending both above and below the molecular plane. Structurally, the solid TTF·TCNQ salt consists of segregated stacks of cations and anions of the donor and the acceptor, respectively. Partial electron transfer ensures a relatively stable bonding between alternating donor–acceptor stacks. Compounds of this type belong therefore to the class of charge-transfer compounds and display semiconductor properties.

The electrocatalytic properties of organic semiconductor salts towards various flavoenzymes was first reported in 1981 [56,57]. An unequivocal interpretation of this effect is not yet available, but it may be associated with the exceptionally strong protein adsorption at the solid surface prompted by electrostatic attraction between the charged protein molecule and the intrinsic electric charge at the electrode surface.

The TTF·TCNQ electrode can be prepared in several ways [58]. The pure compounds can be used either as a single crystal or as pressed pellets. Alternatively, the finely dispersed organic semiconductor mixed with an inert oil yields a conducting paste that can be manipulated in the same way as carbon paste electrodes. Slow evaporation of a salt solution applied to the surface of a graphite electrode (drop coating) is the most straightforward method for preparing conducting salt-modified electrodes.

The potential window of a conducting salt electrode is determined by its electrochemical reactions. In neutral solutions, at a potential more positive than $+0.4$ V vs. SCE, the TCNQ$^{\cdot-}$ is oxidized to neutral TCNQ that is water insoluble and remains on the electrode surface. At potentials more negative than about -0.2 V, TTF and TCNQ$^{\cdot-}$ are

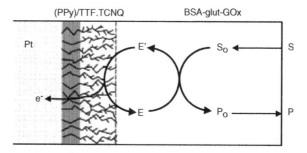

Figure 14.18 Schematics of a disposable glucose electrode based on TTF·TCNQ integrated in an insulating poly(pyrrole) film. Glucose oxidase (GOx) was immobilized by crosslinking with glutaraldehyde and bovine serum albumin (BSA). Reproduced with permission from [60]. Copyright 2002, American Chemical Society.

released. Therefore, the working potential should be adjusted between these limits and not be allowed to drift outside it; otherwise the electrode surface may become completely covered with insoluble decomposition products.

The feasibility of conducting salt electrode-based enzyme sensors was demonstrated with various flavoenzymes such as glucose oxidase, D-amino acid oxidase and L-amino acid oxidase that enables the determination of glucose or amino acids, respectively. The enzyme was immobilized by adsorption onto the electrode surface. A wider linear response range is obtained by coating the sensing layer with a dialysis membrane that imposes a diffusion barrier. Screen printing is a suitable technology for mass fabrication of sensors of this type, which are characterized by very good long-term stability [59].

TTF·TCNQ integrated in a nonconducting poly(pyrrole) film has been used for fabricating a disposable glucose sensor (Figure 14.18) [60]. TTF · TCNQ crystals have been grown through the insulating polymer film yielding a highly branched structure with a very large effective area. Oxygen interference was detected only at glucose concentrations below 1 mM. Electroactive species (such as ascorbic acid, uric acid, cysteine, paracetamol, and acetaminophen) gave very low or undetectable signals. This favorable behavior was ascribed to the protecting effect of the negatively charged polymer that effectively rejects anionic interferents.

Conducting organic salt electrodes allowed the development of amperometric enzyme sensors for various analytes of biological interest [61].

14.5.2 Direct Electron Transfer with FAD-Heme Enzymes

An interesting direct electron-transfer scheme was recently reported for the case of a cellobiose dehydrogenase-based lactose sensor [62]. Structurally, cellobiose dehydrogenase includes two prosthetic groups: a FAD and a heme cytochrome. The first catalytic step involves lactose conversion catalyzed by the FAD moiety (Figure 14.19). FAD is recycled by an internal electron transfer to the heme group that further conveys electrons to the electrode and thus achieves the detection/transduction step. Remarkably, this scheme allows for direct electron transfer to the usual electrode materials such as gold.

The particular structure of cellobiose dehydrogenase has been cleverly exploited in the design of a fuel cell-type glucose biosensor [63]. A fuel cell is a kind of galvanic cell that generates a current by spontaneous electrochemical oxidation of a conventional fuel (such as glucose, in this case). In this glucose biosensor the anode reaction consists of glucose dehydrogenation by a scheme similar to that in Figure 14.19, using cellobiose dehydrogenase as catalyst. The cathode reaction is oxygen reduction catalyzed by bilirubin oxidase. This fuel-cell can be operated either in a synthetic solution (phosphate buffer) or in human serum and is promising for *in vivo* applications.

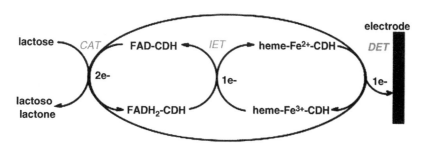

Figure 14.19 Schematics of reactions taking place at cellobiose dehydrogenase/electrode interface in a lactose biosensor. CAT: enzyme-catalyzed reaction; IET: intramolecular electron transfer; DET: electrochemical reaction. Reproduced with permission from. [62]. Copyright 2006, American Chemical Society.

14.5.3 Achieving Direct Electron Transfer by Means of Nanomaterials

Although the mediated electron transfer method proved convenient in performing enzyme communication with the electrode, the geometry of the enzyme–mediator–electrode system at the molecular level is not optimal, which renders the electron transfer relatively inefficient. Close control of the architecture of the biocatalytic layer can be achieved by integrating the enzyme with nanoparticles in order to carry out direct electron transfer between the electrode and the enzyme's active site [64,65]. An application of gold nanoparticles is demonstrated in Figure 14.20. According to path A in this scheme, gold nanoparticles are functionalized with a FAD derivative and this is allowed next to interact with apo-glucose oxidase in order to reconstitute the whole enzyme. Thus, a nanoparticle-enzyme conjugate is obtained. This adduct is further bridged with a dithiol-modified gold electrode via the gold nanoparticle. In an alternative approach (B), nanoparticle–FAD conjugates have first been assembled on the dithiol-modified gold electrode and next the enzyme was reconstituted on the resulting layer. Both methods led to a wired glucose oxidase monolayer that is characterized by a very fast electron transfer and responds linearly to glucose concentration over a broad range in the mM region, with no interference from dissolved oxygen.

Carbon nanotubes have also proved suitable for achieving direct electron transfer between an enzyme and the electrode. For example, Figure 14.21 displays a method for preparing a glucose sensor including carbon naotubes for enzyme wiring [67]. In this approach a gold electrode has been modified by coating with a self-assembled monolayer of cysteamine (A) Next, 120-nm single-walled carbon nanotubes have been covalently attached by amide bonding involving carboxyl terminal groups (B). Further, two different immobilization procedures have been employed for

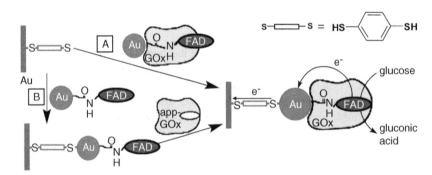

Figure 14.20 Self-assembly of a glucose oxidase–gold nanoparticle conjugate in order to achieve direct electron transfer to the underlying gold electrode. Inset upper right: the dithiol used as linker. Adapted with permission from [66]. Copyright 2003, The American Association for the Advancement of Science.

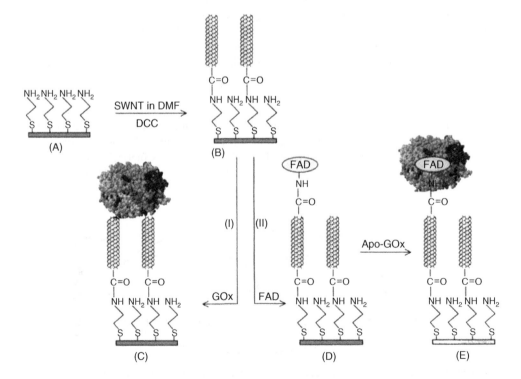

Figure 14.21 Main steps in the fabrication of a modified gold electrode for achieving glucose oxidase wiring via carbon nanotubes. Adapted with permission from [67]. Copyright 2005 Wiley-VCH Verlag GmBH & Co. KGaA.

appending glucose oxidase to the upper end of the nanotubes. In alternative I, glucose oxidase itself was attached to the carbon nanotube layer (C). Procedure II involves attaching the FAD prosthetic group to carbon nanotubes (D), followed by reconstitution of the full enzyme by allowing the apoenzyme to interact with the FAD-modified electrode (E).

Both the above examples illustrate a general principle in enzyme immobilization and conjugation with nano-materials, namely reconstitution of the enzyme by an affinity reaction between the apoenzyme and the catalytic molecular unit [68].

In conclusion, direct electron exchange between enzyme and electrodes can be achieved using different technologies and materials. This approach avoids certain complications that arise with other transduction methods. It is worth mentioning that copper-containing proteins [69] and ligninolytic redox enzymes [70] have been extensively studied from this standpoint and show a remarkable potential for application in amperometric enzyme sensors based on direct electron transfer.

Questions and Exercises (Section 14.5)

1 What kind of electrode material allows performing electron transfer from the active site of a redox enzyme? Review the chemical structures and electrochemical behavior of this material.
2 Review briefly the fabrication methods for electrodes composed of conducting organic salts.
3 What particular features of cellobiose dehydrogenase allow direct electron transfer in sensors based on this enzyme?
4 What kinds of nanomaterial are useful for performing direct electron transfer from the active site of a redox enzyme? Outline the structural features of the enzyme–nanomaterial conjugate that allows this task to be carried out.

14.6 NAD/NADH$^+$ as Mediator in Biosensors

Many dehydrogenases use NAD(P)$^+$/NAD(P)H as a cosubstrate. Hence, general transduction methods in conjunction with such enzymes are based on the detection of these cofactors [71,72]. Electrochemical oxidation of NADH to NAD$^+$ is, at first sight the obvious choice, but in fact at metal or carbon electrodes this reaction occurs with a large overvoltage. However, NADH oxidation occurs with a low overvoltage at organic conductor salt electrodes, which enables the development of convenient amperometric sensors such as, for example, an ethanol sensor [73]. In such a sensor, the NADH/NAD$^+$ couple functions as a mediator that conveys the electron from the enzyme to the electrode by free diffusion.

At the same time, much effort has been devoted to produce sensors based on mediated electron transfer from NADH to metal or carbon electrodes in what can be called a double-mediation system (Figure 14.22). In this approach, the electron exchange between NADH and the electrode is effected by a redox mediator. A large number of mediators have been tested to this end looking for long-term stability and low operating potential. Among the most convenient, Meldola Blue, N-methylphenazinium, TTF and TCNQ can be mentioned.

Enzymatic reactions involving NAD$^+$/NADH have also proved useful for detecting an inorganic cosubstrate such as the ammonium ion. An ammonium sensor has been obtained using the glutamate dehydrogenase enzyme (GLDH), 2-oxoglutaric acid as substrate and Meldola Blue (MB) as mediator [74]. The reaction sequence in this sensor is:

$$2-\text{oxoglutarate} + \text{NH}_4^+ + \text{NADH} \xrightarrow{\text{GLDH}} \text{glutamate} + \text{NAD}^+ + \text{H}_2\text{O} \qquad (14.5)$$

$$\text{NADH} + \text{MB}^+ \rightarrow \text{MBH} + \text{NAD}^+ \qquad (14.6)$$

$$\text{MBH} \rightarrow \text{MB}^+ + 2e^- + \text{H}^+ \qquad (14.7)$$

This sensor has been fabricated by screen printing for the purpose of field analysis applications.

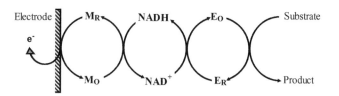

Figure 14.22   Scheme of a mediated NADH/NAD$^+$ sensor.

Questions and Exercises (Section 14.6)

1 What is NAD/NADH$^+$ and what is its role in enzymatic reactions?
2 Write the chemical and electrochemical reactions involved in a NAD/NADH$^+$-mediated enzyme sensor.
3 How can electron exchange between NAD/NADH$^+$ and an electrode be achieved for transduction purposes?

14.7 Summary

Amperometric enzyme sensors are based on redox enzymes that are catalysts in the redox reactions of analyte substrates. In order to perform electrochemical transduction, the electron-transfer process involved in the enzymatic reaction should be coupled in some way with an electrode in order to obtain a current. There are three main approaches to amperometric transduction in amperometric enzyme sensors. If a natural cofactor or a product is electrochemically active, amperometric monitoring of one of these components provides the response current and sensors based on this principle belong to the first generation of amperometric enzyme sensors. In this design, the sensor is dependent on the natural cofactor, whose concentration must be controlled. A reagentless sensor is obtained using an artificial mediator as cofactor. By incorporating the mediator within the sensing layer, the sensor becomes independent of any possible reagent, which makes the operation much more facile. As the mediator is recycled by the electrochemical reaction, it is not consumed during the run. Sensors of this type are called second-generation amperometric enzyme sensors. A great number of organic and inorganic mediators have been tested. Among them, metal complexes such as ferrocenes and osmium complexes have proved to be the most convenient. This is due to the fact that their electrochemical reactions do not involve hydrogen ions and, consequently, are not pH-dependent. The most advanced design in mediated sensors is based on redox polymer layers that incorporate both the enzyme and the mediator, the last being covalently attached to the polymer backbone. Well-structured sensing films can be obtained by self-assembly using the layer-by-layer preparation method.

In both first- and second-generation sensors, electron exchange between electrode and the enzyme active site is achieved indirectly, through a natural or artificial cofactor. A great simplification in the sensor design is achieved if the electron exchange between electrode and the enzyme occurs directly, with no intermediates. This design is a third-generation amperometric enzyme sensor. Direct electron transfer was first achieved using particular organic semiconductors as electrodes. In a more advanced approach, conducting nanoparticles are conjugated with the enzyme to promote direct electron exchange with the electrode. Certain enzymes are able to perform direct electron transfer to conducting polymers deposited at the electrode surface. Therefore, in third-generation sensors, there are no low molecular weight compounds acting as electron conveyors between the enzyme and the electrode.

From the standpoint of mass production, the most convenient technology is that based on screen printing [75,76]. This technology enables fabrication of cheap sensors suitable for single-use applications. As the running time is very short, simple immobilization by adsorption is preferred since the leaking out of the components of the sensing layer is not critical under such conditions. Immobilization by entrapment or electrochemical polymerization is the suitable alternative for fabricating more rugged sensors. Planar biosensors produced by screen printing are suitable for a wide range of applications, from "drop-on assay," which can be useful in field analysis, to automated analysis methods such as flow-injection analysis.

The most impressive application of amperometric enzyme sensors is the glucose sensor that is widely used in diabetes monitoring and also in a series of industrial applications, [77,78]. Amperometric enzyme sensors for other analytes have also been developed for various applications in medicine, biotechnology, food and drinks industry [79] and environment monitoring.

As enzymes are widely used as labels in electrochemical immunosensors and nucleic acid sensors, the basic principles of amperometric enzyme sensors form the basis for the development of various types of sensor based on affinity reactions or nucleic acid hybridization.

References

1. Clark, L.C. and Lyons, C. (1962) Electrode systems for continuous monitoring in cardiovascular surgery. *Ann. N. Y. Acad. Sci.*, **102**, 29–45.
2. Habermuller, L., Mosbach, M., and Schuhmann, W. (2000) Electron-transfer mechanisms in amperometric biosensors. *Fresenius J. Anal. Chem.*, **366**, 560–568.
3. Pingarron, J.M., Yanez-Sedeno, P., and Gonzalez-Cortes, A. (2008) Gold nanoparticle-based electrochemical biosensors. *Electrochim. Acta*, **53**, 5848–5866.

4. Scheller, F.W. and Schubert, F. (1992) *Biosensors*, Elsevier, Amsterdam.
5. Kauffmann, J.M. and Guilbault, G.G. (1992) Enzyme electrode biosensors – theory and applications. *Methods Biochem. Anal.*, **36**, 63–113.
6. Ion, A. and Bănică, F.G. (2002) *Electrochemical Methods in Analytical Chemistry*, Ars Docendi, Bucharest.
7. Ricci, F. and Palleschi, G. (2005) Sensor and biosensor preparation, optimisation and applications of Prussian Blue modified electrodes. *Biosens. Bioelectron.*, **21**, 389–407.
8. Lindgren, A., Ruzgas, T., Gorton, L. *et al.* (2000) Biosensors based on novel peroxidases with improved properties in direct and mediated electron transfer. *Biosens. Bioelectron.*, **15**, 491–497.
9. Yao, Y.L. and Shiu, K.K. (2008) A mediator-free bienzyme amperometric biosensor based on horseradish peroxidase and glucose oxidase immobilized on carbon nanotube modified electrode. *Electroanalysis*, **20**, 2090–2095.
10. Willner, I. and Katz, E. (2000) Integration of layered redox proteins and conductive supports for bioelectronic applications. *Angew. Chem. Int. Edit.*, **39**, 1180–1218.
11. Katz, E. and Willner, I. (2004) Integrated nanoparticle-biomolecule hybrid systems: Synthesis, properties, and applications. *Angew. Chem. -Int. Edit.*, **43**, 6042–6108.
12. Kerman, K., Saito, M., Yamamura, S. *et al.* (2008) Nanomaterial-based electrochemical biosensors for medical applications. *TrAC-Trends Anal. Chem.*, **27**, 585–592.
13. Claussen, J.C., Franklin, A.D., ul Haque, A. *et al.* (2009) Electrochemical biosensor of nanocube-augmented carbon nanotube networks. *ACS Nano*, **3**, 37–44.
14. Chaubey, A. and Malhotra, B.D. (2002) Mediated biosensors. *Biosens. Bioelectron.*, **17**, 441–456.
15. Calvo, E.J. and Danilowicz, C. (1997) Amperometric enzyme electrodes. *J. Braz. Chem. Soc.*, **8**, 563–574.
16. Fultz, M.L. and Durst, R.A. (1982) Mediator compounds for the electrochemical study of biological redox systems – a compilation. *Anal. Chim. Acta*, **140**, 1–18.
17. Martens, N., Hindle, A., and Hall, E.A.H. (1995) An assessment of mediators as oxidants for glucose-oxidase in the presence of oxygen. *Biosens. Bioelectron.*, **10**, 393–403.
18. McMahon, C.P., Rocchitta, G., Kirwan, S.M. *et al.* (2007) Oxygen tolerance of an implantable polymer/enzyme composite glutamate biosensor displaying polycation-enhanced substrate sensitivity. *Biosens. Bioelectron.*, **22**, 1466–1473.
19. Li, G., Xu, H., Huang, W.J. *et al.* (2008) A pyrrole quinoline quinone glucose dehydrogenase biosensor based on screen-printed carbon paste electrodes modified by carbon nanotubes. *Meas. Sci. Technol.*, **19**, 7.
20. Kotzian, P., Brazdilova, P., Kalcher, K. *et al.* (2007) Oxides of platinum metal group as potential catalysts in carbonaceous amperometric biosensors based on oxidases. *Sens. Actuators B-Chem.*, **124**, 297–302.
21. Wang, S.F., Tan, Y.M., Zhao, D.M. *et al.* (2008) Amperometric tyrosinase biosensor based on Fe_3O_4 nanoparticles-chitosan nanocomposite. *Biosens. Bioelectron.*, **23**, 1781–1787.
22. Canevet, D., Salle, M., Zhang, G.X. *et al.* (2009) Tetrathiafulvalene (TTF) derivatives: key building-blocks for switchable processes. *Chem. Commun.*, 2245–2269.
23. Kowalewska, B. and Kulesza, P.J. (2009) Application of tetrathiafulvalene-modified carbon nanotubes to preparation of integrated mediating system for bioelectrocatalytic oxidation of glucose. *Electroanalysis*, **21**, 351–359.
24. Nunes, G.S., Montesinos, T., Marques, P.B.O. *et al.* (2001) Acetylcholine enzyme sensor for determining methamidophos insecticide - Evaluation of some genetically modified acetylcholinesterases from Drosophila melanogaster. *Anal. Chim. Acta*, **434**, 1–8.
25. Skladal, P. (1992) Detection of organophosphate and carbamate pesticides using disposable biosensors based on chemically modified electrodes and immobilized cholinesterase. *Anal. Chim. Acta*, **269**, 281–287.
26. Cass, A.E.G., Davis, G., Francis, G.D. *et al.* (1984) Ferrocene-mediated enzyme electrode for amperometric determination of glucose. *Anal. Chem.*, **56**, 667–671.
27. Higgins, J.J., Hill, H.A.O., and Plotkin, E.V. (1985). US Pat. 4545382.
28. Gorton, L. (1995) Carbon-paste electrodes modified with enzymes, tissues, and cells. *Electroanalysis*, **7**, 23–45.
29. Razumiene, J., Gureviciene, V., Laurinavicius, V. *et al.* (2001) Amperometric detection of glucose and ethanol in beverages using flow cell and immobilised on screen-printed carbon electrode PQQ-dependent glucose or alcohol dehydrogenases. *Sens. Actuators B-Chem.*, **78**, 243–248.
30. Frasconi, M., Deriu, D., D'Annibale, A. *et al.* (2009) Nanostructured materials based on the integration of ferrocenyl-tethered dendrimer and redox proteins on self-assembled monolayers: an efficient biosensor interface. Nanotechnology, **20**, Article nr. 505501.
31. Chen, M. and Diao, G.W. (2009) Electrochemical study of mono-6-thio-beta-cyclodextrin/ferrocene capped on gold nanoparticles: Characterization and application to the design of glucose amperometric biosensor. *Talanta*, **80**, 815–820.
32. Qiu, J.D., Deng, M.Q., Liang, R.P. *et al.* (2008) Ferrocene-modified multiwalled carbon nanotubes as building block for construction of reagentless enzyme-based biosensors. *Sens. Actuators B-Chem.*, **135**, 181–187.
33. Yan, Y.M., Tel-Vered, R., Yehezkeli, O. *et al.* (2008) Biocatalytic growth of Au nanoparticles immobilized on glucose oxidase enhances the ferrocene-mediated bioelectrocatalytic oxidation of glucose. *Adv. Mater.*, **20**, 2365–2370.
34. Heller, A. (1990) Electrical wiring of redox enzymes. *Accounts Chem. Res.*, **23**, 128–134.
35. Boguslavsky, L., Hale, P.D., Geng, L. *et al.* (1993) Applications of redox polymers in biosensors. *Solid State Ion.*, **60**, 189–197.
36. Karan, I.K. (2005) CT Enzyme biosensors containing polymeric electron transfer systems, in *Biosensors and Modern Biospecific Analytical Techniques* (ed. L. Gorton), Elsevier, Amsterdam, pp. 131–178.
37. Katakis, I. and Heller, A. (1997) CT Electron transfer via redox hydrogels between electrodes and enzymes, in *Frontiers in Biosensorics, 1, Fundamental Aspects* (eds F.W. Scheller, F. Schubert, and J. Fedrowitz), Birkhäuser, Basel, pp. 229–241.

38. Coman, V., Gustavsson, T., Finkelsteinas, A. *et al.* (2009) Electrical wiring of live, metabolically enhanced *Bacillus subtilis* cells with flexible osmium-redox polymers. *J. Am. Chem. Soc.*, **131**, 16171–16176.

39. Gooding, J.J. and Hibbert, D.B. (1999) The application of alkanethiol self-assembled monolayers to enzyme electrodes. *TrAC-Trends Anal. Chem.*, **18**, 525–533.

40. Calvo, E.J., Battaglini, F., Danilowicz, C. *et al.* (2000) Layer-by-layer electrostatic deposition of biomolecules on surfaces for molecular recognition, redox mediation and signal generation. *Faraday Discuss.*, **116**, 47–65.

41. Gooding, J.J., Mearns, F., Yang, W.R. *et al.* (2003) Self-assembled monolayers into the 21(st) century: Recent advances and applications. *Electroanalysis*, **15**, 81–96.

42. Hodak, J., Etchenique, R., Calvo, E.J. *et al.* (1997) Layer-by-layer self-assembly of glucose oxidase with a poly(allylamine) ferrocene redox mediator. *Langmuir*, **13**, 2708–2716.

43. Tsai, T.W., Heckert, G., Neves, L.F. *et al.* (2009) Adsorption of glucose oxidase onto single-walled carbon nanotubes and its application in layer-by-layer biosensors. *Anal. Chem.*, **81**, 7917–7925.

44. Csoregi, E., Quinn, C.P., Schmidtke, D.W. *et al.* (1994) Design, characterization, and one-point *in-vivo* calibration of a subcutaneously implanted glucose electrode. *Anal. Chem.*, **66**, 3131–3138.

45. Marko-Varga, G., Emneus, J., Gorton, L. *et al.* (1995) Development of enzyme-based amperometric sensors for the determination of phenolic-compounds. *TrAC-Trends Anal. Chem.*, **14**, 319–328.

46. Lindgren, A., Stoica, L., Ruzgas, T. *et al.* (1999) Development of a cellobiose dehydrogenase modified electrode for amperometric detection of diphenols. *Analyst*, **124**, 527–532.

47. Cosnier, S. and Popescu, I.C. (1996) Poly(amphiphilic pyrrole)-tyrosinase-peroxidase electrode for amplified flow injection-amperometric detection of phenol. *Anal. Chim. Acta*, **319**, 145–151.

48. Coche-Guerente, L., Labbe, P., and Mengeaud, V. (2001) Amplification of amperometric biosensor responses by electrochemical substrate recycling. 3. Theoretical and experimental study of the phenol-polyphenol oxidase system immobilized in Laponite hydrogels and layer-by-layer self-assembled structures. *Anal. Chem.*, **73**, 3206–3218.

49. Hall, G.F., Best, D.J., and Turner, A.P.F. (1988) The determination of *p*-cresol in chloroform with an enzyme electrode used in the organic-phase. *Anal. Chim. Acta*, **213**, 113–119.

50. Saini, S. and Turner, A.P.F. (1995) Multiphase bioelectrochemical sensors. *TrAC-Trends Anal. Chem.*, **14**, 304–310.

51. Ramanavicius, A., Ramanaviciene, A., and Malinauskas, A. (2006) Electrochemical sensors based on conducting polymer-polypyrrole. *Electrochim. Acta*, **51**, 6025–6037.

52. Schuhmann, W. (1995) Conducting polymer based amperometric enzyme electrodes. *Mikrochim. Acta*, **121**, 1–29.

53. Gerard, M., Chaubey, A., and Malhotra, B.D. (2002) Application of conducting polymers to biosensors. *Biosens. Bioelectron.*, **17**, 345–359.

54. Bartlett, P.N. and Cooper, J.M. (1993) A review of the immobilization of enzymes in electropolymerized films. *J. Electroanal. Chem.*, **362**, 1–12.

55. Habermüller, K., Ramanavicius, A., Laurinavicius, V. *et al.* (2000) An oxygen-insensitive reagentless glucose biosensor based on osmium-complex modified polypyrrole. *Electroanalysis*, **12**, 1383–1389.

56. Cenas, N.K. and Kulys, J.J. (1981) Biocatalytic oxidation of glucose on the conductive charge-transfer complexes. *Bioelectrochem. Bioenerg.*, **8**, 103–113.

57. Kulys, J.J. (1986) Enzyme electrodes based on organic metals. *Biosensors*, **2**, 3–13.

58. Bartlett, P.N. (1990) CT Conducting organic salt electrodes, in *Biosensors: A Practical Approach* (ed. A.E.G. Cass), IRL Press, Oxford, pp. 47–95.

59. Khan, G.F. (1997) TTF-TCNQ complex based printed biosensor for long-term operation. *Electroanalysis*, **9**, 325–329.

60. Palmisano, F., Zambonin, P.G., Centonze, D. *et al.* (2002) A disposable, reagentless, third-generation glucose biosensor based on overoxidized poly(pyrrole)/tetrathiafulvalene-tetracyanoquinodimethane composite. *Anal. Chem.*, **74**, 5913–5918.

61. Pauliukaite, R., Malinauskas, A., Zhylyak, G. *et al.* (2007) Conductive organic complex salt TTF-TCNQ as a mediator for biosensors. An overview. *Electroanalysis*, **19**, 2491–2498.

62. Stoica, L., Ludwig, R., Haltrich, D. *et al.* (2006) Third-generation biosensor for lactose based on newly discovered cellobiose dehydrogenase. *Anal. Chem.*, **78**, 393–398.

63. Coman, V., Ludwig, R., Harreither, W. *et al.* (2010) A direct electron transfer-based glucose/oxygen biofuel cell operating in human serum. *Fuel Cells*, **10**, 9–16.

64. Willner, B., Katz, E., and Willner, I. (2006) Electrical contacting of redox proteins by nanotechnological means. *Curr. Opin. Biotechnol.*, **17**, 589–596.

65. Willner, I., Willner, B., and Tel-Vered, R. (2011) Electroanalytical applications of metallic nanoparticles and supramolecular nanostructures. *Electroanalysis*, **23**, 13–28.

66. Xiao, Y., Patolsky, F., Katz, E. *et al.* (2003) "Plugging into enzymes": Nanowiring of redox enzymes by a gold nanoparticle. *Science*, **299**, 1877–1881.

67. Liu, J.Q., Chou, A., Rahmat, W. *et al.* (2005) Achieving direct electrical connection to glucose oxidase using aligned single walled carbon nanotube arrays. *Electroanalysis*, **17**, 38–46.

68. Zayats, M., Willner, B., and Willner, I. (2008) Design of amperometric biosensors and biofuel cells by the reconstitution of electrically contacted enzyme electrodes. *Electroanalysis*, **20**, 583–601.

69. Shleev, S., Tkac, J., Christenson, A. *et al.* (2005) Direct electron transfer between copper-containing proteins and electrodes. *Biosens. Bioelectron.*, **20**, 2517–2554.

70. Christenson, A., Dimcheva, N., Ferapontova, E.E. *et al.* (2004) Direct electron transfer between ligninolytic redox enzymes and electrodes. *Electroanalysis*, **16**, 1074–1092.

71. Gorton, L. and Bartlett, P.N. (2008) NAD(P)-based biosensors, in *Bioelectrochemistry: Fundamentals, Experimental Techniques and Applications* (ed. P.N. Bartlett), Wiley, Chichester, pp. 157–198.

72. Radoi, A. and Compagnone, D. (2009) Recent advances in NADH electrochemical sensing design. *Bioelectrochemistry*, **76**, 126–134.

73. Albery, W.J., Bartlett, P.N., Cass, A.E.G. *et al.* (1987) Amperometric enzyme electrodes. 4. An enzyme electrode for ethanol. *J. Electroanal. Chem.*, **218**, 127–134.

74. Hart, J.P., Serban, S., Jones, L.J. *et al.* (2006) Selective and rapid biosensor integrated into a commercial hand-held instrument for the measurement of ammonium ion in sewage effluent. *Anal. Lett.*, **39**, 1657–1667.

75. Albareda-Sirvent, M., Merkoçi, A., and Alegret, S. (2000) Configurations used in the design of screen-printed enzymatic biosensors. A review. *Sens. Actuators B-Chem.*, **69**, 153–163.

76. Ricci, F., Moscone, D., and Palleschi, G. (2007) CT Mediated enzyme screen-printed electrode probes for clinical, environmental and food analysis, in *Electrochemical Sensor Analysis* (eds S. Alegret and A. Merkoçi), Elsevier, Amsterdam, pp. 559–584.

77. Wang, J. (2008) Electrochemical glucose biosensors. *Chem. Rev.*, **108**, 814–825.

78. Wang, J. (2001) Glucose biosensors: 40 years of advances and challenges. *Electroanalysis*, **13**, 983–988.

79. Prodromidis, M.I. and Karayannis, M.I. (2002) Enzyme based amperometric biosensors for food analysis. *Electroanalysis*, **14**, 241–261.

15

Mathematical Modeling of Mediated Amperometric Enzyme Sensors

A general discussion of enzyme sensor modeling has been presented in Chapter 4 where it has been assumed that the sensor response is proportional to the product concentration at the transducer surface. This condition can be fulfilled in the case of certain first-generation amperometric enzyme sensors. However, the mechanism of mediated amperometric enzyme sensors clearly deviates from the above assumption as the mediator is involved in a series of processes such as diffusion, reaction with the enzyme and the electrochemical reaction. That is why modeling of mediated amperometric enzyme sensors is a distinct problem, which is comprehensively reviewed in the literature [1–4].

This chapter introduces the principles of amperometric enzyme sensor modeling in two limiting cases depending on the location of the diffusion layer, namely the external diffusion and the internal diffusion cases.

In the external diffusion case, substrate diffusion is localized out of the sensing layer, within a membrane or within the diffusion layer formed in the solution when the solution is in direct contact with the enzyme layer. No diffusion processes occur within the enzyme layer.

The internal diffusion case is characterized by the fact that the enzyme reaction produces concentration gradients within the enzyme layer. Hence, diffusion occurs in parallel with the enzymatic reaction. This model assumes that no concentration gradients appear in the solution phase.

15.1 External Diffusion Conditions

This section addresses the modeling of mediated amperometric sensors under the assumption that the sensor functions under external diffusion control.

As discussed in Chapter 5, in the case of an enzyme sensor functioning under external diffusion conditions, the compounds present within the enzyme layer are uniformly distributed and no diffusion occurs within this film. A diffusion barrier is met out of the enzyme layer, within the external membrane or, when the membrane is absent, within the diffusion layer that develops in the solution phase near the sensing film. Such a situation arises when the thickness of the enzyme layer is sufficiently small and the substrate diffusion coefficient within this region is very high. These conditions are best fulfilled when the biocatalytic film consists of only one or a few molecular layers [5–7].

15.1.1 Model Formulation

The physical model used in the present approach is shown in Figure 15.1. The sensing element includes the enzyme and a mediator, both entrapped between an external membrane and the working electrode. The working electrode acts as electron collector and, in conjunction with a counterelectrode, it allows current to flow across the system.

Reactions emphasized in Figure 15.1 can be formulated as follows:

$$S + E_O \underset{k_{-1}}{\overset{k_1}{\rightleftharpoons}} [E_O S] \tag{15.1}$$

$$[E_O S] \xrightarrow{k_2} P + E_R \tag{15.2}$$

$$E_R + M_O \xrightarrow{k'_M} E_O + M_R \tag{15.3}$$

$$M_R \rightleftharpoons M_O + e^- \tag{15.4}$$

In this process, the substrate (S) is converted to the product P by an electron exchange involving the oxidized form of the enzyme (E_O) and going via the intermediate enzyme–substrate complex $[E_O S]$. As a result of this process, a substrate concentration gradient develops across the membrane and prompts the substrate to diffuse from the test

Chemical Sensors and Biosensors: Fundamentals and Applications, First Edition. Florinel-Gabriel Bănică.
© 2012 John Wiley & Sons, Ltd. Published 2012 by John Wiley & Sons, Ltd.

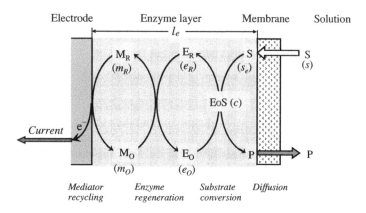

Figure 15.1 Schematic presentation of the processes occurring in a mediated amperometric enzyme sensor under external diffusion conditions. Symbols in parentheses indicate concentrations of relevant species symbolized by capital letters.

solution. It will be assumed that the amount of substrate consumed during the measurement is very low and no noticeable alteration of the bulk concentration occurs. The reduced form of the enzyme (E_R) formed in reaction (15.2) is next recycled to the oxidized form by an electron-transfer reaction involving the oxidized form of the mediator, M_O, (reaction (15.3)). Finally, the reduced mediator (M_R) produced in the above reaction is recycled by an electrochemical reaction at the electrode. This final reaction produces the current that serves as the sensor response. It will be assumed that the mediator is confined within the sensing layer and that no mediator exists in the test solution.

The overall process consists of an electron transfer from the substrate to the working electrode via two intermediate steps with the enzyme and the mediator acting as successive electron conveyors. Denoting the number of electrons by n_e, the overall process can be represented as:

$$S \rightarrow P + n_e \tag{15.5}$$

Therefore, the response current, which is proportional to the absolute value of the substrate flux (J) and the sensor area (A) is given by the following equation:

$$i = n_e F A J \tag{15.6}$$

where F is the Faraday constant.

The goal of the present approach is to derive an equation relating the current to the substrate concentration in solution. As is typical of electrochemical processes, the current is proportional to the overall reaction rate. Therefore, the rate (v) of each step in the sensor process should be considered. The pertinent diffusion fluxes and reaction rates are summarized in Equations (15.7)–(15.10), in which $k_{S,m}$ is the mass-transfer coefficient of the substrate, k_1 and k_{-1} are the rate constants in reaction (15.1), k_2 is the rate constant of reaction (15.2), and k'_M is the rate constant of reaction (15.3). The meaning of concentration symbols is given in Figure 15.1.

$$J = k_{S,m}(s - s_e) \quad \text{(diffusion)} \tag{15.7}$$

$$v_S = k_1 e_O s_e - k_{-1} c \quad \text{(formation of the enzyme – substrate complex)} \tag{15.8}$$

$$v_C = k_2 c \quad \text{(substrate conversion to the product)} \tag{15.9}$$

$$v_M = k'_M m_O e_R \quad \text{(regeneration of the oxidized enzyme)} \tag{15.10}$$

In addition, the mass-balance equation (15.11) for the enzyme should be taken into account. In this equation, e_t is the total concentration of the enzyme in the sensing layer.

$$e_t = e_O + e_R + c \tag{15.11}$$

Equations (15.8) and (15.9) are similar to those used in the Michaelis–Menten kinetic approach and need no comment. Conversely, Equation (15.10) deserves a preliminary discussion. This equation includes the concentration of the mediator in the oxidized state but this concentration depends on the electrode potential according to the Nernst equation:

$$E = E^0 + \frac{RT}{F} \ln \frac{m_O}{m_R} \tag{15.12}$$

where E^0 is the standard electrode potential of the mediator redox couple. Upon introducing the total mediator concentration $m_t = m_O + m_R$, one can derive from Equation (15.12) the following equation relating the M_O concentration to the total mediator concentration and the electrode potential:

$$m_O = m_t f_O \quad \text{where} f_O = \left(1 + e^{-\frac{F}{RT}(E-E^0)}\right)^{-1} \tag{15.13}$$

Accordingly, f_O represents the fraction of the total mediator concentration that is present in the oxidized form.

According to Equation (15.13), when the potential shifts from a very negative value with respect to E^0 to a very positive one, f_O varies from 0 to 1 and m_O changes from 0 to m_t. Keeping in mind this property, the rate Equation (15.10) can be reformulated as follows:

$$v_M = (k'_M m_t f_O)e_R \tag{15.14}$$

As m_t is a constant parameter, it is convenient to introduce the pseudofirst-order rate constant k_M defined as:

$$k_M = k'_M m_t \tag{15.15}$$

Upon substitution in Equation (15.10) the reaction rate of enzyme regeneration assumes the following form:

$$v_M = k_M f_O e_R \tag{15.16}$$

The above equation shows that the reaction rate of enzyme regeneration is a function of electrode potential. This implies that the flux, and consequently the current, are also dependent on potential and can vary from zero to a limiting value as this parameter varies within a specific region around the standard potential. The best sensitivity is achieved when the reaction (15.3) occurs with maximum velocity and this condition is met if E is sufficiently positive with respect to E^0 in order to render $f_O \approx 1$. Under these *limiting conditions*, the velocity of enzyme regeneration assumes the following limiting form:

$$v_{M,l} = k_M e_R \tag{15.17}$$

In the next section, the sensor response under limiting conditions is addressed.

15.1.2 Sensor Response: Limiting Cases

It will be assumed in this section that the electrode potential is sufficiently positive so as to render $f_O \approx 1$ so that the substrate flux assumes its limiting value (J_l). As the current is proportional to the substrate flux, the next discussion focuses on the correlation between the flux and the substrate concentration. To this end, it will be assumed that the sensor functions under *steady-state* conditions, in which the flux equates the reaction rates transposed in a surface-normalized form. As indicated in Section 4.2.2, the surface-normalized reaction rate represents the reaction rate occurring at a sensor of unit surface area such that the reaction volume is numerically equal to the thickness of the enzyme layer, l_e. Under these circumstances, the steady-state condition can be formulated as follows:

$$J_l = l_e v_S = l_e v_C = l_e v_{M,l} \tag{15.18}$$

By substituting appropriate expressions for each reaction rate term one obtains:

$$J_l = l_e(k_1 e_O s_e - k_{-1}c) = l_e k_2 c = l_e k_M e_R \tag{15.19}$$

Equations (15.19), along with Equation (15.11) form a system of equations with J_l, e_O, e_R and c as unknowns. By solving for J_l the following equation is obtained:

$$J_l = l_e e_t \left(\frac{1}{k_2 K_M^{-1} s_e} + \frac{1}{k_2} + \frac{1}{k_M}\right)^{-1} \tag{15.20}$$

where K_M is the Michaelis–Menten constant.

Among the terms in the right-hand side of Equation (15.20), only the first term depends on the diffusion rate via the substrate concentration, the remaining terms being dependent on rate constants only. It is possible therefore to reduce this equation to a simpler form, by assuming that the substrate concentration in the enzyme film is sufficiently low to make the second and third terms negligibly small with respect to the first term. This condition can be fulfilled by securing a very low substrate flux through the external membrane. Under these circumstances, the enzymatic reaction occurs under *first-order kinetic* conditions and Equation (15.20) turns into the following form, in which $J_{1,1}$ stands for the limiting flux under pseudofirst-order kinetics:

$$J_{1,1} = \frac{k_2 e_t l_e}{K_M} s_e \qquad \left(k_2; k_M \gg k_2 K_M^{-1} s_e \right) \tag{15.21}$$

This equation includes the flux-dependent term s_e that can be obtained from Equation (15.7) after substituting $J_{1,1}$ for J:

$$s_e = s - \frac{J_{1,1}}{k_{S,m}} \tag{15.22}$$

From Equations (15.21) and (15.22) one obtains:

$$J_{1,1} = \left(\frac{K_M}{l_e k_2 e_t} + \frac{1}{k_{S,m}} \right)^{-1} s \tag{15.23}$$

Accordingly, under first-order kinetics, the response is directly proportional to the substrate concentration in the solution phase (s), which is suitable for performing substrate determination. In addition, the response is in this case independent of the kinetics of the enzyme-regeneration (step (15.3)). If the sensor is designed so that $l_e k_2 e_t \gg K_M$, the first term in the right-hand side of Equation (15.23) can be disregarded, and the response function turns to the following simple form, where J_d represents the limiting flux observed under first-order kinetics and external diffusion control:

$$\left(J_{1,1} \right)_{l_e k_2 e_t \gg K_M} \equiv J_d = k_{S,m} s \tag{15.24}$$

J_d is the maximum flux achievable at a given substrate concentration and for specific parameters of the external membrane such as permeability and thickness. In other words, the best sensitivity will be achieved when the enzyme concentration is so high that the response becomes independent of the enzyme loading and the kinetic constants. As these parameters may vary as an effect of pH change, action of activators or inhibitors, or enzyme denaturation, the above conditions make the sensor independent of such effects and bring about the best stability of the calibration parameters. On the other hand, Equation (15.24) demonstrates that the sensor sensitivity depends only on the mass-transfer coefficient of the substrate under the above-mentioned conditions.

It is of interest to examine also the opposite situation in which the sensor response is determined by kinetic factors only. Under these conditions, substrate consumption within the enzyme layer is a slow process as compared with diffusion. This could render s_e sufficiently large to make the first term in the right-hand side of Equation (15.20) negligible with respect to the following terms and achieve the *zero-order kinetics*. Consequently, the limiting flux assumes the particular value $J_{1,0}$ that corresponds to the zero-order kinetics, and the response function becomes:

$$J_{1,0} = l_e k_2 (r_T + 1)^{-1} e_t \tag{15.25}$$

where r_T is the quotient of k_2 and k_M reaction rates:

$$r_T = \frac{k_2}{k_M} = \frac{\text{Rapidity of substrate conversion}}{\text{Rapidity of enzyme regeneration}} \tag{15.26}$$

In other words, r_T represents the *turnover number* for the substrate conversion relative to that for the enzyme reoxidation. Clearly, under zero-order kinetics, the response is independent of substrate concentration in the solution phase. In turn, the response is directly proportional to the total enzyme concentration within the enzyme layer. This kind of response is useful when the enzyme is employed as a label tag for an electrochemically inactive analyte, such as an antibody. These conditions are also convenient when the sensor is designed for inhibitor determination. Equations (15.25) and (15.26) prove, at the same time, that the response is also depend on the mediator concentration that is included in the pseudofirst-order constant k_M. However, if this concentration is sufficiently high to make $k_M \gg k_2$, r_T becomes negligible with respect to unity and then Equation (15.25) assumes the following limiting form:

$$\left(J_{1,0} \right)_{r_T \ll 1} = l_e k_2 e_t \tag{15.27}$$

Therefore, at a sufficiently high mediator concentration, the response will not be affected by accidental changes in mediator concentration.

In conclusion, an enzyme amperometric sensor under external diffusion conditions can be designed to respond either to the substrate concentration (first-order kinetics) or the enzyme content within the sensing element (zero-order kinetics). In the second case, the response becomes independent of the mediator concentration if the pseudorate constant k_M is much higher than the k_2 rate constant. This condition can be fulfilled by selecting a mediator characterized by a high k'_M constant and keeping as high as possible the total concentration of the mediator within the sensing layer.

15.1.3 The Dynamic Range and the Limit of Detection

The previous section proved that a shift from the proportional response range to substrate-independent response could, in principle, be achieved simply by increasing the substrate concentration with no alteration in sensor parameters. Consequently, the response is expected to display an approximately linear trend at low substrate concentrations but it will level off above an yet unspecified limit. As the extent of the proportional range is a very important figure of merit, the following discussion will deal with the effect of sensor design parameters (such as the enzyme loading and the mediator concentration) on the upper limit of the proportional response range. In order to solve this problem, it is necessary to derive a general response equation, with no restrictions in enzyme loading and other sensor characteristics. As in the previous section, this problem will be approached under the limiting condition in which all the mediator is present only in the oxidized form ($f_O \approx 1$). The required equation can be obtained by substituting s_e in Equation (15.20) by the following expression derived from Equation (15.7) with J_1 instead of J:

$$s_e = s - \frac{J_1}{k_{S,m}} \tag{15.28}$$

Upon making this substitution and solving for J_1, one obtains:

$$J_1 = k_{S,m}s\left(\frac{l_e k_2 e_t}{k_{S,m} K_M}\right)\left[\frac{s}{K_M}\left(\frac{k_2}{k_M}+1\right)+\left(\frac{l_e k_2 e_t}{k_{S,m} K_M}\right)+1\right]^{-1} \tag{15.29}$$

It is easy to realize that the term $l_e k_2 e_t/k_{S,m} K_M$ in the above equation is the dimensionless substrate modulus for external diffusion (α), which was introduced in Section 4.2.4. As shown there, α is the quotient of the rapidity of the enzyme reaction and the rapidity of substrate diffusion. The calculation can be simplified by introducing the dimensionless concentration of the substrate (S), defined as follows:

$$S = \frac{s}{K_M} \tag{15.30}$$

Upon substitution of S and α in Equation (15.29) the response equation becomes:

$$J_1 = k_{S,m} K_M \alpha S[S(r_T + 1) + (1 + \alpha)]^{-1} \tag{15.31}$$

This is a rather intricate equation in S but, for practical purposes, it can be simplified by assuming certain particular cases. Recalling the definition of α, it is seen that the sensor will operate under diffusion control if this parameter is so great than the following condition is fulfilled:

$$1 + \alpha \gg S(r_T + 1); \quad \text{or } S \gg (r_T + 1)(1 + \alpha)^{-1} \tag{15.32}$$

Under such conditions, the flux assumes a limiting value that is first order in substrate concentration:

$$J_{1,1} = \frac{\alpha}{\alpha + 1} k_{S,m} s \tag{15.33}$$

It is easy to prove that this equation is equivalent to the previously derived Equation (15.23). Equation (15.33) demonstrates that the diffusion-controlled flux, J_d, defined in Equation (15.24), will be obtained when $\alpha \gg 1$. So, in this particular case, the sensitivity reaches its maximum value that depends only on the diffusion coefficient and the thickness of the external membrane.

If $1 + \alpha \ll S(r_T + 1)$, the response function turns into the limiting form (15.25) that is characteristic of enzyme saturation (zero-order kinetics).

Equation (15.32) yields a rough estimation of the limit of the linear range. More accurately, this limit can be defined as the substrate concentration yielding a response that is 95% of that expected in the case of a linear response. Using Equation (15.31) for the true response and Equation (15.33) for the linear one, it is found that the response is linear if:

$$S \leq 0.05(\alpha + 1)(r_T + 1)^{-1} \tag{15.34}$$

Hence, the limit of the linear range increases with α but decreases with increasing the r_T parameter. It is therefore suitable to design the sensor so that $r_T \ll 1$ and $\alpha \gg 1$. Under these circumstances, the limit of the linear range increases almost proportionally to α.

The above treatment proves that, in addition to α, r_T plays an essential role in deciding the sensor characteristics.

The effect of sensor parameters on the response function is best illustrated by plotting the response as a function of substrate concentration. To this end, Equation (15.31) will be set in the following form:

$$\frac{J_1}{J^*} = \frac{\alpha}{(r_T + 1)S + \alpha + 1} S \tag{15.35}$$

where J^* represents the particular value of J_d for $s = K_M$, that is:

$$J^* = k_{S,m} K_M \tag{15.36}$$

As the current is proportional to the flux, Equation (15.35) leads to:

$$\frac{i_1}{i^*} = \frac{J_1}{J^*} = \frac{\alpha}{(r_T + 1)S + \alpha + 1} S \tag{15.37}$$

where i_1 is the actual limiting current and i^* is the particular value of the limiting current recorded at $\alpha \gg 1$ and for $S = 1$, in accordance with Equation (15.36).

Equation (15.37) has been used to generate the response curve under selected conditions in order to show the effect of the dimensionless parameters α and r_T (Figure 15.2). Figure 15.2A displays the shape of the response function for various α values at a very low r_T value. It demonstrates that for any value of α, the response is quasilinear at sufficiently low substrate concentrations, but the linear range expands considerably as the α parameter increases.

The effect of the r_T parameter on the response function at a very high α value is illustrated in Figure 15.2B. This figure proves that an appreciable increase in the span of the linear range can be obtained if k_M is much larger than k_2. Such favorable conditions can be achieved if the reoxidation of the enzyme (reaction (15.3)) is much faster that substrate oxidation (reaction (15.2)). Hence, the mediator concentration in the biocatalytic layer should be as high as possible. On the other hand, the proper selection of the mediator itself and its immobilization procedure can affect favorably the actual rate constant of the enzyme reoxidation (k'_M) that is included in k_M.

In summary, the best sensitivity and response function for the mediated sensor under external diffusion control is achieved when the sensor process occurs under full diffusion control, with no effect of kinetic parameters and enzyme loading. These conditions are met if the sensor is designed so as to secure a sufficiently high value of the substrate modulus α and as low as possible a value of the r_T ratio. In addition, under these conditions, the maximum stability of the calibration parameters is achieved, as any accidental change in enzyme activity or mediator concentration has no effect on the sensor response.

After discussing the criteria for adjusting the upper limit of the linear response range, it is useful to deal also with its lower limit, which is the limit of detection. To this end, it is worth recalling that the true signal is the electrolytic current, which is proportional to the substrate flux, according to Equation (15.6). In addition to the substrate-related current, a background current may arise as a result of secondary processes (such as electrochemical reactions involving concomitant compounds). In agreement with the general definition, the limit of detection is the concentration yielding a current equal to 3 times the standard deviation of the background fluctuations. It is therefore

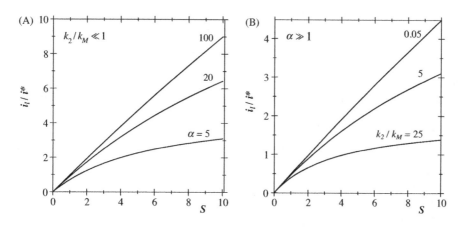

Figure 15.2 (A) Effect of the substrate modulus (α) on the response function at a very low r_T value. α values are indicated on each curve. (B) Effect of the $r_T = k_2/k_M$ ratio on the response function at a constant substrate modulus (α) within the diffusion-control region. r_T values are indicated on each curve.

important to adjust the sensor design and the working conditions such that the background is much lower that the expected response current. This can be achieved by using a semipermeable external membrane that prevents diffusion of interferents.

A simpler method consists of including in the sensor structure a second working electrode coated with the same type of layer as that of the sensing electrode except for the fact that the enzyme is absent. If both electrodes are polarized at the same potential, the second electrode generates only the background current that can be subtracted from the current recorded at the sensing electrode.

15.1.4 Other Theoretical Models

The simple model above enables the derivation of the essential trends in the response dependence on various sensor parameters. A more refined model should consider the stepwise character of reaction (15.2). This reaction may consist of an internal electron transfer between enzyme and substrate within the complex, followed by splitting of the resulting product–enzyme complex to yield the final product and the free enzyme in the reduced form [4,8]. The approach in [8] takes also into account the effect of enzyme inhibition by the product of the reaction. If the mediator is a soluble species that is present in the test solution, its diffusion to the enzyme layer should also be considered [9,10].

A modeling of amperometric enzyme electrodes based on direct electron transfer from the enzyme to the electrode is presented in ref. [4].

The advent of nanomaterial application to enzyme sensor prompted the development of theoretical models of such systems. The problem of amperometric enzyme sensors including carbon nanotubes in the catalytic film is addressed in refs. [11,12]

A fair alternative to the analytical approach to modeling is numerical simulation that can provide relevant information such as concentration profiles and the response value under selected conditions.

15.1.5 Outlook

An amperometric sensor functioning under external diffusion conditions can be built up by assembling a very thin biocatalytic layer including both the enzyme and a mediator. The key parameter of such a sensor is the external modulus factor α, which represents the rapidity of the enzyme reaction relative to that of the diffusion process. In addition, the sensor behavior depends on r_T, the relative turnover of the enzyme reaction with respect to the turnover of the enzyme regeneration. The overall reaction rate depends on the electrode potential, which controls the concentration of the oxidized mediator. In order to maximize sensitivity, the electrode potential should be selected so that the mediator is present only in the oxidized form, which causes the current to assume a limiting, potential-independent value. The limiting current represents the actual sensor response. The limiting current is proportional to the substrate concentration if α is large enough and $r_T \ll 1$. If $\alpha \gg 1$, the response becomes independent of the parameters of the biocatalytic layer and the current–concentration proportionality constant depends only on the diffusion coefficient within the external membrane and the thickness of this, both of them being included in the mass-transfer coefficient. This is the highest sensitivity that can be achieved. In addition, under these conditions the response is independent of the parameters of the biocatalytic layer, which imparts to the sensor a very good resilience.

Questions and Exercises (Section 15.1)

1 Prove that f_O tends to one if $E \geq (E^0 + 0.118)$ V at 25 °C in the absence of any other chemical reaction.
2 List the parameters and constants included in the definition of the α modulus and discuss the effect of each of them on the sensor response under external diffusion control.
3 Prove that Equations (15.23) and (15.33) are similar.
4 A mediated amperometric enzyme sensor operating under external diffusion conditions, includes a 5-μm thick enzyme layer contained between a 2-mm diameter working electrode and a 100-μm thick semipermeable membrane. The diffusion coefficient of the substrate in this membrane is 10^{-10} m^2 s^{-1} and the Michaelis–Menten constant is 10^{-6} mole/ml.

    a. Calculate the substrate modulus for the following values of enzyme specific activity (in μM/min ml): 25; 50; 100; 200; 500.

    b. Calculate the sensitivity of this sensor in each of the above cases (in current/concentration units). Comment on the effect of α on sensitivity.

    c. Consider the cases in which the specific activity is either 25 or 500 μM/min ml and assume that, in each case, the specific activity drops after some time to 80% of its initial value. Calculate the relative variation in sensitivity in each case and comment on the α value effect on the long term stability of the calibration parameters.

d. Calculate the limit of detection for 25 and 500 μM/min ml specific activity, if the standard deviation of the fluctuations in the background current is 0.01 μA.

Hints and answers: (a) Convert all problem data into SI units; in this case, the given specific activity is an approximation of the maximum reaction rate of the enzyme reaction. For an activity of 500 μM/min ml, $\alpha = 41.7$. (b) Sensitivity (i/s) is given by the product of two terms: $nFAk_{S,m}$ and $\alpha/(1+\alpha)$, where A is the electrode surface area. For an activity of 500 μM/min ml, the sensitivity is 0.59 μA/mM. (c) The per cent variation in sensitivity is $100(b_f - b_i)/b_i$, where b_i and b_f stand for the initial and final sensitivities, respectively. (d) For an activity of 500 μM/min ml, the limit of detection is 0.05 mM.

5 For a mediated amperometric enzyme sensor under external diffusion conditions, (a) derive a mathematical expression for the limit of the linear range under the assumption that the maximum accepted deviation from the true concentration is 1%. (b) Assume that $\alpha = 5$ or 20 and $r_T = 10$ or 0.1. Calculate the limit of the linear range for each combination of α and r_T values and comment on the effect of these parameters on the extent of the linear response range.

Hints and answer: The true concentration arises from Equation (15.31), whereas the concentration under the assumption of a linear response function can be derived from Equation (15.33). As the concentration is proportional to substrate flux, the $J_S/J_{S,d}$ ratio yields the ratio of the correct concentration to the concentration including an error due to the deviation from linearity. Derive an expression for the above flux ratio

(answer, $J_S/J_{S,d} = \left[S(r_T + 1)(1 + \alpha)^{-1} + 1 \right]^{-1}$), equate it with 0.99 and solve for S in order to find the

concentration leading to a 1% deviation from the real one (answer: $S_{99\%} = 0.01(\alpha + 1)(r_T + 1)^{-1}$).

15.2 Internal Diffusion Conditions

The internal diffusion functioning regime is characterized by the fact that no significant concentration gradients develop within the test solution as a result of the enzymatic reaction. Conversely, within the enzyme layer the concentration of the substrate is nonuniformly distributed, which also results in spatial variations in the velocity of the enzymatic reaction. Such circumstances arise when the enzyme layer is relatively thick and the diffusion coefficient of the substrate within this layer is very low (see Section 4.3 for details).

The approach in this section follows the general lines in refs. [1,13].

15.2.1 Model Formulation

The structure of the sensor addressed in the present approach is shown in Figure 15.3. Accordingly, the sensor is a three-phase system that consists of a homogeneous biocatalytic layer confined between an electrode and the test solution. Both the enzyme and mediator are entrapped in the catalytic film, with the enzyme firmly bound to the matrix and not free to diffuse. If the mediator is not tightly bound to the matrix, it can travel freely and a true diffusion coefficient D_M expresses its mobility. In this case, electrons are conveyed between the enzyme and electrode by free diffusion of the reduced mediator. However, if the mediator is covalently attached to the matrix (as in the case of redox hydrogel systems), the charge gradient induced by the enzymatic reactions causes electrons to move to the electrode surface by stepwise leaps from a reduced mediator molecule to a neighboring oxidized one. Under such

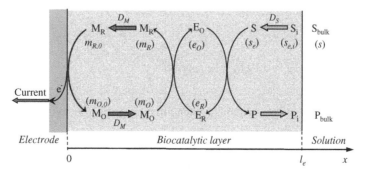

Figure 15.3 Schematics of an amperometric mediated enzyme sensor with internal diffusion. The mediator is entrapped within the biocatalytic layer along with the enzyme. Thick arrows indicate diffusion. Symbols in parentheses indicate concentrations of relevant species symbolized by capital letters.

circumstances, the diffusion coefficient corresponds to the pseudodiffusion of electrons through the matrix rather than to effective displacement of the mediator molecules.

This model does not apply to the case of a conducting entrapment matrix, such as a carbon paste. In this case, electrons are transferred from mediator molecules to conducting particles in order to be further conveyed to the electrode. Such a reaction is heterogeneous in nature and the specific rate constant for mediator–conducting-particle electron transfer should also be introduced.

The chemical reaction sequence consists of substrate conversion followed by enzyme reoxidation according to the two states enzyme mechanism (the ping-pong mechanism):

$$S + E_O \xrightarrow{k_E} P + E_R \tag{15.38}$$

$$M_O + E_R \xrightarrow{k_M} M_R + E_O \tag{15.39}$$

The pseudorate constant for the reaction (15.38) is, according to the Michaelis–Menten kinetics:

$$k_E = \frac{k_2}{K_M + s_e} \tag{15.40}$$

where K_M is the Michaelis–Menten constant. This rate constant turns into the second-order rate constant k_2/K_M if $s_e \ll K_M$. It should be kept in mind that k_2 and K_M represent here the apparent parameters of the immobilized enzyme that can be different from those characteristic of the enzyme in solution.

As a result of the above chemical reactions, electrons are transferred from the substrate to the mediator. In order to keep the process going, the mediator is reconverted to the active form after diffusing to the electrode surface where electron transfer takes place as follows:

$$M_R \rightarrow M_O + ne^- \tag{15.41}$$

This electrochemical reaction forms the transduction stage that provides the response signal as an electric current.

As this model refers to the case of internal diffusion, it is assumed that no noticeable concentration gradients occur in the solution phase owing to a very fast mass transfer. This condition can be met by vigorous stirring and/or by very low substrate diffusivity within the biocatalytic layer. Diffusion is coupled with chemical reactions; hence, a mathematical description of the process can be made by means of Fick's second equation including an additional term that accounts for reactant consumption. In the steady state, the pertinent equations are:

$$D_S \frac{\partial^2 s_e}{\partial x^2} = \frac{k_2 e_O s_e}{K_M + s_e} \tag{15.42}$$

$$D_M \frac{\partial^2 m_O}{\partial x^2} = k_M e_R m_O \tag{15.43}$$

D_S and D_M are diffusion coefficients of the substrate and mediator, respectively. The meaning of the concentration symbols is given in Figure 15.3. Fick's equations (15.42) and (15.43) are mass-balance equations and indicate that the amount of diffusing component (left-hand term) equates to the amount of component converted by the enzymatic reaction (right-hand term).

The response current is proportional to the flux at the electrode surface ($x = 0$) with the proportionality constant $b = nFA$, so:

$$i = bJ_M \tag{15.44}$$

where:

$$J_M = -D_M \left(\frac{dm_O}{dx}\right)_{x=0} \tag{15.45}$$

Therefore, in order to derive the response current, an equation relating the flux to the substrate concentration in the test solution should be derived.

15.2.2 Dimensionless Parameters and Variables

In order to solve this modeling problem, it is necessary to introduce a series of dimensionless parameters and variables that allows the raw differential equations to be converted into more manageable forms. As is commonly the case, dimensionless parameters allow, in addition, revealing some important physical characteristics of the system and a better understanding of the behavior of the sensor in response to alterations in its design parameters and working conditions.

The reaction mechanism previously underlined points to two reacting species: the substrate and the mediator. It is important, therefore, to introduce for each reactant a parameter describing its chemical reactivity relative to its diffusivity. Such a parameter was already introduced for the substrate (Section 4.3.1) in the form of the substrate Thiele modulus ϕ_S, which, in this case is:

$$\phi_S^2 = \mathrm{Da_S} = \frac{l_e^2 k_2 e_t}{D_S K_M} = \frac{\text{Substrate conversion capacity}}{\text{Substrate replenishing capacity}} \tag{15.46}$$

$\mathrm{Da_S}$ stands here for the substrate Damköhler number. In other words, $\mathrm{Da_S}$ indicates the capacity of the enzyme to convert the substrate relative to the substrate replenishment capacity.

A similar parameter, ϕ_M should also be introduced for the mediator. If we assume that the M_O concentration at $x = 0$ equates to the total mediator concentration (m_t) and falls to zero at the limit of the film ($x = l_e$), the M_O flux is, under the assumption of the Nernst diffusion layer, $J_{M,max} = (D_M/l_e)m_t$. At the same time, the rate of the reaction (15.39) attains a maximum value when all the enzyme is present in the reduced form ($e_R = e_t$) and all the mediator is in the oxidized form ($m_O = m_t$). Therefore, the maximum rate of the reaction (15.39) is:

$$v'_{M,max} = l_e k_M m_t e_R \tag{15.47}$$

Note that this reaction rate is surface normalized. By definition, the Damköhler number for the mediator ($\mathrm{Da_M}$) is:

$$\mathrm{Da_M} = \frac{v'_{M,max}}{J_{M,max}} = \frac{l_e^2 k_M e_R}{D_M} \tag{15.48}$$

Therefore, the mediator Thiele modulus is:

$$\phi_M = \sqrt{\mathrm{Da_M}} = l_e \sqrt{\frac{k_M e_t}{D_M}} \tag{15.49}$$

If we keep in mind that the enzyme reoxidation consists of an electron transfer from the enzyme to the mediator and electrons are then conveyed by mediator diffusion, the physical meaning of ϕ_M can be formulated as follows:

$$\phi_M^2 = \mathrm{Da_M} = \frac{\text{Enzyme re-oxidation capacity}}{\text{Electron replenishing capacity}} \tag{15.50}$$

If one of the above ϕ coefficients is very high, it follows that the rate of the pertinent process (enzyme oxidation or substrate conversion) is diffusion controlled. In the opposite case, kinetic control occurs.

As substrate conversion and enzyme reoxidation occur in sequence, it is important to define a parameter that accounts for their relative rapidity. This is the η parameter below:

$$\eta = \frac{\mathrm{Da_M}}{\mathrm{Da_S}} = \frac{k_M}{k_2/K_M} \frac{D_S}{D_M} \tag{15.51}$$

If we substitute $k_M e_t$ for k_M and $k_2 e_t / K_M$ for k_2/K_M, we can realize that η represents the quotient of the pseudofirst-order rate constant for reaction (15.39) and the pseudofirst order rate constant for reaction (15.38), corrected by the ratio of the diffusion coefficients. The meaning of η can therefore be phrased as follows:

$$\eta = \frac{\text{Electron replenishing capacity}}{\text{Substrate conversion capacity}} \tag{15.52}$$

The term "capacity" refers here to both the transport and chemical reaction processes that occur sequentially. Thus, the substrate conversion capacity can be limited by either substrate diffusion or the enzymatic conversion step.

The η parameter compares the conversion and electron-conveying capacities in terms of both rate constants and diffusion coefficients. It disregards the effect of substrate and mediator concentrations that can affect essentially the velocity of chemical reactions. The interplay of these variables can be accounted for by the following dimensionless variable:

$$\gamma = \frac{k_M m_t}{(k_2/K_M)s} \tag{15.53}$$

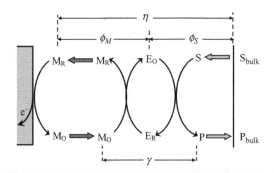

Figure 15.4   Applicability of dimensionless parameters to various step combinations in the overall sensor process.

The numerator in Equation (15.53) is proportional to the limiting reoxidation rate $v_{M,l}$ that occurs when both the M_O supply and substrate conversion are infinitely fast such that $m_O = m_t$ and $e_R = e_t$, that is, $v_{M,l} = k_M m_t e_R$. On the other hand, the denominator is proportional to the limiting conversion rate under first-order conditions, which is apparent when both substrate replenishment is infinitely fast ($s_e = s$) and enzyme reoxidation is extremely rapid ($e_O = e_t$). In this case, the rate of conversion is $v_{S,l} = (k_2 e_t / K_M)s$. Consequently:

$$\gamma = \frac{v_{M,l}}{v_{S,l}} = \frac{\text{Enzyme reoxidation capacity}}{\text{Substrate conversion capacity}} \tag{15.54}$$

So, γ indicates the enzyme reoxidation capacity relative to the substrate conversion capacity in the absence of any diffusion limitation. It depends on experimental conditions via s and on the sensor design characteristics, via m_t.

It is evident that each of the dimensionless parameters introduced above compares the rapidity of two successive steps included in the overall sensor process. The applicability of the above-defined dimensionless parameters to various step combinations in the sensor process is demonstrated in Figure 15.4.

15.2.3   Limiting Conditions

In order to solve the differential equations it is essential to define the limiting conditions, i.e. the physical state at the electrode interface ($x = 0$) and at the external limit of the biocatalytic layer ($x = l_e$). So, at $x = l_e$, the substrate is not yet converted and its concentration depends only on the partition coefficient (k_S^p) as $s_{e,i} = k_S^p s$. In the next treatment, the partition will be neglected for simplicity, hence $s_{e,i} = s$.

At the opposite limit ($x = 0$), reaction (15.41) takes place, allowing electrons to be transferred further to the external electronics to provide the response signal. It will be assumed that the electrochemical reaction proceeds reversibly and, consequently, the Nernst equation with respect to interface concentrations ($m_{O,0}$ and $m_{R,0}$) is obeyed:

$$E = E^0 + \frac{RT}{nF} \ln \frac{m_{O,0}}{m_{R,0}} \tag{15.55}$$

Therefore, the molar fraction of M_O at the interface is:

$$f_O = \frac{m_{O,0}}{m_t} = \left[1 + \exp\left(-\frac{nF}{RT}(E - E^0) \right) \right]^{-1} \tag{15.56}$$

If the mass balance of the mediator is formulated as:

$$m_t = m_O + m_R \tag{15.57}$$

then, the concentration of the oxidized mediator at the electrode surface is given by:

$$m_{O,0} = m_t f_O \tag{15.58}$$

The electrode potential is the driving force of the overall process. It determines the interface concentration of M_O and, therefore, the rate of M_O replenishment. The electrode, by means of the electrode potential, acts as a valve that controls the pace of the overall process between two limits: zero reaction rate if $E \ll E^0$ and the limiting reaction rate at $E \gg E^0$. If the electrochemical reaction of the mediator is fast (reversible), the change between the above limits occurs over a potential span of several tens of millivolts.

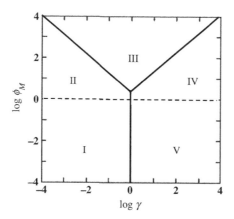

Figure 15.5 Two-dimensional case diagram for $\eta = 1$ (equal conversion and reoxidation capacities), $S \ll 1$ and $f_O = 1$ (limiting response conditions). Adapted with permission from [13]. Copyright 1995 Elsevier.

15.2.4 Solving the Differential Equations. The Case Diagram

Upon substituting dimensionless parameters and variables in Equations (15.42) and (15.43) one obtains new differential equations that are more manageable as they put into evidence very clearly various approximate forms that can be solved analytically. Each approximate case is defined by some constraints imposed on relevant dimensionless parameters. Thus, the constraint $\phi_M \ll 1$ implies that the enzyme reoxidation is very sluggish with respect to mediator diffusion and the overall reaction rate is determined by the backoxidation of the enzyme, that is, the process is *mediator controlled*. This situation occurs if substrate conversion is very rapid as compared with the enzyme reoxidation, which implies $\gamma \ll 1$. The above restrictions identify therefore a *kinetic regime* in which the overall kinetics is mediator controlled, with no kinetic effects from enzyme reoxidation and substrate diffusion processes. However, if $\phi_M \gg 1$, the mediator diffusion is very sluggish with respect to enzyme reoxidation and acts as the rate-determining step if additional restrictions are fulfilled.

Various cases defined as above are displayed on the kinetic case diagram in Figure 15.5. Clearly, the kinetic regime depends on all dimensionless parameters involved in the mathematical model. In order to plot a handy, two-dimensional case diagram, two parameters are selected as variables and the remaining parameters are presumed to be constant. For example, Figure 15.5 was plotted under the assumption that γ and ϕ_M are variable whereas η is constant and $S = s_e / K_M \ll 1$ (in order to secure first-order kinetics for the substrate conversion). Similar diagrams can be plotted for various combinations of constant and variable parameters in order to explore the kinetic behavior of the sensor. As far as the diagram in Figure 15.5 is concerned, any change in either η or S affects both the position and extent of each region.

Among the five kinetic cases displayed in Figure 15.5 several selected cases of particular relevance to sensor applications will be discussed in the following sections.

15.2.5 Kinetic Currents

Cases I and V are clearly seen on the lower part of the case diagram in Figure 15.5, at very low values of both the mediator and substrate Thiele coefficients, i.e. $\phi_M^2 \ll 1$ and $\phi_S^2 \ll 1$. Accordingly, diffusion of both mediator and substrate are very fast relative to the relevant chemical reaction. Taking into account the definition of the ϕ coefficients, it is obvious that these conditions are met if both the enzyme and the mediator loading are very low and the film is very thin. Under these conditions, fast diffusion allows concentrations of both substrate and oxidized mediator to achieve a uniform distribution throughout the film Therefore, $m_O = m_{O,0}$ and $s_e = s$ at any point within the sensing film. As a consequence, the overall reaction rate is dictated by chemical reaction steps and the process occurs under kinetic control. That is why the current generated under these conditions is a *kinetic current*. This case has been termed the *thin-film approximation* for internal diffusion because the distance a substrate molecule can travel before being converted is much greater than the film thickness. The same argument applies to the diffusion of charge. This case should not be confused with the thin-film model for external diffusion conditions, discussed above in Section 15.1.

15.2.6 Diffusion Currents

Diffusion control becomes evident when either ϕ_S or ϕ_M or both are much greater than one, which implies slow diffusion of the substrate or the mediator or both. The last of the above three possibilities, which pertains to *case III* in Figure 15.5, proved to be the most convenient from the standpoint of sensor calibration and will be discussed in detail next.

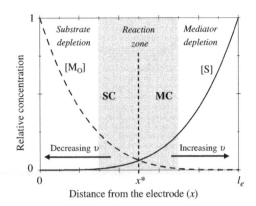

Figure 15.6 Schematics of concentration profiles for substrate (solid line) and mediator (broken line) through the biocatalytic film in case III, for $\gamma = 1$, $\eta = 1$ and limiting current conditions ($m_{O,0} = m_t$). Relative concentrations: $[M_O] = m_O/m_t$; $[S] = s_e/s$ (partition neglected). SC = substrate controlled kinetics; MC = mediator-controlled kinetics. Arrows indicate the shift of the reaction zone in response to γ or η modification.

Concentration profiles in case III are schematically drawn in Figure 15.6 where both the substrate and oxidized mediator (M_O) become depleted within specific boundary regions of the catalytic film. Of course, substrate conversion cannot proceed in such zones, where diffusion is the sole occurring process. Substrate conversion takes place only in the middle zone, where both reactants (that is, the substrate and the oxidized mediator) are present. Reactant concentrations can match each other in the middle of the reaction zone, but at the extremities, one of them falls to negligible values. The lowest-concentration reactant determines the kinetics in the pertinent segment. So, the process is mediator controlled (MC) at the right boundary of the reaction zone but proceeds under substrate control (SC) at the opposite side, whereas mixed control occurs within the midpart of the reaction zone.

Concentration profiles in Figure 15.6 are symmetrical in this particular case in which it is assumed that both γ and η are equal to one. This implies that substrate and mediator flux balance each other (i.e., $\eta = (\phi_M/\phi_S)^2 = 1$). At the same time substrate conversion and enzyme reoxidation balance each other as $\gamma = 1$. In general, the position of the reaction zone is determined by the replenishment factor defined as:

$$v = \frac{\eta}{\gamma f_O} = \frac{D_S s}{D_M m_t f_O} \tag{15.59}$$

Accordingly, the replenishment factor represents the promptness of substrate supply to the reaction zone, relative to the rapidity of the mediator supply.

As demonstrated in [13], the position of the reaction zone (indicated by x^* in Figure 15.4) is determined by the supply factor according to:

$$x^* = \frac{l_e}{1 + v} \tag{15.60}$$

This equation demonstrates that the reaction zone shifts to the limits of the enzyme layer when v deviates from one. If the relative substrate supply is sluggish ($v \ll 1$), the reaction will occur at the external periphery ($x^* \approx l_e$), whereas the remaining part of the catalytic slab can act as a replacement reserve to compensate for gradual decay of the exposed enzyme. However, too sluggish substrate supply causes the kinetics to shift to case IV in Figure 15.5, where the reaction is substrate controlled all across the catalytic layer.

Conversely, with a fast substrate supply ($v \gg 1$), the reaction will occur near the electrode surface and, at very high v, the process becomes mediator controlled all through the film (case II in Figure 15.5). Note that the supply factor depends not only on the s/m_t ratio but also on the electrode potential, via the f_O term. The actual kinetics under limiting-current conditions can be inferred by setting $f_O = 1$.

The diffusion current (i_d) recorded under case III conditions is a linear function of the substrate concentration, as follows [1]:

$$i_d = \frac{nFA}{l_e}(D_M m_t f_O + D_S s) \tag{15.61}$$

So, the sensitivity is inversely proportional to the thickness of the biocatalytic slab and is independent of the rate constants and the enzyme or mediator concentrations. This is especially convenient when performing substrate determinations. However, the response includes also a substrate-independent term that is proportional to the mediator concentration. Therefore, the response is dependent on two calibration parameters. Stringent limits of applicability of this response function are imposed by the conditions determining the occurrence of case III, namely, the supply factor should be located somewhere between 0.1 and 10 in order to avoid the kinetics from shifting the system to cases II or IV. Optimization of the response range depends on the interplay of substrate and mediator concentrations.

If partition cannot be neglected, s in the response function should by corrected by the partition coefficient which means that the sensitivity is affected by partition.

In short, case III is characterized by a convenient response function, but care should be exercised in order to prevent the kinetics from shifting to cases II or IV. In each of these cases, the response is a function of $s^{1/2}$ and depends in addition on rate constants and both enzyme and mediator loading [1]. The main optimization parameters are the substrate and mediator concentrations that should be selected so as to render the supply factor close to one. Consequently, the linear response range cannot be expected to exceed one order of magnitude.

15.2.7 Outlook

This section gave an insight into the inherent complexity of the processes occurring in a mediated amperometric enzyme sensor. Despite this intricacy, theory provides useful guidelines for selecting design parameters and working conditions that secure a convenient response function. Such conditions can be met by operating the sensor under kinetic control, (case V) and nonsaturated enzyme kinetics. The sensitivity in this case can be adjusted by means of the enzyme loading but, at the same time, the response will be affected by enzyme decay or inhibition. A diffusion-limited current (case III) is obtained at high enzyme and mediator loadings, with the additional condition of a supply factor close to unity. The sensitivity depends in this case on the substrate diffusion coefficient and the thickness of the enzyme layer but is free of effects from kinetic parameters and enzyme or mediator concentrations. Consequently, this regime allows the sensor to be almost insensitive to small temperature or pH fluctuations and also not to be affected by partial enzyme inactivation.

Following the treatment in [13], it was assumed here that a mediator molecule can accept two electrons. The case of a single-electron mediator was considered in [14]. On the other hand, it was assumed in this section that the enzyme–mediator reaction is a single-step process. A further refinement can be introduced by assuming that the enzyme–mediator reaction proceeds according to the Michaelis–Menten kinetics as was assumed in ref. [15] for the case for a solution-dissolved mediator.

Although the immobilized mediator design is the most convenient format, the case of the solution-dissolved mediator is also of interest for both mechanistic investigation and for some practical applications. Sensor modeling under these conditions has been reviewed in refs. [16,17] (which also includes sections profiled on immobilized mediator sensors).

Questions and Exercises (Section 15.2)

1 Give an account of the transport processes involved in a mediated amperometric enzyme sensor operating under internal diffusion conditions. Point out the characteristic constants of these processes and comment on their physical meaning.

2 Comment on the chemical reactions involved in the sensor process and outline what are the characteristic constants of these reactions.

3 Each dimensionless parameter introduced in this section compares the maximum possible velocities of two processes occurring in sequence. Put into words the definition of each parameter in terms of maximum velocities.

4 What are the conditions under which the sensor response is a kinetic current or a diffusion current?

5 What parameters determine the position of the reaction zone within the enzyme layer under internal diffusion control conditions? What position is best from the standpoint of sensor response stability?

References

1. Bartlett, P.N., Toh, C.S., Calvo, E.J. *et al.* (2008) Modelling biosensor response, in *Bioelectrochemistry: Fundamentals, Experimental Techniques and Applications* (ed. P.N. Bartlett), John Wiley & Sons, Chichester, pp. 267–325.

2. Eddowes, M.J. (1990) Theoretical methods for analysing biosensor performance, in *Biosensors: A Practical Approach* (ed. A. E.G. Cass), IRL Press, Oxford, pp. 211–263.

3. Bartlett, P.N., Tebbutt, P., and Whitaker, R.G. (1991) Kinetic aspects of the use of modified electrodes and mediators in bioelectrochemistry, in *Progress in Reaction Kinetics* (ed. G. Porter), Pergamon Press, London, pp. 55–155.

4. Albery, W.J. and Craston, D.H. (1987) Amperometric enzyme electrodes: theory and experiment, in *Biosensors: Fundamentals and Applications* (eds I. Karube, G.S. Wilson, and A.P.F. Turner), Oxford University Press, Oxford, pp. 180–210.

5. Gooding, J.J., Hall, E.A.H., and Hibbert, D.B. (1998) From thick films to monolayer recognition layers in amperometric enzyme electrodes. *Electroanalysis.*, **10**, 1130–1136.

6. Gooding, J.J. and Hibbert, D.B. (1999) The application of alkanethiol self-assembled monolayers to enzyme electrodes. *TrAC-Trends Anal. Chem.*, **18**, 525–533.

7. Willner, I. and Katz, E. (2000) Integration of layered redox proteins and conductive supports for bioelectronic applications. *Angew. Chem. Int. Ed.*, **39**, 1180–1218.

8. Albery, W.J. and Bartlett, P.N. (1985) Amperometric enzyme electrodes. 1. Theory. *J. Electroanal. Chem.*, **194**, 211–222.

9. Lyons, M.E.G. (2003) Mediated electron transfer at redox active monolayers. Part 4: Kinetics of redox enzymes coupled with electron mediators. *Sensors*, **3**, 19–42.

10. Baronas, R. and Kulys, J. (2008) Modelling amperometric biosensors based on chemically modified electrodes. *Sensors*, **8**, 4800–4820.

11. Lyons, M.E.G. (2009) Transport and kinetics at carbon nanotube - redox enzyme composite modified electrode biosensors. *Int. J. Electrochem. Sci.*, **4**, 77–103.

12. Lyons, M.E.G. (2009) Transport and kinetics at carbon nanotube -redox enzyme composite modified electrode biosensors Part 2. Redox enzyme dispersed in nanotube mesh of finite thickness. *Int. J. Electrochem. Sci.*, **4**, 1196–1236.

13. Bartlett, P.N. and Pratt, K.F.E. (1995) Theoretical treatment of diffusion and kinetics in amperometric immobilized enzyme electrodes .1. Redox mediator entrapped within the film. *J. Electroanal. Chem.*, **397**, 61–78.

14. delle Noci, S., Frasconi, M., Favero, G. *et al.* (2008) Electrochemical kinetic characterization of redox mediated glucose oxidase reactions: A simplified approach. *Electroanalysis.*, **20**, 163–169.

15. Limoges, B., Moiroux, J., and Saveant, J.M. (2002) Kinetic control by the substrate and/or the cosubstrate in electrochemically monitored redox enzymatic homogeneous systems. Catalytic responses in cyclic voltammetry. *J. Electroanal. Chem.*, **521**, 1–7.

16. Bourdillon, C., Demaille, C., Moiroux, J. *et al.* (1996) From homogeneous electroenzymatic kinetics to antigen-antibody construction and characterization of spatially ordered catalytic enzyme assemblies on electrodes. *Accounts Chem. Res.*, **29**, 529–535.

17. Savéant, J.M. (2006) Enzymatic catalysis of electrochemical reactions, in *Elements of Molecular and Biomolecular Electrochemistry*, Wiley-Interscience, Hoboken, N.J., pp. 298–347.

16

Electrochemical Affinity and Nucleic Acid Sensors

This chapter introduces affinity and nucleic acid sensors based on dynamic electrochemistry transduction methods. Most of the methods reviewed here belong to the amperometric/voltammetric class, but applications of reciprocal derivative chronopotentiometry are also emphasized as this method provides excellent sensitivity in certain transduction processes based on irreversible electrochemical reactions.

In both classes of sensor presented in this chapter, the recognition occurs by formation of molecular complexes involving multiple noncovalent interactions. However, nucleic acid sensors present the particular feature of recognition by hydrogen bonding that involves specific nucleobase pairs.

Another common feature of sensors in both above classes is the extensive use of transduction labels that can be either enzymes or electrochemically active small molecules.

16.1 Amperometric Affinity Sensors

A great deal of research work on amperometric transduction in immunosensors has led to important advances in this field [1–4]. As compounds involved in immunosensing are usually not electrochemically active, amperometric immunosensors rely on labeling with either electrochemically-active compounds or redox enzymes.

Application of synthetic receptors (such as molecularly imprinted polymers) provides opportunities for substituting expensive and unstable antibodies with stable materials for the determination of small molecules. In this area, much progress has been made in the development of sensors for the determination of electrochemically active compounds, although inactive species can also be determined by means of redox probes or by competitive assay.

16.1.1 Redox Labels in Amperometric Immunosensors

In principle, one can resort to either the sandwich or competitive format using immunoreagents tagged with redox labels in order to assess amperometrically the extent of the recognition reaction. However, the sensitivity of this method is often not sufficient for the analysis of real samples. Signal amplification by electrochemical recycling of the redox label can be applied in such instances. Recycling can also be performed by means of an enzyme if the redox probe can act as its substrate. Ferrocenes are among the most common redox labels in such applications.

16.1.2 Enzyme-Linked Amperometric Immunosensors

Enzyme labels allow very sensitive immunosensors to be designed since one single label produces a large amount of detectable product. Both *hydrolases* (alkaline phosphatase and β-galactosidase) and *oxidases* (peroxidase, glucose oxidase and laccase) are suitable for such applications. A common example is alkaline phosphatase that catalyzes the dephosphorylation of *p*-aminophenyl phosphate (PAPP) leading to *p*-aminophenol (PAP) that can be detected by its electrochemical reaction (Figure 16.1). The limit of detection can be improved to a great extent by electrochemical recycling of PAP.

Redox enzyme labels provide various opportunities for developing amperometric immunosensors. For example, Figure 16.2A outlines the principles of an enzyme-linked competitive format immunosensor. As is typical in this format, an enzyme-labeled analyte-analogue is added to the sample containing the target analyte. The enzyme is brought to near the surface of the electrode by the immunoreaction and catalyzes the conversion of the substrate into an electrochemically active product that undergoes an electrochemical reaction to generates the response current.

The principle of the sandwich format applied to amperometric immuno sensors is illustrated in Figure 16.2B. In this format, the analyte interacts with the immobilized receptor at the electrode surface to form a binary complex. Then, an enzyme-labeled secondary antibody binds to the binary complex and, in the presence of the enzyme

Chemical Sensors and Biosensors: Fundamentals and Applications, First Edition. Florinel-Gabriel Bănică.
© 2012 John Wiley & Sons, Ltd. Published 2012 by John Wiley & Sons, Ltd.

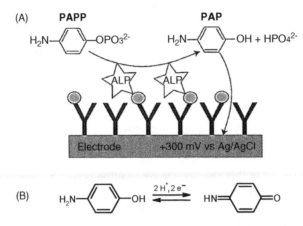

(B)

$$H_2N-\text{\textcircled{}}-OH \underset{}{\overset{2\,H^+,2\,e^-}{\rightleftarrows}} HN=\text{\textcircled{}}=O$$

Figure 16.1 An amperometric immunosensor based on alkaline phosphatase (ALP) as label. (A) Transduction mechanism; (B) electrochemical reaction of PAP. Adapted with permission from [4]. Copyright John Wiley & Sons, Ltd.

substrate, an electrochemically active product forms. The electrochemical reaction of this product provides the sensor response in the form of an electric current.

In the case of an oxidase label, the product can be hydrogen peroxide that form by the reduction of dissolved oxygen. In order to obtain an oxygen-free immunosensor, one can resort to an electron-transfer mediator such as a ferrocene derivative or hydroquinone. After incorporating the enzyme label in the sensing part, the immunosensor acts as an amperometric enzyme sensor composed of an enzyme monolayer. Therefore, it functions under the condition of external diffusion control (see Section 15.1). As the enzyme loading is rather low, the device has the advantage that it functions under a zero-order kinetic regime and generates a response proportional to the amount of enzyme present at the interface. As the amount of enzyme label is determined by the analyte concentration, the response current is itself a function of this concentration.

Screen-printing technology that allows facile fabrication of cheap, disposable sensors, has proved to be very convenient for manufacturing amperometric immunosensors [6]. For example, the device presented in Figure 16.3A has been designed as a sensor for determining the 2,4-dichlorophenoxyacetic acid pesticide. It is built up on the screen-printed transducer strip (I) that includes the working electrode and an Ag/AgCl reference electrode connected to the measuring instrument via silver paste paths coated with an insulator. The capture antibody can be immobilized on the working electrode surface. Alternatively, it can be bound to a nylon mesh disk (II) which is stored and assembled with the transducer just prior to the assay. In order to deal with very small volume samples, a miniature cell (III) is attached over the working area. As this sensor has been designed to work in the competitive format, horseradish peroxidase has been conjugated with analyte molecules and added to the sample along with hydroquinone and hydrogen peroxide. Hydroquinone is catalytically oxidized to *p*-benzoquinone that is then reduced at the electrode back to hydroquinone. The reduction current represents the response signal. As expected for a competitive assay, the response current decreases with increasing analyte concentration as shown in Figure 16.3B.

A similar manufacturing technology has been used to produce a 5-channel multiple sensor on a single strip, which affords simultaneous assay of different samples [7].

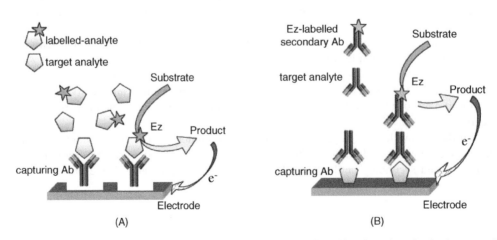

Figure 16.2 Amperometric immunoassay formats using a redox enzyme label (Ez). (A) antigen detection by the competitive immunoassay format. (B) Antibody detection by the sandwich immunoassay format. Reproduced with permission from [5]. Open access.

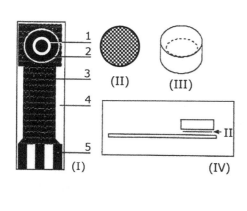 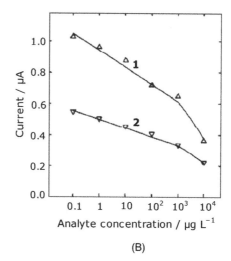

(A)   (B)

Figure 16.3 A screen-printed amperometric immunosensor. (A) sensor configuration. (I) The screen-printed strip. (1) working electrode; (2) reference electrode; (3) insulator layer; (4) ceramic support; (5) electric contacts. (II) Nylon mesh disk with immobilized antibody. (III) A piece of plastic tube forming a minicell. (IV) Assembly of the sensor. (B) Calibration graph for the determination of the 2,4-dichlorophenoxyacetic acid pesticide with the above sensor. (1) Maximum current recorded after adding the sample; (2) steady-state current. Adapted with permission from [6]. Copyright 1995 Elsevier.

16.1.3 Separationless Amperometric Immunosensors

Clearly, the approaches outlined above require the labeled compound to be washed out prior to signal evaluation. This is not a major problem when the assay is run in an automated flow-analysis system, but avoiding separation simplifies the protocol and shortens the analysis time.

The separation step can be eliminated by a proper design of the immunosensor if a redox enzyme is used as label. This is based on the following principle. Redox enzymes perform electron transfer from substrate to the electrode via a mediator that can be immobilized within a redox polymer in order to allow for enzyme wiring (see Section 14.2.5). Hence, the immunosensor sensing layer can be obtained by assembling the capture receptor along with a redox polymer at the electrode surface. Binding of the labeled reagent into the immunocomplex brings the enzyme into the sensing layer where electron transfer is feasible and gives rise to a current proportional to the surface enzyme concentration. The bulk solution enzyme can hardly assume a favorable orientation at the surface and is hence not operative. Consequently, the current can be monitored with no prior separation of the labeled reagent. A redox polymer sensor therefore elicits spatial resolution as it can distinguish between the dissolved labeled species and the same species included in the immunocomplex.

Figure 16.4 presents a competitive format immunosensor based on this principle. The antianalyte antibody is incorporated within a redox polymer while the enzyme-labeled analyte is added to the sample and competes with the analyte itself in the antibody-binding process. Therefore, the amount of enzyme brought to the surface increases with decreasing analyte concentration. Ideally, in this arrangement, only the enzyme included in the immunocomplex is able to perform substrate conversion by electron transfer to the electrode via the redox polymer.

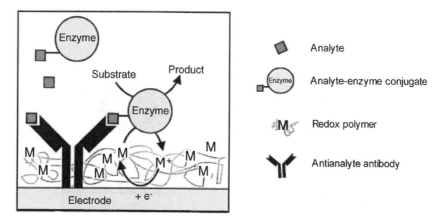

Figure 16.4 An amperometric competitive format immunosensor based on wired enzyme transduction. Adapted with permission from [3]. Copyright 2000 Springer Science + Business Media B.V.

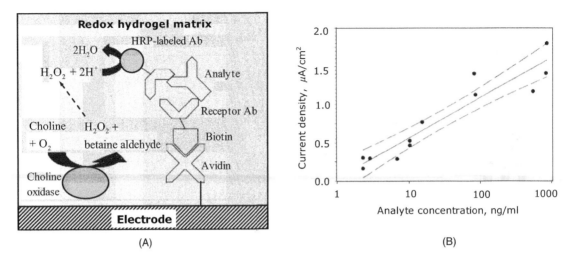

(A)   (B)

Figure 16.5 An amperometric sandwich format immunosensor based on wired enzyme transduction. (A) Sensor configuration; (B) Response function. Broken lines show the 95% confidence interval limits. Adapted with permission from [8]. Copyright 1999 Springer-Verlag.

A separationless sandwich format amperometric immunosensor is shown in Figure 16.5A. Here, a receptor antibody is immobilized within a redox polymer matrix at the electrode surface by avidin–biotin linking. After analyte binding to the receptor, the peroxidase-labeled signaling antibody is added to form the sandwich complex. The enzyme included in this complex catalyzes the reduction of hydrogen peroxide coupled with electron transport via the redox polymer. The contribution of the bulk enzyme to this process is negligible. In this particular example, choline oxidase is integrated with the sensing layer in order to produce hydrogen peroxide needed in the peroxidase reaction.

An electrode modified with avidin, redox polymer and choline oxidase forms a generic platform to which a suitable biotinylated antibody can be added in order to determine a particular analyte. A typical response function for rabbit IgG determination is shown in Figure 16.4B that demonstrates a linear trend on a semilogarithmic plot extending over almost three orders of magnitude.

It should be noted, however, that the contribution of the solution phase enzyme to the total current cannot be totally eliminated. Random interactions between the labeled reagent in the solution and the redox polymer are possible to some extent and give rise to current fluctuations. That is why the points on the graph in Figure 16.5B show a relatively large spread around the average trend. In order to minimize this effect, the labeled antibody to antigen concentration ratio should not exceed 4 : 1.

16.1.4 Nanomaterials Applications in Amperometric Immunosensors

Application of nanomaterials in amperometric immunosensors brought about new opportunities for sensor design and signal amplification [9]. As an example, Figure 16.6 presents a method of signal amplification in a sensor based on a redox probe. In this case, the secondary antibody used in the sandwich format is tagged with liposomes that encapsulate a redox probe solution. After forming the ternary complex (Figure 16.6A), liposomes are broken down by a surfactant and this releases the redox probe. Released redox probe molecules diffuses to the electrode and

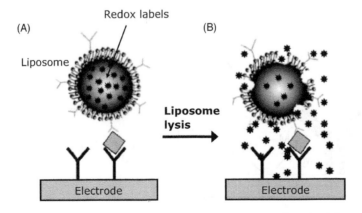

Figure 16.6 Amperometric immunoassay based on a redox probe loaded within liposome labels. Adapted with permission from [9]. Copyright 2007 Elsevier.

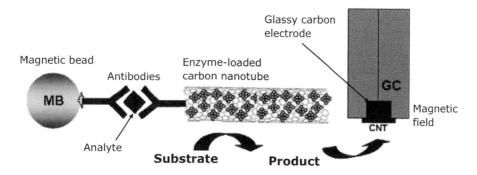

Figure 16.7 An application of carbon nanotubes and magnetic beads in amperometric immunosensors. Adapted with permission from [9] (Copyright 2007 Elsevier B.V.) and [12] (Copyright 2004 American Chemical Society).

undergoes electrochemical conversion. This approach secures a large amount of redox compound which induce a high sensitivity particularly when a sensitive detection methods (such as differential pulse voltammetry) is applied.

Gold nanoparticles used as labels can function as catalysts for certain chemical reactions and, combined with electrochemical recycling, allows achieving detection limits as low as $1\,fg/\mu L$ with the dynamic range stretching over nine orders of magnitude [10].

Carbon nanotubes have been applied as immobilization supports for functional molecules involved with sensor functioning while magnetic beads tags are useful for magnetic accumulation at the electrode surface. Both these principles are outlined in Figure 16.7. In this example, the immunoreaction proceeds in the homogeneous sandwich format, the analyte being coupled to two distinct antibodies. One of them is labelled with an enzyme-loaded carbon nanotube while the second one is tagged with magnetic beads. The ternary complex formed in the solution phase is then collected at the electrode surface by applying a magnetic field. The high enzyme loading achieved by means of the nanotube imparts this sensor with a 100-times enhanced sensitivity with respect to the version based on a single enzyme labeled antibody. On the other hand, magnetic beads integrated with the recognition system allows the immunocomplex to be easily detected without pretreatment steps such as preconcentration or purification, which are implicitly required in standard methods [11].

The enzyme labeling principle illustrated above has also been applied to develop an amperometric immunosensor for prostate specific antigen (PSA), which is a prostate cancer biomarker [13]. As above, the secondary antibody has been attached to carbon nanotubes along with the peroxidase label (Figure 16.8A). The capture antibody has been immobilized to a carbon nanotube forest appended to the underlying electrode. The current signal is generated by the peroxidase-catalyzed reduction of hydrogen peroxide, the electron transfer from the enzyme to the electrode being achieved directly, via the carbon nanotubes. This sensor elicits a linear response within the concentration range of clinical interest (Figure 16.7B). The error bars demonstrate a good reproducibility. Excellent accuracy has been demonstrated by the good agreement with results obtained by the standard ELISA test. If amplification is necessary, it can be obtained by adding hydroquinone as mediator.

16.1.5 Imprinted Polymers in Amperometric Affinity Sensors

Recognition by molecularly imprinted polymers has been used to develop amperometric sensors in which the polymer serves to selectively accumulate the analyte at the electrode surface [14,15]. A facile method for the *in situ* synthesis of imprinted polymers involves electrochemical polymerization at the sensor electrode surface, as demonstrated in Figure 16.9. The polymerization is conducted in a solution containing both the monomer and the template

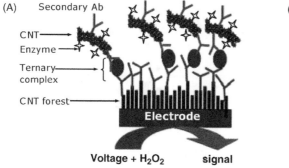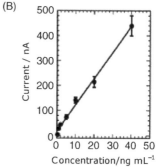

Figure 16.8 Carbon nanotubes as support for antibody immobilization at the electrode surface and support for the enzyme label. (A) Sensor configuration; (B) calibration graph. Adapted with permission from [13]. Copyright 2006 American Chemical Society.

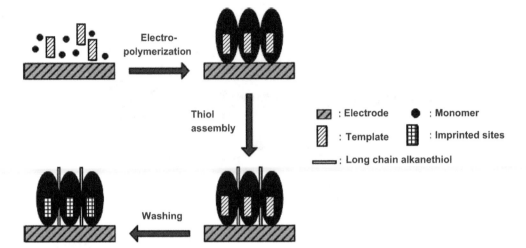

yielding a polymer layer which includes template molecules. In order to block the exposed electrode surface, a long-chain alkanethiol is chemisorbed onto the surface afterwards. Finally, embedded template molecules are washed out, leaving imprinted sites within the polymer layer.

Electrochemically active analytes can be detected by their own electrochemical reaction. In order to achieve a reasonable sensitivity, highly sensitive voltammetric techniques (such as differential pulse voltammetry or square-wave voltammetry) should be applied.

Preformed imprinted polymers can be immobilized as particles embedded in a suitable matrix. Alternatively, they can be cast over the electrode surface to form thin films or can be grown *in situ* by electrochemical polymerization in the presence of the template. Overoxidized polypyrrole proved to be a convenient material for this type of application.

The assay with an imprinted polymer sensor proceeds via three main steps. First, analyte accumulation is performed in a nonaqueous polar solvent such as acetonitrile. The solvent swells the polymer and facilitates access of the analyte to the binding sites within the polymer pores (Figure 16.10). The accumulation time is selected by preliminary trials so as to obtain a sufficiently large response. Next, the sensor is carefully washed with the same solvent in order to remove the adsorbed analyte molecules. Finally, quantitation is carried out in an aqueous solution with the pH sufficiently low to promote the dissociation of the analyte-receptor hydrogen bonds. The analyte is thus released and diffuses within the pores to the electrode surface where the electrochemical reaction takes place. If sensor regeneration is to be attempted, this should be done by means of a solvent that dissolves the reaction products. Often, regeneration of the sensor is hardly achievable. As such sensors are inexpensive and easily manufactured it is suitable therefore to design them as disposable, single-use sensors.

Molecularly imprinted polymers should be prepared so that diffusion of both reactant and product within the polymer pores is possible. Otherwise, the polymer pores become clogged with the product.

An application to the determination the macrocyclic antibiotic rifamycin SV is demonstrated in Figure 16.11. This compound undergoes an anodic reaction at the bare electrode (curve 1) but such a determination can be impaired by the interference from other electroactive compounds. The signal generated after accumulation at an imprinted

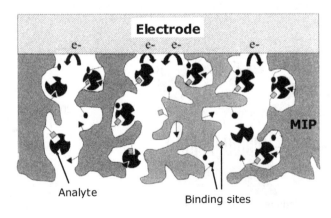

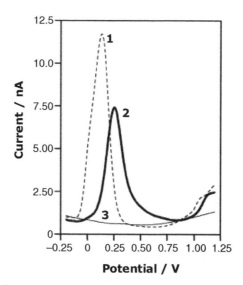

Figure 16.11 Differential pulse voltammograms for the determination of rifamycin SV at a bare carbon electrode (1), at an electrode coated with an imprinted polymer (2) and at a nonimprinted polymer modified electrode (3). 2.5 mM rifamycin in a pH 7.5. phosphate buffer. Reference electrode: Ag|AgCl|KCl(sat.). Adapted with permission from [17]. Copyright 2004 Springer Verlag.

polymer electrode (curve 2) is somewhat lower and shifted towards more positive potentials, but the risk of interference is greatly reduced. Nonspecific accumulation is negligible, as is demonstrated by curve 3 that represents the voltammogram recorded with an electrode coated with the same polymer in the nonimprinted form.

If the analyte itself is not electrochemically active, the response current can be produced by a redox probe whose permeation through the polymer film is modulated by the recognition event. The competitive detection scheme is also applicable with imprinted polymer sensors. In this version, the analyte is first bound to the receptor layer and the sensor is then incubated with an inactive analyte analog. The analyte thus released is finally detected by its electrochemical reaction.

The quality of imprinted polymer sensors can be assessed by means of the response current enhancement factor CE, defined as follows [15]:

$$CE = \frac{I_{MIP}}{I_{NMIP}} \tag{16.1}$$

Here, I_{MIP} is the response of an imprinted polymer coated electrode and I_{NMIP} is the response of a control electrode coated with the same polymer in the nonimprinted form. The selectivity of the sensor can be estimated by comparing CE values for the analyte itself and for possible interferents.

The response time is determined by the rate of analyte diffusion within the pores to the electrode surface. That is why the response time with sensors using MIP particles are longer than that with thin films. Imprinted polymers in the form of nanoparticles can elicit even shorter response time.

Long-term stability of the sensor is determined by the stability of the polymer itself. Best stability has been reported for highly crosslinked polymers such as acrylic or vinyl polymers.

Although much progress has been made in the application of imprinted polymers, it is still important to expand their application to real samples. Also, elimination of nonaqueous solvents in the assay protocol is sought.

16.1.6 Outlook

The design of amperometric immunosensors relies on strategies based on either redox labels or on enzyme labels. In both cases, the capture receptor is immobilized at the surface of an electrode and the label is brought to the electrode surface by suitable immunoreactions, either in the competitive or the sandwich format.

Redox label immunosensors provide satisfactory sensitivity only when combined with a recycling method. On the other hand, use of enzyme labels leads to very sensitive sensors owing to the fact that one single enzyme molecule catalyzes the conversion of a large number of substrate molecules per unit time. High sensitivity is provided by enzymes with a large turnover number. The enzyme-substrate couple is selected so as to produce an electrochemically active compound whose electrochemical reaction generates the response current. Hydrolase or oxidase enzymes are suited as labels in amperometric immunosensors.

Oxidase enzymes offer the possibility of developing separationless immunosensors. In this configuration, the capture receptor is immobilized at the electrode surface within a redox polymer matrix, which provides mediated electron transfer from the enzyme active site to the electrode. Therefore, the enzyme tagged to an immunoreagent present in the solution phase does not contribute to the sensor response and need not be washed out.

Application of nanomaterials is advantageous in that large numbers of labels can be attached to an immunoreagent. Thus, liposomes allow the incorporation of a large number of small-molecule redox probes, whilst carbon nanotubes used as supports for enzyme immobilization are useful for preparing multienzyme molecule tags. Tagging with magnetic beads eliminates the necessity for immobilization of the capture receptor as the immunocomplex formed in the bulk of the solution can be accumulated at the electrode surface by means of a magnetic field.

Affinity amperometric sensors for various compounds can be produced by means of molecularly imprinted polymers. Imprinted polymers for such applications can be readily synthesized by electrochemical polymerization at the sensor electrode surface. If the analyte is electrochemically active, the imprinted polymer imparts selectivity by preferential accumulation of the analyte. In the case of nonelectrochemically inactive analytes, one can resort to a competitive assay, using an electrochemically active analyte analog.

Questions and Exercises (Section 16.1)

1 Sketch diagrammatically the configuration of an amperometric immunosensor based on redox labels. Select some suitable redox labels and suggest procedures for the enzymatic or electrochemical recycling of the label.

2 Write relevant chemical and electrochemical reactions involved with a hydrolase-label amperometric immunosensor.

3 Draw a summary flow chart for the fabrication of an amperometric immunosensor based on horseradish peroxidase as label and write relevant chemical and electrochemical reactions.

4 Select an enzyme–substrate couple and a redox polymer that can be used to produce a separationless immunosensor in the competitive format. Outline the main steps in sensor fabrication and draw schematically the reaction mechanism in this sensor.

5 Sketch diagrammatically the configuration of a separationless amperometric sandwich-type immunosensor using glucose oxidase as peroxide-generating enzyme.

6 Give an overview of possible methods for using nanomaterial as platforms for label-assembly in the design of amperometric immunosensors. What are the advantages of this approach?

7 What are the principles of the magnetic beads applications to amperometric immunosensors and what benefits are obtained by their use?

8 Outline the main steps in the electrochemical preparation of a molecularly imprinted polymer layer at an electrode surface.

9 Give an overview of the main steps in the application of amperometric determinations by means of imprinted polymers. What are the advantages and the drawbacks of such analytical techniques?

10 How can electrochemically inactive compounds be determined by amperometry using imprinted polymer-modified electrodes?

16.2 Electrochemical Nucleic Acid-Based Sensors

Amperometric transduction in nucleic acid sensors can be achieved in several ways [18,19]. First, certain nucleobases are themselves electrochemically active even when included in a nucleic acid sequence. Therefore, nucleic acids containing such nucleobases are self-indicating and need, in principle, no redox labels for amperometric transduction. However, much better sensitivity is achieved when mediated electron transfer to the active nucleobase is used instead. In this case, carefully structured probes are needed in order to prevent the mediator from contributing directly to the overall current. This trouble can be avoided by using redox labels that are either intercalated in the duplex species formed by hybridization or are covalently attached to a species involved with this process. Considerable amplification of the response signal is achieved by using redox enzyme or nonbiotic catalysts as labels. IUPAC recommendations concerning concepts, terms and methodology in electrochemical nucleic acid-based sensors are available in ref. [20].

16.2.1 Electrochemical Reactions of Nucleobases

The most straightforward transduction method is based on the electrochemical reactions of nucleobases residues that are a components of the nucleic acid molecule. The electrochemical activity of nucleic acids was first noted by Paleček using a mercury electrode [21]. Further studies showed that the electrochemical response is due to the reduction of adenine and cytosine residues. The electrochemical activity is sensitive to changes in the nucleic acid

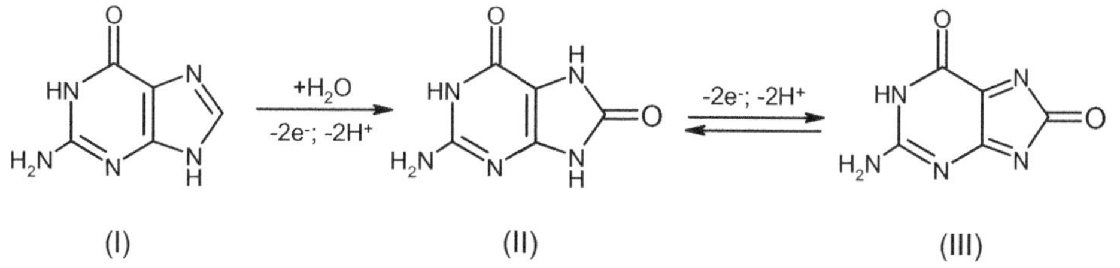

Figure 16.12 Electrochemical oxidation of guanine.

structure and conformation. Therefore, these electrochemical reactions form the basis of a series of methods for assessing DNA hybridization, damage and denaturation [22–25].

Research work based on mercury electrodes laid the fundament of the electrochemistry of nucleic acids and revealed its potential analytical applications. Nevertheless, a liquid mercury electrode is not suitable for developing biosensors and solid electrodes represent the practical alternative for electrochemical transduction. Solid electrodes are best suited for transduction by anodic reactions of nucleobases, particularly guanine and adenine. These compounds, as well as their nucleosides and nucleotides yield well-developed voltammetric signals at graphite electrodes over a broad pH range [26,27]. The anodic reaction of guanine involves two successive two-electron transfer processes, according to the scheme in Figure 16.12. However, the product (III) is not stable and subsequently decays to other final products.

Anodic reactions of guanine and adenine are demonstrated in Figure 16.13 that displays cyclic voltammograms of calf tymus DNA at a bare glassy carbon electrode (curve 1), the same electrode coated with a nonionic surfactant (curve 2) and at same electrode modified with multiwalled carbon nanotubes (MWCNTs) with no prior accumulation (curve 3), and after prior adsorptive accumulation of DNA at the MWCNT-modified surface (curve 4). Curve 4 displays distinct anodic peaks due to the oxidation of guanine and adenine. Figure 16.13 also demonstrates the large background current, which is a common feature of the anodic reaction of nucleobases included in nucleic acid. This characteristic severely impairs the limit of detection [28].

A solution to the above-mentioned background interference was advanced by Wang *et al.* [29] who employed a carbon paste electrode and monitored the electrochemical reaction by reciprocal derivative chronopotentiometry instead of linear scan voltammetry. This method is characterized by an excellent signal/background ratio and allows the determination of DNA at concentrations as low as $0.5 \, mg \, L^{-1}$.

16.2.2 Amperometric Nucleic Acid Sensors Based on Self-Indicating Hybridization

Direct amperometric transduction is based on the electrochemical oxidation of guanine residues in the target after hybridization with the electrode-immobilized nucleic acid probe. However, guanine, if present in the probe, also reacts giving rise to a disturbing background current. This problem can be avoided if the probe is synthesized such as to contain inosine instead of guanine. Inosine is not electrochemically active and the background current in this case is very low. Based on these principles, a method for microfabrication of thick-film hybridization nucleic acid sensors has

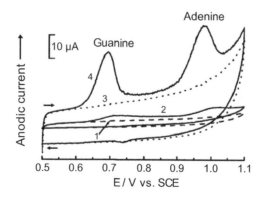

Figure 16.13 Cyclic voltammograms of a $50 \, \mu g \, L^{-1}$ calf tymus DNA solution at a bare and modified glassy carbon electrode. (1) Bare glassy carbon electrode; (2) DHP-coated electrode; (3) MWNT-DHP coated electrode; (4) Same electrode as in (3), after 2 min DNA accumulation at the surface. 0.1 M phosphate buffer (pH 7); potential scan rate, $0.1 \, V \, s^{-1}$. DHP denotes the nonionic surfactant dihexadecyl hydrogen phosphate. Adapted with permission from [30]. Copyright 2003 Springer.

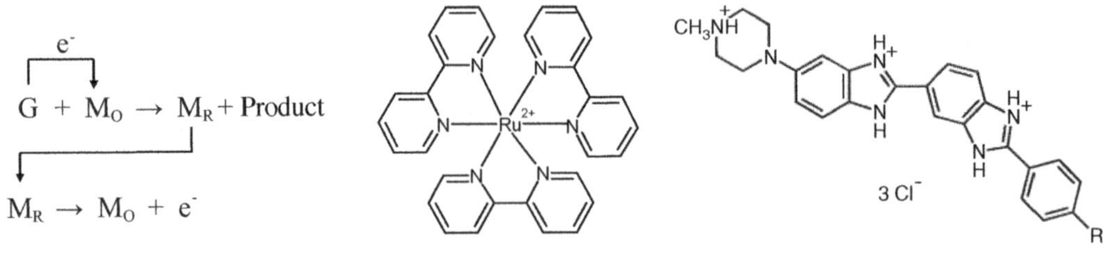

$$
\begin{array}{c}
e^- \\
\downarrow \\
G + M_O \rightarrow M_R + \text{Product} \\
\downarrow \\
M_R \rightarrow M_O + e^-
\end{array}
$$

(A)   (B)   (C)

Figure 16.14 (A) Mediated electrochemical oxidation of guanine (G). M_R and M_O represent the redox mediator in the reduced and oxidized form, respectively. Typical mediators in guanine electrochemical oxidation: (B) tris(bipyridine)ruthenium(II) ($[Ru(bypy)_3]^{2+}$); (C) the staining dye Hoechst 33258 (R = −OH).

been developed [31]. The best quantitation method is reciprocal derivative chronopotentiometry (Section 13.8.8). Such single-use sensors proved suitable for decentralized trace quantitation of DNA in order to detect pathogenic microorganisms.

Nevertheless, when dealing with real samples, various concomitants (such as mismatched and noncomplementary oligomers, chromosomal DNA, RNA and proteins) can undergo electrochemical oxidation along with guanine in the target and lead to a poor sensitivity owing to the enhanced background current. This drawback has been circumvented by linking the probe molecules to magnetic beads and effecting the hybridization in the solution phase. Bead-tagged hybrids are drawn to the electrode surface by a magnet and by this means are separated from the concomitants [32].

The main deficiency of direct guanine detection arises from the slow kinetics of the electrochemical reaction that results in relatively low peak currents and poor sensitivity with standard voltammetric methods. In order to alleviate this limitation, one can resort to a mediated oxidation scheme (Figure 16.14A). In this case, after hybridization, a positively charged redox mediator is attached to the double strand by electrostatic attraction. Suitable mediators are certain metal chelate complexes (such as $[Ru(bypy)_3]^{2+}$[33], $[Co(phen)_3]^{3+}$, $[Cr(bypy)_3]^{3+}$[34] (bypy = 2,2′-bipyridine; phen = ortho-phenanthroline) or the staining dye Hoechst 33258 (Figure 16.14). The mediator undergoes an electrochemical oxidation followed by its reaction with guanine, which is the final electron donor in the process [35]. The actual electrochemical reaction involves only the mediator and is therefore a fast electrochemical reaction. This brings about a considerable enhancement of sensitivity compared with direct guanine oxidation.

Carbon nanotubes offer considerable possibilities for improving the design of amperometric DNA sensors [36]. An example in this respect is represented by the application of carbon nanotubes as probe supports in a DNA sensor based on mediated guanine oxidation, as shown in Figure 16.15 [37]. In this approach, multiwalled carbon nanotubes have been grown on a Cr layer spotted with Ni catalyst sites. In order to impart chemical and mechanical stability to the resulting structure, nanotubes have been embedded in a SiO_2 layer formed by chemical vapor deposition. This dielectric layer allows each nanotube to act as an independent electrode in the array. Mechanical polishing, followed by electrochemical etching has then been applied to shorten and level nanotubes to the same plane as the SiO_2 matrix. As a result, only the end of the nanotube, bearing carboxyl groups, remains exposed. About 70 nanotubes have been thus formed on a $20 \times 20 \, \mu m$ metallic spot. Next, up to 900 probes have then been attached to each nanotube end by amide bonding to carboxyl groups. PCR amplicons of about 300 bases have been used as hybridization

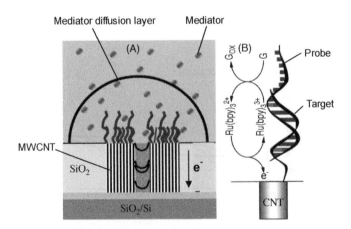

Figure 16.15 Amperometric hybridization sensor based on multiwalled carbon nanoelectrodes with detection by $Ru(bpy)_3^{2+}$ mediated oxidation of guanine. (A) Schematic of the sensor. Due to the relatively large distance between nanotubes, an individual, hemispherical diffusion layer is set up for each of them. (B) Reaction scheme showing the mediated electrochemical oxidation of guanine bases in the hybridized target. Adapted with permission from [37]. Copyright 2004 The Royal Society of Chemistry.

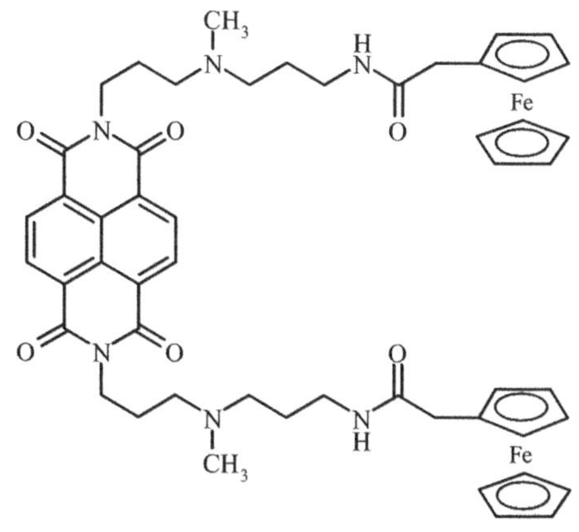

Figure 16.16 Ferrocenyl naphthalene diimide, an intercalating redox probe. Adapted with permission from [39]. Copyright 2003 Wiley-VCH Verlag GmBH & Co. KGaA.

targets. Each target molecule contains about 70 intrinsic guanine bases and thus produces a sufficiently large oxidation current which allows the use of very low nanoelectrode densities. The transduction relies on mediated oxidation of guanine residues using tris(bipyridine)ruthenium(II) as mediator.

The response signal of the sensor in Figure 16.15 has been obtained by multiscan alternating-current voltammetry, which yields a peak-shaped current potential curve. The peak current for the first scan, i_1 results from the joint contribution of two electrochemical processes: mediated guanine oxidation and oxidation of the free mediator. As guanine oxidation is an irreversible process, the second scan yields only the current associated with the free mediator reaction (i_2). The actual response can therefore be obtained as $i_1 - i_2$.

This approach resulted in a very sensitive sensor being obtained by conventional microfabrication technology. It can detect about 300 hybridized targets per spot, close to the detection limit of the fluorescence-based microarray technique. Such sensitivity makes it suitable for direct *in vitro* detection of mRNA.

16.2.3 Intercalating Redox Indicators

Intercalation of a redox indicator represents a simple procedure for labeling a duplex nucleic acid [38,39]. A frequently used indicator of this type is daunomycin that undergoes facile electrochemical oxidation at 0.4–0.5 V vs. Ag|AgCl in neutral solutions. Figure 16.16 shows a purposely synthesized redox indicator, namely, ferrocenyl naphthalene diimide. In this case, the flat, middle part of the molecule binds to DNA by intercalation, while the lateral chains bearing ferrocene active sites extend beyond the duplex structure.

Single-use hybridization sensors based on intercalating indicators have been fabricated by adsorptive immobilization of the probe at carbon screen-printed electrodes [40]. The response has been recorded after daunomycin intercalation by reciprocal derivative chronopotentiometry using the integrated signal (peak area) as analytical signal. In order to avoid a spurious response, the sensor should be carefully rinsed after both hybridization and intercalation. The flow chart of such an assay is shown in Figure 16.17A. Testing of the sensor was carried out with a 21-mer micro-organism oligonucleotide as probe and a complementary target from the same organism as target. A significant response has been noted even in the presence of a noncomplementary target, probably due to nonintercalative binding of daunomycin to the probe layer. However, the response increases clearly with target concentration. In order to correct for the nonspecific response the posthybridization signal should be corrected by subtracting the signal obtained with a noncomplementary target.

A typical response of a redox intercalator-based sensor is shown in Figure 16.17B(a). This sensor has been built up on a screen-printed carbon electrode using $[Co(bypy)_3]^{3+}$ as intercalated indicator. Note the relatively large response for the blank sample (dotted lines) that should be subtracted from the overall signal obtained with target-containing samples (solid lines). Figure 16.17B(b) presents the plot of the corrected response vs. target concentration; this calibration graph is linear over a limited concentration range.

16.2.4 Covalently Bound Redox Indicators in Sandwich Assays

Principles of sandwich DNA sensors have been introduced in Section 7.5.2. This method involves a capture probe and a labeled signaling probe. After the hybridization of the capture probe with the target, the overhanging target fragment is conjugated by hybridization with the signaling probe that is tagged with a ferrocene carboxylic acid residue in order to perform amperometric transduction (Figure 16.18). A sensor thus configured requires the non-hybridized signaling probe in the solution to be washed out prior to the transduction step. This can be avoided by

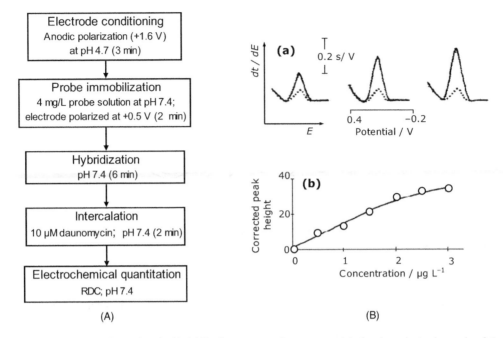

Figure 16.17 (A) Main steps in electrochemical hybridization assays using screen-printed carbon electrodes and an intercalated redox indicator. After each step, the sensor should be carefully washed in order to remove remaining reagents. (B) Reciprocal derivative chronopotentiometry (RDCP) response of a hybridization electrochemical sensor to the concentration of DNA from *Escherichia coli*. Recorded after hybridization (2 min) at a probe-modified screen printed carbon electrode. Indicator: $[Co(bypy)_3]^{3+}$; pH 7; 0.5 M NaCl; (a) RDCP curves with increasing target concentration (1.0 μg/ml steps). Dotted lines: blank response; solid lines: target added. (b) Calibration graph for target determination. (B) Adapted with permission from [34]. Copyright 1997 Elsevier.

careful engineering of the sensing layer, as in the case of the sensor presented in ref. [41]. In this example the capture probe is integrated in a composite self-assembled monolayer at a gold electrode along with molecular wires and insulating alkane molecules terminated by polyethylene glycol moieties. A molecular wire is a molecule that includes an extended conjugated π-orbital system that allows electrons to flow from the gold electrode across the alkane layer. The role of the external polyethylene glycol sheet is to block the access of electroactive species in solution (including free signaling probes) to the terminus of molecular wires. After hybridization, ferrocene tags that are attached to the signaling probe by sufficiently long alkane fragments, protrude from the surface layer and reach the molecular wires that enable electron transfer between ferrocene and the electrode. This is not possible for ferrocenes linked to signaling probe molecules in the solution phase.

The use of a redox probe brings about improved transduction characteristics as the redox probe reaction is fast and occurs at relatively low potentials. This results in a convenient signal/background ratio and allows the detection to be performed by means of very sensitive electrochemical methods, such as differential pulse voltammetry or alternating-current voltammetry. Moreover, the sandwich configuration with a thoroughly engineered sensing layer lessens the risk of interferences. Electrochemical DNA sensors based on sandwich assays are available on a commercial basis as disposable chips.

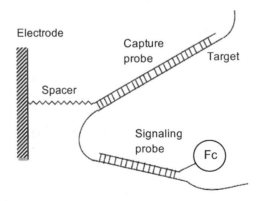

Figure 16.18 An electrochemical DNA sensor based on sandwich assay. Fc stands for ferrocene. Adapted with permission from [38]. Copyright 2001 Chemical Society of Japan.

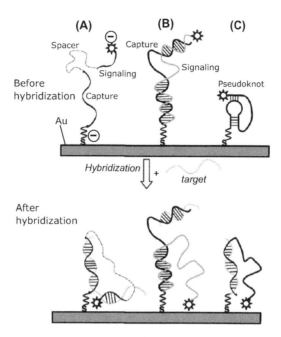

Figure 16.19   Hybridization sensors with a "signal-on" architecture. (A) Sandwich assay using capture and signaling probes connected through a linker; (B) Target induced strand displacement in a discontinuous duplex between capture and signaling probes; (C) target-induced disruption of a pseudoknot. Adapted with permission from [42]. Copyright 2009 Wiley-VCH Verlag GmBH & Co. KGaA.

16.2.5   Covalently Bound Redox Indicators in Spatially Resolved Transduction

An important recognition–transduction approach relies on the hybrid configuration-dependent response (Figure 16.19). Thus, case A in Figure 16.19 corresponds to a sandwich assay in which the capture and signaling probes are connected through a non-nucleotide spacer. The negative redox marker is repelled by the negatively charged probe at the electrode surface. Target hybridization with both capture and signaling probes gives rise to a ternary complex that shifts the label to near the electrode surface and allows the electrochemical reaction to proceed. In case B, the capture and signaling strands form a discontinuous duplex. By hybridization, the target displaces the tagged end of the signaling probe and allows the redox label to approach the electrode surface. Finally, in case C a labeled probe is shaped as a pseudoknot. It changes its configuration by hybridization with the target and the label shifts as a consequence to the electrode surface. All the above methods are of the "signal on" type because the hybridization event displaces the label to the electrode surface from a position far from the electrode. In this way, the response current increases with respect to the background current. Conversely, "signal-off"-type methods are based on label displacement from the electrode surface to a location far from the electrode in response to the hybridization event. As a result, the current decreases after hybridization. "Signal-off" methods are less sensitive owing to limitations imposed by the background current.

16.2.6   Enzyme Labels in Amperometric Nucleic Acid Sensors

Enzymes are widely used as labels in affinity sensors and can therefore also act as transduction labels in nucleic acids sensors [19,39,43]. Inexpensive enzymes with a high turnover number, such as horseradish peroxidase, glucose oxidase and alkaline phosphatase have been used to this end. A proper calibration graph is obtained when the response current is proportional to the concentration of the hybrid-bonded enzyme, that is, when the enzyme assembly operates under kinetic control.

An enzyme-linked DNA sensor can be developed according to the scheme in Figure 16.20. In this approach, an ethanethiol-terminated probe has been attached by self-assembly onto a gold screen-printed electrode while the target has been tagged with biotin at the 5' end. Hybrid labeling has been effected with a streptavidin-conjugated enzyme (alkaline phosphatase) that binds to the hybridized target through streptavidin–biotin affinity reaction. After washing out the solution enzyme, the substrate (α-naphthyl phosphate) is left to undergo enzymatic hydrolysis to the electrochemically active product P_1. Electrochemical oxidation of this product to P_2 provides the response current that has been monitored by differential pulse voltammetry. This scheme needs the target to be tagged with an affinity label. Tagging is effected during the prior polymerase chain reaction by adding a tagged nucleotide to the reaction mixture.

The principles of the mediated enzyme sensor can also be applied for the purpose of transduction in DNA biosensors. An easy way in this respect relies on attaching a mediator (e.g., ferrocene) to the target. By hybridization, the

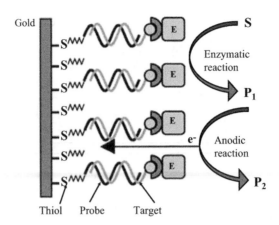

Figure 16.20 Configuration of an electrochemical enzyme-linked DNA sensor based on affinity coupling of the enzyme to the target. E stands for enzyme and S is the substrate. Adapted with permission from [44]. Copyright 2004 Elsevier.

mediator is brought into the surface layer. A suitable enzyme and its substrate are then added to the solution afterwards and the response current is recorded.

Wiring of the enzyme with the electrode through a redox polymer brings about noticeable advantages, particularly due to the fact that the redox polymer protects the electrode surface against the contact with electroactive species in the solution and thus eliminates the need for washing out or separating them before the transduction step. Such a sensor is presented in Figure 16.21. The sensing layer has been formed on a gold ultramicroelectrode (diameter $\leq 100\,\mu m$). First, cysteamine has been self-assembled at the gold electrode surface and then, an -NH$_2$ terminated probe ((5$'$ → 3$'$): NH$_2$-C$_6$-ATTCGACAGGGATAGTTCGA) has been coupled to the -NH$_2$ groups in cysteamine using polyoxyethylene bis(glycidyl ether) as the linker. An osmium redox hydrogel layer was then grown over so as to include the probe. A biotin-tagged sequence from *Listeria monocytogenes* ((5$'$ → 3$'$): biotin-ATTCGACAGGGA-TAGTTCGA) has been used as probe, whereas the control oligonucleotide sequence consisted of (5$'$ → 3$'$): biotin-TCGAACTATCCCTGTCGAAT. The enzyme was glucose oxidase tagged with biotin. The hybridization occurs within the matrix formed by the redox polymer and brings the biotin-tagged target into the surface layer. Subsequently, this tag binds the avidin-tagged enzyme to the hybrid.

In order to assess the hybridization, the substrate (glucose) is then added to the sample solution. Enzymatic oxidation of glucose releases electrons that are transferred to the positively charged electrode through the redox polymer layer and gives rise to the response current. Clearly, the response is a function of the amount of available hybrid and, consequently, of the concentration of the target in the sample.

For calibration and assay, the steady-state current has been recorded at a constant potential ($-0.35\,V$ vs. Ag/AgCl, in samples of pH 7.4 containing 20 mM glucose and 0.15 M NaCl. The calibration graph plotted as current *vs.* the logarithm of the target concentration is linear from 10^{-14} to 10^{-4} M. The very small size of the sensor allows to assay samples of about 10 μL volume.

The previously outlined approaches rely on fully dedicated technologies leading to a device specific to a well-defined target. In order to increase the throughput of the sensor fabrication process, it is advantageous to resort to a generic platform that can be readily adapted to various applications. Such a platform has been constructed

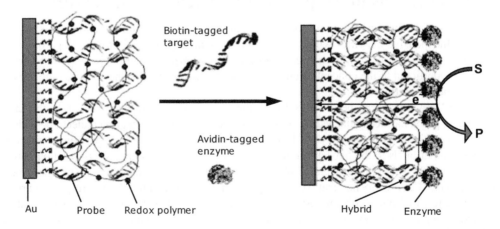

Figure 16.21 A redox hydrogel-based enzyme-linked hybridization sensor. Adapted with permission from [45]. Copyright 2009 Wiley-VCH Verlag GmBH & Co. KGaA.

using an avidin-containing redox polymer layer that has been coated on a carbon electrode [46]. The biotiny-lated capture probe was thereafter attached to the previously formed layer through biotin–avidin binding. Target determination has been carried out by the sandwich assay method, using a peroxidase-tagged signaling probe. In order to perform the assay, the electrode is incubated with the tested solution which contains the analyte nucleic acid, the signaling probe and H_2O_2. The hybridization of the signaling probe sets up an electrical contact between the redox polymer and the peroxidase label. Enzymatic reduction of H_2O_2 proceeds and causes elec-trons to flow through the redox polymer towards the immobilized enzyme. Remarkably, the test could be con-ducted in the presence of extraneous DNA and serum because the surface redox polymer blocks the access of interferents to the electrode surface. A screen-printing fabrication technology based on these principles has been devised [47].

The advantage of this design results from the possibility of readily attaching various biotinylated capture probes to the same platform in order to obtain sensors for various nucleic acids.

In order to enhance the response current, the enzyme has been tagged with several complementary probes [48]. One of them binds to the immobilized target whilst the remaining ones attach by hybridization to other enzyme molecules. The result is a molecular dendritic structure including several enzymes molecules attached to each probe–target hybrid.

Nonbiotic materials can also be used as catalyst labels. Thus, in a sandwich assay, platinum nanoparticles have been attached to the signaling probe. Platinum is a catalyst for electrochemical reduction of H_2O_2 and the resulting current allows the hybridization event to be traced [49].

16.2.7 Electrochemical DNA Arrays

Due to the simplicity of the transduction and, implicitly, of the measuring equipment, as well as to the possibility of performing measurements on turbid samples, electrochemical DNA arrays have been subjected to intensive research effort [50]. Thus, a DNA array has been fabricated by electrochemical polymerization of polypyrrole in the presence of capture-probe-tethered pyrrole at gold microelectrodes arrayed on a silicon chip [51]. This process has been per-formed by means of microcells actuated by a micromanipulator or by scanning electrochemical microscopy. Spots ($50 \times 50 \, \mu m$) have been fabricated in this way on 128 microelectrodes arrays. Amperometric signal transduction has been performed by means of a redox intercalator.

A commercially available microarray platform takes advantage of the enzyme amplification effect. This platform can test up to 12 000 probes in less than 25 s [52]. It is based on a biotin-labeled target sample which after hybridiza-tion is tagged with avidin-conjugated peroxidase. In the presence of H_2O_2, the substrate (3, 3', 5, 5'-tetramethylben-zidene) is converted to a reducible product that is detected by means of its cathodic current.

16.2.8 Nucleic Acids as Recognition Materials for Non-Nucleotide Compounds

As the toxic action of many pollutants (mutagens and carcinogens) is related to their interaction with DNA, it is reasonable to exploit this property by using dsDNA as recognition material in pertinent biosensors for pollution monitoring. To this end, a DNA layer is immobilized at the surface of an electrode. After incubating the sensor with the sample, possible alterations in DNA structure can be noted as modifications in the electrochemical response. The most straightforward method for detecting DNA interaction with the analyte is based on changes in the electrochem-ical signal of guanine oxidation. In this respect, a thorough investigation of various low molecular compounds of analytical concern has been carried out using both ds- and ssDNA [40]. It has been proved that some compounds that interact with DNA by intercalation reduce the guanine signal, whereas other compounds with similar action do enhance it. As expected, the effect is stronger with dsDNA. This implies that a DNA sensor cannot discriminate between various pollutants but can be used as a warning device that indicates the presence of harmful compounds in environmental samples. A preconcentration step of natural water samples is necessary to this end. Compared with microbiological genotoxity tests, the DNA sensor assay is faster and is suitable for field-analysis applications.

Another alternative approach uses the electrochemical reactivity of the analyte itself. Thus, pollutants like aromatic amines or phenols undergo electrochemical oxidation at carbon electrodes after being accumulated at the surface by interaction with a previously attached DNA layer [34]. The response due to the accumulated analyte comes in addition to the guanine signal that is also altered by analyte intercalation. Detection limits close to 0.1 μM have been reported for some polycyclic aromatic amines when the analyte electrochemical signal has been recorded.

16.2.9 Aptamer Amperometric Sensors

Aptamers, that are specific receptors for proteins, exhibit particular advantages with respect to antibodies and have therefore been intensively investigated for the purpose of amperometric biosensor development [42,53–56]. Various

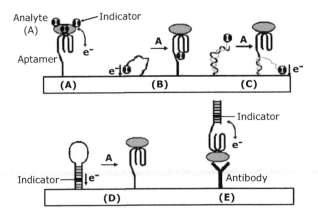

Figure 16.22   Amperometric aptamer sensors using a redox indicator. Adapted with permission from [53]. Copyright 2009 Wiley-VCH Verlag GmBH & Co. KGaA.

amperometric transduction methods for such sensors have been advanced using either redox indicators or redox enzymes.

16.2.9.1 Aptamer Sensors Based on Redox Indicators

In order to perform amperometric transduction, a redox indicator can be attached to the molecule-analyte by various methods. A simple labeling procedure is based on the electrostatic attraction between a negatively charged molecule and a cationic indicator (such as methylene blue). Specific binding of the molecule to the electrode-attached aptamer brings the indicator to a position where it can undergo an electrochemical reaction yielding the response current (Figure 16.22A). Figure 16.22B shows a signal-off approach, in which the indicator is linked to the aptamer that is unfolded in the free form and allows electrons to flow between the redox indicator and the electrode. Analyte binding induces a drastic change in the aptamer conformation that cuts off the electron flow. In Figure 16.22C the aptamer is hybridized with an indicator-bearing oligonucleotide so that the indicator is kept far from the surface. This duplex is partially disrupted by the analyte–aptamer interaction which lets the indicator undergo an electrochemical reaction (signal on approach). An application of a hair pin structured aptamer is shown in Figure 16.22D. In the free form, the duplex moiety of the aptamer is labeled by intercalation with a redox indicator close to the electrode surface. Analyte binding to the aptamer breaks down the duplex and the indicator shifts away by diffusion (signal off). A mixed sandwich approach is shown in Figure 16.22E where the analyte recognition is performed by an antibody, whereas the labeled indicator reacts with the analyte at another site so as to bring the indicator close to the electrode. Care should be taken to prevent a positively-charged indicator from binding to the antibody molecule. This can be achieved by blocking the nonspecific sites on the antibody.

16.2.9.2 Aptamer Sensors Based on Catalytic Reactions

Various strategies for enzyme-linked transduction have been tested with aptamer-based thrombin sensors [57], as emphasized in Figure 16.23. Thus, the method (A) in this figure relies on the ability of thrombin to catalyze the

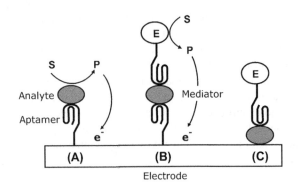

Figure 16.23   Possible configurations of enzyme-linked amperometric aptamer sensors. (A) The analyte (thrombin) catalyzes the formation of an electrochemically active compound; (B) sandwich assay using a second, enzyme-labeled aptamer; (C) the analyte is adsorbed at the electrode surface and then coupled with an enzyme-labeled aptamer. Adapted with permission from [53]. Copyright 2009 Wiley-VCH Verlag GmBH & Co. KGaA.

conversion of peptide-bound *p*-nitroaniline to free *p*-nitroaniline. This compound undergoes electrochemical reduction in both forms, but at different potentials, which allows the detection of the free form by differential pulse voltammetry.

Figure 16.23B shows a sandwich assay involving a second aptamer that is linked to a redox enzyme (for example, peroxidase). In the presence of the substrate and a suitable mediator, the enzymatic reaction gives rise to the response current. In both previous approaches, the aptamer receptor has been attached to a gold electrode by self-assembly via a thiol spacer. Next, this layer has been incubated with mercaptoethanol that adsorbs strongly on the gold free surface and displaces the aptamer molecules randomly adsorbed. In addition, adsorbed mercaptoethanol molecules block any exposed surface site and prevent spurious adsorption of sample components.

In the third alternative (Figure 16.23C) the thrombin analyte has been adsorbed on the mercaptoethanol-modified gold surface and then the resulting layer has been exposed to the enzyme-labeled aptamer. As in case (B), the transduction has been achieved by monitoring the current flowing in the presence of the enzyme substrate and a mediator.

Among these three alternatives, the second one achieved the best sensitivity due to the high turnover number of the label enzyme. In the third alternative, nonspecific adsorption of the labeled aptamer gives rise to a significant background current with a negative effect on the detection limit. Further improvements are needed in order to render such sensors compatible with real biological samples.

16.2.10 Outlook

Hybridization sensors allow simple and fast identification of harmful microorganisms in environment sample and in foods and can be used to monitor biological warfare agents by detecting microorganism nucleic acids contained in such organisms.

Various transduction strategies for amperometric DNA sensors have been developed. The label-free method relies on the anodic reaction of the guanine base included in the nucleic acid sequence. A more general method is based on redox indicators intercalated in the target-probe hybrid. Alternatively, redox indicators can be covalently attached to a signaling probe so as to place them either near to or far from the electrode surface. By hybridization, the indicator is shifted to be far from or near to the electrode, respectively. The first alternative is of the "signal-off" type; the second is a "signal-on" technique.

Much improvement in sensitivity is brought about by utilization of enzyme labels that perform a chemical amplification of the response current. The enzyme produces an electrochemically active compound that generates the response current. Redox enzymes are particularly advantageous because they allow coupling with a redox hydrogel in order to perform wired electron transfer to the electrode. In this way, the enzyme present in the bulk of the solution is not involved in the electrochemical reaction which eliminates the need for washing out before the detection step.

When compared with other transduction methods, electrochemical transduction renders a DNA sensor relatively simple, less costly and more amenable for miniaturization and mass production. A critical issue with such sensors is their relatively poor detection limits. That is why preliminary amplification by the polymerase chain reaction is usually needed [58]. Much improvement in sensitivity is expected from the use of carbon nanotubes [59].

Hand-held analyzers based on nucleic acid sensors have also been developed allowing field analysis to be conducted with obvious advantages [60].

Amperometric aptamer sensors appear as an alternative to standard immunological assay and open new possibilities for detecting proteins. Transduction in amperometric aptamer sensors can be achieved by means of redox indicators, but best sensitivity is achieved by means of enzyme labels.

Questions and Exercises (Section 16.2)

1 In amperometric DNA sensors the hybridization event can be reported directly with no use of labels. What are the electrochemical reactions involved in this process? What are the most suitable electrochemical quantitation methods for use in these sensors?

2 How can the electrochemical response signal be enhanced?

3 What advantages are brought about by the utilization of carbon nanotubes in such sensors?

4 How can the dsDNA intercalation be employed for amperometric transduction? Draw the chemical formula of several redox intercalators and discuss their behavior with reference to the chemical structure.

5 Comment on methods for mass fabrication of DNA sensors based on intercalated redox labels.

6 What are the principles and advantages of the sandwich amperometric assay? Discuss the methods for improving the specificity and the sensitivity of such sensors.

7 The response of a hybridization sensor depends to a large extent on the position of the redox label with respect to the electrode surface. Review several transduction methods based on this principle and point out the advantages of this approach.

8 Amperometric transduction can be achieved by means of a free-diffusing redox probe and an intercalated mediator. Comment on the principles of this method and point out the physicochemical factors that determine the strength of the response current.

9 Why are catalytic labels advantageous in the design of amperometric DNA sensors? What kind of catalysts can be used to this end?

10 Comment on the configuration of an enzyme-linked DNA sensor based on the detection of the product of an enzymatic reaction. What is the effect of the substrate concentration on the response of such a sensor?

11 How can a redox enzyme be used for amperometric transduction in DNA sensors? What advantages does this approach bring about?

12 Review the feasibility and possible advantages of amperometric transduction in DNA microarrays.

13 What kind of non-nucleic acid analytes can be recognized by DNA-based chemical sensors? What are the main applications of chemical sensors based on this principle?

14 What are aptamers and how can they be obtained? What are the main analytical applications of aptamers?

15 How can an aptamer interact with the analyte? Discuss aptamer advantages with respect to other recognition materials employed for the same purpose.

16 Review the strategies for amperometric transduction in aptamer sensors based on redox labels.

17 What are the advantages of enzyme-based transduction in amperometric aptamer sensors? Discuss several methods for enzymatic transduction in such sensors.

References

1. Centi, S., Laschi, S., and Mascini, M. (2009) Strategies for electrochemical detection in immunochemistry. *Bioanalysis*, **1**, 1271–1291.
2. Fowler, J. M., Wong, D. K. Y., Halsall, H. B. *et al.* (2008) Recent developments in electrochemical immunoassays and immunosensors, in *Electrochemical Sensors, Biosensors, and Their Biomedical Applications* (eds X. Zhang, H. Ju and J. Wang) Academic Press, Amsterdam, pp. 115–143.
3. Warsinke, A., Benkert, A., and Scheller, F.W. (2000) Electrochemical immunoassays. *Fresenius J. Anal. Chem.*, **366**, 622–634.
4. Yakovleva, J. and Emneus, J. (2008) Electrochemical immunoassay, in *Bioelectrochemistry: Fundamentals, Experimental Techniques and Applications* (ed. P.N. Bartlett), John Wiley & Sons, Chichester, pp. 377–410.
5. Belluzo, M.S., Ribone, M.E., and Lagier, C.M. (2008) Assembling amperometric biosensors for clinical diagnostics. *Sensors*, **8**, 1366–1399.
6. Kalab, T. and Skladal, P. (1995) A disposable amperometric immunosensor for 2,4-dichlorophenoxyacetic acid. *Anal. Chim. Acta*, **304**, 361–368.
7. Skladal, P. and Kalab, T. (1995) A multichannel immunochemical sensor for determination of 2,4-dichlorophenoxyacetic Acid. *Anal. Chim. Acta*, **316**, 73–78.
8. Campbell, C.N., de Lumley-Woodyear, T., and Heller, A. (1999) Towards immunoassay in whole blood: separationless sandwich-type electrochemical immunoassay based on in-situ generation of the substrate of the labeling enzyme. *Fresenius J. Anal. Chem.*, **364**, 165–169.
9. Liu, G.D. and Lin, Y.H. (2007) Nanomaterial labels in electrochemical immunosensors and immunoassays. *Talanta*, **74**, 308–317.
10. Das, J., Aziz, M.A., and Yang, H. (2006) A nanocatalyst-based assay for proteins: DNA-free ultrasensitive electrochemical detection using catalytic reduction of p-nitrophenol by gold-nanoparticle labels. *J. Am. Chem. Soc.*, **128**, 16022–16023.
11. Kuramitz, H. (2009) Magnetic microbead-based electrochemical immunoassays. *Anal. Bioanal. Chem.*, **394**, 61–69.
12. Wang, J., Liu, G.D., and Jan, M.R. (2004) Ultrasensitive electrical biosensing of proteins and DNA: Carbon-nanotube derived amplification of the recognition and transduction events. *J. Am. Chem. Soc.*, **126**, 3010–3011.
13. Yu, X., Munge, B., Patel, V. *et al.* (2006) Carbon nanotube amplification strategies for highly sensitive immunodetection of cancer biomarkers. *J. Am. Chem. Soc.*, **128**, 11199–11205.
14. Blanco-Lopez, M.C., Lobo-Castanon, M.J., Miranda-Ordieres, A.J. *et al.* (2004) Electrochemical sensors based on molecularly imprinted polymers. *TrAC-Trends Anal. Chem.*, **23**, 36–48.
15. Suryanarayanan, V., Wu, C.T., and Ho, K.C. (2010) Molecularly Imprinted Electrochemical Sensors. *Electroanalysis*, **22**, 1795–1811.
16. Blanco-Lopez, M.C., Lobo-Castanon, M.J., Miranda-Ordieres, A.J. *et al.* (2003) Voltammetric sensor for vanillylmandelic acid based on molecularly imprinted polymer-modified electrodes. *Biosens. Bioelectron.*, **18**, 353–362.
17. Blanco-Lopez, M.C., Gutierrez-Fernandez, S., Lobo-Castanon, M.J. *et al.* (2004) Electrochemical sensing with electrodes modified with molecularly imprinted polymer films. *Anal. Bioanal. Chem.*, **378**, 1922–1928.
18. de-los-Santos-Alvarez, P., Lobo-Castanon, M.J., Miranda-Ordieres, A.J. *et al.* (2004) Current strategies for electrochemical detection of DNA with solid electrodes. *Anal. Bioanal. Chem.*, **378**, 104–118.
19. Sassolas, A., Leca-Bouvier, B.D., and Blum, L.J. (2008) DNA biosensors and microarrays. *Chem. Rev.*, **108**, 109–139.

20. Labuda, J., Brett, A.M.O., Evtugyn, G. *et al.* (2010) Electrochemical nucleic acid-based biosensors: Concepts, terms, and methodology (IUPAC Technical Report). *Pure Appl. Chem.*, **82**, 1161–1187.

21. Paleček, E. (1960) Oscillographic polarography of highly polymerized deoxyribonucleic acid. *Nature*, **188**, 656–657.

22. Paleček, E. and Fojta, M. (2001) Detecting DNA hybridization and damage. *Anal. Chem.*, **73**, 74A–83A.

23. Paleček, E. (2009) Fifty years of nucleic acid electrochemistry. *Electroanalysis*, **21**, 239–251.

24. Paleček, E., and Fojta, M. (2012) *Nucleic Acid Electrochemistry: Basics and Applications*, Springer, Berlin.

25. Oliveira-Brett, A.M. (2008) Electrochemical DNA assays, in *Bioelectrochemistry: Fundamentals, Experimental Techniques and Applications* (ed. P.N. Bartlett), John Wiley & Sons, Chichester, pp. 411–442.

26. Brabec, V., Vetterl, V., and Vrana, O. (1996) Electroanalysis of biomolecules, in *Experimental Techniques in Bioelectrochemistry* (eds V. Brabec, G. Milazzo, and D. Walz), Birkhäuser, Basel, pp. 287–359.

27. Dryhurst, G. (1977) *Electrochemistry of Biological Molecules*, Academic Press, New York.

28. Brabec, V. (1980) Electrochemical oxidation of nucleic acids and proteins at graphite electrode – qualitative aspects. *Bioelectrochem. Bioenerget.*, **7**, 69–82.

29. Wang, J., Bollo, S., Paz, J.L.L. *et al.* (1999) Ultratrace measurements of nucleic acids by baseline-corrected adsorptive stripping square-wave voltammetry. *Anal. Chem.*, **71**, 1910–1913.

30. Wu, K.B., Fei, J.J., Bai, W. *et al.* (2003) Direct electrochemistry of DNA, guanine and adenine at a nanostructured film-modified electrode. *Anal. Bioanal. Chem.*, **376**, 205–209.

31. Wang, J., Cai, X.H., Tian, B.M. *et al.* (1996) Microfabricated thick-film electrochemical sensor for nucleic acid determination. *Analyst*, **121**, 965–969.

32. Wang, J., Kawde, A.N., Erdem, A. *et al.* (2001) Magnetic bead-based label-free electrochemical detection of DNA hybridization. *Analyst*, **126**, 2020–2024.

33. Napier, M.E., Loomis, C.R., Sistare, M.F. *et al.* (1997) Probing biomolecule recognition with electron transfer: Electrochemical sensors for DNA hybridization. *Bioconjugate Chem.*, **8**, 906–913.

34. Wang, J., Rivas, G., Cai, X. *et al.* (1997) DNA electrochemical biosensors for environmental monitoring. A review. *Anal. Chim. Acta*, **347**, 1–8.

35. Popovich, N.D., Eckhardt, A.E., Mikulecky, J.C. *et al.* (2002) Electrochemical sensor for detection of unmodified nucleic acids. *Talanta*, **56**, 821–828.

36. He, P.A., Xu, Y., and Fang, Y.Z. (2006) Applications of carbon nanotubes in electrochemical DNA biosensors. *Microchim. Acta*, **152**, 175–186.

37. Koehne, J., Li, J., Cassell, A.M. *et al.* (2004) The fabrication and electrochemical characterization of carbon nanotube nanoelectrode arrays. *J. Mater. Chem.*, **14**, 676–684.

38. Takenaka, S. (2001) Highly sensitive probe for gene analysis by electrochemical approach. *Bull. Chem. Soc. Jpn.*, **74**, 217–224.

39. Takenaka, S. (2003) Electrochemical detection of DNA with small molecules, in *Small Molecule DNA and RNA Binders: From Synthesis to Nucleic Acid Complexes* (eds M. Demeunynck, C. Bailly, and W. D. Wilson), Wiley-VCH, Weinheim, pp. 224–245.

40. Mascini, M., Palchetti, I., and Marrazza, G. (2001) DNA electrochemical biosensors. *Fresenius J. Anal. Chem.*, **369**, 15–22.

41. Umek, R.M., Lin, S.W., Vielmetter, J. *et al.* (2001) Electronic detection of nucleic acids – A versatile platform for molecular diagnostics. *J. Mol. Diagnos.*, **3**, 74–84.

42. Miranda-Castro, R., de-los-Santos-Alvarez, N., Lobo-Castanon, M.J. *et al.* (2009) Structured nucleic acid probes for electrochemical devices. *Electroanalysis*, **21**, 2077–2090.

43. Willner, I., Shlyahovsky, B., Willner, B. *et al.* (2009) Amplified DNA biosensors, in *Functional Nucleic Acids for Analytical Applications* (eds Y. Li and Y. Lu), Springer, New York, pp. 199–252.

44. Carpini, G., Lucarelli, F., Marrazza, G. *et al.* (2004) Oligonucleotide-modified screen-printed gold electrodes for enzyme-amplified sensing of nucleic acids. *Biosens. Bioelectron.*, **20**, 167–175.

45. Hajdukiewicz, J., Boland, S., Kavanagh, P. *et al.* (2009) Enzyme-amplified amperometric detection of DNA using redox mediating films on gold microelectrodes. *Electroanalysis*, **21**, 342–350.

46. Campbell, C.N., Gal, D., Cristler, N. *et al.* (2002) Enzyme-amplified amperometric sandwich test for RNA and DNA. *Anal. Chem.*, **74**, 158–162.

47. Dequaire, M. and Heller, A. (2002) Screen printing of nucleic acid detecting carbon electrodes. *Anal. Chem.*, **74**, 4370–4377.

48. Dominguez, E., Rincon, O., and Narvaez, A. (2004) Electrochemical DNA sensors based on enzyme dendritic architectures: An approach for enhanced sensitivity. *Anal. Chem.*, **76**, 3132–3138.

49. Polsky, R., Gill, R., Kaganovsky, L. *et al.* (2006) Nucleic acid-functionalized Pt nanoparticles: Catalytic labels for the amplified electrochemical detection of biomolecules. *Anal. Chem.*, **78**, 2268–2271.

50. Dill, K. and Ghindilis, A. (2009) Electrochemical Detection on Microarrays, in *Microarrays* (eds K. Dill, P. Grodzinski, and R.H. Liu), Springer, New York, pp. 25–34.

51. Szunerits, S., Bouffier, L., Calemczuk, R. *et al.* (2005) Comparison of different strategies on DNA chip fabrication and DNA-sensing: Optical and electrochemical approaches. *Electroanalysis*, **17**, 2001–2017.

52. Ghindilis, A. L., Smith, M. W., Schwarzkopf, K. R. *et al.* (2007) CombiMatrix oligonucleotide arrays: Genotyping and gene expression assays employing electrochemical detection. *Biosens. Bioelectron.*, **22**, 1853–1860.

53. Hianik, T. and Wang, J. (2009) Electrochemical aptasensors – recent achievements and perspectives. *Electroanalysis*, **21**, 1223–1235.

54. Sassolas, A., Blum, L.J., and Leca-Bouvier, B.D. (2009) Electrochemical aptasensors. *Electroanalysis*, **21**, 1237–1250.

55. Xu, Y., Cheng, G.F., He, P.G. *et al.* (2009) A review: Electrochemical aptasensors with various detection strategies. *Electroanalysis*, **21**, 1251–1259.

56. Willner, I. and Zayats, M. (2007) Electronic aptamer-based sensors. *Angew. Chem. Int. Ed.*, **46**, 6408–6418.

57. Mir, M., Vreeke, M., and Katakis, L. (2006) Different strategies to develop an electrochemical thrombin aptasensor. *Electrochem. Commun.*, **8**, 505–511.

58. Pedrero, M., Campuzano, S., and Pingarron, J.M., (2011) Electrochemical genosensors based on PCR strategies for microorganisms detection and quantification. *Anal. Methods*, **3**, 780–789.

59. Teles, F.R.R. and Fonseca, L.R. (2008) Trends in DNA biosensors. *Talanta*, **77**, 606–623.

60. Lee, T.M.H. and Hsing, I.M. (2006) DNA-based bioanalytical microsystems for handheld device applications. *Anal. Chim. Acta*, **556**, 26–37.

17

Electrical Impedance-Based Sensors

The flow of direct current (DC) through an electric circuit is impeded only by the electrical resistance presented by the circuit. By contrast, in the case of alternating currents (AC) other mechanisms impede the flow of current, in addition to the resistance. For example, a capacitor cannot be crossed by direct current but it allows alternating current to flow across it by means of alternating charging–discharging processes. Nevertheless, a capacitor presents certain opposition to the AC flow by a mechanism that is fundamentally different from that of the electrical resistance. The physical quantity that quantifies the opposition of a capacitor to the flow of alternating current is called *electrical impedance*. Among other factors, the impedance of a capacitor depends on the dielectric constant of its insulator. This property can be used to develop chemical sensors based on the interaction of the analyte with a suitable insulator material. A sensor of this type is a capacitive sensor and is based on the use of impedance measurements for transduction purposes.

The electrical resistance itself represents a particular case of electrical impedance. Sensors based on resistance monitoring have been developed using materials that can change the electrical resistivity in response to their interaction with an analyte.

Of particular relevance to chemical-sensor science is the electrical impedance of electrochemical cells. As shown in Chapter 13, an electrochemical reaction is a multistep process involving diffusion, ion migration and electron exchange at the electrode/solution interface. In addition, the electrical double layer developed at the electrode/solution interface behaves like a capacitor. This capacitor-like structure provides another path for current flow in addition to the electron transfer process. Impedance measurements allow the characterization of each of the above-mentioned steps in the electrochemical reaction and, in addition, provide information about the characteristics of the electrical double layer. Chemical sensors can be produced by integrating an electrochemical cell with a sensing element so that one of the above-mentioned processes is affected by the interaction with the analyte. The change in the electrochemical impedance of the cell provides a transduction method for such sensors.

This chapter introduces first the definition of electrical impedance and the main properties of this physical quantity. Application of the impedance concept to electrochemical processes is then presented. Further sections introduce various types of sensor based on impedance monitoring. As will be shown below, various recognition mechanisms are suitable for designing impedimetric sensors. Among them, biological or biomimetic recognition processes (such as affinity interactions, nucleic acid hybridization, and enzyme-catalyzed processes) should be first mentioned. Other recognition mechanisms have been exploited in the design of impedimetric sensors for gases and vapors.

17.1 Electrical Impedance: Terms and Definitions

Electrical impedance indicates the opposition that an electrical circuit presents to the passage of an alternating current when an alternating voltage is applied.

As is well known, the flow of direct current through a conductor is governed by Ohm's law that states that the current is the ratio of the applied voltage to the conductor resistance ($I = V/R$). In the case of AC circuits, some circuit components (such as capacitors and inductors) induce a time lapse between the voltage and the current that should be taken into account when assessing how much opposition the current flow experiences. That is why an extension of the concept of resistance to AC circuits has been made by introducing the concept of electrical impedance.

A sine wave AC quantity is characterized by its amplitude (V_m or I_m), frequency (f, in Hz, often indicated by the angular velocity $\omega = 2\pi f$ in rad s^{-1}) and the phase angle (ϕ) with respect to a reference signal (Figure 17.1A). In general, the applied voltage results from the superposition of a bias DC voltage (V_{DC}) and a sine wave voltage V_{AC}:

$$V = V_{DC} + V_{AC} = V_{DC} + V_m(\sin \omega t) \tag{17.1}$$

By convention, it is assumed that the phase angle of the AC voltage is zero; hence, the voltage is the reference signal for defining the phase angle (which is often also termed the phase shift). The response current consists of a

Chemical Sensors and Biosensors: Fundamentals and Applications, First Edition. Florinel-Gabriel Bănică.
© 2012 John Wiley & Sons, Ltd. Published 2012 by John Wiley & Sons, Ltd.

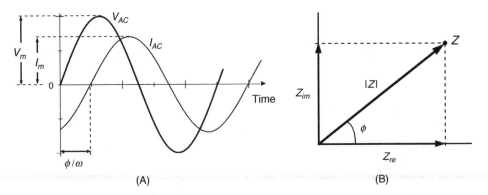

Figure 17.1 (A) Time evolution of sine wave voltage and current. Often, the amplitude is reported as peak-to-peak value ($2V_m$) or as the root mean square value ($V_{rms} = V_m/\sqrt{2}$). (B) Vector (phasor) representation of the impedance (Argand diagram).

DC (I_{DC}) and an AC (I_{AC}) component:

$$I = I_{DC} + I_{AC} = I_{DC} + I_m(\sin \omega t + \phi) \tag{17.2}$$

Time evolution of the AC voltage and current is shown in Figure 17.1A, which also demonstrates the meaning of the phase angle. As already mentioned, the AC current may be shifted with respect to the voltage on the time scale. In this case, these two variables are not in phase. For a capacitor, the current is delayed by $-\phi/\omega$ s, whereas an inductor causes the current to lag behind the voltage by ϕ/ω s. In contrast to the resistance, which is frequency independent, the impedance can depend on this parameter.

In order to derive the impedance definition, it is useful to express the AC voltage and current as complex numbers using the Euler relationships with $j = \sqrt{-1}$:

$$\sin x = \frac{e^{jx} - e^{-jx}}{2}; \quad \cos x = \frac{e^{jx} + e^{-jx}}{2} \text{ or } e^{jx} = \cos x + j\sin x \tag{17.3}$$

Accordingly, one obtains:

$$V_{AC} = V_m \exp(j\omega t) \tag{17.4}$$

$$I_{AC} = I_m \exp(j\omega t + \phi) \tag{17.5}$$

By analogy with Ohm's law, the impedance Z is defined as the ratio of the AC voltage to the AC current:

$$Z = \frac{V_{AC}}{I_{AC}} = |Z|\exp(j\phi) \tag{17.6}$$

where the impedance modulus is $|Z| = V_m/I_m$, for which the SI unit is the ohm (Ω). The above equation demonstrates that the impedance is determined by two parameters, its modulus and the phase angle. A graphical representation of the impedance can be obtained if one resorts again to the Euler relationship in order to express the phase angle dependence by means of trigonometric functions. The result is thus:

$$Z = |Z|(\cos \phi + j\sin \phi) = Z_{re} + j Z_{im} \tag{17.7}$$

According to this equation, the impedance is a complex number with the real numbers Z_{re} and Z_{im} representing the projection of the impedance modulus on the axes of a Cartesian coordinate system. That is why the impedance can be represented in the complex plane by a vector (often called a phasor) as shown in Figure 17.1B. In mathematics, such a representation is termed an Argand diagram. The absolute value (or modulus or magnitude) of the impedance is obtained as:

$$|Z| = \sqrt{(Z_{re} + jZ_{im})(Z_{re} - jZ_{im})} = \sqrt{Z_{re}^2 + Z_{im}^2} \tag{17.8}$$

In general, a circuit includes various components such as resistors, capacitors and inductors and the current flow is determined by the overall impedance. Thus, for a circuit formed of two impedances in series, the equivalent impedance is obtained by vector addition of the components, as demonstrated in Figure 17.2. Analogous to resistance

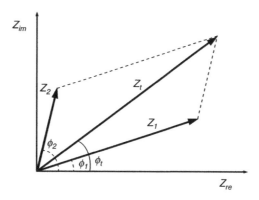

Figure 17.2 Adding up impedances according to the vector rule of addition.

addition, impedance addition is formulated in algebraic terms as:

$$Z_t = \sum_{i=1}^{n} Z_i \tag{17.9}$$

When calculating the equivalent impedance of a parallel circuit it is convenient to resort to *admittance* (*Y*) that is by definition the reciprocal of impedance:

$$Y = \frac{1}{Z} \tag{17.10}$$

Therefore, the admittance can be is expressed as:

$$Y = |Y|e^{-jf} = |Y|(\cos \phi - j \sin \phi) = Y_{re} - j\,Y_{im} \tag{17.11}$$

In the complex plane, the impedance vector is oriented at a $-\phi$ angle with respect to the horizontal axis. The admittance of a parallel circuit is obtained by adding the admittances of the composing components:

$$Y_t = \sum_{i=1}^{n} Y_i \tag{17.12}$$

The SI admittance unit is Ω^{-1}, also called the siemens (S) with usual subunits mS and μS.

Most of this chapter addresses applications of electrochemical impedance measurements for transduction purposes in chemical sensors. In this class of methods, the relevant impedance arises from the current flow through an electrochemical cell. The typical feature of an electrochemical cell is that the circuit involves two types of charge carriers, namely, ions in the electrolyte and electrons in the external circuit. Transition between the two types of electric conduction is performed by electrochemical reactions at the electrode/electrolyte interface.

Another approach based on the concept of impedance is represented by chemical sensors derived from the capacitor design. The capacitor impedance depends on the dielectric constant of its insulator. If this property is affected by interaction with analyte, the change in the capacitor impedance will indicate the analyte concentration.

17.2 Electrochemical Impedance Spectrometry

17.2.1 Basic Concepts and Definitions

Electrochemical impedance spectrometry is an advanced method for investigating the response of an electrochemical system to small perturbations of the equilibrium state [1–4]. In the standard version of this method, the perturbation is applied in the form of a sine wave small-amplitude voltage or current covering a broad frequency spectrum, typically from 10 mHz to 100 kHz. Data analysis allows the determination of characteristic rate constants of the electrochemical reaction and the coupled diffusion process. In addition, this method allows the characterization of the charge distribution at the electrode/solution interface as well the electric resistance of the solution. All these parameters can be influenced by the state of the electrode surface. Recalling that an electrochemical sensor consists of an electrode modified by a sensing layer, it is clear that the recognition event could affect one or more of the above-mentioned factors. Hence, electrochemical impedance spectrometry is a valuable method for monitoring

Table 17.1 Impedance components used to describe electrochemical systems.

	Impedance Element	Impedance Definition	Phase Angle/ Radians	Frequency Dependence
1	Resistor (R)	$Z_R = R$	0	No
2	Capacitor (C)	$Z_C = -\dfrac{j}{\omega C} = \dfrac{1}{j\omega C}; \; |Z_C| = \dfrac{1}{\omega C}$	$-\pi/2$	Yes
3	Constant-phase element (CPE)*	$Z_{CPE} = \dfrac{1}{Y_0(j\omega)^a}; \quad 0 \leq a \leq 1$	$-a\pi/2$	Yes
4	Warburg impedance (W) for semi-infinite linear diffusion**	$Z_W = \dfrac{\sigma}{\sqrt{\omega}}(1-j)$ $\sigma = \dfrac{RT}{n^2 F^2 A \sqrt{2}} \left(\dfrac{1}{c_O \sqrt{D_O}} + \dfrac{1}{c_R \sqrt{D_R}} \right)$	$-\pi/4$	Yes

* Y_0 and a are frequency-independent parameters of the constant-phase element
**The definition of the Warburg impedance corresponds to the case in which the expansion of the diffusion layer is not limited (infinite diffusion). n is the electron number in the electrochemical reaction, F is the Faraday constant, R is the ideal gas constant, T is the absolute temperature, D and c denote the diffusion coefficient and concentration, respectively, of the oxidized (subscript $_O$) and reduced (subscript $_R$) forms of the redox couple, A is the electrode surface area. σ is the Warburg constant.

recognition processes. Its particular advantage resides in the fact that no labels are needed for transduction. Most of the current applications of this method are related to affinity-based recognition processes but other recognition methods are suitable for coupling with impedimetric transduction.

In general, impedance spectrometry measurements provide information about three main parameters of the electrochemical cell: ohmic resistance of the solution, capacitance of the electrode/solution interface and the rate constant of the electrochemical reaction. Each of these parameters can be affected by a recognition event occurring at the electrode surface. Often, an impedimetric sensor is designed so as to provide reliable values for only one of these parameters by minimizing the effect of the other two. Such an approach is met in the cases of conductometric and capacitive sensors. The determination of all three of the above-mentioned parameters is much more laborious and requires more sophisticated equipment. However, such data are very informative and useful in characterizing the sensor surface during the development of an impedimetric sensors or other types of sensors.

It is therefore not surprising that impedance spectrometry aroused considerable interest in chemical-sensor research, as demonstrated by a series of comprehensive reviews published during the last decade [5–9]. A comprehensive review of terminology and nomenclature in electrochemical impedance spectrometry is available in [10].

The main impedance elements in electrochemical cells are summarized in Table 17.1. The ohmic resistance corresponds to charge transport by ion migration in the cell (solution resistance, R_s). It follows the Ohm Law ($V_{AC} = R_s I_{AC}$) and does not cause a phase shift of the current. In standard impedance measurements, R_s depends on the concentration of the supporting electrolyte and if this is between 0.01 and 1 M, it does not contribute significantly to the total impedance. On the other hand, if a recognition process causes important changes in the ion concentration, the solution resistance can act as a suitable response signal (see Section 17.7).

In order to perform electrochemical impedance data analysis, the cell is modeled by a suitable equivalent circuit that responds the AC excitation in the same way than the cell. An equivalent circuit consists of a combination of resistors, capacitors and other characteristic components of an electrochemical cell, as summarized in Table 17.1. Rigorously, the equivalent circuit is derived by mathematical modeling of the electrochemical process. However, due to the intricacy of mathematical models, one resorts often to intuitively derived equivalent circuits.

In principle, AC current flow through the cell can be hindered by processes occurring at both the working and auxiliary electrodes. In order to avoid complications, the surface of the auxiliary electrode is made very large compared to that of the working electrode. Hence, its impedance is negligible and the measured impedance represents the contribution of the working electrode processes alone.

As the frequency is the key parameter, data are represented as impedance module vs. frequency to obtain an impedance spectrum. More informative is a representation of $-Z_{im}$ vs. Z_{re} that is called a complex-plane diagram or a Nyquist plot. An alternative to the complex-plane diagram is the Bode diagram that represents both $\log|Z|$ and $\log\phi$ vs. $\log\omega$. This representation makes the effect of frequency more evident.

In practice, two types of models are used depending on the presence or absence of electrochemical reactions in the system. Non-faradaic models refer to processes consisting only of ions redistribution following the applied sine-wave potential. Such models include only capacitive and resistive components. If an electrochemical reaction occurs, one has to take into account the restricted reaction rate, which introduces a specific, electron-transfer resistance R_{et}.

17.2.2 Non-Faradaic Processes

In the absence of electrochemical reactions, the cell impedance depends only on the ohmic resistance R_s and the charge redistribution at the interface. The ohmic resistance is independent of frequency; its impedance module equals the resistance itself and the phase angle is zero. Therefore, in the Nyquist plot, the resistive impedance is situated on the horizontal axis.

Nonuniform charge distribution at the interface gives rises to the double-layer capacitance C_{dl}. Capacitance is a characteristic parameter of a capacitor that, in the simplest form, consists of two metal plates separated by an insulator. When a voltage is applied, equal electric charges of opposite sign accumulate on each plate. By definition, the capacitance is the Q/V ratio, where Q is the charge accumulated in the capacitor and V is the voltage across it. The SI unit of capacitance is the farad symbolized as F (not to be confused with the Faraday constant). 1 farad is 1 coulomb per volt.

Due to the insulator, a DC current cannot cross the capacitor. However, when an AC voltage is applied, periodic variation in the charge on each plate allows the AC current to flow although no electron transport across the insulator is possible. Nevertheless, the AC current meets some resistance, which is expressed by the capacitor impedance (also termed capacitive reactance) (Table 17.1). Capacitor impedance is inversely proportional to the capacitance and frequency and can become negligible at very high frequencies. In addition, the AC voltage across a capacitor lags the current by a phase angle of $\pi/2$ radians (see also Figure 13. 20A).

At a charged metal electrode, ions of opposite charge accumulate in the solution phase in order to impart charge balance (see Section 13.5). As a result, a double electric layer develops at the interface. It consists of the electric charge at the electrode surface and the opposite charge featured by ions in the nearby solution layer. This system behaves like a capacitor and is characterized by the double-layer capacitance which depends on the thickness of the double layer (δ_{dl}), the electrode surface area (A) and the dielectric constant of the interface layer (ε_{dl}):

$$C_{dl} = \frac{\varepsilon_{dl}A}{\delta_{dl}} \tag{17.13}$$

The double-layer capacitance is potential dependent because the electrode potential determines the counterions charge as well as modification in the dielectric constant at the interface by adsorption of ions or neutral molecules. For a plain polycrystalline gold electrode, the double-layer capacitance could be of a few tens of $\mu F\,cm^{-2}$, depending on the applied potential.

In the absence of electrochemical reactions, the simplest equivalent circuit of the cell consists of the solution resistance in series with the double-layer capacitance (Figure 17.3A). The capacitive impedance is inversely proportional to frequency, whereas the resistance is independent of this parameter. When the frequency is varied, the impedance vector tip shifts along a vertical line. At very high frequencies the capacitive impedance becomes negligible and the overall impedance is determined by the solution resistance alone. Hence, data in a Nyquist plot are represented by a vertical line crossing the horizontal axis at $Z_{re} = R$ (Figure 17.3A). From any point on the plot, the double-layer capacitance can be obtained as $C = 1/Z_{im}\omega$.

Certain processes in electrochemical cells are modeled by a parallel combination of a resistor and a capacitor. This applies for example to an electrode coated with an insulator layer. This system is modeled by the insulator resistance in parallel with the capacitance as shown in Figure 17.3B. The equivalent impedance of this circuit is:

$$|Z| = \left(\frac{1}{R^2} + (\omega C)^2\right)^{-1/2} \tag{17.14}$$

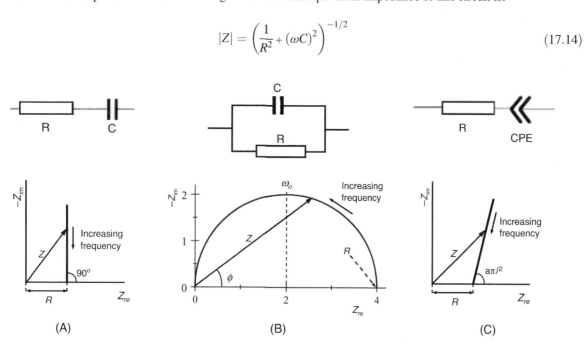

Figure 17.3 Impedance response of non-faradaic processes. Up: equivalent circuits for: (A) electrode–electrolyte solution system; (B) an electrode coated with an insulating layer; (C) an electrode with CPE behavior. Down: Nyquist plots for the equivalent circuits above.

The Nyquist plot for this circuit appears as a semicircle tracked by the tip of the impedance vector Z when varying the frequency. At the very high frequency limit, the capacitor impedance is negligible and the current is opposed only by the ohmic resistance R. The impedance vector then lies on the horizontal axis and the intersection of the semi-circle with this axis gives the R value. The capacitance can be calculated from the frequency ω_0 at which the tip of impedance vector and the center of the semicircle are located on the same vertical line, that is, for $\phi = \pi/4$. The resulting capacitance is then $C = (\omega_0 R)^{-1}$. Actually, the solution resistance should also be included in series with the previous parallel combination. However, as the insulator resistance is usually very high, the solution resistance can be neglected.

The previous approach assumes that the double layer behaves as an ideal capacitor. This applies only in the case of perfectly smooth electrodes such as mercury. Solid electrodes employed in electrochemical sensors deviate more or less from this behavior and are better described by the *constant phase element* (CPE, Figure 17.3C). Its impedance is given in Table 17.1 where Y_0 and a are frequency-independent parameters. Impedance deviation from the ideal behavior is in this case described by the subunit parameter a that determines the phase angle, as shown in the same table as well as in Figure 17.3C. If a is very close to 0, the CPE can be approximated by an ideal capacitor. It is possible to infer the capacitance associated with the CPE using the following relationship [11]:

$$C_{CPE} = \frac{1}{Y_0(\omega_m'')^{a-1}} \tag{17.15}$$

Here, ω_m'' is the frequency at which the imaginary part of the CPE impedance has a maximum. Y_0 value can be obtained by fitting the impedance data to the equivalent circuit in Figure 17.3C.

17.2.3 Faradaic Processes

Electrochemical impedance measurements allow for assessing the kinetic parameters of an electrochemical reaction. As these parameters are strongly dependent on the surface state, impedance investigation is also a useful method for investigating modified electrodes. For this purpose, a redox couple (redox probe) such as $K_3[Fe(CN)_6]/K_4[Fe(CN)_6]$ should be present in the solution. A standard experiment of this type is carried out under the following conditions: (1) both components of the redox couple are soluble species in the solution, preferably at similar concentrations; (2) the redox couple undergoes a reversible electrochemical reaction; (3) the DC bias potential is set at the equilibrium potential of the redox couple (that is, the open-circuit potential); (4) the AC amplitude of the AC voltage is very low (usually ≤ 5 mV) in order to operate within the quasilinear region of the current–voltage curve (see curve 1 in Figure 13.10).

The rate of electron transfer between the electrode and a redox species is intrinsically limited by the relevant rate constant. Therefore, current flow between solution and electrode is more or less thwarted, an effect that is quantified by the *electron-transfer resistance* R_{et} introduced in Section 13.6.5. According to Equation (13.78), R_{et} is inversely proportional to the standard rate constant of the electrochemical reaction and the concentration of the redox probe components in the case in which the oxidized and reduced forms of the redox probe have similar concentrations.

At the same time, diffusion of the redox probe towards the electrode surface could limit the current flow. The effect of diffusion on the total impedance is quantified by the *Warburg impedance* Z_W that depends only on the diffusion coefficients and the concentration of the redox couple (Table 17.1). As diffusion and electron transfer are successive process, R_{et} and Z_W are coupled in series to form the *Faradaic impedance* Z_F.

Modeling of Faradaic processes should also take into account the double-layer capacitance C_{dl} and the solution resistance R_s, as shown in Figure 17.4A. Thus, the AC current can flow along two alternative pathways, the double-layer capacitor and the Faradaic path. That is why in the equivalent circuit these components are coupled in parallel.

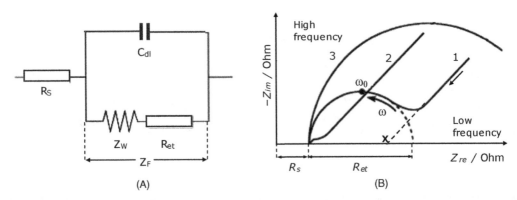

Figure 17.4 Faradaic process-associated impedance. (A) The equivalent Randles circuit; (B) The Nyquist plot for the equivalent circuits in (A). Arrows indicate increasing frequency.

On the other hand, the overall current encounters the solution resistance that is hence in series with the previous combination.

The contribution of each element to the total impedance depends upon its time constant, which is roughly inversely proportional to the characteristic rate constant and is manifest within various frequency regions, as shown in the Nyquist plot in Figure 17.4B. Considering the curve 1 in this plot, it is apparent that the very sluggish diffusion manifests itself only at very low frequencies giving rise to the $\pi/4$ radian (45°) tilted straight line. At higher frequencies, the contribution of Z_W becomes negligible (it is inversely proportional to $\sqrt{\omega}$) as the diffusion cannot follow the voltage oscillation and the electrochemical reaction occurs under diffusionless conditions. The impedance in this case is determined by the combined effect of the double-layer capacitance and the rate of the electron transfer, and the impedance response is analogous to a parallel R–C combination (Figure 17.3B). At very high frequencies, the double-layer impedance becomes negligible and the current flow is now limited only by the solution resistance. Therefore, the Nyquist plot for a Faradaic process consists of a partial semicircle continued by a linear region at very low frequencies. This plot allows the determination of the parameters of the electrochemical process to be effected as follows. R_s and R_{et} result immediately as shown in Figure 17.4B. The double-layer capacitance results from ω_o, the frequency at the maximum point on the semicircle, as:

$$C_{dl} = (\omega_o R_{et})^{-1} \tag{17.16}$$

In order to derive the diffusion coefficient of the redox probe, the straight line region is extrapolated to $Z_{im} = 0$ (point X). The Z_{re} value at this point allows the calculation of the σ parameter in the Warburg impedance from the following equation that applies to semi-infinite linear diffusion:

$$(Z_{re})_X = R_s + R_{et} - 2\sigma C_{dl} \tag{17.17}$$

Curve 1 in Figure 17.4B corresponds to the case in which the plot displays all three characteristic regions. If the electrochemical reaction is very fast, R_{et} is very low and the semicircle is hardly noticeable (Figure 17.4B, curve 2). In the opposite case in which the electrochemical reaction is very slow, R_{et} is very large and the plot displays only a portion of the semicircle with no linear region corresponding to diffusion control. Clearly, the graphical treatment of such data is hardly feasible. It is more suitable to determine the parameters of the equivalent circuit by data fitting to a selected equivalent circuit using a dedicated software package. Data fitting provides more accurate parameter values and, in addition allows the accuracy of the fit to be estimated, which indicates whether or not the selected equivalent circuit is in agreement with the mechanism of the actual electrochemical process. It is recommended to carry out the initial data fitting using a CPE instead of a pure capacitor in the equivalent circuit. In this case, a sizeable a value causes the semicircle in the Nyquist plot to be depressed. The CPE can be replaced by a pure capacitor only if the a parameter is very close to 0.

17.2.4 Probing the Electrode Surface by Electrochemical Impedance Spectrometry

In addition to monitoring the effect of the recognition process, electrochemical impedance spectrometry is a valuable method for investigating the state of the electrode surface after modification [12]. This an important step in developing various types of sensors based on recognition layers at a conductor material surface [7,13].

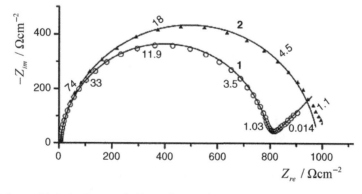

Figure 17.5 Nyquist plots for a gold electrode coated with a self-assembled monolayer consisting of a small molecule (1-(7-selenaoctanoyl)-glycerol, curve 1) and the same compound grafted to a carotenoid (curve 2). Numbers on the curves indicate the frequency in Hz. Electrolyte: 0.1 M KNO$_3$; K$_3$[Fe(CN)$_6$]/K$_4$[Fe(CN)$_6$] (5 mM each); AC potential amplitude, 5 mV. Adapted with permission from [14]. Copyright 2006 Elsevier B.V.

As an example, Figure 17.5 shows Nyquist plots for a gold electrode modified with self-assembled monolayers. Curve 1 data corresponds to an electrode modified with a small molecule (1-(7-selenaoctanoyl)-glycerol) and has been fitted to the Randles circuit with a CPE instead of a capacitor. Clearly, the solution resistance is negligible with respect to the electron-transfer resistance. As the a parameter of the CPE is 0.93, the semicircle is slightly depressed. Curve 2 represents data for a gold electrode modified with a carotenoid derivatized with with 1-(7-selenaoctanoyl)-glycerol. The modifier in this case is a long, hydrophobic molecule that forms a compact layer that is penetrated with difficulty by the redox probe. In this case, the Warburg impedance has been neglected since data at very low frequencies are not reliable and the analysis has been limited to the semicircle region. The CPE a parameter was found to be almost independent of the adsorption time and not affected by certain treatments applied to the surface layer. At the same time, both the electron-transfer resistance and the capacitance (calculated from the CPE parameters by Equation (17.15) increases with the deposition time and also after subjecting the adsorbed carotenoid to electrochemical oxidation. Such effects reflect the alteration of the surface film. Thus, an extended adsorption time results in a higher coverage degree, whereas the anodic reaction grafts hydroxyl groups to the carotenoid backbone leading to a higher dielectric constant.

Often, the sensing layer is build up over an electrode surface layer that acts as an anchor or performs as an insulator. It is useful in such instances to check the integrity of this film by means of electrochemical impedance spectrometry. As shown below, this method allows to estimate the average size of film defects (pinholes) and the average distance between them. It is assumed for simplicity that the pinholes are circular with the uniform radius r_a and are separated by constant intervals $(2r_d)$. If the degree of coverage is very high $(\theta > 0.9)$, it is related to the pinhole parameters as:

$$1 - \theta = r_a^2/r_d^2 \tag{17.18}$$

Pinhole parameters can be determined by analysis of the frequency effect on the components of the Faradaic impedance. Introducing the parameter

$$q = D/0.36r_a^2 \tag{17.19}$$

the following relationships have been derived [15]:

$$Z_{F,re} = \frac{R_{et}}{1 - \theta} + \frac{\sigma}{\omega^{1/2}} + \frac{\sigma}{1 - \theta}\left[\frac{(\omega^2 + q^2)^{1/2} + q}{\omega^2 + q^2}\right]^{1/2} \tag{17.20}$$

$$Z_{F,im} = \frac{\sigma}{\omega^{1/2}} + \frac{\sigma}{1 - \theta}\left[\frac{(\omega^2 + q^2)^{1/2} - q}{\omega^2 + q^2}\right]^{1/2} \tag{17.21}$$

$Z_{F,re}$ and $Z_{F,im}$ represent the real and imaginary part of the Faradaic impedance, respectively. A graphical analysis of the $Z_F - \omega^{-1/2}$ plots according to the above equations allows the determination of the coverage degree and the q parameter that can be used further to calculate the pinhole parameters by means of Equations (17.18) and (17.19).

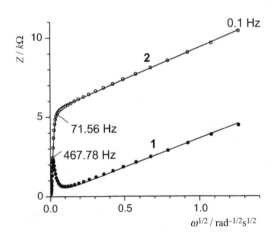

Figure 17.6 Determination of the pinhole parameter by electrochemical impedance spectrometry. Data plot and fitting according to Equations (17.20) and (17.21). Real (1) and imaginary (2) components of the Faradaic impedances recorded at a gold electrode coated with a 1-(7-selenaoctanoyl)-glycerol self-assembled monolayer. Electrolyte: 0.1 M KNO$_3$; K$_3$[Fe(CN)$_6$]/K$_4$[Fe(CN)$_6$] (5 mM each). Frequency span: 20 kHz to 0.1 Hz (40 values equally distributed on a logarithmic scale). $V_m = 5$ mV. Points are experimental data; lines represent the fitting curves according to Equations (17.20) and (17.21), respectively. Adapted with permission from [16] Copyright 2004 Institute of Organic Chemistry and Biochemistry AS CR, v.v.i.

Alternatively one can resort to data fitting to Equations (17.20) and (17.21) as shown in Figure 17.6 that demonstrates the very good fit of these equations with the experimental data. For the system in this figure, such investigations demonstrated that the coverage degree increases with the adsorption time as expected but, at the same time, the defect radius as well as the interdefect distance increases also. Defect coalescence during long deposition times is responsible for this effect. The above method, which has been developed for the case of self-assembled monolayers, is in principle suitable for testing any kind of molecular layer formed at an electrode surface.

Electrochemical impedance spectrometry has also proved useful for assessing the ionization state of surface-adsorbed compounds. Ionization results in a dramatic modification of the electric charge at the surface that is manifested by an change in capacitance. Thus, the pH effect on the capacitance of a gold electrode coated with seleno-methionine allowed the determination of the ionization constants in both amino and carboxyl functionalities of the adsorbed compound [17]. It was thus proved that the logarithm of carboxyl ionization constant in the adsorbed amino acid is about two units greater than the that of the dissolved form. The ionization constant in the amino group was not altered by adsorption.

17.3 Electrochemical Impedance Affinity Sensors

17.3.1 Electrochemical Impedance Transduction in Affinity Sensors

The treatment in the previous section refers mostly to plain electrodes in which the metal surface is in direct contact with the solution. The situation changes dramatically when a sensing layer is present at the electrode surface (Figure 17.7). First, the distance between the electrode and the ion layer increases. Secondly, water (which has a very high dielectric constant) is replaced by an organic film that has a much lower dielectric constant that in turn, results in a reduced capacitance. Thirdly, the voltage drop across the organic layer gives rise to a second capacitor. As a result, the interface capacitance C_i results from the contribution of the ionic layer (C_{dl}) and sensing layer (C_{sl}) capacities connected in series.

$$\frac{1}{C_i} = \frac{1}{C_{dl}} + \frac{1}{C_{sl}} \tag{17.22}$$

Experimental determinations provide only the total interface capacitance. As the recognition process could change radically both the thickness of the recognition layer (δ_{sl}) and the local dielectric constant, each term in the Equation (17.22) can be affected. If the sensing layer shows itself to be resistant to ion transfer, a pertinent resistance should be included in parallel to the C_{sl} element. Hence, it is clear that the capacitance represents a suitable response signal for monitoring affinity reactions at the electrode surface. As a rule, the interface capacitance decreases during the recognition step and the signal should be recorded after the equilibrium state has been attained.

An alternative response signal is the electron-transfer resistance R_{et} as it is very sensitive to the state of the interface. Thus, in the case of an electrode covered with a sensing layer, the recognition process results in an obvious modification of R_{et} compared with the response in the absence of the analyte (Figure 17.7).

It is not possible to make general predictions about the way R_{et} will varies in response to the recognition process. Roughly, it is expected that the change in R_{et} will be more distinct if the analyte is substantially larger than the redox probe or has different physical properties, for example, electric charge. Thus, a charged analyte develops attractive or repulsive forces depending on the charge of the redox probe. It is judicious therefore to compare the R_{et} values

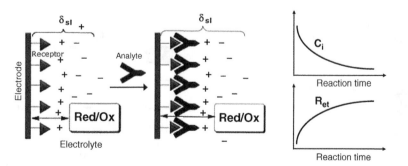

Figure 17.7 Effect of the recognition process on the characteristics of the surface layer. The thickness of the double layer increases causing the capacitance to decrease. The access of the redox probe to the electrode surface is obstructed yielding a higher electron-transfer resistance. Adapted with permission from [7]. Copyright 2003 Wiley-VCH Verlag GmbH & Co. KGaA.

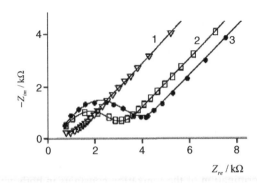

Figure 17.8 Faradaic impedance spectra recorded at a gold electrode coated with a peptide-receptor for the foot-and-mouth-disease antibody. (1) Bare gold electrode; (2) the electrode coated with a peptide monolayer; (3) after binding an antibody to the receptor. Data recorded in the presence of 2 mM $Fe(CN)_6^{4-/3-}$ at pH 7 (phosphate buffer). Adapted with permission from [18]. Copyright 1995, Elsevier B.V.

obtained by means of redox probes with different charges, such as $Fe(CN)_6^{4-/3-}$ which is negatively charged, and $Ru(NH_3)_6^{2+/3+}$ which bears a positive charge.

Taking into account the effect of the analyte size, it could be inferred that large-antigen detection is more sensitive than the detection of small protein molecules by an antibody-recognition layer.

Figure 17.8 shows the typical behavior of an impedimetric affinity sensor in the presence of a redox probe. Curve 1, which corresponds to the plain gold electrode, demonstrates that the electrode process is controlled by diffusion over most of the frequency region. Curve 2, recorded after forming the receptor layer, demonstrates an increased interface capacitance and also an enhanced electron-transfer resistance in agreement with the behavior shown in Figure 17.7. Analyte–receptor binding results in subsequent enhancement each of these parameters (curve 3).

17.3.2 Configuration of Impedimetric Biosensors

Two alternative configurations of an impedimetric sensor are possible, as shown in Figure 17.9. In version (A), the sensing layer is build up at the surface of a working electrode whose potential is controlled with respect to a reference electrode. Version (B) relies on in-plane disposition of twin interdigitated electrodes. In this form, typical electrodes are some 25 μm high, 50 μm wide and placed at 50 μm intervals. The sensing layer is formed either between electrodes, over the electrodes or over the whole surface. When a voltage is applied, electrodes assume equal potentials of opposite sign. Ideally, the current should flow only across the sensing layer where the recognition event brings about substantial impedance modification. Actually, the electric-field lines curve out in the bulk solution, as shown in Figure 17.9B. This renders the response dependent on changes in solution as well. A correction for this effect can be made by using a reference sensor with the recognition receptor replaced by inert molecules.

As the background signal can be rather high, it is best to resort to differential measurements based on a reference sensor. The most convenient operation mode is the two-channel flow-through analysis system including two similar sensor, each of them being installed in one of the channels. The sample flows through the analysis channel while the reference sensor in the second channel is kept in contact with an analyte-free, reference solution.

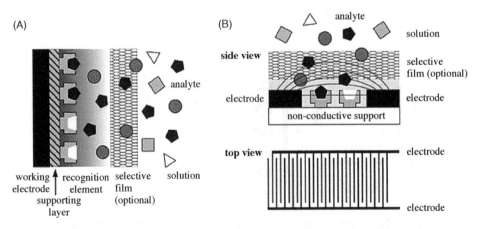

Figure 17.9 Configuration of impedance biosensors. (A) Bulk electrode sensor; (B) planar, interdigitated electrodes configuration. Reproduced with permission from [8]. Copyright 2008 Springer Verlag.

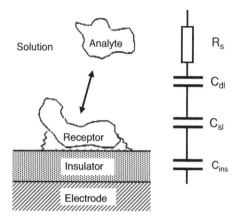

Figure 17.10 Structure and equivalent circuit of the surface layer of an impedimetric biosensor with insulated electrode. Adapted with permission from [19]. Copyright 2001 Wiley-VCH Verlag GmbH & Co. KGaA.

17.3.3 Capacitive Biosensors

Capacitive biosensors are based on the change in interface capacitance as a result of the recognition process. Generally, the interface is represented by a capacitor–resistor series that allows uncomplicated data-processing procedures. Further, no redox probe is needed, which simplifies the read-out equipment and the sensor operation.

Equation (17.22) for the overall capacitance holds only if no ion transfer between the metal electrode and the solution is possible. Otherwise an additional resistor in parallel with the interface capacitance should be included. In order to avoid such complications, a compact insulator layer should be formed at the interface between the metal and the sensing layer (Figure 17.10). However, this adds one more capacitance (C_{ins}) to the equivalent circuit. In order to reduce the contribution of the insulator layer to the total impedance, this layer should be as thin as possible and have a high dielectric constant. Such conditions can be achieved by using an oxide layer on tantalum surface. Receptor immobilization on the oxide layer can be performed via silanization.

Alternatively, one can resort to self-assembled monolayers on gold electrodes. Long-chain alkanethiol layers provide very good insulation and can function also as linkers for the receptor when terminated by a functional group. Defects in the sensing layers can be covered by chemisorption of plain alkanethiols that bind to, or bend over, the exposed metal surface.

As shown in Figure 17.10, the interface capacitance of an insulated electrode sensor consists of three components in series and the total interface capacitance C_i is:

$$\frac{1}{C_i} = \frac{1}{C_{ins}} + \frac{1}{C_{sl}} + \frac{1}{C_{dl}} \tag{17.23}$$

The insulator capacitance is constant, whereas the other components can vary in response to the recognition process.

Figure 17.11 presents the response of an immunocapacitive sensor to a nonspecific analyte (curve 1) and also to that of a specific analyte at increasing concentrations (curves 2 and 3). Clearly, nonspecific interactions also give rise to a certain variation in the capacitance but the analyte effect is discernible. Curves in Figure 17.11 also demonstrate that after washing with buffer solution, the signal does not return to the initial value. This is a clear indication of

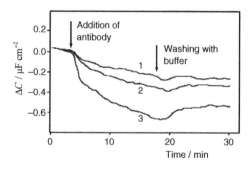

Figure 17.11 Specific capacitance variation in response to an affinity recognition event. The analyte is the foot-and-mouth-disease antibody; the receptor is the respective antigen. Response curves correspond to (1) a nonspecific antibody; (2) 1.74 mg/mL; (3) 17.5 mg/mL specific antibody. Adapted with permission from [20]. Copyright 1996 Elsevier.

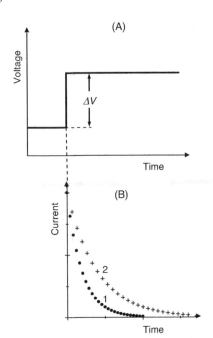

Figure 17.12 Principle of the potential-step method for capacitance determinations. (A) Time evolution of the applied potential; (B) current response. The capacitance in the case of curve 2 is twice as high as that for the curve 1.

incomplete regeneration of the sensing layer. Full regeneration can be obtained by washing with a 6 M urea solution containing 0.1% Tween 20 surfactant. The surfactant also reduces nonspecific adsorption and improves the removal of denatured protein molecules.

There are two main methods for determining the capacitance in the absence of an electrochemical reaction, namely AC measurement and the potential-step method.

AC capacitance measurements rely on measurements at a single frequency. If the interface behavior is modeled by a resistor–capacitor series combination, $Z_{re} = R_s$ and $Z_{im} = C_{sl}$. Thus, the capacitive impedance is represented by the projection of the total impedance on the vertical axis in the phasor representation. If the AC current is fed to a lock-in amplifier tuned at $\phi = -\pi/2$, the output signal represents the capacitive impedance only. This results in appreciable simplification of the equipment. However, substantial deviation from the previously assumed equivalent circuit brings about uncertainties about the relevance of the result. Proper selection of the frequency is an important option. Lower frequencies appear to be more suitable because the capacitive impedance is greater under such conditions. However, some published work demonstrates that a frequency close to 10 kHz could result in a more stable signal.

The potential-step method consists of applying a sudden DC voltage change to the cell (Figure 17.12A). The system responds by producing a redistribution of ions in the double layer in order to match the new charge on the electrode. This, in turn, induces a time-variable current (Figure 17.12B) that depends on the interface capacitance and decays according to the following equation:

$$i(t) = \frac{\Delta V}{R_s} \exp\left(-\frac{t}{R_s C_i}\right) \qquad (17.24)$$

An $\ln i(t)$ vs. t plot of the above equation yields a straight line whose slope and intercept allow the determination of the total capacitance. This equation is based on the simple model of a series combination of a resistor and a capacitor. As this model is very approximate, a straight line plot appears only within a limited time interval. This is due to the fact that the potential step can induce modifications in the thickness and structure of the interface layer. That is why the potential step should not exceed 50 mV. AC impedance measurements make use of very small potential variations and are expected to be less sensitive to such problems.

As demonstrated by Equation (17.24), the current decay rate also depends on the resistance. In order to secure a sufficiently high number of current samples, it is necessary to keep the solution resistance as high as possible in order to obtain a relatively slow current variation. That is why most of the step experiments with capacitive biosensors have been carried out in relatively diluted buffers, typically 10 mM. Faster sampling-rate instruments remove this restriction but are more expensive.

The main advantage of the potential-step method over the AC alternative arises from the operation simplicity and the low cost of the read-out equipment.

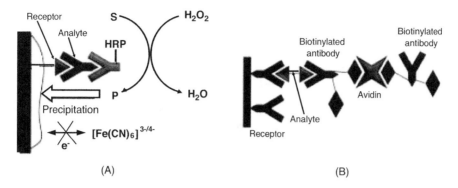

Figure 17.13 Response amplification in impedimetric sensors by postrecognition treatment. (B) Product precipitation by an attached enzyme; (B) growing of a protein layer by biotin–avidin interaction. Adapted with permission from [7]. Copyright 2003 Wiley-VCH Verlag GmbH & Co. KGaA.

17.3.4 Signal Amplification

A major problem with impedimetric immunosensors is the relatively small change in impedance produced by the antibody–antigen binding. Signal variation is therefore insignificant when the surface concentration of the Ab–Ag complex is very far from the saturation value.

Signal amplification can be produced by applying a special approach for immobilizing the recognition element using conducting polymers such as polypyrrole. The recognition event has a marked effect on the conformation of the polymeric network and thus enhances the response. A lipid bilayer with incorporated ion channels also elicits strong dependence of the ion-transfer process in response to antibody–antigen binding.

Alternatively, signal amplification can be achieved by postrecognition processes aimed at enhancing the thickness of the surface layer and altering other relevant properties. This scheme is analogous to the sandwich ELISA procedure. After receptor–analyte binding, an enzyme-labeled antianalyte antibody is added to the system in order to form a sandwich tertiary complex. The enzyme (e.g., horse radish peroxidase) catalyzes the conversion of the substrate that is selected such as to give an insoluble product. As a result, a precipitate layer forms at the electrode surface, thus enhancing dramatically both the capacitive impedance and the electron-transfer resistance. The amplification degree can be controlled by the time interval used for biocatalytic formation of the precipitate.

Another postrecognition enhancement method is based on growing a protein aggregate attached to the initially formed complex (Figure 17.13B). A secondary antibody grafted with several biotin residues is attached to the analyte–receptor complex. Using avidin as a bridge, an additional biotinylated antibody can be added. This sequence can be reproduced several times in order to increase the thickness of the surface layer. By this means, a considerable increase of the impedance is obtained. In general, postrecognition attachment of any charged and/or bulky particles that alter the impedance brings about signal amplification.

A considerable advantage is obtained when the affinity recognition layer is used in conjunction with a field effect transistor that provides electronic signal amplification [21]. Recognition receptors can be attached to the silicon dioxide gate insulator layer by means of aminosiloxane crosslinking.

17.3.5 Synthetic Receptor-Based Impedimetric Sensors

The previous section focused mainly on impedimetric immunosensors. It is important, however, to point out that electrochemical impedance is also suitable for monitoring recognition by synthetic receptors. Thus, a calix[6]arene derivative functionalized with carboxyl groups at the lower rim has been used as receptor for impedimetric detection of epinephrine and dopamine [22]. This receptor was derivatized with branched aliphatic chains at the upper rim, which allows it to be integrated in a phospholipid film supported by a thioalkane self-assembled monolayer. Figure 17.14A presents the impedance response the bare gold electrode. As the electrochemical reaction rate of the redox couple is relatively high, the semicircle is hardly visible and the $\pi/4$ tilted line characteristic of the Warburg impedance is clearly developed. If a phospholipid film is formed over the thiol layer, the electron-transfer resistance increases dramatically, as demonstrated by curve 1 in Figure 17.14B. This effect is much more marked after including calixarene molecules into the phospholipid layer (Figure 17.14B, curve 2). This curve also suggests that the ω_0 frequency is much higher in the case of curve 2 that is caused by a higher double-layer capacitance with respect to that for the electrode coated only with thioalkane (curve 1). The increase in the capacitance is due to the ionized carboxyls on the calixarene.

In order to extract values of relevant parameters, curves like that in Figure 17.14B have been fitted to the equivalent circuit shown in the same figure. Here, C_{sl} and R_{sl} represent the capacitance and the resistance, respectively, associated with the surface layer.

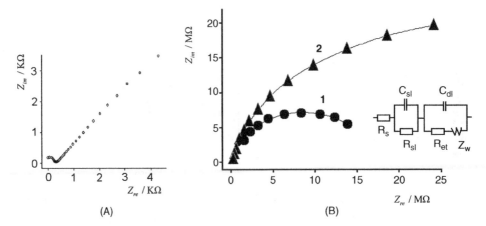

Figure 17.14 Effect of electrode modification on the electrochemical impedance response. (A) Nyquist plot for the $K_3[Fe(CN)_6]/K_4[Fe(CN)_6]$ (5 mM each) couple at a bare gold electrode. (B) The same electrode coated with octadecanethiol self-assembled monolayer (curve 1) and a mixed phospholipid - calixarene film supported on the thiol monolayer (curve 2). McIlvaine buffer (pH = 5); AC amplitude, 5 mV; open circuit potential. Adapted from [22]. Credit T. Hianik and M. A. Mohsin.

C_{sl} proved to be sensitive to the interaction of negatively charged carboxylate ions at calixarene with small organic ions such as epinephrine (adrenaline) and dopamine in the protonated state. The response to epinephrine additions is shown in Figure 17.15 that demonstrates an evident decrease in C_{sl} with increasing concentration of the target compound.

Molecularly imprinted polymers that are excellent biomimetic receptors have also been used in the development of impedimetric sensors. An impedimetric glucose sensor has been developed using a polymer obtained by electro-chemical polymerization of *o*-phenylenediamine at a gold electrode in the presence of glucose as template. After blocking unoccupied sites by chemisorption of 1-dodecanethiol, the electrode impedance behavior is purely capacitive. This sensor responds to glucose binding within the biological concentration range (Figure 17.16) but is insensitive to lactose and ascorbic acid that are structurally related to glucose. The shape of the response curve is similar to that of the Langmuir isotherm, which is characteristic of affinity sensors. Such a sensor gave a stable response for 10 h but the response increased at longer times. Because of the simplicity of the fabrication technology, this approach is suitable for fabricating disposable sensors.

17.3.6 Applications of Impedimetric Affinity Sensors

In principle, impedimetric transduction is suitable for application to any kind of affinity sensor. Thus, imprinted polymers have been employed for impedimetric sensing of some organic compounds. Calixarenes bearing negative substituents at the lower rim and attached to a phospholipid layer are convenient receptors for the impedimetric determination of small, positively charged proteins, such as cytochrome *c* [24].

Metal ions such as Hg^{2+}, Cd^{2+}, Cu^{2+} and Zn^{2+} at femtomolar concentrations can also be determined by capacitance measurement. The ion receptor can be a protein that undergoes conformational change upon metal ion binding to functional groups at side chain residues. Regeneration of such a metal-ion sensor can be done by a strongly chelating reagent such as EDTA. A potassium ion sensor has been produced by including valinomycin into a tethered bilayer lipid membrane [25].

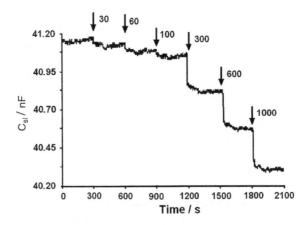

Figure 17.15 Capacitive response of a calixarene-based sensor to epinephrine additions (shown in the figure in μM). Same conditions as in Figure 17.14. Adapted from [22]. Credit T. Hianik and M. A. Mohsin.

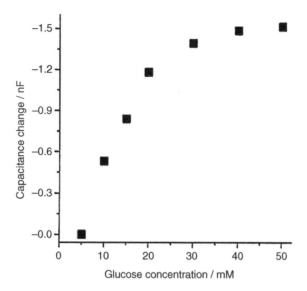

Figure 17.16 Response curve of a capacitive glucose sensor based on a molecularly imprinted polymer receptor. 10 mM Tris buffer; 0.1 M NaCl. The capacitance has been calculated from the imaginary component of the impedance at 10 Hz. Adapted with permission from [23]. Copyright 2001 Elsevier Science B.V.

Simultaneous determination of several analytes in the same sample (multiplexing) is achievable if an array of specifically modified electrodes is assembled on the same support, such that each of them responds to a distinct target compound. Each sensor in the network can be monitored sequentially by a single readout system. However, affinity sensor arrays are vulnerable to nonspecific interactions as each particular element may respond to a greater or lesser extent to several analytes. This limits the possible degree of multiplexing and is particularly troublesome with genuine biological samples. Nevertheless, an array of several immuno sensors has more diagnostic potential than a single sensor responding to a single analyte.

17.4 Biocatalytic Impedimetric Sensors

Although electrochemical impedance spectrometry is mostly associated with affinity reactions, this method is also suitable for performing transduction in enzymatic sensors. For example, signal generation in impedimetric enzyme sensors can be achieved by precipitation of the product of a peroxidase reaction. Hydrogen peroxide involved in this reaction can be produced by an oxidase-catalyzed reaction which allows to determine the substrate of the oxidase enzyme. This principle has been exploited in the glucose bienzyme sensor shown in Figure 17.17. Glucose oxidation catalyzed by glucose oxidase produces hydrogen peroxide. This product is used in the oxidation of the peroxidase substrate to form an insoluble product that accumulates at the electrode surface and causes considerable changes in both the capacitive and faradaic impedances. The response is related to glucose concentration and increases steadily with the incubation time that dictates the thickness of the precipitate layer. Hence, the incubation time allows for the sensitivity of the sensor to be tuned.

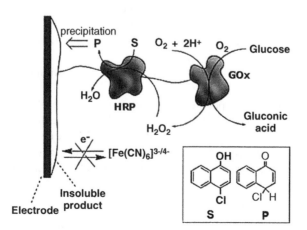

Figure 17.17 An impedimetric glucose sensor based on precipitate formation. Adapted with permission from [7]. Copyright 2003 Wiley-VCH Verlag GmbH & Co. KGaA.

Another approach to biocatalytic impedimetric sensors is based on electrodes covered with biodegradable polymers. The degradation can occur directly under the effect of the enzyme on the polymer backbone or indirectly due to the effect of the products of the enzymatic reaction. The alteration of the film thickness can be monitored by impedance measurements. Determination of an enzyme substrate or inhibitor is feasible in this way. Moreover, this method can be used in immunoassays using enzymes as labels.

17.5 Outlook

Electrochemical impedance spectrometry is a relatively simple transduction method as only electrical measurements are involved. Label-free sensing is the most attractive feature of this method but in many instances amplification is needed in order to achieve suitable sensitivities.

Capacitance-based transduction is very simple, as it can be achieved by AC measurements at a single frequency or by the potential-step method and a redox probe is not required.

In contrast, determination of the electron-transfer resistance can be achieved only with an added redox probe and upon recording the whole impedance spectrum. This can takes at least 10 min and the surface layer can suffer spontaneous modifications during this time interval. In order to reduce the measurement time one can resort to faster methods based on multiple excitation frequencies applied simultaneously. Another fast method is based on excitation with white noise, that is, a randomly variable voltage. An alternative method relies on excitation with a potential step. In this case, the Fourier transform is used to convert the current–time response into an amplitude–frequency spectrum which is similar to that obtained by excitation with a sine voltage at various frequencies. In addition to saving time, short time readout methods avoid uncontrolled impedance variations that may occur during the measurement.

When compared with amperometric methods, impedance transduction presents an advantage resulting from the very small amplitude of the applied voltage. Commonly, the amperometric methods require a large potential deviation from the equilibrium potential. This gives rise to strong electrostatic forces within the surface layer that can harm it. Impedance methods involve very small voltage oscillations around the equilibrium potential, which avoids damaging the sensing layer.

Questions and Exercises (Sections 17.1–17.5)

1 Define the concepts of impedance and admittance. How can these quantities be represented numerically and graphically?

2 What equivalent circuits are used for modeling non-Faradaic processes? How can circuit parameters be derived from impedance data?

3 Comment on the meaning of each component in the Randles circuit. How are the parameters of these components correlated with the processes in an electrochemical cell?

4 How can physicochemical parameters of an electrochemical process be derived from impedance measurements?

5 What are the effects of (a) electrode modification and (b) recognition event at a modified electrode on the cell parameters determining the impedance response?

6 How can the response be measured in capacitive sensors? Review the advantages and disadvantages of each alternative.

7 How can the sensitivity of an impedance sensor be enhanced?

8 How can enzymes be used as receptors in impedance sensors?

9 What are the advantages of molecularly imprinted polymers as substitutes for biological compounds as receptors in impedance sensors?

10 What are the advantages of molecularly imprinted polymers as substitutes for biological compounds as receptors in impedance sensors?

11 A 10 mm diameter electrode has a specific double-layer capacitance of $40\,\mu F/cm^{-2}$. (a) Calculate the capacitive impedance of this electrode at 1 kHz frequency; (b) calculate the total impedance in the absence of any electrochemical reaction if the solution resistance is $20\,k\Omega$; (c) draw the phasor representation of the capacitive, resistive, and total impedance; (d) calculate the phase angle of the total impedance; e) do the same calculations in (a)–(d) for 100 Hz frequency; (f) discuss the frequency effect on the impedance parameters.

Answer: for 1 kH, $Z_{dl} = 50.7\,k\Omega$; $|Z| = 54.5\,k\Omega$; $\phi = 68.5°$.

12 An electrode coated with an insulated layer can be modeled by a parallel capacitor–resistor circuit. Assuming that the electrode has a diameter of 5 mm, the insulator resistance is $100\,k\Omega$, the specific capacitance is

10 $\mu F/cm^{-2}$ and the frequency is 5 kHz, calculate the following parameters: (a) Capacitor and resistor admittance; (b) Total admittance of the circuit; (c) Total impedance and the phase angle. (d) Draw the phasor diagram of this circuit; (e) Calculate the real and imaginary part of the total impedance and plot the corresponding point on the Nyquist diagram. (f) Perform similar calculations as in (a–c) at a frequency of 50 Hz. (g) Comment on the frequency effect on the impedance parameters.

Answer: (a) Capacitor admittance: 6.2 μS; (b) Total admittance: 11.7 μS; (c): Total impedance: 85.1 KΩ; $\phi = 37.1°$.

13 For each curve in Figure 17.5 estimate the parameters of the equivalent Randles circuit assuming that the double layer responds as an ideal capacitor. (Such data are useful as input parameters in data fitting by means of dedicated software.) As data in this figure are reported as specific impedances, assume that the electrode area is 0.11 cm^2.

14 The table below show the current response of a capacitive sensor to a potential step of 40 mV. Plot the natural logarithm of the current vs. time and estimate the interface capacitance and the solution resistance. Alternatively, you can use a computer software to fit an exponential function to the i–t data points and calculate the capacitance from the parameters of the fitting function. Compare the results given by each calculation method and comment on the reason for possible differences.

t/s	0.005	0.01	0.015	0.02	0.025	0.03	0.035	0.04	0.045	0.05
i/μA	15.65	12.06	9.73	7.09	5.91	4.12	3.68	2.34	2.30	1.51

Answer: $R_s = 2.0$ kΩ; $C_i = 9.90$ μF.

17.6 Nucleic Acid Impedimetric Sensors

It is well known that the DNA hybridization reaction involves hydrogen bonding between nucleic acid bases that drives out water from the resulting double helix. This produces a reduction in the amount of water and a similar change in solvated ion concentration within the sensing layer at the electrode surface. Consequently, hybridization at an electrode surface causes the local ionic conduction to diminish, leading to an increase in the non-Faradaic impedance. The marked change in the surface film structure can equally be monitored by means of Faradaic impedance measurements in the presence of a redox probe.

17.6.1 Non-Faradaic Impedimetric DNA Sensors

The change in ion distribution caused by hybridization can be detected simply by assessing the change in the local ohmic resistance. This effect is best exploited by using interdigitated electrode arrays, as shown in Figure 17.18A. In a typical application, electrodes have been fabricated from platinized platinum, whereas the probe was attached by silane coupling to the glass surface between the electrodes [26]. The equivalent circuit (Figure 17.18B) includes the ohmic resistance of the solution (R_s) and two constant-phase elements (CPE) representing the nonideal capacitive contribution of each electrode. The hybridization event is indicated by a decrease in R_s. The measurement can be effected in a buffer solution but a measurement in ultrapure water assures that the sole source of available ions is the immobilized DNA. As the ion distribution at the interface affects both the interface capacitance and the solution

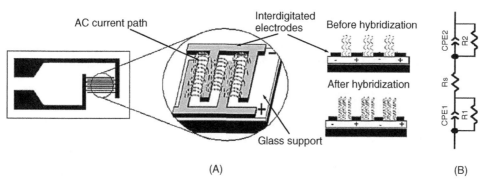

(A)   (B)

Figure 17.18 Configuration of a non-Faradaic impedimetric hybridization sensor (A) and its equivalent circuit (B). Adapted with permission from [26]. Copyright 2004 Elsevier B.V.

resistance, the value of the latter should be extracted by data fitting according to the selected equivalent circuit. The maximum sensitivity was observed at frequencies below 100 Hz.

This approach is suitable for the fabrication by microlithography of DNA arrays consisting of individually addressable electrode pairs [27]. A multiplex impedance analyzer allows the assessment of the solution resistance at each individual site. In order to operate with a three-electrode system a large-area counterelectrode is included in the array. This allows for performing electropolymerization sequentially at each individual site in order to obtain the sensing layer. A single reference electrode consisting of silver-plated platinum is also added to each chip.

Due to the limited number of recognition sites, such an array is suited for detecting a restricted number of genes but not for a complete genome assay. So, this reagentless type of array can be employed for detecting a selected pathogen.

17.6.2 Faradaic Impedimetric DNA Sensors

As the target strand brings a negative charge, the hybridization event causes an enhancement of the negative charge density at the surface. As a result, the electron-transfer resistance for the electrochemical reaction of a negatively charged redox probe (such as $Fe(CN)_6^{3-/4-}$) is also enhanced. A further increase in electron-transfer resistance arises if the hybrid molecule is enlarged by subsequent attachment of a bulky and negatively charged moiety, such as a protein. This principle is illustrated in Figure 17.19A for the case in which avidin is used as enlarging reagent. Curves in Figure 17.19B are Nyquist plots for each of the states depicted in Figure 17.19A. Thus, curve 2 shows an increase in R_{et} with respect to bare gold (curve 1) owing to the negative charge of the probe. Hybridization causes the amount of negative charge to increase and produces a further increase in R_{et} (curve 3). Next, a biotinylated oligonucleotide has been hybridized yielding a sandwich structure that brings about an even higher resistance (curve 4). Finally, avidin coupling to the biotinylated nucleotide results in an appreciable increase in R_{et} due to the enhanced barrier to the redox probe reaction.

Posthybridization modification of the sensing layer can enhance considerably the sensitivity of the impedimetric sensor. As is well known, an avidin molecule contains four biotin binding sites. This property has been exploited for signal amplification by subsequent binding of biotinylated particles (such as liposomes) to the structure 5 in Figure 17.19A.

An alternative amplification strategy is based on hybridization-triggered precipitation of an insoluble product at the sensor surface. To this end, daunomycin is intercalated in the hybrid duplex. Its electrochemical reduction stimulates catalytic electrochemical reduction of dissolved oxygen that generates H_2O_2. In the presence of a peroxidase, H_2O_2 oxidizes 4-chloronaphthol to an insoluble product. Insulation of the electrode by the precipitate results in an enhanced electron-transfer resistance.

A convenient engineering of charge distribution after hybridization is possible when using a molecular beacon-type probe [29]. As shown in Figure 17.20, the sensing layer is prepared by self-assembly of the thiol-tethered probe at the surface of a gold electrode along with mercaptoacetic acid (step 1). This mixed layer allows a sufficient distance between probe molecules in order to avoid steric obstruction in the hybridization event. The probe layer exhibits a large negative charge that prevents a negatively charged redox probe (here, $Fe(CN)_6^{3-/4-}$) from reaching the electrode surface. Consequently, the sensor displays in this state a large electron-transfer resistance. As a consequence of target–probe hybridization (step 2), the conformation of the negative entity changes considerably but negative charges are still present and the electron-transfer resistance suffers only a slight decrease. This situation changes considerably if thionine is added to the sample (step 3). This positively charged compound intercalates in the double-strand hybrid and neutralizes the negative charges of the nucleic acid backbone. As a result, the transfer of the redox probe through the

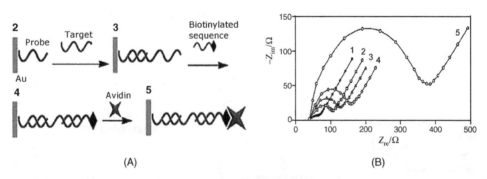

(A) (B)

Figure 17.19 Impedimetric hybridization sensor with amplification by the biotin–avidin reaction. (A) Principle. (B) Nyquist plots: except curve 1, each curve in the diagram corresponds to the state labeled by the same number in (A). (1) The bare Au electrode. (2) The gold electrode with the surface-attached probe. (3) The same, after hybridization with the DNA target. (4) After coupling a biotinylated nucleotide to the probe-target hybrid. (5) After the affinity reaction with avidin. The impedance spectra were recorded in 0.01 M phosphate buffer (pH 7.0) containing 0.01 M $Fe(CN)_6^{3-/4-}$ as the redox probe. Bias potential, 0.17 V; AC amplitude, 5 mV. (A) Adapted with permission from [7]. Copyright 2003 Wiley-VCH Verlag GmBH & Co. KGaA. (B) Adapted with permission from [28]. Copyright 1999 Royal Society of Chemistry.

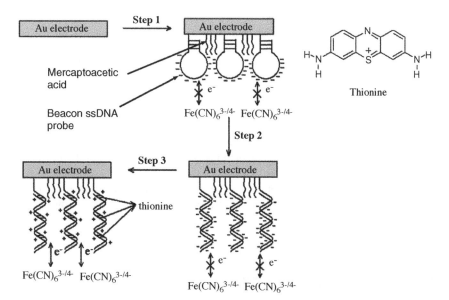

Figure 17.20 An impedimetric DNA sensor based on a molecular beacon probe. Thionine intercalation in the duplex balances out the negative charge and facilitates the transfer of the negative redox probe to the electrode surfaces. Adapted with permission from [29]. Copyright 2006 Wiley-VCH Verlag GmBH & Co. KGaA.

surface layer is facilitated and the electron-transfer resistance is diminished. This resistance difference is proportional to the decimal logarithm of the target concentration over a broad range extending from 10^{-12} to 10^{-6} M. Signal amplification by a positive intercalator is in principle also feasible with a linear probe, but, due to its space-filling capacity, the molecular beacon induces a much larger resistance variation under similar conditions.

17.6.3 Impedimetric Aptasensors

Faradaic inpedance measurements in the presence of a redox probe have been exploited for developing aptamer-based chemical sensors. For example, a thiol-derivatized DNA aptamer has been immobilized at a gold electrode by thiol self-assembly along with 6-mercapto-1-hexanol. The thrombin analyte reacts with the immobilized aptamer and thus increases the electron-transfer resistance for the $Fe(CN)_6^{3-/4-}$ electrochemical reaction. An enhanced resistance has been achieved by denaturation of captured thrombin with guanidine, which results in an increased radius of the thrombin molecule and a more efficient blocking of the redox probe access to the electrode [30]. As shown in Figure 17.21, an increase in the analyte concentration results in an appreciable increase in the electron-transfer resistance. This sensor allows attaining a very low detection limit, close to 0.01 pM. The amplification procedure in this case exploits the outstanding stability of the aptamer against denaturation. Such an approach is not feasible with an immunosensor because the receptor itself can undergo denaturation resulting in the dissociation of the immunocomplex.

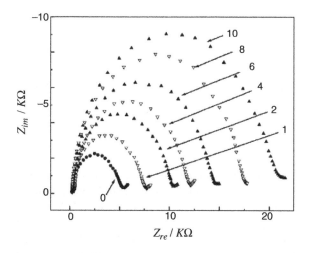

Figure 17.21 Thrombin determination by an impedimetric aptasensor using thrombin denaturation after capture. Nyquist plots for various thrombin concentrations as shown on each curve in 10^{-14} M. Adapted with permission from [30]. Copyright 2006 Wiley-VCH Verlag GmBH & Co. KGaA.

Questions and Exercises (Section 17.6)

1 Why is it that electrochemical impedance spectrometry is particularly suited for assessing nucleic acid hybridization?
2 What are the main approaches for impedimetric monitoring of hybridization?
3 Comment on the configuration of a non-Faradaic impedimetric nucleic acid sensor and applications of this principle in nucleic acid microarrays.
4 How can affinity reactions be used to amplify the sensor response of nucleic acid impedimetric sensors?
5 Discuss some amplification strategies based on posthybridization alteration of the receptor-analyte adduct.
6 How can DNA intercalation be exploited for increasing the sensitivity of an impedimetric DNA sensor?
7 Discuss the principles of aptamer-based impedimetric sensors and possible methods for enhancing its response signal.

17.7 Conductometric Sensors

As is well known, electrolyte solutions conduct electricity, the current flow being achieved by ion movement under the action of the electric field [31–33]. Investigation of charge-transport properties is a key method in the physical chemistry of electrolyte solutions and provides a broad range of analytical applications including applications to chemical sensors [34]. Although monitoring of solution conductivity is the most straightforward method used in conductometric sensors, more advanced methods rely on monitoring variations in the conductivity of conducting polymers (see, for example, [35]). In this chapter the two above approaches are discussed in turn.

17.7.1 Conductivity of Electrolyte Solutions

Conducting properties of electrolyte solutions are indicated by means of the *conductance* (L), which is the reciprocal of the electrical resistance:

$$L = \frac{1}{R} \tag{17.25}$$

Conductance is therefore similar to the resistive admittance. The SI conductance unit is Ω^{-1}, also known as the siemens (S).

For a conductor of uniform area A and length *l*, the resistance depends on the geometrical parameters, and also on a material constant called resistivity (ρ), as follows:

$$R = \rho \frac{l}{A} \tag{17.26}$$

The reciprocal of resistivity is called *conductivity* (λ):

$$\lambda = \frac{1}{\rho} \tag{17.27}$$

Therefore, the conductance depends on the intrinsic conductivity of the conductor, and also on its geometrical parameters:

$$L = \lambda \frac{A}{l} \tag{17.28}$$

In line with the general definition, solution conductance is a measure of the ability of the solution to transport electricity by ion transfer under the effect of an applied electric field. In order to measure the conductance, a conductometric cell is assembled by immersing two electrodes in the test solution. An ideal conductometric cell consists of two similar, parallel electrodes with a solution volume confined between them (Figure 17.22). The meaning of the geometric parameters A and *l* in Equation (17.28) is evident from Figure 17.1. However, in contrast to metal conductors, a network like that in Figure 17.22 involves two different conduction mechanisms: electronic conduction in the external circuit and ionic conduction in the solution. Continuing current flow at the electrode/solution interface occurs by means of electrochemical reactions, such as reduction or oxidation of water.

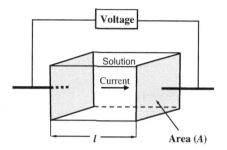

Figure 17.22 Idealized conductometric cell. A voltage is applied between two similar, parallel electrodes that confine the solution.

The definition of conductance in Equation (17.28) also applies to electrolyte solutions. However, practicable conductometric cells deviate from the simple geometry shown in Figure 17.22 and neither the electrode area nor the interelectrode distances are, as a rule, not known accurately. This is why it is usual to express the geometrical characteristics of a conductometric cell by means of a composite parameter termed the *cell constant* ($\sigma = A/l$, in meters). Therefore, the solution conductance is proportional to the intrinsic conductivity with σ being the proportionality constant. This constant is determined by measuring the conductance of standard solutions with known conductivity [36].

Clearly, the conductivity is similar to the conductance of a cell with $\sigma = 1$ m. Hence, the SI conductivity unit is $\Omega^{-1}\,m^{-1}$ but currently, the conductivity is reported usually in $\Omega^{-1}\,cm^{-1}$ (or $S\,cm^{-1}$).

As the conductivity is in general determined by the density of charge carriers, the solution conductivity depends on the concentration of the dissolved salt. In order to obtain a concentration-independent parameter, the *molar conductivity* (Λ) has been introduced as the ratio of the conductivity to the salt concentration c (in $mol\,L^{-1}$):

$$\Lambda = \frac{\lambda}{c} \tag{17.29}$$

The SI molar conductivity unit is $S\,m^2\,mol^{-1}$.

At extremely low concentrations (that is, at infinite dilution) electrostatic interaction between ions in solution is negligible and each ion moves independently of the others. Hence, the infinite dilution molar conductivity is obtaining by addition of the contribution of each type of ion. For an electrolyte with the formula M_mA_n (where M is a metal ion and A is an anion), this rule is formulated as:

$$\Lambda_\infty = m\Lambda_M + n\Lambda_A \tag{17.30}$$

where Λ_∞ is the limiting molar conductivity and Λ_M and Λ_A represent the individual contributions of each type of ion (ionic molar conductivity). For a solution containing N various ions the conductivity at very low concentrations is therefore:

$$\lambda = \sum_{i=1}^{i=N} (c_i\Lambda_i) \tag{17.31}$$

where c_i and Λ_i represent the concentration and molar conductivity of each ion, respectively. At the same time, Λ_i depends on the charge on each ion and on its mobility. Most common inorganic ions have an ionic conductivity between 50 and $100\,S\,cm^2\,mol^{-1}$. An exception is represented by hydrogen and hydroxyl ions that have ionic conductivities of 350 and $199\,S\,cm^2\,mol^{-1}$, respectively. Such large values stem from the particular transport mechanisms of the hydrogen and hydroxyl ions.

Ion mobility, and hence also the solution conductivity, depend on the temperature according to the following equation:

$$\Lambda_{i,T} = \Lambda_{i,T_o}[1 + \alpha(T - T_0)] \tag{17.32}$$

where T and T_0 are the measurement and the calibration temperatures, respectively, and α is the temperature coefficient, which is, for most inorganic ions (except H^+ and OH^-), about $0.02\,K^{-1}$.

As the conductivity is temperature dependent, accurate conductance measurements are carried out in thermostated cells or by measuring the solution temperature and applying temperature corrections.

Equation (17.30), which implies that the molar conductivity is independent of the electrolyte concentration, is reasonably accurate at concentration below 1 mM. At higher concentrations, electrostatic interionic forces cause marked deviations from Equation (17.30) and the empirical Kohlrausch equation is used to indicate the effect of concentration:

$$\Lambda = \Lambda_\infty - ac^{1/2}, \qquad (17.33)$$

where a is a specific constant of the dissolved electrolyte.

In conclusion, solution conductivity depends on concentration, charge and mobility of each ion in the solution. At electrolyte concentrations below 1 mM, the conductivity is roughly proportional to the electrolyte concentration. From the analytical standpoint, direct conductometric determinations are completely lacking in selectivity. However, conductometry is a suitable method for monitoring variations in the total ion concentration, which is exploited in various types of chemical sensors.

17.7.2 Conductance Measurement

In agreement with the method originally developed by Kohlrausch, conductance measurement is carried out as a rule by using AC methods in order to avoid the alteration of the solution composition by electrochemical reactions. Therefore, conductance measurements involve the measurement of the cell impedance under selected conditions that avoid effects arising from processes other than ion transport in solution. The complete equivalent of the conductometric cell (Figure 17.23A) accounts for any possible impedance component in the system. Thus, the double electric layer that develops at the interface of each electrode is modeled by the C_{dl} capacitors. Electrochemical reactions at each electrode give rise to the Faradaic impedances Z_F whereas the current flow through the solution is modeled by the resistor R. In addition, the electrode couple produces a stray capacitance C_{st}. This circuit is rather intricate, and, in order to simplify the measurement, the cell and the operation mode are designed such as to remove certain of the elements in Figure 17.2A. Thus, the effect of the double-layer capacitance is removed by using platinized platinum electrodes with a very large surface area. Due to the very large electrode area, C_{dl} becomes very high and its impedance becomes negligible. As the capacitance is inversely proportional to frequency, a fairly high frequency is used in measurements. The faradaic impedance contribution is made negligible by using small AC amplitudes. Platinum, which is usually employed as the electrode material, exhibits electrocatalytic properties and thus allows electrochemical reactions to occur virtually with no overvoltage. Also, the electrodes are placed in the same plane in order to minimize any stray capacitance. Under such conditions, the whole impedance is determined by the solution resistance alone.

Such optimal conditions can be achieved in a standard conductometric cell but are not always possible in conductometric sensors in which size and shape restrictions determine deviations from the ideal design. It is therefore better to represent a conductometric transducer by the equivalent circuit in Figure 17.23B that differs from the complete one only by the absence of Faradaic impedances. In this network, C_{dl} represent the total capacitance formed by combining the two C_{dl} capacitors in Figure 17.23A.

The impedance spectrum of the circuit in Figure 17.23B often displays a fairly narrow midfrequency range in which the whole impedance equates to the resistance. If such a plateau can be located, it is best to select the measurement frequency to be within it.

A method well suited for use in microelectrode cells is based on the so-called bipolar pulse measurement that minimizes the effect of both series and parallel capacitances [37,38]. As shown in Figure 17.24, two consecutive voltage pulses of equal amplitude and width are applied successively to the cell and the response current is monitored. At the beginning of the first pulse, the stray capacitance is rapidly charged, producing a small spike on the current response. Thereafter, the current decreases slowly as the voltage across the C_{dl} components increases. After

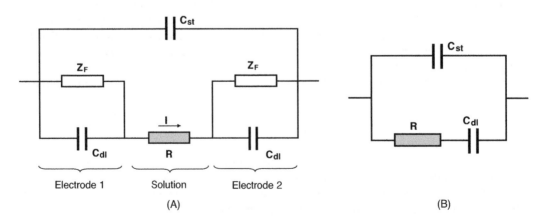

Figure 17.23 The equivalent circuit of a conductometric cell. (A) Complete circuit; (B) the equivalent circuit of a conductometric transducer.

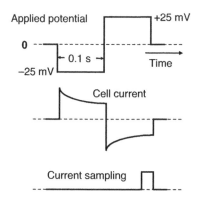

Figure 17.24 Time evolution of relevant parameters in the double-pulse method of conductance measurement.

applying a reverse pulse, a small spike appears again due to charging of the stray capacitance but the cell impedance at the end of the second pulse corresponds to the solution resistance alone. The current that is sampled at the end of the second pulse is proportional to the potential pulse amplitude and inversely proportional to the solution resistance. This method is very fast and works well with electrodes of about $100\,\mu m$ size. Dedicated computer-interfaced electronic equipment can be assembled readily [39].

17.7.3 Conductometric Transducers

Conductometric transducers for biosensor application are designed in the form of small-sized interdigitated electrodes (Figure 17.25). Electrodes are prepared by microfabrication methods in which a metal is deposited on an insulator support (for example, glass). Among the large variety of tested electrode materials, platinum metals perform best. In order to make temperature corrections, a resistance thermometer can be added to the transducer. As far as the equivalent circuit is concerned, the Faradaic impedance cannot be neglected in this case due to the small electrode area.

Miniature-size and facile fabrication are the main advantages of interdigitated transducers. Optimal sensitivity is achieved by decreasing both the electrode working surface and the overall transducer size [40].

17.7.4 Conductometric Enzymatic Sensors

Many enzyme-catalyzed reactions produce ionic species that modify the conductivity of the reaction solution. Hydrogen and hydroxyl ions are commonly involved in enzymatic reactions, but, as the assay is conducted in buffered solutions, such ions are neutralized by buffer components and do not contribute to the transport of electric charge. That is why conductometric transduction works well when the enzymatic reaction produces ionic species other than H^+ or OH^- [34,40]. A number of hydrolysis reactions fulfill this requirement. Thus, enzymatic urea hydrolysis results in ammonium and hydrogencarbonate ions:

$$(NH_2)_2CO + 3H_2O \xrightarrow{\text{Urease}} 2NH_4^+ + HCO_3^- + OH^- \tag{17.34}$$

The amino acid asparagine, which is important in nitrogen metabolism, undergoes enzymatic hydrolysis yielding two oppositely charged ions:

$$\text{L-asparagine} + H_2O \xrightarrow{\text{L-asparaginase}} \text{L-asparatate}^- + NH_4^+ \tag{17.35}$$

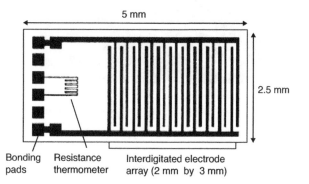

Figure 17.25 An interdigitated microelectrode array conductometric transducer. The resistance thermometer is introduced to perform automatic temperature corrections. Reproduced with permission from [41]. Copyright 1993 American Chemical Society.

Table 17.2 Examples of enzymatic reactions resulting in ionic products for application in conductometric enzymatic sensors.

Enzyme	Reaction examples	Comments
Amidases	$R(CO)\text{-}NH_2 + H_2O \xrightarrow{\text{Amidase}} -RCOO^- + NH_4^+$	Production of ionic groups of opposite charges from a neutral substrate molecule
Dehydratases	$\underset{\text{Serine}}{HOCH_2\text{-}CHNH_2\text{-}COOH} \xrightarrow{\text{Serine dehydratase}} \underset{\text{Pyruvic acid}}{CH_3\text{-}CO\text{-}COO^- + NH_4^+}$	
Decarboxylases	$\underset{\text{Lactic acid}}{CH_3 - CHOH - COOH} + O_2 \xrightarrow{\text{Lactate oxidase}} CH_3COO^- + HCO_3^- + 2H^+$	
Esterases	$\underset{\text{Acetylcholine}}{(CH_3)_3N^+ - (CH_2)_2 - O(CO)CH_3} + H_2O \xrightarrow{\text{Acetylcholinesterase}} \underset{\text{Choline}}{(CH_3)_3N^+ - (CH_2)_2 - OH} + CH_3COO^- + H^+$	Ion generation from a cationic substrate molecule
Kinases	$\underset{\text{ATP}}{R - O(PO_3H)_3^-} + Gluc \xrightarrow{\text{Glucokinase}} \underset{\text{ADP}}{R - O(PO_3H)_2^-} + Gluc - PO_3H^-$	Separation of two anionic species
Phosphatase	$RO - PO_3H^- + H_2O \xrightarrow{\text{Phosphatase}} ROH + H_2PO_4^- + H^+$	Change in the size/charge of ionic groups
Sulfatase	$H_3C - SO_3^- + H^+ \xrightarrow{\text{Sulfatase}} H_3C - OH + SO_4^{2-}$	

ATP and ADP denote adenosine triphospahate and adenosine diphosphate, respectively; Gluc and Gluc-PO$_3$H$^-$ denote D-glucose and D-glucose 6-phosphate, respectively. The state of ionic species corresponds to near-neutral pH.

In some cases, a sequence of enzymatic reactions can be combined in a multienzyme sensor in order to obtain ionic products. Thus, creatinine, which is an indicator of renal function, can be determined by a conductometric sensor which includes creatininase, creatinase and urease, to make use of the following sequence of reactions:

$$\begin{aligned} Creatinine + H_2O &\xrightarrow{\text{Creatininase}} Creatine^- + H^+ \\ Creatine^- + H_2O &\xrightarrow{\text{Creatinase}} Sarcosine^- + Urea + H^+ \end{aligned} \tag{17.36}$$

The urea produced in the previous sequence is finally hydrolyzed according to reaction (17.34) yielding ionic products.

Other examples of enzymatic reactions resulting in a change of the ionic composition are summarized in Table 17.2.

As bulk-solution ions also contribute to the charge transport, it is good to keep their concentration at the minimum possible level. Hence, conductometric sensors are customarily operated in pH buffers with concentrations ≤ 10 mM.

A typical configuration of a conductometric enzymatic sensor is shown in Figure 17.26. It includes platinum electrodes 10 μm wide and 1 mm in length. Enzyme immobilization can be achieved by crosslinking with bovine albumin under the effect of glutaraldehyde vapors. As the background conductance can be rather high, it is best to use

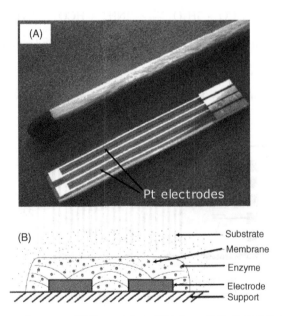

Figure 17.26 A conductometric enzymatic sensor. (A) General view; (B) schematic representation. Adapted with permission from [42]. Copyright 2003 Elsevier B.V.

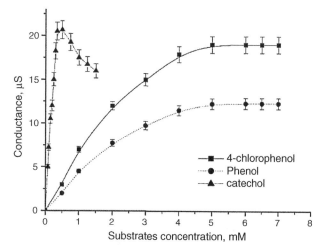

Figure 17.27 Calibration curves of a conductometric tyrosinase sensor for the determination of 4-chlorophenol, phenol and catechol. 5 mM phosphate buffer (pH 6). Adapted with permission from [42]. Copyright 2003 Elsevier B.V.

differential measurements. A reference sensor, which generates the background signal, is obtained similarly, except for the fact that the surface layers consist of crosslinked bovine albumin alone.

The response of a conductometric tyrosinase sensor (like that in Figure 17.26) to several phenolic substrates is shown in Figure 17.27. As the response is proportional to the product concentration within the enzyme layer, optimization of the sensor characteristics should follow the guidelines in Chapter 4. The enzyme loading factor is the key parameter that determines the extent of the linear response range. Modeling of a urea conductometric enzyme sensor is presented in ref. [43].

Conductometric enzymatic sensors are suitable for testing water pollution using enzyme inhibition [40]. As each enzyme is inhibited by a particular class of pollutant, a multisensor based on a certain number different enzymes provides more insight into the degree of pollution. The same task can be performed by algal whole-cell biosensors as the algae metabolism depends on several ion-generating enzymes, each of them featuring inhibition by a specific class of pollutant. Thus, immobilized *Chlorella vulgaris* microalgae causes conductivity modification by alkaline phosphatase and acetylcholinesterase enzymes. The first enzyme is inhibited by heavy-metal ions while the second one responds to carbamates and organophosphorus pesticides. As the differential measurement is more reliable, a reference sensor can be obtained by immobilizing inactive algae. Algal sensors should be operated under daylight conditions, light irradiation being essential to algae metabolism.

A conductometric pH sensor was developed using a responsive hydrogel that shrinks reversibly within a particular pH range [44]. Shrinking causes water elimination and hence a decrease in gel conductivity.

17.7.5 Conductometric Transduction by Chemoresistive Materials

Conductometric sensors presented in the previous section are based on modifications produced in the solution conductivity in response to biocatalytic processes. In such sensors, the change in conductivity occurs into the solution embedded within the sensing element. A limitation of this approach stems from the effect of the pH buffer that prevents detection of conductivity variation due to enzyme-generated hydrogen ions.

A more sophisticated approach rests on monitoring conductivity changes in chemoresistive materials such as conducting polymers. Typical materials suitable for such applications are polyaniline and polyindole [35,45]. For example, emeraldine, a particular form of polyaniline, undergoes a protonation–deprotonation process under the effect of changes in the solution pH (Figure 17.28A). This results in an alteration of the polymer conductance as the conductor form of the polymer turns into an insulator form by deprotonation.

Conductometric conductor polymer-based sensors are usually prepared by depositing a thin layer of a polymer film between two metallic or graphite contacts (Figure 17.28B). Film deposition can be achieved by electrochemical polymerization [46], screen printing with a paste containing polymer microparticles [47], or electrophoretic deposition of polymer nanoparticles [48]. The film resistance can be measured by forcing a small-amplitude AC current through the film and measuring the in-phase voltage in order to obtain the resistance as the voltage to current ratio. Some authors prefer to apply a DC voltage of a few mV between the electrodes and measure the current that is proportional to the film conductance. The second approach needs a bipotentiostat to control individually the potential of each electrode. In both cases, the small magnitude of the excitation signal is intended to prevent electrochemical reactions from occurring at the electrode surfaces.

Conductometric pH sensors represent the most straightforward application of pH-sensitive polymers [49]. When using pure polyaniline, the protonation–deprotonation reaction occurs within pH 4 and 5, which limits the response

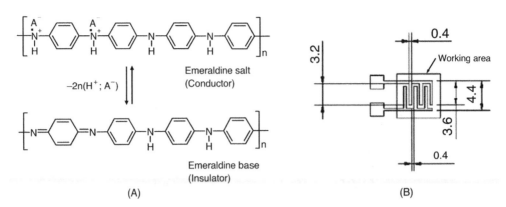

Figure 17.28 (A) The chemical structure and the proton exchange reaction of the emeraldine form of polyaniline. (B) Configuration of a conductometric pH sensor based on polyaniline responsive material. Two interdigitated gold electrodes are coated with a polyaniline film about 0.1 mm thick. All dimensions are in mm. (B) Adapted from [47]. Open access.

range to this domain. A broader response range has been obtained with composite films including emeraldine, polyvinyl butyral (as binder) and a surfactant that prevents the agglomeration of polymer particles [47]. Such a sensor can be fabricated by screen printing in a few mm size (Figure 17.28B). Although ion transport in the adjacent solution contributes to the total conductance, the response is dominated by charge transport within the polymer film. Electrochemical impedance measurements reveal that this sensor behaves as a series R–C circuit [50]. The response to pH changes over the pH range between 4 and 8 is almost linear. At high pH, the charge carriers in the film are depleted and the response becomes independent of pH. After each run, the sensor should be regenerated by immersion in an acidic solution that regenerates the emeraldine salt form. Depending on the sample pH, the response time varies between 1 and 2 min.

Another pH-responsive polymer is poly(*p*-phenylene-vinylene), a polymer that develops a photocurrent under illumination [51]. On a glass support, a 30-nm thin polymer film is intercalated between two aluminum electrodes and the working area is delimited by a silicon layer (Figure 17.29A). The photocurrent is measured under an applied voltage of 1 V. Since the polymer conductivity increases with increasing pH, the photocurrent follows the same trend (Figure 17.29B). The response is linear and reasonably reproducible at pH < 7 and at pH > 8 but a plateau is noticeable between these values. The fabrication processes are compatible with array microfabrication for multiple sample sensing.

Conductometric enzymatic sensors can be obtained simply by coating a conductometric pH sensor with an enzyme layer obtained by glutaraldehyde crosslinking [46,52] or by enzyme entrapment in a polymer. The fabrication method is very simple and suitable for microarray manufacturing [53]. Hydrogen ions produced by the enzymatic reaction diffuse partially into the polyaniline film and alter its conductance.

An alternative to this method involves incorporation of polyaniline tubules in an isoporous polycarbonate membrane, as shown in Figure 17.30 (see also [54]). On each side, the membrane is coated with thin gold layers that act as electrodes. Polyaniline is prepared by electrochemical polymerization so as to fill the membrane pores. During the polymerization process, the enzyme, which is present in the solution, is incorporated within the tubule. The sensitivity of such a sensor is much higher that that obtained with the bilayer version above since diffusion of hydrogen ions

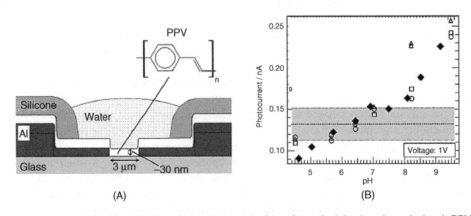

Figure 17.29 (A) A conductometric pH sensor based on a photoconductive polymer (poly(*p*-phenylene vinylene), PPV). The drawing is not to scale. (B) Sensor response to pH. Filled symbols: photocurrents *vs.* pH of the buffer solution for one device. Open symbols: three different series of the same buffer solution are measured using a second device to show reproducibility. The dotted line indicates the average of the deionized water control measurements made in between pH measurements, with its standard deviation indicated by the gray strip. Adapted with permission from [51]. Copyright (2006) Elsevier B.V.

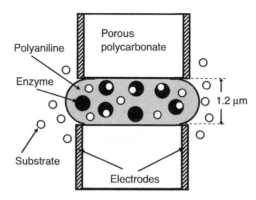

Figure 17.30 A conductometric enzymatic sensor based on polyaniline microtubules.

to the solution occurs to a very low extent. Moreover, this approach is also compatible with array fabrication. To this end, twin electrodes are deposited on a single polycarbonate strip. At each site, a different enzyme can be coimmobilized with polyaniline by electrochemical polymerization.

Living cells can also be entrapped in polyaniline in order to act as a source of enzyme. Thus, an urea sensor has been developed by using the *Brevibacterium ammoniagenes* microorganism as a source of urease [55]. The catalytic action of urease within the cell releases ammonia, thereby causing a local increase in the pH and hence an alteration of the conductivity.

Conductometric immunosensors can be obtained in the sandwich format using a signaling antibody tagged with polyaniline [56]. The recognition layer is flanked by two electrodes that serve to drive the current throughout in order to allow measuring the conductance (Figure 17.31). In this state, the conductance is determined by the ion transport in solution only. After forming the receptor–analyte–signaling antibody complex, a polyaniline layer develops between the electrodes and shunts the solution resistance, causing the conductance to increase. The response function in terms of conductance *vs.* analyte concentration is linear from 0.1 to 5 μg/ml and the limit of detection can be as low as about 0.5 μg/ml. The limit of detection is controlled by two factors: (i) solution resistance in parallel with the polyaniline resistance and (ii) nonspecific interactions. Binding of a nonspecific antibody also causes the conductance to increase to some extent, probably due to the accompanying counterions. However, the response to a nonspecific antibody is low and levels off at concentrations above 1 μg/ml. A competitive assay format with such a sensor works equally well.

Antigen sensors can be designed by using a specific antibody as the capture receptor [57,58]. Moreover, the sensor can be designed in the form of a series of disposable pads assembled with the transducer. In this way, the sensor-regeneration step, which can damage the sensor is eliminated [58].

Other promising chemoresistive materials are phthalocyanines that are suitable for detecting immunocompounds labeled with peroxidases. Such enzymes catalyze the oxidation of the iodide ion to iodine by hydrogen peroxide. Iodine thus formed alters the conductivity of a phthalocyanine film. As the amount of iodine depends on the incubation time, signal amplification can be achieved in accordance with this parameter. The sensing layer is built up on a hydrophobic membrane over the chemoresistive film, which is therefore not exposed to the solution thus preventing interference from the solution resistance. Iodine crosses this membrane by diffusion to reach the chemoresistive film. A limit of detection of 0.2 μg/ml has been reported for such a sensor [60].

Another approach to conductometric transduction in affinity sensors is based on gold nanoparticles as analyte tags. As previously, the sensing layer is assembled within the gap between the two electrodes. By analyte–receptor binding, gold nanoparticles are brought into the gap and are then grown by reductive silver deposition. A metal conductor

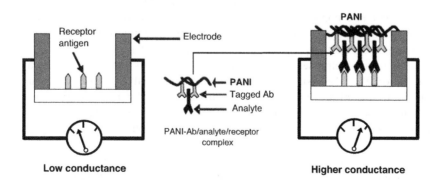

Figure 17.31 Sandwich assay conductometric immunosensor. PANI stands for polyaniline. Adapted from [59] (Open access).

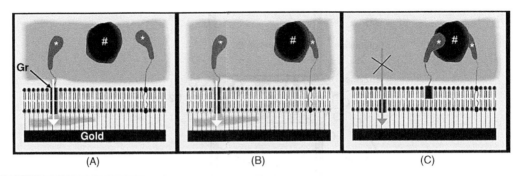

Figure 17.32 An ion-channel-based conductometric sensor. (A) Sensor configuration; (B) initial receptor–analyte binding; (C) two receptor binding accompanied by ion-channel disruption. Symbols: (∗): antibody fragment (Fab); (#): antigen; Gr: gramicidin ion channel. Arrows indicate the ion flux. Adapted with permission from [63]. Copyright 2008, Elsevier B.V.

layer forms in this way between the electrodes and causes a large decrease in the resistance. This approach has been initially suggested for nucleic acid sensors [61] but it can be adapted for use in other affinity sensors.

17.7.6 Ion-Channel-Based Conductometric Sensors

Ion channels consist of ion-permeable molecules embedded in impermeable, hydrophobic membranes (Section 6.6). Ion diffusion through ion channels can be modulated by a properly designed recognition process and the response can be provided by measuring the conductance across the membrane [62].

A conductometric immunosensor based on an artificial ion channel system is presented in Figure 17.32. The ion channel is formed by gramicidin, an antibiotic that enhances the permeability of the bacterial cell membrane to inorganic monovalent cations. This allows cations to travel through unrestricted, thereby destroying the ion gradient between the cytoplasm and the extracellular environment.

The sensing layer consists of a lipid membrane bilayer supported on a gold electrode. The first layer includes raw gramicidin, whereas the second one contains gramicidin molecules tethered to a Fab antibody fragment. Other analyte-specific antibody molecules are attached to the membrane via long lipid molecules in order to capture the antigen (Fig. 17.32B). When the Fab receptor binds to the previously formed complex as shown in (C), gramicidin molecules in the second layer are shifted from their initial position, thus disrupting the ion channel. The recognition event is indicated by a decrease in the electric conductance across the sensing layer.

This ion-channel switch sensor demonstrated very good performance in threshold qualitative test for influenza A virus in raw clinical samples with no crossreactivity to a large number of other viruses. If necessary, the sensitivity can be enhanced by assembling multiple sensors in an array exposed to the same sample. The array also provides opportunities for assessing multiple analytes in a single test.

17.7.7 Outlook

Conductometric sensors can be obtained according to several deign principles. The first one relies on conductivity changes within the electrolyte solution contained within the sensing layer. This is an extremely simple method, but the limit of detection is restricted by the parallel resistance developed in the solution phase. In addition, hydrogen ions cannot be effective in the transduction process as the buffer system keeps their concentration constant. Therefore, conductometric enzyme sensors are restricted to enzymes producing other kinds of ions.

The second alternative is based on conductor materials that display ion-dependent conductivity. Typical materials of this kind are certain conducting polymers. Thus, the emeraldine form of polyaniline features H^+-dependent conductivity. Hence, it is suitable as a transducer material in pH sensors and, implicitly, in sensors based on enzymes producing hydrogen ions. Shunting by the solution resistance, which is a common problem in the previous approach, could be less embarrassing, as conducting polymers feature a relatively low resistivity.

Another approach to conductometric sensors rests on the formation of conductive layers as a consequence of the recognition event. Before that, the system resistance is determined only by ion transport in solution and has a relatively large value. Formation of a conductive layer results in a considerable enhancement of the conductivity. The conductive layer can be formed of conducting polymers or of metal nanoparticles.

Affinity recognition can be used in conductometric sensors in conjunction with ion-channel-based conductometric transduction.

Conductometric sensors are amenable to large-scale production, are compatible with array fabrication, and require nonsophisticated read-out equipment. In contrast to amperometric or potentiometric sensors, conductometric sensors do not require a reference electrode. Conductometric sensors can make use of various biosensing materials such as enzymes, immunocompounds and living cells. The range of applications covers pH determination, the assay of enzyme substrates and of inhibitors, and immunoassays.

Questions and Exercises (Section 17.7)

1 Review the meaning of conductance and conductivity. What factors determine the conductivity of an electrolyte solution?

2 What are the suitable experimental conditions in conductance measurements?

3 What is the cell constant if the conductivity of a 0.1 mole/kg KCl solution at 25 °C is 1.28246 S m^{-1} and the measured conductance of the same solution is 26.3 mS ?

   Answer: 2.05 cm.

4 The conductivity of 0.100 M HCl is 39.4 mS cm^{-1}. What is the molar conductivity of the solution?

   Answer: 0.0394 S mol^{-1} m^2.

5 Check on-line sources for molar ionic conductivity values and estimate the conductivity of the following solutions: (a) 0.5 mM H$^+$, 0.9 mM Na$^+$ and 1.4 mM; HCO$_3^-$; (b). 0.25 mM Ca^{2+}, 0.9 mM Na$^+$ and 1.4 mM; HCO$_3^-$. Comment on the effect of H$^+$ substitution by Ca^{2+}.

   Answer: (a) 0.28; (b) 0.13 mS cm^{-1}.

6 Explain why the molar conductivity of a solution is not directly proportional to the electrolyte concentration.

7 Elaborate on the configuration of conductometric transducers and its implication in the design of conductometric sensors.

8 What kinds of enzyme are suitable for use in conductometric biosensors? What are the experimental conditions under which a conductometric sensor should be operated?

9 What factors determine the width of the linear response range for a conductometric enzyme sensor?

10 What are the advantages of whole-cell-based conductometric sensors?

11 Review the applications of conductometric sensors in monitoring environment pollution.

12 Elaborate on the structure and response mechanism of a conducting polymer-based pH sensor.

13 What is the basic configuration of a conductometric pH sensor and how can its conductance be measured?

14 What is the functioning mechanism of a conducting polymer-based conductometric enzyme sensor?

15 Review possible configurations of above sensors and manufacturing methods for producing them.

16 Using Figure 17.31 as a model, draw schematically the functioning principle of a conductometric immunosensor based on a competition assay.

17 Sketch a possible configuration of a conductometric affinity sensor based on metal nanoparticles as tags.

17.8 Impedimetric Sensors for Gases and Vapors

The quality of normal atmospheric air is determined to a large extent by the water vapor and carbon dioxide content. If the air is polluted, the content of inorganic gases (such as carbon monoxide and nitrogen oxides) or organic vapors is also of interest. The monitoring of indoor air quality requires the real-time determination of humidity and carbon dioxide content, in addition to the air temperature. Good approaches to such analytical problems are provided by electrical impedance-based sensors. As is typical of this kind of sensor, the overall impedance is not measured, but the sensor and the measurement technique are designed so as to provide the response in terms of either electrical resistance or capacitance.

This section focuses mainly on humidity sensors, but it also introduces the principles of capacitive sensors for gases and organic vapors with particular attention being paid to the determination of carbon dioxide in air.

17.8.1 Humidity: Terms and Definitions

As water vapor is a normal component of atmospheric air, the measurement and control of humidity are important not only for human comfort but also for various technological processes. Humidity measurement forms the objective of the scientific branch called *hygrometry* and laboratory methods for the measurement of humidity are well established. However, continuous monitoring of humidity as well as automatic regulation of this parameter can only be effected by using humidity sensors.

The humidity level in a gas is often measured as *the relative humidity* (r_h, or RH) which is defined as:

$$r_h = \frac{p_w}{p_s} \times 100\%, \tag{17.37}$$

where p_w is the partial pressure of water vapor in the gas sample and p_s is the saturation pressure of the same vapor, that is, the partial pressure found when the gas-water system is at thermodynamic equilibrium.

Humidity can also be measure in terms of *absolute humidity* that is the vapor content in grams of water per cubic meter of dry air. A parameter connected to gas humidity is the *dew point* (T_d) that represents the temperature in °C to which the gas must be cooled down (at constant pressure) to reach saturation by water vapor.

Humidity sensors can be constructed using materials whose electrical properties are altered upon water vapor absorption. Typical materials of this kind are ceramic materials and polymer electrolytes. In general, a humidity sensor is constructed by placing a humidity sensitive film between two electrodes. Transduction is performed by measuring either the electrical resistance of the film or the capacitance of the device [64,65].

17.8.2 Resistive Humidity Sensors

Humidity measurements can be carried out using the property of water to dissolve ionic compounds to give rise to conductive ionic solutions. Hence, the name *ionic conductivity sensor* is ascribed to such devices. A simple humidity sensor of this kind is based on a lithium chloride solution impregnated in a porous binder. Water absorption from the atmosphere enhances the conductivity of the device. This sensor cannot withstand high humidity and has a rather slow response.

More advanced humidity sensors rely on porous ceramics and organic polymers sensing materials.

Certain porous *ceramic oxides* undergo a change in their ionic conductivity as a consequence of water vapor absorption of [66]. Water absorption results in the formation of a primary layer consisting of water molecules doubly hydrogen-bonded to two surface hydroxyls. In this ice-like layer, water molecules are rigidly attached to the surface. Next, a series of successive water layers form by means of single hydrogen bonding to the previous layer. In these sublayers water molecules are free to rotate and allow protons to jump from one molecule to an adjacent one. This mechanism, which is known as the Grotthus transport of protons, gives rise to ionic conductivity along the adsorbed water multilayer, and this permits the water vapor content in a gas to be determined. Hence, the term *protonic-type humidity sensor* has been ascribed to devices based on such materials. Protons appear in the water layer upon dissociation of the strongly polarized adsorbed water molecules.

Depending on the humidity level, protonic conductivity can take place through the adsorbed water multilayers (at low humidity), or in capillary-condensed water (at high humidity).

In a porous material, *capillary condensation* can occur if the pore radius is below a certain limited value r_K given by the following Kelvin equation that applies to cylindrical pores with one end closed:

$$r_K = \frac{2\gamma M}{\rho R T \ln(p_s/p_w)} \tag{17.38}$$

Here, γ, ρ and M are the surface tension, density and molecular mass of water, respectively. Due to capillary condensation, liquid water appears in the pore even if the vapor partial pressure is below the saturation value. This occurs because condensation is assisted by the previously indicated hydrogen-bonding mechanism. The smaller the value of r_K or the lower the temperature, the more easily capillary condensation occurs. Equation (17.38) applies to pores open at only one end. When the cylindrical pores are open at both ends, condensation proceeds in pores with a radius below $r_K/2$ while desorption is determined by r_K.

Actual ceramic materials includes both types of pores. In addition, the pores are not cylindrical and pore diameters varies over a broad range. Linear response over a broad humidity range can be expected when pore diameters are asymmetrically distributed over a wide range. The sensitivity of the ceramic oxide is related to the number of water-adsorption sites, which can be increased by the presence of defect lattice sites and unsaturated oxygen atoms at the surface. It is also important to secure a high porosity of the ceramic, ranging from 25 to 50%.

Various mixed oxides have been used to sense humidity, as summarized in [66,67]. In general, mixed-oxide ceramics provide good humidity responses. This class includes spinel solid solutions made of $MgCO_3$ and TiO_2 or TiO_2 mixed with small amounts of V_2O_5 or Nb_2O_5 or a mixture of Al_2O_3, TiO_2 and SiO_2. In general, ceramic-oxide-based sensors exhibit a hysteresis, which indicates that the surface undergoes irreversible modification upon water adsorption. This arises from the formation of hydroxyl groups at the surface, which disturb proton conduction. Hence, the sensor should be fitted with a heater for self-cleaning/drying at about 500 °C in order to achieve sensor recovery.

Addition of alkali ions to oxide ceramics affects the microstructure favorably. In addition, alkali cations can contribute to conduction, in addition to the proton. Such additives enhance the sensitivity and remove the hysteresis.

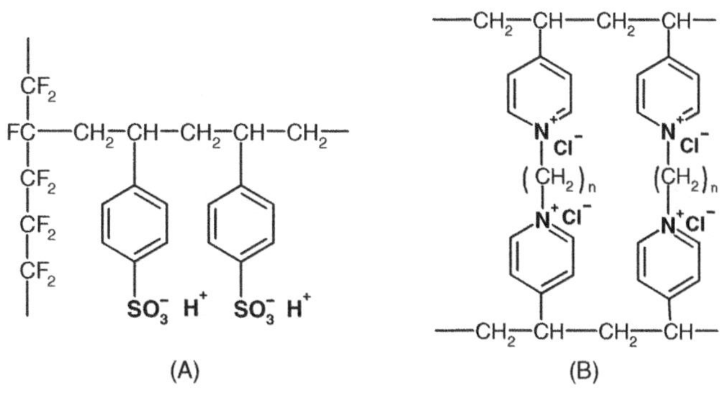

Figure 17.33 Examples of polymer electrolytes used in resistive humidity sensors. (A) PTFE-graft polystyrene sulfonic acid; (B) crosslinked polyvinyl pyridine.

There is currently much interest in the application of oxide nanomaterials (such as TiO_2 or Al_2O_3 nanotubes) to the design of resistive humidity sensors. Nanomaterials allow for a closer control of the structure of the sensing layer with favorable effects on the response characteristics.

Resistive humidity sensors can also be fabricated from *carbon films* modified with nanocrystals of sodium carbonate or acetate [68]. Owing to the presence of oxygen functional groups in both carbon and sodium salt particles, the above-mentioned composite material is strongly hydrophilic. Water absorption leads to partial dissolution of the salt that enhances the ionic conductivity of the film.

Another class of humidity-responsive material is based on *organic polymer electrolytes* that are polymers containing ionized groups attached to the organic backbone together with mobile inorganic counterions in pore water. Typical ionized groups in such polymers are quaternary ammonium and sulfonate groups. Due to the presence of such groups, organic polymer electrolytes are hydrophilic and absorb water proportionately to the atmosphere humidity. Electrical conductivity is imparted to absorbed water by the mobile counterions. So, the conductivity increases with the humidity owing to the increase in the volume of absorbed liquid.

Initially, polymer electrolytes formed by polymerization of an ionic monomer were used in resistive humidity sensors. However, these materials suffer from poor resilience at elevated humidity owing to their solubility in water. For this reason, other polymeric materials, formed from hydrophobic networks to which ionic groups are attached, have been developed [69]. As an example, Figure 17.33A shows a humidity-sensitive polymer produced by grafting sulfonated polystyrene to the hydrophobic backbone of polytetrafluoroethylene (PTFE). Crosslinking of hydrophilic polyvinyl pyridine gives rise to the electrolyte polymer in Figure 17.33B, which is stable in the presence of water owing to the presence of the hydrophobic polyvinyl chains. Another kind of stable humidity-sensitive material consists of intermixed networks of hydrophilic and hydrophobic polymers.

Humidity-sensitive polymer layers can be synthesized *in situ* over a dielectric support fitted with a pair of interdigitated electrodes. Alternatively, polymer sensitive layers can be produced by screen printing or by ink-jet printing techniques.

Conducting polymers are currently being investigated for possible application as humidity-sensing materials [70–72].

No matter what kind of sensing material is used, the sensor displays an intricate frequency response since a capacitive component is always present along with various resistive components arising from the metal contacts and the layer itself. The measurement procedure should be optimized so as to measure the layer resistance with as little as possible contribution from other impedance elements.

17.8.3 Capacitive Humidity Sensors

The electrical capacitance of a capacitor is inversely proportional to the permittivity of the dielectric between the conducting plates. Permittivity indicates the ability of a material to undergo electric polarization upon application of an electric field, and by this means reduce the total electric field inside the material. Thus, permittivity is related to the ability of a material to permit an electric field to develop. It is well known that water has a very large permittivity, while the permittivity of polymers is very low. Based on these principles, humidity-sensitive materials have been developed using polymers such as polyimides, polyethylene glycol (PEG), poly-4-vinyl pyridine, cellulose acetate and other cellulose derivatives [73]. These materials display an appreciable increase in permittivity upon water absorption. This effect can be ascribed to hydrogen bonding of water molecules to polar groups in the polymer, leading to the formation of strongly polar chemical bonds. The change in permittivity can be detected as a change in the electrical capacitance of the device. Hydrophilic polymers (for example, cellulose acetate) have strong basic sites for the hydrogen bonding of water. Less-hydrophilic polymers, such as carboxymethyl cellulose, do not form strong hydrogen bonds with water and the interaction between polymer and water is expected to be weak. As a result, the

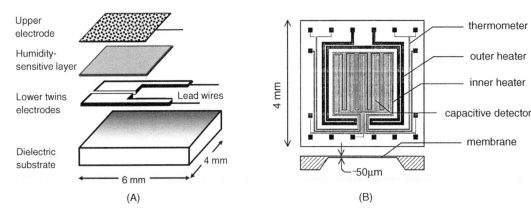

Figure 17.34   (A) Exploded view of a capacitive humidity sensor; (B) Principle of a dew-point hygrometer. (B) Reproduced with permission from [74]. Copyright 2000 Elsevier.

water absorption isotherm for less-hydrophilic polymers is close to a straight line, which secures a wider linear response range.

A capacitive humidity sensor is presented in Figure 17.34A. The device is supported on a dielectric plate of glass, ceramic, or silicon. Twin electrodes are deposited on the substrate by vapor deposition and an approximately 1 μm thick sensitive layer is applied over it. On top of this, a gold upper electrode is formed by vapor deposition to yield an approximately 20-nm thick porous layer that permits rapid vapor diffusion to the sensing film. The upper electrode acts as a common counterelectrode so that the equivalent circuit is represented by two capacitors connected in series.

Certain inorganic materials, such as alumina, silicon dioxide and porous silicon are also suitable as sensing materials in humidity sensors. These materials are very stable under elevated temperature conditions and are applied to humidity sensors operating under such conditions.

Currently, fabrication of humidity sensors relies on micromachining technology that can produce small-sized devices incorporating additional electronics and sensing devices. Thus, devices including a humidity sensor, a thermometer and thermoelectric cooler have been developed for automatic determination of the dew point. The design of such a device is presented in Figure 17.34B. It includes a capacitive humidity sensor formed over a Si-Si$_3$N$_4$-SiO$_2$ structure and also includes a polyimide sensing layer coated with a thin SiO$_2$ or Si$_3$N$_4$ film. The sensing layer is deposited between the interdigitated electrodes. A Pt resistive thermometer is also integrated into the device along with two heaters to evaporate condensed water and clean the surface. The device is cooled by a heat pump and the dew point can be detected optically by a light beam directed at the reflective coating. At the dew point, water condensation over the surface produces a marked decrease in the intensity of reflected light. The temperature value indicated at this moment represents the dew point. Alternatively, the dew point can be detected by the humidity sensor, which indicates a strong decrease in humidity when water vapor condenses. In order to diminish the thermal capacitance of the system and reduce the response time, the device is positioned over a 50-μm thick silicon membrane.

In general, polymers cannot withstand high humidity levels and elevated temperatures for a long time, but the response function is closer to a straight line than that of ceramics-based humidity sensors.

New prospects in the design of capacitive humidity sensors arise from application of nanomaterials. In this respect, promising perspectives are provided by silicon nanowires as humidity-sensing material. Silicon nanowires are produced in an easy way, by chemical etching of a silicon wafer in a solution of hydrogen fluoride and silver nitrate to form a structure like that shown in Figure 17.35A. In order to perform capacitance measurements, a pair of

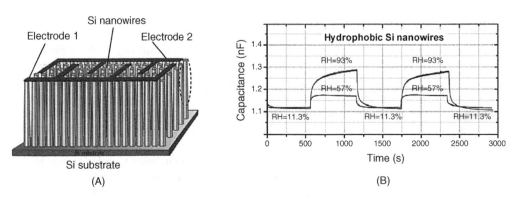

Figure 17.35   A silicon-nanowire capacitive humidity sensor of about 5 × 5 mm in size. (A) Configuration; (B) the capacitance vs. relative humidity response under humidity cycling for a hydrophobic silicon nanowire sensor. Adapted with permission from [75]. Copyright 2011 Elsevier.

interdigitated electrodes is formed by screen printing on top of the nanowire layer. The sensing mechanism is based on water vapor adsorption on the nanowires. The $\log C$ vs. RH response obtained with plain nanowires displays two linear regions with a transition region around RH = 50%. The change in the response slope at high humidity may be due to capillary condensation. However, a linear response over the whole humidity range has been obtained after turning the nanowire surface hydrophobic by silylation. In this way, capillary condensation is prevented [75]. Silylation brings about the additional benefit of a shorter recovery time, which is reduced to about 2 min (Figure 17.35B). Notably, this sensor displays almost no hysteresis; hence it does not need thermal regeneration prior to each run.

17.8.4 Capacitive Gas Sensors

Certain ceramic materials undergo a change in permittivity upon interaction with polar gases such as CO, CO_2 and NO. This property has been exploited in the design of capacitive sensors for determining these gases, which is important in assessing air quality [76].

Molecular sieves have proved particularly effective in this respect. A molecular sieve is a material containing tiny pores of a precise and uniform size. Thus, phosphorus molecular sieves respond preferentially to CO and CO_2 over water vapors. The structure of such sieves consists of four- and six-membered rings of alternating phosphorus and aluminum atoms bridged by oxygen. These rings are arranged so as to form channels 0.73 nm in diameter. Response to the mentioned gases arises from the dipole–dipole interactions between the gas analyte and the solid surface.

Another class of molecular sieve, namely zeolites is suited for sensing layers in capacitive sensors for hydrocarbons, such as butane.

Ceramic material-based CO_2 sensors should be operated at elevated temperature and this implies high power consumption for heating. In a search for normal temperature responsive materials, it was found that polyaminopropyl siloxane responds reversibly to CO_2 at temperatures between 15 to 50 °C [77]. This material is obtained by hydrolysis and polycondensation of 3-aminopropyl trialkoxysilane and it interacts with CO_2 so as to form hydrogen-bonded carboxylate ions. This reaction results in strongly polarized hydrogen bonds that appreciably alter the permittivity.

Polar organic molecules can be sensed by means of polymers that contain polar groups and undergo dipole–dipole interactions with the analyte vapor.

17.8.5 Integrated Impedimetric Gas Sensors and Sensor Arrays

Monitoring of atmosphere quality involves the measurement of a series of parameters, including temperature, humidity, and the concentration of certain gases such as CO_2 and possible pollutants. It is hence best to develop integrated sensors that indicate such parameters in a reliable way, with minimal power consumption and minimal interference from electromagnetic noise. In addition, mass production requires easy and cost-efficient fabrication technology.

A response to such demands is represented by the "plastic chip" platform for capacitive sensors shown in Figure 17.36. It was assembled on flexible polyimide foil, an electrotechnical grade material on which gas or vapor sensitive polymer layers are deposited. Sensing polymer layers (10–30 μm thick) can be produced by drop-, spray-, or spin-coating techniques. Two interdigitated electrode structures have been formed over each sensor area. In addition, a platinum track serves as a resistive thermometer. Metal deposition has been carried out by evaporation using a lift-off technique. The nominal capacitance of an electrode pair without the sensing layer was 10 pF. By proper selection of the polymer coating, one of the two sensors can be fabricated as a reference (that is, analyte-insensitive) device,

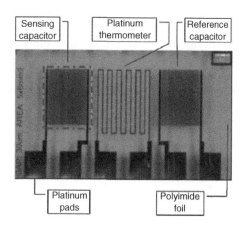

Figure 17.36 A platform for integrated capacitive sensors. It includes a resistive Pt resistance thermometer, a measuring sensor and a reference sensor assembled on plastic foil (commercial polyimide). The chip area is 14 × 25 mm. Reproduced with permission from [78]. Copyright 2009 Elsevier.

whereas the second one is coated with a responsive layer and functions as the actual sensor. This configuration allows for background correction of the overall signal.

A common problem with gas and vapors sensors arises from their unsatisfactory selectivity, a given sensor being able to respond to more than one single analyte in the sample. Typically, this problem is mitigated by arrays of crossreactive sensors, each sensor providing a mixed response being influenced to a greater or lesser extent by each analyte. Optimally, the sensitivity to each analyte should vary widely from sensor to sensor. Multivariate data-processing methods allow the determination of the actual concentration of each analyte from the composite response of the array. A platform such as that shown in Figure 17.36 can be expanded so as to include a number of different sensors, each of them coated with a layer displaying different response characteristics in order to obtain an assembly of crossreactive sensors. The great advantage of capacitive or resistive transduction resides in the simple fabrication technology and easy signal measurement.

17.8.6 Outlook

The concept of electrical impedance can be advantageously employed to develop sensors for water vapors (humidity sensors) and other volatile species. Research and development efforts in this area have been prompted by the high demand for reliable and resilient humidity sensors for application in the control of indoor atmospheres as well as for industrial applications. Two kinds of sensing material have been used mostly in such applications, namely ceramic metal oxides and polymer electrolytes. Porous ceramic materials absorb water vapor forming adsorbed water multilayers that display protonic electrical conductivity which depends on the gas sample humidity. Capillary condensation leading to water-filled pores is an alternative mechanism for use in humidity sensing.

Polymer electrolytes, that are hydrophilic materials, respond to humidity by absorbing water and displaying electrical conductivity by mobile counterions. With both the above materials, transduction is performed by means of the electrical conductance (or resistance) of the sensing film. Being more resilient, ceramic materials are preferred in sensors for high-temperature applications.

An alternative transduction method relies on the modification of the electric permittivity upon water absorption. Typically, hydrophilic polymers including functional groups that are able to form strong hydrogen bonds are employed to this end.

Other materials, such as conducting polymers and carbon films doped with sodium salts have also been investigated for application to humidity sensors. Much expectation is put on applications of nanomaterials that would enable a rigorous control of the sensing layer structure and impart a very high specific area.

In addition to humidity sensing, the above principles can be utilized in the development of sensors for certain gases such as carbon dioxide.

Questions and Exercises (Section 17.8)

1 Review the main physical parameters used to assess the water vapor content in a gas.
2 Give an overview of water vapor interaction with porous ceramic materials and point out the characteristics of ceramic materials used in resistive humidity sensors.
3 What kinds of polymer are useful in resistive humidity sensing and how do they interact with water vapor?
4 What are the principles of humidity determination by capacitive sensors? Outline the structure of a capacitive humidity sensor.
5 Elaborate on the possible effect of the surface hydrophobic layer on the response characteristics (linear range, hysteresis, and recovery time) of a silicon nanowire humidity sensor. *Hint*: take into account the interaction between water molecules and the nanowire surface.
6 Review briefly the materials used in capacitive CO_2 sensors and outline the molecular mechanism leading to a change in the permittivity of the material.
7 Prepare a short essay on possible integration of various sensors and transducer that are suited for testing the air quality from the standpoint of both physical and chemical parameters. Outline the application of micromachining technology in this area.

References

1. Janata, J. (2002) Electrochemical sensors and their impedances: A tutorial. *Crit. Rev. Anal. Chem.*, **32**, 109–120.
2. Orazem, M.E. and Tribollet, B. (2008) *Electrochemical Impedance Spectroscopy*, John Wiley & Sons, Hoboken, N.J.
3. Barsoukov, E. and Macdonald, J.R. (2005) *Impedance Spectroscopy: Theory, Experiment, and Applications*, Wiley-Interscience, Hoboken, N.J.

4. Brett, C.M.A. and Brett, A.M.O. (1993) *In Electrochemistry: Principles, Methods, and Applications*, Oxford University Press, Oxford, pp. 224–252.

5. Daniels, J.S. (2007) Label-free impedance biosensors: Opportunities and challenges. *Electroanalysis*, **19**, 1239–1257.

6. Katz, E. and Willner, I. (2004) Immunosensors and DNA sensors based on impedance spectroscopy, in *Ultrathin Electrochemical Chemo- and Biosensors: Technology and Performance* (ed. V.M. Mirsky), Springer, Berlin, pp. 67–115.

7. Katz, E. (2003) Probing biomolecular interactions at conductive and semiconductive surfaces by impedance spectroscopy: Routes to impedimetric immunosensors, DNA-sensors, and enzyme biosensors. *Electroanalysis*, **15**, 913–947.

8. Lisdat, F. and Schafer, D. (2008) The use of electrochemical impedance spectroscopy for biosensing. *Anal. Bioanal. Chem.*, **391**, 1555–1567.

9. Prodromidis, M.I. (2010) Impedimetric immunosensors-A review. *Electrochim. Acta*, **55**, 4227–4233.

10. Sluyters-Rehbach, M. (1994) Impedances of electrochemical systems - terminology, nomenclature and representation .1. Cells with metal-electrodes and liquid solutions. *Pure Appl. Chem.*, **66**, 1831–1891.

11. Hsu, C.H. and Mansfeld, F. (2001) Technical note: Concerning the conversion of the constant phase element parameter Y0 into a capacitance. *Corrosion*, **57**, 747–748.

12. Finklea, H.O. (1996) Electrochemistry of organized monolayers of thiols and related molecules on electrodes, in *Electroanalytical Chemistry: A Series of Advances* (eds A.J. Bard and I. Rubenstein), Dekker, New York, pp. 109–335.

13. Guan, J.G., Miao, Y. Q. and Zhang, Q. J. (2004) Impedimetric biosensors. *J. Biosci. Bioeng.*, **97**, 219–226.

14. Foss, B.J., Ion, A., Partali, V. *et al.* (2006) Electrochemical and EQCM investigation of a selenium derivatized carotenoid in the self-assembled state at a gold electrode. *J. Electroanal. Chem.*, **593**, 15–28.

15. Finklea, H.O., Snider, D.A., Fedyk, J. *et al.* (1993) Characterization of octadecanethiol-coated gold electrodes as microarray electrodes by cyclic voltammetry and AC-impedance spectroscopy. *Langmuir*, **9**, 3660–3667.

16. Foss, B.J., Ion, A., Partali, V. *et al.* (2004) O-1-[6-(methylselanyl)hexanoyl]glycerol as an anchor for self-assembly of biological compounds at the gold surface. *Collect. Czech. Chem. Commun.*, **69**, 1971–1996.

17. Bănică, A., Culeţu, A., and Bănică, F.G. (2007) Electrochemical and EQCM investigation of L-selenomethionine in adsorbed state at gold electrodes. *J. Electroanal. Chem.*, **599**, 100–110.

18. Knichel, M., Heiduschka, P., Beck, W. *et al.* (1995) Utilization of a self-assembled peptide monolayer for an impedimetric immunosensor. *Sens. Actuators B-Chem.*, **28**, 85–94.

19. Berggren, C., Bjarnason, B., and Johansson, G. (2001) Capacitive biosensors. *Electroanalysis*, **13**, 173–180.

20. Rickert, J., Göpel, W., Beck, W. *et al.* (1996) A mixed self-assembled monolayer for an impedimetric immunosensor. *Biosens. Bioelectron.*, **11**, 757–768.

21. Zayats, M., Raitman, O.A., Chegel, V.I. *et al.* (2002) Probing antigen-antibody binding processes by impedance measurements on ion-sensitive field-effect transistor devices and complementary surface plasmon resonance analyses: Development of cholera toxin sensors. *Anal. Chem.*, **74**, 4763–4773.

22. Mohsin, M.A., Hianik, T., Bănică, F.G. *et al.* (2010) A study of the properties of self-assembled bimolecular layers on the gold electrode with incorporated calixarenes for dopamine and epinephrine detection, in *Sensing in Electroanalysis* (eds K. Vytřas, K. Kalcher, and I. Švancara), University Press Center, Pardubice.

23. Cheng, Z.L., Wang, E.K., and Yang, X.R. (2001) Capacitive detection of glucose using molecularly imprinted polymers. *Biosens. Bioelectron.*, **16**, 179–185.

24. Mohsin, M.A., Bănică, F.G., Oshima, T. *et al.* (2011) Electrochemical impedance spectroscopy for assessing the recognition of cytochrome c by immobilized calixarenes. *Electroanalysis*, **23**, 1229–1235.

25. Knoll, W., Morigaki, K., Naumann, R. *et al.* (2004) Functional tethered bilayer lipid membranes, in *Ultrathin Electrochemical Chemo- and Biosensors: Technology and Performance* (ed. V.M. Mirsky), Springer, Berlin, pp. 239–252.

26. Hang, T.C. and Guiseppi-Elie, A. (2004) Frequency dependent and surface characterization of DNA immobilization and hybridization. *Biosens. Bioelectron.*, **19**, 1537–1548.

27. Guiseppi-Elie, A. and Lingerfelt, L. (2005) Impedimetric detection of DNA hybridization: Towards near-patient DNA diagnostics, in *Immobilisation of DNA on Chips I* (ed. C. Wittmann), Springer-Verlag, Berlin, pp. 161–186.

28. Bardea, A., Patolsky, F., Dagan, A. *et al.* (1999) Sensing and amplification of oligonucleotide-DNA interactions by means of impedance spectroscopy: a route to a Tay-Sachs sensor. *Chem. Commun.*, 21–22.

29. Xu, Y., Yang, L., Ye, X.Y. *et al.* (2006) Impedance-based DNA biosensor employing molecular beacon DNA as probe and thionine as charge neutralizer. *Electroanalysis*, **18**, 873–881.

30. Xu, Y., Yang, L., Ye, X.Y. *et al.* (2006) An aptamer-based protein biosensor by detecting the amplified impedance signal. *Electroanalysis*, **18**, 1449–1456.

31. Pospišil, L. (1996) Electrochemical impedance and related techniques, in *Experimental Techniques in Bioelectrochemistry* (eds V. Brabec, G. Milazzo, and D. Walz), Birkhäuser, Basel, pp. 1–39.

32. Holler, F.J. and Enke, C.G. (1996) Conductivity and conductometry, in *Laboratory Techniques in Electroanalytical Chemistry* (eds P.T. Kissinger and W.R. Heineman), Marcel Dekker, New York, pp. 237–264.

33. Coury, L. (1999) Conductance measurements. *Curr. Separ.*, **18**, 91–96.

34. Sheppard, N.F. Jr. and Guiseppi-Elie, A. (1998) Enzyme sensors based on conductimetric measurement, in *Enzyme and Microbial Biosensors. Techniques and Protocols* (eds A. Mulchandani and K.R. Rogers), Humana Press, Totowa, pp. 157–173.

35. Wei, D. and Ivaska, A. (2006) Electrochemical biosensors based on polyaniline. *Chem. Anal.*, **51**, 839–852.

36. Pratt, K.W., Koch, W.F., Wu, Y.C. *et al.* (2001) Molality-based primary standards of electrolytic conductivity - (IUPAC technical report). *Pure Appl. Chem.*, **73**, 1783–1793.

37. Johnson, D.E. and Enke, C.G. (1970) Bipolar pulse technique for fast conductance measurements. *Anal. Chem.*, **42**, 329–335.

38. LeSuer, R.J., Fan, F.R.F., and Bard, A.J. (2004) Scanning electrochemical microscopy, 52. Bipolar conductance technique at ultramicroelectrodes for resistance measurements. *Anal. Chem.*, **76**, 6894–6901.

39. Papadopoulos, N. and Limniou, M. (2001) A computer-controlled bipolar pulse conductivity apparatus. *J. Chem. Educ.*, **78**, 245–246.

40. Jaffrezic-Renault, N. and Dzyadevych, S.V. (2008) Conductometric microbiosensors for environmental monitoring. *Sensors*, **8**, 2569–2588.

41. Sheppard, N.F., Tucker, R.C., and Wu, C. (1993) Electrical-conductivity measurements using microfabricated interdigitated electrodes. *Anal. Chem.*, **65**, 1199–1202.

42. Anh, T.M., Dzyadevych, S.V., Van, M.C. *et al.* (2004) Conductometric tyrosinase biosensor for the detection of diuron, atrazine and its main metabolites. *Talanta*, **63**, 365–370.

43. Sheppard, N.F., Mears, D.J., and Guiseppi-Elie, A. (1996) Model of an immobilized enzyme conductimetric urea biosensor. *Biosens. Bioelectron.*, **11**, 967–979.

44. Sheppard, N.F., Lesho, M.J., and McNally, P. (1995) Microfabricated conductimetric pH Sensor. *Sens. Actuators B-Chem.*, **28**, 95–102.

45. Lange, U., Roznyatouskaya, N.V., and Mirsky, V.M. (2008) Conducting polymers in chemical sensors and arrays. *Anal. Chim. Acta*, **614**, 1–26.

46. Miwa, Y., Nishizawa, M., Matsue, T. *et al.* (1994) A conductometric glucose sensor-based on a twin-microband electrode coated with a polyaniline thin-film. *Bull. Chem. Soc. Jpn.*, **67**, 2864–2866.

47. Gill, E., Arshak, A., Arshak, K. *et al.* (2007) pH sensitivity of novel PANI/PVB/PS3 composite films. *Sensors*, **7**, 3329–3346.

48. Dhand, C., Sumana, G., Datta, M. *et al.* (2010) Electrophoretically deposited nano-structured polyaniline film for glucose sensing. *Thin Solid Films*, **519**, 1145–1150.

49. Korostynska, O., Arshak, K., Gill, E. *et al.* (2007) Review on state-of-the-art in polymer based pH sensors. *Sensors*, **7**, 3027–3042.

50. Gill, E.I., Arshak, A., Arshak, K. *et al.* (2009) Investigation of thick-film polyaniline-based conductimetric pH sensors for medical applications. *IEEE Sens. J.*, **9**, 555–562.

51. Pistor, P., Chu, V., Prazeres, D.M.F. *et al.* (2007) pH sensitive photoconductor based on poly (para-phenylene-vinylene). *Sens. Actuators B-Chem.*, **123**, 153–157.

52. Hoa, D.T., Kumar, T.N.S., Punekar, N.S. *et al.* (1992) A biosensor based on conducting polymers. *Anal. Chem.*, **64**, 2645–2646.

53. Sangodkar, H., Sukeerthi, S., Srinivasa, R.S. *et al.* (1996) A biosensor array based on polyaniline. *Anal. Chem.*, **68**, 779–783.

54. Sukeerthi, S. and Contractor, A.Q. (1999) Molecular sensors and sensor arrays based on polyaniline microtubules. *Anal. Chem.*, **71**, 2231–2236.

55. Jha, S.K., Kanungo, M., Nath, A. *et al.* (2009) Entrapment of live microbial cells in electropolymerized polyaniline and their use as urea biosensor. *Biosens. Bioelectron.*, **24**, 2637–2642.

56. Sergeyeva, T.A., Lavrik, N.V., Piletsky, S.A. *et al.* (1996) Polyaniline label-based conductometric sensor for IgG detection. *Sens. Actuators B-Chem.*, **34**, 283–288.

57. Mubammad-Tahir, Z. and Alocilja, E.C. (2003) A conductometric biosensor for biosecurity. *Biosens. Bioelectron.*, **18**, 813–819.

58. Muhammad-Tahir, Z. and Alocilja, E.C. (2003) Fabrication of a disposable biosensor for Escherichia coli O157: H7 detection. *IEEE Sens. J.*, **3**, 345–351.

59. Okafor, C., Grooms, D., Alocilja, E. *et al.* (2008) Fabrication of a novel conductometric biosensor for detecting Mycobacterium avium subsp paratuberculosis antibodies. *Sensors*, **8**, 6015–6025.

60. Sergeyeva, T.A., Lavrik, N.V., Rachkov, A.E. *et al.* (1998) An approach to conductometric immunosensor based on phthalocyanine thin film. *Biosens. Bioelectron.*, **13**, 359–369.

61. Park, S.J., Taton, T.A., and Mirkin, C.A. (2002) Array-based electrical detection of DNA with nanoparticle probes. *Science*, **295**, 1503–1506.

62. Cornell, B.A. (2008) Ion channel biosensors, in *Handbook of Biosensors and Biochips* (ed. R.S. Marks), John Wiley & Sons, New York, Chapter 21.

63. Oh, S.Y., Cornell, B., Smith, D. *et al.* (2008) Rapid detection of influenza A virus in clinical samples using an ion channel switch biosensor. *Biosens. Bioelectron.*, **23**, 1161–1165.

64. Yamazoe, N. and Shimizu, Y. (1986) Humidity sensors - Principles and applications. *Sens. Actuators*, **10**, 379–398.

65. Rittersma, Z.M. (2002) Recent achievements in miniaturised humidity sensors - a review of transduction techniques. *Sens. Actuators A-Phys.*, **96**, 196–210.

66. Traversa, E. (1995) Ceramic sensors for humidity detection - the state-of-the-art and future developments. *Sens. Actuators B-Chem.*, **23**, 135–156.

67. Lee, C.Y. and Lee, G.B. (2005) Humidity sensors: A review. *Sens. Lett.*, **3**, 1–15.

68. Lukaszewicz, J.P. (2006) Carbon films for humidity sensors. *Sens. Lett.*, **4**, 281–304.

69. Sakai, Y. (1993) Humidity sensors using chemically-modified polymeric materials. *Sens. Actuators B-Chem.*, **13**, 82–85.

70. Zeng, F.W., Liu, X.X., Diamond, D. *et al.* Humidity sensors based on polyaniline nanofibres. *Sens. Actuators B-Chem.*, **143**, 530–534.

71. Wang, R., Zhang, T., He, Y. *et al.* (2009) Complex impedance analysis of the humidity sensing properties of polypyrrole. *Acta Phys. -Chim. Sin.*, **25**, 327–330.

72. Tu, J.C., Li, N., Yuan, Q. *et al.* (2009) Humidity-sensitive property of Fe^{2+} doped polypyrrole. *Synth. Met.*, **159**, 2469–2473.

73. Sadaoka, Y. (2009) Capacitive-type relative humidity sensor with hydrophobic polymer films, in *Solid State Gas Sensing* (eds E. Comini, G. Faglia, and G. Sberveglieri), Springer Science+Business Media, LLC, Boston, MA, pp. 109–151.

74. Jachowicz, R. and Weremczuk, J. (2000) Sub-cooled water detection in silicon dew point hygrometer. *Sens. Actuators A-Phys.*, **85**, 75–83.

75. Chen, X., Zhang, J., Wang, Z. *et al.* (2011) Humidity sensing behavior of silicon nanowires with hexamethyldisilazane modification. *Sens. Actuators B-Phys.*, **156**, 631–636.

76. Ishihara, T. and Matsubara, S. (1998) Capacitive type gas sensors. *J. Electroceram.*, **2**, 215–228.

77. Patel, S.V., Hobson, S.T., Cemalovic, S. *et al.* (2010) Materials for capacitive carbon dioxide microsensors capable of operating at ambient temperatures. *J. Sol-Gel Sci. Techn.*, **53**, 673–679.

78. Oprea, A., Courbat, J., Bârsan, N. *et al.* (2009) Temperature, humidity and gas sensors integrated on plastic foil for low power applications. *Sens. Actuators B-Chem.*, **140**, 227–232.

18
Optical Sensors – Fundamentals

Analytical chemistry uses light on a large scale to investigate the chemical composition of various kinds of sample. Methods based on light interaction with the sample are known as optical methods of analysis [1]. In many instances, the compound of interest is able to interact with light without any prior chemical modification. However, quite often the analyte cannot be detected directly by an optical method. In such cases, the analyte is converted into a detectable compound by a chemical reaction involving a specific reagent. This procedure involves chemical reactions in the solution phase in which both the analyte and the reagent are free to diffuse.

In order to obtain an optical sensor, the specific reagent is included in a sensing layer. The analyte–reagent reaction is monitored by a light beam that is conveyed by an optical fiber (or other type of waveguide) integrated with the sensing layer. Therefore, waveguiding is a crucial process in optical sensors.

Optical transduction can be achieved by measuring the light power absorbed or emitted by a component of the sensing layer at a specific wavelength. As the dependence of light power on the wavelength represents an optical spectrum, such methods are classified as spectrochemical methods. Application of spectrochemical methods needs a component of the sensing layer to be able to absorb or emit light. If this condition cannot be fulfilled, one must resort to optical signaling labels.

An alternative to spectrochemical transduction is represented by the optical monitoring of a physical property of the sensing layer that varies upon interaction with the analyte. Examples of such properties are the refractive index and the thickness of the sensing layer. This class of transduction techniques does not need an optical label and are denoted label-free methods.

This chapter gives an overview of the fundamental physical and physicochemical principles of optical transduction for the development of optical sensor platforms. Specific applications of optical sensors are introduced in Chapter 19. Further, Chapter 20 deals with applications of nanomaterials in optical sensors.

The field of optical sensors is covered in various texts (e.g., [2] and review papers (e.g., [3–5]).

18.1 Electromagnetic Radiation

Optical sensors rely on the interaction of electromagnetic radiation with the sensing layer. Electromagnetic radiation is characterized by its intensity, propagation direction, frequency (ν) or wavelength (λ), and polarization. According to the wavelength range, electromagnetic radiation shows certain properties that allow a series of spectral regions to be defined, as shown in Figure 18.1. In general, optical sensor function in the ultraviolet (UV) visible (Vis) or infrared (IR) regions. Strictly speaking, the term light denotes radiation in the visible region, but it is commonly applied also to ultraviolet and infrared radiation.

Light propagation is a wave phenomenon and from this standpoint, light is characterized by *frequency* (ν) or *wavelength* (λ). W*ave number* ($\tilde{\nu}$) is usually employed with reference to infrared radiation. The relationship between the above wave parameters is:

$$\lambda = \frac{v_p}{\nu} = \frac{1}{\tilde{\nu}} \tag{18.1}$$

where v_p is phase velocity of the wave in the propagation medium.

When electromagnetic radiation exchanges energy with a substance, energy transfer occurs in the form of discrete energy packets (called *photons*). According to the Plank equation, the photon energy is directly proportional to the wave frequency, the proportionality constant being the Plank constant, h. Therefore, photon energy (E_{ph}) is related to the wave parameters as follows:

$$E_{ph} = h\nu = \frac{hc}{\lambda} = hc\tilde{\nu} \tag{18.2}$$

where c is the light velocity in vacuum. Hence, photon energy increases proportionally with the frequency and the wave number but is inversely proportional to the wavelength.

Chemical Sensors and Biosensors: Fundamentals and Applications, First Edition. Florinel-Gabriel Bănică.
© 2012 John Wiley & Sons, Ltd. Published 2012 by John Wiley & Sons, Ltd.

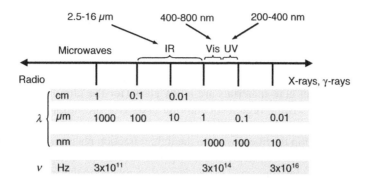

Figure 18.1 Spectral regions in the spectrum of electromagnetic radiation. Adapted with permission from [6]. Copyright 1996 John Wiley & Sons Limited.

Electromagnetic radiation applies electrical and magnetic forces on subatomic particles, such as electrons, which are endowed with electric charge and magnetic moment. The two forces are oriented perpendicularly to each other. As the strength of the electromagnetic field is time-variable, free electrons experience oscillations when subject to an electromagnetic field and, in this way, an exchange of energy between the substance and the field can take place.

Electrons involved in chemical bonds behave in a particular way because, owing to quantum-mechanical restrictions, they are allowed to adopt only particular, quantized energy levels. Energy exchange between electromagnetic radiation and bound electrons is allowed only if the photon energy matches the energy difference between two molecular orbitals. In this way, chemically bonded atom systems can absorb or emit electromagnetic radiation. Energy transferred to the system by absorption promotes an electron from a filled orbital to a higher, empty orbital. On the other hand, when an electron shifts from a filled orbital to a lower, empty orbital, electromagnetic radiation is emitted. In addition to the energy-matching conditions, such energy transitions are restricted by a series of quantum mechanics selection rules.

An optical sensor should be integrated with additional optical devices such as a wavelength-selection device and optoelectronic components such as light sources and light detectors [7–9]. Common light sources are lasers and light-emitting diodes.

Wavelength selection can be performed by an optical filter, which selects light within a defined spectral region that is extremely narrow in the case of interference filters. More advanced wavelength selection is achieved by means of a monochromator. This apparatus is fitted with an optical prism or a diffraction grating that perform spatial separation of light according to its wavelength. Therefore, the monochromator allows various components of a polychromatic beam to be inspected. A monochromator integrated with a light detector forms a spectrometer. Currently, miniature spectrometers of a few centimeters in sizes are commercially available.

The light detector provides an electrical signal that is proportional to the intensity of the incident light. A very sensitive light detector is the photomultiplier tube. Photodiodes and phototransistors are small and inexpensive light detectors.

Semiconductor-based detectors can be integrated in detector arrays in which a large number of detectors are uniformly distributed over an area of about a square centimeter. The charge-coupled device, which is also found in common electronic photocameras, is an example of this kind. A one-dimensional array of hundreds or thousands of photodiodes can be used in applications that do not require two-dimensional imaging.

The light beam of interest can be interfered with background light. In order to avoid this, it is usual to modulate the incident light beam, which means that one of its parameters varies periodically with time according to a selected function. Simple modulators are based on a rotating half-disk ("chopper") placed in the path of the beam. It converts the constant-intensity beam produced by the source into a pulsed beam. Suitable electronics connected to the detector allows the background constant signal to be rejected.

The commonly used term "light intensity" is somewhat elusive. Accurately defined quantities are *irradiance*, and *radiant emittance*, that denote the power of electromagnetic radiation per unit area at a surface. Irradiance is used when the electromagnetic radiation is incident on the surface. Radiant emittance is used when the radiation is emerging from the surface. The SI units for these quantities are watts per square meter (W/m^2). In practical applications, light intensity is often reported in arbitrary units representing the response of the light detector.

18.2 Optical Waveguides in Chemical Sensors

Optical sensors rely to a large extent on waveguides that are optical devices capable of letting light propagate along selected directions that can deviate from the straight line. Waveguides can be constructed in the form of optical fibers, in a planar configuration or according to other geometries.

18.2.1 Optical Fibers: Structure and Light Propagation

An optical fiber is a flexible, transparent and very thin fiber made of silica glass or plastic that acts as a waveguide, or light pipe, to transmit light between the two ends of the fiber even if the fiber is bent.

An optical fiber (Figure 18.2A) is made of a core and a cladding that has a slightly lower refractive index compared with that of the core. Typical core materials are silica glass (for the ultraviolet, visible and near-infrared regions), plastic materials (for the visible region) or fluoride glass (for the infrared region). For protection and reinforcement, an external plastic layer is coated over the fiber.

Light propagation is confined within the fiber owing to the light refraction phenomenon in which the direction of a light beam changes at the interface of two isotropic and transparent media with different refractive indexes n_1 and n_2 (Figure 18.2B). According to Snell's law, the angle of incidence θ_1 is related to the angle of refraction θ_2 as follows:

$$\frac{\sin \theta_1}{\sin \theta_2} = \frac{n_2}{n_1} \tag{18.3}$$

If the angle θ_1 is increased gradually, the angle θ_2 increases also and a situation arises where the refracted beam is directed along the interface, that is, at $\theta_2 = 90°$. At even higher values of θ_1, the angle θ_2 becomes greater than 90°, which implies that the light beam does not cross the interface but returns back to the initial medium. This is *total internal reflection*, a phenomenon that should be not confused with simple (specular) reflection. In specular reflection, the incidence and reflection angles are equal to each other, while in total internal reflection the same angles are correlated by Snell's law.

The critical incidence angle θ_c which gives $\theta_2 = 90°$ results from Snell's law as $\theta_c = n_2/n_1$ and is therefore constant for a particular pair of transparent media.

In an optical fiber, refraction occurs when the beam enters the fiber and then each time the beam reaches the core/cladding interface. In order for light propagation to be confined within the fiber, total internal reflection should take place at this interface. Assume that beam (1) in Figure 18.2A is directed so that the incidence angle at the core/cladding interface is the critical angle for total internal reflection. In order to achieve this, the beam should enter the fiber at a particular angle θ_0 that results from Snell's law as:

$$\sin \theta_0 = \frac{\sqrt{n_1^2 - n_2^2}}{n_0} = \text{Numerical aperture} \tag{18.4}$$

This angle determines the *cone of acceptance* of the optical fiber. A beam that comes from outside of this cone cannot propagate along the fiber. Conversely, any beam that enters the fiber through the cone of acceptance does propagate further (see, e.g., beam (2) in Figure 18.2A). This beam experiences total internal reflection each time it reaches the cladding edge and can propagate along to the end of the fiber even if the fiber is winding. At the exit, the beam direction is circumscribed to a *cone of illumination*, similar to the cone of acceptance. A cone of illumination forms only in cases of incoherent light, such as that produced by common lamps. However, coherent light produced by lasers leaves the fiber in the form of a columnated beam.

Equation (18.4)) indicates that the *numerical aperture* depends on the refractive index (n_0) of the medium from which the light enters the fiber or into which it leaves the fiber. n_0 is close to 1 for air but is 1.33 for water.

The fiber in Figure 18.2A is a *step-index fiber* because the refractive index changes suddenly at the core/cladding interface. A *gradient-index fiber* is an optical fiber whose core has a refractive index that decreases with increasing radial distance from the fiber axis.

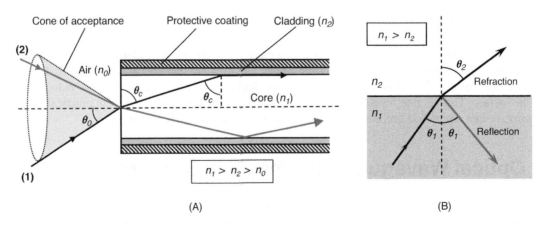

Figure 18.2 (A) The optical fiber: structure and mechanism of light transmission. (B) Light reflection and refraction at the interface of two isotropic media with different refractive indices n_1 and n_2.

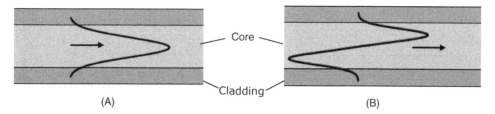

Figure 18.3   Propagation modes in waveguides. Curves indicate the transversal distribution of the wave amplitude. (A) single-mode waveguide; (B) double-mode waveguide.

Light propagates as a transversal wave. Its propagation in a waveguide is affect by the overlapping of rays following different pathways. As a result, stationary oscillations with the amplitude varying along the fiber axis appear, forming *propagation modes*. Depending on the core-cladding materials and the fiber radius, light can propagate as a single mode or as multiple modes, as shown in Figure 18.3. Remarkably, the tails of the amplitude distribution curves expand beyond the core/cladding interface. This feature is essential in certain types of optical sensor and is discussed in detail in Section 18.2.3.

In conclusion, in order to let light propagate along an optical fiber, light should enter the fiber as a convergent beam with the convergence angle smaller than that corresponding to the numerical aperture. With incoherent light, the exit beam is a divergent beam circumscribed to the cone of illumination that is symmetrical to the cone of acceptance.

In optical sensors, the optical fiber can play either a passive or an active role. As a passive component, the optical fiber serves only to convey light from the source to the sensing layer and then from the sensing layer to the light detector. Conversely, in an active fiber sensor, the fiber is intrinsically involved in the transduction process.

18.2.2   Passive Fiber Optic Sensor Platforms

A possible fiber optic sensor configuration is based on a sensing layer deposited on a flat support. A tiny amount of sample is deposited on the sensing layer and allowed to undergo recognition. Upon completion of the reaction, the resulting change in the optical properties of the layer are assessed by means of a fiber bundle (Figure 18.4A) that delivers the interrogation beam through the central fiber and collects the response beam by the other fibers. This arrangement is suitable for both reflectance and fluorescence transduction. A bifurcated fiber bundle can be used in the same fashion (Figure 18.4B).

An alternative design relies on a sensing layer integrated with the optical fiber to obtain tip-sensitive devices. The sensing layer can be added as a membrane at the fiber end (Figure 18.4C) or can be attached in a groove etched at the tip (Figure 18.4D). Fiber optic sensors with the sensing layer installed at the tip of the fiber are often called *optodes*.

18.2.3   Active Fiber Optic Sensor Platforms

Interaction between the light beam and the sensing layer can be achieved by attaching a sensing layer directly over the core after removing the cladding over a short distance (Figure 18.5A(a)). The sensing layer can be installed in the middle of the fiber (b) or at the fiber end (c).

Light can interact with the sensing layer if the refractive index of the core is greater than that of the added layer. As indicated earlier (Figure 18.3), at the interface between the core and the sensing layer light energy is manifest over a

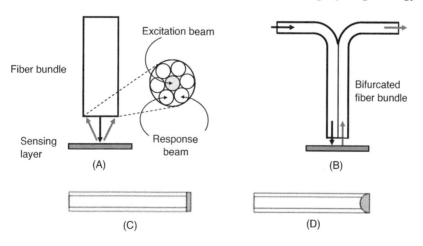

Figure 18.4   Passive optical fiber sensor platforms. (A) Fiber bundle; (B) bifurcated fiber bundle; (C) and (D): tip-based sensors: (C) modified end-face; (D) etched tip. Reprinted with permission from [10]. Copyright 2008 American Chemical Society.

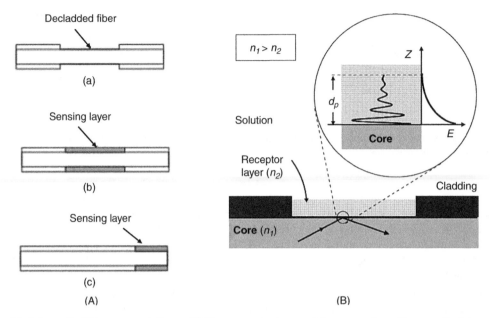

Figure 18.5 (A) Active optical fiber sensor platforms. (B) The evanescent wave at the core surface. The magnified view shows schematically the decay of the wave amplitude (*E*) with the distance from the core surface. (A) Reprinted with permission from [10]. Copyright 2008 American Chemical Society.

short distance beyond the core in the form of an *evanescent wave* (Figure 18.5B). The amplitude of the evanescent wave decreases exponentially with the distance from the core surface (*z*) and the wave practically vanishes at a distance called the *depth of penetration* (d_p). As the depth of penetration may be 100–200 nm, the light beam crossing the fiber interacts only with particles that are in close proximity of the core but not with particles in solution. This highly localized interaction is the strength of active waveguide sensors, also known as *intrinsic optical sensors*.

If the coating layer is optically transparent, the evanescent wave travels back to the core with no loss of energy. However, backcoupled light changes certain parameters depending on the thickness and the refractive index of the coating. Such changes are exploited in label-free optical sensors. If light-absorbing species are present within the sensing layer, then the evanescent wave loses energy by absorption and thereby its intensity is attenuated. This is the basis of the *attenuated total internal reflection* analytical method, which can serve as a transduction method in optical sensors.

On the other hand, the evanescent wave may perform fluorescence excitation within the sensing layer. Light emitted by fluorescence couples back to the fiber and can be detected at the fiber end.

Another application of the evanescent wave is light transfer between two waveguides. This can be accomplished by placing two waveguides close together so that the evanescent field generated by one of them does not decay much before it reaches the other one. If the receiving waveguide can support modes of the appropriate frequency, the evanescent field gives rise to propagating-wave modes, thereby connecting (or coupling) the wave from one waveguide to the second one.

18.2.4 Planar Waveguides

A planar waveguide is formed from a dielectric slab core sandwiched between two cladding layers with lower refractive indexes [11]. Light propagation along the core is secured by total internal reflection, as in the case of the optical fiber.

Planar waveguides are particularly suitable for obtaining intrinsic sensors, as shown in Figure 18.6. The receptor layer is formed over the waveguide core and its components interact with the evanescent wave produced at the core surface provided that the $n_2 > n_1 > n_3 \geq n_0$ condition is fulfilled.

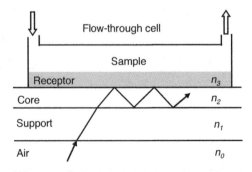

Figure 18.6 Configuration of an intrinsic planar waveguide sensor. The receptor layer is formed over the core.

The planar waveguide configuration imparts to the optical sensor a rugged geometry and a large contact surface with the sample.

Large-scale fabrication of planar waveguide sensors is possible using *integrated optical circuit* technology that allows the waveguide to be integrated with other optical components such as power splitters, optical amplifiers, optical modulators, filters, lasers and detectors. As in the case of electronic integrated circuits, integrated optical circuit technology makes use of photolithography to pattern wafers by etching and material deposition. Arrays of planar waveguide sensors can be produced in this way. At the same time, planar waveguides are compatible with miniature integrated analytical systems known as microfluidic analytical systems [12,13].

18.2.5 Capillary Waveguides

Optical waveguides can be prepared in the form of a capillary tube through which the sample is allowed to flow [14]. Light is coupled into and out of the capillary through small diffraction gratings fabricated at the capillary ends. The sensing layer can be formed on the inside capillary wall in order to be put in direct contact with the sample solution. In this case, the waveguide is in the passive configuration. Fluorescence excitation through the evanescent wave is an alternative design. Capillary waveguide sensors are suitable for application in microfluidic analytical systems [15].

18.2.6 Outlook

Waveguides are obtained by integrating two transparent materials with different refractive indexes. The higher-index material forms the core, while the lower-index coating forms the cladding. Light propagation along the waveguide is possible only if the incident beam is focused on the input end at an angle that is smaller than the angle that defines the numerical aperture.

Waveguiding is an essential feature of optical sensors. First, waveguiding allows for conveying light from the light source to the sensing element and then to the optical devices that perform signal processing in order to deliver the analytical response. This is the typical format of passive waveguide-based optical sensors. Moreover, the evanescent wave that expands beyond the core limit can interact with a sensing layer over the core surface. This interaction affects the parameters of the propagating wave and can be used to monitor the recognition event. This is the active (or intrinsic) waveguide sensor format.

Waveguide optical sensors can be designed according to various geometrical configurations. Thus, integration of an optical fiber with a sensing layer leads to very small sensors of the dip-in form. Planar and capillary waveguides are particularly suited for flow-through analytical applications including microfluidic analytical systems.

Questions and Exercises (Section 18.2)

1 What is the optical phenomenon that governs light propagation in a waveguide?
2 What is the mechanism of light propagation in a waveguide? What geometrical condition should be fulfilled in order to achieve light propagation along the waveguide?
3 What is a passive waveguide sensor format? Review possible configurations of passive fiber optic sensors.
4 What is an evanescent wave and what are its applications in chemical sensors?
5 How are planar waveguide sensors structured? What are the advantages of this format?

18.3 Spectrochemical Transduction Methods

Spectrochemical methods of analysis rely on light absorption or emission by sample compound's molecules [1]. Such processes are connected with transitions between molecular orbitals and are possible only if the energy difference between orbitals involved matches the photon energy. Therefore, light absorption or emission occurs within particular spectral bands centered on the nominal wavelength.

Very often, neither the receptor nor the analyte have specific absorption or emission bands. In order to perform transduction, an optical label that performs either light absorption or emission is included in the recognition system. Common labels are organic dyes and certain metal complexes.

18.3.1 Light Absorption

Light absorption occurs when the photon energy in the incident beam fits the energy difference between a filled and an empty orbital in the absorbing molecular species. Owing to the interplay of electronic and vibrational energy

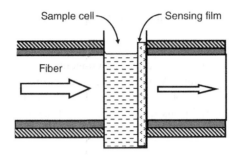

Figure 18.7   Configuration of an absorptiometric optical sensor. The sensing film should be optically clear.

levels, light absorption occurs within a broad wavelength interval in the UV-VIS region but is restricted to relatively narrow spectral bands in the infrared.

In light-absorption measurements, the power of the beam transmitted through the sample (P) is compared with that of a reference beam (P_0) that crosses a reference sample which contains no analyte but is similar to the sample itself in all other respects. According to the Lambert–Beer law, the degree of absorption in the sample is expressed by the *absorbance* (*A*) as follows:

$$A = log \frac{P_0}{P} = log \frac{1}{T} = abc \tag{18.5}$$

where *T* is the *transmittance*, *a* is the molar absorptivity of the analyte, *b* is the thickness of the absorbing layer and *c* is the concentration of the analyte. Therefore, the absorbance is directly proportional to the concentration and the sensitivity of the method depends on the thickness *b* and the characteristic absorptivity. As the absorptivity depends on the wavelength, the wavelength is adjusted so as to obtain maximum sensitivity.

As $P_0 > P$, absorbance is a positive number. Good accuracy is obtained if *P* is commensured with P_0 so that the absorbance assumes values between 0.1 and 1. This limits the working range of the method to about one order of magnitude.

If the sample contains several absorbing species, the total absorbance at a given wavelength is the sum of the individual absorbances of each species:

$$A = b(a_1 c_1 + a_2 c_2 + \cdots a_n c_n) \tag{18.6}$$

where $a_1 \ldots a_n$ indicates the molar absorptivity of each absorbing species and $c_1 \ldots c_n$ represent individual concentrations. Absorbance additivity allows corrections to be made to the required signal when it is slightly interfered with the absorbance produced by another compound.

Light-absorption-based sensors can be obtained by mounting a sample cell between two aligned optical fibers, as shown in Figure 18.7. The sensing film can be formed on plastic strips that are installed in the cell prior to the run. Owing to constructive restrictions, the reference signal (P_0) cannot obtained at the measurement wavelength, as is done in standard spectrophotometry. In practice, the reference signal is obtained at a different wavelength where the analytical response is negligible.

18.3.2 Diffuse Reflectance Spectrometry

Reflection represents the change in direction of a light beam at an interface between two different media such that the beam returns into the medium from which it originated. Two distinct types of reflection are possible. The first is *specular* (or mirror-type) reflection in which the angle at which the wave is incident on the surface equals the angle at which it is reflected.

The second type of reflection, called *diffuse reflection*, represents the reflection of light from a surface such that an incident ray is reflected at many angles rather than at just one angle as is the case of specular reflection. This occurs when light strikes the surface of a nonhomogeneous, nonmetallic material. As shown in Figure 18.8A, due to multiple reflection and refraction at the microscopic irregularities inside the material, light bounces off in all directions. In other words, diffuse reflection consists of light scattering at the surface of a nonhomogeneous specimen. Diffuse reflection takes place with polycrystalline materials, powders, and organic materials composed of cells or fibers (such as paper).

Analytical application of diffuse reflection is based on light absorption by light absorbing compounds within the specimen. The correlation between reflectance and the concentration of the absorbing species is provided by the Kubelka–Munk equation:

$$f(R) = \frac{(1 - R)^2}{2R} = \frac{ac}{s} \tag{18.7}$$

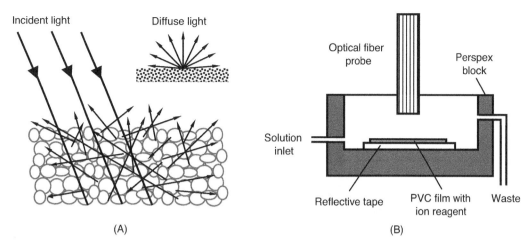

Figure 18.8 (A) Diffuse reflection from the surface of a granular solid (refraction phenomena are not represented); (B) A strip-type reflectance based sensor platform. (B) Adapted with permission from [16]. Copyright 1998 Elsevier.

In this equation, R is the reflected light power, $f(R)$ is the *remission function*, a is the molar absorptivity of the light absorbing component, c is the concentration of this species and s is the scattering coefficient of the material. Hence, the remission function is directly proportional to the concentration of the substance of interest. Through the molar absorptivity term, the remission function is a function of the wavelength, which requires the wavelength of the incident light to match the absorption band of the analyte.

Reflectance spectrometry is a suitable method for the analysis of solids. For example, the content of protein, starch and water in wheat can be determined by reflectance spectrometry in the near-infrared region. Diffuse reflectance is also suitable for noninvasive analysis of cutaneous blood.

Dip-type reflectance sensors can be obtained by attaching a sensing membrane at the end of a bifurcated optical fiber. For flow-analysis purposes, reflectance sensors are obtained by receptor immobilization on a strip or on beads contained in a flow-through cell. A strip-type reflectance-based sensor platform is shown in Figure 18.8B. In order to determine toxic metal ions, a colored reagent is incorporated in a PVC membrane to form the sensing part. Reflectance measurement at a suitable wavelength indicates the ion concentration in the sample. Similarly, pH sensors can be obtained by incorporation of pH color indicators in the sensing membrane.

Disposable test strips are widely used in enzymatic assay of glucose in blood. A part on the strip is impregnated with glucose oxidase and a cosubstrate that develops a color caused by the enzymatic oxidation of glucose. Measurement of color intensity and conversion in glucose concentration is done by a hand-held meter.

18.3.3 Luminescence

Luminescence (from Latin *lumin-, lumen* = light) is a form of cold body radiation emission by a substance. It is connected to electronic transitions between molecular orbitals of excited molecular species and involves two main steps, namely *excitation*, which means production of a labile, excited state, followed by *relaxation* to the ground state by light emission.

Depending on the excitation mechanism, one can distinguish different kinds of luminescence. Luminescence produced by light excitation is called *photoluminescence* and can be of the *fluorescence* or *phosphorescence* type, depending on the relaxation pathway.

Excited molecules can be produced by certain chemical reactions that give rise in this way to *chemiluminescence*.

Certain living organisms, such as the firefly, produce light by enzyme-catalyzed chemical reactions. This particular kind of chemiluminescence is called *bioluminescenece*. Bioluminescence-producing enzymes extracted from living organisms are used in analytical applications.

Luminescence that accompanies certain electrochemical reactions is another type of chemiluminescence that is known as *electrochemiluminescence*.

Compounds that are able to produce luminescence are called *luminophores* and are commonly employed as optical labels in chemical sensors.

Photoluminescence requires the sample to be irradiated with an excitation light beam that can induce fluorescence of the matrix compounds in the sample giving thus background light. As a consequence, the limit of detection is restricted to values higher than that expected for a background-free response. In contrast, chemiluminescence occurs in the absence of an excitation beam and the background interference problem is less critical. As will be shown later, certain methods for fluorescence measurement can reduce considerably the effect of the background radiation.

The next five sections introduce fluorescence and related phenomena as optical transduction methods. The last two sections present chemiluminescence and electrochemiluminescence.

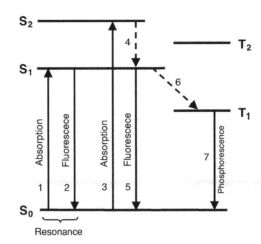

Figure 18.9 Simplified energy level diagram (Jablonski diagram) showing electronic states of a molecule and transitions between them in photoluminescence phenomena. Full arrows indicate radiative transitions (light absorption or emission); broken arrows denote nonradiative transitions (for example, dissipation of energy from the molecule to its surroundings). S denotes singlet states; T denotes triplet states.

18.3.4 Fluorescence Spectrometry

Fluorescence consists of light emission by molecules previously excited through light absorption. Fluorescence provides extremely sensitive transduction methods and is widely used in optical sensors [17–19].

The mechanism of fluorescence is illustrated in Figure 18.9. Molecular excitation is performed by light absorption at a wavelength that promotes molecule transition from the ground electronic level S_0 to a higher, empty level S_1 (1). This can be followed by molecule relaxation back to the ground level (2), when the excitation energy is released in the form of light with a similar frequency. This kind of fluorescence, which is called *resonance fluorescence*, is seldom met in molecular compounds. However, excitation can promotes the electron to a higher level, S_2 (3). By exchange of kinetic energy with the surroundings, the energy state of the molecule shifts then to the level S_1 through a nonradiative transition (4) that alters the vibration state of the molecule. This is followed by a transition back to the ground level (5) accompanied by light emission. However, as a fraction of the incoming photon energy was lost in the nonradiative transition, emitted light has a lower frequency. The difference in frequency (or wavelength) between incident and emitted light is called the *Stokes shift*.

Another possible mechanism of relaxation involves a nonradiative transition from the S_2 level to a metastable T_1 level (transition 6). This state may have a rather long lifetime, from milliseconds to hours, depending on the compound involved. Therefore, the molecule relaxes to the ground state with a relatively long delay (7). Light emission through this mechanism is called *phosphorescence*.

Both fluorescence and phosphorescence are included in the class of *photoluminescence* phenomena. The relaxation mechanism in photoluminescence depends on the energy-level distribution in the involved compound and on quantum-mechanics restrictions.

Some compounds, such as the amino acid tryptophan and the nicotinamide adenine dinucleotide (NAD^+) coenzyme, are fluorescent. If the analyte is not fluorescent it can be labeled with a fluorescent tag (called a *fluorophore*), which could be an organic dye, a metal complex or a luminescent nanoparticle.

The efficiency of photoluminescence is given by the *quantum yield* (QY) which is defined as follows:

$$QY = \frac{\text{Number of emitted photons}}{\text{Number of absorbed photons}} \qquad (18.8)$$

The quantum yield is always smaller than one because both radiative and nonradiative relaxation occurs simultaneously. It is 0.90 for fluorescein in 0.1 M NaOH, but only 0.03 for reduced nicotinamide adenine dinucleotide (NADH) in water. A high quantum yield secures a good sensitivity in photoluminescence chemical analysis.

It should also be borne in mind that both the quantum yield and the emission wavelength are sensitive to the environment of the molecule such as the solvent, solution components or the other molecular fragment to which the fluorophore group is appended.

Two critical parameters should be selected in fluorimetry, namely the excitation and the emission wavelength. These parameters are selected by means of the absorption (Figure 18.10A) and emission (Figure 18.10B) spectra of the compound of interest, which is in this example the $[Ru(bipy)_3]^{2+}$ complex (bipy denotes 2,2′-bipyridine). Two maxima are evident on the absorption spectrum and each maximum wavelength is suitable for excitation. The emission spectrum recorded with excitation at 450 nm has a maximum at 620 nm; this wavelength has to be selected for fluorescence intensity measurements.

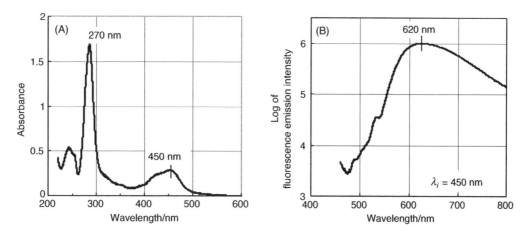

Figure 18.10 Absorption **(A)** and emission **(B)** spectra of the fluorescent $[Ru(bipy)_3]^{2+}$ complex.

18.3.5 Steady-State Fluorescence Measurements

In steady-state fluorescence, the sample is irradiated with a constant power beam with a frequency matching the energy difference between the ground and the excited levels. The response function can be derived with reference to Figure 18.11A by taking into account the direct proportionality relationship between the emitted power (F) and the absorbed power ($P_0 - P$):

$$F = k_f(P_0 - P) \tag{18.9}$$

The constant k_f depends upon the quantum yield, the geometry and physical parameters of the measuring system. On the other hand, the Lambert–Beer law (Equation (18.5)) gives:

$$P = P_0 10^{-abc} \tag{18.10}$$

From the above equations one obtains:

$$F = k_f P_0 (1 - 10^{-abc}) = k_f P_0 (1 - 10^{-A})$$

where $A = abc$ is the absorbance. If $A < 0.05$, the above equation turns into the following limiting form:

$$F = (2.303 k_f P_0 ab)c \tag{18.11}$$

Therefore, the power of the fluorescence beam is directly proportional to the concentration of the fluorescent compound provided that the absorbance, and hence the concentration, is very low. This conclusion is illustrated in Figure 18.11B. The sensitivity of the method depends on two specific constants of the fluorophore, namely the molar

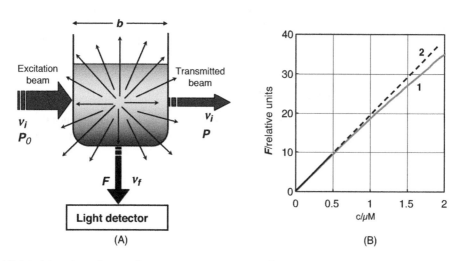

Figure 18.11 (A) Principles of steady-state fluorescence measurement. (B) Response function in steady-state fluorescence measurements (curve 1) and the limiting response at very low concentrations (curve 2).

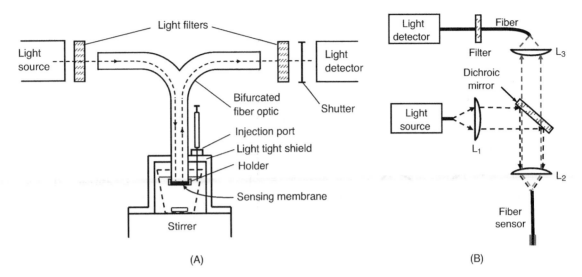

Figure 18.12 Typical configurations of fluorescence optical fiber sensors. (A) Bifurcated fiber optic sensor; (B) Fiber optic senor using a dichroic beam splitter. L denotes lenses. (A) Reprinted with permission from [22]. Copyright 1982 American Chemical Society.

absorptivity and the fluorescence quantum yield (which is included in the k_f constant). The product of these two quantities is called the *brightness of the fluorescent molecule*. On the other hand, fluorescence sensitivity is proportional to the power of the excitation beam and, therefore, the sensitivity can be enhanced by increasing the incident power. However, fluorophores can decompose under strong irradiation (*photobleaching*). Therefore, stability under irradiation is an important criterion in the selection of the fluorophore.

Fluorescence is a very sensitive analytical method. Its limit of detection is determined as a rule by background radiation arising from the fluorescence of other compounds in the sample. For examples, proteins produce a background radiation owing to the fluorescence of their aromatic amino acids.

Fluorescence is widely used as a label-based transduction method in immunosensors and nucleic acid sensors. Common fluorescent labels are organic dyes and certain metal complexes. Nanoparticles, such as semiconductor nanoparticles and carbon nanotubes, are superior in many respects to organic dyes as fluorescent labels (see Chapter 20).

As background fluorescence of biological samples is limited to the visible region, fluorescence in the infrared region is more advantageous from the standpoint of the detection limit [20,21]. Near-infrared fluorescent labels can be organic dyes, semiconductor nanoparticles or single-walled carbon nanotubes.

Two possible configurations of fluorescence-based sensors are shown in Figure 18.12. The sensor in Figure 18.12A uses a bifurcated optical fiber bundle to convey excitation and fluorescence radiation to and from the sensing membrane attached at the distal end of the optical fiber. The optical filter in front of the light source selects the excitation wavelength while the emitted light is passed through a second filter that selects the fluorescence wavelength.

In the configuration shown in Figure 18.12B, a *dichroic mirror* is used as a beam splitter in order to separate the fluorescence light from the excitation beam. A dichroic mirror reflects the light with the wavelength under a certain wavelength threshold but is transparent to the radiation with the wavelength over a second threshold. Owing to the Stokes shift, fluorescence radiation has a greater wavelength than the excitation radiation and a dichroic mirror should be selected so as to reflect the excitation beam and transmit the fluorescence radiation. As the dichroic mirror functions as a band-pass filter, a narrow-band filter is placed in front of the light detector in order to select the specific fluorescence wavelength.

18.3.6 Time-Resolved Fluorimetry

In time-resolved fluorimetry, a very short irradiation pulse is applied and then the fluorescence decay is recorded as a time function in the absence of the excitation radiation. During the irradiation, a fraction of analyte molecules are excited and then relax according to the kinetics of a first-order reaction:

$$F = k_d c^*(t) \tag{18.12}$$

where $c^*(t)$ is the time-variable concentration of the excited state and k_d is the decay rate constant. Hence, fluorescence intensity varies as:

$$F = F_0 \exp(-k_d t) = F_0 \exp\left(-\frac{t}{\tau_0}\right) \tag{18.13}$$

where F_0 is the power at the initial moment and τ_0 is the *fluorescence lifetime* ($\tau_0 = k_d^{-1}$), that is, the time needed for the fluorescence intensity to decrease to $1/e = 0.378$ of its initial value. The fluorescence lifetime can be determined from the slope of the ln F vs. time plot.

As the light-intensity measurement is conducted in the absence of the excitation radiation, the response does not suffer interference from the scattered excitation light. If constant background radiation is present, time-resolved fluorimetry allows the actual lifetime of the fluorophore to be inferred by appropriate data processing. The effect of the background fluorescence can be eliminated by using fluorophores with a lifetime much longer than that of the background fluorescence. Intensity sampling after a sufficiently long time delay provides a response signal that is free from background interference.

Clearly, time-resolved fluorimetry provides a better limit of detection compared to steady-state measurements. However, time-resolved fluorimetry requires very fast optoelectronic components.

In an alternative technique (*phase-resolved fluorimetry*), the sample is excited by a radiation beam with the intensity modulated according to a sine function with frequency f. In response to this excitation, fluorescence radiation is also a sine function, but it is shifted on the time axis because relaxation does not occur instantly. The correlation between the lifetime and the phase shift ϕ is:

$$\tau_0 = \frac{\tan \phi}{2\pi f} \tag{18.14}$$

Phase-modulation fluorescence measurements allow *dual lifetime self-referencing sensors* to be developed [23]. To this end, two different fluorophores are included in the sensing part. A short-lifetime fluorophore serves as indicator while an analyte insensitive long-lifetime fluorophore produces the reference signal. The two fluorophores are selected such that their absorption and emission bands overlap, which allows for excitation with the same light source and fluorescence measurement at the same wavelength. The excitation source is modulated at a frequency compatible with the lifetime of the indicator. Owing to the very short lifetime, the indicator signal has a negligible phase shift ($\phi_{Ind} \approx 0$), whereas the phase shift of the reference signal ϕ_{Ref} assumes a significant but constant value. A change in the analyte concentration modifies the intensity of the indicator fluorescence but the reference intensity remains constant. The phase shift of the mixed signal (ϕ_m) depends on the amplitude of both the indicator signal (A_{Ind}) and the constant reference signal (A_{Ref}) as follows:

$$\cot \phi_m = \cot \phi_{Ref} + \frac{1}{\sin \phi_{Ref}} \frac{A_{Ind}}{A_{Ref}} \tag{18.15}$$

This equation demonstrates a linear relationship between $\cot \phi_m$ and the amplitude ratio, which, in turn, depends on the analyte concentration through the A_{Ind} term.

This method allows the effects of temperature variations and fluctuations in the excitation power to be removed by a ratiometric approach and brings about a better accuracy when compared with measurements based solely on indicator fluorescence measurements.

A generic platform for fluorometric sensing by dual lifetime referencing is presented in Figure 18.13. Light arising from a blue light-emitting diode (LED) undergoes total internal reflection at each face of the polymer reflective

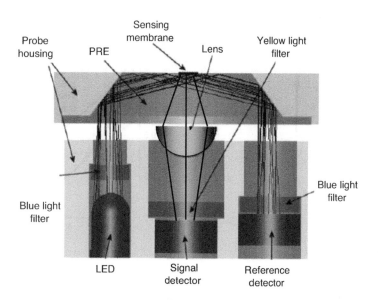

Figure 18.13 A generic platform for dual lifetime referencing fluorometric sensors. LED: light-emitting diode; PRE: polymer reflective element. Reprinted with permission from [24]. Copyright 2006 Elsevier.

element (PRE). At the upper edge, it excites fluorescence of both indicator and reference fluorophores in the sensing membrane by the evanescent wave to produce yellow light. Fluorescence light is collected and focused onto the signal light detector by a half-ball lens. A yellow filter is positioned in front of the signal detector to reject scattered blue light. A blue light filter is included in front of the reference detector in order to eliminate the longer wavelengths arising from the LED emission and sensing layer fluorescence. In response to the sine modulated excitation light, the sensing membrane generates a sine response that is phase shifted offset from the reference signal. The phase shift of the signal with respect to the reference, along with amplitude values, allows the analyte concentration to be calculated by means of Equation (18.15).

18.3.7 Fluorescence Quenching

Quenching refers to any process that lessens the fluorescence intensity of a given substance. This can occur by various mechanisms. The process presented in Figure 18.14A is characteristic of *collisional dynamic quenching*. Under the effect of a steady excitation beam, molecules L are excited to L^* and can decay by two concurrent routes. One of these involves relaxation back to the ground level accompanied by fluorescence light emission. A second route is possible in the presence of quencher A that can abstract excess energy from the excited species and allow it to relax without light emission. This produces a decrease in the fluorescence intensity that depends on the quencher concentration according to the Stern–Volmer equation:

$$\frac{F_0}{F} = 1 + k_{SV}c_Q \tag{18.16}$$

In this equation, F_0 and F are the fluorescence power in the absence and in the presence of the quencher, respectively, c_Q is the quencher concentration and k_{SV} is the Stern–Volmer constant ($k_{SV} = k_q/k_d$). The Stern–Volmer equation allows the quencher concentration to be determined by means of a linear calibration function. This equation applies to the *steady-state fluorescence quenching* that implies that the concentration of the excited form remains constant during the measurement.

Steady-state measurement of fluorescence quenching is impaired by background radiation that affects the limit of detection. Time-resolved measurements, which are more advantageous in this respect, are based on the determination of the time-dependent fluorescence intensity ($F(t)$) normalized to the initial intensity (F_0). This quotient decreases exponentially with the time as:

$$\Psi_Q = \frac{F(t)}{F_0} = \exp\left[-(k_d + k_q c_Q)t\right] = \exp\left[-\frac{t}{\tau}\right] \tag{18.17}$$

Here, τ is the fluorescence lifetime in the presence of the quencher which is given by the following equation:

$$\tau = (k_d + k_q c_Q)^{-1} \tag{18.18}$$

As shown in Figure 18.14B, the Ψ_Q parameter decreases faster in the presence of the quencher owing to the occurrence of two relaxation routes. Consequently, the half-lifetime measured in the presence of a quencher is smaller than

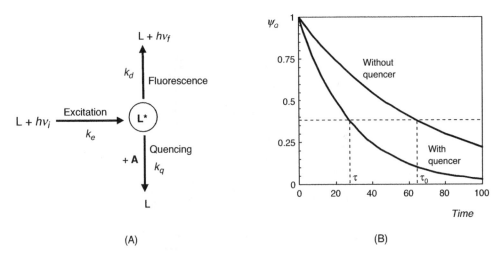

(A)   (B)

Figure 18.14 (A) Mechanism of dynamic fluorescence quenching. L is a luminophore and A is a quencher. k_e, k_d, and k_q are the rate constants for excitation, emission and quenching, respectively (B) Time evolution of the response in time-resolved measurement of fluorescence quenching.

that obtained without quencher, in accordance with Equation (18.18). Taking into account the definition of the half-life-time in the absence of the quencher (τ_0), one obtains the following equation that is formally similar to Equation (18.16):

$$\frac{\tau_0}{\tau} = 1 + k_{SV}c_Q = 1 + \tau_0 k_q c_Q \qquad (18.19)$$

Equation (18.19) allows the quencher concentration to be determined from lifetime measurements.

Fluorescence quenching is a very sensitive method for the determination of quenchers such as oxygen (Section 19.3). Monitoring of the chloride ion as fluorescence quencher has been used to develop an optical sensor for seawater salinity [23].

18.3.8 Resonance Energy Transfer

Resonance energy transfer (RET, also referred to by the acronym FRET which stands for fluorescence resonance energy transfer) consists of energy transfer from an excited species (donor, D) to an acceptor (A) through a nonradiative process (Figure 18.15A). As a result, the acceptor shifts to an excited state and can relax by fluorescence emission. This process is critically dependent on the distance between the two species. In order for RET to occur, the absorption spectrum of the acceptor should overlap to a large extent the emission spectrum of the donor.

The distance at which RET is 50% efficient, called the *Förster distance*, is typically in the range of 2–9 nm. Therefore, the rate constant of an RET process decreases with r^6, where r is the distance between the donor and the acceptor.

The mechanism in Figure 18.15A is an oversimplification. Actually, the excited donor D^* can also experience fluorescence or nonradiative relaxation. On the other hand, the excited acceptor A^* can relax by nonradiative transitions that reduces or eliminates completely the fluorescence of the acceptor. In this case, RET produces only quenching of the donor fluorescence. For the simple case in Figure 18.15A, fluorescence intensity generated by the acceptor is proportional to the donor concentration, c_A:

$$F_A = k' c_{D^*} c_A \qquad (18.20)$$

The proportionality constant k' includes the energy transfer rate constant (k_{et}) and the response depends also on the concentration of the excited donor (c_{D^*}).

The strong dependence of the energy transfer efficiency on the donor–acceptor distance allows for spatially resolved transduction in affinity sensors, as demonstrated in Figure 18.15B. In this example, the immobilized receptor is tagged with an energy donor while the analyte is labeled with an acceptor. Energy transfer occurs after the recognition event, when the donor and the acceptor come in close proximity of each other. However, the labeled analyte in the solution phase cannot take part in the energy transfer and does not contribute to the response signal. Therefore, RET is a spatially resolved transduction method, which avoids washing-out of the solution analyte prior to the measurement.

18.3.9 Chemiluminescence and Bioluminescence

Chemiluminescence consits of light emission by excited molecules produced by chemical reactions [25–27]. In general, the chemiluminescence processes involves the oxidation of a compound L_{ox} to an excited species L_{ox}^* that decays to the ground state L_{ox} according to the following reaction scheme:

$$A + L_{red} \xrightarrow[\text{chemical reaction}]{\text{Biochemical or}} P + L_{ox}^* \longrightarrow L_{ox} + h\nu \qquad (18.21)$$

In this scheme, A is a coreactant or cosubstrate of the reaction, and P is a reaction product. As the second step in the scheme (18.21) is extremely fast, the intensity of the emitted light is proportional to the reaction rate of the first step.

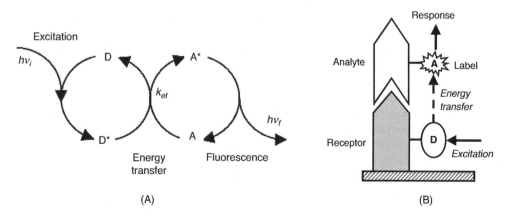

(A) (B)

Figure 18.15 (A) Mechanism of resonance energy transfer. (B) Spatially resolved transduction based on RET. D: donor; A: acceptor.

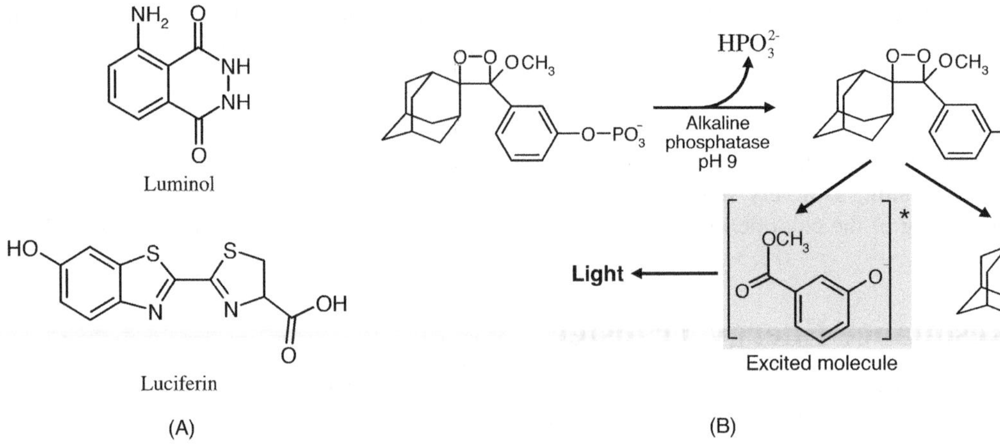

Figure 18.16 Reagents and reactions in chemiluminescence. (A) Chemical formulas of luminol and luciferin; (B) Mechanism of alkaline phosphatase promoted luminescence of adamantyl 1,2-dioxetane phenyl phosphate.

The quantum yield of a chemiluminescence process is defined as follows:

$$\phi_{CL} = \frac{\text{Number of emitted photons}}{\text{Number of reacting molecules}} \qquad (18.22)$$

A common chemiluminescence label is *luminol* (Figure 18.16A) that is oxidized by hydrogen peroxide in the presence of the horse radish peroxidase enzyme (HRP) to yield an excited product molecule that emits light [28]:

$$2H_2O_2 + \text{Luminol} \xrightarrow{\text{HRP}} \text{3-aminophtalate} + N_2 + 3H_2O + h\nu(\lambda = 430\,\text{nm}) \qquad (18.23)$$

Reaction (18.23) can be coupled with oxidase-catalyzed reactions that produce hydrogen peroxide and allows the oxidase substrate to be determined in this way.

Much stronger chemiluminescence is produced by enzymatic hydrolysis of adamantyl 1,2-dioxetane phenyl phosphate catalyzed by alkaline phosphatase (Figure 18.16B). The hydrolysis product is unstable and decomposes into two products, one of which forms in an electronically excited state that emits light.

Various living species (such as the firefly) are capable of emitting light by *bioluminescence*, a process in which excited molecules are produced by a biological process. Enzymes isolated from such organisms are available for analytical applications. For example, the *luciferase* enzyme that is extracted from the *Photinus pyralis* firefly uses adenosine-5′-triphosphate (ATP) as cosubstrates in the oxidation of luciferin (Figure 18.16A) by oxygen. This reaction results in light emission and is used for extremely sensitive determination of ATP, which is an important coenzyme involved in metabolism.

Chemiluminescence sensors do not need an excitation light source and are therefore less complicated than fluorescence-based sensors. Absence of the excitation beam removes the possibility of scattered background light appearing, which allows very low detection limits to be attained.

18.3.10 Electrochemically Generated Chemiluminescence

Electrochemically generated chemiluminescence (ECL), also known as *electrochemiluminescence*, is a chemiluminescence process produced directly or indirectly as a result of an electrochemical reaction. The fundamentals of ECL are covered in refs. [29,30] and analytical applications are amply reviewed in refs. [31–33].

Among the wide range of reagents known to generate ECL, only a few (for example, luminol and certain metal complexes) are currently used in sensor applications.

Luminol electrochemiluminescence can be observed under anodic polarization in an alkaline solution containing hydrogen peroxide. Under these conditions, luminol is electrochemically oxidized to a radical anion. This product is further oxidized by the hydrogen peroxide anion (HOO⁻) or by the superoxide radical ($O_2^{-\bullet}$) that form by electrochemical oxidation of hydrogen peroxide. The product of the above reaction is an excited molecule that emits blue light at the characteristic wavelength of luminol chemiluminescence. This reaction allows very sensitive detection of hydrogen peroxide and can also be used as the transduction method in oxidase based biosensors in which hydrogen peroxide is produced upon substrate oxidation by oxygen. However, the alkaline pH required in luminol ECL is not always compatible with the enzyme.

Various inorganic compounds, complexes and clusters are ECL active but tris(2,2′-bipyridine) ruthenium (II) ($[\text{Ru(bipy)}_3]^{2+}$) is the most frequently used inorganic ECL reagent to date [34].

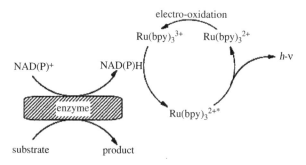

Figure 18.17 Principle of ECL sensors based on NAD$^+$-linked dehydrogenases. Reprinted with permission from [32]. Copyright 2001 Elsevier.

There are various reaction sequences leading to the ECL of $[Ru(bipy)_3]^{2+}$. One of them consists of oxidation to $[Ru(bipy)_3]^{3+}$ and reduction to $[Ru(bipy)_3]^+$ at a pair of interdigitated electrodes polarized at different potentials. These two species react in solution by electron exchange, yielding the Ru(II) complex in the excited state. Relaxation of the excited molecule results in light emission:

$$[Ru(bipy)_3]^{3+} + [Ru(bipy)_3]^+ \rightarrow [Ru(bipy)_3]^{2+} + [Ru(bipy)_3]^{2+*} \tag{18.24}$$

$$[Ru(bypy)_3]^{2+}* \rightarrow [Ru(bypy)_3]^{2+} + h\nu\ (620\,nm) \tag{18.25}$$

Alternatively, electrochemistry can be used to generate only the oxidized form while the reduced form is produced by chemical reaction with a reducing agent. Various amine derivatives, such as amino acids, proteins and various pharmaceuticals can be determined with high sensitivity using this method. Tripropylamine is one of the most efficient reagents in this reaction.

Another ECL mechanism relies on oxidation of $[Ru(bipy)_3]^{2+}$ to $[Ru(bipy)_3]^{3+}$ by the peroxodisulfate ion $(S_2O_8^{2-})$ while $[Ru(bipy)_3]^+$ is produced by electrochemical reduction.

In sensor development, $[Ru(bipy)_3]^{2+}$ is immobilized at the electrode surface by various methods such as the Langmuir–Blodgett technique, gel entrapment or incorporation in a cation-exchanger polymer (Nafion).

The principle of $[Ru(bipy)_3]^{2+}$ application to enzymatic sensors based on dehydrogenases is presented in Figure 18.17. The enzymatic reaction involves the NAD(P)$^+$/NAD(P)H cofactor. The reduced form (NAD(P)H) that results in the enzymatic reaction reduces the electrogenerated $[Ru(bipy)_3]^{3+}$ to the excited form $[Ru(bipy)_3]^{2+*}$, which relaxes by light emission. Sensors for glucose, ethanol, and L-lactate have been produced in this way.

ECL reagents are also used as luminescent labels in immunosensors and in nucleic acid sensors.

ECL has the advantage of providing light from the electrode surface area only, which allows more efficient collection of emitted light. In addition, ECL can be shut on/off by changing the electrode potential or can be modulated in the same way in order to discriminate against the continuum background light. The absence of a scattered light background, which is a general advantage of chemiluminescence, applies also to ECL and brings about very low detection limits.

18.3.11 Raman Spectrometry

Raman spectra are obtained by irradiating the sample with a monochromatic laser beam. Light scattered by the sample is collected at 90° and analyzed by means of a spectrometer.

The characteristic features of the Raman spectrum are introduced in Figure 18.18. Rayleigh scattering produces a very strong signal at the excitation frequency. Signals placed symmetrically with respect to the excitation frequency arise from transitions involving virtual states and vibrational states in the ground electronic level. Stokes Raman lines are placed at frequencies lower than that of the excitation frequency owing to the ΔE_S difference in energy between the absorbed and scattered photon. Anti-Stokes Raman lines are connected with transitions from a higher vibration level in the ground state followed by relaxation to the ground level. Therefore, the emitted photon is more energetic than the absorbed one and the additional energy ΔE_{AS} results in a higher frequency of scattered light with respect to the excitation light. Raman spectra are plotted as scattered light intensity vs. the Raman shift ($\Delta \tilde{\nu}_R$), which is the difference between the scattered wave number and the excitation wave number.

Raman spectral lines are extremely weak when recorded in homogeneous systems but are tremendously enhanced if the substance of interest is adsorbed at a rough silver or gold surface. This method, which is called *surface-enhanced Raman spectrometry(SERS)* [35] is used in sensors for detecting the analyte itself or a label attached to the analyte [36]. As the Raman spectrum includes a series of specific lines, it can serve as a fingerprint that allows the target compound to be identified and quantified with good accuracy.

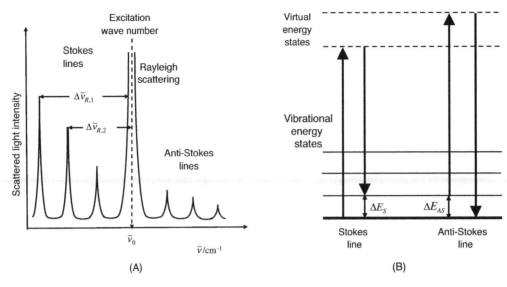

Figure 18.18 (A) Schematic presentation of a Raman spectrum. (B) Energy level diagram showing energy states and transitions involved in Raman scattering.

The intensity of Raman lines is directly proportional to the analyte concentration. However, the Raman response can be swamped by fluorescence light from the analyte, a problem that can be alleviated by using excitation in the near-infrared region.

18.3.12 Outlook

A series of different transduction methods based on light absorption, emission, and scattering are available. Light absorption measurement provides a response that is proportional to the concentration of the absorbing species, but the working range is limited to about one order of magnitude. Diffuse reflectance is accompanied by light absorption at a specific wavelength and represents an alternative to standard absorptiometry. As diffuse reflectance allows solid surfaces to be interrogated, it is very advantageous from the standpoint of sensor design. Both diffuse reflectance and light absorption measurements provide only moderate limits of detection, but can still be useful in applications in which this parameter is not critical.

When a very low limit of detection is essential, luminescence-based transduction methods are more suitable. Fluorescence transduction can be implemented by means of relatively simple schemes and do not need complicated auxiliary equipment. On the other hand, time-resolved fluorimetry provides much better limits of detection but requires more complicated optoelectronic instrumentation. In particular, the dual lifetime referencing method is of a particular interest because this ratiometric measurement method delivers more accurate analytical results.

Fluorescence quenching is an appropriate method for determining compounds that produce dynamic fluorescence quenching. In addition, fluorescence quenching can also be accomplished by resonance energy transfer. Moreover, resonance energy transfer can prompt the fluorescence of compounds that cannot be excited directly by light absorption. Being strongly localized, resonance energy-transfer processes allow for spatially-resolved transduction.

Chemiluminescence, bioluminescence and electrochemiluminescence are processes that do not need an excitation source and, therefore, are not impaired by a background produced by light scattering and nonspecific fluorescence of sample components. In addition, electrochemiluminescence offers the possibility of controlling light emission by means of the electrode potential.

Raman scattering provides a spectral signature in the infrared spectral region and can be extremely sensitive when used in the surface-enhanced Raman spectrometry mode.

Very often, neither the analyte nor the recognition material is endowed with suitable properties for spectrochemical transduction. In this case, spectrochemical transduction is based on luminescent labels and the recognition process is designed so as to result in a modification of the label concentration within the sensing layer.

Questions and Exercises (Section 18.3)

1 Review briefly the molecular processes involved in spectrochemical transduction.
2 What law governs light absorption and what is the meaning of each term in the pertinent equation?
3 Comment on a possible configuration of sensors based on light absorption. How can the reference beam be obtained?

4 What are the differences between specular reflection and diffuse reflection from the standpoint of (a) the ray's path, and (b) light interaction with the reflecting surface?

5 How can the concentration of a light-absorbing compound be estimated by diffuse reflectance measurements?

6 Give a short overview of diffuse reflectance-based chemical sensors.

7 What is luminescence and what methods are used to produce excited molecules that can emit light?

8 Point out the differences between fluorescence and phosphorescence as far as the molecular mechanism of light emission is concerned.

9 What specific parameter determines the sensitivity in the detection of a photoluminescent species?

10 Using data in Figure 18.10, calculate the Stokes shift for the $[Ru(bypy)_3]^{2+}$ complex.

11 What is the response function in steady-state fluorescence measurements and what determines its limit of application?

12 Review the parameters that determine the sensitivity in fluorometric determinations.

13 Give an overview of the main components of a fluorescence-based optical sensor and its auxiliary devices.

14 What are the principles of time-resolved spectrometry? Derive the light intensity–time function in this method. What are the advantages of time-resolved fluorimetry?

15 How can the lifetime of an excited species be determined by means of a sinusoidally modulated excitation beam?

16 Give an overview of the principles and required instrumentation in a ratiometric measurement method based on phase-resolved fluorimetry.

17 What is the mechanism of the fluorescence dynamic quenching?

18 Give a derivation of the Stern–Volmer equation using the chemical kinetics approach to the mechanism in Figure 18.14A and assuming that the concentration of the excited state is constant.

19 Demonstrate that the pre-c_Q terms in Equations (18.16) and (18.19) are similar.

20 Give a derivation of Equation (18.20) which correlates the intensity of the fluorescence radiation produced by RET and the acceptor concentration in the case of the mechanism shown in Figure 18.15A.

21 What characteristics of the resonance energy transfer make it suitable for transduction in optical affinity sensors?

22 Draw a scheme of an oxidase-based enzymatic sensor that uses chemiluminescence transduction.

23 Assume a molecular beacon nucleic acid probe with a RET acceptor and donor attached at the extremities of the probe molecule. Draw schematically the configuration change upon binding to the target nucleic acid. Draw a graph showing the signal variation with the target concentration if (a) the acceptor quenches the donor fluorescence; (b) the acceptor emits light upon resonance energy transfer from the donor.

24 Draw a simplified scheme to illustrate the structure and functioning of an alcohol sensor based on chemiluminescence transduction.

25 Draw the reaction scheme for the application of luminol in electrochemiluminescence transduction.

26 Draw schematically the mechanism of electrochemiluminescence generation by $[Ru(bipy)_3]^{2+}$ at interdigitated electrodes.

27 Write schematically the chemical and electrochemical reaction involved in a sensor based on NAD^+-dependent dehydrogenase using electrochemiluminescence of $[Ru(bipy)_3]^{2+}$ for transduction.

28 What are the advantages of chemiluminescence and electrochemiluminescence over fluorescence transduction?

29 What is the mechanism of light emission in Raman spectrometry? How can the sensitivity of the Raman response be enhanced?

18.4 Transduction Schemes in Spectrochemical Sensors

If the recognition process is at equilibrium, two main strategies can be distinguished in spectrochemical transduction, namely direct and indirect transduction [3]. Direct transduction makes use of the optical signal produced by a species directly involved in the recognition process; this species could be either the receptor itself or the receptor–analyte complex. Indirect transduction is achieved by competition between the analyte and an optically detectable analyte-analog.

18.4.1 Direct Transduction

A reversible transduction reaction at equilibrium involves the analyte A and the receptor R confined at the sensor surface that form an association complex AR (reaction (18.26) with the equilibrium constant given by

Equation (18.27):

$$A + R \rightleftarrows AR \qquad (18.26)$$

$$K_e = \frac{c_{AR}}{c_A c_R} = \frac{1}{K_d} \qquad (18.27)$$

In Equation (18.27), c_A is the concentration of the analyte in the solution; c_R and c_{AR} represent the density of free and combined receptor sites within the sensing element and K_d is the dissociation constant of AR. Taking also into account the mass balance for the receptor ($c_{R,t} = c_R + c_{AR}$), one obtains the following equations relating the concentration of detectable species (R or AR) to the analyte concentration:

$$c_{AR} = \frac{K_e c_A}{1 + K_e c_A} c_{R,t} \qquad (18.28)$$

$$c_R = \frac{1}{1 + K_e c_A} c_{R,t} \qquad (18.29)$$

Assuming that either R or AR or both R and AR can be detected by an optical method, and the signal is proportional to the concentration of the pertinent species, one can distinguish three possible response functions.

i. If AR is the detectable species, Equation (18.28) gives by rearrangement the following forms:

$$c_{AR} = \frac{c_{R,t}}{K_e^{-1} + c_A} c_A = \frac{c_{R,t}}{K_d + c_A} c_A \qquad (18.30)$$

The response is shown schematically by curve 1 in Figure 18.19A, where, the analyte concentration is represented by the dimensionless variable c_A/K_d. Clearly, this is a nonlinear response. However, assuming that $c_A \ll K_e^{-1}$ (or $c_A \ll K_d$), Equation (18.30) can be put in a limiting linear form as follows:

$$c_{AR} = K_e c_{R,t} c_A = K_d^{-1} c_{R,t} c_A \qquad (18.31)$$

This linear response is shown in Figure 18.19A by line 2, which matches the actual response as long as $c_A/K_d \ll 1$. The sensor sensitivity is determined by the equilibrium constant and the total receptor site density (see Equation 18.31). Hence, contradictory requirements as far as the dissociation constant are evident. High sensitivity demands a low dissociation constant, whereas a wide linear response range is obtained if the dissociation constant is high.

ii. A second case to be considered is that in which the detectable species is the receptor itself whereas the complex is inactive. If the signal is proportional to the density of free receptor sites, Equation (18.29) predicts a nonlinear response function, which is plotted in Figure 18.19B. As expected, the signal

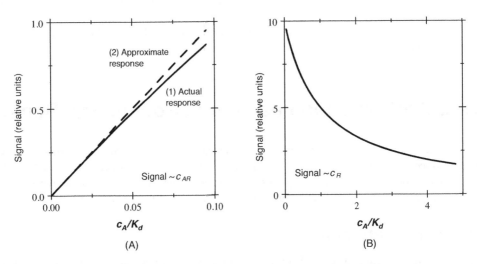

Figure 18.19 Response of a direct equilibrium sensor. **(A)** Detection of the analyte–receptor complex; **(B)** detection of the free receptor.

decreases asymptotically with increasing analyte concentration. However, Equation (18.29) can be rearranged to give the following linear form:

$$\frac{1}{c_R} = c_{R,t}^{-1} + K_e c_{R,t}^{-1} c_A \tag{18.32}$$

Therefore, if the response is proportional to the receptor concentration, the reciprocal of the signal is a linear function on analyte concentration. Response sensitivity increases with the formation constant K_e but decreases with the density of the receptor sites. The upper limit of the linear response range is about $c_A = 5K_d$; at higher concentrations, the response variation with the analyte concentration is too small.

iii. A third possible case is that in which both the receptor and the complex produce an optical signal at a distinct wavelength such that they can be assessed individually. Using Equations (18.29) and (18.30) one obtains:

$$\frac{c_{AR}}{c_R} = K_e c_A \tag{18.33}$$

Therefore, if one measures the signal of both AR and R, by taking their ratio one obtains a measurable quantity which is proportional to the analyte concentration. This is a *ratiometric response*, which is very convenient because possible fluctuations in instrument parameters affect in the same way both terms in the ratio and the fluctuations cancel out. However, a reasonable accuracy is obtained only if the terms in the ratio have comparable values so that the ratio value is between 1 and 10. Therefore, the working range of a ratiometric method is limited to about an order of magnitude.

Direct transduction is used in pH sensors, in which a pH indicator undergoes a color change by protonation. It is also suitable for use in optical ion sensor in which ion binding to the receptor produces a complex with a specific absorption or emission wavelength. Resonance energy-transfer-based sensors are also appropriate for application of the direct transduction method.

18.4.2 Indirect (Competitive-Binding) Transduction

The receptor in a direct sensor should be at the same time analyte-selective and able to provide a signal response either in the free or bound state. If such a receptor is not available, one can resort to indirect transduction, in which an analyte-analog (L) endowed with detection capabilities is available. The analog can be added to the sample solution or can be incorporated in the sensing element so as to be free to diffuse. There is in this case a competition between the analyte and the analog for binding sites, described by the following equations and equilibrium constants:

$$A + R \rightleftarrows AR \quad K_A = \frac{c_{AR}}{c_A c_R} \tag{18.34}$$

$$L + R \rightleftarrows LR \quad K_L = \frac{c_{LR}}{c_L c_R} \tag{18.35}$$

Using the equilibrium constant equations, the analyte concentration results as follows:

$$c_A = \frac{c_{AR} c_L}{c_{LR}} \frac{K_L}{K_A} \tag{18.36}$$

This equation can be further modified by using the mass-balance conditions for both the analyte and the analog. Assuming that the equilibrium constants K_A and K_L are sufficiently large to yield $c_R \ll c_{LR}$ and $c_R \ll c_{AR}$, one obtains:

$$c_A = \frac{(c_{R,t} - c_{LR}) c_L}{c_{LR}} \frac{K_L}{K_A} \tag{18.37}$$

The ratio of equilibrium constants in this equation reflects the fact that the response depends on the relative affinity of analyte and analog for the receptor.

Further processing of this equation allows to derive the response function for the cases in which either the free analog or the analog complex is detectable. In the first case, the signal increases with the analyte concentration and attains a limit when receptor sites are saturated with the analyte. In the second case, the signal decreases asymptotically with the analyte concentration.

Indirect transduction permits the employment of very selective receptors without any alteration of the receptor molecule. An example of this type is represented by competitive binding of an analog that can perform resonance energy transfer to the receptor which results in either fluorescence emission or fluorescence quenching. This strategy is used in affinity sensors and also in certain types of ion sensor.

18.4.3 Outlook

If the recognition process is at equilibrium, optical transduction can be approached in a direct or indirect way.

Direct transduction relies on monitoring the optical signal produced by the free receptor or the analyte–receptor complex. Linear response functions can be obtained within a concentration range determined by the equilibrium constant of the recognition reaction. Ratiometric transduction is possible when both the free receptor and the recognition product give rise to discernible optical signals.

Indirect transduction makes use of an analyte-analog that competes with the analyte for receptor sites. In this case, the receptor is not detectable but the analog is selected such as to be able to produce an optical signal either in the free state or in the form of a complex with the receptor.

Questions and Exercises (Section 18.4)

1 Comment on the effect of the dissociation constant and the receptor site density in the case of a direct sensor based on the monitoring of the optical signal of the recognition product.
2 What parameters determine the sensitivity of a direct sensor based on the optical signal of the receptor in the free state?
3 Discuss the advantages and possible drawbacks of the direct ratiometric transduction method taking into account the response sensitivity, the width of the linear response range and the required instrumentation.
4 Comment on possible applications of direct transduction in sensors based on resonance energy transfer.
5 What are the principles of the indirect transduction scheme?
6 What are the required properties of the analyte-analog used in an indirect transduction scheme?
7 Draw schematically the response curve of an indirect senor assuming that either the free analyte-analog or its combination with the receptor provides the response signal.

18.5 Fiber Optic Sensor Arrays

Optical fiber bundles can be used as platforms for assembling arrays of individually addressable optical sensors [37]. A bundle composed of 6000 to 50 000 individual fibers makes it possible to obtain about 2×10^7 sensors per cm^2. Individual sensors can be formed either at the fiber tip (passive fiber configuration) or onto the fiber core (active fiber configuration).

Individually addressable sensors at a fiber bundle can be formed by photopolymerization. To this end, the fiber end is first derivatized to incorporate functional groups compatible with the polymer. For example, if a methacrylate polymer matrix is to be formed, the fiber is treated with 3-(trimethoxysilyl)propyl methacrylate for copolymerization of the matrix-forming monomer. Polymerization is next conducted in a solution containing the monomer, a polymerization initiator, a crosslinker and a fluorescent indicator. Initiation of the photopolymerization process is produced by delivering light through the fiber to its distal face. By exposing distinct regions of the bundle surface to different photopolymers, an array of different sensors can be produced onto a single imaging bundle. The polymer can serve as immobilization support for the capture receptor. Multianalyte fiber optic sensors can be formed in this way. The multiple responses are obtained by means of a charge-coupled device arrays.

The same procedure can also be used to attach a distinct sensor to each individual fiber in a bundle. This kind of multiple sensor allows high-resolution imaging to be performed. For example, visualization of the pH distribution in the section of a biological specimen can be achieved in this way.

Fiber optic sensor arrays are useful for various applications. First, integrated multiple sensors can be obtained and applied for the determination of a series of different analytes, such as pH, oxygen and carbon dioxide in biological samples. Arrays of similar sensors make possible visualization of the spatial distribution of a particular analyte.

Arrays of crossreactive optical sensors are suitable for performing artificial olfaction, that is, identification of particular gas or vapor mixtures by devices of the artificial-nose type (see Section 1.7). Use is made in such devices of the *solvatochromic effect*, which consists of changes in the fluorescence parameters in response to modifications in the solvent polarity. An optical electronic nose is obtained by forming an array of optical sensors, each individual sensor including a fluorescent indicator embedded in a particular polymer matrix. When exposed to an organic vapor mixture, each sensor responds by a change in the fluorescence wavelength under the combined effect of all components of the mixture. However, as each component has a specific affinity for the polymer matrix, it affects in a specific way the fluorescence of each individual sensor. Using suitable algorithms, vapor mixtures can be identified using arrays composed of a relatively small number of crossreactive optical sensors.

18.6 Label-Free Transduction in Optical Sensors

The previous section introduced a series of label-based transduction methods used in optical sensors. As the introduction of the optical label brings about certain complications, it becomes desirable in many applications to implement label-free transduction methods that rely on changes in certain intrinsic properties of the sensing element as a result of its interaction with the analyte. One of these properties is the refractive index.

A phenomenological definition of the refractive index has been presented before in connection with the refraction phenomenon (Section 18.2.1). The general meaning of this quantity can be rationalized by considering the propagation of electromagnetic waves through material media. Light velocity in material media is smaller than that in vacuum and, consequently, the light wavelength in a material medium is lesser that the wavelength in vacuum at the same frequency. These relationships are formulated in Equation (18.38) in which n is the refractive index of the considered medium, c and λ_0 are light velocity and wavelength, respectively, in vacuum, and v_p and λ are phase velocity and wavelength of the light wave, respectively, in the considered medium.

$$n = \frac{c}{v_p} = \frac{\lambda_0}{\lambda} \tag{18.38}$$

Therefore, the refractive index is an essential parameter in light propagation through material media. Notably, the refractive index is wavelength dependent, a property that is demonstrated by light dispersion produced by an optical prism. As light is an electromagnetic wave phenomenon, the refractive index is connected to the electrical and magnetic properties of the propagation medium.

Very small variations of the refractive index can be measured by the different methods presented below. As the response depends only on an intrinsic property of the sensing layer, such methods are categorized as label-free transduction methods [38–41].

18.6.1 Surface Plasmon Resonance Spectrometry

When free electrons in plasma experience forces applied by a variable electromagnetic field with a particular wavelength, they can oscillate in a coherent way. Large groups of electrons in the oscillation state are called *plasmons* [42].

Plasmons appear also at the surface of certain metals under light irradiation forming *surface plasmons*, this phenomenon being known as *surface plasmon resonance (SPR)* [43-45]. In a very thin metal film the plasmon region extends over the whole film thickness.

As shown in Figure 18.20A, SPR form in a thin gold layer at the surface of an optical prism provided that the angle of incidence θ is selected so that total internal reflection occurs at the interface. Under these conditions, free electrons in the gold film oscillate as plasmons and capture a fraction of the energy of the incident radiation. At the same time, electron oscillations generate an electromagnetic wave that extends beyond the film surface as an evanescent wave.

The occurrence of plasmon resonance in a metal layer is critically dependent on three factors: the angle of incidence, the wavelength of the incident radiation, and the refractive index (n) of the sample in which the evanescent wave is extending. For example, if the wavelength is held constant and the angle of incidence is varied, the intensity of reflected light attains a minimum at a particular angle called the *surface plasmon resonance angle* (Figure 18.20B). This minimum indicates strong absorption of radiant energy by plasmons. However, the minimum shifts to a different angle if a minute change in the refractive index of the sample occurs in response to the recognition event.

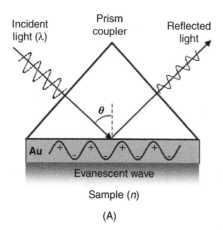
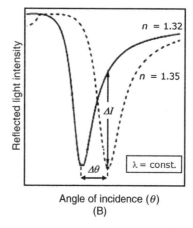

Figure 18.20 Surface plasmon resonance. (A) Plasmon excitation via a prism coupler (Kretschmann prism setup). Incident light should be *p*-polarized, that is, oscillations should occur in a plane parallel to the plane of the figure. (B) Intensity of reflected light as a function of angle of incidence at a fixed wavelength and two samples with different refractive indices.(B) Adapted with permission from [51] Copyright 2005 Elsevier.

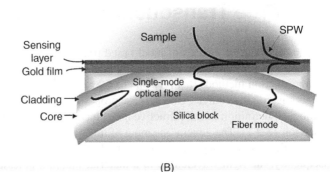

Figure 18.21 SPR fiber optic sensors. (A) Dip-probe configuration; (B) Structure of a SPR biosensor based on a single-mode optical fiber. (A) Reprinted with permission from [44]. Copyright 1998 John Wiley & Sons, Ltd. (B) Reprinted with permission from [52]. Copyright 2001 Elsevier.

A similar dependence can be noticed if the angle of incidence is held constant and the wavelength of the incident radiation is varied. The wavelength at which the intensity of reflected light is at a minimum (called the *plasmon wavelength*) is also sensitive to the change in the refractive index of the sample.

An alternative method for SPR coupling relies on grating-shaped metal surfaces [46].

SPR provide a straightforward transduction method in sensors based on affinity reactions such as immunosensors and nucleic acid sensors [45, 47–50]. A SPR immunosensor can be obtained by immobilization of a capture receptor at the surface of the gold layer in the Kretschmann prism setup (Figure 18.20A). The capture of the target analyte modifies the refractive index and produces a change in the characteristics of the reflected light. In order to perform automatic assay, the coupler and the gold layer coated with the receptor layer are placed in a flow-through cell. The analytical response can be either the surface plasmon resonance angle at constant wavelength, the plasmon wavelength shift at constant angle of incidence or the change in the reflected light intensity at a given wavelength or angle of incidence (see Figure 18.20).

SPR transduction can also be implemented in fiber optic sensors designed in the active fiber configuration. In the example in Figure 18.21A, the cladding at the tip of the optical fiber was removed and a thin gold film was deposited over the core to serve as SPR support. A receptor layer was then formed over the gold film. The silver layer mirror at the end of the fiber reflects back the light, thus allowing the radiation to excite plasmon resonance each way (forth and back).

Alternatively, a decladded portion of the fiber was polished flat and a gold layer was deposited over this surface (Figure 18.21B). In order to detect the staphylococcal enterotoxin B (SEB), a double-layer of anti-SEB antibody was attached by glutaraldehyde crosslinking. The sensor was operated using a flow-injection system and a mild acid solution was used to regenerate the surface for replicate analyses. The total time for an analysis cycle was approximately 9 min and the detection limit was 4 ng mL^{-1}, which compare favorably with conventional immunoassay methods.

In conclusion, SPR measurement allows label-free transduction to be performed in sensors based on affinity reactions that lead to changes in the refractive index of the sensing layer.

18.6.2 Interferometric Transduction

Light interference is produced by superposition of two or more coherent waves arising at a given location. A one dimensional traveling wave is described by the following equation:

$$\psi(x, t) = A \cos\left(\frac{2\pi}{\lambda}x - 2\pi f + \phi\right) \tag{18.39}$$

where $\psi(x, t)$ is the strength of the electrical or magnetic field, A is the amplitude, λ is the wavelength, f is the frequency and ϕ is the phase angle.

Wave interference is demonstrated in Figure 18.22 for the simple case of the superposition of two waves with the same wavelength, plotted as thin curves. Depending on the phase difference ($\Delta\phi$) between the superposed waves, the

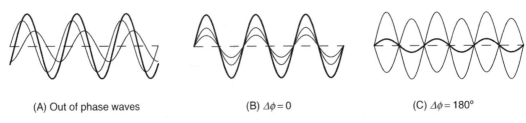

(A) Out of phase waves   (B) $\Delta\phi = 0$ (C) $\Delta\phi = 180°$

Figure 18.22 Interference of two waves of similar frequency. (A) General case (out-of-phase waves); (B) in-phase waves-constructive interference; (C) waves in antiphase-destructive interference. Thin lines represent superposed waves; thick lines indicate the resulting wave.

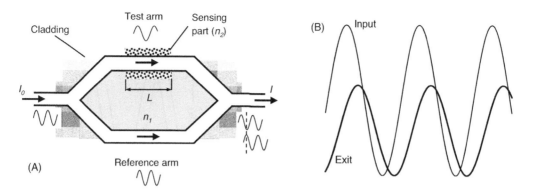

Figure 18.23 Mach–Zehnder-type wave guide interferometer. (A) Principle; (B) Waveforms for the input and exit beams.

amplitude of the resulting wave (thick lines) can be greater or smaller than the amplitude of each superposed wave (Figure 18.22A). The resulting wave is phase shifted with respect to both waves.

A particular situation arises when the superposed waves are in phase, that is, $\Delta\phi = 2\pi m$, where m is zero or an integer number. In this case, the amplitude of the resulting wave is $A_t = A_1 + A_2$, where A_1, and A_2 are the amplitudes of the superposed waves (Figure 18.22B). This is a *constructive interference* phenomenon. If the phase difference is $\Delta\phi = m\pi$ radians (with m being an odd number) the waves are in antiphase and the resulting amplitude is $A_t = A_1 - A_2$ (Figure 18.22C). Light extinction through *destructive interference* occurs when two superposed waves with similar amplitude are in antiphase.

Application of interferometry in optical sensor relies on the interference of light waves that propagate along two paths with different refractive indexes. Initially, the two waves are in-phase, but the emergent waves are out of phase and can interfere. The resulting wave has a lower intensity and a different phase angle with the respect to the incident ones. In sensor applications, one of the two beams serves as reference while the second one interacts with the receptor layer and undergoes a delay determined by the refractive index of this layer [53].

The above principles can be implemented by means of various kinds of optical devices. A typical sensor based on the *Mach–Zehnder interferometer* is presented in Figure 18.23A. In this sensor, a coherent beam produced by a laser is split in two beams by means of waveguides that are of the single-polarization and single-mode type. The two beams meet at the exit waveguide and are sent to a light detector. If the two paths have the same lengths and optical properties, there is no phase shift and constructive interference results. However, if on a portion of the test path the cladding is removed and replaced with a sensing layer, the test beam interacts with this layer by means of the evanescent wave. As a result, the propagating light experiences the effect of the refractive index of the sensing layer. As a result, the test beam is phase shifted with respect to the reference beam. Consequently, the exit light intensity is smaller than that of the incident light and the resulting wave is shifted with the angle $\Delta\phi$ offset of the reference beam. Owing to the interference of the two waves, the intensity of the exit beam depends on the propagation parameters of the test arm as follows [54]:

$$\frac{I(\Delta\phi)}{I_o} = \frac{1}{2}\left[1 + \cos\left(\frac{2\pi}{\lambda}L\Delta n_{\text{eff}}\right)\right] \tag{18.40}$$

Here, n_{eff} is the effective refractive index that quantifies the phase delay per unit length in a waveguide, relative to the phase delay in vacuum. Δn_{eff} is the difference between effective refractive indexes in the measuring and test arm and L is the length of the measuring window. I_o and $I(\Delta\phi)$ are the intensities of the incident and exit beams, respectively. Hence, a very low variation in the refractive index of the sensing part can be detected by measuring the $I(\Delta\phi)/I_0$ ratio. This allows for monitoring affinity reactions occurring at the sensing part.

An alternative setup is based on the principle of the *Young interferometer* (Figure 18.24A). In this device, two similar light beams pass through the parallel reference and sensing waveguides. As in the previous case, a sensing layer is formed over the core surface of the sensing waveguide and thereby modifies the phase angle of the sensing beam. Divergent beams at the exit of each channel interfere, producing interference fringes on the detector screen (a charge-coupled device array). According to ref. [55], for $d \ll f_s$ the spatial distribution of the intensity $I(x)$ along the detector screen is:

$$I(x) = 1 + \cos\left(\frac{2\pi n d}{\lambda_0 f_s}x - \delta\right) \tag{18.41}$$

where n is the refractive index of the cladding d is the distance between the two waveguide outputs, λ_0 is the wavelength in vacuum and f_s is the distance between the detector screen the optic device. The first term under cosine in

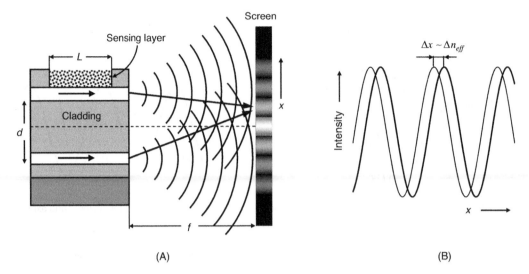

Figure 18.24 A Young-type waveguide interferometer. **(A)** Functioning principle; **(B)** Shift of interference fringe as a result of the change in refractive index.

Equation (18.41) indicates the phase difference in the case in which both waveguides are cladded whereas δ represents the additional phase shift produced by the difference between the refractive indexes on each path, Δn_{eff}: According to Equation (18.41), the light intensity on the screen varies with the distance according to a periodic function, which describes the sequence of fringes in Figure 18.24A. The δ term in Equation 18.41 is given by the following expression:

$$\delta = \frac{2\pi L}{\lambda_0} \Delta n_{\text{eff}} \qquad (18.42)$$

Therefore, if a particular intensity on the screen is selected, it shifts on the screen surface with a distance Δx in response to a change in Δn_{eff} as follows:

$$\Delta x = \frac{f_s L}{d} \Delta n_{\text{eff}} \qquad (18.43)$$

Accordingly, Δx represents the response signal of a Young interferometer sensor, as demonstrated in Figure 18.24B. The fringe shift is proportional to the variation of the effective refractive index on the test arm.

Besides the above interferometric devices, which are based on two waveguide arms, simple and robust devices have been developed using reflectometric interference spectrometry [39]. In this approach, interference is caused by light reflection at two parallel interfaces of a thin layer. The principle behind this device is similar to that which produces colors at an oil film on water and is used in the design of the Fabry–Pérot interferometer.

Summing up, interferometric transduction exploits the tiny change in the refractive index of the sensing part as a result of its interaction with the analyte. It can be coupled with various recognition schemes such as gas absorption in polymers, affinity reactions and nucleic acid hybridization.

18.6.3 The Resonant Mirror

As shown in Figure 18.25A, the resonant mirror is an assembly formed of a coupling prism (3) coated with a low refractive index layer (2) over which a very thin, high refractive index layer 1 is formed. The sensing film (for example, an antibody layer) is deposited over the layer 1. Under irradiation with a laser beam, total internal reflection occurs at the prism/layer 2 interface giving rise to an evanescent wave in the overlayer 2. This layer is sufficiently thin to allow light coupling into the layer 1 that acts as a waveguide with claddings composed of the layer 2 and the sensing layer. This arrangement gives rise to a leaky waveguide mode as the mode in the layer 1 expands into the layer 2 as an evanescent wave and further into the prism, as a constant-amplitude wave (Figure 18.25B). Therefore, light is not totally confined within the layer 1 but it leaks steadily into the prism producing the reflected beam. This optical phenomenon is called *frustrated total internal reflection*.

Because the resonant mirror has very minor losses, light will be reflected at any angle of incidence. However, a phase shift of π radians occurs in the reflected light when it is in resonance with the dielectric modes, yielding the most efficient coupling. This occurs at a particular incidence angle θ. A change in the refractive index of the sensing layer as a result of the recognition event produces a change in the incidence angle that produces resonance. Therefore, the response signal of a resonance mirror sensor is the incidence angle that gives rise to maximal reflected intensity. The response depends on both the thickness and the refractive index of the sensing layer.

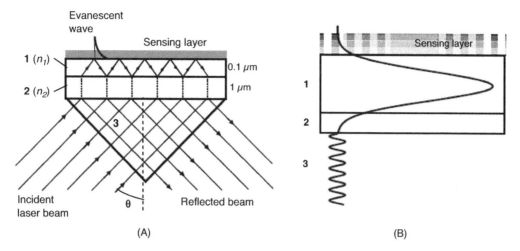

Figure 18.25 (A) Principle of the resonant mirror sensing. (1) High refractive index resonant layer (titanium dioxide); (2) low refractive index coupling layer (silica); (3) equilateral prism. $n_1 > n_2$. Layer thicknesses are not at scale. (B) Leaky mode at the resonant mirror.

18.6.4 Resonant Waveguide Grating

Recently, a reflectometric sensor design based on subwavelength structured surfaces has been advanced [56,57]. As shown in Figure 18.26, the active part is a grating formed of alternating regions of different refractive indexes, n_1 and n_2. The grating periodicity (d) should be small compared with the wavelength of the incident light. This device behaves as a narrowband filter if the effective refractive indices satisfy the following relations: $n_1 > n_2$; $n_2 > n_{substrate}$. If this structure is irradiated with polychromatic light, it reflects only light with the resonance frequency that depends on the grating period and the effective refractive index. An added biolayer (for example, an antibody) increases the optical path length of the incident radiation through the structure and this modifies the wavelength at which the maximum reflectance is observed. The wavelength shift is sensitive to a recognition event that results in an increase of the thickness of the biolayer (t_{bio}). A detection limit of 0.016 nM protein has been reported for such a device [58]. Remarkably, this device is compatible with the microtiter plate design that allows for parallel assay of a great number of samples.

18.6.5 Outlook

In label-free transduction methods the sensing layer is formed over the core of a waveguide. Guided light interacts with the sensing layer through the evanescent wave that extends beyond the core limit. This interaction is sensitive to changes in the refractive index or in the thickness of the sensing layer produced by the recognition event. Indirect interaction between the guided light and the receptor layer can be achieved by means of surface plasmon resonance devices.

Compared with label-based methods, label-free transduction is advantageous in several respects. First, the interaction is strictly localized at the sensing layer zone and no response is produced by species in the bulk of the solution. Moreover, as the light does not pass through the solution, the measurement can be performed in turbid or light-absorbing samples. However, these advantages are obtained at the expense of a more intricate sensor design.

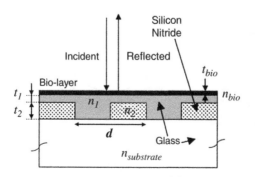

Figure 18.26 Reflective biosensor based on a subwavelength structured surface. Reprinted with permission from [58]. Copyright 2002 Elsevier.

18.7 Transduction by Photonic Devices

Development of optical telecommunication systems and the prospects of high-speed optical computing led to the development of new devices for processing optical signals. Some of these devices open up new prospects in the design of optical sensors and are currently the object extensive research efforts. As in the devices introduced in the previous section, transduction is based on a perturbation in guided-light propagation produced by the sample. However, photonic devices presented in this section appear as much more sensitive and better suited for running extremely low amounts of sample.

18.7.1 Optical Microresonators

An optical resonator is a system that supports an electromagnetic wave with a particular wavelength (called the resonant wavelength) with much greater amplitude than at other wavelengths. Optical microresonators can be shaped as microrings, microspheres or cylindrical microcavities. Such structures are termed *optical microcavities*.

An *optical ring resonator* consists of a waveguide in a closed loop coupled to one or more input/output waveguides (Figure 18.27A). When light of the appropriate wavelength is coupled to the loop by the input waveguide, the light undergoes total internal reflection at the waveguide edge and returns back on itself in phase, undergoing constructive interference. As a result, it becomes trapped within the waveguide in the form of *whispering-gallery modes* (*WGM*), as shown in Figure 18.27B. In order for resonance to occur, the ring circumference should be an integer multiple of the wavelength so as to satisfy the following condition:

$$\lambda_r = \frac{L}{m} n_{eff} \qquad (18.44)$$

where L is the round-trip length, m is the cavity mode order (an integer number), λ_r is the resonance frequency, and n_{eff} is the effective refractive index. At resonance, light is trapped producing thereby a sharp dip on the transmittance–wavelength curve. The resonator quality factor (Q) is defined as:

$$Q = \frac{\delta\lambda_r}{\lambda_r} \qquad (18.45)$$

where $\delta\lambda_r$ is the full width at half-maximum of the absorption band. The quality factor can be around 10^6, which is extremely high.

Sensor applications of microresonators rely on the well-known interaction of the evanescent wave with a sensing layer formed over a waveguide. However, in contrast to the straight waveguide, the effective light–analyte interaction length of a ring resonator (L_{eff}) is determined by the number of revolutions of the trapped light supported by the resonator. In turn, L_{eff} depends on the quality factor and the refractive index of the resonator (n) as:

$$L_{eff} = \frac{Q\lambda_r}{2\pi n} \qquad (18.46)$$

Typically, L_{eff} can be as long as 10–15 cm, despite the small physical size of the resonator that is in the micrometer region. A high quality factor is essential for resolving a small change in λ_r produced by the binding of a few molecules.

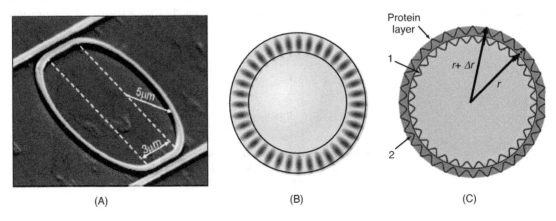

(A)   (B)   (C)

Figure 18.27   Optical microresonators. (A) Optical ring resonator. Input and output waveguides are shown on each side. (B) Distribution of the electromagnetic field in the whispering-gallery mode in a ring resonator; (C) whispering-gallery modes in a microsphere resonator without (1) and with a surface protein layer (2). (A) Reprinted with permission from [59]. Copyright 2007, The Optical Society (OSA). (C) Reprinted by permission from [60]. Copyright 2008 Macmillan Publishers Ltd. (Nature Methods).

Whispering-gallery modes can also obtained with dielectric microspheres, as shown in Figure 18.27C [60]. In this case, the whispering-gallery wave circumnavigates along the sphere equator. Due to the relative large size (about 100 μm), the microsphere has a lower refractive index resolution compared with the ring resonator. On the other hand, the microsphere resonator displays a better detection limit owing to a greater quality factor.

Silicon microspheres are particularly suitable for protein sensing by immunoreactions because incidentally, silicon and proteins have similar refractive indexes. So, binding of a Δr thick protein layer over the r radius sphere expands its radius to $r + \Delta r$ without alteration in the refractive index (Figure 18.27C). As the round-trip length increases upon analyte binding, the resonance wavelength also varies as $\Delta\lambda_r/\lambda_r = \Delta r/r$. This allows detection of extremely small variations in the thickness of the binding layer in response to the recognition event.

Another possible WGM resonator geometry is the cylindrical optical microcavity, which is a thin capillary of 15–150 μm diameter. Microcapillary resonators are readily compatible with microfluidic systems in which the microcavity acts as a flow-through cell with the receptor film formed over the inner surface.

Optical microresonators provide unparalleled label-free methods for very sensitive transduction in protein, nucleic acid and micro-organism sensing [60,61]. Remarkably, detection of single antigen particles by microsphere resonators has been reported.

18.7.2 Photonic Crystals

A photonic crystal is composed of a periodic arrangement of two dielectric materials of different refractive indexes, as shown in Figure 18.28A that displays a one-dimensional photonic crystal. Here, dimensionality implies the number of directions along which the periodicity is manifested.

The analogy of photonic crystals with true crystals becomes evident when the propagation of electromagnetic waves is examined. When irradiated with an X-ray beam, a true crystal produces a reflection at each atom layer (Figure 18.28B). Reflected beams focused on a detector interfere in accordance with the distance between the atom layers (d) and the angle θ. Constructive interference occurs when the phase difference is a multiple of 2π, which leads to Bragg's law:

$$m\lambda = 2d \sin \theta \tag{18.47}$$

where m is an integer (diffraction order). Interference occurs only if the interplane distance (*lattice period*) is on the scale of the optical wavelength.

Light propagation in a photonic crystal (like that in Figure 18.28) follows a similar trend. A light wave is reflected at each interface and the interference of reflected waves interfere with each other The condition for constructive interference in the photonic crystal is:

$$m\lambda = 2n_{\text{eff}}d \tag{18.48}$$

where the integer number m is the diffraction order, n_{eff} is the effective refractive index in the periodic structure and d is the lattice period of the photonic crystal in the direction of light propagation. When light is incident on the photonic crystal structure, at certain wavelengths light will get reflected according to Equation (18.48) depending on the period and the effective refractive index. This peak in reflection will be associated with a trough in transmission. Therefore, light interference leads to the formation of a *photonic bandgap* that is a wavelength interval within which light propagation through the photonic crystal is not allowed.

A two-dimensional photonic crystal can be formed as a periodical network of pores in a slab, as shown in Figure 18.29A. Three-dimensional photonic crystals display periodical variation of the refractive index along all the

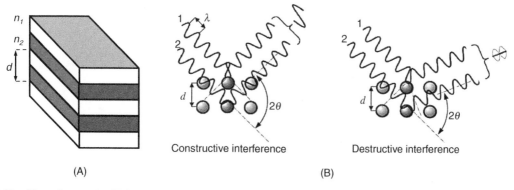

(A) (B)

Figure 18.28 Photonic crystals. (A) A one-dimensional photonic crystal. (B). Diffraction and interference of electromagnetic waves at a true crystal formed of a periodic network of atoms. d is the distance between two successive atom layers in the crystal. (B) Adapted from http://commons.wikimedia.org/wiki/File:Loi_de_bragg.png Last accessed 17/05/2012.

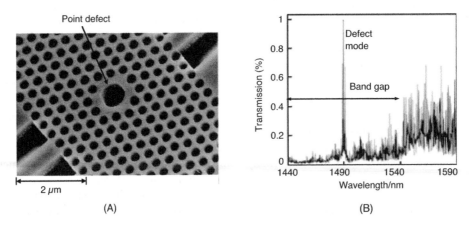

(A)   (B)

Figure 18.29 Photonic defects in photonic crystals. **(A)** A silicon two-dimensional photonic crystal with a point defect. Pore diameter, 240 nm; lattice constant $d = 400$ nm). Input and output light coupling is achieved by lateral waveguides (not shown). **(B)** Transmission spectrum for the photonic crystal showing the defect mode within the photonic bandgap. Reprinted with permission from [62]. Copyright 2007 The Optical Society (OSA).

three axes. An example of this type is a periodic arrangement of microspheres embedded in a background of different refractive index. Colloidal photonic crystal films can be fabricated by the self-assembly of monodisperse colloidal nano-particles into an ordered, crystal-like structure.

The behavior of the photonic crystal is altered if a *photonic defect* is introduced in the form of a different size element in the periodic structure (Figure 18.29A). This produces a defect mode that allows light propagation at the *defect resonance wavelength*. Consequently, on the transmission or reflection spectrum, the defect mode appears as a sharp peak within the bandgap (Figure 18.29B).

Optical transduction by photonic crystals is essentially different from that based on the evanescent wave. The evanescent wave is only a tiny tail of the electromagnetic field in the waveguide. Conversely, photonic crystals local-ize and reinforce the electromagnetic field in the low refractive index pore region. Therefore, modification of the inner pore surface with biomacromolecules can be monitored with extremely high sensitivity. If an affinity reaction occurs at the surface, changes in the surface layer thickness and its refractive index result in a shift of the defect resonance wavelength. Although the wavelength shift is in the nanometer region, it can be still detected with very good resolution [63]. Various label-free sensing schemes based on photonic crystals are feasible [64].

A particular kind of two-dimensional photonic crystal is the *photonic crystal fiber* [65]. This is an optical fiber with a periodical network of holes running along the entire length of the fiber (Figure 18.30). A defect along the fiber axis creates a core-region that acts as a waveguide at the particular frequency of the defect mode.

Photonic crystal fibers are promising platforms for label-free optical sensors [66,67]. If the recognition process is conducted at the hole surface, light propagation is affected by the change in the effective refractive indexes of the core and cladding and a shift in the defect resonance wavelength will be noticed. In addition, label-based transduc-tion can also be effected by means of hollow-core fibers. In this case, light-absorbing labels should be present at the walls of the holes around the defect. The evanescent wave interacts with the label, according to the Lambert–Beer law [68]. Fluorescence transduction is also possible upon label excitation by the evanescent wave. Surface-enhanced Raman spectrometry at hollow-core photonic crystal fibers is another promising transduction method [69].

Holes in photonic crystal fibers can act as fluidic channel to deliver the sample solution and thus allow photonic crystal fibers to be integrated in microfluidics analytical systems. At the same time, the fiber structure provides a relatively large contact area with the sample, which results in a very sensitive detection.

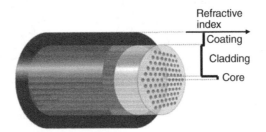

Figure 18.30 Triangular photonic crystal fiber. The refractive index of the core is greater than that of the surrounding region. The defect is a missing hole along the fiber axis. The sketch to the right indicates the variation of the refractive index with the distance from the fiber axis. Reprinted with permission from http://www.photonics.com/Article.aspx?AID=25277. Credit Kim P. Hansen and NKT Photonics. Last Accessed 14/05/2012.

18.7.3 Outlook

The advent of photonic devices brings about new opportunities for improving the performances of optical sensor both in the label-free and the label-based detection modes. Photonic devices allow for advanced miniaturization and multiplexing, are suited for operation in microfluidic systems, and attain remarkably low detection limits.

References

1. Ingle, J.D. and Crouch, S.R. (1988) *Spectrochemical Analysis*, Prentice Hall, Englewood Cliffs, NJ.
2. Zourob, M. and Lakhtakia, A. (eds) (2010) *Optical Guided-wave Chemical and Biosensors*, Springer-Verlag, Berlin (2 vols.).
3. Seitz, W.R. (1984) Chemical sensors based on fiber optics. *Anal. Chem.*, **56**, A16–A34.
4. Seitz, W.R. (1988) Chemical sensors based on immobilized indicators and fiber optics. *CRC Crit. Rev. Anal. Chem.*, **19**, 135–173.
5. Potyrailo, R.A., Hobbs, S.E., and Hieftje, G.M. (1998) Optical waveguide sensors in analytical chemistry: Today's instrumentation, applications and trends for future development. *Fresen. J. Anal. Chem.*, **362**, 349–373.
6. Eggins, B.R. (1996) *Biosensors: An Introduction*, Wiley, Chichester.
7. Wilson, J. and Hawkes, J.F.B. (1998) *Optoelectronics: An Introduction*, Prentice Hall, London.
8. Boisdé, G. and Harmer, A. (1996) *Chemical and Biochemical Sensing with Optical Fibers and Waveguides*, Artech House, Boston.
9. Dybko, A. (2006) Fundamentals of optoelectronics, in *Optical Chemical Sensors* (eds F. Baldini, A.N. Chester, J. Homola, and S. Martellucci) Springer, Dordrecht, pp. 47–58.
10. McDonagh, C., Burke, C.S., and MacCraith, B.D. (2008) Optical chemical sensors. *Chem. Rev.*, **108**, 400–422.
11. Herron, J.N., Wang, H.K., Tan, L. *et al.* (2006) Planar waveguide biosensors for point-of-care clinical and molecular diagnostic, in *Fluorescence Sensors and Biosensors* (ed. R.B. Thompson) CRC Press, Boca Raton, pp. 283–332.
12. Schmidt, H. and Hawkins, A.R. (2008) Optofluidic waveguides: I. Concepts and implementations. *Microfluid. Nanofluid.*, **4**, 3–16.
13. Hawkins, A.R. and Schmidt, H. (2008) Optofluidic waveguides: II. Fabrication and structures. *Microfluid. Nanofluid.*, **4**, 17–32.
14. Wolfbeis, O.S. (1996) Capillary waveguide sensors. *TrAC-Trends Anal. Chem.*, **15**, 225–232.
15. Borecki, M., Korwin-Pawlowski, M.L., Beblowska, M. *et al.* (2010) Optoelectronic capillary sensors in microfluidic and point-of-care instrumentation. *Sensors*, **10**, 3771–3797.
16. Vaughan, A.A. and Narayanaswamy, R. (1998) Optical fibre reflectance sensors for the detection of heavy metal ions based on immobilised Br-PADAP. *Sens. Actuators B-Chem.*, **51**, 368–376.
17. Valeur, B. (2002) *Molecular Fluorescence: Principles and Applications*, Wiley-VCH, Weinheim.
18. Demchenko, A.P. (2009) *Introduction to Fluorescence Sensing*, Springer, Dordrecht.
19. Thompson, R.B. (2006) *Fluorescence Sensors and Biosensors*, CRC Press, Boca Raton.
20. Mizaikoff, B. (2003) Mid-IR fiber-optic sensors. *Anal. Chem.*, **75**, 258A–267A.
21. Amiot, C.L., Xu, S.P., Liang, S. *et al.* (2008) Near-infrared fluorescent materials for sensing of biological targets. *Sensors*, **8**, 3082–3105.
22. Saari, L.A. and Seitz, W.R. (1982) pH sensor based on immobilized fluoresceinamine. *Ana. Chem.*, **54**, 821–823.
23. Huber, C., Klimant, I., Krause, C. *et al.* (2000) Optical sensor for seawater salinity. *Fresenius J. Anal. Chem.*, **368**, 196–202.
24. Burke, C.S., Markey, A., Nooney, R.I. *et al.* (2006) Development of an optical sensor probe for the detection of dissolved carbon dioxide. *Sens. Actuators B-Chem.*, **119**, 288–294.
25. Dodeigne, C., Thunus, L., and Lejeune, R. (2000) Chemiluminescence as a diagnostic tool. A review. *Talanta*, **51**, 415–439.
26. Roda, A., Pasini, P., Guardigli, M. *et al.* (2000) Bio- and chemiluminescence in bioanalysis. *Fresenius J. Anal. Chem.*, **366**, 752–759.
27. Blum, L.J. (1997) *Bio- and Chemi-Luminescent Sensors*, World Scientific, Singapore.
28. Marquette, C.A. and Blum, L.J. (2006) Applications of the luminol chemiluminescent reaction in analytical chemistry. *Anal. Bioanal. Chem.*, **385**, 546–554.
29. Bard, A.J. and Faulkner, L.R. (2001) *Electrochemical Methods: Fundamentals and Applications*, Wiley, New York.
30. Richter, M.M. (2004) Electrochemiluminescence (ECL). *Chem. Rev.*, **104**, 3003–3036.
31. Marquette, C.A. and Blum, L.J. (2008) Electro-chemiluminescent biosensing. *Anal. Bioanal. Chem.*, **390**, 155–168.
32. Fähnrich, A., Pravda, M., and Guilbault, G.G. (2001) Recent applications of electrogenerated chemiluminescence in chemical analysis. *Talanta*, **54**, 531–559.
33. Knight, A.W. (1999) A review of recent trends in analytical applications of electrogenerated chemiluminescence. *TrAC Trends Anal. Chem.*, **18**, 47–62.
34. Yin, X.- B., Dong, S., and Wang, E. (2004) Analytical applications of the electrochemiluminescence of tris (2,2'-bipyridyl) ruthenium and its derivatives. *TrAC Trends Anal. Chem.*, **23**, 432–441.
35. Campion, A. and Kambhampati, P. (1998) Surface-enhanced Raman scattering. *Chem. Soc. Rev.*, **27**, 241–250.
36. Storhoff, J.J., Marla, S.S., Garimella, V. *et al.* (2005) Labels and detection methods, in *Microarray Technology and its Applications* (eds R. Müller and V. Nicolau), Springer, Berlin, pp. 147–179.
37. Epstein, J.R. and Walt, D.R. (2003) Fluorescence-based fibre optic arrays: a universal platform for sensing. *Chem. Soc. Rev.*, **32**, 203–214.

38. Brecht, A. and Gauglitz, G. (1995) Optical probes and transducers. *Biosens. Bioelectron.*, **10**, 923–936.

39. Gauglitz, G. (2005) Direct optical sensors: principles and selected applications. *Anal. Bioanal. Chem.*, **381**, 141–155.

40. Gauglitz, G. and Proll, G. (2008) Strategies for label-free optical detection, in *Biosensing for the 21st Century* (eds R. Renneberg, F. Lisdat, and D. Andresen), Springer, Berlin pp. 395–432.

41. Fan, X.D., White, I.M., Shopoua, S.I. *et al.* (2008) Sensitive optical biosensors for unlabeled targets: A review. *Anal. Chim. Acta*, **620**, 8–26.

42. Campbell, D.J. and Xia, Y.N. (2007) Plasmons: Why should we care? *J. Chem. Educ.*, **84**, 91–96.

43. Homola, J. (2008) Surface plasmon resonance sensors for detection of chemical and biological species. *Chem. Rev.*, **108**, 462–493.

44. Earp, R.L. and Dessy, R.E. (1998) Surface plasmon resonance, in *Commercial Biosensors: Applications to Clinical, Bioprocess, and Environmental Samples* (ed. G. Ramsay), J. Wiley, New York, pp. 99–164.

45. Homola, J. (ed.) (2006) *Surface Plasmon Resonance Based Sensors*, Springer-Verlag, Berlin.

46. Voros, J., Ramsden, J.J., Csucs, G. *et al.* (2002) Optical grating coupler biosensors. *Biomaterials*, **23**, 3699–3710.

47. Scarano, S., Mascini, M., Turner, A.P.F. *et al.* (2010) Surface plasmon resonance imaging for affinity-based biosensors. *Biosens. Bioelectron.*, **25**, 957–966.

48. Homola, J., Yee, S.S., and Gauglitz, G. (1999) Surface plasmon resonance sensors: review. *Sens. Actuators B-Chem.*, **54**, 3–15.

49. Homola, J. (2003) Present and future of surface plasmon resonance biosensors. *Anal. Bioanal. Chem.*, **377**, 528–539.

50. Mullett, W.M., Lai, E.P.C., and Yeung, J.M. (2000) Surface plasmon resonance-based immunoassays. *Methods*, **22**, 77–91.

51. Homola, J., Vaisocherova, H., Dostalek, J. *et al.* (2005) Multi-analyte surface plasmon resonance biosensing. *Methods*, **37**, 26–36.

52. Slavik, R., Homola, J., Ctyroky, J. *et al.* (2001) Novel spectral fiber optic sensor based on surface plasmon resonance. *Sens. Actuators B-Chem.*, **74**, 106–111.

53. Gauglitz, G. (2006) Interferometry in bio-and chemosensing, in *Optical Chemical Sensors* (eds F. Baldini, A.N. Chester, J. Homola, and S. Martellucci), Springer, Dordrecht, pp. 217–237.

54. Drapp, B., Piehler, J., Brecht, A. *et al.* (1997) Integrated optical Mach-Zehnder interferometers as simazine immunoprobes. *Sens. Actuators B-Chem.*, **39**, 277–282.

55. Brandenburg, A. (1997) Differential refractometry by an integrated-optical Young interferometer. *Sens. Actuators B-Chem.*, **39**, 266–271.

56. Shamah, S.M. and Cunningham, B.T. (2011) Label-free cell-based assays using photonic crystal optical biosensors. *Analyst*, **136**, 1090–1102.

57. Fang, Y. (2010) Resonant waveguide grating biosensor for microarrays, in *Optical Guided-Wave Chemical And Biosensors II* (eds M. Zourob and A. Lakhtakia), Springer-Verlag, Berlin, pp. 27–42.

58. Cunningham, B., Li, P., Lin, B. *et al.* (2002) Colorimetric resonant reflection as a direct biochemical assay technique. *Sens. Actuators B-Chem.*, **81**, 316–328.

59. De Vos, K., Bartolozzi, I., Schacht, E. *et al.* (2007) Silicon-on-Insulator microring resonator for sensitive and label-free biosensing. *Opt. Express*, **15**, 7610–7615.

60. Vollmer, F. and Arnold, S. (2008) Whispering-gallery-mode biosensing: label-free detection down to single molecules. *Nature Methods*, **5**, 591–596.

61. Sun, Y.Z. and Fan, X.D. (2011) Optical ring resonators for biochemical and chemical sensing. *Anal. Bioanal. Chem.*, **399**, 205–211.

62. Lee, M.R. and Fauchet, P.M. (2007) Nanoscale microcavity sensor for single particle detection. *Opt. Lett.*, **32**, 3284–3286.

63. Lee, M. and Fauchet, P.M. (2007) Two-dimensional silicon photonic crystal based biosensing platform for protein detection. *Opt. Express*, **15**, 4530–4535.

64. Zhao, Y.J., Zhao, X.W., and Gu, Z.Z. (2010) Photonic crystals in bioassays. *Adv. Funct. Mater.*, **20**, 2970–2988.

65. Hansen, K.P. and Iman, H. (2010) Photonic crystal fiber. *Optik & Photonik*, **5**, 37–41.

66. Skibina, Y.S., Tuchin, V.V., Beloglazov, V.I. *et al.* (2011) Photonic crystal fibres in biomedical investigations. *Quantum Electron.*, **41**, 284–301.

67. Malinin, A.V., Skibina, Y.S., Tuchin, V.V. *et al.* (2011) The use of hollow-core photonic crystal fibres as biological sensors. *Quantum Electron.*, **41**, 302–307.

68. Jensen, J.B., Pedersen, L.H., Hoiby, P.E. *et al.* (2004) Photonic crystal fiber based evanescent-wave sensor for detection of biomolecules in aqueous solutions. *Opt. Lett.*, **29**, 1974–1976.

69. Yang, X., Shi, C., Newhouse, R. *et al.* (2011) Hollow-core photonic crystal fibers for surface-enhanced Raman scattering probes. *Int. J. Opt.*, Article ID 754610.

19

Optical Sensors – Applications

As indicated in the previous chapter, there are a broad range of optical transduction methods that allow the development of a great diversity of optical sensor platforms. In order to develop a particular kind of optical sensor, one has to select a suitable recognition method and integrate the recognition element with an appropriate optical sensor platform.

This chapter presents typical strategies for the development of optical sensors for various classes of applications focusing on recognition methods for particular kinds of analytes. Included are optical sensors for inorganic species, such as the hydrogen ion, metal ions, and certain gaseous compounds. Optical sensors based on enzymes, affinity recognition receptors, and nucleic acids are also presented.

19.1 Optical Sensors Based on Acid–Base Indicators

19.1.1 Optical pH Sensors

Optical pH sensors make use of *acid–base indicators* (also called pH indicators) that are organic compounds capable of undergoing acid dissociation accompanied by a change in color:

$$HIn(\lambda_1) \rightleftarrows In^-(\lambda_2) + H^+; \quad K_i = \frac{[In^-][H^+]}{[HIn]} \tag{19.1}$$

Symbols in square parentheses represent activities of pertinent species. The equilibrium constant in Equation (19.1) is the *indicator constant*. An absorbance measurement at the specific wavelength of one of indicator form (λ_1 or λ_2) permits the solution pH to be determined.

An absorbance–pH relationship can be derived by combining the indicator constant expression (19.1) and the mass balance equation for the indicator ($c_t = [In^-] + [HIn]$, where c_t is the total concentration of the indicator). Assuming that only the basic species [In$^-$] absorbs light and taking into account the pH definition (pH $= -\log a_{H^+} \approx -\log[H^+]$) the concentration of the basic form of the indicator depends on pH as:

$$[In^-] = \frac{K_i}{10^{-pH} + K_i} c_t \tag{19.2}$$

A plot of [In$^-$] absorbance vs. pH is shown in Figure 19.1A for the hypothetical case of an indicator with $pK_i = -\log K_i = 3$. The symmetrical, S-shaped curve shows a portion centered around pH $= pK_i$ where the absorbance variation is approximately linear. This is the indicator transition pH range, in brief the *pH range*. Therefore, pH determination based on a single indicator is limited to a narrow pH range of about 2–3 pH units, with strong deviation from linearity beyond this region. It is possible to broaden the pH range by incorporating in the sensing layer several pH indicators with pK_i value selected so as to give, by absorbance superposition, an extended and pseudolinear response region.

As the indicator constant depends on the ionic strength of the solution and temperature, these parameters have to be kept under control.

Acid–base indicators can be immobilized by covalent binding or entrapment in cellulose membranes or polymeric materials, or by photochemical coupling to the fiber end. Sol-gel chemistry is also used for indicator immobilization by entrapment. Indicator immobilization can modify the absorption spectrum of the indicator, and, in favorable situations, expands the transition pH range.

An alternative to conventional acid–base indicators are conducting polymers (such as polyaniline or polypyrrole), that change their absorptivity in both the visible and near-infrared regions in response to pH change. The advantage of conducting polymer pH sensors is a wide response range, because these materials are polyelectrolytes with multiple acid dissociation constants. However, they are susceptible to interference by other ions and show a longer response time.

Diffuse reflectance provides essentially the same kind of response as light-absorption measurement but is more convenient from the standpoint of sensor fabrication. Figure 19.1B shows a miniature diffuse reflectance-based pH sensor for *in vivo* gastric pH monitoring. It is composed of two measuring channel, each of them fitted with a

Chemical Sensors and Biosensors: Fundamentals and Applications, First Edition. Florinel-Gabriel Bănică.
© 2012 John Wiley & Sons, Ltd. Published 2012 by John Wiley & Sons, Ltd.

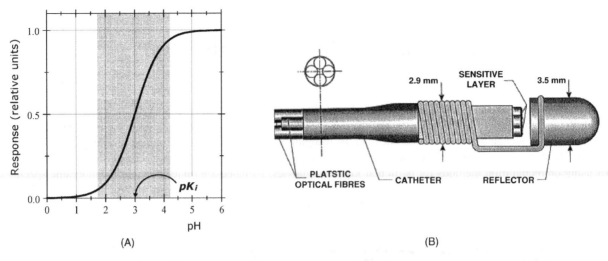

Figure 19.1 pH optical sensors. (A) pH response of a single-color indicator pH sensor. The gray area indicates the response range. (B) A miniature pH optical sensor for *in vivo* determination of pH in foregut. (B) Reprinted with permission from [1]. Copyright 1995 Elsevier.

different pH indicator, namely thymol blue (pH range 1.2–2.8) and bromophenol blue (pH range 3–4.6). The pH indicators are covalently immobilized on controlled-pore glass beads attached to the end of a plastic optical fiber by heating. In each channel, one fiber carries the light from the source to the probe and the other fiber carries the reflected light from the probe to the light detector. A Teflon reflector placed in front of the fibers provides a better coupling of the modulated light. The reference signal is obtained at a wavelength far from the maximum absorption wavelength of each indicator. This sensor has a response range sufficiently wide for *in vivo* determination of pH in the digestive tract.

Fluorescent acid–base indicators display fluorescence intensity depending on pH similar to that shown in Figure 19.1A. As in the case of absorption- or reflectance-based pH sensors, the response range can be expanded by using combinations of indicators with different pK_i values. Common fluorescent pH indicators are fluorescein derivatives and 8-hydroxypyrene-1,3,6-trisulfonic acid (HPTS) (Figure 19.2). Fluoresceinamine can be immobilized at hydroxyl-containing membranes by covalent coupling with cyanuric chloride. HPTS is a strong polyacid and can be incorporated in an anion exchanger or in hydrophobic polymer as an ion pair with an organic hydrophobic cation. Its indicator constant ($pK_i \approx 7.5$) makes it suitable for pH determinations in the physiological pH range.

The ratiometric measurement method can be implemented in pH sensors if both the acid and base form of the indicator absorb light or fluoresce at specific wavelengths. The ratiometric pH response is obtained by putting the indicator constant expression in (19.1) in the following form:

$$pH = pK_i + \log \frac{[In^-]}{[HIn]} \tag{19.3}$$

Compared with the pH glass electrode, which provides a very wide response range, the response range of optical pH sensors is limited to a narrow range around the pK_i value. However as already mentioned, the response range of optical pH sensors can be expanded by means of sensing layers including combinations of a series of properly selected indicators. Optical pH sensors have the advantages of easy miniaturization and reliable response in samples with a very low ion content, that are troublesome when using the pH glass electrode. An overview of pH optical sensor is presented in ref. [2].

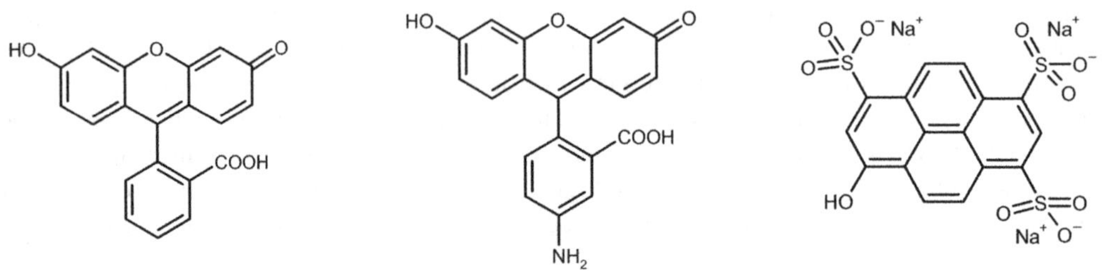

Fluorescein   Fluoresceinamine   HPTS

Figure 19.2 Fluorescent indicators used in pH sensors.

19.1.2 Optical Sensors for Acidic and Basic Gases

Acidic gases (e.g., carbon dioxide) and basic gases (e.g., ammonia) impart on dissolution in water a particular pH that depends on the concentration in the gas sample. This property is exploited in the potentiometric sensor (Section 10.16) in which the pH change is monitored by a pH glass electrode. Similarly, optical gas sensors are obtained by means of optical pH probes. The indicator dye should be selected so as to match its indicator constant to the ionization constant of the dissolved gas. A pH buffer should be incorporated in the sensing layer in order to prevent complete protonation (or dissociation) of the indicator, which would significantly limit the response range of the sensor.

As in the case of potentiometric gas sensors, optical gas sensors are suitable for testing both liquid and gas samples with interferences arising only from other acidic or basic gases.

The response range of single indicator sensors is limited to about one order of magnitude, but it can be extended using an optimized mixture of indicators instead of a single one.

A common application is the optical carbon dioxide sensor, which is of interest to clinical chemistry, chemistry of aquatic systems and also to the food and beverage industry [3]. The sensing layer in an optical CO_2 sensor is composed of an aqueous solution containing hydrogen carbonate as the pH buffer and a light-absorbing indicator dye. This solution is entrapped at the end of an optical fiber bundle by means of a $\sim 10\,\mu m$ thick Teflon membrane that is permeable to CO_2 but not to ions. Sample CO_2 passes through the membrane and dissolves in the internal solution, affecting its pH and hence the optical response produced by the pH indicator.

More rugged and reliable CO_2 sensors are obtained by incorporating an anionic indicator dye in a plasticized polymer membrane as an ion pair with a hydrophobic cation. The ion pair binds also several water molecules that allow acid–base equilibria to occur as in the aqueous solution.

An ammonia sensor has been developed using two indicators: chlorophenol red ($pK_i = 6.25$, $\lambda_1 = 578$ nm) and bromothymol blue ($pK_i = 7.30$; $\lambda_2 = 618$ nm) [4]. The sensing element in this sensor is a very thin layer of ammonium chloride buffer entrapped behind a microporous Teflon membrane. Light of a selected wavelength is conveyed to the indicator solution by two fibers. After crossing the solution, light is scattered back by the Teflon membrane, collected by a third fiber and transmitted to the light detector. The absorbance response is nonlinear and the response range stretches from 0.2 to 30 μM ammonia. Response and recovery times are of about 10–20 min owing to the slow diffusion thorough the membrane.

Much shorter response times and a more rugged configuration are obtained by incorporation of the indicator dye in sol-gel materials or in a gas-permeable polymer matrix.

Fluorescence-based gas sensors have also been produced. For example, a carbon dioxide sensor has been developed using the HPTS fluorescent indicator incorporated in a silicone matrix as an emulsion of a room-temperature ionic liquid [5]. The response function of this sensor can be put in a linear form as follows:

$$\frac{F_o}{F} = 1 + kc_{CO_2} \tag{19.4}$$

where F_o and F are fluorescence intensities in the absence and in the presence of CO_2, respectively and k is a constant depending upon the first dissociation constant of carbonic acid, the indicator constant and the total concentration of bicarbonate in the buffer solution incorporated in the sensing film. The dual lifetime referencing method has also been applied by incorporating in the sensing part a CO_2-insensitive, long-lifetime fluorophore in order to obtain a ratiometric response.

Phase-resolved fluorimetry combined with the dual lifetime referencing method (Section 18.3.6) has been applied to obtain a carbon dioxide sensor [6]. The HPTS fluorescent indicator was incorporated in a poly(dimethylsiloxane) matrix as an ion pair with a hydrophobic quaternary ammonium base. A ruthenium complex incorporated into silica microparticles and included in the immobilization matrix acts as the reference fluorophore. The measured phase shift is correlated with carbon dioxide partial pressure according to the following nonlinear function:

$$\cot \phi_m = \cot \phi_{max} + \frac{\cot \phi_0 - \cot \phi_{max}}{1 + k_A p_{CO_2}} \tag{19.5}$$

where ϕ_m is the phase angle at a particular CO_2 concentration, ϕ_0 the phase angle measured in the absence of CO_2, and ϕ_{max} the phase angle measured at saturation, when all indicator sites are in the protonation form. The k_A constant depends on the indicator constant and the basicity of the buffer used in the sensing membrane. The layout of such a sensor is presented in Figure 18.13.

The main advantage of optical gas sensors over the potentiometric ones arises from the possibility of preparing the sensing layer as a solid membrane and not as an entrapped solution. This avoids serious drawbacks related to the difference between the osmotic pressure in the sample and that in the internal solution of the sensor. This difference produces water transfer to or from the internal solution, which affects the response of a potentiometric gas sensor.

Questions and Exercises (Section 19.1)

1 What are the properties of acid–base indicators? What factor determines the pH response range of these indicators? What parameters determine the sensitivity of the response to pH?

2 Derive the response function of an optical sensor in the case in which the acid form of the indicator is detected. Plot it as a $[In^-]/c_t$ vs. pH graph and compare this curve with the curve in Figure 19.1A.

3 Derive an equation that relates the pH to the ratio of the acidic and basic indicator forms assuming that the absorption band of these forms overlaps partially and the total absorbance measured at each wavelength is the sum of the absorbances of each form. What advantages brings about this transduction scheme? *Hint*: use Equation (19.3) as starting point.

4 How can the response range of an optical pH sensor be expanded?

5 Draw a simplified scheme of the pH sensor in Figure 19.1B showing light paths for the interrogating and response light beams.

6 Discuss briefly the methods for the immobilization of pH indicators. *Hint*: Refer to Chapter 5.

7 Compare optical pH sensors with the potentiometric pH sensor as far as advantages and drawbacks are concerned.

8 Write the chemical reactions involved in a carbon dioxide optical sensor and the relevant equilibrium constants. Define criteria for the selection of the pH indicator for this kind of sensor.

9 Sketch the possible configuration of an optical carbon dioxide sensor based on (a) an aqueous solution as sensing element; (b) a plasticized polymer membrane as sensing element. In each case, mention the components of the sensing phase and possible fabrication schemes.

10 Write the chemical reactions involved in an ammonia optical sensor and the relevant equilibrium constants. Define criteria for the selection of the pH indicator for this kind of sensor.

11 Draw schematically a possible structure of an optical ammonia sensor based on two pH indicators.

12 Refer to Figure 18.13 and comment on the functioning principle of a dual lifetime referencing carbon dioxide sensor. Discuss the composition of the sensing membrane and the response function.

19.2 Optical Ion Sensors

Optical ion sensors (also called *ion optodes*) can be developed as either direct or indirect sensors [7–9]. Selective ion recognition can be achieved by means of supramolecular chemical interactions.

19.2.1 Direct Optical Ion Sensors

Direct ion sensors are feasible when the receptor itself is a chromophore or a luminophore and experiences a change in optical properties upon binding to an analyte ion. Such a receptor should fulfill two critical conditions: (i) to display selectivity to the analyte ion and (ii) to change the absorption or fluorescence properties upon binding the analyte. Such double-function receptors can be obtained by tagging a selective ion receptor with a chromogenic or fluorogenic fragment. Ion binding brings an electric charge that shifts the absorption or fluorescence wavelength. Hence, at a fixed wavelength one can notice either an increase or a decrease of the optical signal. This principle has been applied to develop molecular ion sensors.

Sensing an alkali metal ion by a macrocyclic receptor tagged with a fluorescent moiety is demonstrated in Figure 19.3. In the absence of the metal ion, fluorescence is quenched by electron transfer from the nitrogen atom at the

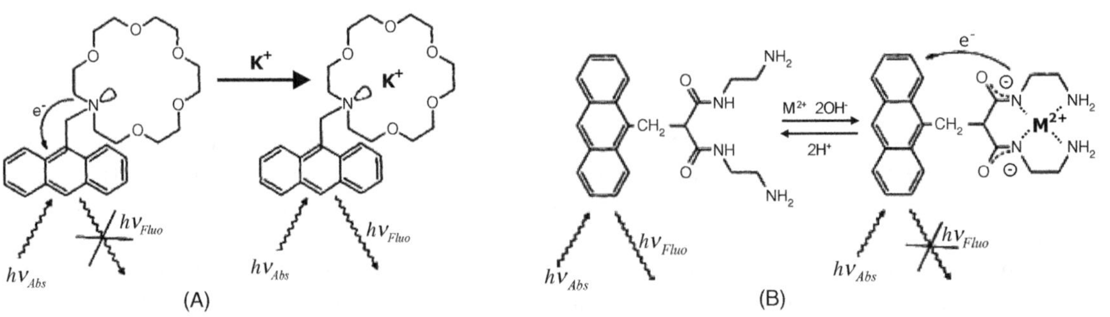

Figure 19.3 Ion sensing by photoinduced electron transfer. (A) Sensing of the potassium ion by a macrocyclic ligand tagged with a fluorescent moiety (anthracene); (B) sensing of a transition-metal ion (copper or nickel) by coordination to a polydentate ligand with an appended fluorescent moiety. Reprinted with permission from [10]. Copyright 1995 The Royal Society of Chemistry.

macrocycle to the excited state of anthracene (*photoinduced electron transfer*). In order for quenching to occur the energy level of the donor orbital should be higher than that of the acceptor orbital. If the metal ion is bound to the receptor macrocycle, its charge lowers the energy level of the electron pair at the nitrogen atom and prevents the electron transfer. Consequently fluorescence intensity increases with increasing ion concentration. In the case of an alkali metal ion, the photoinduced electron transfer is shut off owing to the ion–dipole interaction between the ion and the electron donor site.

Transition-metal ions can be detected by a mechanism involving formation of donor sites in the receptor molecule after ion binding, as shown in Figure 19.3B. In this case, the fluorophore tag is bound to a carbon atom at the ion receptor and thereby photoinduced electron transfer cannot take place. Metal-ion binding by coordination leads to the ionization of amide nitrogens and creates negative sites that act as electron donors in photoinduced electron transfer. Therefore, ion binding results in fluorescence quenching.

Other mechanisms for direct sensing on metal ions are described in the literature [10–12]. Anion-sensing schemes based on photoinduced electron transfer have also been developed [13].

19.2.2 Indirect Optical Ion Sensors

The indirect approach to optical ion sensing relies on well-established selective ion receptors but avoiding their chemical modification that can impair the selectivity. As typical of indirect sensing, not the receptor or the ion-receptor complex is detectable and the transduction is performed by means of a signaling reagent involved in the sensing process.

A general indirect transduction scheme is based on the charge-balance restriction within the sensing membrane. Sensing membranes are similar to those used in potentiometric ion sensors, that is, they contains the receptor and additives entrapped in a plasticized polymer matrix [14–16]. For optical transduction purposes, a lipophilic pH indicator is included in the membrane as signaling reagent.

Figure 19.4 shows two possible indirect transduction schemes coupled with ion recognition by a neutral ion receptor (R). For cation sensing (Figure 19.4A), a hydrophobic indicator is incorporated as the protonated, neutral molecule (HIn). Owing to the very high stability constant of the MR metal complex, the cation M^+ is extracted into the membrane but, for charge-balance reasons, a hydrogen ion is evicted leaving the indicator in the anionic form. Indicator ionization modifies its optical properties as in function of the analyte concentration.

An anion-sensing scheme is shown in Figure 19.4B. In this case, the indicator is present in the membrane as an anion (In^-) forming an ion pair with a hydrophobic counterion X^+. The analyte anion A^- is extracted into the membrane by strong coupling with the receptor and thus introduces an additional negative charge. For charge-balance reasons, the anionic indicator is protonated by extraction of a hydrogen ion from the solution. Protonation of the indicator alters its optical properties and thereby allows for the analyte concentration in the sample to be measured by means of an optical signal.

Depending on the optical properties of the indicator, either light absorption or fluorescence emission can be used to obtain the optical response.

Analogous transduction schemes have been developed for the case of ion recognition by charged receptors [15].

The merits of indirect optical ion sensors have been discussed in detail in ref. [15]. The response function has a sigmoid shape and the measured signal depends on the solution pH, which requires this parameter to be controlled by means of a pH buffer system.

A pH-independent transduction scheme is based on *solvatochromism*, which means solvent-dependent light absorption or emission. A sensing scheme based on this principle is shown in Figure 19.5. As shown in this Figure, a cationic fluorescent dye (In^+), which is tagged with a long hydrocarbon tail, is incorporated in the sensing membrane along

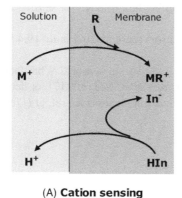

(A) **Cation sensing**   (B) **Anion sensing**

Figure 19.4   Indirect transduction in optical ion sensors based on neutral receptors (R) and a pH indicator. (A) Cation sensing by ion exchange involving the hydrogen ion. (B) Anion sensing based on coextraction of the hydrogen ions and the analyte. The receptor, the indicator, and the counterion (X^+) are hydrophobic species that cannot be extracted into the solution.

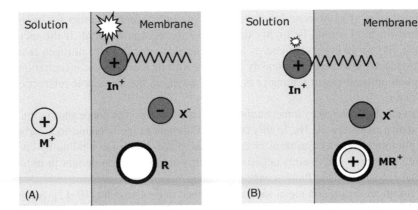

Figure 19.5 Indirect transduction in optical ion sensors based on the solvatochromic effect on the fluorescence of a signaling dye. (A) Before ion extraction (strong fluorescence); (B) after ion extraction (lessened fluorescence).

with the ion receptor (R) and a hydrophobic counterion (X^-). Before ion extraction, the dye in the nonpolar membrane environment fluoresces strongly (Figure 19.5A). Ion extraction leads to the formation of a positively charged complex, which repels the charged head of the dye into the solution phase. In this polar environment, fluorescence is very weak. Therefore, fluorescence intensity decreases with increasing ion concentration in the solution. A potassium optical sensor based on this principle has been developed using valinomycin as receptor and alkyl-Acridine Orange as indicator [17]. The fluorescence response of this sensor decreases with analyte concentration as a linear function.

As potentiometric and indirect optical ion sensors are based on similar recognition receptors, it is useful to compare the performances of these to kinds of sensors [15]. Optode *selectivity* can be modeled by an equation similar to the Nikolskii-Eisenman equation used in potentiometry. The *dynamic range* of optical sensors stretches over 2–4 order of magnitudes. It is narrower than that of potentiometric sensors, which can be of 5–9 orders of magnitude. On the other hand, the *detection limit* of optical ion sensors can reach subnanomolar levels, whilst the limit of detection of potentiometric sensors is, in general, in the micromolar region. The better detection limit of optical sensors is due to the fact that the analyte ion is not present within the sensor system and cannot contaminate the sample, as occurs with potentiometric sensors. As far as the lifetime of the sensors is concerned, it is limited in both cases by slow leaching of membrane components into the sample. In contrast to potentiometric sensors, the ion optode response depends directly on the concentration of active components in the membrane and is therefore more sensitive to leaching effects.

An advantage of ion optodes arises from the insensitivity of the response to mechanical damage of the membrane. In contrast, a physical hole in the potentiometric sensor membrane produces an electrical short between the internal and external solutions and, therefore, causes complete breakdown. The greater ruggedness of optodes allows for a wider variety of shapes and sizes of sensor to be used, as well as extreme miniaturization. Last, but not least, a potentiometric ion sensor depends on a reliable reference electrode, which complicates the sensor structure. Such a complication is absent in the case of ion optodes. However, it should not be forgotten that the readout instrumentation for potentiometric ion sensor is simpler.

Questions and Exercises (Section 19.2)

1. Comment on the structure of a molecular ion sensor and the role of each distinct molecular module in the sensing process.
2. What is photoinduced electron transfer? How can this effect be exploited for sensing alkali metal ions or transition-metal ions?
3. Write schematically the reactions involved in the ion sensors presented in Figure 19.4 pointing out the ion transfer between the solution and the membrane phase.
4. What are the underlying principles of pH-independent indirect ion sensors? What factor determines the limit of detection and the upper limit of the linear response range? See ref. [17] for details.
5. Give a comparative overview of potentiometric and optical ion sensors (see ref. [17] for details).

19.3 Optical Oxygen Sensors

The property of oxygen to produce dynamic fluorescence quenching forms the basis of oxygen optical sensors [18–21] (see also Section 18.3.7). Common fluorophores used in oxygen sensors are ruthenium (II) complexes with 2,2′-bipyridine (bipyridyl), 1,10 phenanthroline and related compounds (Figure 19.6). Analogous complexes of other platinum group metals (Os^{2+}, Ir^{3+}) as well as certain organic dyes and polycyclic aromatic hydrocarbons have also been used.

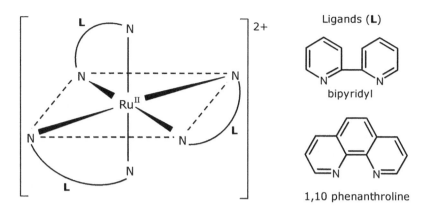

Figure 19.6 Structure of Ru(II) complexes used in optical oxygen sensors.

Luminophores can be immobilized within polymers such as cellulose acetate or polyvinyl chloride along with a plasticizer that enhances oxygen permeation [22]. Sol-gel entrapment is also a suitable immobilization method.

The response of oxygen sensors deviates from the Stern–Volmer equation (Equation (18.16)) owing to the effect of the immobilization matrix. Upon immobilization, the luminophores can be found in various types of site, each of them having a particular quenching constant. On the other hand, the response depends on the concentration of absorbed oxygen in the matrix, which is a nonlinear function of oxygen concentration in the sample. In many cases, Equation (19.6) (in which the a coefficients are empirical constants and c_{O_2} is oxygen concentration in the sample) fits well the experimental data [18].

$$\frac{F_0}{F} = 1 + a_1 c_{O_2} + \frac{a_2 c_{O_2}}{1 + a_3 c_{O_2}} \tag{19.6}$$

As in Equation (18.16), F_0 and F represent the fluorescence power in the absence and in the presence of the quencher, respectively. Good sensitivity is favored by long-lived excited states, high oxygen solubility and high rate of diffusion of oxygen in the immobilization matrix. In this respect, phosphorescent platinum and palladium porphyrin complexes are suited for very sensitive oxygen sensors because of their much longer lifetime of the excited state.

Nanomaterial application as luminophores opens up new prospects in the field of optical oxygen sensors. Thus, the fluorescent C_{70} fullerene has been incorporated in a sol-gel based on organically modified precursors [23]. It allows for oxygen sensing and imaging down to the ppb level.

Fiber optic oxygen sensors can be designed in either the passive or the active (intrinsic) fiber configuration. An intrinsic sensor has been produced by forming a methylene blue layer on a decladded plastic optical fiber using the sol-gel method [24]. Dye excitation is performed by the evanescent wave, which brings about a more rapid response and very short recovery time. The response range stretches from 0.6 to 20.9%, which makes it applicable in the diagnosis of oxygen-deficiency diseases.

Oxygen microsensors can be obtained by photolithography. For example, the platinum octaethylporphyrin luminophore has been integrated in photoresist plastic material along with a plasticizer. The mixture has been deposited over a fused silica substrate and then patterned by ultraviolet irradiation in order to obtain arrays of 3 μm oxygen sensors for monitoring oxygen consumption rate in single cells [25].

Optical oxygen sensors function equally well in gas and aqueous solution analysis. They can be produced at a very small size for *in vivo* applications. Also, oxygen optical sensors can be integrated with optical pH and CO_2 sensors to obtain multianalyte probes for *in vivo* determinations.

Moreover, an optical oxygen sensor can function as transduction element in oxygen-linked enzyme sensors.

Compared with the electrochemical oxygen sensor (Section 13.10.1), the optical one is more compact and rugged as it is an all-solid device. The optical oxygen sensor can be easily miniaturized and display shorter response and recovery times. On the other hand, the optical sensor is more expensive than the electrochemical one owing to the included optical and optoelectronic components.

Questions and Exercises (Section 19.3)

1 What methods are used for performing immobilization of a fluorescent indicator in optical oxygen sensors? What condition should the immobilization matrix fulfill?

2 What are the effects of fluorophore immobilization on the response of an optical oxygen sensor?

3 What property of the fluorophore imparts high sensitivity of an optical oxygen sensor? Give a review of fluorophores commonly used in optical oxygen sensors.

4 Comment on possible configurations of passive and active optical fiber oxygen sensors.

19.4 Optical Enzymatic Sensors

19.4.1 Principles and Design

The functioning of an optical enzymatic sensor relies on the change in the optical properties of a fluorescent indicator in response of its interaction with a reactant or product involved in the enzymatic reaction. Common transduction procedures are based on optical monitoring of pH change or oxygen consumption. A typical configuration of a fiber optic enzymatic sensor is shown in Figure 19.7. For transduction purposes, a fluorescent indicator layer (2) is formed over a transparent support (3) at the end of the optical fiber. This layer is sensitive to the change in the concentration of a cosubstrate or a product involved in the enzymatic reaction. The recognition layer consisting of immobilized enzyme (1) interacts with the substrate, and the products or an unconsumed reactant diffuses further to the indicator layer and affect the fluorescence intensity. The indicator layer can include a pH indicator in order to monitor the production or consumption of hydrogen ions by the enzymatic reactions. In the case of an oxidase enzyme-base sensor, oxygen consumption can be monitored by fluorescence quenching. An alternative optical transduction scheme is based on the spectrochemical properties of the NAD^+ coenzyme.

19.4.2 Optical Monitoring of Reactants or Products

Common reactants or products involved in enzymatic reactions are hydrogen ions, oxygen, hydrogen peroxide, carbon dioxide, ammonia, and ammonium ions. In principle, an optical enzymatic sensor can be obtained by integrating an enzyme layer with a sensor for one of the above-mentioned species.

A common scheme relies on the pH-linked transduction method. Enzymes belonging to the oxidase or hydrolase classes are suitable for such applications. As is typical of pH-linked enzymatic sensors, proper functioning requires the presence of a low buffer capacity pH buffer. As sensor functioning is based on pH variation, care should be exercised to avoid drastic effects of pH change on enzyme activity which can affect unfavorably the response function.

Despite such problems, various applications of pH-linked optical transduction have been reported. For example, pH-based monitoring of acetylcholine esterase activity has proved useful in the development of inhibition-based sensors for pesticides and warfare agents.

Figure 19.8 shows a glucose optical sensor based on fluorescence quenching by residual oxygen in the glucose oxidase-catalyzed reaction. This sensor is designed in the catheter format and has been used for either *in vitro* or *in vivo* determination of glucose in fish blood. Glucose oxidase was immobilized within a polymer film over an ultra-thin dialysis membrane (3). The rolled enzyme membrane was place in a metal needle (1) with lateral holes that allow the sample to get in contact with the enzyme membrane. For transduction purposes, a fiber optic oxygen probe (4) is installed inside the membrane roll. Before measurement, the sensor is filled with an oxygen-saturated neutral buffer solution and then the sensor is applied to the sample. Owing to the gradual oxygen consumption, the oxygen probe response decreases steadily. That is why the readout is measured after a predetermined time interval.

Hydrogen peroxide is a product of reactions catalyzed by oxidase-type enzymes. Therefore, monitoring of this product by chemiluminescence or electrochemiluminescence provides alternative transduction methods in enzymatic optical sensors.

Electrochemiluminescence has been applied to the development of an array of six different enzymatic sensors [28]. This array is composed of six sensing spots formed at the surface of a carbon electrode. Each spot contains a particular oxidase enzyme for the determination of its substrate. Enzymes and luminol are loaded on anion-exchanger beads that are then entrapped in a polymer matrix. Chemiluminescence intensity produced at each spot is proportional to the amount of hydrogen peroxide resulting from the enzymatic oxidation of the substrate. Such a

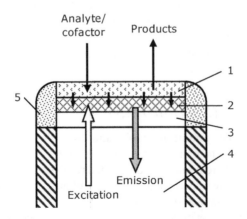

Figure 19.7 Schematics of a fiber optic enzymatic sensor. (1) Enzyme layer; (2) indicator layer; (3) transparent support; (4) optical fiber; (5) glue. Redrawn with permission from [26]. Copyright 2008 American Chemical Society.

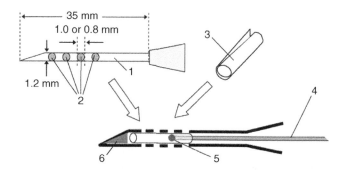

Figure 19.8 A catheter-type optical glucose sensor based on fluorescence quenching by oxygen. (1) Needle-type hollow container; (2) holes; (3) immobilized enzyme membrane; (4) optical fiber; (5) optical oxygen probe; (6) epoxy filling. Reprinted with permission from [27]. Copyright 2006 Elsevier.

sensor array allows a series of analytes to be determined in parallel. Tests conducted with blood serum gave satisfactory results in the determination of glucose, lactate, and uric acid.

Ammonia monitoring by a pH indicator-based optical probe represents a viable transduction method in optical urea sensor based on the urease enzyme [29]. As ammonia is in equilibrium with the ammonium ion, an ammonium optode is suitable for performing transduction in urea sensors [30,31].

19.4.3 Coenzyme-Based Optical Transduction

Nicotinamide adenine dinucleotide (NAD^+) which functions as a coenzyme for certain dehydrogenases, displays characteristic absorption (350 nm) and fluorescence (450 nm) bands. This allows NAD^+ to act as an optical marker in optical enzymatic sensors. In the initial stages of development, NAD^+ was added to the sample solution. In a more advanced approach, a self-contained sensor has been produced by using NAD^+-tagged poly(ethylene glycol) in order to prevent NAD^+ from crossing dialysis membranes [32]. This modified coenzyme along with two enzymes was entrapped within the sensing compartment between a dialysis membrane and the fiber optic tip. One of the enzymes (e.g., alcohol dehydrogenase) catalyzes substrate conversion using NAD^+ as cofactor, whilst the second enzyme (lactate dehydrogenase) serves to regenerate NAD^+ by reduction of NADH with pyruvate.

A two-enzymes system (an oxidase and a luciferase) extracted from the marine bacteria *Vibrio harveyi* or *Vibrio fischeri* has been employed in the determination of NADH or NADPH coenzymes by chemiluminescence produced in a two-step reaction. In the first step (reaction (19.7)), which is catalyzed by an oxidoreductase, NAD(P)H is oxidized by flavin mononucleotide (FMN) giving reduced flavin mononucleotide ($FMNH_2$). In the next step (reaction (19.8)), $FMNH_2$ acts as cosubstrate in the oxidation of a long-chain aldehyde by luciferase, a reaction that produces chemiluminescence.

$$NAD(P)H + H^+ + FMN \xrightarrow{\text{Oxidoreductase}} NAD(P)^+ + FMNH_2 \tag{19.7}$$

$$FMNH_2 + R - CHO + O_2 \xrightarrow{\text{Luciferase}} FMN + R - COOH + H_2O + h\nu(\lambda = 490\,nm) \tag{19.8}$$

The above sequence of reactions provide an optical transduction method in sensors based on $NAD(P)^+$-dependent dehydrogenases according to the mechanism in Figure 19.9. This transduction mechanism involving coenzyme recycling provides a limit of detection close to 0.4 μM, which is 1 or 2 orders of magnitude better than that obtained by direct detection of NADH.

19.4.4 Outlook

Optical enzymatic sensors can be obtained by combining an enzyme layer with an optical system in order to monitor reactants or products involved in the enzymatic reaction. Depending on the mechanism of the enzymatic reaction,

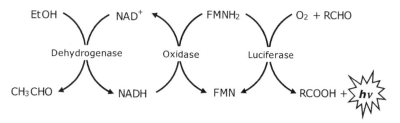

Figure 19.9 Chemiluminescence ethanol assay by means of reactions (19.7) and (19.8).

either hydrogen ions, oxygen, hydrogen peroxide, certain gases (ammonia and carbon dioxide) or the ammonium ion can serve for transduction by means of optical probes for such species.

Recent advances in enzymatic optical sensors are surveyed in refs. [26,33]. The field of optical glucose sensors is reviewed in ref. [34].

Questions and Exercises (Section 19.4)

1 Review briefly the classes of enzyme that are suitable for producing optical enzyme sensors. For each class, indicate the reactant or product that can be detected by optical methods.
2 What are the main components of an optic fiber enzymatic sensor?
3 Comment on the role of each component of the sensor shown in Figure 19.8.
4 Draw a sketch of an electrochemiluminescence-based optical enzyme sensor. See ref. [28] for orientation.
5 Sketch possible configurations of optical urea sensors.
6 Discuss the configuration and the functioning mode of a chemiluminescence-based sensor using a dehydrogenase enzyme as the sensing element.

19.5 Optical Affinity Sensors

19.5.1 Optical Immunosensors

Optical immunosensors can be designed in either the competitive or sandwich format. Owing to its outstanding sensitivity, fluorescence is the preferred optical transduction method. Common fluorescent labels in immunoassay are cyanine dyes (Figure 19.10). Depending on structure details, cyanine dyes can fluoresce over a wide wavelength region, from the ultraviolet to the near-infrared. For labeling purposes, reactive groups are attached to the dye molecule in order to allow covalent bonding to proteins. Luminescent semiconductor nanocrystals (quantum dots) represent a promising alternative to organic dyes, as shown in detail in Chapter 20.

Figure 19.11A shows an intrinsic optical immunosensor in which the sensing layer is formed at the end of a tapered optical fiber. In order to perform flow analysis, the sensor is installed in a flow-through cell. Antibodies can be immobilized at the silica fiber surface by bifunctional silane coupling. The fluorescence light is separated from the excitation light by a dichroic mirror.

In order to perform a competitive binding assay, the antibody layer is first saturated with the fluorescence-labeled analyte. Next, the sensor is incubated with the sample solution that also contains the labeled analyte at a predetermined concentration. Competition between the analyte and the labeled analyte leads to a partial replacement of the

Cy3   Cy5

Figure 19.10 Cyanine fluorescent labels.

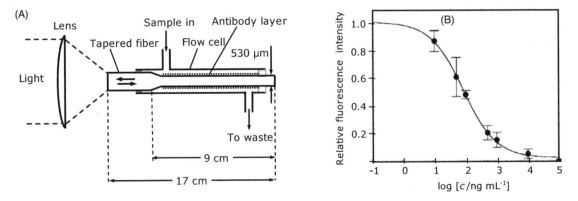

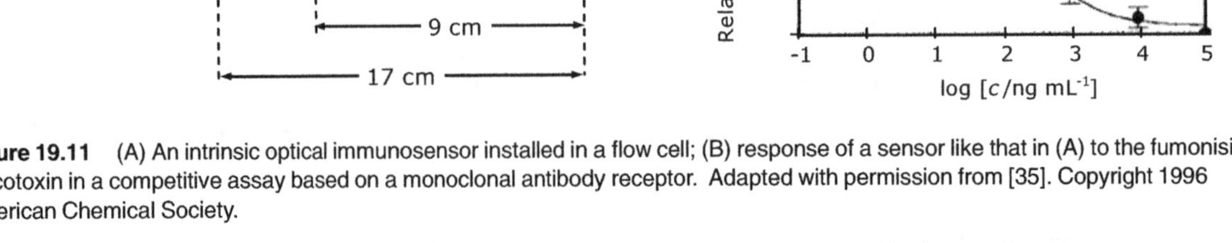

Figure 19.11 (A) An intrinsic optical immunosensor installed in a flow cell; (B) response of a sensor like that in (A) to the fumonisin B$_1$ mycotoxin in a competitive assay based on a monoclonal antibody receptor. Adapted with permission from [35]. Copyright 1996 American Chemical Society.

antibody-bound labeled analyte and thus reduces the fluorescence intensity. Sensor regeneration is performed by means of a solution of the labeled analyte that restores the saturation of the sensing layer.

As fluorescence excitation occurs through the evanescent wave, labels in the solution cannot be excited, which makes any washing-out step unnecessary.

A sensor of this type has been designed for the determination of the fumonisin B_1 mycotoxin in corn extracts [35]. Figure 19.11B shows the response of this sensor and demonstrates that the working range extends from 10 to 1000 ng/ mL. This sensor is highly selective towards fumonisin B_1 and does not crossreact with structurally related compounds.

Besides fluorescence, resonance energy transfer is a very convenient transduction method in immunosensors (see Section 18.3.8).

Optical immunosensors have been developed for the determination of haptene-type analytes such as toxins, drugs, and pesticides. Antigens, such as bacterial cells can be sensed by means of immobilized antibodies. Alternative recognition elements for bacteria are antibacterial peptides which are produced by living organisms for protection against harmful microbes.

Immobilized aptamers, which can act as antibody substitutes, have also been exploited in the development of affinity optical sensors [36].

Label-free transduction in immunosensors and protein arrays represents a practical alternative to the schemes based on fluorescent labels. Surface plasmon resonance spectrometry, interferometry, as well as other methods presented in Sections 18.6 and 18.7 allow the labeling process to be precluded at the expense of a more intricate design of the sensor platform. Application of such methods is favored by the significant variation in the refractive index in response to the recognition process.

19.5.2 Optical Sensors Based on Biological Receptors

Optical glucose sensing can be achieved by means of the natural receptor Concanavalin A (Con A), which displays affinity to various carbohydrate compounds. An indirect, competitive binding sensor for glucose based on this receptor is shown in Figure 19.12. The receptor is immobilized at the surface of a glucose-permeable tubular membrane at the end of an optical fiber. The membrane-entrapped solution contains a labeled analyte-analog such as dextran. Dextran is a glucose-based polymer and can function as a glucose analog because glucose residues in its structure form affinity complexes with Con A. Being a bulky molecule, dextran cannot pass through the membrane. In contact with a glucose-containing sample, the following equilibrium is established:

$$Dex^*(Con A) + Gluc(sample) \rightleftarrows Gluc(Con A) + Dex^*(internal solution) \qquad (19.9)$$

where Dex^* is dextran tagged with a fluorescent label and Gluc is glucose.

Owing to the geometry of the cone of illuminance, labels bound to the receptors at the membrane surface are not excited and the fluorescence intensity depends only on the label concentration in the illumination region. In contact with the sample, labeled dextran is displaced by glucose molecules, which increases label concentration within the illumination zone and hence enhances the fluorescence intensity. As a result, fluorescence intensity increases nonlinearly with the analyte concentration.

The above sensing scheme has been further improved by using labeled Con A immobilized at Sephadex microparticles as shown in Figure 19.13A. Sephadex is a crosslinked dextran gel that displays a large specific surface area which is available for affinity interaction with Con A. Hence, Con A tagged with a fluorescent label has been attached to

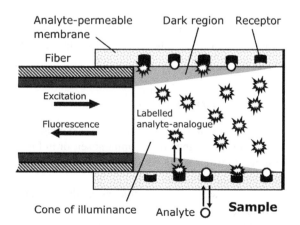

Figure 19.12   A reversible affinity optical sensor based on competitive binding. The membrane is a hollow dialysis fiber that is permeable to glucose but not to the large dextran molecule. Redrawn with permission from [37]. Copyright 1987 John Wiley and Sons.

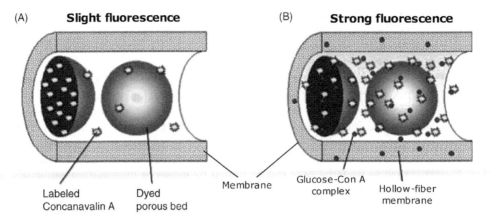

(A) **Slight fluorescence** (B) **Strong fluorescence**

Labeled Concanavalin A

Dyed porous bed

Membrane

Glucose-Con A complex

Hollow-fiber membrane

Figure 19.13 A competitive-binding glucose sensor based on dye modified Sephadex. Adapted with permission from [39]. Copyright 2000 American Chemical Society.

Sephadex. In addition, Sephadex particles have been modified with two organic dyes that absorb both the excitation and the fluorescence light. Modified Sephadex particles have been entrapped in a hollow-fiber dialysis membrane attached at the end of an optical fiber in the same way as in Figure 19.12. Owing to the included dyes, fluorescence intensity in this configuration is very low. In contact with a glucose-containing sample, glucose molecules pass through the membrane and form glucose–Con A associations that are free to diffuse into the internal solution. This process can be represented by the following reaction scheme in which $(Con A)^*$ indicate labelled Con A (see also Figure 19.13B):

$$[(Con A)^*(Sephadex)] + Gluc(sample) \rightleftarrows [(Con A)^*(Gluc)](internal\ solution) \qquad (19.10)$$

This competitive binding process releases the labeled receptor into the internal solution where excitation and fluorescence emission are possible. Consequently, fluorescence intensity increases as a function of the glucose concentration.

This sensing scheme is characterized by a very large density of receptor sites within the active region. Consequently, small variations in glucose concentration result in large variations in the response signal, which enhances the accuracy and expands the dynamic range of the sensor. This sensor can be produced in a very small size ($0.5 \times 0.2\,mm$) and is therefore suitable for transdermal glucose monitoring. By using a near-infrared fluorescent label, interference from the fluorescence background produced by the skin tissue is avoided [38].

A major drawback of Con A is its lack of selectivity. Better selectivity towards glucose has been obtained by the glucose oxidase enzyme in the inactive form (apoenzyme) that has been used in a competitive binding format sensor based on resonance energy transfer transduction [40]. Another selective glucose receptor is the *Escherichia coli* glucose-binding protein in [41].

19.5.3 Outlook

Luminescence is widely used to perform transduction in immunosensors. Although fluorescence is the most widely used transduction process, chemiluminescence and electrochemiluminescence are also of interest as these methods do no need an excitation light source. Fluorescence resonance energy transfer is another method of choice in optical immunosensors. Label-free transduction methods represent a promising alternative to label-based schemes owing to the simplification of the assay protocol.

Affinity reactions based on natural receptors allows developing optical sensors for small molecules such as glucose. Such sensors have a relatively simple structure and can be easily fabricated at a small size that suits *in vivo* applications.

Questions and Exercises (Section 19.5)

1 Draw the flow chart for the fabrication of an immunosensor based on an active optical fiber.
2 Draw a possible optical setup for running the optical immunosensor in Figure 19.11A.
3 Design possible optical immunosensors using transduction by resonance energy transfer in the direct assay format and the competitive assay format.
4 (a) Design an optical affinity sensor with the same structure as that in Figure 19.12 but using the analyte analog immobilized at the membrane inner surface and the labeled receptor free to diffuse within the internal solution; (b) using the reaction scheme (19.9) as model, write the recognition reaction for the sensor in (a); (c) comment on the functioning mode and draw schematically the response function of this sensor.

5 The sensing scheme developed in the previous exercise can be improved by using dextran microparticles as the immobilization support for labeled Con A. Discuss the construction details of this kind of sensor and its advantages over previous affinity-based glucose sensors.
6 Discuss the selectivity of Con A-based optical glucose sensors and the possibilities for improving the selectivity of optical affinity glucose sensors.

19.6 Optical DNA Sensors and Arrays

Optical transduction is widely used in both nucleic acid sensors and in nucleic acid arrays [42–44]. Transduction based on luminescent labels is a simple method suited to both self-contained nucleic acid sensors and DNA microarrays [45,46]. Label-free transduction can be achieved by a series of methods previously reviewed in Section 18.6 and 18.7. Application of label-free sensors simplifies the assay protocol but demands a more intricate fabrication technology.

19.6.1 Fluorescence Transduction in Nucleic Acid Sensors

The high sensitivity of fluorescence spectrometry is exploited on a large scale in the design of nucleic acids sensors and arrays. Fluorescence labels can be appended to nucleic acids in two ways: by covalent binding or by intercalation. Covalently bound labels can be included in a nucleic acid strand during the polymerase chain reaction by providing a fluorophore-tagged primer. Cyanine dyes are common fluorescent labels used in optical nucleic acid assays.

An alternative labeling method is based on fluorescent intercalating and groove binding of planar molecule dyes (see Section 7.3.2). When such a label is present in an aqueous solution, fluorescence is quenched by water molecules. After intercalation in a double-strand nucleic acid, fluorescence is strongly enhanced owing to the hydrophobic environment between the base pairs. Usual intercalating labels are shown in Figure 19.14.

Typical sensing strategies in nucleic acid optical sensors are summarized in Figure 19.15 (see also refs. [26], [43]). The panel (A) demonstrates the transduction by a fluorescent label attached to the target strand.

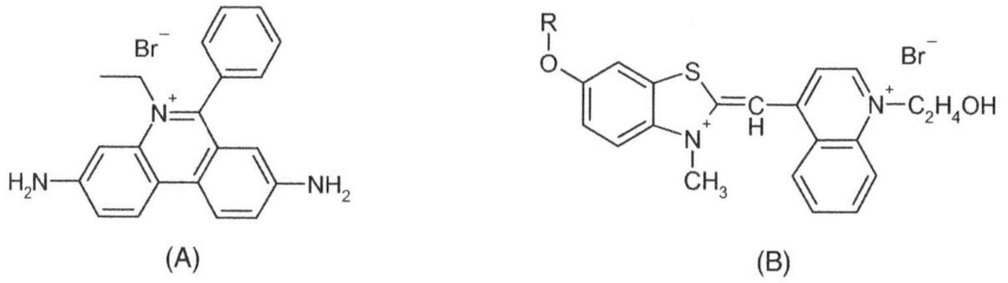

Figure 19.14 Fluorescent dyes for DNA labeling by intercalation. (A) Ethidium bromide; (B) thiazole orange dyes. R is a methyl group in thiazole orange methoxyhydroxyethyl (TOMEHE).

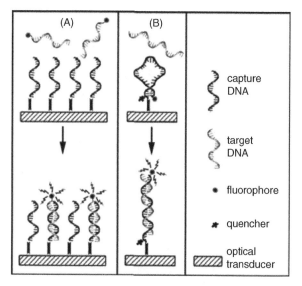

Figure 19.15 Typical sensing schemes in optical DNA sensors. (A) Conventional DNA sensor using a fluorescently labeled complementary strand. (B) molecular beacon sensor. Reprinted with permission from [26]. Copyright 2008 American Chemical Society.

Target labeling can be avoided by resorting to an intercalating fluorescent dye that binds to the double strand formed upon hybridization.

The molecular beacon approach is illustrated in Figure 19.15B. The molecular beacon capture probe is designed so as to include a fluorophore and a fluorescence quencher in close proximity. Probe–target hybridization shifts the fluorophore far from the quencher and results in fluorescence enhancement. This approach precludes the labeling of the target analyte.

Another method for avoiding the labeling of the target is based on the competitive hybridization approach. In this method, a labeled analyte analog is initially hybridized to the capture probe so as to achieve saturation. Next, the saturated receptor layer is allowed to interact with the sample. In this stage of the process, the sample analyte displaces a fraction of the labeled analog, leading to a decrease in the fluorescence.

Alternatively, transduction in optical nucleic acid sensors can be carried out in the sandwich format. In this approach, the probe is designed so that after hybridization with the target, an overhanging target fragment remains in the single-strand form. This fragment is then used to bind by hybridization a signaling probe tagged with a fluorescent label.

19.6.2 Fiber Optic Nucleic Acid Sensors

As shown in Figure 19.16, an intrinsic fiber optic DNA sensor has been obtained by immobilization of the capture probe onto a piece of decladded 400-μm diameter quartz fiber [47]. To this end, the fiber surface is first activated with a long-chain aliphatic spacer terminated in 5′-O-dimethoxytrityl-2′-deoxyribonucleoside. Starting with this molecular unit, solid-phase DNA synthesis has been performed at the fiber surface in order to obtain the capture probe.

The transduction setup (Figure 19.16) includes the excitation source (an Ar laser) and a photomultiplier light detector. Discrimination of excitation and emission beams is achieved by a dichroic mirror.

Hybridization is conducted in a small plastic cell that contains the sample diluted with the hybridization buffer. After hybridization the cell is flushed with the buffer and the staining of the resulting duplex is achieved by incubation with the ethidium bromide intercalating dye. Finally, unreacted dye is flushed away by a stream of fresh buffer and the fluorescence signal is measured. In this step, the intercalated dye is excited by the evanescent wave. Emitted light is coupled back to the fiber and conveyed through the objective to the photomultiplier. The DNA-bound dye has an emission maximum at 595 nm that is well beyond the cut-off wavelength of the dichroic mirror (495 nm).

After each hybridization assay, the sensor should be regenerated by flushing with hot (85 °C) hybridization buffer, which promotes duplex denaturation.

The background-corrected fluorescence intensity serves as the analytical signal. The response is linear with respect to the target concentration up to 800 ng/mL and the reported limit of detection is 90 ng/mL.

The robustness of the quartz fiber and the nucleic acid probe, as well as the strength of the covalent probe binding, impart this sensor with excellent stability for long-term storage and for stringent cleaning conditions.

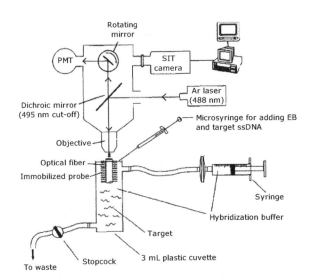

Figure 19.16 Setup of an optical fiber DNA sensor. The sensing component is a piece of optical fiber onto which the capture probe is immobilized. The upper part of the figure shows a fluorescence microscope that includes the excitation source (an Ar laser), the photomultiplier light detector (PMT) and a dichroic mirror that resolves excitation and fluorescence beams. Sample solutions, staining dye (ethidium bromide-EB) and hybridization buffer are delivered by syringes to the hybridization cell. Adapted with permission from [47]. Copyright 1994 Elsevier.

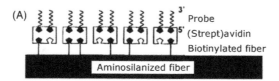

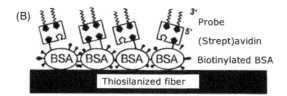

Figure 19.17 Possible structures of the sensing part of an intrinsic fiber optic DNA sensor. (A) The biotinylated probe is immobilized through avidin bridges to a biotin layer covalently attached to an aminosilanized quartz fiber. (B) Biotinylated bovine serum albumin adsorbed on a thiosilanized fiber is used to attach the probe by biotin–avidin bridges. Reprinted with permission from [48]. Copyright 1996 American Chemical Society.

The next example refers to nucleic acid optical sensors produced by immobilization of preformed capture probes on the decladded surface of a quartz optical fiber in order to obtain intrinsic DNA sensors. Two immobilization methods are presented in Figure 19.17. In both cases, the first step consists of surface modification by silanization with reagents containing either amine (A) or thiol (B) terminal groups. In case (A), biotin was then covalently linked to the amine function and the biotinylated probe was attached finally to the previous layer by means of avidin bridges. In case (B), biotinylated bovine serum albumin (BSA) was adsorbed on the primary thiol layer and served as a linker for attaching the biotinylated probe by means of avidin. In order to detect the hybridization event, the target was labeled with fluorescein. This sensor has been included in a flow-through cell that allows the assay sequence to be run under automatic control.

The fluorescent label is excited by the evanescent wave and emitted light couples back to the optical fiber and is detected in order to assess the response signal. The evolution of the sensor signal during an analysis cycle is shown in Figure 19.18 for both the target (curve 1) and a noncomplementary oligonucleotide (curve 2). Note that the last one gives rise to a negligible signal, which proves the absence of nonspecific interactions.

This sensor was used in two assay formats. The first format makes use of direct detection of the labeled target and produces the calibration curve shown in Figure 19.19A. In this case, the signal increases with the target concentration up to saturation at about 10 nM. The second version is based on a competitive assay. This consists of incubation with both the nonlabeled target and a tracer composed of a similar oligonucleotide tagged with a fluorophore. As is typical for competition assays, the target and the tracer compete for the same binding sites and, at a constant tracer concentration, the response decreases with increasing target concentration (Figure 19.19B). The detection limit is better in the first format, but the second format has the advantage of precluding the labeling of the target itself. Detection limits in Figure 19.19 have been achieved for 32.9 °C hybridization temperature and 60-min hybridization time.

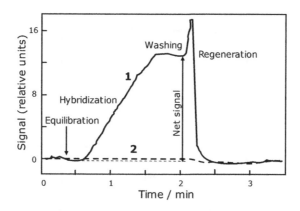

Figure 19.18 Time evolution of the response signal during a DNA assay using the sensor in Figure 19.17(A). Curve 1: target; curve 2: noncomplementary oligonucleotide. Equilibration and washing was done using the hybridization buffer (pH 7). Chemical regeneration was performed by means of a 50% (w/w) urea solution at the hybridization temperature (26.7 °C). Target: (fluorescein-5′-GTTGTGTGGAATTGTG-3′); noncomplementary oligonucleotide: fluorescein-5′- CTGCAACACCTGACAAACCT -3′), each at 10 nM concentration. Adapted with permission from [48]. Copyright 1996 American Chemical Society.

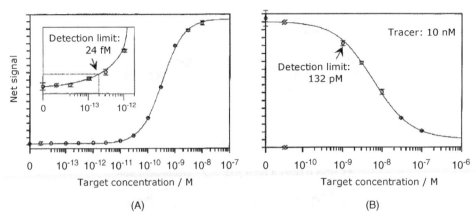

Figure 19.19 Response curves for direct (A) and competitive hybridization assays (B) by the sensor in Figure 19.17B. The error bar corresponds to the standard deviation of three measurements. Reprinted with permission from [48]. Copyright 1996 American Chemical Society.

A critical issue is the sensor stability during a long sequence of assays. At a constant target concentration, the successive response values decay slowly according to an exponential law. In order to account for this drift, the corrected signal value (y_{corr}) should be calculated according to the following equation, where y_m is the measured signal, t is the utilization time and k_d is the decay constant that should be determined prior the analysis:

$$y_{corr} = y_m/e^{k_d t} \qquad (19.11)$$

19.6.3 Fiber Optic Nucleic Acid Arrays

Nucleic acid arrays can be formed at the distal end of optical fiber bundles by the methods introduced in Section 19.6.2.

More control over the array design is achieved if the capture probe is immobilized on microspheres with diameters equal to the fiber core diameter [49,50]. In order to couple microspheres to individual fibers, fiber ends are chemically etched to form microwells. Owing to the size complementarity, each microwell accommodates a single microsphere modified with a particular capture probe (Figure 19.20). As the microspheres are randomly distributed at the bundle surface, an optical decoding is required in order to identify the microsphere coupled to each individual fiber. Identification is achieved by encoding each sensing element with multiple fluorescent dyes prior to the array fabrication. The identity of each microsphere is ascertained from the wavelength and intensity of the fluorescence produced by the encoding dyes. Clearly, encoding dyes must have emission spectra distinct from those of the fluorophores used for sensing. This technique allows for coupling of about 100 different microspheres to a single bundle.

Fiber optic DNA arrays allow for reliable identification of a pathogen microorganism by simultaneous detection of a series of characteristic DNA fragments. Moreover, this design is suitable for manufacturing of multipathogen detection arrays. The high number of individually addressable sensing elements makes this approach appropriate for genomics applications.

Figure 19.20 Atomic force micrograph of etched optical fiber bundle with 3-μm wells accommodating microspheres derivatized with recognition receptors. Reprinted with permission from [50]. Copyright 2003 The Royal Society of Chemistry.

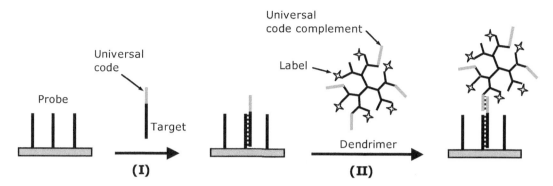

Figure 19.21 DNA detection using fluorophore labeled dendrimers.

19.6.4 Optical DNA Microarrays

Optical detection by fluorescence is widely used in standard DNA microarrays that consists of a collections of microscopic DNA-probe spots attached to a solid surface [44,51]. Each spot identifies by hybridization a particular target in a sample. Signal reading is carried out by means of an optical scanner that includes both laser excitation sources and very sensitive light detectors. The detection limit thus achieved is close to 1 fluorophore/μm^2. However, owing to the small amount of available material, it is often necessary to perform prior DNA amplification by means of the polymerase chain reaction.

In order to avoid DNA amplification, various methods for enhancing the fluorescence signal have been advanced [52,53]. For example, Figure 19.21 presents a sandwich format labeling method based on a dendrimer-tagged signaling oligonucleotide. Each dendrimer tag includes about 250 fluorescent molecules. In order to bind the dendrimer label to the hybrid, cDNA targets are coded with a single universal sequence tag that is complementary to the dendrimer-tagged oligonucleotide. After hybridization (step I), the hybrid is stained by coupling the overhanging universal code sequence with the complementary oligonucleotide attached to the dendrimer (step II). This procedure results in about a 16-fold improvement in the detection limit that allows the assay to be performed with much less starting RNA material in gene expression analysis.

Signal amplification in fluorescence-based DNA microarrays can also be performed by means of nanomaterial tags such as semiconductor nanoparticles (quantum dots) or metal nanoparticles (see Chapter 20 for details).

Surface-enhanced Raman spectrometry is another optical method for ultrasensitive hybridization monitoring in DNA microarrays [53]. Application of Raman spectrometry in DNA microarrays can be approached in two ways. In one approach, the capture probe is immobilized at a nanoscale roughened silver layer. This layer induces an enhancement of the Raman signal produced by a label such as a Rhodamine dye [54]. A second approach is based on metal nanoparticle tags modified with a Raman label dye. The metal nanoparticle is attached to the hybrid in the sandwich format and induces a strong enhancement of the Raman signal produced by the dye label. Compared with the previous silver-layer-based method, the metal nanoparticle approach is characterized by a much lower scatter background because Raman scattering is produced only by nanoparticle tags. More details on metal nanoparticle applications in Raman spectrometry transduction are available in Chapter 20.

Label-free detection of the hybridization on microarrays can be achieved by surface plasmon resonance (SPR) [53,55]. Detection limits achieved by standard SPR instruments are in the nanomolar range, which requires DNA amplification prior to the analysis. However, a limit of 54 nM has been attained with a high-resolution SPR instrument that discriminates an angle shift of 10^{-4}.

19.6.5 Outlook

The design of optical nucleic acid sensors can be based on either label- or label-free transduction. Whilst label-based sensors are characterized by a simple design and facile multiplexing, the label-free approach involves a more intricate fabrication technology but presents a simpler assay protocol.

Luminescent labels can be introduced in the target molecule during the polymerase chain reaction. Alternatively, an intercalating luminescent label can be attached to the double helix formed upon target-probe hybridization. The sandwich format provides an additional method for labeling the probe-target hybrid. Another transduction procedure relies on the application of labeled capture probes in the molecular beacon configuration.

Label-based optical detection is widely used in standard DNA microarrays. In order to avoid the need to use DNA amplification using the polymerase chain reaction, one can resort to nanoparticles tagged with multiple signaling luminophore molecules. Label-free detection in DNA microarrays can be achieved by surface plasmon resonance spectrometry and other methods introduced in Section 18.6.

Questions and Exercises (Section 19.6)

1 Review the methods for fluorescent labeling of single-strand and double-strand nucleic acids. Comment on the particular structural features of intercalating fluorescent labels.

2 Give an overview of recognition–transduction strategies in optical DNA sensors.

3 Draw a summary flow chart for the fabrication of an optical sensor for nucleic acids. Comment on the details of the assay sequence.

4 Give a short overview of methods for the immobilization of a preformed capture probe at the surface of a decladded optical fiber.

5 Comment on the assay procedure, reproducibility, limit of detection, and the extent of the quasilinear response range of the DNA sensor shown in Figure 19.17.

6 Give an overview of the possible methods for the fabrication of fiber optic DNA arrays.

7 Draw a summary flow chart for the fabrication and operation of a DNA microarray based on labeling with multiple fluorophore markers in the sandwich format.

8 Draw schematically a possible configuration of a surface plasmon resonance-based DNA array.

References

1. Baldini, F., Bechi, P., Bracci, S. *et al.* (1995) *In-vivo* optical-fiber pH sensor for gastroesophageal measurements. *Sens. Actuators B-Chem.*, **29**, 164–168.

2. Lin, J. (2000) Recent development and applications of optical and fiber-optic pH sensors. *TrAC Trends Anal. Chem.*, **19**, 541–552.

3. Mills, A. (2009) Optical sensors for carbon dioxide and their applications, in *Sensors for Environment, Health and Security: Advanced Materials and Technologies* (ed. M.I. Baraton), Springer, Dordrecht, pp. 347–370.

4. Rhines, T.D., and Arnold, M.A. (1988) Simplex optimization of a fiber-optic ammonia sensor based on multiple indicators. *Anal. Chem.*, **60**, 76–81.

5. Borisov, S.M., Waldhier, M.C., Klimant, I. *et al.* (2007) Optical carbon dioxide sensors based on silicone-encapsulated room-temperature ionic liquids. *Chem. Mater.*, **19**, 6187–6194.

6. Burke, C.S., Markey, A., Nooney, R.I. *et al.* (2006) Development of an optical sensor probe for the detection of dissolved carbon dioxide. *Sens. Actuators B-Chem.*, **119**, 288–294.

7. Oehme, I., and Wolfbeis, O.S. (1997) Optical sensors for determination of heavy metal ions. *Mikrochim. Acta*, **126**, 177–192.

8. Langry, K. and Rambabu, B. (1999) Ionic optodes: role in fiber optic chemical sensor technology. *J. Chem. Technol. Biotechnol.*, **74**, 717–732.

9. Bakker, E.B.E., Crespo, G., Grygolowicz-Pawlak, E. *et al.* (2011) Advancing membrane electrodes and optical ion sensors. *Chimia*, **65**, 141–149.

10. Fabbrizzi, L. and Poggi, A. (1995) Sensors and Switches from Supramolecular Chemistry. *Chem. Soc. Rev.*, **24**, 197–202.

11. Valeur, B. and Leray, I. (2000) Design principles of fluorescent molecular sensors for cation recognition. *Coord. Chem. Rev.*, **205**, 3–40.

12. Callan, J.F., de Silva, A.P., and Magri, D.C. (2005) Luminescent sensors and switches in the early 21st century. *Tetrahedron*, **61**, 8551–8588.

13. Fabbrizzi, L., Licchelli, M., Rabaioli, G. *et al.* (2000) The design of luminescent sensors for anions and ionisable analytes. *Coord. Chem. Rev.*, **205**, 85–108.

14. Johnson, R.D. and Bachas, L.G. (2003) Ionophore-based ion-selective potentiometric and optical sensors. *Anal. Bioanal. Chem.*, **376**, 328–341.

15. Bakker, E., Buhlmann, P., and Pretsch, E. (1997) Carrier-based ion-selective electrodes and bulk optodes. 1. General characteristics. *Chem. Rev.*, **97**, 3083–3132.

16. Buhlmann, P., Bakker, E. and Pretsch, E. (1998) Carrier-based ion-selective electrodes and bulk optodes. 2. Ionophores for potentiometric and optical sensors. *Chem. Rev.*, **98**, 1593–1687.

17. Kawabata, Y., Tahara, R., Kamichika, T. *et al.* (1990) Fiber optic potassium ion sensor using alkyl-acridine orange in plasticized poly(vinyl chloride) membrane. *Anal. Chem.*, **62**, 1528–1531.

18. Mills, A. (1997) Optical oxygen sensors. *Platinum Metals Rev.*, **41**, 115–127.

19. Wang, X.-D., Chen, H.-X., Zhao, Y. *et al.* (2010) Optical oxygen sensors move towards colorimetric determination. *Trac-Trends Anal. Chem.*, **29**, 319–338.

20. Mills, A. (2009) Oxygen indicators in food packaging, in *Sensors for Environment, Health and Security: Advanced Materials and Technologies* (ed. M.I. Baraton), Springer, Dordrecht, pp. 371–388.

21. Grist, S.M., Chrostowski, L., and Cheung, K.C. (2010) Optical oxygen sensors for applications in microfluidic cell culture. *Sensors*, **10**, 9286–9316.

22. Amao, Y. (2003) Probes and polymers for optical sensing of oxygen. *Microchim. Acta*, **143**, 1–12.

23. Nagl, S., Baleizao, C., Borisov, S.M. *et al.* (2007) Optical sensing and imaging of trace oxygen with record response. *Angew. Chem. Int. Ed.*, **46**, 2317–2319.

24. Cao, W. and Duan, Y. (2006) Optical fiber evanescent wave sensor for oxygen deficiency detection. *Sens. Actuators B-Chem.*, **119**, 363–369.
25. Zhu, H., Tian, Y., Bhushan, S. *et al.* (2010) High throughput micropatterning of optical oxygen sensor. IEEE Sensors 2010 Conference, pp. 2053–2056.
26. Borisov, S.M. and Wolfbeis, O.S. (2008) Optical biosensors. *Chem. Rev.*, **108**, 423–461.
27. Endo, H., Yonemori, Y., Musiya, K. *et al.* (2006) A needle-type optical enzyme sensor system for determining glucose levels in fish blood. *Anal. Chim. Acta*, **573–574**, 117–124.
28. Marquette, C.A., Degiuli, A., and Blum, L.J. (2003) Electrochemiluminescent biosensors array for the concomitant detection of choline, glucose, glutamate, lactate, lysine and urate. *Biosens. Bioelectron.*, **19**, 433–439.
29. Xie, X.F., Suleiman, A.A., and Guilbault, G.G. (1990) A urea fiber optic biosensor based on absorption measurement. *Anal. Lett.*, **23**, 2143–2153.
30. Wolfbeis, O.S. and Li, H. (1993) Fluorescence optical urea biosensor with an ammonium optrode as transducer. *Biosens. Bioelectron.*, **8**, 161–166.
31. Sansubrino, A. and Mascini, M. (1994) Development of an optical fiber sensor for ammonia, urea, urease and IgG. *Biosens. Bioelectron.*, **9**, 207–216.
32. Schelp, C., Scheper, T., Buckmann, F. *et al.* (1991) Two fiberoptic sensors with confined enzymes and coenzymes - Development and application. *Anal. Chim. Acta*, **255**, 223–229.
33. Choi, M.M.F. (2004) Progress in enzyme-based biosensors using optical transducers. *Microchim. Acta*, **148**, 107–132.
34. Pickup, J.C., Hussain, F., Evans, N.D. *et al.* (2005) Fluorescence-based glucose sensors. *Biosens. Bioelectron.*, **20**, 2555–2565.
35. Thompson, V.S. and Maragos, C.M. (1996) Fiber-optic immunosensor for the detection of fumonisin B$_1$. *J. Agric. Food Chem.*, **44**, 1041–1046.
36. Sassolas, A., Blum, L.J., and Leca-Bouvier, B.D. (2011) Optical detection systems using immobilized aptamers. *Biosens. Bioelectron.*, **26**, 3725–3736.
37. Schultz, J.S. (1987) Sensitivity and dynamics of bioreceptor-based biosensors. *Ann. N. Y. Acad. Sci.*, **506**, 406–414.
38. Ballerstadt, R., Polak, A., Beuhler, A. *et al.* (2004) *In vitro* long-term performance study of a near-infrared fluorescence affinity sensor for glucose monitoring. *Biosen. Bioelectron.*, **19**, 905–914.
39. Ballerstadt, R. and Schultz, J.S. (2000) A fluorescence affinity hollow fiber sensor for continuous transdermal glucose monitoring. *Anal. Chem.*, **72**, 4185–4192.
40. Chinnayelka, S. and McShane, M.J. (2005) Microcapsule biosensors using competitive binding resonance energy transfer assays based on apoenzymes. *Anal. Chem.*, **77**, 5501–5511.
41. Ge, X.D., Tolosa, L., Simpson, J. *et al.* (2003) Genetically engineered binding proteins as biosensors for fermentation and cell culture. *Biotechnol. Bioeng.*, **84**, 723–731.
42. Kuswandi, B., Tombelli, S., Marazza, G. *et al.* (2005) Recent advances in optical DNA biosensors technology. *Chimia*, **59**, 236–242.
43. Yang, M.S. McGovern, M.E., and Thompson, M. (1997) Genosensor technology and the detection of interfacial nucleic acid chemistry. *Anal. Chim. Acta*, **346**, 259–275.
44. Sassolas, A., Leca-Bouvier, B.D., and Blum, L.J. (2008) DNA biosensors and microarrays. *Chem. Rev.*, **108**, 109–139.
45. Müller, H.J. and Röder, T. (2006) *Microarrays*, Elsevier, San Diego.
46. Baldi, P. and Hatfield, G.W. (2002) *DNA Microarrays and gene Expression: From Experiments to Data Analysis and Modeling*, Cambridge University Press, Cambridge.
47. Piunno, P.A.E., Krull, U.J., Hudson, R.H.E. *et al.* (1994) Fiber optic biosensor for fluorometric detection of DNA Hybridization. *Anal. Chim. Acta*, **288**, 205–214.
48. Abel, A.P., Weller, M.G., Duveneck, G.L. *et al.* (1996) Fiber-optic evanescent wave biosensor for the detection of oligonucleotides. *Anal. Chem.*, **68**, 2905–2912.
49. Epstein, J.R., Leung, A.P.K., Lee, K.H. *et al.* (2003) High-density, microsphere-based fiber optic DNA microarrays. *Biosens. Bioelectron.*, **18**, 541–546.
50. Epstein, J.R. and Walt, D.R. (2003) Fluorescence-based fibre optic arrays: a universal platform for sensing. *Chem. Soc. Rev.*, **32**, 203–214.
51. Teles, F.R.R. and Fonseca, L.R. (2008) Trends in DNA biosensors. *Talanta*, **77**, 606–623.
52. Willner, I., Shlyahovsky, B., Willner, B. *et al.* (2009) Amplified DNA biosensors, in *Functional Nucleic Acids for Analytical Applications* (eds Y. Li and Y. Lu), Springer New York, New York, pp. 199–252.
53. Storhoff, J.J., Marla, S.S., Garimella, V. *et al.* (2005) Labels and detection methods, in *Microarray Technology and Its Applications* (eds R. Müller and V. Nicolau), Springer, Berlin, pp. 147–179.
54. Allain, L.R. and Vo-Dinh, T. (2002) Surface-enhanced Raman scattering detection of the breast cancer susceptibility gene BRCA1 using a silver-coated microarray platform. *Anal. Chim. Acta*, **469**, 149–154.
55. Scarano, S., Mascini, M., Turner, A.P.F. *et al.* (2010) Surface plasmon resonance imaging for affinity-based biosensors. *Biosens. Bioelectron.*, **25**, 957–966.

20

Nanomaterial Applications in Optical Transduction

A new and impressive impetus in optical sensing technology has been brought about by the introduction of nano-materials with outstanding photophysical properties such as semiconductor nanocrystals, carbon nanotubes and metal nanoparticles. In many respects, such materials are superior to the traditional molecular luminophores and are expected to replace them in many applications. Although applications reported so far refer mostly to molecular probes, such results are also useful for possible further integration with optical waveguides in order to develop true chemical sensors.

20.1 Semiconductor Nanocrystals (Quantum Dots)

20.1.1 Quantum Dots: Structure and Properties

Quantum dots (QDs, see Section 8.6) are semiconductor nanocrystals typically between 2 and 10 nm diameter (that is, 10–50 atoms in diameter). From the chemical standpoint, QDs are compounds of the elements from the periodic groups II-VI [1], III-V [2] or Ib-VI [3]. Common QD materials are of the MX type, where M is Cd, Zn or Pb and X is S, Se or Te. QDs have as a rule a roughly spherical shape but QDs with other shapes (cubes, spheres, pyramids, etc.) have also been synthesized. QDs feature outstanding photophysical properties [4] that brought about a revolution in bioimaging and optical sensing methods [5–8].

Electronic and optical properties of QDs derive from their semiconducting nature [9]. In order to understand these properties it is useful to keep in mind that under normal conditions all valence electrons in a semiconductor are localized in covalent bonds and the valence electrons levels are grouped into the *valence band* (Figure 20.1, see also Section 11.1.2). If such a material is excited by electromagnetic radiation, valence electrons can be promoted to a higher energy level located in the *conduction band*. There is an energy interval E_G between the valence and conduction bands (*bandgap*) where no electron levels exists. Excitation can therefore only occur if the incident photon energy overcomes the bandgap. The excited electron leaves behind a positive vacancy (*hole*) at its initial position and a Coulomb attraction force between the hole and the leaving electron develops. Due to the confined available space, this pair remains in a bound state forming a virtual particle called *exciton*. From the quantum-mechanical standpoint, this state is described by a wave function similar to that of the hydrogen atom and displays a set of discrete energy levels. This is not surprising because both the hydrogen atom and the exciton consist of an electron interacting with a positive charge (hence the name of artificial atoms ascribed to QDs by some authors). However, the size of the exciton (1–10 nm) is much larger that that of a simple atom due to charge screening by lattice atoms. As shown in Figure 20.1, the excited state relaxes first by partial conversion of exciton energy into lattice vibration energy until the electron reaches the edge of the bandgap. As no energy levels are available within the bandgap the relaxation terminates by the electron jumping back to the valence band accompanied by electron–hole recombination. This results in the emission of a photon whose energy equals the bandgap energy difference, this process being analogous to the fluorescence of molecular compounds.

As the emitted photon energy depends on the bandgap, it is important to realize that this factor depends first of all on the chemical composition. In addition, the particle size plays a crucial role because this parameter determines the electron–hole distance, which is termed as the *Bohr radius*. The size effect is illustrated in Figure 20.2 that shows the absorption and fluorescence spectra of CdSe nanoparticles of various sizes. Figure 20.2A demonstrates that for a given size, light absorption occurs over a broad wavelength region over a threshold dictated by the particle size. Accordingly, the absorption threshold shifts to higher wavelengths (that is, lower photon energies) with increasing particle size.

Light emission itself is restricted to a narrow band with the peak wavelength depending on the bandgap (Figure 20.2B). Consequently, the emission band shifts to higher wavelengths as the particle size increases. This means that the fluorescence wavelength can be tuned by means of the particle size so that various fluorescent labels can be obtained using the same material. Note that for each particle size the maximum fluorescence intensity occurs in the region of the absorption threshold. This contrasts with the behavior of molecular luminophores that display as a rule a distinct interval between the adsorption and emission bands (Stokes shift).

The above-mentioned features of QDs fluorescence have a considerable impact on QDs applications as fluorescence labels. Thus, the sharpness of the emission bands allows facile discrimination of various QD fluorophores in

Chemical Sensors and Biosensors: Fundamentals and Applications, First Edition. Florinel-Gabriel Bănică.
© 2012 John Wiley & Sons, Ltd. Published 2012 by John Wiley & Sons, Ltd.

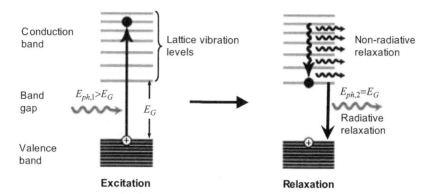

Figure 20.1 Mechanism of light emission by quantum-dot fluorescence. Electrons excited to energies greater than the band edge efficiently relax first to the band edge through the release of phonons, that is, quanta of lattice vibration. Finally, electrons relaxes to the valence band with light emission. $E_{ph,1}$ and $E_{ph,2}$ indicate the energy of the adsorbed and emitted photon, respectively; E_G is the bandgap energy difference. Adapted with permission from [8]. Copyright 2010 American Chemical Society.

imaging applications. On the other hand, due to the broadness of the absorption band, a set of different QD fluorophores can be excited by a single monochromatic beam provided its wavelength is below the cutoff limit of the smallest particle.

Such properties impart QDs superior characteristics as florescent labels when compared with molecular luminophores. Molecular luminophores excited within limited wavelength regions and display broad emission bands due to the interplay of electronic, vibrational and rotational energy levels. When using molecular luminophores, a specific excitation beam must be available for each label, which restricts the number of detectable labeled analytes. That is why multiplex assays are hard to implement with conventional luminophores but easily achievable using QD labels.

In addition, in contrast to molecular luminophores, QDs have a very high quantum yield (20–70%), which allows achieving much better sensitivity. Furthermore, QDs are very stable under the conditions of a prolonged irradiation and allow intensive excitation energy in contrast to organic luminophores that experience photochemical decay

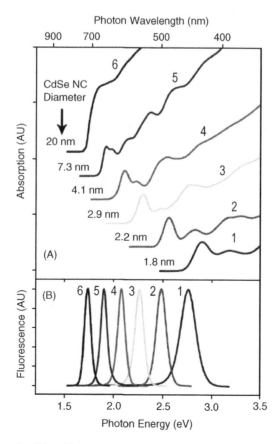

Figure 20.2 Size effect on the absorption (A) and fluorescence (B) spectra of CdSe semiconductor nanoparticles. Nanoparticle diameter increases from curve 1 to 6. Note that depending on size, semiconductor nanoparticles can fluoresce at various wavelengths within the visible and near-infrared regions. Adapted with permission from [8]. Copyright 2010 American Chemical Society.

(photobleaching). Moreover, QD fluorescence is not dependent on pH as occurs with organic dyes. However, QDs can be pH sensitive if the capping ligand contains an ionizable acid functionality. In such a case, the electric charge resulting from ligand ionization can affect the bandgap and the quantum yield.

Another advantage of QDs arises from the relatively long decay time of the excited state (up to several tens of nanoseconds). This characteristic facilitates implementation of time-resolved measurements that is characterized by an advanced signal–background discrimination.

All of these characteristics render QDs superior to molecular fluorophores in optical sensing and imaging applications.

The quantum yield of a QD is lowered by nonradiative recombination that largely occurs at the particle surface and is therefore greatly influenced by the surface state. Capping a core quantum dot with several atomic shells reduces nonradiative recombination and results in brighter emission, provided that the shell is a semiconductor with a wider bandgap than the core material [10]. Thus, (core)shell quantum dots such as (CdSe)ZnS and (CdSe)CdS display efficient and stable fluorescence. Capping with silica, a material with a very broad bandgap, is another alternative for improving the quantum yield.

In conclusion, QDs are readily available materials with advanced photophysical properties that allow attaining superior performances in imaging and optical sensing. Some applications of QDs in this area are introduced in the next section.

20.1.2 Applications of Quantum Dots in Chemical Sensing

Despite the fact that the advent of the QDs in the field of biosensor technology is relatively recent, a large number of applications, such as optical labels in bioimaging and sensing, have already been reported [11–15]. QDs can be used not only as passive optical labels but can also be integrated in more elaborate transduction schemes based on resonance energy transfer, fluorescence quenching or chemiluminescence.

20.1.2.1 Determination of Ions and Small Molecules

The luminescence of QDs can be very sensitive to the chemical state of the capping layer and this makes such nanoparticles amenable to sensing application by chemical tailoring of the surface properties in order to impart selectivity.

Chemical sensing systems for ions and small molecules can be developed by making use of the fluorescence changes induced by adsorption or interactions with an ion complexing capping ligand. Depending on the properties of the analyte, various mechanisms can define the QD–analyte interaction [16]. For example, fluorescence enhancement occurs when ions such as Zn^{2+} or Cd^{2+} interact with the QD surface and passivates surface trap states. A similar effect has been observed when Zn^{2+} interacts with cysteine-modified QDs. In this case, several QDs aggregate around the metal ion through cysteine–Zn^{2+} coordination, thereby activating the surface states that are responsible for enhanced emission.

In some instances, the analyte–QD interaction quenches the fluorescence. This effect can be due to electron transfer from the analyte (e.g., Cu^{2+}) to the capping ligand (for example, thioglycerol) to give Cu^+. This ion quenches the fluorescence by facilitating nonradiative recombination of electrons and holes in the excited QD. A different quenching mechanism occurs when Fe^{3+} combines with cysteine-capped QDs. The resulting Fe^{3+} complex partially absorbs QD emitted light and functions therefore as an inner filter. Quenching has also been noted when Ag^+ adsorbs on the CdS QD surface and displaces Cd^{2+}, yielding Ag_2S that facilitates electron–hole recombination. Quenching by electron transfer from the analyte to the excited QD valence band has been noted in the interaction with certain small molecules such as dopamine.

It should be borne in mind that any detection scheme based on metal-ion complexation is strongly influenced by the pH that modulates by protonation-deprotonation the binding capacity of the ligand.

Optical pH determination has been accomplished with QDs capped with thiocarboxilic acids, such as mercaptopropionic or mercaptosuccinic acid that modulate the QD fluorescence by proton dissociation.

A more advanced approach to the determination of ions and small molecules is based on supramolecular recognition using QDs functionalized with specific receptors, such as macrocyclic ligands, cyclodextrin, calixarenes or porphyrin [17].

20.1.2.2 Fluorescent Labels in Biosensors

In the previous examples, the QD capping layer is directly involved in the analyte-recognition process. QDs can also be used as passive optical labels in immunoassay and nucleic acid-based sensing with substantial advantages as compared with conventional dyes fluorophores.

For example, an application in optical immunoassay has been developed using antibodies attached to CdSe-ZnS core-shell QDs in order to form the signaling conjugate in a sandwich assay on a microtiter plate. Several antibodies

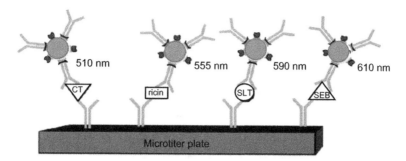

Figure 20.3 Parallel optical immunodetection of pathogen toxins (1 μg mL^{-1} each) in a sandwich assay using QDs of different sizes as labels. Emission wavelength (in nm) is shown for each QD. Antigens: cholera toxin (CT), ricin, shiga-like toxin 1 (SLT), staphylococcal enterotoxin B (SEB). Adapted with permission from [13]. Copyright 2008 Wiley-VCH Verlag GmBH & Co. KGaA.

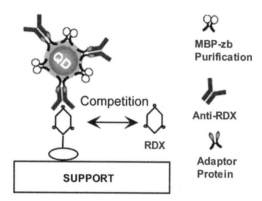

Figure 20.4 Competition/displacement assay for immunodetection of the RDX explosive. MBP-zb (maltose-binding protein with an electrostatic leucine-zipper attachment domain) was attached in the process of purification of antibody-modified QDs. Adapted with permission from [18]. Copyright 2004 Elsevier and the Association for Laboratory Automation.

can be detected simultaneously using QDs with a specific size for labeling each analyte (Figure 20.3). Although some overlapping of emission bands occurs, individual bands can be resolved by spectrum deconvolution. This example illustrates the possibility of performing multiplex assay using different size QDs as labels and a single excitation source.

A competitive immunodetection scheme for a hapten (the RDX explosive) is presented in Figure 20.4. In this example, the biotinylated anti-RDX receptor is attached to the surface of the QD through an adapter protein (MBP-zb) and this assembly is immobilized by affinity binding of the antibody to the RDX modified support. By competition, sample RDX displaces the support-attached QD conjugates and thus causes the fluorescence intensity to decrease in accordance with analyte concentration. RDX determination can be thus achieved in a concentration interval extending from 0.5 to 100 ng/ml.

Signal amplification and multiplexed analysis have been accomplished by QDs attached to or incorporated in polymer beads, with each bead acting as an enhanced luminescence label [13].

20.1.2.3 *Resonance Energy Transfer*

As is well known, an essential condition for resonance energy transfer (RET) to occur is the matching of the emission band of the energy donor with the absorption band of the energy receptor. This can be easily accomplished with QDs as donors because the emission wavelength of such a species is size-tunable. It is therefore not surprising that QDs have been used in various RET-based sensing schemes.

RET proved useful for developing pH-sensitive QD–dye conjugates [19]. Thus, squaraine (a pH-sensitive dye) has been covalently attached to the polymer encapsulating layer of QDs. Energy transfer occurs from the excited QD to the dyes whose acceptor efficiency increases with increasing pH, while the QD behavior is independent of this parameter. Consequently, dye emission increases with increasing pH, whereas QD emission undergoes an opposite variation owing to increasing RET efficiency. Solution pH can be inferred from the ratio of the QD emission to the dye emission with all the advantages of the ratiometric approach.

QD RET has been advantageously exploited for devising affinity sensors. A substantial improvement in the energy-transfer efficiency has been achieved by attaching several receptor dyes to the same QD by affinity-based assembly [18]. In this configuration, a single donor can interact with several acceptors placed at equal distance from

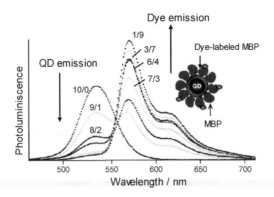

Figure 20.5 Enhancement of RET efficiency in immunosensors using multiple acceptors (here the Cy3 dye) in a QD-protein conjugate. Both dye-labeled and crude maltose-binding protein molecules (MBP) are attached by affinity binding to the QD. Total number of MBP is constant. Spectra pertains to various MBP/MBP-dye molar ratios as indicated on each curve. Adapted with permission from [18]. Copyright 2004 Elsevier.

the conjugate center, as shown in Figure 20.5. This figure demonstrates that fluorescence intensity decreases with increasing the number of quenching dye molecule conjugated with a QD. This change is accompanied by an increase in dye fluorescence intensity. As demonstrated by this example, a QD can act not only as a fluorophore but also as a scaffold for assembling other components of the transduction system.

20.1.2.4 Fluorescence Quenching

A molecular quencher can capture the excess energy of an excited QD causing the fluorescence intensity to lessen. Size tunability of the QD fluorescence wavelength allows facile fulfilling of the above-mentioned energy-transfer condition. Several applications of this effect are further envisaged.

Aptamer-capped QDs have been used for developing a detection scheme for a protein-analyte that also functions as a fluorescence quencher [20]. To this end, near-infrared PbS QDs have been grown in an aqueous solution in the presence of 15-mer thrombin-specific aptamer that functions as both QD stabilizer and analyte receptor (Figure 20.6). Thrombin combines selectively with the aptamer via either the heparin or the fibrinogen binding sites and in the resulting product it performs energy exchange with the excited QD. Consequently, photoluminescence quenching occurs in proportion to the thrombin concentration with a detection limit of about 1 nM. Other proteins can adsorb nonspecifically to the QD but do not act as quenchers. Interference has been noted only with prostate-specific antigen that also has a heparin-binding exosite.

Quenching QD fluorescence is a convenient approach for designing DNA sensors based on the molecular-beacon configuration. In such an approach, CdS/ZnS were linked to the 5′ end of a hairpin capture probe that incorporates at its 3′ end the quencher molecule 4-(4′-dimethylaminophenylazo)benzoic acid (Dabcyl) [21]. In this state, QD fluorescence is quenched by the nearby quencher. After hybridization with the target, in the resulting hybrid the fluorophore and the quencher are located at a long distance from each other, which results in an enhanced fluorescence signal. Multiplex analysis can easily be implemented using a set of different probes, each of them labeled with a QD of specific emission wavelength.

QDs have also been employed for devising fiber optic sensors [15]. An example of this kind is the affinity sensor shown in Figure 20.7. The receptor in this case is Protein A that shows affinity to immunoglobulin G used as the model analyte. QD-modified protein A was immobilized by glutaraldehyde crosslinking on an aminosilanized glass

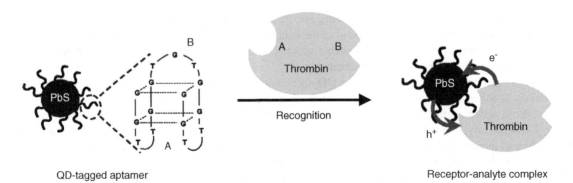

Figure 20.6 Aptamer-based determination of thrombin by QD fluorescence quenching. A is the heparin-biding exosite and B is the fibrinogen binding exosite in thrombin. Adapted with permission from [20]. Copyright 2006 American Chemical Society.

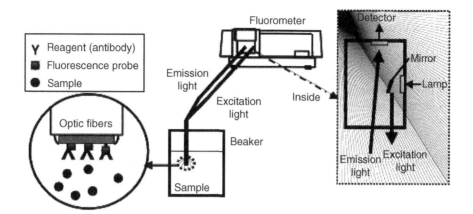

Figure 20.7   Design of a fiber optic fluorescent immunosensor using QDs as labels. Adapted with permission from [22]. Copyright 2005 Elsevier.

slide of 1 mm diameter that was next attached to the distal end of a bifurcated optical fiber. The fiber allows the excitation light to reach the sensing layer and, on the other hand, conveys fluorescence light to the light detector. Receptor–analyte interaction results in fluorescence quenching in relation with analyte concentration and allows this one to be determined at concentrations between 0.5 and 5 µg/ml. The main advantage of employing a QD as fluorophore is the great intensity of the fluorescence light that permits using a long optical fiber without considerable loss in intensity.

20.1.2.5   Chemiluminescence

As envisaged above, QDs photoluminescence appears when radiant energy absorption forms an electron–hole pair. Alternatively, excited QDs can be produced by means of charge-transfer processes that involve suitable electron donors and acceptors [23]. For example, in an alkaline solution, H_2O_2 decomposes to yield $OH \cdot$ radicals and superoxide ions (O_2^-). The radical can inject a hole, whilst the superoxide ion injects an electron into a QD:

$$OH \cdot + QD \rightarrow QD(h^+) + OH^- \tag{20.1}$$

$$O_2^- + QD \rightarrow QD(e^-) + O_2 \tag{20.2}$$

Next, a charge-transfer process involving charge-injected QDs results in an excited QD that relaxes by light emission:

$$QD(h^+) + QD(e^-) \rightarrow QD^* + QD; \quad QD^* \rightarrow QD + Light \tag{20.3}$$

Therefore, chemiluminescence of QDs is triggered by the chemical reaction of a coreactant that produces electron donors and electron acceptors. Charge injection from such reactants into QDs is conditioned by the relative position of the energy levels in the donor/acceptor and the QD (Figure 20.8). Thus, hole injection occurs if the redox potential of the acceptor is below the hole level in the QD. On the other hand, electron injection is possible if the redox

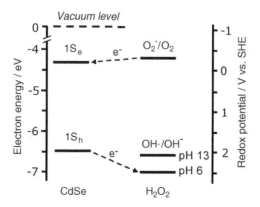

Figure 20.8   Energy level diagram of CdSe QDs and an electron donor (O_2^-) and acceptor (OH·) involved in QD chemiluminescence. SHE means standard hydrogen electrode. Adapted with permission from [24]. Copyright (2004) American Chemical Society.

potential of the donor is above the electron level in the QD conduction band. As the redox potential can be pH dependent, attention should be paid to this parameter.

As pointed out, radicals are involved in the mechanism of QDs chemiluminescence. Consequently, any radical scavenger can in principle be determined by its inhibiting effect on chemiluminescence. Such scavengers are small organic molecules or biomolecules containing $-OH$, $-SH$ or $-NH_2$ groups [23]. On the other hand, ionic species interacting with the surface-protecting layer of the QD affect the chemiluminescence behavior. It was thus demonstrated that metal ions such as Pb^{2+}, Hg^{2+}, Ag^+, Fe^{3+} can be determined by means of mercaptoacetic-covered CdS QDs [23].

20.1.2.6 *Electrochemically Generated Chemiluminescence (ECL)*

As pointed out before, QD chemiluminescence is triggered by suitable electron acceptor/donors that can inject holes or electrons in the QD. If such species result from en electrochemical reaction, the term ECL is used to denote the pertinent luminescence process. An example of this kind is the interaction of QDs with the $SO_4^-\cdot$ radical-anion formed by the electrochemical reduction of peroxodisulfate. $SO_4^-\cdot$ injects a hole in a CdSe QD and hole recombination with an electron from the conduction band results in light emission. Alternatively, ECL can be triggered by the electrochemical reaction of a QD which injects an electron into the conduction band giving rise to the $QD(e^-)$ form. If oxygen is also present in the solution, it is electrochemically reduced to form the OOH^- ion that abstracts an electron from the valence band of $QD(e^-)$ and turns it into the excited form QD^* that emits light [25]. A similar process occurs in the presence of H_2O_2, but this compound is a less-efficient coreactant. Such processes have been noted with QDs immobilized on Pt or paraffin-impregnated graphite electrodes.

Analytical applications of QD-ECL are mostly similar to the above-mentioned applications of QD chemiluminescence with the advantage that the process can be triggered electrically. As oxygen and H_2O_2 are involved in reactions catalyzed by oxidases, QD-ECL appears as a possible transduction procedures in biosensors base on such enzymes.

20.1.2.7 *Quantum Dots as Photoelectrochemical Labels*

Photoexcitation of a QD yields an electron–hole pair. Trapping a conduction-band electron in surface states imparts to the pair a sufficiently long lifetime to permit the transfer of the trapped electron to a positively charged electrode giving rise to an anodic photocurrent (Figure 20.9A). If an electron donor (D) is present, the overall process will consist of D oxidation mediated by the photoexcited QD. The current can be reversed by rendering the electrode negative with respect to the valence-band edge and a photomediated reduction of a suitable electron acceptor A takes place in this instance (Figure 20.9B). Clearly, such an approach enables photoelectrochemical detection of a QD-tagged biocompound in the presence of a sacrificial electron donor or acceptor.

An application of QDs as a photoelectrochemical label for monitoring nucleic acid hybridization is presented in Figure 20.10. In this approach, the electrode and CdS QDs surfaces are functionalized by thiolated oligonucleotides 1 and 3, respectively. The oligonucleotide 1 is complementary to the 5′ end of the target 2, while the oligonucleotide 3 is complementary to the 3′ ends of the target. After the target–probe recognition (step I), the modified QD is bound by hybridization with the overhanging moiety of the target (step II). In step III, QDs tagged with a 2/1 oligonucleotide hybrid are attached by hybridization to the oligonucleotide 3 that is present in the QD layer produced in step II. By repeating the above sequence a layered aggregate including a high number of QDs can be grown sequentially. This strategy brings about a considerable enhancement of the sensitivity.

Upon photoexcitation, QDs convey electrons from the triethanolamine donor to the electrode giving rise to the response photocurrent that is in relation with the target concentration. However, direct electron uptake from distant

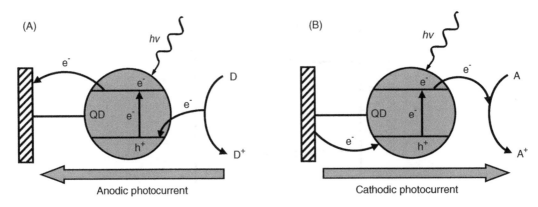

Figure 20.9 Photocurrents generated by quantum dots associated with electrodes. (A) anodic photocurrent; (B) cathodic photocurrent. Reproduced with permission from [13]. Copyright 2008 Wiley-VCH Verlag GmBH & Co. KGaA.

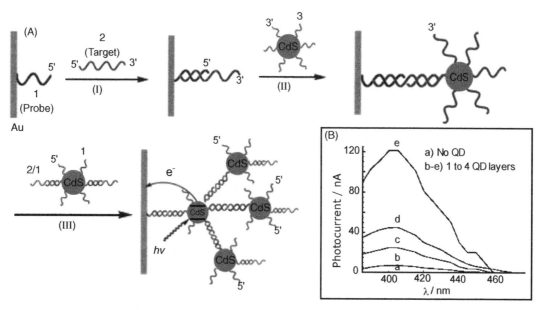

Figure 20.10   QD photoelectrochemical monitoring of DNA hybridization. (A) Recognition and QD labeling process; (B) photocurrent action spectra (current–wavelength curves) for numbers of QD layers. Electron donor: triethanolamine (20 mM). Adapted with permission from [26]. Copyright 2001 Wiley-VCH Verlag GmBH & Co. KGaA.

QDs is hardly probable and, in order to achieve this process, $[Ru(NH_3)_6]^{3+}$ has been used as mediator. Alternatively, the electron transfer from excited QDs to the electrode is conveniently effected by intercalating daunomycin (as an electron relay) in the DNA duplex. As shown in Figure 20.10B, the response photocurrent increases with the number of QD layers.

QD photoelectrochemistry has been successfully applied for the development of enzyme-inhibition sensing schemes. For example, Figure 20.11A, shows the configuration of an acetylcholine esterase-based inhibition sensor. Here, an enzyme–QD conjugate is covalently attached to a gold electrode. Under the catalytic effect of the enzyme, the acetyl thiocholine substrate (1) is hydrolyzed, giving thiocholine (2) that undergoes then an electrochemical oxidation mediated by the excited QD. As the product of the enzyme reaction acts as an electron donor, no sacrificial donor is needed here. Figure 20.11B shows the effect of the excitation wavelength on the photocurrent and demonstrates that the current reaches a plateau below the absorption threshold of the QD. The plateau photocurrent increases with substrate concentration in a way reminiscent of Michaelis–Menten curve. At a constant substrate concentration and in the presence of an AChE inhibitor, the plateau photocurrent decreases with increasing inhibitor concentration, as is usual with enzyme-inhibition sensors.

20.1.3   Outlook

QDs are suited as labels in various luminescence-based transduction methods and appear to be convenient substitutes to classical molecular labels owing to their stability, facile tuning of the optical properties and straightforward

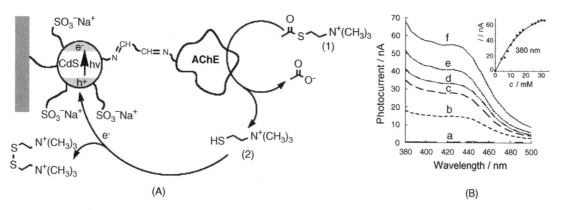

Figure 20.11   QD-based photoelectrochemical detection of acetylcholine esterase inhibitors. (A) sensor configuration. (B) Photocurrent action spectra recorded in the presence of various substrate concentrations (mM) as follows: (a) 0; (b) 6; (c) 10; (d) 12; (e) 16; (f) 30. Inset: photocurrent at 380 nm vs. substrate concentration. 0.1 M phosphate buffer, pH = 8.1. Adapted with permission from [27]. Copyright 2003 American Chemical Society.

functionalization. QDs can function as luminescent labels in a wide range of applications, from sensing small molecules and inorganic ions to monitoring affinity interactions, nucleic acid hybridization and enzymatic reactions.

Various applications of QDs in chemical sensing are comprehensively surveyed in refs. [13,28,29].

Questions and Exercises (Section 20.1)

1. What kind of material displays quantum dot characteristics in the nanometer size region?
2. What are the properties of the electron–hole pair generated by photoexcitation in a QD?
3. Give a short account of the fluorescence mechanism in a QD.
4. How can the fluorescence wavelength of a QD be tuned?
5. Review concisely the advantages of QD as optical labels with respect to molecular optical labels.
6. What kinds of interaction allow ion determination by means of QD fluorescence?
7. Draw a flow chart for the preparation and operation of the affinity sensor in Figure 20.3.
8. Do the same as above for the affinity sensor in Figure 20.4.
9. Sketch the sequence of chemical and photophysical process that occurs in pH determination by means of a QD–pH-sensitive dye conjugate.
10. Comment on possible methods for signal amplification in immunosensors by means of QDs.
11. Describe the functioning principle of a sensor based on quenching of QD fluorescence by the analyte.
12. Draw a simple picture of the recognition–transduction mechanism in a QD-based nucleic acid sensor based on the molecular-beacon approach in the signal-on and signal-off configurations.
13. What condition should be fulfilled in order to achieve chemical excitation of QD fluorescence? Write the mechanism of QD chemiluminescence involving hydrogen peroxide as a source of electron donors and acceptors.
14. How can QD chemiluminescence be applied for the determination of small-molecule compounds and metal ions?
15. Write schematically the mechanism of an enzyme sensor based on QD electrochemically generated chemiluminescence.
16. Give a schematic representation of the mechanism of photoelectrochemical reactions of QDs.
17. Review in brief the application of QD electrochemically generated chemiluminescence in biosensors.

20.2 Carbon Nanotubes as Optical Labels

Carbon nanotubes (CNTs), already introduced in Section 8.3.1, display outstanding optical properties such as light absorption, photoluminescence and Raman scattering that form the basis of their applications in optical sensing [30].

20.2.1 Light Absorption and Emission by CNTs

Optical properties of CNTs result from the particular distribution of electron energy levels in these nanomaterials. As is well known, a macroscopic conductor (including graphite) exhibits continuous valence and conduction bands. These bands overlap in a metal conductor, thereby permitting spontaneous promotion of valence electrons into the conduction band which imparts high electric conductivity. In a semiconductor, the conduction and valence bands are separated by a bandgap and, as a result, only electrons with energy overcoming the bandgap can reach the conduction band. As only a very small fraction of valence electrons fulfill this condition, pure semiconductors are characterized by a very low conductivity at ambient temperature (see Section 11.1.2).

By contrast, the density of energy states in single-walled carbon nanotubes (SWCNTs) is distributed in a discrete form, displaying marked spikes at specific energy values where the density of states tends to infinity (Figure 20.12). Such patterns are called *van Hove singularities* [31]. In a metallic SWCNT, the valence and conduction bands overlap partially, whereas in a semiconductor SWCNT these bands are separated by a bandgap where no allowed energy states exist. The HOMO (highest occupied molecular orbital) and LUMO (lowest unoccupied molecular orbital) of semiconducting SWCNTs correspond to the first van Hove singularities in the valence and conduction band (v_1 and c_1), respectively.

In order for light absorption to occur, the incoming photon energy must match the difference between two valence and conduction van Hove singularities with similar indices. Hence, $v_1 - c_1$ (E_{11}) and $v_2 - c_2$ (E_{22}) transitions are allowed but $v_2 - c_1$ and $v_1 - c_2$ transitions are forbidden. The transitions are labeled as in Figure 20.12 but sometimes, the symbol E is replaced by M (metallic) or S (semiconductor) in order to indicate the nanotube character.

In semiconductor SWCNTs, photon absorption gives rise to a quantum-confined electron–hole pair (exciton) with a Bohr radius of approximately 2.5 nm. SWCNTs luminescence occurs as a result of an E_{22}-type excitation (Figure 20.12). After excitation both electron and hole quickly relax to v_1 and c_1 states, respectively. Then, electron–hole

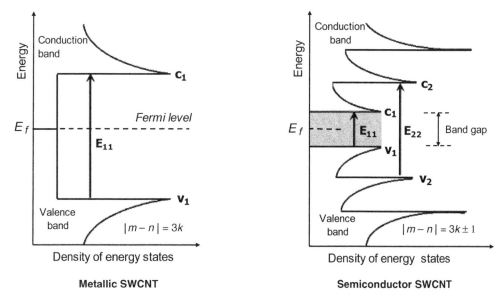

Figure 20.12 Density of electronic energy states in SWCNTs.

recombination takes place and results in light emission with a relatively large Stokes shift caused by previous relaxation processes. It is important to note that $v_1 - c_1$ transitions fall in the near-infrared region (800–2000 nm), a spectral region where no fluorescence background from biological components appears. The quantum yield is very low (typically around 0.01%) but values up to 20% have been attained by proper optimization of the procedure for isolating individual nanotubes in solution. The low quantum yield is due to luminescence quenching that can be induced by interactions between nanotubes themselves or nanotubes and other materials, including the immobilization support in the case of supported CNTs. Oxygen is a strong quencher, so, brightly fluorescent CNTs have been obtained upon coating them with an oxygen-excluding surfactant [32].

Remarkably, the energy separation of the van Hove singularities depends on the nanotube structure and surface modification by adsorption or chemical functionalization. This opens up many possibilities for tuning the optical properties of SWCNTs. For example, Figure 20.13 shows the absorption and emission spectra of a mixture of SWCNTs in suspension. The large difference between the spectra of CNTs modified by a simple surfactant (curve 1) or by dextran (curve 2) is obvious. Note that each nanotube type (identified by the *n, m* indices) gives rise to a distinct spectral band in both absorption and fluorescence spectra. Band intensities are, however, very different, in accordance with the adsorbate capacity to prevent quenching by ambient water and oxygen.

CNTs are characterized by excellent photostability, which is an important property in sensor applications. It is also worth noting that CNTs, as opposed to other fluorophores including QDs, do not exhibit blinking. This characteristic is significant when tracing the displacement of a CNT-labeled biocompound. The fact that fluorescence bands of CNTs are located in the near-infrared region results in a minimal interference by background light, which is characterized by lower wavelength.

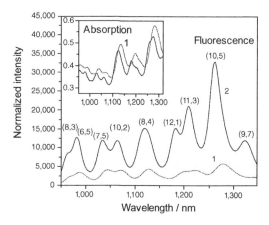

Figure 20.13 Absorption and fluorescence spectra of a water suspension of SWCNTs coated with sodium cholate (curves 1) or 3,4-diaminophenyl-functionalized dextran (curve 2). CNT *n, m* indices are indicated on each band in the fluorescence spectrum. Reprinted by permission from Macmillan Publishers Ltd: Nature Chemistry [33], Copyright 2009.

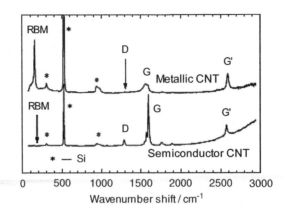

Figure 20.14 Raman spectra of isolated SWCNTs immobilized on an oxidized silica substrate. Star labels indicate contributions of the oxidized silica. Adapted with permission from [34]. Copyright 2005 Elsevier.

20.2.2 Raman Scattering by CNTs

Raman scattering is another optical method for characterizing CNTs [34]. In contrast to photoluminescence, which is exhibited only by semiconductor nanotubes, Raman scattering occurs with both metallic and semiconductor forms. During the Raman scattering event, the exciting photon promotes an electron from the valence band to the conduction band. The excited electron undergoes further exchanges of energy with quantified lattice vibrations (phonons). Finally, the excited electron relaxes to the valence band by photon emission. The energy of the emitted photon differs from that of the incoming photon by the amount of energy exchanged with phonons. This energy difference is detected as a shift in frequency with respect to the frequency of the incoming photon (Figure 20.14). The frequency shift is proportional to the energy difference between the vibration states. The above mechanism is characteristic of *resonance Raman scattering*; it occurs when the energy of either the incoming or emitted photon matches the difference between two electronic levels in the nanotube. *Nonresonant Raman scattering* occurs when the excited electron is promoted to a virtual state but such a process is much less intensive as compared to the previous one.

As typical in Raman spectrometry, the Raman spectrum of a carbon nanotube is related to the characteristic molecular vibration modes. Owing to its particular geometry, a SWCNT exhibits the specific *radial breathing mode* (RBM) that corresponds to radial expansion–contraction. The RBM frequency is inversely proportional to the nanotube diameter and therefore allows this parameter to be inferred from Raman spectrometry data. This specific band permits distinguishing SWCNTs from other graphite forms.

In addition to the RBM mode, CNTs display a series of Raman bands common to most graphite forms as shown in Figure 20.14. Thus, the G band (G for graphite) is due to vibrations along tangents to the nanotube surface. Its fine structure exhibits several signals with the two strongest component being labeled G$^+$ and G$^-$ bands. The CNT G bands are slightly shifted with respect to that of graphene. The D band originates from structural defects (disorder-induced mode), whereas the G$'$ band represents the second overtone of the D band. Due to different selection rules, the G$'$ band is stronger than D.

In contrast to photoluminescence, which is limited to semiconductor SWCNTS, Raman scattering occurs with both metallic and semiconductor nanotubes. CNTs Raman bands are relatively insensitive to the environment and proved consequently useful in CNT application as optical labels in bioassay. In addition, the Raman signal exhibits an excellent signal to background ratio and hence outstanding detection limits when using CNTs as Raman labels.

Fluorescence and Raman spectrometry are widely used for characterizing SWCNTs. In addition, these spectrometric properties lay the basis for CNTs application in bioimaging and biosensing. The main advantage of CNTs arises from the position of the characteristic bands in the near-infrared region, which minimizes the interference of the background radiation. In addition, the excellent resistance of CNTs to photobleaching renders such materials particularly advantageous in imaging and chemical sensing.

20.2.3 CNT Optical Sensors and Arrays

Due to the sensitivity of SWCNTs fluorescence to the local environment, this kind of material has been used to develop quenching-based sensing scheme, as emphasized next in several examples.

Nitric oxide (NO), a messenger in biological signaling, can be determined by means of its ability to quench the fluorescence of SWCNTs by electron withdrawal from the valence band of the excited nanotube. Other reactive

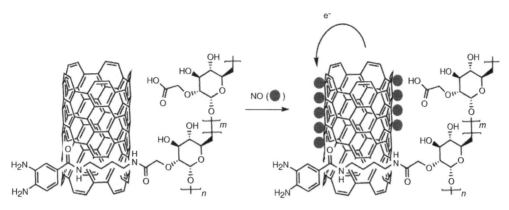

Figure 20.15 Nitrogen oxide detection by means of a 3,4-diaminophenyl-functionalized dextran modified SWCNT. Reprinted by permission from Macmillan Publishers Ltd: Nature Chemistry [33], Copyright 2009.

nitrogen and oxygen species behave similarly and selectivity should therefore be imparted by modifying the CNT with a NO selective receptor such as 3,4-diaminophenyl-functionalized dextran [33,35]. The principle of selective NO sensing by dextran-functionalized SWCNTs is illustrated in Figure 20.15. This principle has been applied in a the design of a nanosensor for reversible, real-time, and spatially resolved detection of nitric oxide in macrophage cells and exhibits potential applications for *in vivo* assays.

A CNT glucose-detection scheme has been developed using the property of this enzyme to produce hydrogen peroxide by substrate oxidation [36]. In this approach, the enzyme is immobilized by adsorption on the CNT wall and CNT fluorescence in the near-infrared region is excited with light from a laser pointer (635 nm). For the purpose of transduction, $Fe(CN)_6^{3-}$ is added to the sample solution because this anion adsorbs irreversibly onto the nanotube wall and quenches fluorescence by electron withdrawing. Hydrogen peroxide produced by the enzyme reaction reduces $Fe(CN)_6^{3-}$ to $Fe(CN)_6^{4-}$, which is not a quencher. As a result, the fluorescence intensity is enhanced, making it possible to determine the concentration of glucose in the millimolar range. A near-infrared glucose probe can be obtained by entrapping a modified nanotube suspension into a sealed dialysis capillary along with the enzyme. The dialysis membrane is permeable to small molecules but not to nanotubes.

CNTs have been used as Raman labels in proteins arrays based on immunoassay in the sandwich format (Figure 20.16). The array support consists of gold nanoclusters that have been covered with branched PEG-COOH by condensation with self-assembled cysteamine ($H_2N-(CH_2)_2-SH$) (Figure 20.16A). Various protein-antigens have been then covalently linked to the previous layer as receptors for the target antibodies. In order to obtain SWCNT-tagged secondary antibodies, nanotubes have been modified by adsorption of a mixture of two polyethylene glycol-modified phospholipids (DSPE-3PEO and DSPE-PEG-NH$_2$) (Figure 20.16B). A secondary antibody (goat antimouse immunoglobulin G – GaM-IgG) was linked to this layer and served for specific binding of SWCNTs tags to the target antibody in the sandwich format (Figure 20.16C). In this configuration, the CNT G^+ signal (which was used as a response signal) is strongly amplified as a result of the local field being enhanced by gold nanoclusters. This makes it possible to detect the analyte (aHSA) down to 1 pM level, which is about 100 times lower than the detection limit achieved by fluorescence using the molecular Cy3 labels. Multiplexing has been achieved by using isotopically pure C^{13} and C^{14} nanotube; each of them exhibiting a distinct G band.

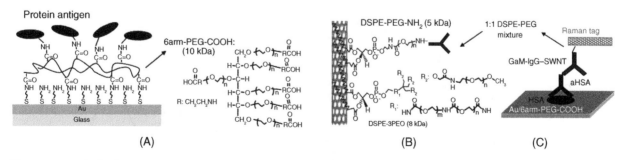

Figure 20.16 Antibody determination in the sandwich format using CNTs as Raman labels. (A) Structure of the sensing layer; (B) carbon nanotubes modified with a secondary antibody; (C) structure of the receptor-analyte-tagged antibody ternary complex. Reprinted by permission from Macmillan Publishers Ltd: Nature Biotechnology [37], Copyright 2008.

20.2.4 Outlook

SWCNTs are very convenient materials for applications as fluorescent or Raman labels in optical sensing mostly due to their extraordinary photostability that allows them to withstand very strong excitation beams. Also important is the near-infrared location of SWCNTs emission bands that render them practical for detection in biological media with no interference from background fluorescence. Future progress is expected to arise with the advance in the synthesis methods as far as selective production of CNTs with predefined properties is concerned.

Questions and Exercises (Section 20.2)

1 What are the differences between metallic and semiconductor SWCNTs as far as the distribution of the energy states is concerned?
2 How is the fluorescence influenced by the particular distribution of energy states in semiconductor SWCNTs?
3 Sketch the mechanism of Raman scattering by SWCNTs using a simplified scheme of the energy-level distribution.
4 What kind of spectral bands can be identified in the Raman spectrum of a SWCNT?
5 What are the advantages of SWCNTs Raman scattering in chemical-sensor applications?
6 Draw a summary flow chart for sensing layer preparation and immunoassay using SWCNTs as Raman scattering labels, in accordance with Figure 20.16.

20.3 Metal Nanoparticle in Optical Sensing

20.3.1 Optical Properties of Metal Nanoparticles

Metal nanoparticles have been introduced in Section 8.2. It has long been known that colloidal solutions of gold or silver feature a distinct, bright color. When a light beam crosses such a solution, it is partially extinct along the propagation path but, at the same time, light arising from the solution can be detected along all possible directions. Despite the apparent similarity with fluorescence, such an optical phenomenon is caused by a fundamentally different mechanism. As is well known, fluorescence is due to electron transitions between valence orbitals in molecules. In the case of colloidal solutions of metals, light extinction and scattering is caused by collective oscillation of free electrons under the effect of the forces applied by the oscillating electromagnetic field. Under irradiation, a spherical nanoparticle behaves as an oscillating dipole and hence becomes a light source (Figure 20.17A). This is why light is scattered and gives a bright appearance of the solution. At the same time, light scattering attenuates the incident beam that is manifested by an increased absorbance over a specific spectral region (graph 1 in Figure 20.17).

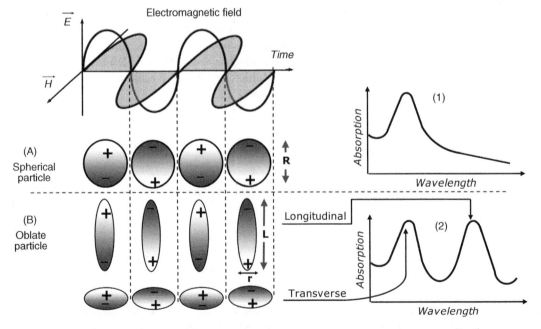

Figure 20.17 Plasmon resonance in metal nanoparticles under the influence of an electromagnetic field (light). $\vec{E}$ represents the strength of the electric field; $\vec{H}$ is the strength of the magnetic field. Graphs: absorption spectrum of spherical (1) and oblate (2) colloidal metal particles. Adapted with permission from [40]. Copyright 2009 Wiley-VCH Verlag GmBH & Co. KGaA.

Collective oscillation of conduction electrons is known as *plasmon resonance* (shortened, *plasmon*). The term plasmon describes a particular state of conduction electrons that undergo oscillation under the effect of an electromagnetic field (see Section 18.6.1). Figure 20.17 illustrates the apparition of bulk plasmons that involve electrons in the whole particle. Simultaneously, surface-localized plasmons, consisting of electrons oscillating along the metal surface, can form. Metal structures often support both types of plasmons simultaneously. The *plasmon wavelength* is the incident light wavelength at which plasmon scattering occurs with maximum intensity. The plasmon wavelength depends on the particle material, particle size and also on the dielectric constant of the surrounding medium. As the dielectric constant is related to the refractive index, this parameter is often used to express the effect of the local environment. This property is the basis of analytical applications of metal nanoparticles.

In contrast to spherical nanoparticles, nonspherical particles (such as nanorods) show more than one absorption band because surface plasmons created along various directions are energetically different. Thus, an oblate nanoparticle supports two oscillation modes, a longitudinal and a transversal one, each of them characterized by a specific plasmon wavelength. Interestingly, the resonance wavelength of nanorods can be tuned by changing the *aspect ratio* (that is, the length/radius ratio, *L/r*). This feature offers additional opportunities for analytical applications [38,39].

Gold is the ideal material for such applications due to its chemical inertness and compatibility with various chemical modification methods but silver displays similar optical properties. Both these metals form nanoparticles with distinct absorption bands in the visible spectral region, in contrast to other metals that absorb over very broad bands and are therefore of little use for analytical application. It is important to mention that metal nanoparticles feature extremely high light absorption coefficients, allowing detection of such species at extremely low concentrations.

20.3.2 Optical Detection Based on Metal Nanoparticles

As mentioned before, metal nanoparticles induce light scattering that depends on the properties of the local environment that affect the plasmon wavelength. The analyte concentration can be inferred from the shift of the plasmon band. There are two possible approaches for detecting the effect of plasmon resonance. The first one relies on recording the absorption spectrum of the colloidal solution containing both nanoparticles and the analyte. Alternatively, light scattered by nanoparticles attached to a solid surface can be inspected by suitable methods such as dark-field microscopy.

The special properties of metal nanoparticles can be exploited in several ways as follows: (i) aggregation-dependent shift of the plasmon frequency; (ii) shift of the plasmon frequency owing to local change in the dielectric constant; (iii) surface-enhanced Raman scattering; and (iv) enhancement or quenching of the fluorescence of molecular luminophores [38,39]. In order to impart selectivity to the analyte–nanoparticle interaction, the surface of the particle should be modified with suitable recognition receptors.

i. *Nanoparticle aggregation* results in a color change that depends on the specific size and shape of the resulting assembly. For analytical applications, several specific receptor entities are appended to the nanoparticles to promote aggregation via analyte bridging (Figure 20.18A). The absorption band of the aggregate is broadened and red-shifted with respect to the single-nanoparticle band. This scheme has been employed for detecting toxic metal ions, DNA, proteins, antibodies and also some small molecules. For example, gold nanospheres modified with a lithium-specific ligand demonstrate a shift of the absorption maximum which is proportional to the lithium concentration. This allows performing lithium determination between 10 and 100 mM using the resonance shift as analytical signal (Figure 20.18B).

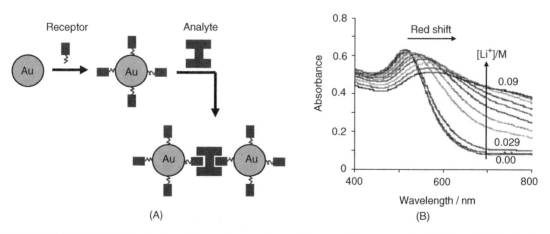

(A)   (B)

Figure 20.18   (A) A general scheme for chemical sensing based on metal nanoparticle aggregation. (B) Effect of Li$^+$-induced aggregation of gold nanoparticles modified with a Li$^+$ ligand-receptor. Li$^+$ concentration has been varied between 0 and 0.09 M, as shown on selected curves. Adapted with permission from [39]. Copyright 2008 The Royal Society of Chemistry.

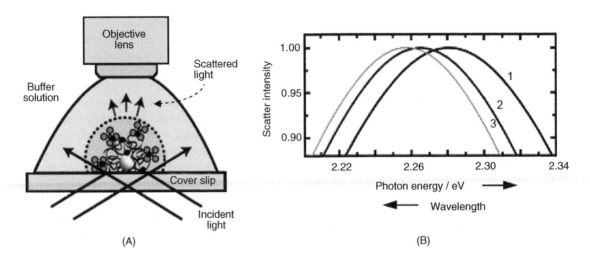

Figure 20.19 An affinity biosensor using biotin-modified gold nanoparticles to detect streptavidin by means of light scattered by plasmon resonance. (A) Sensor configuration. (B) Calculated scattering spectrum for (1) naked particles; (2) biotin modified particles; (3) after streptavidin binding to biotin. Adapted with permission from [42]. Copyright 2003 American Chemical Society.

ii. *The local dielectric constant* at the nanoparticle surface affects the plasmon wavelength. As the dielectric constant is connected to the refractive index, this last tem can also be used as an environment indicator.

Owing to the above-mentioned effect, the absorption of an analyte at a nanoparticle surface causes the plasmon band to shift in accordance with surface coverage. Since the degree of coverage is dependent on the analyte concentration, this parameter can be inferred from the band shift.

Based on these principles, the response of biotinylated Au nanoparticles to streptavidin binding has been investigated by dark-field microscopy that allows the intensity of scattered light to be monitored at various wavelengths, as shown in Figure 20.19A. The calculated plasmon scattering spectrum is shown in Figure 20.19B for the naked nanoparticle (1), the biotinylated particle (2) and after avidin–biotin coupling (3). It is apparent that changes in the local refractive index caused by biotinylation and subsequent streptavidin binding significantly alter the plasmon resonance wavelength. The shift of curve 3 with respect to curve 2 allows streptavidin to be detected at concentrations as low as 1 μM.

A modeling of the response of an immunosensor revealed that it is best to consider the normalized plasmon band shift ($\Delta R/\Delta R_{max}$), where ΔR is the band shift at incomplete coverage and ΔR_{max} is the band shift at the maximum analyte coverage. The $\Delta R/\Delta R_{max}$ parameter is related to the surface concentration of the antibody–antigen complex [AB] by a Langmuir-type function as follows [41]:

$$\frac{\Delta R}{\Delta R_{max}} = \theta = \frac{K_{a,surf}[AB]}{1 + K_{a,surf}[AB]} \tag{20.4}$$

where θ is the surface coverage by the antibody–antigen complex (AB), [AB] is the surface concentration of the complex, and $K_{a,surf}$ is the affinity constant of the surface-confined immunoreaction.

iii. *Enhanced Raman scattering* (Section 18.3.11) is a common spectrometric method for investigating organic compounds adsorbed at metal surfaces. Enhancement of Raman scattering also occurs with molecules adsorbed on metal nanoparticles as a result of the very strong local electromagnetic field caused by plasmon resonance. In this way, the vibrational spectrum of the adsorbate can be recorded at extremely low concentrations. Nonspherical particles give rise to a stronger Raman scattering relative to spherical ones. This behavior is particularly evident for sharp-edged particles such as cubes and blocks.

iv. *Quenching or enhancement of fluorescence* by metal nanoparticles has been found to occur with the usual molecular fluorophores [38,43]. On the one hand, fluorescence enhancement arises from the local amplification of the excitation light intensity owing to plasmon resonance. On the other hand, fluorescence quenching occurs when the emission band overlaps the plasmon absorption band allowing resonance energy transfer to occur. Both quenching and enhancement depend on the shape of the metal particles and also on the orientation of dipole moment of the luminophores with respect to the particle surface. Gold nanoparticles are efficient quenchers in nucleic acid sensors based on the molecular-beacon approach.

20.3.3 Metal Nanoparticles in Optical Sensing

Most of the applications reported so far rely on preformed nanoparticles used either as a colloidal solution or attached to a solid support in a relatively uncontrollable way (such as electrostatic attraction). In the process of

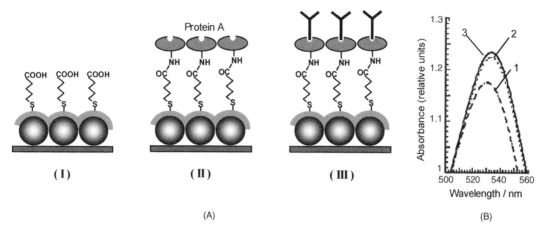

Figure 20.20 A plasmon resonance immunosensor based on gold-capped silica nanoparticles. (A) Main steps in the fabrication of the sensing element (B) absorption spectra for (1) gold capped nanoparticles (I); (2) anti-casein antibody-modified surface (III); (3) after casein binding to the sensing antibody. Adapted with permission from [44]. Copyright 2007 Institute of Physics Publishing Ltd.

immobilization, care should be taken to avoid damaging the optical properties of the nanoparticle. However, in order to develop an optical sensor, nanostructured materials should be firmly and reliably linked to a solid support. This seems to be best achieved by preparing such a material directly on the selected support. An example of this kind is a casein immunosensor based on gold-capped silica nanoparticles that is shown in Figure 20.20. In this example, silica nanoparticles are first attached to a gold-plated glass slide using siloxane chemistry. Next, the gold capping is formed by thermal evaporation and carboxyl groups are attached by thiol chemisorption to obtain the structure (I) in Figure 20.20A. Further, Protein A is appended by crosslinking with carbodiimide (structure (II) and the casein–antibody receptor is linked by affinity in order to complete the fabrication of the sensing element (structure III). Light transmitted through this assembly is attenuated by plasmon resonance. Transmitted light is collected by an optic fiber probe and conveyed to a spectrometer for assessing the shift in the plasmon wavelength (Figure 20.20B). This sensor allows casein concentration between 10^{-8} and 10^{-4} M to be determined.

Coupling the evanescent wave with gold nanoparticles has been applied to hybridization monitoring by means of scattered light in microarray assays [45,46]. Figure 20.21 illustrates the principles of this method. The capture probe is immobilized on a glass slide (A) and the probe–target hybrid is labeled by a gold nanoparticle-tagged oligonucleotide in a sandwich assay format (B). The glass slide functions as a planar waveguide that confines the excitation light beam thus preventing it from reaching the light detector by scattering. However, the evanescent wave causes electron oscillations in each nanoparticle, which results in light emission. So, each nanoparticle attached to the surface turns into a light source by plasmon resonance light scattering. The intensity of the scattered light can be enhanced by growing a silver layer on each gold nanoparticle via chemical reduction of Ag^+ with hydroquinone (C).

Outstanding performances have been obtained with this configuration. Thus, the detection limit is three orders of magnitude better than in fluorescent detection based on Cy3 and approaches 50 fM. Moreover, the dynamic range is very broad, covering three orders of magnitudes. An additional advantage arises from the high melting temperature and sharp melting transition of the nanoparticle-labeled hybrid. Whereas this transition spans over an 18 °C region for a dye-labeled hybrid, it encompasses only 3 °C with gold-particle labels. Remarkably, such characteristics enable single-base mismatch determination at a concentration as low as 100 attomolar.

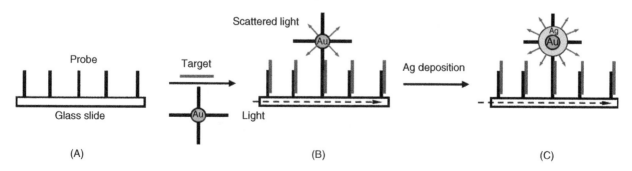

Figure 20.21 Light scattering by metal nanoparticles for hybridization imaging.

The previous discussion has been confined to distinct nanoparticles as plasmon sites. However, plasmon-supporting structures can also be fabricated by nanofabrication on solid surfaces [47].

So far, various applications of metal nanoparticles as optical probes in batch chemical analysis and imaging have been reported [38,39,48,49]. Currently, available knowledge in this area lays the basis for future applications in the field of optical chemical sensors.

Questions and Exercises (Section 20.3)

1 What are the optical phenomena that occur when a light beam crosses a colloidal solution of metal nanoparticles?
2 What is the mechanism of light absorption and light emission by spherical nanoparticles?
3 What are the particular features of light absorption and emission by nonspherical particles?
4 Draw the scheme of a possible nucleic acid hybridization sensor based on gold-nanoparticle plasmon resonance.
5 Draw schematically the structure and the response of a nucleic acid hybridization sensor based on the effect of the dielectric constant on the gold nanoparticle plasmon resonance. How can the aggregation effect be minimized?
6 Briefly review the principles of surface-enhanced Raman spectrometry with particular reference to the application of metal nanoparticles.
7 Draw a simple sketch that demonstrates fluorescence quenching or enhancement by metal nanoparticles.
8 Using Figure 20.20 as a guide, draw schematically the structure of a possible nucleic acid sensor based on light scattering by metal nanoparticles.
9 Draw a possible structure of an immunosensor based on light emission by Au nanoparticles combined with a planar waveguide (use Figure 20.21 as a guide).

20.4 Porous Silicon

Porous silicon is a form of the chemical element silicon that is produced by chemical, electrochemical or photochemical etching of monocrystalline silicon in the presence of hydrofluoric acid. The resulting structure is composed of aligned pores and silicon columns perpendicular to the surface. The freshly prepared surface contains metastable Si–H sites that react with water, oxygen or other oxidizing reagents (Figure 20.22). In this way, oxygen is incorporated in the lattice forming various oxygen functionalities. Hydroxyl groups formed in this way allow for surface functionalization by condensation reactions with alkoxy- or chlorosilanes [50].

Application of porous silicon relies either on its luminescence properties or light-interference phenomena [50,51].

Porous silicon is a luminescent material owing to nanoscale regions of crystalline silicon embedded in the porous columnar skeleton. Such regions display typical properties of quantum dots and allow for optical transduction by luminescence quenching. Quenching of porous silicon fluorescence makes possible the development of optical gas sensors as well as immunosensors or nucleic acid sensors. However, fluorescence transduction is particularly vulnerable to interference from the sample components.

Interferometric transduction (Section 18.6.2) is possible when using porous silicon as an immobilization support for the receptor in affinity-based chemical sensors. Porous silicon produces light reflection both at the upper surface and at the bulk silicon surface. Owing to the difference in the optical path length, reflected beams produce interference patterns. When the analyte binds to a surface-appended receptor, the resulting change in the optical thickness produces a shift of the interference fringes depending on the analyte concentration.

Controlled etching can form single-dimensional photonic crystal structure at the surface of a silicon support (Section 18.7.2). Such a device can perform label-free transduction based on the modification of the refractive index at the silicon surface.

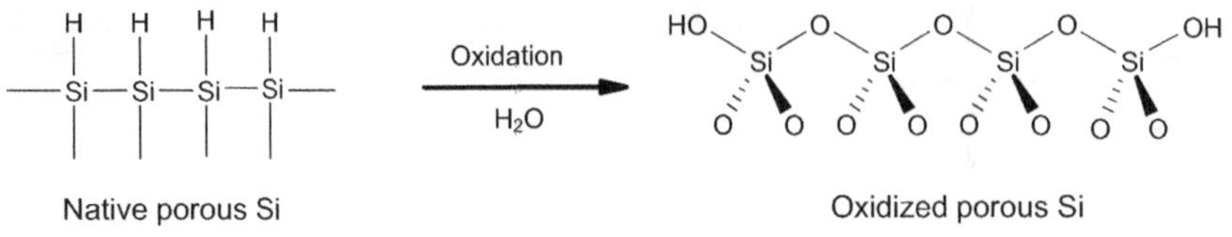

Figure 20.22 Oxidation of native porous silicon.

20.5 Luminescent Lanthanide Compound Nanomaterials

Gd_2O_3 nanoparticles doped with Tb or Eu are highly photostable luminescent materials characterized by narrow emission bands, large Stokes shift and a relatively long lifetime [52]. Suitable encapsulation imparts water compatibility and provides functionalization opportunity via reactive groups at the surface layer. Thus, a polysiloxane shell has been used for covering Tb: Gd_2O_3 nanoparticles. This brings the additional advantage of an enhanced luminescence that has been assigned to the removal of Tb-OH surface groups that are detrimental for luminescence. Eu: Gd_2O_3 nanoparticles covered with poly (L-lysine) have been used as reporter labels in a competitive fluorescence microassay for phenoxybenzoic acid, a generic biomarker for human exposure to pyrethroid insecticides.

New research results on luminescent lanthanide-doped nanoparticles are being reported at a high pace [53] and lay the ground for future applications of such materials as fluorescent probes in biosensor technology, immunoassay and high-throughput screening. Hybrid materials containing fluorescent lanthanide nanoparticles or fluorescent lanthanide complexes show promising prospects for application in this area.

20.6 Summary

Nanomaterials have brought about new opportunities for improving the figures of merit of optical sensing. Nanomaterials of particular relevance to this field are semiconductor nanocrystals (quantum dots) carbon nanotubes, metal nanoparticles and luminescent lanthanide compounds.

Photophysically active nanomaterials can be used simply as substitutes for molecular luminescent labels used in the early stages of development of optical sensors. In this alternative, nanomaterials proved to be superior as far as stability, signal intensity and multiplexing possibilities are concerned. On the other hand, certain nanomaterials, such as metal nanoparticles and carbon nanotubes, developing new transduction schemes based on optical methods such as Raman spectrometry and plasmon resonance spectrometry.

References

1. Eychmüller, A. (2010) Synthesis and characterization of II-VI nanoparticles, in *Nanoparticles: From Theory to Application* (ed. G. Schmid), Wiley-VCH, Weinheim, pp. 69–100.
2. Banin, U. (2010) Synthesis and characterization of III-V semiconductor nanoparticles, in *Nanoparticles: From Theory to Application* (ed. G. Schmid), Wiley-VCH, Weinheim, pp. 101–127.
3. Dehnen, S., Eichhöffer, A., Corrigan, J.F. *et al.* (2010) Synthesis and characterization of Ib-VI nanoclusters, in *Nanoparticles: From Theory to Application* (ed. G. Schmid), Wiley-VCH, Weinheim, pp. 127–239.
4. Schmid, G. (2004) *Nanoparticles: From Theory to Application*, Wiley-VCH, Weinheim.
5. Comparelli, R., Curri, M.L., Cozzoli, P.D. *et al.* (2007) Optical biosensing based on metal and semiconductor nanocrystals, in *Nanomaterials for Biosensors* (ed. C.S.S.R. Kumar), Wiley-VCH, Weinheim, pp. 123–174.
6. Bakalova, R., Zhelev, Z., Ohba, H. *et al.* (2007) Quantum dot-based nanohybrids for fluorescent detection of molecular and cellular biological targets, in *Nanomaterials for Biosensors* (ed. C.S.S.R. Kumar), Wiley-VCH, Weinheim, pp. 175–207.
7. Murphy, C.J. (2002) Optical sensing with quantum dots. *Anal. Chem.*, **74**, 520A–526A.
8. Smith, A.M. and Nie, S. (2010) Semiconductor nanocrystals: structure, properties, and band gap engineering. *Acc. Chem. Res.*, **43**, 190–200.
9. Banin, U. and Oded, M. (2010) Optical and electronic properties of semiconductor nanocrystals, in *Nanoparticles: From Theory to Application* (ed. G. Schmid), Wiley-VCH, Weinheim, pp. 371–392.
10. Dorfs, D. and Eychmüller, A. (2008) Multishell semiconductor nanocrystals, in *Semiconductor Nanocrystal Quantum Dots: Synthesis, Assembly, Spectroscopy and Applications* (ed. A.L. Rogach), Springer, Vienna, pp. 101–117.
11. Shi, J.J., Zhu, Y.F., Zhang, X.R. *et al.* (2004) Recent developments in nanomaterial optical sensors. *TrAC Trends Anal. Chem.*, **23**, 351–360.
12. Zayats, M. and Willner, I. (2008) Photoelectrochemical and optical applications of semiconductor quantum dots for bioanalysis, in *Biosensing for the 21st Century*, Springer-Verlag, Berlin, pp. 255–283.
13. Gill, R., Zayats, M., and Willner, I. (2008) Semiconductor quantum dots for bioanalysis. *Angew. Chem. Int. Edit.*, **47**, 7602–7625.
14. Frasco, M.F. and Chaniotakis, N. (2009) Semiconductor quantum dots in chemical sensors and biosensors. *Sensors*, **9**, 7266–7286.
15. Jorge, P.A.S., Martins, M.A., Trindade, T. *et al.* (2007) Optical fiber sensing using quantum dots. *Sensors*, **7**, 3489–3534.
16. Wang, X.J., Ruedas-Rama, M.J., and Hall, E.A.H. (2007) The emerging use of quantum dots in analysis. *Anal. Lett.*, **40**, 1497–1520.
17. Han, C.P. and Li, H.B., (2010) Host-molecule-coated quantum dots as fluorescent sensors. *Anal. Bioanal. Chem.*, **397**, 1437–1444.
18. Mattoussi, H., Medintz, I.L., Clapp, A.R. *et al.* (2004) Luminescent quantum dot-bioconjugates in immunoassays, FRET, biosensing, and imaging applications. *J. Assoc. Lab. Autom.*, **9**, 28–32.
19. Snee, P.T., Somers, R.C., Nair, G. *et al.* (2006) A ratiometric CdSe/ZnS nanocrystal pH sensor. *J. Am. Chem. Soc.*, **128**, 13320–13321.

20. Choi, J.H., Chen, K.H., and Strano, M.S. (2006) Aptamer-capped nanocrystal quantum dots: A new method for label-free protein detection. *J. Am. Chem. Soc.*, **128**, 15584–15585.

21. Kim, J.H., Chaudhary, S., and Ozkan, M. (2007) Multicolour hybrid nanoprobes of molecular beacon conjugated quantum dots: FRET and gel electrophoresis assisted target DNA detection. *Nanotechnology*, **18**, Article Number: 195105.

22. Aoyagi, S. and Kudo, M. (2005) Development of fluorescence change-based, reagent-less optic immunosensor. *Biosens. Bioelectron.*, **20**, 1680–1684.

23. Li, Y.X., Yang, P., Wang, P. *et al.* (2007) CdS nanocrystal induced chemiluminescence: reaction mechanism and applications. *Nanotechnology*, **18**, 8, Article Number: 225602.

24. Poznyak, S.K., Talapin, D.V., Shevchenko, E.V. *et al.* (2004) Quantum dot chemiluminescence. *Nano Lett.*, **4**, 693–698.

25. Zou, G.Z. and Ju, H.X. (2004) Electrogenerated chemiluminescence from a CdSe nanocrystal film and its sensing application in aqueous solution. *Anal. Chem.*, **76**, 6871–6876.

26. Willner, I., Patolsky, F., and Wasserman, J. (2001) Photoelectrochemistry with controlled DNA-cross-linked CdS nanoparticle arrays. *Angew. Chem. Int. Ed.*, **40**, 1861–1864.

27. Pardo-Yissar, V., Katz, E., Wasserman, J. *et al.* (2003) Acetylcholine esterase-labeled CdS nanoparticles on electrodes: Photo-electrochemical sensing of the enzyme inhibitors. *J. Am. Chem. Soc.*, **125**, 622–623.

28. Freeman, R., Xu, J.P., and Willner, I. (2010) Semiconductor quantum dots for analytical and bioanalytical applications, in *Nanoparticles: From Theory to Application* (ed. G. Schmid), Wiley-VCH, Weinheim, pp. 455–511.

29. Costa-Fernandez, J.M., Pereiro, R., and Sanz-Medel, A. (2006) The use of luminescent quantum dots for optical sensing. *TrAC Trend. Anal. Chem.*, **25**, 207–218.

30. Sgobba, V. and Guldi, D.M. (2009) Carbon nanotubes-electronic/electrochemical properties and application for nanoelectronics and photonics. *Chem. Soc. Rev.*, **38**, 165–184.

31. Joselevich, E. (2004) Electronic structure and chemical reactivity of carbon nanotubes: A chemist's view. *Phys. Chem. Chem. Phys.*, **5**, 619–624.

32. Ju, S.Y., Kopcha, W.P., and Papadimitrakopoulos, F. (2009) Brightly fluorescent single-walled carbon nanotubes via an oxygen-excluding surfactant organization. *Science*, **323**, 1319–1323.

33. Kim, J.H., Heller, D.A., Jin, H. *et al.* (2009) The rational design of nitric oxide selectivity in single-walled carbon nanotube near-infrared fluorescence sensors for biological detection. *Nature Chem.*, **1**, 473–481.

34. Dresselhaus, M.S., Dresselhaus, G., Saito, R. *et al.* (2005) Raman spectroscopy of carbon nanotubes. *Phys. Rep.-Rev. Sec. Phys. Lett.*, **409**, 47–99.

35. Strano, M.S., Boghossian, A.A., Kim, W.J. *et al.* (2009) The chemistry of single-walled nanotubes. *MRS Bull.*, **34**, 950–961.

36. Barone, P.W., Baik, S., Heller, D.A. *et al.* (2005) Near-infrared optical sensors based on single-walled carbon nanotubes. *Nature Mater.*, **4**, 86–92.

37. Chen, Z., Tabakman, S.M., Goodwin, A.P. *et al.* (2008) Protein microarrays with carbon nanotubes as multicolor Raman labels. *Nature Biotechnol.*, **26**, 1285–1292.

38. Sau, T.K., Rogach, A.L., Jackel, F. *et al.* (2010) Properties and applications of colloidal nonspherical noble metal nanoparticles. *Adv. Mater.*, **22**, 1805–1825.

39. Murphy, C.J., Gole, A.M., Hunyadi, S.E. *et al.* (2008) Chemical sensing and imaging with metallic nanorods. *Chem. Commun.*, 544–557.

40. Cademartiri, L. and Ozin, G.A. (2009) *Concepts of Nanochemistry*, Wiley-VCH, Weinheim.

41. Haes, A.J. and Van Duyne, R.P. (2002) A nanoscale optical biosensor: Sensitivity and selectivity of an approach based on the localized surface plasmon resonance spectroscopy of triangular silver nanoparticles. *J. Am. Chem. Soc.*, **124**, 10596–10604.

42. Raschke, G., Kowarik, S., Franzl, T. *et al.* (2003) Biomolecular recognition based on single gold nanoparticle light scattering. *Nano Lett.*, **3**, 935–938.

43. Ray, K., Chowdhury, M.H., Zhang, J. *et al.* (2009) Plasmon-controlled fluorescence towards high-sensitivity optical sensing. In *Optical Sensor Systems in Biotechnology* (ed. G. Rao), Springer-Verlag, Berlin, pp. 29–72.

44. Hiep, H.M., Endo, T., Kerman, K. *et al.* (2007) A localized surface plasmon resonance based immunosensor for the detection of casein in milk. *Sci. Technol. Adv. Mater.*, **8**, 331–338.

45. Taton, T.A., Mirkin, C.A., and Letsinger, R.L. (2000) Scanometric DNA array detection with nanoparticle probes. *Science*, **289**, 1757–1760.

46. Willner, I., Shlyahovsky, B., Willner, B. *et al.* (2009) Amplified DNA biosensors, in *Functional Nucleic Acids for Analytical Applications* (eds Y. Li and Y. Lu), Springer, New York, pp. 199–252.

47. Stewart, M.E., Anderton, C.R., Thompson, L.B. *et al.* (2008) Nanostructured plasmonic sensors. *Chem. Rev.*, **108**, 494–521.

48. Mayer, K.M. and Hafner, J.H. (2011) Localized surface plasmon resonance sensors. *Chem. Rev.*, **111**, 3828–3857.

49. Aslan, K., Lakowicz, J.R., and Geddes, C.D. (2005) Plasmon light scattering in biology and medicine: new sensing approaches, visions and perspectives. *Curr. Opin. Chem. Biol.*, **9**, 538–544.

50. Stewart, M.P. and Buriak, J.M. (2000) Chemical and biological applications of porous silicon technology. *Adv. Mater.*, **12**, 859–869.

51. Jane, A., Dronov, R., Hodges, A. *et al.* (2009) Porous silicon biosensors on the advance. *Trends Biotechnol.*, **27**, 230–239.

52. de Dios, A.S. and Diaz-Garcia, M.E. (2010) Multifunctional nanoparticles: analytical prospects. *Anal. Chim. Acta*, **666**, 1–22.

53. Binnemans, K. (2009) Lanthanide-based luminescent hybrid materials. *Chem. Rev.*, **109**, 4283–4374.

Acoustic-Wave Sensors

Mass change is the most straightforward modification caused by the recognition event in a chemical sensor. Mass-change assessment appears therefore as a very advantageous transduction method, particularly as it does not involve labels. However, extremely sensitive mass transducers are required for this purpose as the mass change can be below the μg level. Such ultrasensitive transducers are based on vibrating piezoelectric crystals.

Acoustics is the study of mechanical vibrations in gases, liquids, and solids. As vibrations are perceived as sound, the name acoustic-wave sensor is ascribed to the devices presented in this chapter An acoustic wave can develop either simply at the surface of a vibrating device or it can expand to its whole volume. The first case refers to *surface-acoustic waves*, whereas the second refers to *bulk acoustic waves*.

Applications of acoustic waves in chemical sensors rely on the interaction of the analyte with an adjacent recognition layer, with resulting modifications in the wave frequency and other wave parameters.

As the key physical phenomenon in an acoustic-wave sensor is *piezoelectricity*, the first part of this chapter introduces the basic concept of this physical effect. The second part presents the principles of the thickness–shear mode vibrating piezoelectric crystal that is currently the most widely used mass transducer in chemical sensors. Subsequent sections address a series of applications of this transducer in various kinds of chemical sensors such as gas sensors, affinity sensors and nucleic acid sensors.

Commonly, a transducer or a sensor based on a thickness–shear vibrating piezoelectric crystal is referred to as *quartz crystal microbalance (QCM)* due to the extreme sensitivity to mass loading. The QCM is a *bulk acoustic-wave device*. Despite its name, the QCM is much more than a microbalance. It is also a valuable tool for investigating elasticity and viscosity of various materials including synthetic and biological polymers currently employed in the development of chemical sensors. Therefore, the QCM is an unparalleled tool for assessing physicochemical properties of thin layers used in chemical sensors. Rational use of the wide potential of QCM makes available a wealth of information on the physicochemical properties of the sensing element and also allows for rational design and operation of various types of chemical sensors.

Basic terms and definitions related to piezoelectric materials, sensing devices and analytical procedures with piezoelectric chemical sensors are available in ref. [1]. Various aspects of QCM principles, theory and applications are extensively addressed in collective volumes [2,3] and reviews [4–6].

21.1 The Piezoelectric Effect

The piezoelectric effect, which was discovered in 1880 by the brothers Pierre and Jacques Curie, refers to a particular property of certain crystals that develops electrical polarization under mechanical stress. The term piezoelectric is derived from the Greek *piezein* that means to press. This effect is noticed with crystals that have unique anisotropic symmetry axis, the most common material of this type being crystalline quartz. Other piezoelectric materials commonly used are zinc oxide, lithium niobate and lithium tantalate.

The development of electric polarization in a piezoelectric quartz crystal is demonstrated in Figure 21.1. Due to the different electronegativity, each oxygen atom assumes a partial negative charge while each silicon atom assumes an opposite charge. At equilibrium (unloaded crystal, A) the atoms are positioned such that opposite charges balance each other and no overall charge can be noticed at the macroscopic scale. If the crystal is longitudinally loaded (B), the hexagonal frame is distorted such that an excess negative charge builds up at the upper part, whereas a positive charge appears at the other side. Although the overall charge of the system is still zero, a distinct electric charge appears at each crystal side. In order to detect the charge, the crystal is embedded between two metal plates (electrodes). Electrons at the metal surface are attracted or repelled by electrostatic forces, giving rise to a charge opposite to that at the crystal surface. If the crystal is transversely compressed, the distortion of the crystal lattice results in an opposite electric polarization (C). In this way, crystal distortion along particular directions results in an electrical potential difference between the metal plates. This voltage depends on the distortion degree, that is, on the applied stress. The piezoelectric effect is reversible as the voltage vanishes if the loading is removed.

The above effect is termed the *direct piezoelectric effect*. It is applied to perform conversion of an acoustic signal into an electrical one in the case of the piezoelectric microphone. A *converse piezoelectric effect* is equally possible. It consists of crystal distortion under the effect of a voltage applied to the metal plates. Ultrasound generators make

Chemical Sensors and Biosensors: Fundamentals and Applications, First Edition. Florinel-Gabriel Bănică.
© 2012 John Wiley & Sons, Ltd. Published 2012 by John Wiley & Sons, Ltd.

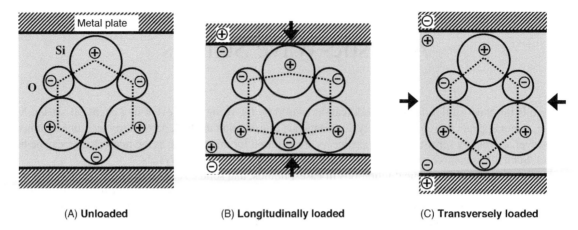

| (A) **Unloaded** | (B) **Longitudinally loaded** | (C) **Transversely loaded** |

Figure 21.1   Effect of the mechanical stress on charge distribution in a piezoelectric SiO$_2$ crystal.

use of this effect as they are based on the vibration of a piezoelectric crystal under the effect of an applied AC voltage. Therefore, a piezoelectric crystal can perform interconversion of electrical and mechanical energy and can act as a stress–electrical voltage transducer.

21.2 The Thickness–Shear Mode Piezoelectric Resonator

21.2.1 The Quartz Crystal Microbalance

Piezoelectric quartz crystals can be cut from a quartz crystal at various orientations with respect to the crystal optical axis. The most advantageous is the AT cut, depicted in Figure 21.2A because the thermal expansion coefficient of the crystal thus obtained is negligible. A shear-deformation mode crystal is shown in Figure 21.2B. It consists of a very thin (about 0.2 mm) AT-cut quartz crystal disk plated on each side with a metal layer (Au, Ag, Pt or others). The active area is that between the electrodes and this may have a diameter of 5 to 20 mm. Electrical contacts are available at the periphery of the plate.

An AT-cut quartz wafer undergoes *thickness–shear deformation* under a vertical load applied to its surface. Figure 21.3 presents the converse piezoelectric effect experienced by such a crystal. When the electrodes are not charged, the plate is at equilibrium (A) but once a voltage is applied, each end of the molecular dipole in the crystal is subject to electrostatic forces that shift laterally the plane in the crystal lattice and distort the crystal as shown in (B). If the voltage is reversed, the crystal is distorted in the opposite direction.

Figure 21.1B shows the shear deformation of a piezoelectric crystal under the effect of an applied DC voltage. The applications of such devices are based on crystal oscillation under the effect of an applied AC voltage so as to form a *piezoelectric resonator*. If an AC voltage is applied to the device, it is expected to undergo periodic shear-distortion following the voltage oscillation (Figure 21.3C). This actually occurs only within a very narrow frequency range

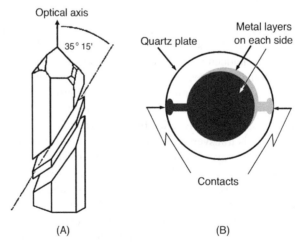

| (A) | (B) |

Figure 21.2   The thickness–shear piezoelectric crystal. (A) AT-cut of a α-quartz wafer. The quartz plate is cut at an angle of 35° 15′ with respect to the optical axis. (B) A thickness–shear mode resonator. The quartz wafer (some 0.2 mm thick and 15–30 mm diameter) is partially sandwiched between two thin metal electrodes fitted with contact pads at the periphery. The vibrating zone is that contained between the electrodes. (A) was reproduced with permission from [6]. Copyright 2000 Wiley-VCH Verlag GmBH & Co. KGaA.

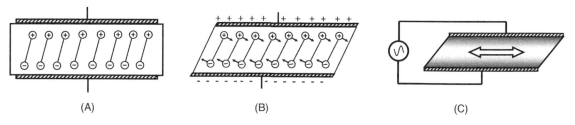

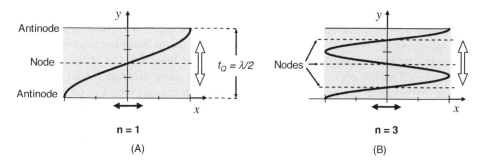

Figure 21.3 The converse piezoelectric effect in the case of the thickness–shear deformation. (A) No voltage applied; (B) DC voltage applied; (C) AC voltage applied. The arrow shows the vibration direction. (A) and (B) were adapted with permission from [4]. Copyright 1992 American Chemical Society.

Figure 21.4 The profile of a standing acoustic wave across the piezoelectric crystal in the fundamental mode (A) and third overtone (B) mode. Black arrows indicate the direction of particle displacement; white arrows indicate the direction of wave propagation. Nodes correspond to zero displacement and antinodes correspond to maximum displacement.

around the specific vibration frequency of the crystal itself. The frequency at which the vibration amplitude reaches a maximum is termed the *resonant frequency* f_0. This depends on the crystal thickness and also on the density and elasticity of the piezoelectric material. As the applied voltage increases, the shear-deformation amplitude increases as well. Due to the elastic deformation, potential energy is stored, as is also the case in a compressed spring. A voltage decrease causes the stress to reduce and due to its elasticity the crystal tends to return to the undistorted position. During the opposite half-period, the crystal undergoes distortion in the opposite direction. The resonance occurs when the voltage frequency matches the intrinsic vibration frequency of the crystal; under these conditions, the vibration amplitude is at a maximum. The amplitude of lateral displacement rarely exceeds a nanometer. It is maximum at the center but decreases gradually to zero at the periphery of the active zone.

The radial displacement profile at resonance is presented in Figure 21.4A for the so-called fundamental mode. In this case, the wavelength is twice the crystal thickness, t_Q. A standing wave forms at resonance, this implying that there is no displacement at the node plane and that the maximum displacement occurs at the lateral faces of the crystal ($y = 0$ and $y = t_Q$). The resonators can be operated at a number of overtones (harmonics), typically indexed by the number of nodal planes, n, parallel to the crystal surfaces. Only odd harmonics can be excited ($n = 1, 3, 5, \ldots$). The vibration amplitude is inversely proportional to the square of the overtone number. The vibration expands over the whole crystal volume; which is why the name bulk acoustic-wave transducer is assigned to such a device. Due to the deformation type, this device is better described as a *thickness–shear mode* (*TSM*) piezoelectric crystal.

As will be shown below, the key parameter in analytical applications of such devices is the resonant frequency. That is why the resonator is integrated with an excitation AC voltage source and a frequency counter (Figure 21.5). Note that a DC current cannot flow across the piezoelectric crystal because it consists of an insulator material. However, the electrodes–crystal system behaves somewhat as a capacitor and therefore allows an AC current to flow along the left-hand loop. The current amplitude is at a maximum when the AC voltage frequency matches the crystal resonant frequency.

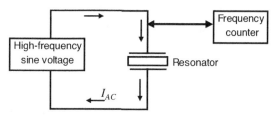

Figure 21.5 Schematics of electrical wiring of the piezoelectric resonator.

21.2.2 The Unperturbed Resonator

In order to understand the functioning principles of QCM sensors it is important to get a clear insight into the physical principle governing the functioning of the unperturbed resonator.

Piezoelectric crystals are electromechanical devices that perform interconversion of electrical and mechanical energy. That is why the resonator can be modeled either by a mechanical layout or an electrical equivalent circuit, as shown in Figure 21.6. The mechanical model (A) consists of a mass M attached to a spring (which indicates energy accumulation by the elastic deformation of the crystal) and a piston that represents energy dissipation by internal friction and damping from the crystal mounting. Once disturbed, this system starts oscillating as shown by the black arrow, but the oscillation dampens due to energy dissipation in the piston. If energy is periodically provided the oscillation is maintained for an undefined time. The spring accumulates energy by extension or compression and its electrical analog is from this standpoint the capacitor C_1. Energy dissipation in the equivalent circuit (Figure 21.6B) is accounted for by the resistor R_1 and the inertial element M in (A) (which accumulates potential energy) is represented by the inductor L_1 in (B). The series combination of R_1, C_1, and L_1 represents the *motional (or acoustic) branch* of the network because AC current flows along it only when the crystal vibrates at resonance. C_0 on the electrical branch is the static capacitance that arises from the electrodes located on the opposite sides of the crystal. A more accurate model includes in addition a parasitic capacitance that arises in the test fixture and is coupled in parallel with C_0. Of course, components of the motional branch depend on the geometrical and physical properties of the piezoelectric material, such as thickness, density and the shear stiffness.

As energy dissipation occurs under crystal vibration, the power loss is compensated by the driving AC voltage supply.

According to Figure 21.6B, the motional impedance of the unloaded resonator is (with $j = \sqrt{-1}$) [7]:

$$Z_{m,1} = R_1 + j\omega L_1 + \frac{1}{j\omega C_1} \tag{21.1}$$

Here, ω is the angular frequency ($\omega = 2\pi f$) R_1 is the resistance, L_1 is the inductance and C_1 is the capacitance of the pertinent component in Figure 21.6. The last two terms in the above equation account for reactances in the equivalent circuit. Reactance quantifies the opposition of a circuit element to a change of current, caused by the build-up of electric or magnetic fields in the element. Those fields act to produce a countervoltage that is proportional to either the derivative of the current (capacitive reactance, phase shift, $-90°$), or the integral of the current (inductive reactance, phase shift, $90°$). The absolute value of an inductive reactance is $X_L = \omega L$ where L represents the inductance. For a capacitor of capacitance C, the absolute value of the reactance is $X_C = 1/\omega C$. Note that a reactance introduces a phase shift, in contrast to a pure resistance. Therefore, $Z' = R_1$ is the real part of the motional impedance, whereas the imaginary part is $Z'' = X_L + X_C$.

The resonance occurs when the impedance of the motional branch is at a minimum, that is, when $Z'' = 0$. At frequencies far from the resonant value, $Z_{m,1}$ is very high and the AC current flows along the electrical branch only. However, near the resonant frequency, the motional impedance becomes very low as a standing acoustic wave develops within the crystal. In this case, the contribution of the motional branch to the total impedance is negligible and current flows mostly along this branch.

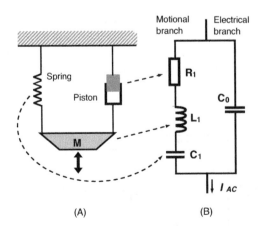

(A)   (B)

Figure 21.6 The mechanical model (A) and the equivalent circuit of the TSM unperturbed resonator (B). The total electrical impedance of the equivalent circuit is similar to that of the resonator itself.

21.2.3 QCM Loading by a Rigid Overlayer. The Sauerbrey Equation

The most common application of the QCM is microweighing. It is therefore essential to consider the effect of mass loading on the resonance frequency.

The simplest case to be considered is the loading of one of the crystal surfaces by a thin rigid layer with uniform thickness and perfect adherence to the surface. Experiments proved that crystal loading causes the resonant frequency to decrease in a measurable amount, Δf_m. This case was first addressed in 1959 by Sauerbrey who derived and tested a relationship between frequency change and the loading mass [8]. The derivation of the Sauerbrey equation is straightforward and will be given next in order to provide a better understanding of the functioning principle of the QCM as well as the limitations of this equation. Referring to Figure 21.4A, it is clear that the standing wave can form only when the wavelength (λ) is twice the crystal thickness (t_Q), that is:

$$t_Q = \frac{\lambda}{2} \tag{21.2}$$

For the unloaded resonator, the wavelength is related to the resonant frequency (f_0) as $\lambda = v_{tr}/f_0$, where v_{tr} is the propagation velocity of the transverse wave. Hence, the resonance condition is:

$$f_0 = \frac{v_{tr}}{2t_Q} \tag{21.3}$$

Coating with a rigid layer expands the thickness of the vibrating system by Δt and Equation (21.3) assumes in this case the following form:

$$f_0 + \Delta f_m = \frac{v_{tr}}{2(t_Q + \Delta t)} \tag{21.4}$$

Assuming that Δf_m is negligible with respect to f_0, the above equations leads to:

$$\frac{\Delta f_m}{f_0} = -\frac{\Delta t}{t_Q} \tag{21.5}$$

In order to account for mass variation, the crystal thickness will be expressed as a function of the crystal density, ρ_Q, volume, V_Q, the area, A, and mass of the vibrating zone, m_Q, as follows:

$$t_Q = \frac{V_Q}{A} = \frac{m_Q}{A\rho_Q} \tag{21.6}$$

Analogously, the thickness of the coating layer is:

$$\Delta t = \frac{\Delta m}{A\rho_L} \tag{21.7}$$

where Δm and ρ_L represent the mass and density of the coating layer, respectively. By inserting in Equation (21.4) the above thickness expressions and assuming that $\rho_Q \approx \rho_L$ one obtains the Sauerbrey equation in the following form:

$$\frac{\Delta f_m}{f_0} = -\frac{\Delta m}{m_Q} \tag{21.8}$$

This equation proves that the frequency decreases proportionally to the mass change. In order to put into evidence the effect of the physical properties of the crystal, m_Q will be substituted by $V_Q\rho_Q = At_Q\rho_Q$. On the other hand, the crystal thickness and the resonant frequency are related by Equation (21.3) that includes the wave velocity. This latter depends on material density (which represents the inertial opposition to displacement) and its shear elastic modulus (μ_Q) that account for the accumulation of potential energy, as is the case for the deformation of a spring. Acoustics theory proves that the wave velocity along the y-axis in Figure 21.4A is:

$$v_{tr} = \sqrt{\frac{\mu_Q}{\rho_Q}} \tag{21.9}$$

Therefore, the thickness–frequency relationship is:

$$t_Q = \frac{1}{2f_0}\sqrt{\frac{\mu_Q}{\rho_Q}} \tag{21.10}$$

By substituting in Equation (21.8) the pertinent expression for m_Q and then using Equation (21.10) one obtains the Sauerbrey equation in the following widely used form:

$$\Delta f_m = -\frac{2f_0^2}{\sqrt{\mu_Q \rho_Q}}\frac{\Delta m}{A} = -C_f \Delta m \qquad (21.11)$$

This equation proves that the resonant frequency decreases as the *mass loading per surface unit* increases. For a particular type of crystal, the mass sensitivity constant C_f can be introduced. This constant is proportional to the squared resonant frequency. That is why 5- to 10-MHz resonators are commonly employed as mass transducers in order to achieve good mass sensitivity. As the QCM responds to mass loading per surface unit, the sensitivity is independent of the crystal diameter. However, in order to avoid edge effects, resonators with the diameter less than 5 mm are seldom used.

The previous approach addressed only the fundamental frequency case. If a higher overtone is considered (for example, $n = 3$), then the standing wave develops when the crystal thickness is $3\lambda/2$ (Figure 21.4B). Denoting by f_n the overtone frequency ($f_n = nf_0$, $n = 1, 3, 5 \ldots$) and using the same reasoning line as before, the Sauerbrey equation is obtained in the following form:

$$\Delta f_m = -\frac{2}{n\sqrt{\rho_Q \mu_Q}} f_n^2 \frac{\Delta m}{A} \qquad (21.12)$$

Because the sensitivity increases as the squared frequency, the sensitivity is enhanced when the resonator is operated at a higher overtone.

In order to illustrate the outstanding mass sensitivity of the QCM, a resonator of 5 MHz resonant frequency operated at the fundamental mode will be considered. Using numerical values of the constants in Equation (21.11) one obtains $C'_f = C_f/A = 56.6\,\text{Hz}\,\text{cm}^2\,\mu\text{g}^{-1}$. That means that the resonant frequency changes by 56.6 Hz when a mass load of 1 μg/cm^2 is applied. As the frequency resolution can be as low as 1 Hz, it can be seen that that a mass load of about 20 ng cm^{-2} can be accurately measured with such a device. The exceptional mass sensitivity of this transducer fully justifies its assigned name (QCM); some authors prefer to call it a *quartz crystal nanobalance* (QCN).

The above derivation of the Sauerbrey equation involved a series of simplifying assumptions that limit to some extent its applicability. In this respect, it should be remembered that the loading film is treated as an extension of crystal thickness. Therefore, this equation applies only to systems in which the following three conditions are met: the deposited mass must be rigid, evenly distributed, and the mass change must be less than 2% of the crystal mass (that is, $\Delta f_m/f_0 < 0.02$). These conditions are best fulfilled by solid layers such as metals and oxides. Molecularly thin films behave also as a rigid layer since the energy dissipated is negligible [9]. In addition this equation holds accurately only if the resonator is operated in vacuum or in a gaseous atmosphere such that the mechanical energy is not dissipated by friction. Under these conditions, the mass sensitivity of the QCM can be calculated from its resonant frequency and the intrinsic properties of the piezoelectric material. In addition, the mass sensitivity is independent of the physical parameters of the deposited material, except its mass. Therefore, no preliminary calibration is required in gas phase applications provided the deposited material covers evenly the active vibrating area. The Sauerbrey equation has been firmly validated using various solid films and the relative error in the film mass determination by means of this equation did not exceed 2%.

21.2.4 The QCM in Contact with Liquids

The QCM is very often operated in contact with a liquid that is entrained in vibration by the resonator surface. This implies that a part of the mechanical energy is dissipated within the fluid, which brings about drastic changes in the vibration parameters. Vibrational coupling with the fluid depends on the fluid rheological properties that determine its deformation behavior under applied stress. From the rheology standpoint we can distinguish Newtonian liquids (such as water and dilute aqueous solutions) and non-Newtonian fluids (e.g., a dense suspension of corn flour in water). The behavior of Newtonian liquids is illustrated in Figure 21.7 for the case in which it is in contact with a moving plate. Due to friction with the plate wall, the plate movement gives rise to a shear stress T_{xy} (drag) that causes the fluid to move in the same direction. The stress represents the force per unit area along the movement direction. It decreases with the distance y due to the friction between successive liquid layers. As a consequence, the flow rate varies along the vertical axis giving rise to a velocity gradient $\partial v/\partial y$. Internal friction is quantified by the *dynamic viscosity*, η, which is defined by the following equation:

$$T_{xy} = \eta \frac{\partial v}{\partial y} \qquad (21.13)$$

The viscosity of a Newtonian liquid is constant and independent of the rate of the stress application. An elastic material recovers its initial shape after removing the stress, because of the potential energy stored under deformation. Newtonian liquids do not behave like that because the applied energy is dissipated by internal friction.

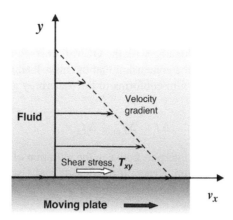

Figure 21.7 Drag of a Newtonian liquid by a moving plate. The fluid velocity decreases as the distance from the surface increases.

Non-Newtonian fluids display a more intricate behavior under stress. Among them, the viscoelastic fluids are particularly relevant in this context because biopolymer gels used as sensing layers in a number of QCM biosensor behave as viscoelastic fluids. A viscoelastic fluid displays both viscous and elastic properties. When the stress is removed, they do not return completely to the initial shape (as an elastic material would do) because energy is partially dissipated by internal friction.

21.2.5 The QCM in Contact with a Newtonian Liquid

If a face of the resonator is immersed in a liquid, the liquid dampens the vibration by friction and causes the resonant frequency to lessen. At the same time, the adjacent liquid layer is subject to shear stress due to the crystal vibration and is itself entrained in periodical shear deformations. The stress in the liquid is maximal at the contact with the resonator surface but it reduces quickly with the distance from the surface due to friction within the oscillating liquid. So, under the effect of the resonator oscillation, a damped wave propagates in the liquid, and decays over a very short distance (Figure 21.8A). The frequency change due to the liquid contact, Δf_L, depends on liquid density ρ_L (inertial parameter) and viscosity η_L (dissipative parameter) according to the following equation [10]:

$$\Delta f_L = -f_0^{3/2} \sqrt{\frac{\eta_L \rho_L}{\pi \mu_Q \rho_Q}} \tag{21.14}$$

This equation holds when the thickness of the liquid layer is much greater than the wave decay length ($\delta_L = \sqrt{\eta_L/\pi f_0 \rho_L}$) that is of some 250 nm for a 5 MHz resonator in water. Therefore, the transfer of the QCM from air to a viscous liquid causes the resonant frequency to decrease although no rigid layer forms at its surface. The frequency change depends on liquid density and its viscosity and is independent of the resonator area. It is equivalent to that produced by a hypothetical rigid liquid film of thickness δ_L.

Equation (21.14) has been derived under the assumption that the liquid film does not slip at the contact with the resonator. This is only an approximation; in fact, film slippage does occur and this enhances the energy dissipation. In addition, due to the surface rugosity, a small amount of liquid at the interface vibrates almost like a rigid layer. Consequently, the real frequency change can be much higher than that predicted by Equation (21.14) (that is, some

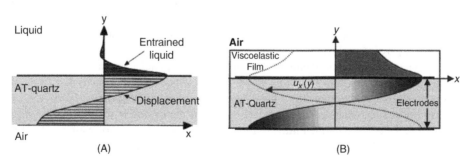

Figure 21.8 Vibration profile of a TSM resonator in contact with a liquid. (A) Semi-infinite Newtonian liquid (the liquid thickness is much greater than the wave decay length); (B) finite thickness viscoelastic film. The thickness of the film is exaggerated relative to that of the quartz disk; $u_x(y)$ represents the shear displacement. (A) Reproduced with permission from [11]. Copyright 1993 American Chemical Society. (B) Reproduced with permission from [12]. Copyright 2000 American Chemical Society.

650 Hz in water). Equation (21.14) has been retrieved as a limiting case of a more general equation including the coefficient of friction between the liquid film and the resonator surface [9].

It is often of interest to measure the mass loading with the QCM immersed in a liquid. A theoretical approach of this problem under nonslip conditions led to the conclusion that the mass loading and the liquid contact produces an additive effect as long as the rigid overlayer film conforms to the Sauerbrey assumptions [13]. Therefore, the total frequency change Δf can be formulated as:

$$\Delta f = \Delta f_{\mathrm{m}} + \Delta f_{\mathrm{L}} \tag{21.15}$$

The additive behavior implies that there is no interplay between the vibration of the rigid film and the wave propagating into the liquid except for energy dissipation. Hence, if the density and viscosity of the liquid are invariable, the frequency change reports only on the change in the mass loading.

21.2.6 The QCM in Contact with a Viscoelastic Fluid

From the standpoint of chemical sensor applications, it is of interest to examine the case in which a finite-thickness viscoelastic film is coated on top of the resonator. This is because in many instances, the recognition layer consists of biopolymers that behave as viscoelastic liquids. The vibration profile of such a system is shown in Figure 21.8B which demonstrates that a standing wave also develops in the overlayer. However, it is dampened as a result of power dissipation and, in addition its phase is shifted with respect to the standing wave in the resonator. That is why the frequency shift induced by a viscoelastic film depends on its mass in an intricate way. A more detailed description of this system can be made by considering its complex impedance, as shown in the next section.

21.2.7 Modeling the Loaded TSM Resonator

21.2.7.1 Basic Principles

The basic physical process in TSM resonators is the propagation of acoustic waves induced and sustained by an applied AC voltage. Therefore, such a system has a composite character as it includes both mechanical and electrical components. The behavior of this system can be analyzed in a more practical way if it is represented only by electrical circuit elements such as resistors, capacitors, and inductors. Such an approach is also used to model electrochemical cells and has been introduced in the chapter dealing with electrochemical impedance spectrometry (Chapter 17) where more details on electrical modeling are available. The advantage of the electrical modeling is twofold. First, it has a sound theoretical basis developed in the frame of the theory of electric network. Secondly, it relies on electrical measurements only and makes use of standard instruments for the analysis of electrical networks. Such instruments are known as *network analyzers*.

Electrical modeling of acoustic-wave devices is of a particular interest when a deeper insight into the functioning principles of acoustic-wave sensors is sought. Such sensors includes, as a rule, a sensing layer that can consist of synthetic polymers or biopolymers that display an intricate behavior under the mechanical stress induced by vibration. Hence, a careful investigation of such properties is essential in the design and operation of acoustic-wave sensors. On the other hand, modeling may suggest particular data processing algorithms that enhance the analytical power of the sensor.

Impedance analysis reveals the effect of the vibrating frequency on the electric impedance of the system. The impedance Z represents the opposition of the system to current flow and is a complex quantity. It is therefore represented by two numbers: the real part Z' and the imaginary part Z''. The complex impedance is hence represented as:

$$Z = Z' + jZ'' \tag{21.16}$$

An important parameter associated with the impedance is the phase angle $\phi = \tan(Z''/Z')$; it represents the delay of a periodical quantity with respect to a reference periodic signal. In the case of acoustic-wave sensors, the phase angle quantifies the loss of energy by internal friction.

An acoustic-wave device vibrates with maximum amplitude at the resonant frequency and the amplitude decays abruptly if the applied frequency deviates from this value. That is why impedance analysis of such devices is confined to a very narrow frequency range (typically 1 kHz) around the resonant frequency.

In order to perform impedance analysis an equivalent circuit of the device is inferred either intuitively or by a theoretical approach. Impedance data are then fitted to this model using dedicated software and finally the fit quality is checked. A good fit is not always a sound indication of the correctness of the model. It should be combined with a careful test of the physical fit of the model to the physical system.

Two types of electrical models of the acoustic-wave devices are commonly employed: the *transmission line model* (Mason) and the *lumped element model* often quoted as the Butterworth–Van Dyke model (Figure 21.6). The second model is less accurate but more handy.

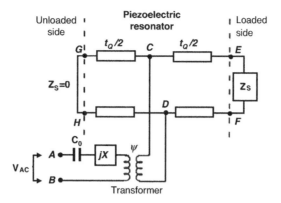

Figure 21.9 Mason transmission line model of the TSM piezoelectric resonator.

In the transmission line model a composite resonator is represented as an arrangement of the respective number of transmission lines in series [7,14]. Energy supply is represented by an electrical transformer with the secondary coil connected to the electrical port at A and B. Both the transformer turn ratio ψ and the element jX depend only on the physical and geometrical parameters of the resonator. Each transmission line has two acoustic ports (EF and GH) each representing one face of the resonator (Figure 21.9). The acoustic ports are connected to the electrical ports by a transmission line that represents the phase shift and the power loss experienced by an acoustic wave propagating across the resonator thickness. Mechanical loading of the resonator is represented by the mechanical impedance Z_s that is by definition the ratio of the surface stress to the particle velocity at the resonator surface (Figure 21.7):

$$Z_s = \frac{T_{xy}}{v_x}\bigg|_{y=0} \tag{21.17}$$

For an unloaded resonator, both sides can be considered as stress-free edges that give $Z_s = 0$, as in the case of a short circuit at both acoustic ports. Commonly, in sensor applications one of the sides is unloaded ($Z_s = 0$) whereas the loading layer is represented by a nonzero mechanical impedance Z_s (Figure 21.9). The total impedance can be represented by the motional impedance Z_m in parallel with the static capacitance C_0.

Although the transmission line model is very accurate the less-accurate lumped-element model is often preferred because it is easier to visualize and often allows for an easier derivation of characteristic parameters. Theoretically, if the Z_s/Z_{m1} ratio does not exceed 0.1, the second model predicts a response that is very close to that given by the first. The approach in next Section is based on the lumped-element model.

21.2.7.2 The Unloaded Resonator

As shown in Figure 21.6B the unloaded resonator can be modeled by an equivalent circuit composed of a resistor, a capacitor and an inductor that account, respectively, for energy dissipation, energy storage and inertial behavior. The impedance of this series combination yields the motional impedance of the unloaded resonator Z_{m1} that, for AT-cut quartz, has an absolute value of $8.8 \times 10^6 \, \text{kg m}^{-2}\,\text{s}^{-1}$.

When dealing with equivalent-circuit analysis, one makes a distinction between the *series* and the *parallel resonant frequency*. The first is the resonance frequency of the motional branch alone, whereas the second applies to the parallel combination of both branches and is reported by the measuring instrument.

21.2.7.3 The Loaded Resonator

Loading effect can be represented by adding the surface motional impedance Z_{m2} in series with the motional impedance of the unperturbed resonator (Figure 21.10A). For the case in which the coating layer is much thinner than the crystal, the additional impedance consists of a resistor R_2 and an inductor L_2 in series. The first element accounts for energy dissipation in the overlayer, while the second represents the mass load. Assuming that the coating is rigid and does not experience any shear deformation under vibration, energy dispersion in this layer is negligible ($R_2 \approx 0$), whereas the reactance of the inductive element L_2 is proportional to the density of the coating layer. The overall mass of the vibrating system is therefore increased to some extent, which causes the resonant frequency to decrease as:

$$\frac{\Delta f}{f_0} = -\frac{2f_0}{Z_{m1}}\frac{\Delta m}{A} \tag{21.18}$$

Here, Δf is the frequency change, f_0 is the resonant frequency of the unloaded crystal, Δm is the mass of the coating (evenly distributed), A is the area of the active surface, and Z_{m1} is a constant determined by the physical properties of the piezoelectric resonator. Equation (21.18) is in fact an alternative form of the Sauerbrey equation.

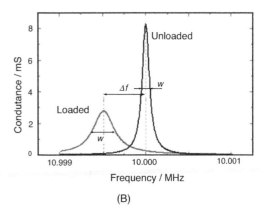

Figure 21.10   Effect of mass loading on the piezoelectric TSM resonator. (A) The equivalent circuit; (B) Conductance spectra of the unloaded and loaded resonator with a resonant frequency of 12 MHz. Conductance is the real part of the complex admittance.

According to this model, the total motional impedance $Z_{m,t}$ of the loaded resonator is:

$$Z_{m,t} = Z_{m1} + Z_{m2} = (R_1 + R_2) + j\omega(L_1 + L_2) + \frac{1}{j\omega C_1} \tag{21.19}$$

Impedance is a complex number that can be separated into a real and an imaginary component. The imaginary component displays a zero phase shift with respect to the applied AC voltage and its reciprocal is termed conductance. If the conductance is plotted against the frequency around the resonant frequency, one obtains a sharp peak, as shown in Figure 21.10B, for the unloaded resonator. The peak frequency is the *series resonant frequency*. The sharpness of the peak is quantified by the full width at half-maximum w. A key parameter related to the bandwidth is the *quality factor Q* that is by definition the quotient of the stored electrical energy and the energy dissipated in one period:

$$Q = 2\pi \frac{E_{\text{stored}}}{E_{\text{dissipated}}} \tag{21.20}$$

The 2π factor is included for convenience. For very high Q values, Q is related to the bandwidth as follows:

$$Q = \frac{f_0}{w} \tag{21.21}$$

For the unloaded resonator, the quality factor is:

$$Q = \frac{X_{L1}}{R_1} = \frac{\omega L_1}{R_1} \tag{21.22}$$

As R_1 is very low, the quality factor of the unloaded resonator is about 105 and the bandwidth is very low. Consequently, the resonant frequency can be measured very accurately.

If a load is applied, it can bring about supplementary dissipation in addition to the inertial contribution. Therefore, the resonant frequency shifts to lower values but, at the same time, the bandwidth increases and the quality factor decreases (Figure 21.10B). The quality factor is in relation to the phase shift ϕ of the acoustic wave, that is, the phase difference of the acoustic wave between the crystal surface and the outer surface of the overlayer. When the quality factor becomes sufficiently low, the AC generator circuit can no longer sustain the oscillation and the QCM ceases to function.

Generally, the overlayer motional impedance is a complex quantity:

$$Z_{m2} = Z'_{m2} + jZ''_{m2} \tag{21.23}$$

The imaginary part Z''_{m2} determines the change in the series resonant frequency Δf_s as follows:

$$\frac{\Delta f_s}{f_0} \propto -Z''_{m2} \tag{21.24}$$

At the same time, the real part Z'_{m2} describes the energy loss within the overlayer:

$$\Delta R \propto Z'_{m2} \tag{21.25}$$

Table 21.1 Functioning regimes of the QCM sensors. M and V are defined in Equations (21.26) and (21.27); ΔM and ΔV are, respectively, the change in the M and V factors induced by resonator loading [15].

1	Gravimetric sensor	$M \to M + \Delta M$	$V = 1$
2	Mass and material effect sensor	$M \to M + \Delta M$	$V \to V \pm \Delta V$
3	Film-properties sensor	$M \approx$ constant	$V \to V \pm \Delta V$

A clear description of the resonator response to loading can be formulated in terms of two factors: the mass factor M and the acoustic (vibrational) factor V [15]:

$$M = \omega \rho_1 t_1; \quad V = \frac{\tan \phi}{\phi} \tag{21.26}$$

where ρ_1 and t_1 are, respectively, the density and the thickness of the overlayer, and ϕ is the phase shift. With these parameters, the motional impedance associated with the overlayer is:

$$Z_{m2} = jMV \tag{21.27}$$

It is apparent from the previous discussion that the QCM parameters could vary in response to both mass loading (the M factor) and energy loss in the overlayer (the V factor). Therefore, the QCM is not only a microbalance but also provides information on other physical parameters of the loading layer. Accordingly, one can distinguish various functioning regimes of the QCM according to the properties of the overlayer as shown in Table 21.1 [15].

In the first case in this table, the QCM functions as a true gravimetric sensor and reports on the mass change in the overlayer. In the second cases, the response is not purely gravimetric. Case 2 corresponds to a composite response that depends on both mass change and changes in the viscoelastic properties of the overlayer. In the third case, the QCM responds only to the change of the viscoelastic properties of the overlayer under constant-mass conditions. An example of this type is represented by the QCM in contact with a liquid.

21.2.7.4 The Ideal Mass Loading

This is represented by a rigid overlayer that moves synchronously with the resonator surface. The transverse wave propagates across this layer with a negligible phase shift. In this limiting case, no energy dissipation occurs and hence $R_2 = 0$. The reactance X_2 $(= \omega L_2)$ of the inductive element represents the imaginary component of Z_{m2} and in the fundamental mode it is [12]:

$$X_2 = j\omega \Delta m \tag{21.28}$$

This is the case 1 in Table 21.1, that is, the case in which the QCM elicits a pure gravimetric response and obeys the Sauerbrey equation. Both the phase shift and the motional resistance R_2 are negligible, which implies an absence of energy dissipation in the overlayer.

21.2.7.5 Loading by a Newtonian Liquid

If one side of the resonator is in contact with a thick Newtonian liquid layer, R_2 and L_2 in Figure 21.10A represent, respectively, the energy dissipated into the contacting liquid and the kinetic energy of the entrained liquid layer. In this case, equal amounts of power are distributed to each component, as shown by the following equation that applies under resonance conditions [7]:

$$R_2 = X_2 = \frac{\sqrt{\pi f \rho_L \eta_L}}{8K^2 C_0 Z_{m1} f_0} \tag{21.29}$$

The liquid contact translates the peak of the admittance magnitude to lower frequency (in accordance with Equation (21.14)) and, at the same time, reduces the quality factor due to power dissipation (Figure 21.10B). The identity of R_2 and X_2 (and implicitly a phase shift of $\pi/4$) proves that the liquid that contacts the resonator is a Newtonian one. It should be kept in mind that Equation (21.29) has been derived under the assumption that no film slip occurs. If this is not the case, some deviations from the above conclusions would occur.

21.2.7.6 Viscoelastic Overlayer

The next discussion addresses the case of a finite-thickness viscoelastic film deposited on top of the resonator and in contact with air on the opposite side [12]. Therefore, no power dissipation occurs at the external surface of this system.

A viscoelastic material exhibits both elastic and viscous behavior. That is why the shear modulus G is a complex quantity:

$$G = G' + jG'' \tag{21.30}$$

The real part G' represents the stress component in phase with strain, giving rise to energy storage (elastic component). The imaginary part G'' represents the stress component $\pi/2$ radians out of phase, which causes power dissipation and accounts for the viscosity effect. The film can vibrate using power transferred from the resonator. That is why the resonator–viscoelastic film system can be treated as two coupled resonators [16]. A standing wave develops across the film due to the interference of the direct wave and the wave reflected at the film/air surface. As both elastic and viscous properties are manifest, the acoustic wave experiences both a phase shift and attenuation while traversing the film. When the phase shift reaches $n'\pi/2$ (where n' is an odd integer) interference becomes constructive, resulting in enhanced absorption of the acoustic power by the film and leading to maximum damping of the crystal resonance. Under such conditions, the film experiences resonance vibration. For a given resonant frequency of the resonator, film resonance is obtained at a specific film thickness.

Viscoelastic coatings could behave as in the cases 2 and 3 in Table 21.1, that is, the QCM response is not purely gravimetric. In case 2, which is the most common, the response is affected by both mass change and energy dissipation. In case 3, the response can change even if the mass does not change appreciably. Hence, the response modification is due mostly to certain changes in the structural properties of the film leading to enhanced dissipation.

An equivalent circuit consisting of a parallel R_2, L_2, C_2 network has been advanced for modeling the motional impedance of a viscoelastic film of finite thickness coated atop of a resonator oscillating in air [12] (Figure 21.11).

Near the resonant frequency, the components of the equivalent circuit are related to the physical properties of the film as follows [12]:

$$R_2 = \frac{2A|G|^2}{\omega t_1 G'} \tag{21.31}$$

$$C_2 = \frac{t_1 G'}{2A|G|^2} \tag{21.32}$$

$$L_2 = \frac{8}{(n'\pi)^2} \Delta m_c \tag{21.33}$$

where t_1 is the coating film thickness and Δm_c is the film mass. This model suggests that near-resonance impedance analysis allows inferring the mass of the viscoelastic film from the reactance of the L_2 element.

In a more advanced approach the resonant frequency is considered as a complex quantity $\tilde{f}$ [17,18]. In this case, the real part of the complex frequency represents the resonant frequency of the series equivalent circuit, whereas the imaginary part is the half-band-half-width Γ ($\Gamma = w/2$, Figure 21.10B). Therefore, the complex frequency is:

$$\tilde{f} = f + j\Gamma \tag{21.34}$$

and the complex frequency change is:

$$\Delta\tilde{f} = \Delta f + j\Delta\Gamma \tag{21.35}$$

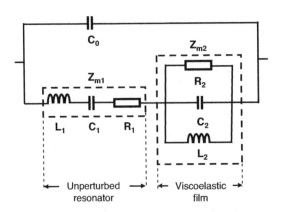

Figure 21.11   Equivalent circuit of a TSM resonator coated with a viscoelastic film.

The change in the complex resonant frequency is related to the load impedance by the following basic equation:

$$\frac{\Delta \tilde{f}}{f_f} \approx j \frac{Z_{m2}}{\pi Z_{m1}} = j \frac{Z_{m2}}{2\pi m_Q} \tag{21.36}$$

where f_f is a parameter approximately equal to the resonant frequency of the fundamental mode and m_Q is the areal mass density of the crystal (that is, mass per surface area unit). This equation applies only if the frequency change is very small compared with the resonant frequency.

For a viscoelastic film, the above equation assumes the following form:

$$\frac{\Delta \tilde{f}}{f_f} = -\frac{Z_{m2}}{\pi \sqrt{\rho_Q \mu_Q}} \tan\left(\frac{\omega}{Z_{m2}} m_f\right) \tag{21.37}$$

where m_f is the areal mass density of the film. This equation shows that the acoustic properties of the overlayer are fully specified by two parameters: its acoustic impedance and its mass per unit area. Therefore, it is not possible to derive simultaneously the thickness, density and viscoelastic parameters of a film from acoustic impedance measurements alone.

21.2.7.7 Multiple Loading

In many applications of the QCM, the surface is coated by multiple layers. Thus, a thioalkane chemisorbed film can serve for immobilization of a protein layer by crosslinking in order to obtain the sensing layer. The first layer could behave as an ideal mass load, whereas the second layer could display viscoelastic properties. Under these conditions, a linear combination of film impedances is a reasonable approximation. This implies that no power dissipation occurs across the first layer and the second one experiences at its lower side the same stress as in the case in which it was coated directly on the resonator. Simple addition of individual motional impedances is also applicable in the case of an ideal mass load in contact with a Newtonian fluid. In fact, this property led to Equation (21.15). However, in general, multiple layers do not add algebraically but combine in a nonlinear way. The most common nonlinear system is that of a viscoelastic film with a Newtonian fluid on top (Figure 21.12). As the wave propagates away from the rigid layer, it experiences a considerable decay. Typical decay lengths are 0, 2 and 0.2 μm for the rigid layer, the viscoelastic film and the aqueous phase, respectively. Due to the phase shift across the viscoelastic film the total motional impedance does not equal the sum of individual impedances of each layer. Instead, the total motional impedance depends in a more intricate fashion on the properties of each layer [7].

The application of the concept of complex frequency brought about a more accurate picture of the TSM resonator with both viscoelastic and Newtonian liquid loading [17,18].

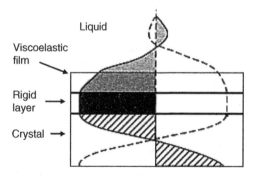

Figure 21.12 Propagation of the acoustic shear wave induced by a TSM resonator coated with a viscoelastic layer exposed to a liquid. Adapted with permission from [19]. Copyright 2003 Wiley-VCH Verlag GmBH & Co. KGaA.

21.2.8 The Quartz Crystal Microbalance with Dissipation Monitoring (QCM-D)

Previous sections proved that in the case of a soft material loading, both the resonance frequency and the quality factor change. Both these parameters can be inferred from impedance spectrometry data by a relatively laborious approach. A particular method of operating the QCM allows straightforward and fast extraction of both parameters [20]. This method, known as the quartz crystal microbalance with dissipation monitoring (QCM-D, [20]), proved to be very practical and is widely used currently particularly for investigating thin soft material films and biorecognition processes [21]. In this method, the resonator is excited at the resonant frequency for about 10 ms, then the power

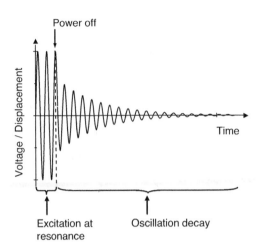

Figure 21.13   Principle of the QCM-D method. An excitation voltage is applied for about 10 ms, then the driving power supply is switched off and the decaying amplitude of the free-oscillating system is monitored. Typically, the decay time is of a few ms.

supply is disconnected and the system is left to oscillate freely (Figure 21.13). Due to energy dissipation, the oscillation amplitude decays exponentially as:

$$V(t) = V_0 e^{t/\tau} \sin(2\pi f_0 t + \phi) \tag{21.38}$$

where τ is the decay time constant and the time, t, is measured from the moment when power is switched off. The duration of the overall run, including data transfer is about 25 ms. Moreover, the system can be operated at multiple overtones in order to assess the effect of the frequency. Data processing allows the extraction of two parameters: the resonant frequency and the decay time constant. The decay time constant is related to the quality factor as:

$$Q = \pi f_0 \tau \tag{21.39}$$

By definition, the dissipation is the reciprocal of the quality factor:

$$D = \frac{1}{Q} \tag{21.40}$$

As the dissipation is related to the bandwidth, it results that this method reports on both terms of the complex frequency included in Equation (21.34).

The QCM-D method has found various applications in fundamental research of thin soft materials layers including biomaterials. Thus, it allows straightforward assessment of the validity of the Sauerbrey equation for particular systems. If the load is purely inertial (rigid overlayer) then the dissipation is very low. As a soft material overlayer can undergo hydration and swelling, the QCM-D method affords facile appraisal of such phenomena and provides insight into structural changes.

21.2.9   Operation of QCM Sensors

As an analytical device, the QCM can be used with either gas or liquid samples. In the second case, the QCM is used as a transducer in affinity sensors. In order to avoid electrical shorting via electrolyte solutions, the resonator is mounted such that only one face is exposed to the sample.

The resonant frequency depends on the temperature; that is why it is recommended to operate the sensor under thermostated conditions. In order to avoid interference due to ambient electromagnetic noise, the QCM should be installed in a Faraday cage.

As demonstrated before, the solution contact itself brings about a considerable change in the resonance frequency. That is why it is preferred in some applications to resort to the "dip-and-dry" operation method. In this method, the resonant frequency of the unaltered sensor is first measured in air and then the sensor is dipped into the sample and left to attain the equilibrium state indicated by an invariable frequency. Next, the sensor is removed from the sample, carefully washed to remove nonspecifically bound sample components and gently dried (e.g., by a stream of nitrogen). Finally, the frequency is measured again with the sensor in air. The frequency change reports on the mass change during the interaction with the analyte in the sample.

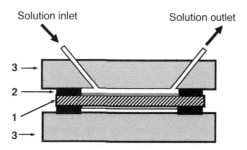

Figure 21.14 Flow cell for QCM sensors. (1) TSM resonator with the sensing layer at the upper side; (2) elastic O-ring; (3) plastic plate.

Affinity assays require a series of washing steps and also sensor regeneration by means of a solution containing immunocomplex-disrupting reagents. That is why it is convenient to install the sensor in a flow cell integrated with an automatic flow-analysis system. Such a cell is presented in Figure 21.14. The quartz crystal is fastened between two plastic material blocks by means of an elastic O-ring. Solutions are pumped through the upper chamber with a volume of about 50 μL and put in contact with the receptor-modified gold layer. Note that only one side of the resonator is exposed to the aqueous sample. In order to avoid gas bubble accumulation at the sensor surface, preliminary sample degassing is recommended.

A QCM flow-analysis system is shown in Figure 21.15. It includes a pump and a series of computer-controlled valves that allow flushing of the sensor with various solutions such as the sample, the regeneration solution and the washing buffer. Data supplied by the frequency counter are stored by the computer for further processing.

The flow-analysis system can be operated in two modes. In the flow-through mode, the fluid sample crosses the cell steadily at constant flow rate until the recognition process attains equilibrium. The frequency change is then read out and the valves are switched such as to allow the regeneration solution and then the washing buffer to flow along the sensor surface.

Alternatively, the analyzer can be operated under the flow-injection analysis regime. In this case, a stream of buffer solution flows steadily through the cell. A minute amount of sample is inserted in this stream by means of a sample-introduction valve and is carried as a plug to the cell. The frequency changes first in response to the recognition reaction but decays back to the background level as the sample plug is flushed out from the cell. Therefore, the response has a transient character and the maximum frequency change is used to derive the analyte concentration. This method is faster, but the contact time is relatively short and the recognition process cannot always reach the equilibrium state. This can in some instances impair the sensitivity and also the reliability of the results.

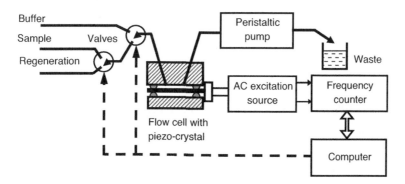

Figure 21.15 A semiautomatic setup for measurements with piezoelectric immunosensors.

21.2.10 Calibration of the QCM

In many analytical applications the frequency change is correlated with the analyte concentration by a calibration graph and the absolute mass of the overlayer is overlooked. However, it is often of interest to assess the loading mass of a loading layer. Even if many workers make use of the Sauerbrey constant derived from quartz properties, it could be useful to perform QCM calibration in solution in order to account for various effects of the liquid contact. Thus, in addition to energy loss by wave propagation in the liquid, deviations might occur due to surface roughness and sensor sealing.

The calibration can be performed by recording simultaneously the current and the frequency change during the electrodeposition of silver on one of the crystal side under cyclic voltammetry conditions. The mass change due to silver deposition/dissolution can be calculated from the associated electric charge by means of the Faraday electrolysis law. The electric charge is obtained as the time integral of the current. A plot of the recorded Δf_m vs. charge yields a straight line and the C_f constant can be derived from the line slope [19,22].

21.2.11 Outlook

The QCM was initially developed as an ultrasensitive mass transducer. This feature has been exploited for developing chemical sensors based on mass change in response to the recognition effect. The label-free character of these sensors made them very attractive for various applications in both gas and liquid sample analysis. Such sensors are manufactured by coating a piezoelectric resonator with a thin recognition layer.

However, the QCM is a true mass transducer only under severely limited conditions. The main limitation is the absence of acoustic energy dissipation within the vibrating system. Hence, the coating layer should be very thin and, to a good approximation it should behave as a rigid layer, both before and after the recognition event. This condition is met by solid overlayers and by molecularly thin layers of organic compounds. More generally, the Sauerbrey equation holds whenever the overlayer thickness is negligible with respect to the wavelength of the acoustic wave. QCM gas sensors fulfill this condition and generally obey the Sauerbrey equation.

If the QCM is put in contact with a Newtonian liquid, acoustic energy is dissipated into this liquid that results in an appreciable change of the resonant frequency and a reduction of the quality factor. However, if the overlayer has a rigid character, its mass change can be assessed as the liquid contact effect is constant at constant viscosity and density of the liquid.

The situation is much more intricate when the overlayer is a non-Newtonian liquid. In contact with such a material, the resonator experiences both resonant frequency change and large energy dissipation. The frequency change in this case is no longer a simple result of the mass change but is also influenced to a great extent by the viscoelastic properties of the coating. This results in a mass sensitivity much lower that that predicted by the Sauerbrey equation. Such a situation is typical for affinity sensor based on biomacromolecules as receptors. In many instances, such a sensor cannot achieve the expected sensitivity and signal amplification by postrecognition expansion of the mass of layer is needed. Although this procedure proved successful, it requires additional reagents and expands the duration of the assay.

The QCM is more than just a microbalance but is also a valuable method of investigation of viscoelastic properties of soft materials. In fact, the term "balance" makes sense even for nongravimetric applications if it is understood in the sense of a *force balance*. This terminology is justified by the fact that the force exerted by the crystal upon the overlayer is balanced by a force originating from the shear gradient inside this layer.

Sensor applications of the QCM still rely mostly on measurement of frequency change. However, recognition methods involving macromolecules or cells bring about many other effects in addition to the mass change, such as modification of the layer thickness and shear modulus. Mass variation may result not only from the addition of supplementary material by the recognition event. This event can be accompanied by water exclusion/inclusion or similar processes involving hydrated ions. That is why a more detailed picture is obtained by impedance measurements. Despite the cost and intricacy of this method, possible applications in sensor transduction may be expected. The QCM-D method, which provides essentially the same kind of information, appears as a convenient substitute for impedance analysis.

Questions and Exercises (Sections 21.1; 21.2)

1 What is the piezoelectric effect? Comment on the reversibility of this effect and its practical applications.
2 Review the physical and constructive features of the TSM piezoelectric crystal.
3 Draw schematically the effect of an applied DC or AC voltage on the TSM crystal. What conditions should be fulfilled by the AC voltage in order to affect the crystal?
4 What are the characteristic features of a TSM acoustic wave and what conditions should be fulfilled by the wave in order to obtain resonant vibration?
5 Comment on the meaning of each component in the mechanical and electrical model of the resonator.
6 What is the main application of the TSM resonator in analytical chemistry? Compare its performance with that of other devices used for similar purposes.
7 What is the output signal of the QCM, and how does it depend on the mass load?
8 How can the mass sensitivity of the QCM be enhanced? What is the effect of the crystal thickness? Does the surface area have an effect on the sensitivity?
9 Review the limitations of the Sauerbrey equation.
10 For a 0.2-mm thin resonator, assess whether or not the Sauerbrey equation holds for the following loads uniformly distributed over the active area:
   a) a 0.5-μm silver layer;
   b) a short-chain thioalkane layer;

c) a 10-μm thick rubber layer;

d) a multilayer formed of large-molecule proteins.

11 Review the behavior of the QCM in contact with water. What are the practical consequences of this behavior on the QCM applications to sensing in liquid samples?

12 What are the effects of a biomacromolecule load on the behavior and the response of the QCM?

13 Review advanced QCM operation methods and comment on their possible advantages in biosensor applications.

14 Review the methods of operation of QCM-based sensors. What precautions should be taken in order to obtain reliable results and maximum sensitivity?

15 Using the Sauerbrey equation, calculate the C_f constant (in Hz/μg) for a resonator with the active area of 5 mm diameter and resonant frequency of 10 MHz using the following quartz constants: $\rho_Q = 2.648\,\mathrm{g}$ cm^{-3}, $\mu_Q = 2.947 \times 10^{11}\,\mathrm{g\,cm}^{-1}\,\mathrm{s}^{-2}$.

Answer: 44.45 Hz/μg.

16 What is the lowest detectable mass by the above device if the frequency resolution is 1 Hz?

Answer: 22.5 ng

17 What is the value of the C_f constant for the same sensor if it is operated at the third overtone?

Answer: 133.4 Hz/μg.

21.3 QCM Gas and Vapor Sensors

The gas phase provides the most convenient operating conditions for the QCM. It is therefore not surprising that gas analysis has been among the first applications of QCM sensors [23–25]. Such sensors can be manufactured by coating a recognition layer on one or both sides of the resonator. The sensor can be exposed directly to the gas sample but in some instances, when the sample contains dust or aerosol, it is better to install the sensor in a gas flow cell to which the sample is pumped through a filter.

In general, QCM gas and vapor sensors elicit good reversibility. Due to the analyte volatility, QCM gas sensors can be regenerated by flushing with a pure gas, eventually accompanied by heating.

The simplest applications rely on using the metal layer on the resonator as the sensing element. Thus, mercury vapor interacts with the surface of gold electrodes forming amalgams that allows the determination of mercury vapors in atmosphere down to 1 ng/L [26]. Sensor regeneration can be achieved by heating in a pure gas atmosphere. As mercury vapors accumulate steadily under contact with the gas, this device can function as a dosimeter for assessing the mercury vapor exposure.

Palladium is known to absorb strongly hydrogen and deuterium. Hence, a QCM with palladium electrodes can function as a sensor for both of the above gases or for assessing the deuterium content in a hydrogen–deuterium gas mixture [27].

Coating with a specific sensing layer permits the determination of various gas compounds. For example, aliphatic amine coatings allow detecting acid gases such as hydrochloric acid or sulfur dioxide. Aromatic hydrocarbons in a hydrocarbon mixture can be detected by means of certain metal complexes that form coordinate bonds with π-electrons in aromatics but that do not interact with aliphatic compounds.

Coating with a specific biomaterial receptor allows selective determination of certain organic compounds. Thus, a formaldehyde vapor QCM sensor has been manufactured by coating the resonator with formaldehyde dehydrogenase and its cofactors (NAD$^+$ and reduced glutathione) [28]. The products of the enzymatic reaction accumulate at the sensor surface and cause the resonant frequency to decrease proportionally to the formaldehyde concentration. Parathion in water samples has been determined in the gas phase by means of an antiparathion antibody-coated QCM [29]. The sample is flushed by a gas stream that carries out the volatile analyte to a QCM gas flow cell.

Better sensitivity in gas analysis can be achieved by means of surface-acoustic wave transducers (Section 21.6) that, in addition, allow for miniaturization and microarraying.

21.4 QCM Affinity Sensors

In principle, any kind of recognition process based on affinity interactions can be monitored by means of a QCM if a suitable receptor layer is formed over the surface of the resonator crystal. Much research effort has been devoted to the development of QCM immunosensors owing to the label-free character of this approach [30–32]. However, the

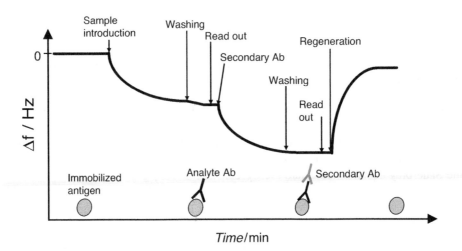

Figure 21.16 Operational sequence in a quartz crystal sensor immunoassay.

viscoelastic characteristic of the sensing layer causes most of the vibrational energy to be dissipated, and the frequency change response can be very small despite the large mass change produced by the recognition event. That is why it is often best to resort to signal amplification by postrecognition modification with high-mass particles.

In addition to immunocompounds, great interest is also shown in using synthetic materials for the preparation of sensing layers in affinity QCM sensors.

21.4.1 QCM Immunosensors

QCM immunosensors are best operated under flow-analysis conditions so as to automate the process. The main steps in a flow-analysis QCM immunoassay are reviewed in Figure 21.16, which refers to the case in which an antibody is determined by means of an antigen-modified quartz crystal. Before sample introduction, it is important to wait until the frequency stabilizes. After sample introduction, the receptor–analyte binding gradually lowers the frequency and when the frequency reaches a stable value the sample is washed out by a buffer solution in order to remove non-specifically bound compounds. Then, the signal is recorded and the sensor is regenerated by disrupting the analyte–receptor complex.

If amplification is sought, a secondary antibody is added after the washing step, as shown in Figure 21.16. Formation of a ternary complex brings about an additional frequency shift.

The previous scheme represents a direct assay, and the amplification by a secondary antibody can be viewed as sandwich format labeling. Competitive immunoassay by QCM immunosensors is also achievable. In this case, an analyte-analog is added to the solution along with the analyte itself. If the mass of the analyte-analog is much lower than that of the analyte, the frequency decreases with increasing analyte concentration.

An example of a QCM immunosensor developed for the diagnosis of the viral infection caused by the African Swine Fever disease (ASF) is shown in Figure 21.17A. The ASF-specific antibody can be detected by means of a sensing layer formed from a protein that is found in the outer membrane of the virus and that functions as an antigen. This protein has been immobilized by irreversible adsorption on a gold-electrode surface. The course of the assay is

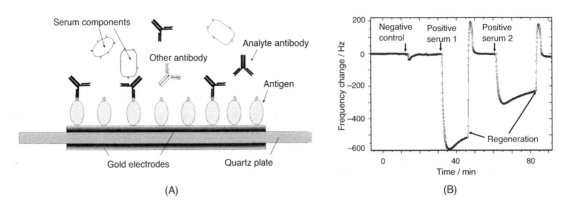

Figure 21.17 (A) Configuration of a piezoelectric immunosensor for the diagnostic of African Swine Fever disease; (B) The course of an immunoassay test with the sensor in (A). Each sample was diluted 10-fold with the pH buffer solution. Adapted with permission from [33]. Copyright 1998 Elsevier.

shown in Figure 21.17B. First, a negative control sample is applied, and then the unknown serum is assessed. No effect due to the blank serum is observed but large frequency shifts are caused by the sera containing the ASV antibody secreted by the organism in response to infection (positive sera). Subsequent regenerations by a pH 11 buffer solution is successful as demonstrated by the return of frequency to the baseline. Such a sensor proved to be reliable for 10 successive runs.

21.4.2 Amplification in QCM Immunosensors

The amplification principles and benefits are demonstrated next by the case of the prostate specific antigen (PSA) determination (Figure 21.18). The PSA sensor has been obtained by amine coupling of the specific antibody to a thiocarboxylic acid initially assembled on the gold electrode surface. As shown in Figure 21.18A, when in contact with the analyte solution, the frequency attains a stable value after a certain time interval that depends on the velocity of the binding interaction and, hence, on the analyte concentration. The effect of nonspecific interactions have been tested with a control sensor formed by the immobilization of a nonspecific antibody. As demonstrated by curve 5, which was recorded with the control sensor in the presence of a large analyte concentration, nonspecific binding is negligible. However, the working range of this sensor is well above the normal PSA level in blood serum (2 ng/mL). Therefore, signal amplification is needed. Line 1 in Figure 21.18B shows the sensor response after amplification by postrecognition binding of a secondary antibody as demonstrated in Figure 21.16. Although the response is enhanced in this way, the working range is still above the normal PSA level. A sufficiently large signal has been obtained by using a secondary antibody conjugated with gold nanoparticles (line 2). The large mass of the nanoparticle brings about a considerable change in frequency and allows reliable determinations of PSA to be performed within the normal concentration range.

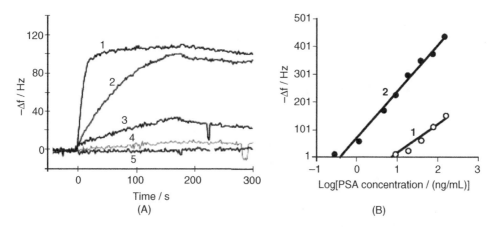

Figure 21.18   Prostate-specific antigen determination in 75% human serum by a QCM immunosensor. (A) Sensor response to various PSA concentrations (in ng/mL): (1) 5000; (2) 312; (3) 78; (4) 21.8. Curve 5 represents the response of the control sensor to 5000 ng/mL PSA. (B) The calibration graph for PSA determination with signal amplification. (1) Amplification by a secondary antibody; (2) amplification by gold nanoparticles. Adapted with permission from [34]. Copyright 2010 Elsevier.

Further signal enhancement is possible by growing the nanoparticles attached to the ternary complex formed at the sensor surface. As shown in Figure 21.19, this can be achieved by the chemical reduction of $HAuCl_4$ under the action of hydroxylamine as the reducing agent. The initial, small gold nanoparticle functions as a crystallization center and grows considerably by silver deposition thus enhancing the mass load. Particle growth can also be accomplished by chemical deposition of silver.

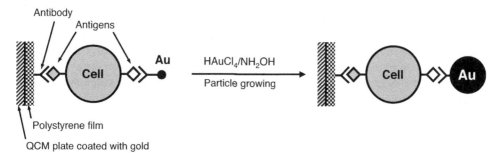

Figure 21.19   Signal amplification by means of gold nanoparticles in QCM immunoassay of human lung carcinoma cell in the sandwich format. The secondary antibody is tagged with gold nanoparticles that are further grown by chemical deposition of gold [35]. Reproduced with permission of The Royal Society of Chemistry (RSC) for the Center National de la Recherche Scientifique.

A substantial simplification of sensor manufacturing and test protocol is obtained if the assay is conducted in the solution phase using magnetic nanoparticles conjugated with the immunocomplex [36,37]. This allows for simple accumulation of this complex at the transducer surface by means of a magnetic field, thus avoiding the laborious immobilization step. Regeneration of the immunosensor in this case can be accomplished by removing the magnetic field.

21.4.3 Determination of Small Molecules Using Natural Receptors

Direct determination of small organic molecules (haptens) by properly tailored antibodies is also achievable by means of QCM sensors. However, the low molecular mass, combined with the viscoelastic effect in the immobilized antibody layer can limit severely the sensitivity of such a sensor. A more convenient approach relies on the competitive assay in which the sensing layer is formed from an analyte-analog (X) immobilized at the resonator surface. To the sample containing the analyte A, a heavy receptor R, which can bind to both the analyte and the analog, is added. Hence, competition between the analyte and an immobilized analog for binding to the receptor R will occur as follows:

$$\begin{cases} A(\text{solution}) + R(\text{solution}) \rightleftharpoons AR(\text{solution}) \\ X(\text{surface layer}) + R(\text{solution}) \rightleftharpoons XR(\text{surface layer}) \end{cases} \tag{21.41}$$

The mass variation is therefore an inverse function of the analyte concentration.

This principle has been applied for the detection of organophosphorus pesticides (A) in water using the cholinesterase enzyme as receptor (R) [38]. The catalytic activity is not addressed at all; instead one makes use of the property of the enzyme to bind reversibly to certain molecules that normally act as inhibitors. Therefore, the recognition interaction in this sensor is of the affinity type. The sensing layer consists of a cocaine derivative (X) immobilized at the surface of the gold electrode. This sensor was shown to respond to pesticide concentrations below the accepted limit for drinking water.

21.4.4 QCM Sensors Based on Molecularly Imprinted Polymers

Among synthetic receptors, molecularly imprinted polymers are the most frequently applied to developing piezoelectric sensors [39]. Most of the published work in this area addresses the detection of small molecules [41]. Despite the low molecular mass, such compounds can lead to clear mass variation due to their capacity to diffuse inside the pores of the imprinted polymer. In this way, the analyte will be accumulated in a sufficiently large amount. Therefore, bulk accumulation occurs in this case, in contrast to the immunosensors that are characterized by surface-localized analyte–receptor binding.

An example of an imprinted polymer application is the piezoelectric bilirubin sensor described in [40]. Bilirubin content in urine is an indicator of liver malfunction. In order to produce the sensor, the gold electrode surface has been coated with allyl mercaptan (2-propene-1-thiol) in the form of a self-assembly monolayer, and benzophenone, as the polymerization initiator, has been then deposited over it. Polymerization has been effected next by UV irradiation in a solution layer over the previously formed film using 4-vinylpyridine as monomer, divinyl benzene as crosslinker and bilirubin as template. After completing the polymerization (about 150 nm film thickness), the template was washed out with 10 mM NaOH. This manufacturing method demonstrated good batch-to-batch reproducibility.

Electropolymerization is another practical method for including receptor sites in a coating layer. Thus, a piezoelectric dopamine sensor has been obtained by electrochemical copolymerization of two pyrrole derivative monomers shown in Figure 21.20A. The first monomer includes a macrocyclic polyether that binds to the positively charge ammonium group in dopamine by ion–dipole interactions. The second monomer imparts to the imprinted site hydroxyl groups that can form hydrogen bonds with the analyte. This sensor has been operated under flow-injection conditions. In this method, a sample plug is carried by a buffer solution stream and flows across the sensor surface. As shown in Figure 21.20B, the response displays a transient character, with a maximum corresponding to the moment when the plug center reaches the sensor. The calibration graph is obtained by plotting the peak response as a function of the analyte concentration. Dopamine concentrations between 0.1 and 130 mM have been determined with negligible interference from compounds such as ascorbic acid, histamine or 2-phenylethylamine. The particular feature of this approach is the inclusion of a host receptor (here, a macrocyclic ether) within the imprinted site.

Molecular imprinting technology has also been applied to the development of piezoelectric sensors for bulky species such as cells, viruses and enzymes [43–45]. However, such species cannot diffuse into the polymer pores and therefore the imprinting technology should be designed so as to produce surface-imprinted polymers. This can be achieved by imprinting the template on a pretreated polymer mixture as shown in Figure 21.21. The polymer mixture is first deposited on the QCM surface and then the printing is performed by means of a stamp with micro-organisms attached to its surface. After polymer curing, the stamp is removed leaving the surface imprinted with a honeycomb-like structure as shown in Figure 21.22A.

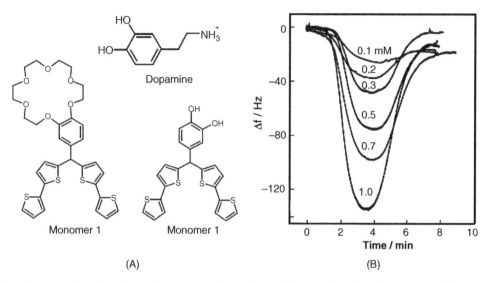

(A)   (B)

Figure 21.20 A dopamine piezoelectric senor based on an electrochemically polymerized imprinted polymer. (A) Structure of dopamine and the monomers used in the synthesis of the molecularly imprinted polymer; (B) Flow-injection analysis response in 0.2 M phosphate buffer (pH = 7.0); sample volume, 100 μL. The analyte concentration is shown on each curve. (B) was adapted with permission from [42]. Copyright 2010 Elsevier B.V.

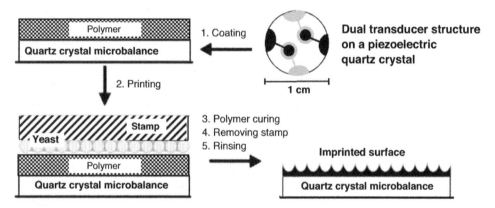

Figure 21.21 A molecularly imprinted polymer as recognition element in a piezoelectric yeast sensor. Adapted with permission from [44]. Copyright 2001 Wiley-VCH Verlag GmBH & Co. KGaA.

In order to assess the specificity of cell recognition two sensors have been assembled on the same piezoelectric crystal (Figure 21.21). One of them, which functions as the working sensor has been coated with the imprinted polymer, whereas the second one (the control sensor) has been covered by the same polymer in the nonimprinted form. The response of each sensor is plotted in Figure 21.22B, which demonstrates the excellent selectivity of the sensing layer.

Similar technology has been used to produce piezoelectric sensors for bacterial pathogens, viruses, and proteins. This opens the way to the development of straightforward and inexpensive methods for the determination of such species without the need to use biochemical reagents and procedures.

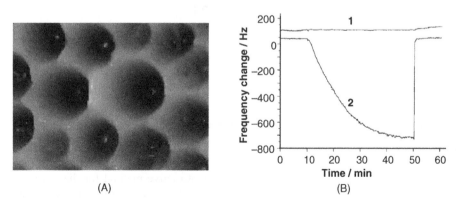

(A)   (B)

Figure 21.22 (A) Atomic force microscopy image of a yeast surface imprinted polymer. (B) The response of the control sensor (nonimprinted polymer, curve 1) and the working sensor (curve 2) to 0.1 mg/ml yeast (*Saccharomyces cerevisiae*). (A) Adapted from [46]. (B) Adapted with permission from [44]. Copyright 2001 Wiley-VCH Verlag GmBH & Co. KGaA.

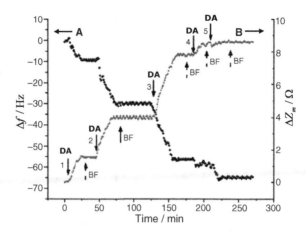

Figure 21.23 The response of a dopamine piezoelectric sensor to a sequential increase in analyte concentration. (A) Resonance frequency; (B) motional impedance. Arrows indicate dopamine addition (DA), and washing buffer addition (BF). Dopamine concentration (in nM): (1) 0.053; (2) 0.28; (3) 0.53; (4) 2.8; (5) 5.3. Reproduced by permission from [47]. Copyright 2010 Elsevier B.V.

21.4.5 QCM Sensors Based on Small Synthetic Receptors

Among the receptors in this class, calixarenes have aroused much interest due to the broad range of their interactions with small molecules and the great flexibility of the immobilization methods. An application of a calixarene to mass-sensitive sensors is illustrated by a dopamine sensor that is based on calix-4-arene functionalized with long-chain thiol moieties at the lower rim [47]. This allows for immobilization along with a long-chain thioalkane by sulfur chemisorption at the gold-electrode surface. The thioalkane layer provides surface protection against contact with the solution. The recognition occurs by the electrostatic interaction of the dopamine cation (Figure 21.20) with anionic carboxylate groups on the upper rim of the calixarene. Despite the low molecular mass of the analyte, track A in Figure 21.23 demonstrates a clear frequency change in response to dopamine addition. This effect has been studied further by recording also the change in the motional impedance that is shown by the track B in Figure 21.23. As the change in the resonant frequency is accompanied by a proportional variation of the motional impedance, it results that the viscoelastic effect alone is responsible for the sensor response. It is thus proved that small molecules that cannot induce detectable mass variations can still be determined by means of the modification in the viscosity of the surface layer (see also Table 21.1).

Various types of sugar sensors rely on phenylboronic acid derivatives as receptors. At a suitable pH, phenylboronic acid forms spontaneously and reversibly a cyclic ester by interacting with hydroxyl groups in sugar molecules, as shown in Figure 21.24. This kind of interaction has been employed to devise a piezoelectric sensor for a glycated hemoglobin, a sugar derivative of hemoglobin, which is of interest in diabetes mellitus monitoring [48,49]. 3-Aminophenylboronic acid has been appended to an inert protein layer previously formed at the surface of the gold electrode, thus providing a high density of binding groups. After each run, sensor regeneration is carried out with 0.2 M HCl and is completed in only 2 min. The sensitivity remained unchanged after a sequence of 15 successive runs. Such a sensor responds, in principle, to any sugar in the sample, including glucose. However, glucose does not affect the response because its mass is negligible with respect to that of the analyte. Glucose competition in the analyte binding can be made negligible by securing a high density of binding sites at the sensor surface.

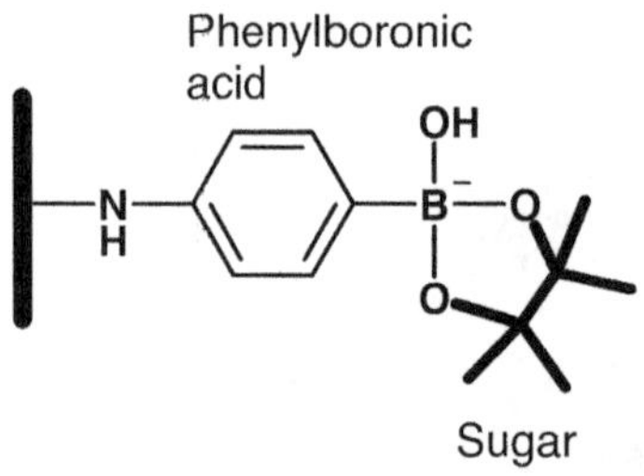

Figure 21.24 The cyclic ester product of phenylboronic interaction with a sugar molecule.

21.4.6 Outlook

The label-free character of QCM transduction makes it a very attractive method for the development of affinity sensors based on either immunocompounds or synthetic receptors. When very large molecules are involved, QCM transduction is impaired by the viscoelastic effect that reduces the sensitivity. This problem can be mitigated by resorting to postrecognition amplification, which can enhance considerably the sensitivity. However, amplification techniques involve processes similar to that in label-based methods.

Very promising perspectives are opened up by application of synthetic receptors that eliminate the use of biochemical compounds and biochemical operations. Thus, molecularly imprinted polymers prove to be effective in the determination of both small molecules and large entities such as proteins, viruses and micro-organisms. Molecular synthetic receptors are also of great interest.

Questions and Exercises (Section 21.4)

1 What kind of problems are anticipated when running a QCM immunosensor and how can such problems be mitigated?
2 Give an overview of nanomaterial applications in QCM affinity sensors.
3 How can the selectivity of a QCM affinity sensor be tested?
4 What characteristics of molecularly imprinted polymers impart selectivity to a QCM affinity sensor? Discuss the details of the recognition processes in sensors of this type.
5 Draw a sketch of the sensor used to obtain the data in Figure 21.23.

21.5 QCM Nucleic Acid Sensors

21.5.1 Hybridization Sensors

The label-free character of the piezoelectric transduction has proved advantageous for constructing DNA hybridization sensors [50,51].

Probe immobilization should be performed so as to obtain an optimum probe density, avoid nonspecific interactions at the sensor surface, and secure good stability and resilience in the regeneration step. Immobilization can be achieved simply by chemisorption of a thiol-tethered oligonucleotide to the gold coating over the quartz wafer. More elaborate methods, such as avidin–biotin-based linking, impart superior characteristics to the recognition layer [52].

A piezoelectric sensor for the detection of genetically modified organisms (GMOs) has been fabricated using a 25-unit oligomer DNA probe that is complementary to a characteristic sequence in the GMO's DNA [53]. The biotinylated probe was immobilized by affinity coupling to the gold layer that was modified first with a mercaptoalcohol $(HS-(CH_2)_{11}-OH)$ and streptavidin-modified dextran. The readout was adjusted to zero with the sensor immersed in the hybridization buffer and then the sample was added (see Figure 21.25). After 20 min incubation, the frequency reached a constant value that was recorded. Finally sensor was washed with the buffer solution in order to remove unbound oligonucleotides. The analytical signal y is the frequency difference between levels A and B, as shown in Figure 21.25.

In order to perform the next assay, the single-strand probe was regenerated by a one-minute treatment with dilute hydrochloric acid. Regeneration could be performed satisfactorily up to 10 times.

An assay with a noncomplementary oligonucleotide yielded no measurable frequency shift proving the absence of non-specific interactions. Tests with PCR-amplified DNA samples from a transgenic soya-bean flour certified reference material gave an almost linear response from 0.01 to $20\,\mu M$ target concentration.

Similarly, detection of bacterial pathogens in environmental samples has been effected by means of sensing layers including capture probes specific to a particular sequence in the DNA of the target micro-organism [54].

The sensitivity of a direct piezoelectric sensor is too low for detecting the target in a biological sample and amplification by PCR should be performed prior to the determination. This time-consuming step can be avoided by means of gold nanoparticle attached to the primary hybridization product [55,56]. This principle is illustrated in Figure 21.26.

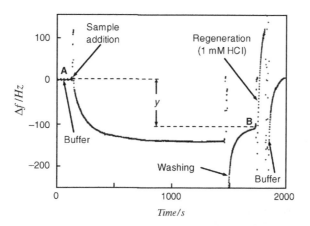

Figure 21.25 Time evolution of Δf during a hybridization assay using a piezoelectric DNA sensor. Adapted with permission from [53]. Copyright 2001 Springer Science+Business Media B.V.

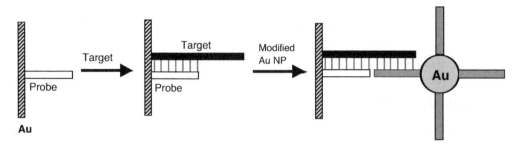

Figure 21.26 Amplification of the piezoelectric sensor response by attaching gold nanoparticles to the probe–target hybrid.

First, the target is hybridized with the capture probe attached to the piezocrystal surface to form the primary hybrid. Next, a gold nanoparticle modified with a sequence complementary to the overhanging moiety of the target is attached to the primary hybrid in a sandwich arrangement. Subsequent amplification can be achieved by growing a dendrimer aggregate starting from this structure and using nanoparticles modified with suitable sequences. Mass amplification by this technique allows the determination of target concentrations down to 10^{-10} M.

Mass amplification can be further enhanced by chemical deposition of a metal on nanoparticles previously attached to the probe–target hybrid. As indicated before, in the chemical reduction of $AuCl_4^-$ to gold by hydroxyl-amine, gold nanoparticle act as crystallization centers and their mass is augmented by gold deposition. This allowed an outstanding detection limit close to 10^{-15} M to be reached.

Signal amplification by means of gold nanoparticles is intricate and time consuming but this drawback is compensated by the possibility of detecting the target in the sample with no prior DNA amplification.

21.5.2 Piezoelectric Aptasensors

Various strategies for protein recognition in piezoelectric aptasensors have been elaborated, as demonstrated in Figure 21.27. In this figure, BF represents a simple aptamer that can bind to one specific site in the target protein. BFA is similar to BF as far as the recognition behavior is concerned. However, in order to keep the aptamer position as far as possible from the surface, the linkage fragment in BFA is composed of a rigid double strand. BFF is composed of two aptamers similar to BF joined together by means of complementary fragments. This receptor, which is an aptabody, includes two similar binding sites and yields a greater response because each receptor binds two analyte molecules. BFH is also a composite receptor but each aptamer unit in its structure binds to a different site in the analyte molecule.

A piezoelectric aptasensor for human α-thrombin has been described in ref. [58]. The recognition layer consists of a mixture of carbon nanotubes and positively charged methylene blue deposited on the gold surface. Aptamers were attached to this primer by electrostatic interaction. Two versions of this sensor have been tested. In the first one, the receptor was a thrombin aptamer. In the second one, an aptabody, formed by conjugation of two aptamers with complementary oligonucleotide tails (like BFF in Figure 21.27) has been used. The aptabody sensor yields a detection limit of 0.3 nM that is three times better that that demonstrated by the single-aptamer version. A common protein (human serum albumin, HSA) had almost no effect on the sensor response, proving the absence of nonspecific interactions. The absence of interferences was also demonstrated by the similarity of calibration curves obtained with thrombin in a pure buffer, in a 3 mg/ml HSA containing buffer and in 10 times diluted blood serum.

As is common with large protein layers, the viscoelastic effect contributes to the overall frequency shift, and, therefore, the resonance frequency and the motional resistance needs to be extracted from the total response [57].

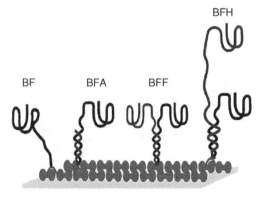

Figure 21.27 Various types of aptamers used for thrombin recognition in QCM sensors. BF: conventional single-stranded aptamer; BFA: similar to BF but containing a double-strand linker fragment that provides a more rigid orientation; BFF: two BF aptamers connected through complementary fragments; BFH: two aptamers, each of them being specific to a particular site in the thrombin molecule [57]. Reproduced with permission of the Royal Society of Chemistry.

A similar aptamer sensor for the determination of prion (an infectious agent composed of protein in a misfolded form) has also been developed [59].

Considerable improvement in detection limit has been achieved by means of postrecognition attachment of gold nanoparticles conjugated with a second aptamer to form a sandwich ternary complex [60]. Further amplification has been obtained by growing the nanoparticles by gold depositions from a $AuCl_4^-$ solution containing 1,4-dihydronicotinamide adenine dinucleotide (NADH) as reducing agent.

21.5.3 Outlook

The QCM is a versatile transduction technique in nucleic acid sensors owing to its label-free character. Superior sensitivity can be achieved by means of posthybridization treatment aimed at increasing the overall mass load by appending gold nanoparticles. Further sensitivity enhancement is achieved by chemical deposition of gold on the appended particles.

Aptamer application in QCM protein sensors represents a promising alternative to the QCM immunoassay.

Questions and Exercises (Section 21.5)

1 What are the main advantages and drawbacks of typical piezoelectric sensors for nucleic acids?
2 Describe a possible method for constructing a QCM DNA sensor.
3 How can the detection limit of a piezoelectric DNA sensor be improved?
4 What is the typical configuration of an aptamer-based piezoelectric protein sensor?
5 What are the precautions required when assessing the response of an aptamer-based piezoelectric protein sensor?
6 Review the fields of application of QCM nucleic acid sensors.

21.6 Surface-Launched Acoustic-Wave Sensors

21.6.1 Principles

The quartz crystal microbalance, which was covered in the previous sections, relies on thickness shear vibration induced by electrodes placed on each side of the piezoelectric wafer. In contrast, surface-acoustic-wave devices are based on acoustic excitation by means of two electrodes placed on the same surface. The electrodes are configured as an *interdigitated transducer* (IDT). The acoustic wave induced by an IDT is propagated in an extremely thin layer at the surface of the piezoelectric wafer. The wave motion is detected by the same or an additional transducer on the same surface. This configuration is typical of *surface-launched acoustic-wave devices (SLAW)*.

Commonly used SLAW devices are shown in Figure 21.28. Such devices are utilized as transducers in chemical sensors for gas or liquid phase analysis. SLAW functioning principles and sensor applications have been reviewed in

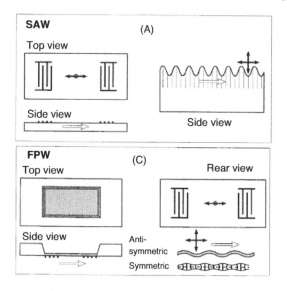
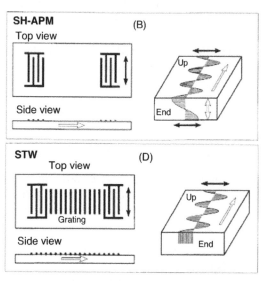

Figure 21.28 SLAW devices and acoustic wave propagation modes. Black arrows indicate particle displacement; white arrows indicate wave propagation. For each type, the typical material is further indicated. (A) SAW: surface acoustic wave or Rayleigh wave (ST-cut quartz). (B) SH-APM: shear horizontal acoustic plate mode (AT-quartz; lithium niobate, $LiNbO_3$). (C) FPW: flexural plate-wave device – Lamb wave (ZnO on silicon nitride). (D) STW: surface transverse wave device – Love wave (AT-cut quartz; lithium tantalate ($LiTaO_3$)). Adapted with permission from [61]. Copyright 2000 Springer Science+Business Media B.V.

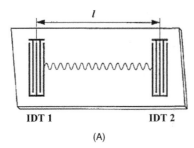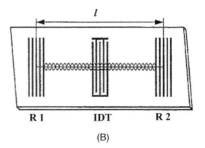

Figure 21.29 Two alternative types of SAW device. (A) Two-port delay-line SAW device; (B) one-port SAW device. The gratings (R1 and R2) are totally reflecting at the resonant frequency. The wafer is laterally cut slightly off the long axis to prevent interference of the back-reflected wave with the main wave. Alternatively, this problem can be mitigated by coating the ridges with a wave-absorber material. Adapted with permission from [63]. Copyright 1995 Elsevier Science S.A.

refs. [11,62–64]. This section introduces first the physical principles of SLAW devices and next presents their sensor applications.

21.6.2 The Surface Acoustic Wave

The *surface acoustic wave* (SAW) also known as the Rayleigh wave (Figure 21.28A) involves particle displacement along two directions: a longitudinal-compressional component (that is, back and forth, parallel to the surface) and shear vertical component (that is, up and down). The superposition of these two components gives rise to particle trajectories that follow an elliptical path around their rest positions. The thickness of the layer involved in vibration is of one or two acoustic wavelengths.

The configuration of SAW devices is shown in Figure 21.29. Version A is a two-port delay line; its name is in accordance with its function in electronic equipment. Two IDTs are formed on the surface of the piezoelectric wafer. The first one (transmitter) is connected to a high-frequency AC voltage supply and excites surface vibration by the converse piezoelectric effect. The acoustic wave is detected by the second IDT (receiver) that generates an AC voltage proportional to the wave amplitude. The time delay t_d is given by the l/v ratio where l is the distance between the IDTs and v is the wave velocity. The maximum energy transfer between IDTs occurs at a characteristic frequency $f_0 = v/\lambda_0$, where λ_0 is the corresponding wavelength. As the first IDT radiates in both directions, there is significant energy loss. However, if the delay line is integrated in the feedback of an electronic oscillator network, the losses can be compensated and the device performs as a resonator based on a resonant traveling wave. As in the case of a vibrating string, there are many close resonance states with closely spaced resonant frequencies that may complicate the device operation.

The device in Figure 21.29B is a single-port resonator. In this case, the vibration is excited by an IDT placed between two gratings that are totally reflecting at the characteristic frequency. A standing wave develops between the reflectors giving rise to a resonator that vibrates at various overtones with frequencies given by $f_n = nv/2l$, where n is the overtone number. The resonant frequencies depend on the properties of the piezoelectric material (density and elasticity) and on the angle of cut. However, the quality factor is much lower than in the case of the QCM.

The wavelength (and hence the operating frequency) is determined by the spacing between the electrodes. Using this parameter the frequency can be adjusted between 150 and 400 MHz. Quartz is a suitable piezoelectric material for such devices, but LiNbO$_3$ displays better electromechanical coupling. As the SAW is strongly damped by liquids its analytical application is limited to gas sensors.

21.6.3 Plate-Mode SLAW Devices

In the case of the SAW device presented above, the vibration is localized at the surface of a piezoelectric plate with the thickness much greater than the wavelength. This section introduces a series of SLAW devices in which the vibration expands over the whole plate thickness although it is excited on only one side. In such devices, the vibrating plate is thinner than the acoustic wavelength.

a. *The flexural plate wave* (FPW or *Lamb wave* device) is shown in Figure 21.28C. As in the case of the SAW, the particle displacement is elliptical. Two vibration modes are possible: the antisymmetric one (flexural) and the symmetric one (dilatational). The sensitivity of the FPW device is approximately 10 times higher than that of an equivalent SAW. Hence, the FPW can be operated at lower frequencies using less-complicated electronics. Another advantage is that the sensing layer can be deposited on the rear side of the wafer therefore avoiding exposure of the IDTs to the sample. FPW devices can be used for both gas-phase and liquid-phase sensor applications. However, the plate thickness is much less than 1 μm, which results in difficult manipulation.

b. *A shear horizontal acoustic plate* (SH-APM, Figure 21.28B) is obtained from an AT-cut quartz plate of some 200 μm thickness on which two IDTs are deposited in order to operate it as a two-port device with the resonant frequency between 100 and 150 MHz. The particle motion is similar to that known for the QCM with the notable difference that the exciting electrodes occupy a negligible area on a single side, whereas the second side is left free. As in the case of the FPW device, several overtones are possible. A particular overtone of interest can be selected by adjusting the plate thickness.

SH-APM sensors can be used in both gas- and liquid-phase sensing. The second kind of application relies on the fact that the motion of surface particles is parallel to the contact surface. Even if wave dampening by a liquid does occur (as in the case of the QCM), vibration can still be sustained.

c. *The surface transverse wave device* (STW), also termed the *Love wave* device (Figure 21.28D), was designed in an attempt to enhance some of the advantages of the devices presented previously. In the STW, the shear-horizontal acoustic wave is trapped in the surface region by a metal grating that is positioned between the transducers. The whole surface is coated with a 50-nm thin silicon dioxide layer (or another material) to provide protection against metal corrosion. This assembly is commonly referred to as an acoustic-wave guide and the wave thus produced is a guided wave. In contrast to the SH-APM device, particle movement is confined to the surface in a pure shear-horizontal mode. This particular feature brings about greater analytical sensitivity.

Wave guiding can be obtained not only by a metal grating but also by a compact layer made from a material in which the wave propagates with a lower velocity than in the piezoelectric substrate. Gold, silicon dioxide, silicon nitride or synthetic polymers can be used to this end, provided that the previous velocity condition is fulfilled. A polymer guiding layer can function itself as the sensing element for analyte recognition by sorption. Each of the above-mentioned materials can function as the support for receptor immobilization.

The interaction of the shear wave in the STW device with the sensing layer arises from the fact that the wave energy is manifested over a very short distance perpendicular to the propagation direction. The signal decays exponentially with the distance from the surface (acoustic evanescent wave) but the depth of penetration is sufficient for interaction with a submicrometer thin film to occur.

21.6.4 SLAW Gas and Vapor Sensors

Among the various available SLAW devices, the SAW one has been mostly employed in gas-phase sensing [65,66]. Surface-wave interaction with the sample alters various wave parameters such as the amplitude, the phase angle, the velocity, and the harmonic content. The two-port SAW delay line is used as the transducer in SAW devices by coating a recognition layer over the vibrating area (Figure 21.30). Therefore, the wave parameters are modified in accordance with the effect of analyte sorption. Most often, the sensor response is represented by the frequency change produced by the recognition event. The frequency change (Δf) decreases proportionally to the mass load per unit area [11]. However, this simple relationship holds only if the analyte sorption does not produce modifications in the physical properties of the sensing layer. As the resonant frequency is much higher than that of a typical QCM, much better sensitivity is achieved by SAW sensors.

Besides frequency monitoring, the response can be assessed by measuring the variation in both wave amplitude and phase angle using a vector voltmeter connected between the two IDTs. The phase variation depends on to the mass load and also on other types of modifications in the sensing layer.

The most advanced monitoring method for SAW sensors is based on network analyzers that reports on both the magnitude and the phase angle of the electrical impedance as a function of frequency.

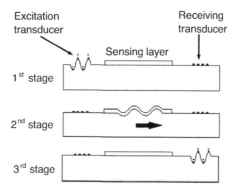

Figure 21.30 Side view of a SAW sensor showing three distinct stages: (1) excitation of the acoustic wave; (2) wave propagation along the coated surface; (3) transduction of the acoustic signal into an electrical one by the receiver. Adapted with permission from [67]. Copyright 2004 Springer Science+Business Media B.V.

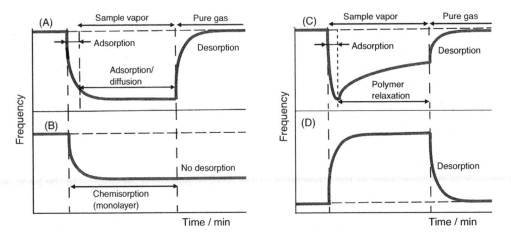

Figure 21.31 Possible response profiles of a SAW vapor sensor. (A) Ideal response; (B) Irreversible chemisorption; (C) combined mass and viscoelastic effects; (D) Large change in polymer properties for a small loading mass. Adapted with permission from [62] Copyright 1997 John Wiley & Sons, Inc.

The sensing layer is selected such as to interact with the analyte either by physical or chemical sorption. Although the SAW device is capable of detecting monolayer adsorption, greater sensitivity can be obtained by means of a thicker film that allows the analyte to diffuse through and interact with a larger number of sorption sites.

Sensing by *chemical sorption* involves relatively strong interaction and provides reasonable selectivity. However, such a sensor is irreversible under normal conditions and will be most useful in dosimetry applications. The dosimeter provides an integrated signal related to the concentration and time of exposure to the target compound. The large dynamic range and low cost of SAW devices renders this alternative a practical one.

Physical sorption-based recognition has the advantage of being a reversible process that allows for easy regeneration of the sensor. However, the selectivity in this case is rather poor. This problem is alleviated by operating arrays consisting of cross-selective sensors. In view of the small size and low cost of the SAW device, this approach is very practical.

Sensor response depends mainly on the mechanism of the analyte–receptor interaction (Figure 21.31). The simplest case is represented by physical adsorption (A) that causes an abrupt change in the signal initially, when only adsorption at the surface is taking place. Subsequently, the analyte diffuses into the sensing layer giving rise to a further slower, signal change until the equilibrium with the gas phase is attained. As this sensor is reversible, regeneration can be achieved by flushing it with a pure gas. Such a sensor is temperature dependent as both the diffusion rate and the equilibrium concentration of the adsorbed analyte are affected by temperature. At higher temperatures the diffusion is faster, producing a shorter response time. On the other hand, a higher temperature reduces the analyte concentration in the sensing layer. Hence, sensitivity can be enhanced by lowering the device temperature but only at the expense of a longer response time.

If the analyte undergoes chemical absorption, the response is irreversible and the original value of the signal will not be restored by purging with a pure gas (Figure 21.31B).

A more intricate behavior can be noted when the properties of the sensing layer are altered by analyte sorption. Thus, Figure 21.31C shows a two-step response evolution. Fast sorption during the first step gives rise to a sudden frequency change due to the mass loading, Further changes in the properties of the analyte-permeated sensing layer causes modifications of the viscoelastic properties that also affect the response signal. As shown in Figure 21.31D, the second effect can in some cases be the more dominant and an increase (or decrease) in the resonant frequency could occur due mostly to modifications of the film properties in response to analyte sorption. In the case of a polymer film, sorption of the analyte in the polymer causes a slow conformational rearrangement until a thermodynamically stable conformation is attained. Relaxation of the polymer matrix can lead to modifications in the viscoelastic properties, the dielectric constant or the electric conductance, each of these effects causing a modification of the response signal.

By far the most common gas sensing materials are synthetic polymers deposited by solvent evaporation. The Langmuir–Blodgett method is also convenient for sensing-layer preparation. Another method exploits the presence of hydroxyl groups at the quartz surface in order to prepare self-assembled monolayers capable of interacting with the analyte. A number of particular applications rely on metal coatings such as palladium in hydrogen and platinum in ammonia sensors.

It should be noted that SAW devices are sensitive to temperature and pressure variations. That is why it is practical to combine the sensor itself with a reference sensor that does not responds to the analyte. The differential signal obtained in this way is free from effects arising from spurious influences. Both the working and reference sensor can be formed on the same piezoelectric wafer.

The limit of detection of SAW sensors is determined as usually by the signal-to-noise ratio. As the noise increases proportionally with frequency whereas the signal increases with the square of the frequency the signal-to-noise ratio will increase proportionately with frequency. Noise reduction can also be achieved by reducing the length of the delay line.

The design and manufacturing of a sensor for a particular analyte in a complex gas sample is a difficult task. However, in certain applications it is not necessary to detect each component but to detect together (as one) a complex set of compounds that impart the mixture a particular property such as odor or aroma. Such a task can be performed by an assembly of sensors fabricated such that each of them responds to a number of compounds in the sample. Although none of the sensors is ideally specific, data processing by statistical methods allows a set of numerical parameters to be ascribed to the mixture that allows it to be characterized sufficiently. This approach forms the physical basis of the artificial nose [67] (see also Section 1.7). The low cost and easy fabrication technology of SLAWs makes them very suitable for such applications [68].

21.6.5 Liquid-Phase SLAW Sensing

The most suitable device for liquid-phase sensing is the STW (Love) device due to its enhanced sensitivity to environment effects and to its high quality factor, high frequency (over 100 MHz), robustness and small size [69,70]. As the device surface is coated with a wave-guiding layer, this layer can function as a support for receptor immobilization.

A typical configuration of a STW device sensor is shown in Figure 21.32. A sensing layer including suitable recognition receptors is deposited on top of the guiding layer. The transmitter IDT that is connected to a high-frequency voltage supply excites the thickness–shear vibration and the guided wave propagates along the device surface. Due to total internal reflection, the wave energy can cross the guiding layer along the vertical direction as an evanescent wave. Acoustic energy is thus transferred to the recognition material on top of the guiding layer. As shown by the graph inserted into Figure 21.32, the wave decays exponentially with distance and its effect is mostly confined to the thin recognition layer. The parameters of the wave reaching the receiver IDT are therefore affected by wave interaction with the sensing film. In order to compensate for spurious effects, a reference sensor is installed on the same piezoelectric plate and the response is corrected by subtracting the reference signal from that of the working sensor. As in the case of the QCMs, STWs are best operated when incorporated in a flow-analysis system. In order to avoid perturbation caused by pressure variation, gravitational flow or syringe pumps are suitable for delivering the sample.

Figure 21.33 presents the time evolution of the response of a STW xylene sensor to the analyte concentration in water. As can be noted, two different response signals are available: the frequency change and the variation of the dissipated power. Power loss (L_{dB}, in decibels, dB) can be determined by the following formula in which P means power, A is the wave amplitude, and the subscripts "out" and "in" indicate output and input value, respectively:

$$L_{dB} = 10 \log_{10}\left(\frac{P_{out}}{P_{in}}\right) = 20 \log_{10}\left(\frac{A_{out}}{A_{in}}\right) \tag{21.42}$$

Similar information can be derived from phase-angle measurements.

The frequency change response in Figure 21.33A reflects both mass loading and energy dissipation in the viscoelastic polymer film. Energy dissipation is, however, unambiguously revealed by data in Figure 21.33B. In both cases, the response is in a linear dependence on the analyte concentration over a limited range that is narrower in the

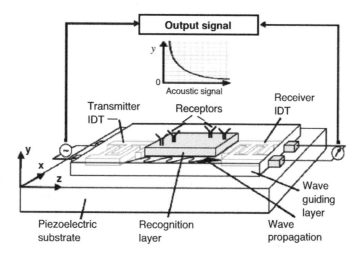

Figure 21.32 Configuration of a SLAW biosensor. The inset graph demonstrates the exponential decay of the wave amplitude along the vertical direction. Adapted with permission from [70]. Copyright 2007 Elsevier B.V.

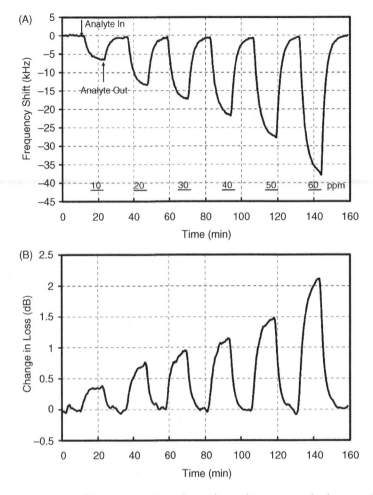

Figure 21.33 Flow-analysis response of a STW xylene sensor to increasing analyte concentration between 10 and 60 ppm. (A) Frequency change response; (B) wave power loss response. The sensing layer was a 0.8-μm thick poly(isobutylene) film. After each sample, the sensor was flushed with pure water. Reproduced with permission from [71]. Copyright 2005 American Chemical Society.

case of loss measurements. So, two distinct response signals can be generated by a SLAW sensor. This feature is of relevance to the design and data processing in sensor arrays.

The previous example refers to the sensing of small molecules by sorption into a polymer coating. However, the field of application also includes the sensing of large biological entities such as proteins, nucleic acids, bacterial toxins, micro-organisms, and whole cells [70]. Sensors for proteins or pathogens make use of antibodies as recognition reagents. Application of aptamers as recognition receptors for proteins has also been demonstrated. Separation of mass loading and viscoelastic effects can be useful for performing transduction even in the absence of significant mass variation.

An important feature of the STW sensor is the possibility of adapting the functioning mode to the size of the detected entity by tuning the penetration depth of the evanescent wave into the sensing overlayer. Thus, a high frequency results in a small penetration depth that is suitable for sensing small molecules. The opposite case fits best the requirements imposed by the detection of large entities such as proteins or cells.

It was recently pointed out that the number of reported applications of SLAW devices in biosensors is relatively low despite the great potential of such devices [70]. This is most probably due to the intricacy of the theoretical modeling and the complexity of readout procedures that require interdisciplinary teams. Recent advances in this area are surveyed in ref. [72].

21.6.6 Outlook

Currently, various SLAW devices with well-established manufacturing technologies and fairly good theoretical descriptions are available. The broadest field of sensing application is still the development of gas sensors, but the advent of the STW (Love) device has opened up very promising perspectives to applications in liquid-phase sensing.

It is important to note that both SLAW and QCM devices provide essentially the same kind of information, namely the mass load and the viscoelastic properties of the coating layer. The advantage of the QCM is its easy fabrication, the small temperature sensitivity, and the low noise level. On the other hand, SLAW sensors use higher frequencies that lead to enhanced mass sensitivity. In addition, the small size of the SLAW sensors, which is in the millimeter range, makes them very suitable for developing sensor arrays.

The performances of piezoelectric sensors (including both QCMs and SLAWs) are often compared with that of an optical method for interface sensing based on *surface plasmon resonance* (SPR, see Section 18.6.1). SPR is generally more sensitive than the QCM but SLAW sensors approach the sensitivity of the SPR method. However, due to device characteristics, the immobilization support in the SPR is restricted to gold. In contrast, the STW (Love) devices provide many more alternatives as the guiding layer (which acts as an immobilization support) can be fabricated from various organic or inorganic materials. Moreover, the SPR method provides no information on the viscoelastic properties of the sensing layer, as the acoustic-wave devices do.

> ## Questions and Exercises (Section 21.6)
>
> 1 What are the general features of SLAW devices?
> 2 Draw the shape of the surface acoustic wave for each of the devices in Figure 21.28 paying attention to the wavelength/plate thickness ratio.
> 3 What are the possible configurations of SLAW devices?
> 4 Review (i) the recognition mechanism and (ii) the time evolution of the response signal in SLAW gas sensors.
> 5 What are the advantages and limitations of SLAW applications in gas sensors? How can such limitations be overcome?
> 6 What SLAW devices are suitable for sensing applications to liquid samples? Which of them performs best and why?
> 7 Give a short account of the immobilization methods suitable for building up the recognition layer on top of the guiding layer in the STW device. Hint: refer to immobilization on gold, SiO_2, and polymers.
> 8 Review the SLAW parameters that reflect the changes produced in the sensing layer by interaction with the analyte. How are such quantities related to the changes in the properties of the sensing layer in response to the recognition event?

21.7 Summary

Acoustic-wave transduction is based on coupling the solid device vibration with a recognition layer whose characteristics change in response to the interaction with the analyte. As a consequence of this interaction, the vibration frequency is altered as a function of the analyte concentration.

Vibrating devices are made of piezoelectric materials and the vibration is maintained by means of an AC voltage applied through metal electrodes. A piezoelectric oscillator is characterized by its resonant frequency which depends on geometrical factors (such as device thickness) and the physical properties of the piezoelectric materials (such as density and elasticity).

Ideally, the change in resonance frequency is determined only by the additional mass loaded through the recognition event. This occurs when the added film adheres firmly to the resonator surface, is rigid, and much thinner than the wavelength of the acoustic wave.

Two kinds of acoustic-wave device can be distinguished, according to the quotient of the device thickness and the depth of penetration of the acoustic wave into the vibrating material. Bulk acoustic-wave devices are characterized by the fact that the vibration encompasses the whole volume of the piezoelectric material. In contrast, in surface-launched acoustic-wave devices, the vibration is limited to a very thin layer at the device surface.

The typical bulk acoustic-wave device is the quartz crystal microbalance, in which the piezoelectric crystal vibrates in the thickness–shear mode. For thin, rigid layers, the frequency change is proportional to the mass variation according to the Sauerbrey equation. Owing to the absence of energy dissipation, the QCM gas sensor functions close to these conditions. When the QCM is operated in a liquid phase, energy dissipation into the liquid brings about a constant frequency change that can be subtracted from the overall response However, when macromolecules are involved, energy dissipation gives rise to strong deviations from the Sauerbrey equation and causes the frequency variation to be much lower than that expected from the mass change. This drawback can be mitigated by signal amplification based on postrecognition modification of the surface layer with gold nanoparticles.

Compared with the QCM, surface-launched acoustic-wave devices are characterized by a much higher resonance frequency, and, hence, provide a better sensitivity. The small size of SLAW devices allows for miniaturization and the fabrication of sensor arrays. SLAW devices in which the vibration is oriented perpendicularly to the resonator surface are useful in gas and vapor sensors. For application in liquid-phase sensing, devices in which the vibration is parallel to the resonator surface are suitable owing to the very low energy dissipation.

In addition to chemical-sensing applications, acoustic-wave devices are valuable tools for investigating thin layers as far as viscoelastic properties are concerned, providing information that is useful in the development of various sensors based on thin layers attached to metal surfaces.

References

1. Buck, R.P., Lindner, E., Kutner, W., and Inzelt, G. (2004) Piezoelectric chemical sensors - (IUPAC Technical Report). *Pure Appl. Chem.*, **76**, 1139–1160.
2. Steinem, C. and Janshoff, A. (eds.) (2007) *Piezoelectric Sensors*, Springer-Verlag, Berlin.
3. Ballantine, D.S., Martin, S.J., Ricco, A.J. *et al.* (1997) *Acoustic Wave Sensors: Theory, Design, and Physico-Chemical Applications*, Academic Press, San Diego.
4. Buttry, D.A. and Ward, M.D. (1992) Measurement of interfacial processes at electrode surfaces with the electrochemical quartz crystal microbalance. *Chem. Rev.*, **92**, 1355–1379.
5. Marx, K.A. (2003) Quartz crystal microbalance: A useful tool for studying thin polymer films and complex biomolecular systems at the solution-surface interface. *Biomacromolecules*, **4**, 1099–1120.
6. Janshoff, A., Galla, H.J., and Steinem, C. (2000) Piezoelectric mass-sensing devices as biosensors - An alternative to optical biosensors? *Angew. Chem. Int. Edit.*, **39**, 4004–4032.
7. Bandey, H.L., Martin, S.J., Cernosek, R.W., and Hillman, A.R. (1999) Modeling the responses of thickness-shear mode resonators under various loading conditions. *Anal. Chem.*, **71**, 2205–2214.
8. Sauerbrey, G. (1959) Use of vibrating quartz for thin film weighing and microweighing (in German). *Z. Phys.*, **155**, 206–222.
9. Rodahl, M. and Kasemo, B. (1996) On the measurement of thin liquid overlayers with the quartz-crystal microbalance. *Sens. Actuators A-Phys.*, **54**, 448–456.
10. Kanazawa, K.K. and Gordon, J.G. (1985) The oscillation frequency of a quartz resonator in contact with a liquid. *Anal. Chim. Acta*, **175**, 99–105.
11. Grate, J.W., Martin, S.J., and White, R.M. (1993) Acoustic-wave microsensors .1. *Anal. Chem.*, **65**, A940–A948.
12. Martin, S.J., Bandey, H.L., Cernosek, R.W. *et al.* (2000) Equivalent-circuit model for the thickness-shear mode resonator with a viscoelastic film near film resonance. *Anal. Chem.*, **72**, 141–149.
13. Martin, S.J., Granstaff, V.E., and Frye, G.C. (1991) Characterization of a quartz crystal microbalance with simultaneous mass and liquid loading. *Anal. Chem.*, **63**, 2272–2281.
14. Lucklum, R., Behling, C., Cernosek, R.W., and Martin, S.J. (1997) Determination of complex shear modulus with thickness shear mode resonators. *J. Phys. D-Appl. Phys.*, **30**, 346–356.
15. Lucklum, R. and Hauptmann, P. (2006) Acoustic microsensors-the challenge behind microgravimmetry. *Anal. Bioanal. Chem.*, **384**, 667–682.
16. Mecea, V.M. (1994) Loaded vibrating quartz sensors. *Sens. Actuators A-Phys.*, **40**, 1–27.
17. Johannsmann, D. (2008) Viscoelastic, mechanical, and dielectric measurements on complex samples with the quartz crystal microbalance. *Phys. Chem. Chem. Phys.*, **10**, 4516–4534.
18. Johannsmann, D. (2007) Studies of viscoelasticity with the QCM, in *Piezoelectric Sensors* (eds C. Steinem and A. Janshoff), Springer-Verlag, Berlin, pp. 49–109.
19. Hillman, A.R. (2003) The electrochemical quartz crystal microbalance, in *Encyclopedia of Electrochemistry* (eds A.J. Bard, M. Stratmann, and P.R. Unwin), Wiley-VCH, Weinheim, pp. 230–289.
20. Rodahl, M. and Kasemo, B. (1996) A simple setup to simultaneously measure the resonant frequency and the absolute dissipation factor of a quartz crystal microbalance. *Rev. Sci. Instrum.*, **67**, 3238–3241.
21. Höök, F. and Kasemo, B. (2007) The QCM-D technique for probing biomacromolecular recognition reactions, in *Piezoelectric Sensors* (eds C. Steinem and A. Janshoff) Springer-Verlag, Berlin, pp. 425–447.
22. Hepel, M. (1999) Electrode-solution interface studied with electrochemical quartz crystal nanobalance, in *Interfacial Electrochemistry: Theory, Experiment, and Applications* (ed. A. Wieckowski) Marcel Dekker, New York, pp. 599–630.
23. Ngeh-Ngwainbi, J., Suleiman, A.A., and Guilbault, G.G. (1990) Piezoelectric crystal biosensors. *Biosens. Bioelectron.*, **5**, 13–26.
24. Guilbault, G.G. and Jordan, J.M. (1988) Analytical uses of piezoelectric crystals - a review. *CRC Crit. Rev. Anal. Chem.*, **19**, 1–28.
25. Alder, J.F. and McCallum, J.J. (1983) Piezoelectric crystals for mass and chemical measurements - a review. *Analyst*, **108**, 1169–1189.
26. Mirsky, M.M. and Vasjari, M. (2007) Chemical sensors for mercury vapours, in *Electrochemical Sensor Analysis* (eds S. Alegret and A. Merkoçi), Elsevier, Amsterdam, pp. 235–251.
27. Bucur, R.V. (1974) Piezoelectric quartz crystal microbalance used as detector for gaseous hydrogen (deuterium). *Rev. Roum. Phys.*, **19**, 779–786.
28. Guilbault, G.G. (1983) Determination of formaldehyde with an enzyme-coated piezoelectric crystal detector. *Anal. Chem.*, **55**, 1682–1684.
29. Ngehngwainbi, J., Foley, P.H., Kuan, S.S., and Guilbault, G.G. (1986) Parathion antibodies on piezoelectric crystals. *J. Am. Chem. Soc.*, **108**, 5444–5447.
30. Mascini, M., Minunni, M., Guilbault, G., and Carr, R. (1998) Immunosensors based on piezoelectric crystal device, in *Affinity Biosensors: Techniques and Protocols* (eds K.R. Rogers and A. Mulchandani), Humana Press, Totowa, pp. 55–76.
31. Skladal, P. (2003) Piezoelectric quartz crystal sensors applied for bioanalytical assays and characterization of affinity interactions. *J. Braz. Chem. Soc.*, **14**, 491–502.
32. Vaughan, R.D. and Guilbault, G.G. (2007) Piezoelectric immunosensors, in *Piezoelectric Sensors* (eds C. Steinem and A. Janshoff), Springer-Verlag, Berlin, pp. 237–280.
33. Uttenthaler, E., Kosslinger, C., and Drost, S. (1998) Quartz crystal biosensor for detection of the African Swine Fever disease. *Anal. Chim. Acta*, **362**, 91–100.

34. Uludag, Y. and Tothill, I.E. (2010) Development of a sensitive detection method of cancer biomarkers in human serum (75%) using a quartz crystal microbalance sensor and nanoparticles amplification system. *Talanta*, **82**, 277–282.

35. Ma, Z.F., Wu, J.L., Zhou, T.H. *et al.* (2002) Detection of human lung carcinoma cell using quartz crystal microbalance amplified by enlarging Au nanoparticles in vitro. *New J. Chem.*, **26**, 1795–1798.

36. Li, J., He, X., Wu, Z. *et al.* (2003) Piezoelectric immunosensor based on magnetic nanoparticles with simple immobilization procedures. *Anal. Chim. Acta*, **481**, 191–198.

37. Zhang, Y., Wang, H., Yan, B. *et al.* (2008) A reusable piezoelectric immunosensor using antibody-adsorbed magnetic nanocomposite. *J. Immunol. Methods*, **332**, 103–111.

38. Halamek, J., Pribyl, J., Makower, A. *et al.* (2005) Sensitive detection of organophosphates in river water by means of a piezoelectric biosensor. *Anal. Bioanal. Chem.*, **382**, 1904–1911.

39. Uludag, Y., Piletsky, S.A., Turner, A.P.F., and Cooper, M.A. (2007) Piezoelectric sensors based on molecular imprinted polymers for detection of low molecular mass analytes. *FEBS J.*, **274**, 5471–5480.

40. Avila, M., Zougagh, M., Escarpa, A., and Rios, A. (2008) Molecularly imprinted polymers for selective piezoelectric sensing of small molecules. *TrAC-Trends Anal. Chem.*, **27**, 54–65.

41. Wu, A.H. and Syu, M.J. (2006) Synthesis of bilirubin imprinted polymer thin film for the continuous detection of bilirubin in an MIP/QCM/FIA system. *Biosens. Bioelectron.*, **21**, 2345–2353.

42. Pietrzyk, A., Suriyanarayanan, S., Kutner, W. *et al.* (2010) Molecularly imprinted poly[bis(2,2′-bithienyl)methane] film with built-in molecular recognition sites for a piezoelectric microgravimetry chemosensor for selective determination of dopamine. *Bioelectrochemistry*, **80**, 62–72.

43. Dickert, F.L. and Lieberzeit, P.A. (2007) Imprinted polymers in chemical recognition for mass-sensitive devices, in *Piezoelectric Sensors* (eds C. Steinem and A. Janshoff), Springer-Verlag, Berlin, pp. 173–210.

44. Hayden, O. and Dickert, F.L. (2001) Selective microorganism detection with cell surface imprinted polymers. *Adv. Mater.*, **13**, 1480–1483.

45. Dickert, F.L., Lieberzeit, P.A., and Hayden, O. (2004) Molecularly imprinted polymers for mass sensitive sensors – from cells to viruses and enzymes, in *Molecular Imprinting of Polymers* (eds S. Piletsky and A.P.F. Turner), Landes Bioscience, Austin, pp. 50–64.

46. Dickert, F.L., Lieberzeit, P.A., Hayden, O. *et al.* (2003) Chemical sensors - from molecules, complex mixtures to cells - supramolecular imprinting strategies. *Sensors*, **3**, 381–392.

47. Snejdarkova, M., Poturnayova, A., Rybar, P. *et al.* (2010) High sensitive calixarene-based sensor for detection of dopamine by electrochemical and acoustic method. *Bioelectrochemistry*, **80**, 55–61.

48. Pribyl, J. and Skladal, P. (2005) Quartz crystal biosensor for detection of sugars and glycated hemoglobin. *Anal. Chim. Acta*, **530**, 75–84.

49. Pribyl, J. and Skladal, P. (2006) Development of a combined setup for simultaneous detection of total and glycated haemoglobin content in blood samples. *Biosens. Bioelectron.*, **21**, 1952–1959.

50. Teles, F.R.R. and Fonseca, L.R. (2008) Trends in DNA biosensors. *Talanta*, **77**, 606–623.

51. Sassolas, A., Leca-Bouvier, B.D., and Blum, L.J. (2008) DNA biosensors and microarrays. *Chem. Rev.*, **108**, 109–139.

52. Tombelli, S. and Mascini, M. (2000) Piezoelectric quartz crystal biosensors: Recent immobilisation schemes. *Anal. Lett.*, **33**, 2129–2151.

53. Minunni, M., Tombelli, S., Mariotti, E. *et al.* (2001) Biosensors as new analytical tool for detection of Genetically Modified Organisms (GMOs). *Fresenius J. Anal. Chem.*, **369**, 589–593.

54. Tombelli, S., Mascini, M., Sacco, C., and Turner, A.P.F. (2000) A DNA piezoelectric biosensor assay coupled with a polymerase chain reaction for bacterial toxicity determination in environmental samples. *Anal. Chim. Acta*, **418**, 1–9.

55. Willner, I., Shlyahovsky, B., Willner, B., and Zayatz, M. (2009) Amplified DNA biosensors, in *Functional Nucleic Acids for Analytical Applications* (eds Y. Li and Y. Lu), Springer, New York, pp. 199–252.

56. Patolsky, F., Ranjit, K.T., Lichtenstein, A., and Willner, I. (2000) Dendritic amplification of DNA analysis by oligonucleotide-functionalized Au-nanoparticles. *Chem. Commun.*, 1025–1026.

57. Hianik, T., Grman, I., and Karpisova, I. (2009) The effect of DNA aptamer configuration on the sensitivity of detection thrombin at surface by acoustic method. *Chem. Commun.*, 6303–6305.

58. Hianik, T., Porfireva, A., Grman, I., and Evtugyn, G. (2008) Aptabodies - New type of artificial receptors for detection proteins. *Protein Pept. Lett.*, **15**, 799–805.

59. Hianik, T., Porfireva, A., Grman, I., and Evtugyn, G. (2009) EQCM biosensors based on DNA aptamers and antibodies for rapid detection of prions. *Protein Pept. Lett.*, **16**, 363–367.

60. Willner, I. and Zayats, M. (2007) Electronic aptamer-based sensors. *Angew. Chem. Int. Edit.*, **46**, 6408–6418.

61. Kaspar, M., Stadler, H., Weiss, T., and Ziegler, C. (2000) Thickness shear mode resonators ("mass-sensitive devices") in bioanalysis. *Fresenius J. Anal. Chem.*, **366**, 602–610.

62. Thompson, M. and Stone, D.C. (1997) *Surface-Launched Acoustic Wave Sensors: Chemical Sensing and Thin-Film Characterization*, Wiley, New York.

63. Benes, E., Groschl, M., Burger, W., and Schmid, M. (1995) Sensors based on piezoelectric resonators. *Sens. Actuators A-Phys.*, **48**, 1–21.

64. Grate, J.W., Martin, S.J., and White, R.M. (1993) Acoustic-wave microsensors. 2. *Anal. Chem.*, **65**, A987–A996.

65. Caliendo, C., Verardi, P., Verona, E. *et al.* (1997) Advances in SAW-based gas sensors. *Smart Mater. Struct.*, **6**, 689–699.

66. Wohltjen, H. (1984) Mechanism of operation and design considerations for surface acoustic-wave device vapor sensors. *Sens. Actuators*, **5**, 307–325.

67. James, D., Scott, S.M., Ali, Z., and O'Hare, W.T. (2005) Chemical sensors for electronic nose systems. *Microchim. Acta*, **149**, 1–17.

68. Grate, J.W. (2000) Acoustic wave microsensor arrays for vapor sensing. *Chem. Rev.*, **100**, 2627–2648.
69. Hossenlopp, J.M. (2006) Applications of acoustic wave devices for sensing in liquid environments. *Appl. Spectrosc. Rev.*, **41**, 151–164.
70. Gronewold, T.M.A. (2007) Surface acoustic wave sensors in the bioanalytical field: Recent trends and challenges. *Anal. Chim. Acta*, **603**, 119–128.
71. Li, Z.H., Jones, Y., Hossenlopp, J. *et al.* (2005) Analysis of liquid-phase chemical detection using guided shear horizontal-surface acoustic wave sensors. *Anal. Chem.*, **77**, 4595–4603.
72. Lange, K., Rapp, B.E., and Rapp, M. (2008) Surface acoustic wave biosensors: a review. *Anal. Bioanal. Chem.*, **391**, 1509–1519.

22

Microcantilever Sensors

Microcantilevers belong to the class of microelectromechanical devices that encompass all those devices produced by microfabrication except integrated circuits [1]. When integrated with a sensing element, a microcantilever responds to the recognition event by mechanical deformation. An alternative mechanical transduction method utilizes cantilevers that vibrate at the resonance frequency. In this case, the mass change produced by the recognition event induces a change in the resonance frequency.

Microcantilevers with a sharp tip were originally introduced as probes for exploring solid surface topography by atomic force microscopy in the early 1980s. It was soon realized that microcantilevers are sensitive to physical parameter such as temperature and humidity. By the mid-1990s, it was demonstrated that cantilevers modified with chemically sensitive layers respond selectively to particular chemical compounds. This opened the way to the development of a new class of chemical sensor in which the transduction is based on the mechanics of elastic solids. Despite its relatively recent introduction, this field of research has already produced very promising results. Early results obtained in the microcantilever sensor area have been summarized in refs. [2,3]. A good introduction to the field is given in ref. [4]. The mechanical principles of microcantilever operation are reviewed in [5,6], while comprehensive overviews of fabrication methods and recent advances are available in refs. [7–9]. Finally, applications of microcantilever sensors in the determination of ions, small molecules and biomacromolecules are surveyed in refs. [9–11].

22.1 Principles of Microcantilever Transduction

22.1.1 The Microcantilever

A microcantilever is a submillimeter beam of solid material, as shown in Figure 22.1A, which displays a rectangular microcantilever. V-shaped, U-shaped or T-shaped microcantilevers have also been employed in developing sensors. Owing to their small size, cantilevers can be integrated in arrays, as demonstrated in Figure 22.1B.

Microcantilevers are produced from metals or silicon materials (e.g., single-crystal silicon, polysilicon or silicon nitride) by micromachining, a technology based on the formation of sequences of thin films by deposition, patterning or etching (Section 5.13.3). Certain polymer materials are also used in microcantilever fabrication, particularly the SU-8 epoxy-based polymer that was originally developed as a negative photoresist mask for the microelectronics industry [12]. Polymer cantilevers are fabricated by injection molding.

Chemical sensing with a microcantilever is achieved by letting the analyte bind to the upper surface. This event induces a surface stress that modifies the plate area. As the length is much greater than the width, the most prominent effect is a modification of the length of the interacting surface which produces a *tensile stress*. If this stress is not compensated at the other side of the thin beam, the whole structure will bend. Figure 22.2A demonstrates the cantilever bending due to an increase in the surface stress (*compressive stress*) which causes the cantilever to bend downwards. If analyte interaction with the surface reduces the surface stress, which produces a *tensile stress*, this results in upward bending of the microcantilever. As the cantilever remains immobile after attaining an equilibrium state, this operational mode is called the *static deformation mode*.

Alternatively, the microcantilever sensor can be operated in the *dynamic mode* by letting it vibrate at the resonance frequency. An increase in mass results in a change in the resonance frequency, as in the case of the quartz crystal microbalance. By this means, the cantilever resonator functions as an extremely sensitive mass transducer with mass resolution in the picogram region. A resonator microcantilever can be fabricated in the single-clamped (Figure 22.2B) or doubly clamped (bridge) format (Figure 22.2C). Compared with the quartz crystal microbalance, beam resonators may be several orders of magnitude more sensitive. In the bridge format, the beam resonator displays mass resolution well below the picogram level.

There are various techniques for measuring the cantilever deflection, the most common being the optical and the piezoresistive methods. Optical transduction is based on deflection of a light beam at the cantilever surface. The deflected beam is detected by a photoelectric position detector. In the piezoresistive method beam deflection is indicated by the change in the resistance of an incorporated resistor.

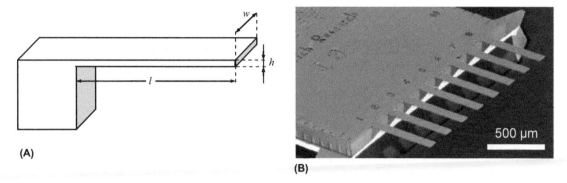

Figure 22.1 The microcantilever. (A) Conformation and geometrical parameters. Typical dimensions: $l = 100\text{–}350\,\mu m$; $w = 20\text{–}40\,\mu m$; $h = 0.6\text{–}1\,\mu m$. (B) Scanning electron micrograph of a cantilever array. (B) Reprinted with permission from [13]. Copyright 2001 Elsevier.

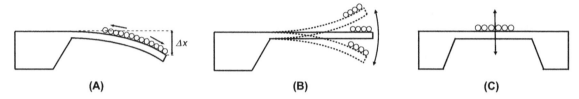

Figure 22.2 Transduction principles in microcantilever sensors. The cantilever is seen from the side. (A) Static deformation mode. The surface stress induces a slight elongation of the upper surface that results in downward beam bending. (B) Dynamic mode (resonator cantilever); (C) Doubly clamped beam resonator (bridge configuration).

In order to promote selective binding of the analyte, the cantilever surface is coated with a recognition receptor film. If the cantilever is to be operated in the static deformation mode, the receptor layer should be formed only on one side, typically on the upper surface. The other surface should be coated with a passive film to prevent any interaction with the analyte. This precaution is necessary because analyte interaction with both surfaces leads to an intricate response. Microcantilever sensors operated in the dynamic mode can be coated on both sides, but if only one side is coated, it is possible to measure both the deflection and the resonance frequency.

Bimaterial cantilevers, obtained by coating a nonmetallic cantilever with a metal film, are utilized as temperature transducers. The difference in the expansion coefficients of the two layers causes the cantilever to bend as a function of the ambient temperature. In this way, the bimaterial cantilever is employed as a microcalorimeter in order to monitor the temperature changes caused by chemical reactions or phase-transition processes occurring at the cantilever surface. Temperature variations down to $10^{-5}\,K$ can be detected in this way.

Microcantilevers can be affected in a spurious way by the thermal drift or by interactions with the environment, particularly when operated in a liquid. In addition, nonspecific binding of sample concomitants or physical adsorption of the analyte may also contribute to the total response. That is why it is essential to use a reference sensor integrated with the working sensor in the same array. The corrected response, which is obtained as the difference between the sensor and reference signals, is much more reliable than the uncorrected response. That is why microcantilever sensors are typically produced and operated in the array format. Operation in an array configuration brings about an additional advantage, namely the possibility of multiplexing or parallelization of the assay.

22.1.2 Static Deformation Transduction

As mentioned above, the static deformation mode relies on beam bending as a result of analyte interaction with the sensor layer. The energy needed to bend the cantilever originates from the change in the Gibbs free energy that accompanies the interaction of the analyte with the upper surface. The energy contribution arises from the analyte binding energy, repulsive or attractive forces between bound molecules, and surface stress induced by direct interaction of an adsorbate with the cantilever surface. Entropy variation in the response to the recognition event can also contribute to the change in the Gibbs surface free energy.

Figure 22.3 demonstrates two types of molecular interactions causing cantilever binding. If the surface interaction generates ionic groups, electrostatic repulsion produces compressive stress and the cantilever bends down (Figure 22.3A). Tensile stress and thereby up bending might be noticed with DNA hybridization sensors (Figure 22.3B). In this case, the single-strand DNA probe assumes a random coiled structure upon immobilization. After hybridization with the target, the stiff double-strand hybrid assumes a near-perpendicular orientation to the surface that relaxes the repulsive interactions and induces tensile deformation.

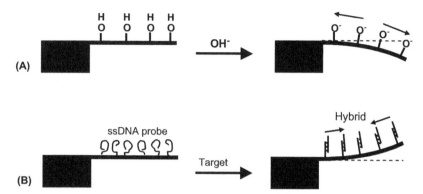

Figure 22.3 Cantilever deflection in response to surface interactions. (A) Ionization of a surface acidic group; (B) DNA hybridization. Reprinted with permission from [4]. Copyright 2008 The Royal Society of Chemistry.

As already mentioned, the sensor response in the static deformation mode is the cantilever deflection (Δx, Figure 22.2A), which is caused by the difference in surface stress ($\Delta\sigma$) between the responsive and the passive surfaces of the cantilever. The bending radius of the cantilever (r) depends on the surface stress difference according to Stoney's equation:

$$\frac{1}{r} = \frac{6(1 - v)}{Eh^2} \Delta\sigma \tag{22.1}$$

where E is the Young modulus (which describes tensile elasticity), h is the thickness of the cantilever and v is the Poisson ratio (which equals the ratio of lateral strain to axial strain). Assuming that $r \gg l$ (where l is the length of the cantilever), Stoney's equation leads to the following equation that gives the cantilever deflection as a function of the surface stress difference:

$$\Delta x = \frac{3(1 - v)}{E} \left(\frac{l}{h}\right)^2 \Delta\sigma \tag{22.2}$$

Equation (22.2) demonstrates that the sensitivity in terms of $\Delta x / \Delta\sigma$ can be enhanced by increasing the length and decreasing the thickness of the cantilever. Sensitivity depends also on the cantilever material through the E and v constants. The Young's modulus is an indicator of the elasticity of the material and a lower value of the Young modulus brings about a higher sensitivity. For example, the polymer SU-8 has a Young modulus of about 40 times lower than that of silicon and hence can provide a greater sensitivity.

22.1.3 Resonance-Mode Transduction

A microcantilever that vibrates at its resonance frequency can act as an extremely sensitive mass transducer. Actuation of a microcantilever resonator can be achieved by means of a piezoelectric resonator integrated with the microcantilever. Typically, PZT (Pb ($Zr_{0.52}Ti_{0.48}$)O_3) or aluminum nitride are employed as piezoelectric materials for actuation. Thermal actuation is possible with bimetal-type cantilevers by periodic supply of heat by means of an external laser or an integrated resistive heater. Electrostatic or magnetomotive actuation methods have also been developed.

A cantilever behaves as a single-dimensional harmonic oscillator with the fundamental resonance frequency (f_0) depending on the mass of the beam (m) and on the spring constant of the beam material (k):

$$f_0 = \frac{1}{2\pi} \sqrt{\frac{k}{nm}} \tag{22.3}$$

where n is a geometrical parameter for the fundamental vibration mode of a rectangular cantilever beam. The product $nm = m^*$ represents the effective mass of the vibrating system. The elastic properties of the beam material are indicated by the spring constant, which is the quotient of strain and stress under a longitudinally applied stress and depends on the Young's modulus and the geometrical parameters as:

$$k = \frac{Ew}{4} \left(\frac{h}{l}\right)^3 \tag{22.4}$$

Using the above two equations, the resonance frequency of a rectangular silicon microcantilever of $500 \times 100 \times 1$ µm was found to be of 6 kHz [4]. Upon combining Equations (22.3) and (22.4), it can be found that the resonance frequency increases with decrease in the length and increase of the thickness of the cantilever.

The resonance frequency decreases if an additional mass is loaded over the cantilever resonator. If the mass load is uniformly distributed over the entire beam surface, the mass change can be found by means of the following equation in which f_m is the frequency of the loaded beam:

$$\Delta m = -\frac{k}{4\pi^2 n} \left(\frac{1}{f_0^2} - \frac{1}{f_m^2} \right) \tag{22.5}$$

At very small changes in frequency the above equation can be converted into the following approximate form that indicates the mass sensitivity of the cantilever resonator:

$$\frac{\Delta f}{\Delta m} \approx \frac{1}{2} \frac{f_0}{m} \tag{22.6}$$

According to this equation, mass sensitivity increases with increasing fundamental resonance frequency, which is obtained by having a large Young's modulus, low density and small dimensions. Greater sensitivity is obtained by operating the resonator at a higher vibration mode (overtone).

The above treatment assumes that the spring constant is not affected by the analyte interaction with the cantilever surface. Actually, this constant can vary as a result of fusion of materials (change in the Young's modulus) and change in cantilever thickness. Such complications are avoided if the sensing layer is localized at the terminal end of the beam, as shown in Figure 22.2B. In this case, the changes in resonance frequency can be mostly attributed to mass loading.

Mass sensitivity depends to a large extent on the quality factor of the resonator (Q), which is defined in the following equation:

$$Q = 2\pi \frac{W_0}{\Delta W} \tag{22.7}$$

where W_0 is the stored vibrational energy and ΔW is the energy dissipated per cycle. The minimum detectable frequency shift of a cantilever is proportional to $\sqrt{1/Q}$. Hence, the minimum detectable frequency, as well as mass sensitivity, increases with increasing quality factor of the cantilever. For proper operation, the quality factor should be at least 100.

As indicated in Equation (22.7) a good quality factor is obtained when energy dissipation is very low. Energy dissipation is produced by intrinsic factors associated with the resonator (internal dissipation) and by energy loss to the ambient medium (external damping). External damping increases with the viscosity and density of the surrounding medium, this effect being much more pronounced in liquids than in gases.

The quality factor can be increased by *active feedback*, a process in which energy is supplied to the resonator in order to compensate for dissipation. This principle is similar to that employed in the quartz crystal microbalance.

Compared with the quartz crystal microbalance, which typically displays mass resolution in the nanogram (10^{-9} g) region, the mass resolution of a resonator microcantilever is in the picogram (10^{-12} g) region, that is, about 100–1000 times better.

Much better mass sensitivity is obtained with doubly clamped slab resonators (Figure 22.2C), as the resonance frequency in the bridge format is about 6 times greater than that of a singly clamped beam of the same dimension. At the same time, the bridge-type resonator shows a higher quality factor. Very high quality factors have been reported for string-like (stretched) doubly clamped beams that allow for mass resolution below the picogram level.

Very high mass sensitivity has been achieved by reducing the size of doubly clamped beams to the nanometer region to produce nanoelectromechanical systems [14,15]. Nanometer-sized resonators have an exceptionally small mass and thereby a very high resonance frequency, which could reach 100 MHz. At the same time, the quality factor of a nanoresonator can attain tremendously great values (up to 10^5). The combined effect of very low mass and a high quality factor leads to mass resolution in the attogram region (10^{-18} g) which makes possible the detection of individual macromolecules.

22.2 Measurement of Cantilever Deflection

22.2.1 Optical Measurement of Cantilever Deflection

In the optical lever technique, a laser beam is impinged onto the cantilever end and the reflected beam is allowed to fall on an array of light microdetectors that act as a position-sensitive detector (Figure 22.4A). Cantilever bending

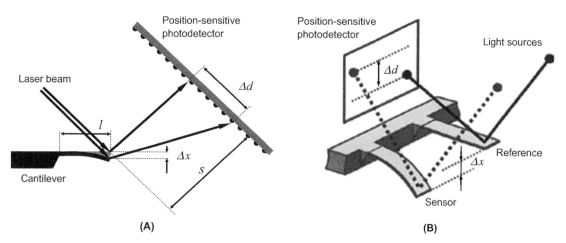

Figure 22.4 Optical measurement of beam deflection. (A) Principle of the optical lever technique. A laser beam is reflected by the cantilever and impinges onto the screen of the position-sensitive detector. Beam deflection produces a measurable Δd shift of the light spot. (B) Measurement of the sensor beam deflection with respect to a reference beam. (B) Reprinted with permission from [4]. Copyright 2008 The Royal Society of Chemistry.

causes the reflected light spot to shift on the detector screen. Using Stoney's equation, the deflection at the end of the cantilever is obtained as:

$$\Delta x = \frac{l}{4s}\Delta d \tag{22.8}$$

where l is the length of the cantilever, s is the distance between the cantilever and the screen of the position-sensitive detector, and Δd is the spot displacement. Figure 22.4B demonstrates the measurement of a sensor cantilever deflection with respect to a reference cantilever.

The optical lever technique allows a vertical deflection of 0.1 nm to be measured routinely. The mean disadvantage of this read-out technique is that it requires frequent optical alignment and calibration. In addition, the optical read-out is susceptible to errors due to light absorption or scattering in the sensor environment. When applied to cantilever arrays, an individual light source should be available for each cantilever. In a simpler configuration, a single laser beam is scanned sequentially across all the beams in the array. Another possibility in array operation is to illuminate all the cantilevers with a single collimated beam. Individual reflected beams are projected onto the image plane of a charge-coupled device camera to assess the deflection of each cantilever.

The optical lever techniques allow for simultaneous measurement of beam deflection, resonance frequency and vibration amplitude.

Optical measurement of cantilever deflection can also be performed by means of light interference. In this technique, coherent light waves reflected by a reference and a measuring cantilever interfere when there is a difference in deflection amplitude. Interference is caused by the difference between the path lengths of each wave (see Section 18.6.2). The deflection difference is indicated by the change in the intensity of the selected diffracted order. This technique is extremely sensitive as it is capable of detecting deflections as small as 0.01 Å, but the technique has a very narrow dynamic range.

A more compact optical read-out system has been developed by forming a SiO_2 grating over a silicon cantilever [16]. By means of the grating, light is coupled in the cantilever, which acts as a waveguide. A second waveguide, which is placed in front of the cantilever, collects guided light. Any displacement of the cantilever disturbs the alignment of the above waveguides and thereby modifies the intensity of the collected light. This system needs no preliminary adjustment except for a relatively rough alignment for light coupling.

22.2.2 Electrical Measurement of Cantilever Deflection

The second most common transduction technique in microcantilever sensors is based on the *piezoresistive effect* that represents the change in the electrical resistance of a conductor under the effect of a mechanical strain. As the electrical resistance is proportional to the resistor length, any change in length results in a modification of the resistance. The strain sensitivity of a piezoresistive transducer is indicated by the *gauge factor* (GF) defined as:

$$GF = \frac{\Delta R/R}{\Delta l/l} \tag{22.9}$$

where ΔR is the change in resistance, R is the initial resistance, Δl is the change in the length and l is the initial length.

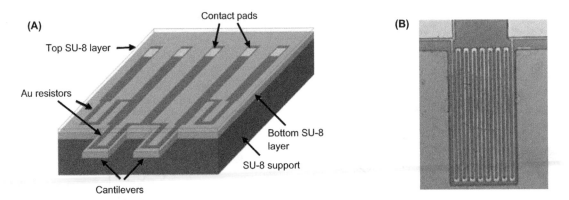

Figure 22.5   Principle of piezoresistive measurement of cantilever deflection using metal-film piezoresistors. (A) Device configuration. Cantilevers are produced from SU-8 photoresist polymer. Two resistors on the chip along with the resistors on each cantilever form a Wheatstone bridge. (B) A close-up view of the cantilever with integrated meander-type gold resistor. Reprinted with permission from [17]. Copyright 2002 IOP Publishing Ltd.

Piezoresistive gauges can be made of metal or semiconductor materials and can be included in the cantilever or attached to its surface. A semiconductor strain gauge is smaller, less expensive, and more sensitive than a metal film gauge. On the other hand, semiconductor gauges are sensitive to temperature variations and are susceptible to drift.

The configuration of a piezoresistive microcantilever transducer is shown in Figure 22.5A. It is formed of two cantilevers, one of them acting as reference. On each cantilever, a gold resistor is formed by vapor deposition in the form of a meander track, as shown in the close-up view in Figure 22.5B. Two resistors are formed in the same way on the chip and connections are provided so as to form a Wheatstone bridge along with cantilever resistors. The sensing process induces unequal variations in beam resistance and thereby an inbalance of the Wheatstone bridge. The output voltage of the bridge indicates the deflection difference.

When operated in liquid samples, all resistors should be coated with an insulator layer in order to prevent short circuiting.

An analysis of the response of the piezoresistive microcantilever led to the conclusion that the sensitivity increases with a decrease in the total thickness of the device [18]. The width and the length of the device also affect the sensitivity, although to a lesser extent.

Compared with optical transduction, electrical transduction is advantageous in that optical components are not required and neither is laser alignment. The response is not affected by optical properties of the surrounding medium. In addition, the read-out electronics can be integrated on the same chip to form a self-contained sensing–measuring device.

One additional advantage results from the possibility of using the resistor on the cantilever as a heating element to control the temperature of the sensing layer. This characteristic can be employed for thermal breaking of the target–probe duplex in DNA sensors in order to perform sensor regeneration after a DNA assay. Heating can also be used to sustain polymerase chain reactions at the cantilever surface.

An *electrostatic* transduction method is based on capacitance measurements. In this method, a capacitor is formed of a metalized cantilever and a fixed electrode placed underneath. As the cantilever deflects, the distance between the two elements changes and thereby modifies the capacitance of the system. Sensitivity in capacitive transduction is enhanced by forming a square pad at the end of the cantilever beam because the capacitance is proportional to the surface area of the movable plate. Capacitive transduction has been employed in gas sensors, since in such cases there is no risk of the ambient fluid short circuiting the plates. This method is advantageous due to the simplicity of the associated electronics and its low power consumption.

22.3 Functionalization of Microcantilevers

In order to impart selectivity, recognition receptors are attached to the cantilever surface. Receptor immobilization can be conducted on silicon material surfaces upon activation to form hydroxyl groups, followed by silanization. Gold-coated cantilevers can be functionalized by forming first a thiocarboxylic acid self-assembled monolayer. In each case, the first layer acts as a linker to which recognition receptor proteins are grafted by covalent immobilization. Thiolated nucleotides can be attached directly to the gold film via thiol chemisorption. Electrochemical polymerization can also be utilized to form a polymer layer containing functional groups to which the receptor is further attached.

The inactive area of the microcantilever should be passivated by coating with a material that does not interact with the analyte or other sample components. Typically, the hydrophilic polymer polyethylene glycol is utilized for passivation against biomolecules.

The reproducibility in the fabrication of the sensing layer is poor at the cantilever edge; therefore, a greater width of the cantilever beam brings about an improvement in batch reproducibility.

As microcantilevers are commonly used in the array format, it is important to utilize techniques that allow the modifying reagents to be delivered sequentially or in parallel to each individual cantilever [19]. Drop deposition can be performed by inkjet printing or by microcontact printing. Such methods achieve modification of the upper side of the cantilever only. If both sides have to be modified, arrays of glass microcapillaries are utilized. Each cantilever is introduced in a microcapillary that guides the solution of the needed reagent.

22.4 Microcantilever Gas and Vapor Sensors

In order to detect gas or vapors, a microcantilever is coated with a thin layer of material that interacts more or less selectively with the analyte [20]. In general, gas-sensitive coatings are similar to those used in other types of gas sensor.

Metal coatings allow certain gases and vapors to be sensed. For example, a palladium coating is used in hydrogen sensing whilst a gold coating is used for trace mercury determination. Gold coating also imparts specificity to mercaptans that adsorb on gold via the thiol group.

Water vapors can be sensed by cantilevers coated with hydrophilic polymers such as polyvinyl alcohol, carboxymethyl cellulose or polyvinylpyrrolidone.

Detection of organic vapors is achieved by means of polymer coatings. Common polymers such as polystyrene, polyurethane, poly(methyl methacrylate) and their blends can be used as coatings in organic vapor sensors. Because the selectivity of such coatings is poor, they are used to fabricate arrays of cross-sensitive sensors for applications in artificial olfaction. As is typical in this technique, each cantilever in the array is coated with a different polymer material so that it responds with a particular sensitivity to each component of the gas sample.

Selectivity to particular classes of organic vapors is imparted by chromatographic stationary phase beads (about 0.1 μm diameter) incorporated in a gel at the cantilever surface. More advanced selectivity to particular analytes is obtained by means of supramolecular chemistry receptors such as cyclodextrins.

Owing to the exceptional sensitivity of microcantilever sensors, detection of the vapors of explosives in the search for explosives is a promising application in the fight against terrorism. Explosive vapors can be detected by adsorption at a silicon surface, or, more generally, by means of specific sensing coatings. TNT (2,4,6-trinitrotoluene) can be detected by deflagration on a heated cantilever, the deflagration producing a bump in the cantilever bending signal [21].

22.5 Microcantilever Affinity Sensors

22.5.1 General Aspects

Microcantilever devices are suitable platforms for developing label-free sensors based on various kinds of affinity reactions. Most of the reported applications in this area make use of antibodies as recognition receptors. Such applications encompass determination of proteins, haptene molecules, bacteria or viruses [10]. An immunosensor can be obtained by coating the cantilever with a selected antibody layer (Figure 22.6) that functions as a selective receptor

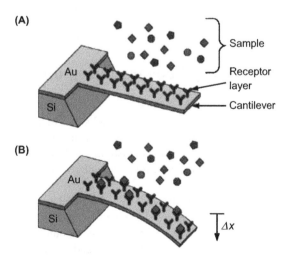

Figure 22.6   Functioning principle of a microcantilever affinity sensor. (A) Before analyte binding; (B) after analyte binding. Reproduced with permission from [22]. Copyright 2006 Elsevier.

for the analyte. Analyte binding induces compressive stress that leads to beam bending in proportion with the analyte concentration.

The response of the microcantilever immunosensor can be amplified if a second antibody is attached to the initial immunocomplex to form a sandwich ternary complex [23]. When operating the sensor in the static deformation mode, formation of the ternary complex produces additional surface stress that enhances the beam deflection. With a sensor operated as a resonator, the frequency shift is enhanced if the second antibody is conjugated with silicon nanoparticles that increase the overall mass of the cantilever.

Exceptional sensitivity has been reported for microchannel resonators operating in high vacuum [24,25]. In this approach, the microcantilever includes a microchannel through which the sample is allowed to flow in order to perform recognition by a receptor layer at the microchannel surface. Operation in vacuum brings about a very high quality factor, which, in conjunction with the very low mass of the resonator, allows a single biomolecule or a single cell to be detected.

Microcantilever immunosensors have been developed for various types of target analytes such as proteins, viruses and bacteria. An overview of the underlying design principles is given in ref. [18].

22.5.2 Microcantilever Protein Sensors

An example of a protein assay by a microcantilever sensor is the determination of creatine kinase and myoglobin in blood [27]. These proteins are biomarkers that help diagnose or rule out a heart attack. Microcantilever sensors for each of the above proteins have been obtained by using the corresponding antibody layer over a microcantilever. A very sensitive resonator microcantilever myoglobin sensor has been produced using antibody immobilization by biotin–streptavidin interaction [28]. If the antibody orientation at the surface is random, a fraction of the present antibodies is not accessible to the analyte. In order to maximize the availability of the antibody to analyte binding, antibody biotinylation has been conducted so as to let all binding sites be exposed to the sample analyte. This immobilization method results in a 10-fold greater sensitivity compared with a randomly oriented antibody layer.

Another biomarker of interest in clinical chemistry is the prostate-specific antigen (PSA), a glycoprotein that is used for the detection of prostate cancer. Using an anti-PSA antibody, a cantilever operated in the static deformation mode allows free-PSA to be detected at concentration from $0.2 \, \mathrm{ng \, mL^{-1}}$ to $60 \, \mathrm{\mu g \, mL^{-1}}$, which includes the clinical relevant PSA concentration range.

Among chemical warfare agents there are certain protein toxins that could be detected by microcantilever immunosensors. An example of this kind is ricin, a potent cytotoxin that prevents living cells from producing essential proteins. A microcantilever sensor including a ricin antibody has allowed a limit of detection of 60 ppt of ricin to be attained [28].

22.5.3 Microcantilever Pathogen Sensors

Detection of pathogenic organisms such as viruses and bacteria has also been approached by microcantilever immunosensors. Pathogen recognition is achieved by means of a corresponding antibody grafted to the microcantilever surface. Identification of bacteria spores is of great importance to the food and beverage industry as well as in the assessment of drinking water quality. Accordingly, sensors for spores of *Bacillus anthracis*, which is the pathogen of the acute Anthrax disease, have been developed. A microcantilever sensor for the spores of the potential biological warfare agent *Francisella tularensis* (which causes tularemia disease) has been developed using a commercial antibody [28]. The reported limit of detection of *Francisella tularensis* is 1000 organisms per ml when the sensor signal is corrected by the signal of a reference cantilever.

Detection of viruses with antibody-modified microcantilevers has also been demonstrated.

22.5.4 Microcantilever Affinity Sensors Based on Other Recognition Receptors

Besides immunosensors, microcantilever sensors based on other molecular receptors have also been developed. For example, a short peptide with a selected amino acid sequence can act as the receptor for the identification of *Bacillus subtilis* spores [29]. Detection of fungal spores has been demonstrated with *Aspergillus niger* that is a spoilage agent and a pathogen to humans. The Concanavalin A protein, which recognizes saccharides in the cell wall, has been used as a recognition receptor [30].

Phenylboronic acid is a suitable receptor for saccharides. In order to perform fructose determination, a thiolated phenylboronic acid derivative was immobilized over a gold-coated microcantilever [31]. Phenylboronic acid interacts with two vicinal hydroxyl groups in fructose to form a boronate compound (Figure 22.7). Depending on the pH, this boronate compound can be present as a neutral or negatively charged species. Thereby, beam deflection has been ascribed to the electrostatic repulsion of charged molecules in the surface molecular layer. As the molar fraction of the charged form increases with the pH, the response signal also increases with this parameter. When operated in the

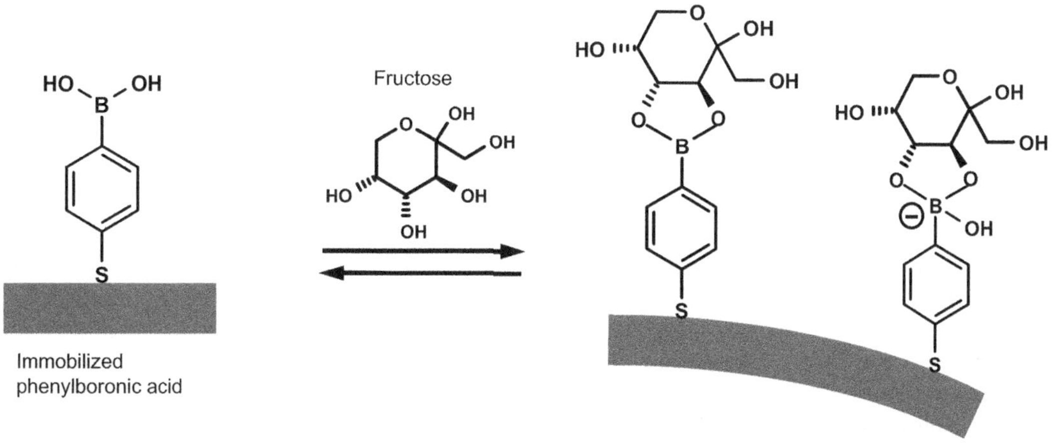

Figure 22.7 Fructose determination by means of a phenylboronic acid coated cantilever. Phenylboronic acid is immobilized by thiol chemisorption on a gold coating. Adapted with permission from [31]. Copyright 2008 American Chemical Society.

static deformation mode at pH 7, the surface stress was linear in analyte concentration over the 0–25 mM range, allowing fructose to be readily detected at physiologically meaningful concentrations.

22.6 Enzyme Assay by Microcantilever Sensors

Enzyme assays aim at measuring enzymatic activity in biological specimens. In principle, an enzyme activity cantilever sensor is obtained upon immobilization of an enzyme substrate on the microcantilever [32]. Substrate conversion under the catalytic effect of the enzyme leads to changes in mass, conformation, and charge that modify the cantilever deflection and allow the enzyme activity to be determined. This approach is also suitable for inhibitor determination, with the enzyme–inhibitor interaction proceeding in the solution phase.

Several methods for signal amplification in enzyme activity sensors have been advanced. In one approach, the immobilized substrate is conjugated with microbeads at its upper end [33]. Under the effect of the target enzyme, the substrate molecule is split, leading to the removal of the bead and, thereby, to a considerable loss of resonator mass.

Signal amplification in static deformation enzyme sensors has been achieved by using an immobilized substrate conjugated with magnetic nanoparticles [34]. Microcantilever deflection is produced in this state by an external magnetic field. The target enzyme breaks down the substrate and the nanoparticle-tagged fragment is disconnected from the cantilever. As a result, cantilever deflection decreases and the deflection variation is in relation to the enzyme activity in the sample.

22.7 Microcantilever Nucleic Acid Sensors

Application of microcantilevers as transducers in DNA hybridization sensors has been demonstrated in ref. [35] using cantilever arrays operated in the static deformation mode. Single-base mismatch detection was possible in this way. Cantilever deflection upon hybridization depends on a series of factors such as surface coverage with the capture probe, degree of disorder at the probe-modified surface, degree of hydration and ionic strength. Depending on the experimental conditions, either compressive or tensile stress can be produced by hybridization.

Detection of nucleic acid hybridization requires reference cantilevers functionalized with an oligonucleotide that is not complementary to the target DNA.

Appreciable signal amplification in microcantilever DNA sensors has been obtained by incorporation of various kinds of nanoparticles.

Magnetic nanoparticles incorporated in the sensing layer upon recognition have been utilized to amplify the microcantilever response in the static deformation mode [36]. The cantilever deflection is enhanced by an external magnetic field that allows determination of DNA with an outstanding limit of detection.

As demonstrated in ref. [37], signal amplification in resonator-type DNA sensors can be obtained by means of metal nanoparticles incorporated into the sensing layer. In this method, the target polynucleotide is hybridized with a shorter capture probe. Next, a complementary probe tagged with gold nanoparticles is hybridized with the overhanging tail of the target. Nanoparticles contribute substantially to the mass increase upon recognition. Additional response enhancement is obtained by chemical deposition of silver onto gold nanoparticles. This is a transposition of a similar amplification protocol originally developed with the quartz crystal microbalance. The above microcantilever sensor produced a linear response in terms of frequency shift vs. logarithm of target concentration over a concentration rage extending from 0.05 to 10 mM.

22.8 Outlook

Microcantilever sensors represent a viable approach to the development of label-free sensors for various kinds of species including gases, small molecules, biomacromolecules and pathogens. From the standpoint of the application field, microcantilever sensors are in competition with well-established sensing methods such as the quartz crystal microbalance, surface acoustic wave and surface plasmon resonance devices. In order to achieve performances comparable with the above techniques, resonant microcantilevers should be produced so as to display a very high resonance frequency, preferably over 1 MHz. Alternatively, high sensitivity can be obtained by operating conventional beam resonators at higher overtones.

Nanoparticle application has opened up a new way to sensitivity enhancement for both static and dynamic mode microcantilever sensors. However, nanoparticles act as a kind of label and the inclusion of the nanoparticle into the sensing layer complicates the assay protocol.

Further progress in this field is expected to be resulting in increased sensitivity and reduced response time. Incorporation of microcantilever arrays into microfluidic cartridges would be one way of shortening the response time and increasing throughput.

Questions and Exercises

1 Comment on microcantilever shape, size, materials and integration in arrays.
2 How can the surface stress be induced in the microcantilever surface? What is the effect of surface stress and how can surface stress be measured?
3 Draw a sketch showing the heat effect on a bimaterial cantilever.
4 What cantilever parameters determine the amplitude of beam deflection in the static deformation mode?
5 What are the particular features of the dynamic mode operation of microcantilever sensors?
6 Suppose that a thiocarboxylic acid (HS-(CH$_2$)$_n$-COOH) is grafted to the surface of a gold-coated microcantilever. Comment on the effect of pH on the cantilever shape.
7 What cantilever characteristics determine the mass sensitivity in the dynamic mode operation?
8 Compare Equation (22.6) with the Sauerbrey equation and comment on why a microcantilever resonator could be more sensitive to mass change compared with the quartz crystal microbalance.
9 What parameter indicates the effect of the surrounding environment on the sensitivity of a microcantilever resonator? What microcantilever configuration provides superior sensitivity and why?
10 What are the particular characteristics of nanosized resonator beams and how do these characteristics affect the mass sensitivity?
11 Discuss the principles, advantages and drawbacks of the optical cantilever technique for the measurement of beam deflection.
12 Comment on certain optical transduction methods that overcome the drawbacks of the optical cantilever technique.
13 Draw the scheme of a Wheatstone bridge and indicate the position of each resistor on a piezoresistive microcantilever couple.
14 What parameters determine the sensitivity of a piezoresistive response to surface stress?
15 Comment on possible cantilever coatings utilized in the sensing of gases and vapors.
16 Describe some methods for signal amplification in microcantilever immunosensors.
17 What kinds of proteins are of interest in the application of microcantilever immunosensors?
18 Give an overview of nonimmunologic recognition receptors utilized in affinity immunosensors.
19 How can enzyme activity be assayed by means of microcantilever devices? Describe certain amplification methods utilized in such sensors.
20 Describe some methods for signal amplification in microcantilever nucleic acid sensors.

References

1. Judy, J.W. (2001) Microelectromechanical systems (MEMS): fabrication, design and applications. *Smart Mater. Struct.*, **10**, 1115–1134.
2. Raiteri, R., Grattarola, M., Butt, H.J. *et al.* (2001) Micromechanical cantilever-based biosensors. *Sens. Actuators B-Chem.*, **79**, 115–126.

3. Sepaniak, M., Datskos, P., Lavrik, N. *et al.* (2002) Microcantilever transducers: A new approach to sensor technology. *Anal. Chem.*, **74**, 568A–575A.

4. Fritz, J. (2008) Cantilever biosensors. *Analyst.*, **133**, 855–863.

5. Ziegler, C. (2004) Cantilever-based biosensors. *Anal. Bioanal. Chem.*, **379**, 946–959.

6. Lang, H.P. and Gerber, C. (2008) Microcantilever sensors. *Top. Curr. Chem.*, **285**, 1–27.

7. Boisen, A., Dohn, S., Keller, S.S. *et al.* (2011) Cantilever-like micromechanical sensors. *Rep. Prog. Phys.*, **74**, Article Nr. 036101.

8. Alvarez, M. and Lechuga, L.M. (2010) Microcantilever-based platforms as biosensing tools. *Analyst.*, **135**, 827–836.

9. Goeders, K.M., Colton, J.S., and Bottomley, L.A. (2008) Microcantilevers: Sensing chemical interactions via mechanical motion. *Chem. Rev.*, **108**, 522–542.

10. Buchapudi, K.R., Huang, X., Yang, X. *et al.* (2011) Microcantilever biosensors for chemicals and bioorganisms. *Analyst.*, **136**, 1539–1556.

11. Wang, C.Y., Wang, D.Y., Mao, Y.D. *et al.* (2007) Ultrasensitive biochemical sensors based on microcantilevers of atomic force microscope. *Anal. Biochem.*, **363**, 1–11.

12. Nordstrom, M., Keller, S., Lillemose, M. *et al.* (2008) SU-8 cantilevers for bio/chemical sensing; Fabrication, characterisation and development of novel read-out methods. *Sensors*, **8**, 1595–1612.

13. Battiston, F.M., Ramseyer, J.P., Lang, H.P. *et al.* (2001) A chemical sensor based on a microfabricated cantilever array with simultaneous resonance-frequency and bending readout. *Sens. Actuators B-Chem.*, **77**, 122–131.

14. Ekinci, K.L. and Roukes, M.L. (2005) Nanoelectromechanical systems. *Rev. Sci. Instrum.*, **76**, Article Nr. 061101.

15. Eom, K., Park, H.S., Yoon, D.S. *et al.* (2011) Nanomechanical resonators and their applications in biological/chemical detection: Nanomechanics principles. *Phys. Rep.-Rev. Sec. Phys. Lett.*, **503**, 115–163.

16. Zinoviev, K., Dominguez, C., Plaza, J.A. *et al.* (2006) A novel optical waveguide microcantilever sensor for the detection of nanomechanical forces. *J. Lightwave Technol.*, **24**, 2132–2138.

17. Thaysen, J., Yalcinkaya, A.D., Vettiger, P. *et al.* (2002) Polymer-based stress sensor with integrated readout. *J. Phys. D-Appl. Phys.*, **35**, 2698–2703.

18. Joshi, M., Gandhi, P.S., Lal, R. *et al.* (2011) Modeling, simulation, and design guidelines for piezoresistive affinity cantilevers. *J. Microelectromech. S.*, **20**, 774–784.

19. Bietsch, A., Zhang, J.Y., Hegner, M. *et al.* (2004) Rapid functionalization of cantilever array sensors by inkjet printing. *Nanotechnology.*, **15**, 873–880.

20. Lang, H.P. (2009) Cantilever-based gas sensing, in *Solid State Gas Sensing* (eds E. Comini, G. Faglia, and G. Sberveglieri), Springer Science+Business Media, LLC, Boston, MA, pp. 305–328.

21. Pinnaduwage, L.A., Wig, A., Hedden, D.L. *et al.* (2004) Detection of trinitrotoluene via deflagration on a microcantilever. *J. Appl. Phys.*, **95**, 5871–5875.

22. Lim, S.H., Raorane, D., Satyanarayana, S. *et al.* (2006) Nano-chemo-mechanical sensor array platform for high-throughput chemical analysis. *Sens. Actuators B-Chem.*, **119**, 466–474.

23. Lee, S.M., Hwang, K.S., Yoon, H.J. *et al.* (2009) Sensitivity enhancement of a dynamic mode microcantilever by stress inducer and mass inducer to detect PSA at low picogram levels. *Lab Chip.*, **9**, 2683–2690.

24. Burg, T.P., Mirza, A.R., Milovic, N. *et al.* (2006) Vacuum-packaged suspended microchannel resonant mass sensor for biomolecular detection. *J. Microelectromech. S.*, **15**, 1466–1476.

25. Burg, T.P., Godin, M., Knudsen, S.M. *et al.* (2007) Weighing of biomolecules, single cells and single nanoparticles in fluid. *Nature.*, **446**, 1066–1069.

26. Arntz, Y., Seelig, J.D., Lang, H.P. *et al.* (2003) Label-free protein assay based on a nanomechanical cantilever array. *Nanotechnology*, **14**, 86–90.

27. Grogan, C., Raiteri, R., O'Connor, G.M. *et al.* (2002) Characterisation of an antibody coated microcantilever as a potential immuno-based biosensor. *Biosens. Bioelectron.*, **17**, 201–207.

28. Ji, H.F., Yan, X.D., Zhang, J. *et al.* (2004) Molecular recognition of biowarfare agents using micromechanical sensors. *Expert Rev. Mol. Diagn.*, **4**, 859–866.

29. Dhayal, B., Henne, W.A., Doorneweerd, D.D. *et al.* (2006) Detection of Bacillus subtilis spores using peptide-functionalized cantilever arrays. *J. Am. Chem. Soc.*, **128**, 3716–3721.

30. Nugaeva, N., Gfeller, K.Y., Backmann, N. *et al.* (2005) Micromechanical cantilever array sensors for selective fungal immobilization and fast growth detection. *Biosens. Bioelectron.*, **21**, 849–856.

31. Baker, G.A., Desikan, R., and Thundat, T. (2008) Label-free sugar detection using phenylboronic acid-functionalized piezoresistive microcantilevers. *Anal. Chem.*, **80**, 4860–4865.

32. Bottomley, L.A., Ghosh, M., Shen, S. *et al.* (2006) Microcantilever apparatus and methods for detection of enzymes, enzyme substrates, and enzyme effectors. U.S. Patent 7,141,385.

33. Liu, W., Montana, V., Chapman, E.R. *et al.* (2003) Botulinum toxin type B micromechanosensor. *Proc. Natl. Acad. Sci. USA*, **100**, 13621–13625.

34. Weizmann, Y., Elnathan, R., Lioubashevski, O. *et al.* (2005) Magnetomechanical detection of the specific activities of endonucleases by cantilevers. *Nano Lett.*, **5**, 741–744.

35. Fritz, J., Baller, M.K., Lang, H.P. *et al.* (2000) Translating biomolecular recognition into nanomechanics. *Science.*, **288**, 316–318.

36. Weizmann, Y., Patolsky, F., Lioubashevski, O. *et al.* (2004) Magneto-mechanical detection of nucleic acids and telomerase activity in cancer cells. *J. Am. Chem. Soc.*, **126**, 1073–1080.

37. Su, M., Li, S.U., and Dravid, V.P. (2003) Microcantilever resonance-based DNA detection with nanoparticle probes. *Appl. Phys. Lett.*, **82**, 3562–3564.

23
Chemical Sensors Based on Microorganisms, Living Cells and Tissues

23.1 Living Material Biosensors: General Principles

Biocatalytic sensors were initially developed using enzymes that have been isolated from biological media or organisms and that catalyze the conversion of a substrate. Signal transduction in enzymatic sensors can be achieved by monitoring the concentration of a reactant or product that is involved in the enzymatic reaction.

Instead of using enzymes that have been isolated it is possible to use microorganisms, living cells or living tissues that contain analyte-specific enzymes. By metabolic processes, the analyte undergoes conversion to particular products, this process involving consumption of oxygen or another reactant. Monitoring of reactant consumption or product formation provides a means of signal transduction in metabolism-based biosensors.

Comprehensive overviews of living material-based biosensors are available in several recent books [1–3] and reviews [4–7]. Mathematical modeling of whole-cell biosensors is discussed in ref. [8].

Due to the similarity in the functioning principles, the design of metabolism-based sensors is essentially similar to that of enzymatic sensors. Thus, a layer containing the selected living material is integrated with a transducer device that indicates the concentration of a reactant or product involved in the metabolism of the sensing entity. In this way, the isolation of the relevant enzyme is avoided and the enzyme functions in its natural environment. The natural environment secures an optimal enzymatic activity and, in addition, provides necessary coenzymes and activators. Moreover, a living entity contains a great number of different enzymes that allows multienzyme sensing schemes to be developed.

Counteracting these advantages, there are several drawbacks that arise from using living materials in sensors. First, reactants and products have to cross the cell membrane and this slows down the diffusion process. Therefore, the response time is longer than that normally observed with sensors based on isolated enzymes. Secondly, the large number of enzymes present in a living entity is a source of interferences as various compounds in the sample may be converted to form various products to which the transducer is sensitive. Methods for improving the selectivity of living-material-based sensors are discussed in Section 23.6.

Various kinds of living material can be used in the design of biosensors. The most common are *living cells* of various origins. Unicellular organisms such as bacteria, algae [7] or yeast are widely used as sensing elements in *microbial biosensors*. Also useful are cells originating from multicellular organisms such as molds, lichens, or cells isolated from higher organisms such as plant or mammalian tissues [9]. Cells can be readily obtained by cell culture and can be conveniently immobilized at the transducer surface by various methods presented in Section 23.3.

Higher plant or animal material can be used in the form of thin slices integrated with the transducer device to form *tissue biosensors*.

In making biosensors, intact and viable biomaterial is normally used. However, because enzymatic activity is preserved for some time after biomaterial becomes nonviable, nonviable biological materials can also be used to construct biosensors.

The discussion above focused on the application of spontaneous metabolic processes that occur when the substrate and certain additional reactants are supplied to the sensing living entity. In such cases, it is the substrate that undergoes metabolic conversion.

There are, however, other sensing mechanisms in which the substrate is not metabolized, but where certain metabolic processes that result in detectable compounds are induced upon cell stimulation by the analyte.

23.2 Sensing Strategies in Living-Material-Based Sensors

23.2.1 Biocatalytic Sensors

Transduction in living material biosensors can be achieved by monitoring the concentration of the product of an enzymatic reaction, such as hydrogen ions, ammonia or carbon dioxide.

Chemical Sensors and Biosensors: Fundamentals and Applications, First Edition. Florinel-Gabriel Bănică.
© 2012 John Wiley & Sons, Ltd. Published 2012 by John Wiley & Sons, Ltd.

If the relevant enzyme in the sensing cell is an oxidase, transduction can be performed by monitoring the consumption of oxygen. Oxygen, which functions as an electron acceptor in the enzymatic reaction, can be replaced by an artificial electron acceptor such as hexacyanoferrate(III) or another redox mediator (see Section 14.2). In this case, transduction is performed by monitoring the reduced form of the electron acceptor.

Oxygen monitoring forms the basis of so-called *respiratory sensors*. Using microorganisms capable of metabolizing a broad variety of organic compounds, this approach allows water pollution with organic compounds to be assessed. In water chemistry, an indicator of the total content of organic material is called the *biological oxygen demand* (BOD) [10]. The biosensor approach is a convenient alternative to the standard laboratory method for determining the BOD. Certain vitamins can be assessed using their stimulating effect on the metabolism of the microorganism.

Respiratory sensors are also suitable for assessing the content of enzyme inhibitors such as toxic metal ions, antibiotics or organic pollutants. As inhibition is, as a rule, irreversible, the sensing membrane must be renewed after each assay taking care to preserve the density and metabolic status of the microbial cells.

Multistep sequences of enzymatic reactions can be implemented if a single-enzyme reaction does not provide a detectable compound. For example, a nitrate ion biosensor has been produced using *Azotobacter vinelandii* that initiates the reaction sequence (23.1)–(23.2), which results in ammonia, which can be detected by various methods:

$$NO_3^- + NADH + H^+ \xrightarrow{\text{Nitrate reductase}} NO_2^- + NAD^+ + H_2O \tag{23.1}$$

$$NO_2^- + 3NADH + 4H^+ \xrightarrow{\text{Nitrite reductase}} NH_3 + 3NAD^+ + 2H_2O \tag{23.2}$$

Multistep reaction sensing can also be performed by associating an isolated enzyme and living cells to form what is called a *mixed biosensor*.

23.2.2 External-Stimuli-Based Biosensors

Some substances can be assessed by exploiting their effect on the permeability of cellular membranes. The cells used for such assays are usually modified by incorporation of an ion that is not naturally occurring and that acts as a marker. The analyte affects in some way the permeability of the cell membrane for this marker that is monitored after its release into the extracellular environment. Thus, the antibiotic Nystatin which can act as an ionophore, modifies the permeability of cell membranes to alkali metal ions. Using Rb^+ as marker, Nystatin promotes the release of this ion from the yeast cells and allows transduction to be performed by means of an Rb^+ sensor.

An alternative approach relies on the breaking down of cells (lysis) under the effect of the analyte. This method has been applied to the assay of the lysozyme enzyme that produces the lysis of polysaccharide present in cell walls of bacteria. Under the effect of this enzyme, an atypical ion incorporated in the microorganism is released and its concentration indicates the lysozyme concentration in the sample.

Another methodology is based on the initiation of metabolism under the influence of the analyte as a stimulus for some physiological process. In this case, the analyte interacts with a specific cell receptor located at the plasma membrane. Formation of the analyte–receptor complex activates a metabolic process that gives rise to a detectable product.

23.3 Immobilization of Living Cells and Microorganisms

Immobilization of microorganisms and cells can be achieved by aggregation–flocculation, adsorption or adhesion, entrapment in polymeric networks or crosslinking with glutaraldehyde [5,11]. As a rule, gentle techniques should be used in order to preserve the viability of cells. That is why covalent immobilization should be used with caution because of the possible damaging effects of chemical reagents. Covalent immobilization at gold surfaces can be achieved by means of primary self-assembled monolayers containing reactive end groups that allow for conjugation with functional groups present at the cell wall.

Passive immobilization of cells into the pores, or adhesion onto the surface, of cellulose or other materials allows direct contact of the cells with the liquid sample. Therefore, no additional diffusion barrier is formed by immobilization, in contrast to gel entrapment. Adhesion is secured by coating the cells, the support or both with polyethyleneimine. Polyethyleneimine is a cationic polymer that imparts adhesion by electrostatic interaction with the negatively charged outer surface of cells.

Microorganism immobilization can be effected by affinity interaction with lectins (which are sugar-binding proteins). To this end, a lectin layer is first attached to the support in order to form a platform onto which microorganisms are further immobilized by interactions with saccharides present in the cell-wall membrane. As the immobilization process is reversible, a sensor formed in this way can be regenerated by eluting the used microorganism layer in order to load the transducer surface with a fresh batch of the microorganism.

Entrapment in poly(vinyl alcohol) or ionotropic gels (such as alginate) is a suitable method for immobilization of viable cells. A suspension of cells in a sodium alginate solution is added drop wise to a calcium chloride solution. As the alginate macromolecule is a polyanion, the divalent calcium ions promote gelation by forming ionic bridges between the macromolecules. In this way, gel particles several millimeters in size are obtained. The alginate gel is not stable in the presence of Ca^{2+}-binding compounds and certain cations that can substitute Ca^{2+} in the gel network. Gel stabilization can be achieved by crosslinking.

Other common entrapment matrices used in whole-cell biosensors are silica or organically modified silica (Ormosil) prepared by the sol-gel method [12]. In order to preserve the viability of the microorganism, inorganic precursors are preferred over organic ones that release toxic alcohols as byproducts. A typical entrapment protocol starts with mixing a cell suspension with glycerol. To this mixture is then added hydrochloric acid and sodium silicate solution in order to promote the sol-gel process [13]. In this preparation, the incorporation of glycerol leads to the formation of a well-defined cavity around the cell that prevents direct contact between cell and matrix. As a result, cell membrane permeability is not affected. In addition, the cell is not subject to mechanical stress that could impair cell viability.

Whole-cell biosensor arrays are produced for parallel assays of multiple samples by forming individually addressable cell spots of micrometer size. This process, which is commonly known as cell patterning, can be conducted in different ways [14]. In such techniques, cells are immobilized in the form of monolayers on solid supports by adhesion. Adhesion is a natural process in which cells bind to an adhesive extracellular matrix. Binding is achieved by the attraction of specific peptide regions in an adhesive extracellular matrix protein (integrin) that forms transmembrane receptors on the plasma membrane. Following this coupling, supramolecular complexes form to provide structural links between the cytoskeleton and the extracellular membrane [15]. In order to promote cell adhesion to inorganic surfaces, a protein layer should be formed by standard protein immobilization methods. This layer then acts as an adhesive extracellular matrix.

Cell patterning can be achieved by elastomer stenciling. In this technique, a micromachined thin sheet (stencil) containing patterned holes is applied over the support. This mask acts as a template during the cell-seeding process.

Microcontact printing (soft lithography) is used to create a pattern of protein spots to which the cells are then allowed to adhere.

Ink-jet patterning is a patterning method that has the advantage of allowing different cell strains to be patterned on the same substrate surface.

An elegant cell-patterning method is based on photolithography coupled with surface-specific chemical reactions [16]. Thus, a gold-electrode array is first formed by photolithography over a silicon dioxide film. This is followed by self-assembly of a long-chain thiocarboxylic acid monolayer over the gold surface. Protein or peptides are finally conjugated by covalent binding through the carboxyl groups in order to form an adhesive extracellular matrix for cell adhesion. In order to prevent cell adhesion to the exposed silicon dioxide insulator film, the silicon dioxide is coated with a protecting layer by silanisation.

23.4 Electrochemical Microbial Biosensors

Electrochemical transduction in whole-cell biosensor exploits the formation or consumption of electrochemically detectable species [17]. Amperometric methods are used for monitoring oxygen consumption in respiratory sensors as well as in sensors based on the inhibition of photosynthetic activity. Amperometry can also be used in conjunction with whole-cell biosensors including an enzyme that converts an artificial substrate into an electrochemically active compound. Ion formation by cell enzymes can be monitored in a selective way by potentiometric ion sensors or in a non-selective way by means of conductance measurements.

23.4.1 Amperometric Microbial Biosensors

The structure of an amperometric microbial biosensor is similar to that of its enzymatic counterpart (Section 14.1). Accordingly, a whole-cell layer is formed at the surface of an amperometric probe that functions as the transducer.

The most extensively studied sensor of this kind is the biological oxygen demand (BOD) sensor. Biochemical oxygen demand represents the amount of dissolved oxygen needed by aerobic microorganisms to break down organic material present in a given water sample at a given temperature over a specific time period. BOD is an indicator of the amount of organic material in the water. The BOD value is most commonly expressed in milligrams of oxygen consumed per liter of sample during 5 days of incubation at 20 °C and is a standard method for the assay of organic pollution of water.

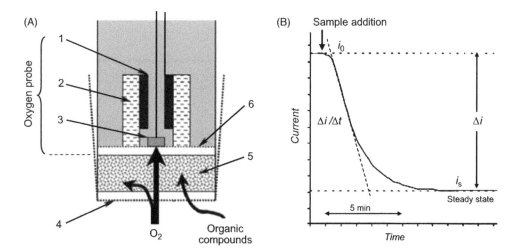

Figure 23.1 The amperometric BOD sensor. (A) Sensor configuration. (1) Ag/AgCl reference electrode; (2) electrolyte; (3) Pt cathode; (4) dialysis membrane; (5) immobilized whole cell layer; (6) oxygen-permeable membrane. (B) Response curve of an amperometric BOD sensor. Adapted with permission from [18]. Copyright 2002 Elsevier.

The above-mentioned BOD determination method (denoted as BOD_5) takes five days. The assay duration can be reduced considerably by using BOD amperometric sensors, which are obtained by assembling an amperometric oxygen probe with a film of microorganisms that are able to metabolize organic compounds in water with oxygen consumption (Figure 23.1A). In the absence of organic compounds, dissolved oxygen diffuses through the microorganism layer and is partially consumed in microorganism respiration. The oxygen probe generates a constant response current (i_0), as shown on the initial portion of the curve in Figure 23.1B. When a sample is added, organic compounds are metabolized by the microorganism and the oxygen consumption increases, leading to a gradual decrease in the response current. A constant, steady-state current (i_s) is obtained after about 10 min. The content of organic compound is indicated by the current difference $\Delta i = i_0 - i_s$. This difference increases proportionally with the BOD over a certain range (e.g., 5–100 mg O_2/L [19]) but levels off at higher BOD values where the limit of the metabolism rate is attained. Clearly, in the BOD assay the water sample should be saturated with air.

An alternative measurement method utilizes the rate of signal variation with time ($\Delta i/\Delta t$) that is indicated by the slope of the quasilinear portion of the current–time curve, as shown in Figure 23.1B. Compared with the steady-state measurement, the rate-base method provides the response after a shorter time lag.

In order to be suitable for application in a BOD sensor, the microorganism should be able to metabolize a broad spectrum of substrates and to maintain its metabolic activity even in the presence of adverse factors such as toxic metal ions or a high salt concentration (when used in the testing of sea water).

Various microbial species have been proposed as recognition elements in BOD sensors. The sensor can include a single-strain culture (e.g., *Bacillus subtilis* or *Trichosporon cutaneum*), a mixture of two identified microbial strains (e.g., *Bacillus subtilis* and *Bacillus licheniformis*) or activated sludge from wastewater processing plants, which include a mixture of bacteria and protozoan microorganisms.

Typically, single-strain BOD sensors display stable characteristics over a reasonable time period but their application is limited by the narrow substrate spectrum of the sensing strain. Good precision and operational stability, as well as a broader substrate spectrum is provided by mixtures of two identifiable strains. Activated sludge responds to a very broad range of substrates but they are not suitable for long-term operation owing to the instability of the composition of the microorganism mixture over time.

Instead of measuring the change in the oxygen concentration, a BOD sensor can be designed to perform transduction by means of a redox mediator such as the ferricyanide ion [20,21]. The mediator accepts electrons from the microorganism cells and after diffusion to the anode undergoes oxidation, thereby generating the response current. In this way, it is possible to determine the BOD of water samples with low oxygen contents without the need for additional supply of air.

In the amperometric BOD sensor, the current is produced under the influence of a constant negative potential applied to the working electrode. An alternative design is based on the *fuel cell* principle, as shown in Figure 23.2 [22]. The cell is composed of two compartments separated by a cation-exchanger membrane. The microorganism is selected so as to be able to perform direct electron transfer to the anode. Therefore, organic material is processed by the microbial film and electrons released in the oxidation reactions are released to the anode. Electrons pass through the external circuit and reduce dissolved oxygen supplied to the cathode chamber. The current indicated by the ammeter depends on the content of organic compounds in the sample supplied to the anode chamber. The first BOD sensor that was reported by Karube *et al.* in 1977 was of the fuel-cell type [23].

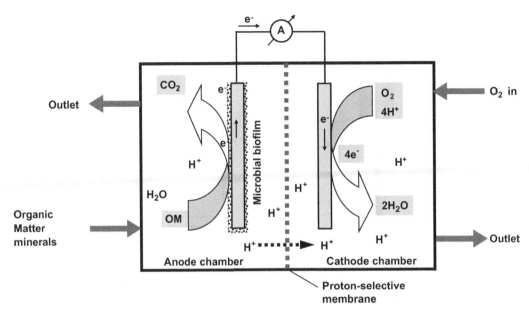

Figure 23.2 A microbial fuel cell. OM indicates organic material. Reprinted with permission from [22]. Copyright 2010 Elsevier.

Electrochemically inactive microorganisms can also be used in microbial fuel cells but in such cases an electron-transfer mediator should be present in the anode chamber. As the mediator has to cross the lipophilic cell membrane, it should be itself lipophilic. Suitable lipophilic mediators are menadione and benzoamines. Ferricyanide ion is widely used as the mediator either alone or in conjunction with menadione.

The application of BOD sensors reduces considerably the duration of the assay, from 5 days in the standard method to about 15 min. More advanced BOD sensor designs are based on a planar configuration that is suitable for multiplexing and allows for parallel assay of multiple samples.

Amperometric respiratory sensors have also been developed for various compounds of interest to the food and beverage industry (e.g., ethanol or glucose) or water pollutants such as phenols or surfactants.

Amperometric transduction can be achieved not only by monitoring the respiration of the microorganism but also by selecting a substrate that forms on oxidation an electrochemically active compound. For example, *p*-nitrophenol can be sensed by means of microorganisms that promote its oxidation to benzoquinone.

Sensors discussed above are based on the metabolic processing of the analyte by microorganisms. An alternative sensing strategy relies on the change of the enzyme activity under the effect of an analyte that act either as an inhibitor or as an activator. In general, enzyme inhibition is a poorly selective process. Improved selectivity can be achieved by means of genetically modified microorganisms in which a selected enzyme is produced in order to impart selectivity to a particular kind of inhibitor.

23.4.2 Potentiometric Microbial Biosensors

In the presence of a selected substrate, microorganism metabolism can result in ions (e.g., H^+) or gases (e.g., ammonia or carbon dioxide) that can be detected by a potentiometric sensor integrated with the microbiological sensing element. Thus, the first reported potentiometric microbial sensor was obtained by combining a potentiometric ammonia sensor with a layer of *Streptococcus faecium* [24]. This microorganism metabolizes the amino acid L-arginine forming ammonia. As a result, the potential-response of the ammonia sensor is a linear function of the logarithm of arginine concentration. Using a similar design, microbial biosensors for various amino acids and pollutants have been developed [25].

In the above approach the signaling ion or gas is produced by the metabolic conversion of the analyte. An alternative sensing mechanism involves the interaction of the analyte with a cell receptor leading to a metabolic process that changes the extracellular ion concentration [26]. The mechanism of induced metabolic response is presented in Figure 23.3. Glucose and oxygen are consumed to produce adenosine triphosphate (ATP) that is the source of energy in various metabolic processes indicated in this figure by the arrows. The products of the metabolic processes are lactic acid and carbon dioxide (or the hydrogen carbonate ion). Excretion of the waste products induces acidification of the extracellular medium that can be detected by a suitable transducer. The rate of these metabolic processes can be enhanced if the analyte is a ligand that interacts with a cell receptor and thereby enhances the metabolic activity. A particular activation mechanism is the activation of the sodium-potassium pump located in the cell membrane. Using energy provided by ATP hydrolysis, this molecular pump drives sodium and potassium ions against their concentration gradients. This results in an increase of the sodium ion concentration in the extracellular fluid accompanied by a decrease of the potassium ion concentration in the same medium. Monitoring of pH or alkali ion

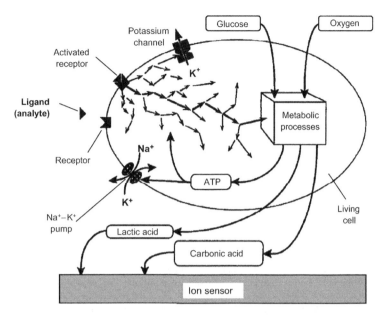

Figure 23.3 Schematic presentation of the cellular metabolism and its relationship with cell receptor activation by an external chemical stimulus. Adapted with permission from [26]. Copyright Elsevier 1992.

concentration by means of suitable potentiometric sensors provide a response signal that is dependent on the concentration of the ligand-analyte.

The most straightforward transduction method is based on pH change produced by excreted acidic waste products. In order to obtain a quasilinear response, the pH should be adjusted to a value as close as possible to the pK value of the buffer system. Typically, natural acid-base compound present in the cell environment (such as carbonate or phosphate ions or amino acids) provide a suitable buffer capacity.

Both ion-selective electrodes and ion-selective field effect devices have been used as transducers in whole cell biosensors [27]. The response in each case is linearly dependent on the logarithm of the analyte concentration. Facile miniaturization and compatibility with microfabrication technology makes potentiometric sensors well suited for integration in sensor arrays and microfluidic analytical systems.

Ion-selective field effect devices have been used to develop sensor arrays for multiple-parameter monitoring of the cell metabolism. This can be accomplished by measuring the concentration of a series of extruded ions such as H^+, K^+ and Ca^{2+} by means of light-addressable potentiometric sensors [28]. Additional sensors for oxygen and organic compounds of physiological relevance can be included in the array in order to provide a more comprehensive assessment of the cell metabolism. Such a device, which is termed a *microphysiometer*, can be used in fundamental research of cell physiology as well as in the assay of antibiotics or toxic species which affect the cell metabolism.

23.4.3 Conductometric Microbial Sensors

Many microbe-catalyzed reactions result in a change of the local ion concentration, which can be monitored in a nonselective way by conductance measurements. The configuration of conductometric cell biosensors is similar to that of conductometric enzymatic sensors (Section 17.7.4). In short, a living-cell layer is formed between two electrodes and the electric conductance of this layer is monitored by a conductometer. For example, a sensor for acrylonitrile has been developed using the *Rhodococcus ruber* microorganism that converts acrylonitrile into the ammonium acrylate ion [29]. A compact design is obtained by means of interdigitated electrode pairs.

Carbamates and organophosphorus pesticides can be assessed by inhibition of the acetylcholinesterase enzyme in the *Chlorella vulgaris* microalgae [30]. Using methylumbelliferylphosphate as substrate, a quaternary ammonium compound forms. The same microorganism contains alkaline phosphatase that is inhibited by Cd^{2+} and Zn^{2+} ions. The substrate of this enzyme (*p*-nitrophenylphosphate) produces by hydrolysis phosphate ions that alter the local conductance. Therefore, this sensor is useful in the assay of water pollution by either pesticides or toxic metal ions.

The response signal in conductometric biosensors is overlapped by a significant background produced by ions in the pH buffer system. Background correction can be performed by subtracting from the overall signal the signal produced by a reference sensor. This sensor includes a strain of inactive microorganism.

23.4.4 Electrical Impedance Transduction

Electrical impedance measurement is suitable for monitoring electrodes coated by living-cell monolayers immobilized by adhesion. Impedance measurements (Sections 17.1 and 17.2) provide a wealth of information about the cell

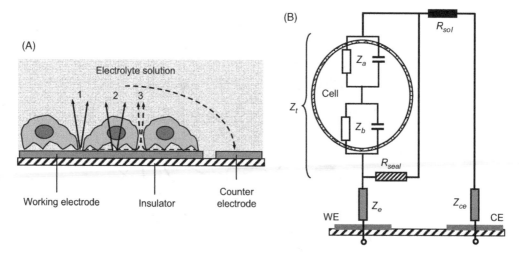

Figure 23.4 Principle of impedimetric assay of whole cells. (A) Current flow at an electrode coated with a cell monolayer. (B) The equivalent circuit of a system formed of a living cell attached to an electrode surface. WE: working electrode; CE: counterelectrode; Z_e and Z_{ce}: impedances at the direct contact of electrodes with the solution; R_{seal}: the resistance at the cell-substrate contact; Z_a and Z_b: apical and basal cell membrane impedances, respectively. (A) Adapted with permission from [14]. Copyright 2007 The Royal Society of Chemistry. (B) Adapted with permission from [32]. Copyright 2009 Elsevier.

physiology, adhesion propensity, shape, spreading and motility [14,31]. Typically, mammalian or higher eukaryotic whole cells are used in this kind of sensor in order to substitute cell cultures for whole animals in the investigation of the effects of toxins, drugs, viruses and bacteria on living organisms [32]. Cell adhesion to the electrode is promoted by prior coating of the electrode with a specific protein.

Current distribution at a whole-cell monolayer coated on a metal electrode is illustrated in Figure 23.4A. As can be seen, ion current can flow at parts of the electrode exposed to the solution (path 1), by crossing the cells (path 2), or by passing through the solution channels between the cell and the electrode surface (path 3). The equivalent circuit of an electrode coated with a cell monolayer is shown in Figure 23.4B. In this circuit, the Z_e impedance corresponds to path 1, while path 3 meets the seal resistance (R_{seal}) which is the resistance at the cell–substrate contact. The parallel track 2 leads the current across the cell where there are two impedances (Z_b and Z_a) generated by the cell membrane. Each of them is composed of a parallel combination of a capacitor and a resistor. The capacitor accounts for the insulating properties of the membrane whilst the resistor represents the parallel combination of the ion channels available for the transport of various ions through the membrane. The counterelectrode has a large surface area exposed to the solution and thereby its impedance (Z_{ce}) is meaningless.

The cell impedance is therefore represented by the series sum of basal and apical cell impedances:

$$Z_{cell} = Z_a + Z_b \tag{23.3}$$

The total impedance associated with the cell (Z_t) includes also the effect of the sealing resistance:

$$Z_t = \frac{Z_{cell} R_{seal}}{Z_{cell} + R_{seal}} \tag{23.4}$$

If $Z_{cell} \gg R_{seal}$, the total impedance approximates to seal resistance, whilst in the opposite case ($Z_{cell} \ll R_{seal}$) the total impedance is mostly determined by the cell impedance. The impedance at the electrode–electrolyte surface is of no relevance and should be kept as high as possible by patterning the cells in a tightly bound, well-spread conformation.

Cell impedance measurements are utilized in cell biology to assess in real time the cell response to physical or chemical stimuli, such as temperature, pH and various chemical compounds. Electrical impedance whole-cell biosensors have been developed for the assay of organic or inorganic toxic species. Arrays of such sensors have been produced for parallel toxicity assays of multiple water samples.

23.5 Optical Whole-Cell Sensors

23.5.1 Optical Respiratory Biosensors

Microorganism respiration can be monitored by means of optical oxygen sensors. Hence, optical analogs of the amperometric microbial sensor (Section 23.4.1) have been developed.

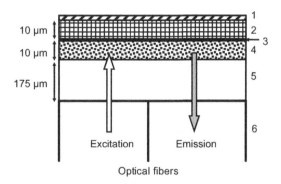

Figure 23.5 Cross section of an optical BOD sensor. (1) porous polycarbonate cover; (2) layer of yeast (*Trichosporon cutaneum*) immobilized in poly (vinyl alcohol); (3) 1 μm layer of charcoal (optical insulator); (4) oxygen-sensitive fluorescent layer; (5) inert and gas-impermeable polyester layer; (6) optical fiber bundle. Redrawn with permission from [34]. Copyright 1994 American Chemical Society.

Fluorescent BOD sensors have been constructed by forming a culture of microorganisms at the end surface of a fiber optic oxygen sensor (Figure 23.5). As oxygen sensing is based on fluorescence quenching, the fluorescence response intensity increases with the oxygen consumption rate and the response function is similar to that of an amperometric BOD sensor. *Trichosporon cutaneum* or *Bacillus subtilis* microorganisms have been used in the sensing layer.

As lower selectivity of the microorganism layer allows the monitoring of more possible pollutants, an optical BOD sensor has been constructed using a combination of three microorganisms (*Bacillus licheniformis*, *Dietzia maris* and *Marinobacter marinus*) [33]. Compared with the sensor that makes use of *Bacillus licheniformis* only, the three-microorganism sensor allows the limit of detection to be decreased from 0.9 to 0.3 mg/L and the response time from 30 to 3.2 min.

Optical respiratory biosensors are also useful in the assay of toxic chemical ionic species that cause respiratory activity to decrease.

23.5.2 External-Stimuli-Based Optical Sensors

In addition to the metabolic function, other biological processes in whole cells can be exploited in the design of optical sensors. For example, photosynthesis, which is an essential process in plant cells, has been used to develop herbicide or pesticide sensors based on the inhibition of an electron-transfer step occurring during photosynthesis [13,35]. As a result, the chlorophyll concentration in the cell increases and this modification can be detected by means of chlorophyll fluorescence at 682 nm. A sensor for the herbicide diuron based on photosynthesis inhibition is shown in Figure 23.6A. The sensing membrane in this sensor has been obtained by entrapment of the *Chlorella vulgaris* alga in a silica matrix prepared by sol-gel chemistry. The membrane was positioned in a flow-through cell and a bifurcated optical fiber cable was used to provide excitation and collect the fluorescent light. The response is obtained as the time derivative of the fluorescence intensity measured 5 min after sample introduction. A typical response curve is presented in Figure 23.6B. The detection limit of this sensor is $1\,\mu g\,L^{-1}$ diuron, which is better than the value reported for high-performance liquid chromatography.

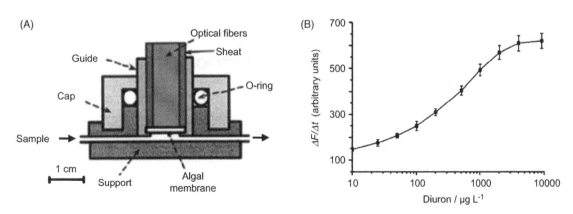

Figure 23.6 Optical alga biosensor for the herbicide Diuron. (A) Sensor configuration; (B) response function. Adapted with permission from [35]. Copyright 2007 Elsevier.

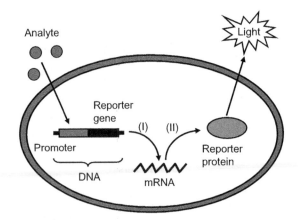

Figure 23.7 Mechanism of bioreporter response to a chemical stimulus. Step (I): transcription; step (II): translation.

23.5.3 Bioreporters

A very promising approach in biosensing is represented by induced synthesis of certain proteins that can be detected by means of luminescence. Living cells performing this function are known as *bioreporters*.

Bioreporters are intact living cells that produce a measurable signal in response to a specific chemical or physical agent in their environment [36–38]. Typically, genetically modified organisms are suitable for such applications. Bioreporter cells contain two specific genetic elements, a promoter sequence (which promotes the gene transcription into RNA) and a reporter gene (Figure 23.7). In a genetically modified organism, the reporter gene is substituted for a natural gene that is involved in cell regulatory functions. When the analyte (or one of its metabolites) is recognized by specific receptors (regulatory protein), a natural regulatory process is triggered. In this process, the reporter gene code is first transcribed into messenger RNA (mRNA) and then translated into a luminescent reporter protein that can be detected (Figure 23.7). In a cell culture, the amount of reporter protein depends on the analyte concentration in the sample.

In order to perform signal transduction by fluorescence, the reporter gene is selected such as to be expressed into green fluorescent protein (GFP) that is encoded in the *gfp* gene. GFP can be modified to emit light at various wavelengths (cyan, red, and yellow), which allows for multianalyte detection. Alternatively, bioluminescence can be produced if the reporter protein is an enzyme of the luciferase class, which is encoded in a *lux* gene. Firefly luciferase is particularly suitable owing to its high quantum yield but bacterial luciferase is also utilized. In order for bioluminescence to occur, an exogenous substrate (luciferin) should be supplied.

In the above mechanism, the light signal is enhanced under the effect of the analyte. An alternative sensing mechanism is based on the reduction of the light signal upon expression of a repressor protein that restrains the expression of the reporter gene.

Bioreporters can be designed for various applications in environment pollution control. These applications address determinations of organic compounds, metal and nonmetal toxic species, and the assay of sample toxicity or stress conditions. Of particular interest is the utilization of bioreporters in the assay of bioavailable toxic metal ions such as Co^{2+} and Ni^{2+}. As the toxic effect is exerted only by bioavailable ion species, the concentration of these species is more enlightening than the total ion concentration. Because the response of a bioreporter cell depends on intricate biochemical reactions, the response time is relatively long and can vary between 30 min to 5 h.

23.6 Improving the Selectivity of Microorganism Biosensors

Biosensors based on living cells are reputed to be poorly selective for various reasons. Typical interferences arise from the presence of a high number of different enzymes in the same cell. Hence, various metabolic pathways are possible and some of them can give rise to the same signaling product starting from different substrates. Therefore, the sensor response would be higher than that expected for the target analyte. Another source of interferences in microbial sensor is represented by contamination with other bacterial species that give rise by metabolism to products that interfere with the transduction process.

Selective microbial sensors can be obtained if the analyte is essential for the growth of the selected microorganism. This approach has been applied in the development of biosensors for vitamins.

Selectivity can also be achieved by proper selection of the operating conditions such as the pH or the concentration of dissolved of oxygen. In certain cases, the interfering metabolic pathway can be repressed by means of enzyme inhibitors.

Good selectivity can be obtained upon combining an isolated enzyme with a microorganism. Thus, urea determination can be selectively performed by a hybrid sensor composed of urease and nitrifying bacteria. Ammonia produced by the enzymatic reaction is further metabolized by the bacteria with oxygen consumption and the measurement of the decrease in the oxygen specific signal is correlated with the substrate concentration. The selectivity of this hybrid sensor originates from the fact that nitrifying bacteria can use only ammonia as an energy source.

It should be pointed that in certain applications, poor selectivity is an advantage rather than a drawback. This applies to BOD sensors or sensors designed for toxicity assay.

23.7 Conclusions

Living cell-based biosensor utilize living biological cells and monitor physiological modifications produced by chemical stimuli. There are two particular strategies for biosensing with living cells. The first relies on the metabolic conversion of the target species accompanied by consumption or production of a detectable chemical species. From this standpoint, the whole cell biosensor behaves as an enzymatic sensor in which the active enzyme functions in its natural environment. The second alternative relies on alteration of the cell physiology or morphology in response to a chemical stimulus.

As the receptor species are included in the cell's natural environment, they are not exposed to degradation or inactivation. This feature brings about a reduced manufacturing cost since the isolation of recognition receptors is avoided.

In general, metabolism-based biosensors display rather poor selectivity owing to the occurrence of multiple metabolic pathways. At first sight, the poor selectivity of such sensors appears to be a drawback, particularly when a specific chemical species is targeted. However, the poor selectivity becomes an advantage when a whole class of target compounds (such as dissolved organic material or toxic species) is being addressed.

Although tissues of higher organisms are suitable for certain applications, whole-cell biosensors are commonly preferred owing to easier fabrication and the high degree of control over the characteristics of the biological material. Whole-cell biosensors make use of either microorganisms or higher-organism cell cultures.

The application of whole-cell biosensors encompasses a broad area, including environment monitoring, medicine and clinical chemistry [28,39], the pharmaceutical industry, food and fermentation industries, and biosecurity. Of great interest is the capacity of whole-cell biosensors to respond to various kinds of toxic species. This characteristic is employed in the assay of ecotoxicity [21,32] and the detection of chemical-warfare agents.

Questions and Exercises

1 What kinds of living materials can be used for analyte recognition in biosensors?
2 Give a synopsis of biocatalytic sensors based on living material with a focus on underlying principles, applications, advantages and shortcomings.
3 Comment on possible sensing methods for target analytes that can act as external stimuli to living cells.
4 Discuss the immobilization methods for living cells with emphasis on (i) passive immobilization; (ii) affinity-based immobilization; (iii) gel entrapment.
5 Give a short overview of cell-patterning methods with applications in cell-based sensor arrays.
6 Give a comparative overview of amperometric and optical BOD sensors.
7 Discuss the similitude and difference between BOD fuel-cell sensors and BOD amperometric sensors driven by an applied electrical voltage.
8 How can a potentiometric gas sensor be used in the development of microbial biosensors?
9 Certain compounds can activate metabolic pathways in living cells upon interaction with cell receptors present at the surface of the cell. Discuss possible transduction methods utilized in biosensors based on cell stimulation.
10 How can electrical impedance measurements be used to assess the response of living cells to physical and chemical stimuli?
11 Comment on the application of the photosynthesis process to optical biosensors.
12 What are the relevant properties of bioreporters and how can bioreceptors be used in the design of whole-cell optical biosensors?
13 Comment on the selectivity of whole-cell biosensors and the implications of the selectivity issue in the design of this kind of sensors.

References

1. Racek, J. (1995) *Cell-Based Biosensors*, Technomic Publishing Company, Inc., Lancaster.
2. Zourob, M., Elwary, S., and Turner, A. (eds) (2008) *Principles of Bacterial Detection: Biosensors, Recognition Receptors and Microsystems*, Springer, New York.
3. Wang, P. and Liu, Q. (eds) (2010) *Cell-Based Biosensors: Principles and Applications*, Artech House, Norwood.
4. Nakamura, H., Shimomura-Shimizu, M., and Karube, I. (2008) Development of microbial sensors and their application. *Adv. Biochem. Eng./Biotechnol.*, **109**, 351–394.
5. D'Souza, S.F. (2001) Microbial biosensors. *Biosens. Bioelect.*, **16**, 337–353.
6. Su, L., Jia, W., Hou, C. *et al.* (2011) Microbial biosensors: A review. *Biosen. Bioelec.*, **26**, 1788–1799.
7. Lei, Y., Chen, W., and Mulchandani, A. (2006) Microbial biosensors. *Anal. Chim. Acta.*, **568**, 200–210.
8. Baronas, R., Kulys, J., and Ivanauskas, F. (2009) *Mathematical Modeling of Biosensors: An Introduction for Chemists and Mathematicians*, Springer Netherlands, Dordrecht.
9. Rechnitz, G.A. (1988) Bioselective membrane electrodes using tissue materials as biocatalysts. *Method Enzymol.*, **137**, 138–152.
10. Nomura, Y., Chee, G.J., and Karube, I. (1998) Biosensor technology for determination of BOD. *Field Anal. Chem. Technol.*, **2**, 333–340.
11. Buchholz, K., Kasche, V., and Bornscheuer, U.T. (2005) *Biocatalysts and Enzyme Technology*, Wiley-VCH, Weinheim.
12. Depagne, C., Roux, C.C., and Coradin, T. (2011) How to design cell-based biosensors using the sole-gel process. *Anal. Bioanal. Chem.*, **400**, 965–976.
13. Pena-Vazquez, E., Maneiro, E., Perez-Conde, C. *et al.* (2009) Microalgae fiber optic biosensors for herbicide monitoring using sol-gel technology. *Biosens. Bioelectron.*, **24**, 3538–3543.
14. Asphahani, F. and Zhang, M. (2007) Cellular impedance biosensors for drug screening and toxin detection. *Analyst.*, **132**, 835–841.
15. Garcia, A.J. (2006) Interfaces to control cell-biomaterial adhesive interactions, in *Polymers for Regenerative Medicine* (ed. C. Werner), Springer, Berlin, pp. 171–190.
16. Veiseh, M., Zareie, M.H., and Zhang, M.Q. (2002) Highly selective protein patterning on gold-silicon substrates for biosensor applications. *Langmuir.*, **18**, 6671–6678.
17. Lagarde, F. and Jaffrezic-Renault, N. (2011) Cell-based electrochemical biosensors for water quality assessment. *Anal. Bioanal. Chem.*, **400**, 947–964.
18. Liu, J. and Mattiasson, B. (2002) Microbial BOD sensors for wastewater analysis. *Water Res.*, **36**, 3786–3802.
19. Riedel, K., Lange, K.P., Stein, H.J. *et al.* (1990) A microbial sensor for BOD. *Water Res.*, **24**, 883–887.
20. Yoshida, N., Yano, K., Morita, T. *et al.* (2000) A mediator-type biosensor as a new approach to biochemical oxygen demand estimation. *Analyst.*, **125**, 2280–2284.
21. Pasco, N.F., Weld, R.J., Hay, J.M. *et al.* (2011) Development and applications of whole cell biosensors for ecotoxicity testing. *Anal. Bioanal. Chem.*, **400**, 931–945.
22. Namour, P. and Jaffrezic-Renault, N. (2010) Sensors for measuring biodegradable and total organic matter in water. *TrAC-Trends Anal. Chem.*, **29**, 848–857.
23. Karube, I., Matsunaga, T., Mitsuda, S. *et al.* (1977) Microbial electrode BOD sensors. *Biotechnol. Bioeng.*, **19**, 1535–1547.
24. Rechnitz, G.A., Kobos, R.K., Riechel, S.J. *et al.* (1977) Bio-selective membrane electrode prepared with living bacterial cells. *Anal. Chim. Acta.*, **94**, 357–365.
25. Corcoran, C.A. and Rechnitz, G.A. (1985) Cell based biosensors. *Trends Biotechnol.*, **3**, 92–96.
26. Owicki, J.C. and Wallace Parce, J. (1992) Biosensors based on the energy metabolism of living cells: The physical chemistry and cell biology of extracellular acidification. *Biosens. Bioelect.*, **7**, 255–272.
27. Poghossian, A., Ingebrandt, S., Offenhausser, A. *et al.* (2009) Field-effect devices for detecting cellular signals. *Semin. Cell Dev. Biol.*, **20**, 41–48.
28. Wang, P., Xu, G.X., Qin, L.F. *et al.* (2005) Cell-based biosensors and its application in biomedicine. *Sens. Actuators B-Chem.*, **108**, 576–584.
29. Roach, P.C.J., Ramsden, D.K., Hughes, J. *et al.* (2003) Development of a conductimetric biosensor using immobilised Rhodococcus ruber whole cells for the detection and quantification of acrylonitrile. *Biosens. Bioelectron.*, **19**, 73–78.
30. Chouteau, C., Dzyadevych, S., Durrieu, C. *et al.* (2005) A bi-enzymatic whole cell conductometric biosensor for heavy metal ions and pesticides detection in water samples. *Biosens. Bioelectron.*, **21**, 273–281.
31. Gu, W.W. and Zhao, Y. (2010) Cellular electrical impedance spectroscopy: an emerging technology of microscale biosensors. *Expert Rev. Med. Devices.*, **7**, 767–779.
32. Banerjee, P. and Bhunia, A.K. (2009) Mammalian cell-based biosensors for pathogens and toxins. *Trends Biotechnol.*, **27**, 179–188.
33. Lin, L., Xiao, L.L., Huang, S. *et al.* (2006) Novel BOD optical fiber biosensor based on co-immobilized microorganisms in ormosils matrix. *Biosens. Bioelectron.*, **21**, 1703–1709.
34. Preininger, C., Klimant, I., and Wolfbeis, O.S. (1994) Optical fiber sensor for biological oxygen demand. *Anal. Chem.*, **66**, 1841–1846.
35. Nguyen-Ngoc, H. and Tran-Minh, C. (2007) Fluorescent biosensor using whole cells in an inorganic translucent matrix. *Anal. Chim. Acta.*, **583**, 161–165.

36. van der Meer, J.R., Tropel, D., and Jaspers, M. (2004) Illuminating the detection chain of bacterial bioreporters. *Environ. Microbiol.*, **6**, 1005–1020.
37. Belkin, S. (2003) Microbial whole-cell sensing systems of environmental pollutants. *Curr. Opin. Microbiol.*, **6**, 206–212.
38. Van der Meer, J.R. (2011) *Bacterial Sensors: Synthetic Design and Application Principles*, Morgan & Claypool, San Francisco.
39. Kintzios, S.E. (2007) Cell-based biosensors in clinical chemistry. *Mini-Rev. Med. Chem.*, **7**, 1019–1026.

Index

Note: Page numbers in *italics* refer to Figures; those in **bold** to Tables

Printed and bound by CPI Group (UK) Ltd, Croydon, CR0 4YY

27/02/2024

14460322-0001